Modeling Sociocultural Influences on Decision Making

Human Factors and Ergonomics Series

PUBLISHED TITLES

Around the Patient Bed: Human Factors and Safety in Health Care
Y. Donchin and D. Gopher

Cognitive Neuroscience of Human Systems: Work and Everyday Life
C. Forsythe, H. Liao, M. Trumbo, and R. E. Cardona-Rivera

Conceptual Foundations of Human Factors Measurement
D. Meister

Content Preparation Guidelines for the Web and Information Appliances: Cross-Cultural Comparisons
H. Liao, Y. Guo, A. Savoy, and G. Salvendy

Cross-Cultural Design for IT Products and Services
P. Rau, T. Plocher and Y. Choong

Data Mining: Theories, Algorithms, and Examples
Nong Ye

Designing for Accessibility: A Business Guide to Countering Design Exclusion
S. Keates

Ergonomics in Design: Methods and Techniques
Marcelo M. Soares and Francisco Rebelo

Ergonomic Workplace Design for Health, Wellness, and Productivity
Alan Hedge

Handbook of Cognitive Task Design
E. Hollnagel

The Handbook of Data Mining
N. Ye

Handbook of Digital Human Modeling: Research for Applied Ergonomics and Human Factors Engineering
V. G. Duffy

Handbook of Human Factors and Ergonomics in Health Care and Patient Safety Second Edition
P. Carayon

Handbook of Human Factors in Web Design, Second Edition
K. Vu and R. Proctor

Handbook of Occupational Safety and Health
D. Koradecka

Handbook of Standards and Guidelines in Ergonomics and Human Factors
W. Karwowski

Handbook of Virtual Environments: Design, Implementation, and Applications, Second Edition,
K. S. Hale and K M. Stanney

Handbook of Warnings
M. Wogalter

Human–Computer Interaction: Designing for Diverse Users and Domains
A. Sears and J. A. Jacko

Human–Computer Interaction: Design Issues, Solutions, and Applications
A. Sears and J. A. Jacko

PUBLISHED TITLES (CONTINUED)

Human–Computer Interaction: Development Process
A. Sears and J. A. Jacko

Human–Computer Interaction: Fundamentals
A. Sears and J. A. Jacko

The Human–Computer Interaction Handbook: Fundamentals
Evolving Technologies, and Emerging Applications, Third Edition
J. A. Jacko

Human Factors in System Design, Development, and Testing
D. Meister and T. Enderwick

Introduction to Human Factors and Ergonomics for Engineers, Second Edition
M. R. Lehto

Macroergonomics: Theory, Methods and Applications
H. Hendrick and B. Kleiner

Modeling Sociocultural Influences on Decision Making:
Understanding Conflict, Enabling Stability
Joseph V. Cohn, Sae Schatz, Hannah Freeman, and David J. Y. Combs

Practical Speech User Interface Design
James R. Lewis

The Science of Footwear
R. S. Goonetilleke

Skill Training in Multimodal Virtual Environments
M. Bergamsco, B. Bardy, and D. Gopher

Smart Clothing: Technology and Applications
Gilsoo Cho

Theories and Practice in Interaction Design
S. Bagnara and G. Crampton-Smith

The Universal Access Handbook
C. Stephanidis

Usability and Internationalization of Information Technology
N. Aykin

User Interfaces for All: Concepts, Methods, and Tools
C. Stephanidis

Variability in Human Performance
T. Smith, R. Henning, and M. Wade

Modeling Sociocultural Influences on Decision Making

UNDERSTANDING CONFLICT, ENABLING STABILITY

Edited by

Joseph V. Cohn • Sae Schatz

Hannah Freeman • David J. Y. Combs

CRC Press
Taylor & Francis Group
Boca Raton London New York

CRC Press is an imprint of the
Taylor & Francis Group, an **informa** business

CRC Press
Taylor & Francis Group
6000 Broken Sound Parkway NW, Suite 300
Boca Raton, FL 33487-2742

First issued in paperback 2019

ISBN-13: 978-1-4987-3669-5 (hbk)
ISBN-13: 978-0-367-87407-0 (pbk)

Library of Congress Cataloging-in-Publication Data

Names: Cohn, Joseph V., editor.
Title: Modeling sociocultural influences on decision making : understanding conflict, enabling stability / editors, Joseph V. Cohn, Sae Schatz, Hannah Freeman, David J.Y. Combs.
Description: Boca Raton, FL : CRC Press, 2016. | Series: Human factors and ergonomics series | Includes bibliographical references and index.
Identifiers: LCCN 2016011243 | ISBN 9781498736695 (alk. paper)
Subjects: LCSH: Conflict management--Social aspects. | Social conflict. | Culture conflict. | Decision making--Social aspects. | Applied sociology.
Classification: LCC HM1126 .M64 2016 | DDC 303.6/9--dc23
LC record available at https://lccn.loc.gov/2016011243

Visit the Taylor & Francis Web site at
http://www.taylorandfrancis.com

and the CRC Press Web site at
http://www.crcpress.com

This book is dedicated to our families, friends, and colleagues whose guidance, support, and mentorship provided us with the inspiration and motivation to see this project through to completion.

Contents

Foreword xiii
Editors xix
Contributors xxi
Introduction xxv

Section I Building Theories

1. **Expeditionary Modeling for Megacities and Other Dense Urban Areas** 5
   Brian P. Kettler, Rachel G. Hingst, and Mark A. Hoffman

2. **More than Just a Story: Narrative Insights into Comprehension, Ideology, and Decision Making** 27
   Scott W. Ruston

3. **The Spread of Information via Social Media** 43
   Brian M. Fairlie

4. **The Spread of Opinions in Societies** 61
   Boleslaw K. Szymanski, Omar Lizardo, Casey Doyle, Panagiotis D. Karampourniotis, Pramesh Singh, Gyorgy Korniss, and Jonathan Z. Bakdash

5. **Culture's Influences on Cognitive Reflection** 85
   Vladimíra Čavojová and Róbert Hanák

6. **Cultural Influences on Cognitive Biases in Judgment and Decision Making: On the Need for New Theory and Models for Accidents and Safety** 103
   Atsuo Murata

Section II Collecting and Analyzing Data

7. **The Sixty Percent Mission: An Introduction to High-Risk Ethnography Process and Protocol in Support of the US Army's Civil Affairs Humanitarian Mission** 115
   Tracy Saint Benoit, Clarissa Graffeo, Mark A. Carter, and Col. Richard Swisher (Ret)

8. **Challenges in Connecting with the Disconnected: An Introduction to Connecting with Communities Disconnected from the Rest of the World** 135
   Christina S. Kang

9. **They Are Beyond WEIRD: Helpful Frameworks for Conducting Non-WEIRD Research** 163
David J. Y. Combs, Sarai Blincoe, Christopher P. Garris, and Eric S. Vorm

10. **Deciphering the Emic Perspective in Data in Order to Assess Threat** 181
Laurie Fenstermacher and Lawrence A. Kuznar

11. **Collecting Data and Semantic Content via Mobile Devices** 205
Alper Caglayan and Laura Cassani

12. **Measuring Changes in Attitudes: Using Factor Analysis to Track Population Attitudes Spatially and Temporally** 221
Joseph Maddux and Jeffrey Appleget

13. **Gaining Insight by Applying Geographical Modeling** 243
Erman Çakıt and Waldemar Karwowski

Section III Building and Validating Sociocultural Models

14. **Sociocultural Capability Requirements across All Phases of Military Operations** 269
Walter L. Perry

15. **Methods to Characterize and Manage Uncertainty for Sociocultural Applications** 289
Perakath Benjamin, Kalyan Vadakkeveedu, and Satheesh Ramachandran

16. **Validating Causal and Predictive Claims in Sociocultural Models** 315
Amy Sliva, Scott Neil Reilly, John Chamberlain, and Randy Casstevens

17. **Rapid Generation of Political Conflict Simulations for Scenarios around the World** 335
Barry G. Silverman, David Q. Sun, Nathan Weyer, and Gnana K. Bharathy

18. **Detailed Model Development Case Study: The Peace Game** 361
Matthew J. Powers

Section IV Applying Sociocultural Models to Gain Insight into Conflict and Instability

19. **Using the Social Framework Model of Trust to Better Understand Trust in Government** 397
Walter W. Kulzy, David J. Y. Combs, and Ronald D. Fricker, Jr.

20. **Understanding Public Opinion toward Violent Extremists** 423
Lewis A. Anderson and Ronald D. Fricker, Jr.

21. Modeling Sociocultural Influences on Decision Making: Assessing Conflict and Stability 449
Michael L. Bernard, George A. Backus, Asmeret Bier Naugle, Robert F. Jeffers, and Regan W. Damron

22. Modeling Social System Resiliency: An Agent-Based Multiscale Approach 473
Steven B. Hall, Curtis L. Blais, and Ryan G. Baird

23. Applying Modeling and Simulation to Foreign Policy: An Afghan Example 495
Corey Lofdahl

Subject Index 511

Author Index 535

Foreword

Why Is This Topic Important?

Our world is increasingly interconnected in terms of communications and commerce. Global interdependence is generally a stabilizing factor. Yet at the same time, our global commons are increasingly congested, competitive, and contested, which destabilizes our planet (Irving 2015; Puşcaş 2010). With nearly $2 trillion in global research and development, the rate of technological change will continue to accelerate, which will likely bring nations and their peoples closer and more intimately connected and codependent. This will further raise the potential for both transnational threats and conflict as well as transnational opportunities for peace and prosperity. One of the key dimensions of these societal tectonic movements is culture. Unique languages, art, cuisine, histories, and traditions provide a rich tapestry to our planet but also sow the seeds for potential culture clash. Understanding beliefs, attitudes, and behaviors across multiple cultures in this increasingly wired world is an essential skill to navigating and shaping this human terrain to ensure a more stable and prosperous world.

Background

Since the emergence of early descriptions of societies, humankind has had a thirst for understanding social, cultural, and behavioral aspects of humans and groups. One of the largest modern reviews of scientific sociocultural research was the Department of Defense's Human Social Cultural Behavioral (HSCB) modeling program. One of the program's goals was to create a global and persistent indications and warnings capability that would complement and enhance conventional sensors by leveraging massive open source data at a global scale to support improved situation awareness, understanding, and decision making. An integrating vision of this program was the notion of a social radar, an ability to sense and track events in the human domain. Metaphorically, the vision for social radar was motivated by sensing science, including passive and active methods. For example, in 1490, Leonardo da Vinci inserted a tube in water to detect vessels. Centuries later, in 1914, sonar (sound navigation and ranging) was invented to penetrate or see through water, using sound propagation to navigate, communicate, or detect objects. In 1941, radar (radio detection and ranging) was invented to use radio waves to see objects (range, angle, and velocity) through air. Soon after astronomer William Herschel discovered nonvisible infrared radiation in 1800, detectors and imagers were invented, and eventually humans were provided the capacity to see with little or no (visible) light. An analogous sensing modality would be valuable to detect human perceptions, attitudes, beliefs, and behaviors as well as to geolocate and track these to support smart engagement with multicultural populations in support of defense, diplomacy, and development (Costa and Boiney 2012).

Analogous to classical radar or sonar and drawn from this legacy, social radar would ideally provide persistent, real-time, anonymous, global access to multilingual, multicultural, multimodal information about geolocated perceptions, beliefs, attitudes, and behavior in a manner that is secure and privacy preserving. Taking this analogy one step further, HSCB brought to operational utility the Defense Advanced Research Projects Agency-created worldwide integrated crisis early warning system for the US Southern Command, US Pacific Command, and others, which could monitor and forecast, in near real-time, destabilizing events in a commander's area of responsibility and identify key drivers to instability. These early systems effort to provide actionable links between social sensing and engagement should enable users to

- Understand: Capabilities grounded in social and behavioral sciences to support the perception and the comprehension of sociocultural features and dynamics in an operational environment
- Detect: Capabilities to discover, distinguish, and locate operationally relevant sociocultural signatures through collection, processing, and analysis of sociocultural behavior data
- Forecast: Capabilities for tracking and forecasting change in entities and phenomena of interest along multiple dimensions (time, space, social networks, types of behavior) through persistent sensing and modeling of the environment
- Mitigate: Capabilities to develop, order/prioritize, execute, and measure courses of action grounded in the social and behavioral sciences that are intended to influence entities and phenomena of interest

Of course social sensing is of value well beyond defense applications. For example, one of the themes explored in this collection is how social media data can be exploited to understand how different cultures make decisions. For example, Massachusetts Institute of Technology (MIT) Media Lab's Sandy Pentland, author of *Honest Signals*, collected and analyzed data from mobile sensors to model human networks and predict human behavior using features such as proximity, friends, and colocation. Pentland was interested in understanding how social exposure predicts behavior. Studying 65 young families over a year, he found that he could predict with 45% accuracy the applications individuals would download based on the behaviors of those in their networks. A subsequent electricity incentive study in Switzerland (Mani, Rahwan, and Pentland 2013) found that social network incentives (of local friends) were four times as efficient as standard incentives. Wang, Abdelzaher, and Kaplan (2015) provide a recent overview of social sensing.

Social sensing can be useful not only for advertising but also for enhancing health. For example, recognizing that 68% of Americans were overweight and costing over $150 billion a year, $62 billion in Medicare cost alone, demonstrates the need for intervention programs to reduce future healthcare costs. In their research on persistent health assessment tools (PHAT), researchers Meredith Keybl and John Henderson (2015) created a model to provide timely predictions of obesity rates at the state level from tweets. PHAT provides tailored obesity demographic and regional groups for public health policy makers (MITRE 2014). By providing finer grained and faster (city, monthly) reporting and decisions on the success of obesity prevention programs beyond traditional surveillance, PHAT enables program managers to allocate resources more efficiently. An interactive dashboard enables public health policy makers to (1) browse, cluster, and search tweets and (2) visualize behavioral risk factor surveillance system data to finer timeline scales.

More broadly, we might seek to apply social radar to not only improve the well-being of particular individuals or sectors such as energy and healthcare, but also change entire societies. For example, while we commonly focus on enhancing productivity and gross national product, an equally important measure is happiness. If the pursuit of happiness is an end state, the gross national happiness or the gross global happiness is perhaps as important to measure (Dodds et al. 2011). Some researchers have noted that happiness is not necessarily causally related to financial success. However, we can begin to measure happiness on a large scale (with small geospatial and temporal intervals). Consider the hedonometer (www.hedonometer.org), which senses local and global happiness. The system performs both offline and real-time analysis of emotional assessments. Researchers initially used a mechanical turk to capture five million human scores of the degree of positive valence of 100,000 words in 10 languages across 24 corpora including books, web, news, social media, movies, and even songs (Dodds et al. 2015). Words like *laughter, happiness*, and *love* appear at the top. Spanish corpora turn out to be the most positive. Coupled with Google Earth visualizations, this enables tracking happiness globally as well as temporally. The hedonometer computational back end processes and parses any digitized text (e.g., Twitter, the *New York Times*) and localization data. Its back end can run on cloud computing solutions (e.g., Amazon Web Services) to compute daily and historical hedonometric values for a variety of languages and corpora, to include the analysis of real-time feeds.

These examples in energy, health, and happiness illustrate the broad importance of sociocultural sensing. The chapters in this book provide a broad set of experiences and perspectives, providing important insights for the future of human interaction, both positive and negative, and some level of understanding into how to shape more opportunity-based interactions for more resilient and prosperous societies.

Why Is This Book Important?

This collection of chapters in an increasingly important area contributes to the scientific foundation for better understanding cross-cultural behavior. It is carefully organized to increase the reader's awareness of theories, data and analysis that support or refine those theories and associated models, creation and validation of sociocultural models, and application of these models to conflict and instability. Culture-based behavior models promise more resilient insights into societies across the globe. Those insights promise to help humans better anticipate conflict and cooperation. This situational and predictive intelligence can foster wiser decisions and increased collective progress.

What Core Questions Are Explored in the Book?

Representing a multiplicity of perspectives, this book provides a unique collection of insights into some fundamental questions including the following:

- How sensitive and comprehensive are current cross-cultural theories of human behavior? Where are the important gaps?

- What are the most important entities, attitudes, social relations, and behaviors to capture in cross-cultural models?
- What are the best methods and tools to capture and represent individual and cross-cultural models?
- What types and sources of data would help create or validate cross-cultural models?
- What classes of global societal needs are these theories and models effective for (e.g., economic forecasting, stability and conflict prediction, global policy, global policing, disease management, urban planning, sociocultural econometrics, critical infrastructure forecasting, attitude awareness, social policy development)?
- How will modeling and simulation for multiscale sociocultural analysis change given the rapid growth of the Internet of Things, mobile platforms, cognitive/emotional computing, and other emerging advanced technologies?
- How can we collect data, potentially from existing platforms (e.g., financial, medical, telecommunications, critical infrastructure, social media) to provide deeper and broader insight into cross-cultural human behavior?
- How do we counter denial and deception to get genuine understanding of security and stability in nations?
- How do we overcome our individual, social, cultural, and national biases to gain clearer understanding of other nations and their interests and intents?
- How confident are we in the validity and the applicability of cross-cultural models to various environments?

Conclusion

We are blessed to live in an age of abundance, yet we are faced with challenges, some persistent—poverty, ignorance, and injustice—as well as some contemporary ones such as earnings inequality gaps, lure of the sound bite, environmental instability, and resistance, either reasoned or violent, to many components of globalization. The promise of cross-cultural understanding in an era of big data is the possibility for personalized matchmaking in work, life, and love; detection and mitigation of deception in global communications; reduction and offsetting of biases ranging from personal to national; emotionally sensitive interfaces; tailored and culturally sensitive group facilitation and dispute arbitration; and more lasting and stable global governance. As I write this on the dawn of a new year, I hope the reader will find that this collection provides fresh insights into bringing to life the promise of cross-cultural understanding.

Mark T. Maybury
Vice President
Chief Technology Officer and Chief Security Officer
The MITRE Corporation
Bedford, Massachusetts

References

Costa, B., and Boiney, J. 2012. Social radar. https://www.mitre.org/sites/default/files/pdf/12_0581.pdf.

Dodds, P. S. et al. 2011. Temporal patterns of happiness and information in a global social network: Hedonometrics and Twitter. *PloS one* 6.12: e26752. http://journals.plos.org/plosone/article?id=10.1371/journal.pone.0026752.

Dodds, P. S. et al. 2015. Human language reveals a universal positivity bias. Proceedings of the National Academy of Science vol. 112 no. 8., 2389–2394. http://www.pnas.org/content/112/8/2389.abstract.

Irving, C. 2015, August 21. Interdependence day: Contending with a new global order. *RAND Review*. http://www.rand.org/blog/rand-review/2015/08/interdependence-day-contending-with-a-new-global-order.html.

Keybl, M., Henderson, J., Zarrella, G., Gibson, J., and Kluchnik, M. 2015. Supplementing obesity-related surveillance with persistent health assessment tools. *Online Journal of Public Health Informatics* 7(1): e86. http://www.ncbi.nlm.nih.gov/pmc/articles/PMC4512335/.

Mani, A., Rahwan, I., and Pentland, S. 2013. Including Peer Pressure to Promote Cooperation. *Scientific Reports* 3: 1735. doi:10.1038/srep01735.

MITRE. 2014, December. Does this tweet make me look PHAT? Tracking obesity trends with social media. http://www.mitre.org/publications/project-stories/does-this-tweet-make-me-look-phat-tracking-obesity-trends-with-social.

Puşcaş, V. 2010, February. Management of post-crisis global interdependencies. *Proceedings of the International Economics Congress on an Interdisciplinary Analysis of the Roles of Global Politics and Civil Society in International Economics, Berlin*. http://www.culturaldiplomacy.org/academy/content/articles/biec/speakers/speakers-pages/files/Managing_the_Post_Crisis_Global_Economic_Interdependence_-_Dr._Puscas.pdf.

Wang, D., Abdelzaher, T., and Kaplan, L. 2015. *Social Sensing: Building Reliable Systems on Unreliable Data*. New York: Elsevier.

Acknowledgments

Special thanks to Kristin Heckman, Lisa Costa, Frank Stech, Barry Costa, Jill Drury, Brian Tivnan, Kerry Buckley, and Mike Cenkl for their feedback on this Foreword.

Editors

Joseph V. Cohn, PhD, earned his BS in biology from the University of Illinois–Urbana Champaign in 1993. He earned his PhD in neuroscience from Brandeis University in 1998. He has coauthored more than 100 publications and presented talks to national and international professional conferences on a host of topics, ranging from foundational neuroscience to the nature of technical innovation. He has coedited a three-volume book series focusing on all aspects of training system development and a single-volume book on enhancing human performance in high-risk environments. He has received industry and professional society awards for his research across the spectrum of the human performance and biomedical sciences. He cochaired the Applied Human Factors and Ergonomics Association's Cross-Cultural Decision Making Conference from 2013 to 2015 and is a fellow of the American Psychological Association and the Society of Military Psychologists, as well as an associate fellow of the Aerospace Medical Association.

Sae Schatz, PhD, is an applied human-systems researcher, learning science professional, and cognitive scientist. She has headed an array of applied research efforts, authored more than 50 peer-reviewed scholarly publications, led the development of three military textbooks, and received industry awards for both her publications and her research efforts. Schatz cochairs the Applied Human Factors and Ergonomics Association's Cross-Cultural Decision Making Conference (2013–), is a service principal for the Interservice/Industry Training, Simulation and Education Conference (2015–), and supports international science and technology working groups associated with the North Atlantic Treaty Organization, Partnership for Peace, and The Technical Cooperation Program. Schatz also maintains close ties with her alma mater, the University of Central Florida, where she earned her PhD in human systems in modeling and simulation in 2008.

Hannah Freeman earned her Bachelor of Arts degrees in international studies (Russian and Eastern European studies) and Hispanic studies from Illinois Wesleyan University, where she was awarded Phi Beta Kappa, in 2012. Freeman earned her Master of Science degree in comparative politics (conflict studies) from The London School of Economics and Political Science in 2013, where she also received a Russian Language Certificate. Her research interests include the former Soviet Union and Soviet bloc, post-Soviet Russia, human rights, national and ethnic conflict, radicalization, political violence, and terrorism.

David J. Y. Combs, PhD, earned his BA in psychology from Simpson University in 2003. He earned his PhD in experimental social psychology from the University of Kentucky in 2010. He has also completed certifications in political psychology (Stanford University), analysis of incomplete data sets (University of Michigan), and Afghanistan–Pakistan regional expertise. He has completed additional coursework (applied survey sampling) with The George Washington University. He has coauthored dozens of papers, conference presentations, and book chapters on social psychological topics such as trust, attitude change, experience of humiliation, and emotions resulting from political events. He is especially interested in applying social psychological theory and methods to understanding cross-cultural interactions within the irregular warfare context.

Editors

Contributors

Lewis A. Anderson
Special Operations Command Korea (SOCKOR)
Seoul, South Korea

Jeffrey Appleget
Naval Postgraduate School
Monterey, California

George A. Backus
Sandia National Laboratories
Albuquerque, New Mexico

Ryan G. Baird
Joint Warfare Analysis Center
King George, Virginia

Jonathan Z. Bakdash
US Army Research Laboratory
Aberdeen, Maryland

Perakath Benjamin
Knowledge Based Systems, Inc.
College Station, Texas

Michael L. Bernard
Sandia National Laboratories
Albuquerque, New Mexico

Gnana K. Bharathy
University of Pennsylvania
Philadelphia, Pennsylvania

Curtis L. Blais
Naval Postgraduate School
Monterey, California

Sarai Blincoe
Longwood University
Farmville, Virginia

Alper Caglayan
Milcord LLC
Waltham, Massachusetts

Erman Çakıt
Department of Industrial Engineering
Aksaray University
Aksaray, Turkey

Mark A. Carter
Intelligent Software Solutions, Inc.
Independence, Missouri

Laura Cassani
Milcord LLC
Waltham, Massachusetts

Randy Casstevens
OpenWare
Herndon, Virginia

Vladimíra Čavojová
Centre for Social and Psychological Sciences
Institute of Experimental Psychology
Slovak Academy of Sciences
Bratislava, Slovakia

John Chamberlain
Charles River Analytics
Cambridge, Massachusetts

David J. Y. Combs
Washington, D.C.

Regan W. Damron
Booz Allen Hamilton GmbH
Stuttgart, Germany

Casey Doyle
Network Science Technology Center
Rensselaer Polytechnic Institute
Troy, New York

Brian M. Fairlie
Washington, D.C.

Laurie Fenstermacher
US Air Force Research Laboratory
Wright-Patterson, Ohio

Ronald D. Fricker, Jr.
Statistics Department
Virginia Polytechnic Institute and State University
Blacksburg, Virginia

Christopher P. Garris
Metropolitan State University of Denver
Denver, Colorado

Clarissa Graffeo
College of Education and Human Performance
University of Central Florida
Orlando, Florida

Steven B. Hall
Naval Postgraduate School
Monterey, California

Róbert Hanák
Centre for Social and Psychological Sciences
Institute of Experimental Psychology
Slovak Academy of Sciences
Bratislava, Slovakia

Rachel G. Hingst
Lockheed Martin Advanced Technology Laboratories
Arlington, Virginia

Mark A. Hoffman
Lockheed Martin Advanced Technology Laboratories
Kennesaw, Georgia

Robert F. Jeffers
Sandia National Laboratories
Albuquerque, New Mexico

Christina S. Kang
California

Panagiotis D. Karampourniotis
Network Science Technology Center
Rensselaer Polytechnic Institute
Troy, New York

Waldemar Karwowski
Department of Industrial Engineering and Management Systems
University of Central Florida
Orlando, Florida

Brian P. Kettler
Lockheed Martin Advanced Technology Laboratories
Arlington, Virginia

Gyorgy Korniss
Network Science Technology Center
Rensselaer Polytechnic Institute
Troy, New York

Walter W. Kulzy
United States Central Command
Model and Simulation
Tampa, Florida

Lawrence A. Kuznar
Indiana University–Purdue University at Fort Wayne
Fort Wayne, Indiana

Omar Lizardo
Department of Sociology
University of Norte Dame
Notre Dame, Indiana

Corey Lofdahl
Charles River Analytics
Cambridge, Massachusetts

Joseph Maddux
Operations Analysis Directorate
United States Marine Corps
Quantico, Virginia

Atsuo Murata
Graduate School of Natural Science and Technology
Okayama University
Okayama, Japan

Asmeret Bier Naugle
Sandia National Laboratories
Albuquerque, New Mexico

Walter L. Perry
RAND Corporation
Arlington, Virginia

Matthew J. Powers
Operations Research Analyst
The Joint Center for International Security Force Assistance
Fort Leavenworth, Kansas

Satheesh Ramachandran
Knowledge Based Systems, Inc.
College Station, Texas

Scott Neil Reilly
Charles River Analytics
Cambridge, Massachusetts

Scott W. Ruston
Center for Strategic Communication
Arizona State University
Tempe, Arizona

Tracy Saint Benoit
College of Education and Human Performance
University of Central Florida
Orlando, Florida

Barry G. Silverman
University of Pennsylvania
Philadelphia, Pennsylvania

Pramesh Singh
Northwestern Institute on Complex Systems
Northwestern University
Evanston, Illinois

Amy Sliva
Charles River Analytics
Cambridge, Massachusetts

David Q. Sun
University of Pennsylvania
Philadelphia, Pennsylvania

Col. Richard Swisher (Ret)
College of Education and Human Performance
University of Central Florida
Orlando, Florida

Boleslaw K. Szymanski
Network Science Technology Center
Rensselaer Polytechnic Institute
Troy, New York

Kalyan Vadakkeveedu
Knowledge Based Systems, Inc.
College Station, Texas

Eric S. Vorm
Indiana University
Bloomington, Indiana

Nathan Weyer
University of Pennsylvania
Philadelphia, Pennsylvania

Introduction

It is impossible, in principle, to explain any pattern by invoking a single quantity.

Gregory Bateson (1979, p. 58)

Humans, by their very nature, are sense makers (Klein, Moon, and Hoffman 2006). We are primed to receive vast amounts of data, through our sensory systems, and build narratives to make sense of that information, with the ultimate goal of using those narratives to guide decision making and using those decisions to take (or, to not take) specific actions (Leedom 2001). Oftentimes, the resultant stories are at odds with reality. In the best of cases, these mismatches simply result in socially awkward actions. In the worst of cases, these mismatches lead to tragedy.

A critical component of a successful social interaction is the ability of individuals to build these narratives in order to understand and predict each other's behaviors across a range of conditions, situations, and cues. This predictive ability requires that the individuals who are interacting share a common history of experiences, often known as transactive memory, to establish a shared context (Liang, Moreland, and Argote 1995; Moreland, Argote, and Krishnan 1996). In humans, the brain has developed in such a way that this understanding naturally arises through repeated training, shared experiences, and other types of social interaction (Dunbar 1998; Moreland and Myaskovsky 2000; Adolphs 2009). This ability is what allows humans to predict, understand, and modify their social interactions in real time (Liang, Moreland, and Argote 1995).

An increasingly important line of investigation focuses not on how humans establish social interactions with one another but rather on how social interactions influence how humans perceive and analyze information, build their narratives, and make decisions to take action based on these narratives (Dehghani et al. 2009). While the neural elements of our decision making processes are determined through our genetic makeup, environmental factors play a key role in how these structures develop and respond (Glimcher et al. 2008). The sociocultural environment, which includes group and individual behaviors such as customs, legal, social, and ethical standards, religious beliefs, and traditions, is a core aspect of the decision making environment. The foundational question to be addressed is to what extent does the sociocultural environment in which an individual develops influence their decision making processes?

In our increasingly globally interconnected world, understanding and appreciating the sociocultural context within which individuals make their decisions is critical to developing successful partnerships. In the past, overcoming cross-cultural barriers required significant investments in developing a shared sociocultural understanding. This process, in turn, forced individuals to adopt, to some degree, the other culture's perspective, leading to a deeper appreciation of the context within which decisions were made. Advances in information and social media technologies have forced a dissolution of these cultural barriers, which have notably altered the way in which shared sociocultural understanding naturally develops.

The consequences of this technology-enabled shift are profound. Sociocultural changes, which may once have occurred at a local level and remained relatively isolated, can now be shared instantaneously and morph organically into a movement that may have deep consequences. As one striking example, the short-lived Arab Spring is often attributed to a single action: the self-immolation of a Tunisian street vendor, which, through social media, turned into a movement that led to the unseating of four reigning leaders across the Arab world, not to mention protests and civil uprisings across more than a dozen countries.

Modeling Sociocultural Influences on Decision Making: Understanding Conflict, Enabling Stability explores how advances in information and social media technologies, combined with social science and modeling and simulation tools, can be leveraged to better understand sociocultural decision making processes, providing an alternative means for gaining a shared understanding in today's high-tech and globally interconnected world. In doing so, the book explores how culture contributes to decision making processes; how these cultural influences may be characterized and modeled, how these models may be harnessed into tools and technologies to facilitate group decision making across cultural contexts; and what challenges remain in further breaking down the barriers to understanding decision making across cultures.

The consequences of understanding how the sociocultural environment influences decision making are profound. From an economic perspective, sociocultural factors play a crucial role in influencing consumers' economic decisions (e.g., the study by Maheswaran and Shavitt [2000]). Marketing approaches and techniques that are positively perceived in one culture may be negatively perceived by another culture. As cultures become more interconnected and reliant on one another to maintain effective commercial relationships, it is imperative to understand how each culture's unique characteristics influence their members' decision making processes. From a global security perspective, the changing nature of the political landscape, which includes constrained military—operations in which actions are bounded by factors such as political and collateral damage considerations—and increasing interactions between state/nonstate actors, government and nongovernment organizations, as well as more direct interactions with indigenous populations, raises unique challenges. Determining, in advance of any action, the potential impact and the consequences that such actions may have on a population—from a cultural perspective—is critical to ensuring that the appropriate courses of actions are implemented and security is achieved. These perspectives and others are addressed throughout this volume.

Modeling Sociocultural Influences on Decision Making: Understanding Conflict, Enabling Stability is divided into four sections, with each successive section building on the previous ones. The sections consist of a series of invited chapters from leading experts from academia, industry, and government. These four sections are as follows:

- Section I: "Building Theories"—This section highlights connections between cross-cultural research and culture-based behavior models.
- Section II: "Collecting and Analyzing Data"—This section discusses some of the challenges, and unique solutions, associated with cross-cultural data collection and its effective analysis.
- Section III: "Building and Validating Sociocultural Models"—This section reviews ways to move from simple data analysis into validated models for use in forecasting and decision making.
- Section IV: "Applying Sociocultural Models to Gain Insight into Conflict and Instability"—This final section provides examples and use cases that underscore the approaches discussed in earlier sections.

Embedded within these four sections are the following features:

- Insights into cross-cultural decision making based on recent international events, with grounding in an historical context;
- Discussions of cutting-edge modeling techniques used today by professionals across multiple organizations and agencies;
- Descriptions of specific cross-cultural decision making tools designed for use by laypeople and professionals; and
- Case studies on the role of cross-cultural decision making, grounded in current events and (in many cases) military applications.

This volume introduces three unique devices to enhance the reader's experience and to assist the reader in gaining a deeper insight into this unique research area. First, contributing authors applied their key findings and conclusions to a use case focusing on a fictional megacity. The use case, provided at the end of this introduction, provides a common thread to bind together the chapters comprising this volume. The use case also serves as a focal point to help readers easily conceptualize the impact that the described theories, data, and tools may have on a realistic challenge. Second, each chapter includes a series of callout boxes that highlight key concepts. Lastly, each chapter concludes with a series of discussion questions, intended to stimulate further reflection and prompt conversation on the ideas presented throughout that chapter.

Recent events highlight the degree to which cross-cultural understanding can facilitate and sustain stability within a geopolitically defined region—but they also underscore the degree to which the absence of this understanding can lead to instability and conflict. It is our fervent hope that this book will lead to a deeper appreciation of the basis for cross-cultural differences in decision making and help laypeople, students, and practitioners facilitate more effective cross-cultural exchanges.

Joseph V. Cohn, PhD
Sae Schatz, PhD
Hannah Freeman, MSc
David J. Y. Combs, PhD

References

Adolphs, R. (2009). The social brain: Neural basis of social knowledge. *Annual Review of Psychology* 60: 693.

Bateson, G. (1979). *Mind and Nature: A Necessary Unity (Advances in Systems Theory, Complexity, and the Human Sciences)*. New York: Hampton Press.

Dehghani, M., Sachdeva, S., Ekhtiari, H., Gentner, D., and Forbus, K. (2009). The Role of Cultural Narratives in Moral Decision Making. *Proceedings of the 31st Annual Conference of the Cognitive Science Society (CogSci), Washington, D.C.* Retrieved 10 May 2016 from http://www.qrg.northwestern.edu/papers/files/qrg_dist_files/qrg_2009/narratives-cogsci09-md-19.pdf.

Dunbar, R. I. (1998). The social brain hypothesis. *Brain*. 9: 10.

Glimcher, P. W., Camerer, C., Poldrack, R. A., and Fehr, E. (2008). *Neuroeconomics: Decision Making and the Brain*. Cambridge: Academic Press.

Klein, G., Moon, B., and Hoffman, R. F. (2006). Making sense of sensemaking I: Alternative perspectives. *IEEE Intelligent Systems* 21(4): 70–73.

Leedom, D. K. (2001). Final report: Sensemaking symposium. Technical Report prepared under contract for Office of Assistant Secretary of Defense for Command, Control, Communications and Intelligence. Vienna, VA: Evidence Based Research. Inc. Retrieved January 1, 2016 from http://www.dodccrp.org/files/sensemaking_final_report.pdf.

Liang, D. W., Moreland, R., and Argote, L. (1995). Group versus individual training and group performance: The mediating role of transactive memory. *Personality and Social Psychology Bulletin* 21(4): 384–393.

Maheswaran, D., and Shavitt, S. (2000). Issues and new directions in global consumer psychology. *Journal of Consumer Psychology* 9(2): 59–66.

Moreland, R. L., and Myaskovsky, L. (2000). Exploring the performance benefits of group training: Transactive memory or improved communication? *Organizational Behavior and Human Decision Processes* 82:117–133.

Moreland, R. L., Argote, L., and Krishnan, R. (1996). Socially shared cognition at work: Transactive memory and group performance. In *What's Social About Social Cognition? Research on Socially Shared Cognition in Small Groups.* Nye, J. L., and Brower, A. M., editors. Thousand Oaks, CA: Sage, 57–84.

Fictional Megacity Use Case

A recent United Nations (UN) report suggests that urban populations will continue to grow at a staggering rate, leading to greater urbanization, larger concentrations of people in smaller areas, and an increasing number of high-population density zones or megacities around the world. By definition, megacities are those urban settlements, such as Tokyo and Delhi, with more than 10 million inhabitants. The UN's *World Urbanization Prospects* estimates that 28 megacities already existed as of 2014, and by 2030, that number will increase by nearly 50% (UN 2014). This urbanization has direct implications for social, financial, environmental, health, and security sectors.

Some worry that the swelling urban populations will lead to increased social unrest, greater disparity between the haves and the have nots, worsening environmental conditions, and serious security challenges (e.g., the studies by Harris et al. [2014] and Kohlhase [2013]). Others offer more positive forecasts. UN Secretary General Ban Ki-moon, for example, acknowledged the potential inherent in megacities in his introduction to the *2012/2013 State of the World's Cities Report*: "…cities can generate and equitably distribute the benefits and opportunities associated with prosperity, ensuring economic well-being, social cohesion, environmental sustainability and a better quality of life in general" (Habitat 2013). Enabling this positive future development of megacities, however, will require innovation in governance, global relations, and our collective understanding of societal level actions.

Because of their growing importance—and footprint on the world—we chose megacities as the focus for the volume's collective use case. The fictional narrative below serves as a common sandbox for exploring ideas in this book. The concepts in the use case are based upon real phenomena, but the locations and the cultures described are fictional. We believe that grounding the theories and the techniques from each chapter into this narrative will help inform practical understanding of population behaviors, clarify the described model and forecasting methods, and encourage reflection on megacities, in general.

Megacity Xanadu

City: Megacity Xanadu
Country: Çawaila
Time: 25 years in the future

Several decades ago, refugees came from Bursuka, fleeing the failed civil war designed to overthrow the ruling despot. At first, the megacity's state government welcomed the refugees, and then later, it turned a blind eye to the thousands of migrants who snuck into the refugee camps despite the determined quota. The borders of megacity Xanadu teemed with migrant workers.

Next came the migrants from North Razinia. When that country's 80-year monarchy abruptly fell, an avalanche of migrants poured across the borders. Megacity Xanadu, unprepared for the onslaught, was ill-equipped to prevent the migration. Many Razinis died in the harsh journey across an unforgiving landscape, and even more were deported by the state government when they arrived. Still, thousands of Razinis successfully evaded nature and the border police to take up residence in the aging refugee camps or cramped warrens that passed as domiciles in the megacity.

Today, megacity Xanadu seethes with underlying chords of racial and classist tension. The North Razinis, newly released into the world beyond their closed societal borders, struggle to mesh with the society. The Bursuks fare little better, and unemployment among them hovers around 40%.

A few of the most fortunate refugees work long hours in factories, making marginal pay. The immigrants compete against local workers (who enjoy clear preference) as well as the other citizens of the local country who travel from the countryside for work in the megacity. The rural workers are treated as secondary citizens but still better than the immigrants.

A thriving black market now seethes throughout the city, connected by various centers of power in each of the city wards. In the northern wards, where the North Razinis have settled, organized criminal groups wasted little time finding ways to exploit the desperate and naive refugees. Sweatshops, illicit drug makers, and human trafficking have been reported. The international community has yet to make serious comment, and the state government has few resources to spare. So long as the northern wards remain stable, they overlook most of the mafia activities.

In the moldering refugee camps to the west, the Bursuka families fare a little better. Their religious clerics form the centers of gravity. The religious sect, only tolerated by the secular local culture, has steadily gained popularity among the downtrodden expatriates. After years of experience as third-class citizens, the Bursukan culture now functions as a society within a society.

Megacity Xanadu sits at the precipice, teetering between unrest and progress. If the tensions in the megacity ignite, the state government would be powerless to quell the resulting violence. Similarly, if health and environmental concerns remain unaddressed, widespread disease and contamination could truncate many lives and strain the country's already saturated healthcare system. However, if the megacity's leaders can better understand the city's growing population and devise appropriate governance approaches, they could turn the population's scale and diversity into an asset—fostering the city's rich social landscape, creating a competitive edge in the global marketplace, and enhancing the quality of life for all of its citizens. The outcome depends, in part, upon understanding the sociocultural context, analyzing how to best maintain stability, and deciding upon those actions that will ultimately encourage prosperity.

References

Habitat. 2013. U. N. State of the World's Cities 2012/2013: Prosperity of Cities. Abingdon: Routledge.

Harris, M., Dixon, R., Melin, N., Hendrex, D., Russo, R., and Bailey, M. 2014, June. *Megacities and the United States Army: Preparing for a Complex and Uncertain Future.* Chief of Staff of the Army, Strategic Studies Group.

Kohlhase, J. E. 2013. The New Urban World 2050: Perspectives, Prospects and Problems. *Regional Science Policy & Practice*, 5(2), 153–165.

United Nations (UN). 2014. World urbanization prospects: The 2014 revision, highlights (ST/ESA/SER.A/352). Geneva: Department of Economic and Social Affairs, Population Division, United Nations.

Section I

Building Theories

Building and expanding theory in the social sciences is a notoriously difficult task. But, difficult or not, the authors in this section have taken on the challenge in exciting ways. This section is not designed to provide researchers with the practical steps to developing and building a theory (e.g., conducting a literature review, developing path diagrams); rather, this section provides theorists and researchers with important theoretical matters to consider when building theories. No single volume, much less one section of a volume, can provide a comprehensive guide to the multitude of issues that researchers should consider when developing new theories in the sociocultural domain. However, while this section is not and cannot be comprehensive, it does have something for everyone and takes special consideration of the megacity theme woven throughout this volume.

In order to orient readers, this introduction to the Section I will sketch out an overarching theory of culture (i.e., Markus and Conner's culture cycle) and will then briefly describe how the individual chapters found within this section fit within that broader framework.

The Culture Cycle

There are dozens (perhaps hundreds) of ways to define, think about, and organize research regarding culture. Likewise, there are a number of ways to think about and organize the diverse set of concepts tackled in this section. One helpful way (although there are admittedly others) to think about culture, and the topics written about in this section, is to consider culture within the context of Markus and Conner's (2013) culture cycle (see Chapter 9, this volume, for a description and a citation). This helpful framework suggests that there are four interlocking and interrelated levels of culture—the *individual level* (culture at the level of the individual person), the *interaction level* (the person as he or she interacts with the world around them), the *institution level* (intuitions that influence a person and are influenced by people), and the *cultural conceptual idea* level (big broad cultural ideas and ideals that are often unseen but bind a culture together in the same way that things like gravity bind together the physical world). Markus and Conner suggest that truly understanding culture means to examine culture at all these levels (although, clearly, whole careers can be made by examining small elements of each level). In this section, the authors each contribute theoretical and research perspectives that can be utilized by other researchers at each level of culture (see the following for more details). Essentially, as I read and organized this section, I could not help but notice that the authors, quite on their own, organized their thinking and writing within this helpful framework.

Section Contents

Individual Level Perspectives

As noted, the individual level of culture encompasses the individual person. Within this section of the volume, Vladimira Čavojová and Hanak discuss research and perspective at this level of the culture cycle. In particular, they lay out a series of research findings regarding how culture does or does not have an impact on the ability of individuals to suppress reflexive thinking in favor of more critical and effortful thinking. As they describe, multiple theories in the social sciences demonstrate that people commonly rely on these two modes of thinking (or a hybrid of the two) for decision making. Their analysis, using a host of non-American and American samples, demonstrates that there is little difference across cultures in the ability to override reflexive automatic thinking in favor of effortful thinking when needed. In fact, her findings indicate that, to an extent, regardless of culture or national origin, most people do not appear to actively override automatic thinking in favor of more effortful thinking, even when it might appear clear that such an override is warranted. Troubling findings indeed. Designing theories about individuals will often incorporate perspectives on how people process information. When future research endeavors to build new theories in the social cultural domain (especially at the individual level of analysis), the findings, and implications of those findings, from this chapter will be a critical pillar in those efforts.

Interaction Level Perspectives

Markus and Conner's interaction level of culture, as would be anticipated, is the level of culture that people experience the most. Our interactions, how we think, feel, and act in association with people and objects, are guided by implicit and explicit cultural norms or expectations. Two chapters from this section of the volume represent this level of culture. Specifically, Brian Fairlie examines a relatively new domain of the interaction level (and cultural modeling broadly)—social media. While there is much recent scholarly activity in this domain, Fairlie provides theorists and practitioners with guidance on building a theory, and perhaps even more importantly, he takes readers through the critical steps in testing the social media theory practically. His chapter helpfully blends current perspectives in social media research with practical examples of how information spreads through social media. Any new social media analyst should have this chapter printed and ready next to their computer.

In addition, Bolek Szymanski et al. present quantifiable perspectives on how public opinion spreads through interactions within societies, and they examine the role of sociocultural factors in this process. They build on two prominent theoretical frameworks of opinion dynamics (i.e., the naming game and the threshold model) and demonstrate the effects of factors such as race, gender, and class on opinion formation, change, and relationship generation. This fascinating chapter describes theory and research that not only has applicability to the spread of information in societies, but also has major implications for understanding thresholds associated with societal tipping points. Their empirical work relies on the interesting data source of incoming freshman college students (as a proxy for the general public—and even more interesting, as a sort of proxy for incoming refugees into a nation). Their perspectives provide not only an innovative way to understand public opinion, but also an interesting case example of how his perspectives could be applied in the Xanadu megacity example.

Institution Level Perspectives

The next level of culture is the institution level. Every interaction (described earlier) occurs within the institution level of culture. Within this section, Atsuro Murata describes the theoretical issues in play within institutions—and the cross-cultural interactions between people within institutions (such as industrial organizations and airlines). He makes a compelling case that cultural factors and nuances probably play a major role in industrial accidents and aviation mishaps. His practical thinking in this chapter helps readers understand how culture can have a pervasive and often unseen impact on judgment and decision making within institutions. He places a special emphasis on the need for professionals in the safety domain to generate new models and theories of safety that incorporate cultural factors as a primary driver of accidents and mishaps. He paints a frightening picture using case studies of tragic accidents that could have been likely averted had models and theories of accidents simply accounted for cultural misunderstandings and nuance.

Cultural Conceptual Ideas Perspectives

The final layer of Markus and Conner's culture cycle is the conceptual ideas level. They suggest that these conceptual ideas, like gravity or quantum mechanics in the physical sciences, are a sort of unseen force that binds cultures together in recognizable ways.

Scott Ruston presents perspectives that fit within this level of culture. He describes overarching cultural narratives and how narratives permeate societies, influence how people think, respond to threats, and interpret the world around them. In many ways, new social cultural theories will be incomplete without a strong grounding and understanding of the social cultural narratives running through a society. As both theorists and practitioners attempt to generate practical theory to help model the behavior of individuals, interactions, institutions, nations, regions, etc., they must infuse their theoretical perspectives with an understanding of the narratives that bind people together.

Overarching Perspectives

Brian Kettler et al. set the stage for this volume overall. Their chapter interweaves the elements of each level of the culture cycle—and how each level is critical for modeling within the context of megacities. They provide readers with a remarkably comprehensive guide to developing theory and challenges inherent to understanding megacities. They discuss the overall challenges from both a theoretical and a practical perspective. They explain that perhaps the greatest challenge in developing theory and research for understanding megacities is the simple lack of data (at multiple levels) that researchers and practitioners will need in order to fully understand the complex megacity of the future. Given the burgeoning focus on megacities in multiple domains (academic, government, etc.), any researcher or practitioner who plans to build theory and research on this subject will be remiss if they do not rely heavily on this fascinating chapter.

Moving Forward

As noted at the beginning of this introduction, building theory in the social cultural sciences is a difficult endeavor. Regardless of the level of culture a researcher is working within, this section provides important groundwork for future theory and research. I hope you will enjoy reading these chapters as much as I have.

Reference

Markus, H., and Conner, A. 2013. *Clash! 8 Cultural Conflicts That Make Us Who We Are*. New York: Hudson Street Press.

1

Expeditionary Modeling for Megacities and Other Dense Urban Areas

Brian P. Kettler, Rachel G. Hingst, and Mark A. Hoffman

CONTENTS

Introduction 5
Urban Operation Challenges 7
The Expeditionary Modeling Vision 8
An Expeditionary Modeling Scenario 11
Expeditionary Modeling Requirements: Data 12
Expeditionary Modeling Requirements: Models 18
Modeling Urban Environments 18
Modeling Populations 19
Example: Athena 20
Composing Models 21
Conclusion 22
References 23

ABSTRACT Fueled by globalization and urbanization, megacities are extreme exemplars of complex operational environments by virtue of their size, density, complexity, and dynamism. These likely epicenters of future instability make them probable locations of future military operations by the United States and coalition forces for missions such as humanitarian assistance and counterinsurgency. A new vision, expeditionary modeling, rapidly brings to bear dynamic, heterogeneous ensembles of data, computational models, and subject matter expertise to understand and predict the (in)direct effects of courses of action on the urban and human terrains of dense urban areas. This chapter presents the benefits, the challenges, and supporting multidisciplinary research for this vision.

Introduction

Megacities are extreme exemplars of complex military operational environments in general and dense urban areas in particular.* Megacities are expected to double in number from more than 20 cities today to in excess of 40 cities by 2025 (US Army Strategic Studies

* Megacities are defined as those having more than 10 million inhabitants. An operational environment includes the city or other regions in which the crisis/military operations is taking place.

MEGACITIES TODAY—AND TOMORROW

There are over 20 megacities today with more than 10 million inhabitants and over 800 dense urban areas of more than 500,000 people. By 2030, there will be over 40 megacities with urban areas having over 60% of the world's population and 70% of the world's gross domestic product.

Group [SSG] Megacities Concept Team, 2014). At the extreme end of the spectrum, China plans a megacity, Hing-hin-ji, that will encompass several large cities with populations of more than 130 million people, 25 times the size of London (Benedictus, 2015). By 2030, cities will have over 60% of the world's population and 70% of the world's gross domestic product as masses move from rural areas to cities.

Megacities are increasingly located in developing regions with fragile stability (e.g., Africa and Asia). Thus, they are likely epicenters of instability. Crises in these areas can have global implications, such as displacing large populations or even contributing to failed states. Military operations in these urban areas by the United States and coalition forces can span a wide range of missions including humanitarian assistance, counterinsurgency operations, major combat operations, or some combination of these. In the late 1990s, Marine Corps General Charles Krulak (1999) described a potential *Three Block War* in which a squad of soldiers might be involved in simultaneous combat, peacekeeping, and humanitarian operations within close proximity in a city. A recent variant of this postulates a four-floor war in which the Marines could be evacuating friendly forces off rooftops, detaining targets on one floor, engaging in combat on another floor, and pursuing insurgents through the streets and tunnels below (Sanborn, 2015).

Dense urban areas are a growing area of concern for the US government. They are of particular interest to the US Army, which would play a major role in urban operations abroad. With this in mind, the Army SSG (2014) recently identified a number of shortfalls in the current understanding, doctrine, and planning scenarios for operations in megacities.

The scale of operations in megacities will be unprecedented. For example, Hurricane Katrina in August 2005 displaced upward of 1 million people across multiple US states with the mobilization of over 72,000 military forces, 1.7 million gallons of water, 3.6 million meals, 11.5 million pounds of ice, and 1.3 million gallons of fuel to provide evacuation support, shelter, and emergency medical care for thousands of people (Felix and Wong, 2015). A rough extrapolation of the total populations of the areas in the four states Katrina affected or partially affected is roughly 17 million people, comparable to a large megacity, although far less dense. For comparison, a megacity is, by definition, at least 30 times larger than Fallujah, Iraq, and 1.5 times the size of Baghdad, Iraq (US Army Capabilities Integration Center, 2015).

This chapter posits that bringing together diverse theories and models from social science and urban science can provide the understanding needed to support military operations in dense urban areas. According to military theorist David Kilcullen (2013), "We need to understand how a city's subsystems and subdistricts fit together. . . . It's partly a matter of taking models that already exist in other disciplines, bringing them into the national security field, building on them, plugging in new variables, and looking closely and creatively at the results. . . ." We believe this understanding should be brought to bear rapidly in the form of new computational aids—based on multidisciplinary models of the operational environment and powered by dynamic data—to help planners mitigate evolving crises. We call the integrated application of the enabling tools and processes *expeditionary modeling* (ExMo).

The second section ("Urban Operation Challenges") briefly describes some of the many challenges of urban operations. The third section ("The Expeditionary Modeling Vision") presents the ExMo vision in detail including an illustrative scenario for the fictional megacity Xanadu. The fourth and fifth sections ("Expeditionary Modeling Requirements: Data" and "Expeditionary Modeling Requirements: Models," respectively) look at the data and model requirements, respectively, and the ways they might be addressed.

Urban Operation Challenges

The four megatrends of population growth, urbanization, littoralization, and connectedness are causing the likely environment for future military conflict to shift toward dense urban areas (Kilcullen, 2013). Besides the megacities, there are currently more than 800 cities with populations in excess of 500,000 people. These big cities are increasingly connected to one another and the rest of the world by vast and largely uncharted political, economic, and communications networks and interdependencies that make it hard to predict and isolate the effects of individual events and actions.

The urban operations triad, shown in Figure 1.1, highlights the three urban foci and their challenges, i.e., complex, dense urban terrain; dense, heterogeneous population; and fragile infrastructure (US Department of Defense, 2013). An urban terrain consists of diverse zones and neighborhoods with multiple potential domains of operation including subsurface (subterranean tunnels, basements), surface (streets, parks), elevated areas (buildings, towers), airspace, cyberspace, and, for coastal cities, undersea and surface. Urban populations may be very diverse in terms of age, ethnicity, religion, education, and economic opportunity. Between urban populations and urban terrain is the varied infrastructure

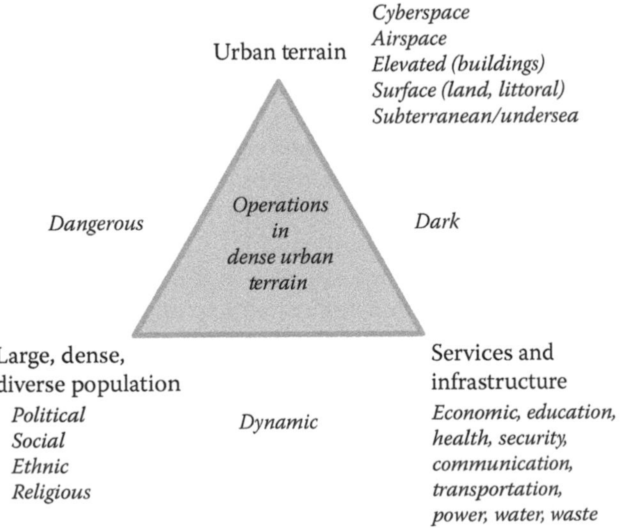

FIGURE 1.1
Urban operations triad.

(e.g., power, water, sanitation, transportation, communications) and the services it enables (e.g., commerce, healthcare, security, education).

Dense urban areas are dynamic, dark, and dangerous. The dynamicity is largely a result of constant population growth and migration and the strains they cause; local and broader political, social, and economic trends; and factors external to the city. The urban terrain itself is constantly changing given new development, disasters, or damage from conflict.

Much of the urban and human terrain is dark, meaning poorly charted. This includes rampant undocumented land ownership (Ellis and Sisco, 2014); poorly documented infrastructure (outdated maps, illegal taps into infrastructure, etc.); activities obscured by or within buildings; informal political, economic, and social networks; and criminal and insurgent activities that may be largely in the shadows. Many complex interconnections and interactions between these networks are also unknown or even deliberately concealed.

Rapidly rising social tensions can make these areas dangerous powder kegs. Potential tensions include increasing gaps between economic classes, mingling of disparate cultures as they are brought into close proximity, growing numbers of youth, and increased strain on already limited resources (food, water, electricity, and other infrastructure/services). Rapidly growing urban areas may exceed their capacities for employment, food, electric power, water, sanitation, security, and other services.

Coalition forces will face high uncertainty about the urban terrain, often unpredictable populations, and limited tactical visibility and maneuverability. They will also face limitations in weapon employment (due to concerns about collateral damage) and degraded communications that hinder command and control. Cities may harbor adversarial actors ranging from criminals to terrorists seeking a safe haven in urban concealment.

Today, understanding the complex dimensions of unique dense urban areas often requires many human experts from different organizations pooling their knowledge and whatever data they have. The automation to support the collection and the application of data, information, and knowledge is often lacking, and the processes are frequently ad hoc. The next section describes some future concepts that address this.

The Expeditionary Modeling Vision

The time available between the decision to intervene in a crisis and the beginning of operations to plan and prepare military operations may be very limited. A 2011 RAND Corporation study showed that 70% of US joint task forces had less than 5 weeks between a warning order and the start of operations (RAND Corporation, 2011). Consequently, during the intelligence preparation of the battlespace (IPB)* process, only a few friendly (blue) courses of action (COAs) can be developed and evaluated against anticipated enemy (red) COAs, those typically deemed mostly likely or most dangerous. Blue COAs may be war-gamed, time permitting. Today, this is typically done by a group of operators and analysts with few automated tools and often lacking detailed knowledge of the specific operational environment. This not only limits the number of COAs evaluated but also introduces human cognitive biases.

Computational models have been shown to mitigate human biases and complement human decision making. For example, statistical models have outperformed human analysts in predicting events (Tetlock, 2006). Such models, trained on data over the longer

* The IPB is also known as the intelligence preparation of the operational environment.

A VISION OF EXPEDITIONARY MODELING

The ExMo approach would rapidly bring together dynamic, heterogeneous ensembles of computational models, dynamic data sources, and deep subject matter expertise to understand the complex interdependencies across the urban and human terrains and predict the outcome of a COA.

term, are not unduly influenced by recent events or human cognitive or cultural biases. These models can complement human analysts, especially when diverse models are brought to bear together, as shown in weather forecasting, polling (Silver, 2012), and geopolitical event forecasting (e.g., the study by O'Brien [2010]).

We envision ExMo as encompassing changing ensembles of diverse models that are powered by dynamic data sources and embedded in decision support tools to aid users in understanding complex operational environments and evaluating COAs for them. As Figure 1.2 shows, decision support aids in future ExMo environments could build on today's geographic information systems (GISs). Today's systems present layers of data on maps, where each layer is stored in a database. For a dense urban area such as a megacity, layers from the bottom to the top would include terrain and climate; services and infrastructures; political, ethnic/religious, civic, commercial, and criminal groups/

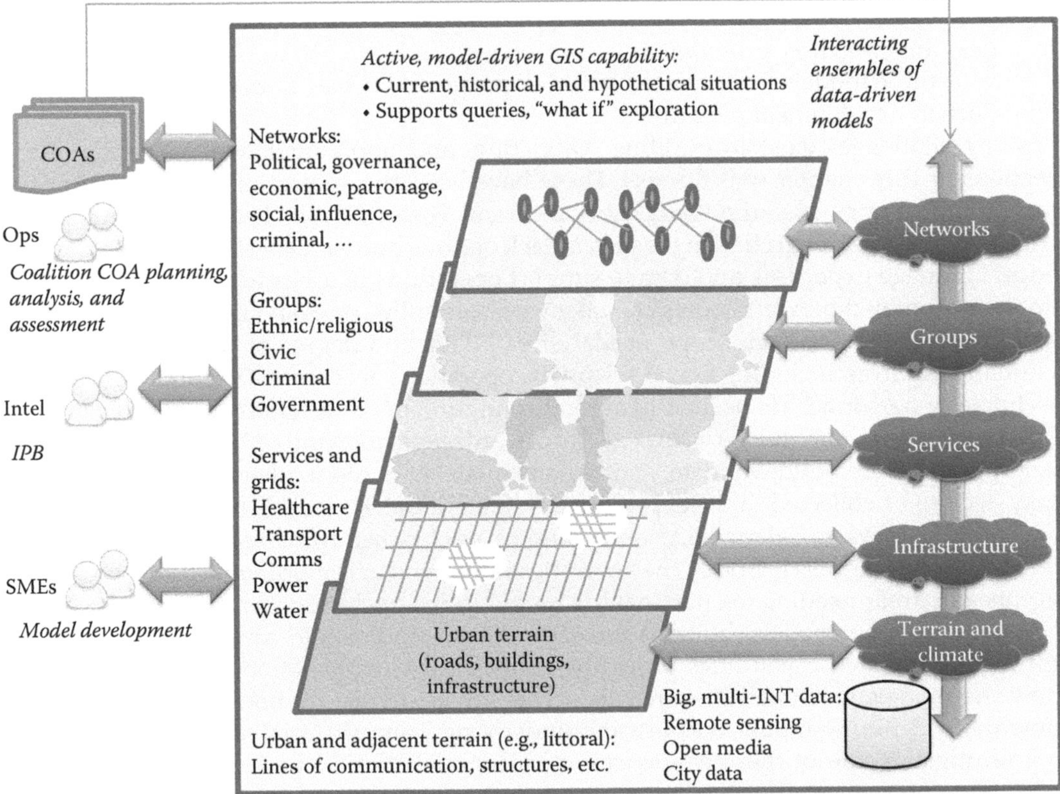

FIGURE 1.2
The ExMo vision: a dynamic model-based capability for exploring the operational environment.

organizations (and associated key individuals); and social, power, and economic networks interconnecting them.

Rather than just being static data in a database that describes the current state of the operational environment, imagine an active version of a GIS in which each layer is generated by an ensemble of interconnected models, the output of which is shown on a zoomable map. Changes within a layer propagate across that layer, and up and down to other layers. Users can thus explore alternative situations, not just the current situation. Users can manipulate a time slider that allows the models to be run forward or backward in simulated time.

As a user interacts with a layer or new data are received, the models in the layer's ensemble output potentially different new results for that layer. Thus, a change to a social group, for example, could cause changes to other groups in the layer. Furthermore, a model ensemble in one layer could be connected to ensembles in other layers. For example, the power infrastructure layer model might receive input that a major substation has been damaged (by a natural disaster, insurgent attack, collateral damage from coalition operations, or mechanical failure). The change in power service is reflected at the services layer. This might in turn change the input to a quality-of-life model at the group level that determines a group's happiness and security.

To enable this future vision, ensembles of models are needed to span the urban triad of terrain, infrastructure/services, and a heterogeneous population. The latter includes applying human social cultural behavioral models based on social science theory and empirical data. Urban terrain models address phenomena such as land use, traffic, and pedestrian behavior. Infrastructure models include those representing physical networks, facilities, and services provided (e.g., power grid, water system, healthcare systems). Typically, these models were created for other purposes such as understanding and managing urban development over time.

Many challenges exist in building, validating, and applying models, as subsequent sections of this chapter will discuss. These have been the subject of recent government research programs (e.g., the study by Boiney and Foster [2013]). The expeditionary setting presents additional challenges such as lack of time, human expertise (regional knowledge, language expertise), and data to support operations in areas where a coalition may not have operated before. Models can take weeks, months, or even years to develop from scratch depending on their scope, availability of data, and degree of validation required. Human resources including social scientists, operations research staff, and information technology personnel are needed to design, configure, operate, and maintain the models. These personnel may be very limited in forward-deployed units. As will be described below, the data to build, validate, and operate models are often a key limiting factor and may need to be collected in the expeditionary operational environment, mined from open sources (e.g., news, social media, country studies) or obtained from mission command systems. Once the models are built, they must be maintained and modified as both real-world events and areas needing deeper analysis unfold in the operational environment.

When applying models, humans must be able to understand and trust their results through automated capabilities providing transparency, explanation, and exploration of how the models generated their results across the ensemble of models used, each with potentially differing inputs, outputs, and underlying computational formalisms.

To mitigate some of these challenges, extant models would be leveraged where possible. This presents challenges in bringing together models not designed to interoperate, each with very different inputs and outputs, at both a syntactic and a semantic level. At the start of planning, relevant models must be quickly identified, located, and tailored to

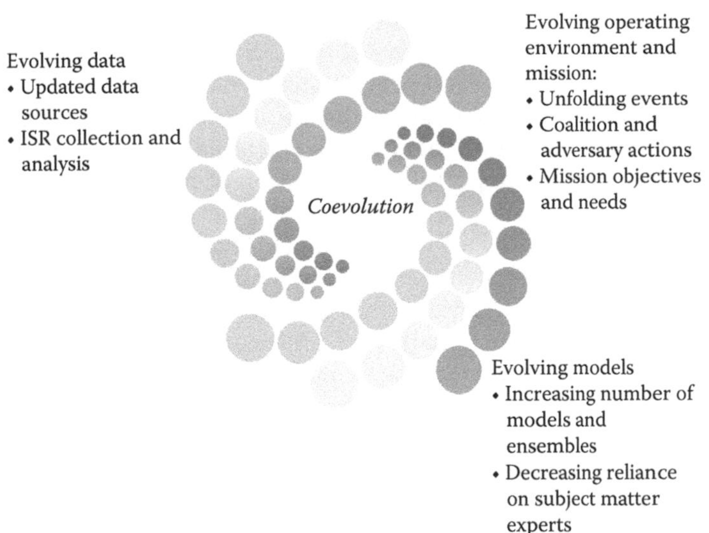

FIGURE 1.3
Coevolution of the models, data, and operating environment/mission.

the operational environment at hand, using extant or collected data. Individual models must then be integrated into a computational modeling and data framework and composed as needed to produce relevant outputs. As the triple helix in Figure 1.3 shows, we anticipate that models, data, and the mission/operating environment will coevolve. As time progresses, new data (from new sources or existing sources as events unfold) will enable applying new models. New models may be driven by changing mission needs, also a result of unfolding events and actions taken.

An Expeditionary Modeling Scenario

As a future use example, consider that in 2045, the megacity Xanadu in the country of Çawaila experiences an earthquake causing mass casualties and damage to key infrastructure. Xanadu has long struggled with a population larger than it is able to fully support, and the Çawaila government is unable to respond effectively to the crisis. The United States establishes Joint Task Force Rapid Relief. The joint force land component commander (JFLCC) and his staff must come up to speed quickly on the operational environment and its unique challenges. In addition to traditional approaches to IPB, the JFLCC G2 and G3 recommend using ExMo capabilities for COA generation and analysis (war-gaming).

Although no models have been specifically built for Xanadu, the staff is able to specify the city's key features (megacity, decaying infrastructure, large immigrant population, race and class tension, etc.) and operational mission objectives (humanitarian assistance and security). An initial suite of relevant models is automatically recommended from a new joint urban modeling library. These models will be automatically refined as Xanadu-specific data are collected and intelligence analysts update key assumptions as they conduct IPB.

Using the recommended models, an ExMo tool prompts the intelligence staff to specify known attributes of the key actors. At the outset, it is known that the Bursuks have strong religious ties and are distrustful of the national government, the North Razinis have a strong criminal element (including black market, drugs, and human trafficking), and the local Çawailas are generally trustful of the government but not of the other two groups. In addition to the information provided by G2 staff, ExMo models use the data from other classified (e.g., order of battle data) and open sources for the models.

Within hours, the JFLCC staff is ready to start some initial COA development and wargaming. An ExMo model has highlighted which earthquake-damaged areas will require immediate reconstruction to support the distribution of needed goods. Other areas are highlighted as likely areas where the majority of displaced persons are likely to relocate and several areas are highlighted as likely targets of increased criminal activity. A COA is entered including the deployment of US Army engineers to rebuild key areas, while US security forces patrol areas identified as high risk, and the local Çawaila government begins the distribution of internationally provided goods. ExMo models show that this COA will result in only a fraction of the population (the native Çawailas) benefitting from the humanitarian assistance, resulting in more Bursuk hostility and a large portion of the goods ending up on the black market, increasing the operating capabilities of the Razini criminal elements. The JFLCC staff is able to rapidly understand that the primary COA attributes that have created this result are the lack of US military presence along the distribution routes and using the local Çawaila government to distribute supplies to the non-Çawaila population. Based on the information provided by ExMo models, the staff changes the COA to include more security along critical distribution routes and providing goods to the Bursuks via their religious leaders and to the Razinis by US forces.

A COA brief is prepared for the commander within hours, and as the preparation for deployment and operations get underway, the ExMo models continue to update with the most current data being collected in the operational environment, providing the planning staff at each echelon with increasingly rich models and data for COA generation and wargaming and facilitating the rapid understanding of the changes in the operational environment and operational assessment of results.

As the ExMo capabilities are repeatedly deployed, the number of models, data, and situations impacted will increase, providing ever-growing base capabilities for the next time.

Expeditionary Modeling Requirements: Data

This section focuses on the challenges of capturing evolving understanding and data about a dense urban area. For illustrative purposes, we will use Lagos, Nigeria, as an example of a megacity with economic opportunity yet fragile stability. According to a recent megacity use case, Lagos has a population of over 20 million today, two orders of magnitude greater than its population in the 1950s; spans over 3400 sq. km. (mainland and several islands), and it is a key deepwater port for most of Nigeria's oil exports (US Army SSG Megacities Concept Team, 2014). Attracted by economic opportunity, the population growth has far outpaced the capacities of the city's infrastructure and governance. Boko Haram, while currently mostly active in northeastern Nigeria, presents a looming security threat. An informal network of patrons (*okas*) presents alternative governance, and local gangs (Area Boys) span most of the city (Packer, 2006).

Characterizing dense urban areas in general is a prerequisite to modeling specific cities or collecting data about them. Cities can be described as "at once a social organism, a human-made physical form and an economic system" (Evans, 2015). The joint integrating

CRITICAL DATA

Data are critical to realize the ExMo vision. Dynamic, diverse data from open sources and sensor collection must be integrated, exploited, and used to provision model-based analysis and forecasting. These data must be organized in a framework, taxonomies, and ontologies that are understandable to humans and to computational models. The data widely vary in completeness, depth, and reliability across different megacities.

concept for joint urban operations states that "urban areas are complex living systems with a wide range of structures, processes, and functions that have evolved to sustain concentrated human societies in confined space" (US Department of Defense, 2007). It describes a city as like "a living organism with structures, processes and functions; supported and protected by a physical infrastructure, and maintaining a steady state of continuously fluxing matter-energy and information" and the enemy as "hostile growth embedded and diffused within the urban system, subsisting off its host." A key challenge then is defeating the growth without killing the host. Kilcullen (2013) uses operations in Baghdad during 2006–2007 as an example of pacifying a city by mostly shutting it down, effectively committing urbicide.

The military has developed several conceptual frameworks for understanding complex operational environments as part of capturing doctrine, methodologies, and best practices for joint and service-specific urban operations (e.g., US Department of Defense, 2013). For example, the army operations process seeks to identify the factors describing an operational environment in the design of military missions (US Army, 2012). Table 1.1 shows operational variables using the widely used political, military, economic, social, infrastructure, and information (PMESII) characterization* for a notional city. Each PMESII variable is broken down into several dimensions: areas, structures, capabilities, organization, people, and events (ASCOPE). A related taxonomy focuses on infrastructure and services in urban environments and includes sewer, water, electricity, academics, trash, medical, safety, and other services. The army megacity study proposed characterizing megacities in terms of their context (historical, cultural, local, regional, international), unique scale, density (population, structures, and electronic signals), connectedness (physical, information), threats, and flows (energy and other inflows, waste and other outflows) (US Army SSG Megacities Concept Team, 2014).

There is a need to unify these conceptual frameworks and taxonomies to both facilitate the exchange of information about these kinds of operational environments and organize the relevant, collected data. Given such a framework, key data challenges are the sourcing of relevant, accurate, and timely data about new or changing operational environments, organizing the data into a usable framework and using it to quickly develop and evaluate blue COAs against likely adversary actions. The expeditionary aspect of the problem limits the time to identify, locate, collect, process, analyze, and exploit these data.

Data collection can mine a range of data sources including various kinds of intelligence including imagery, signals, human, and open source (OSINT). The amount and

* Variants of PMESII include PMESII-PT, where *PT* stands for physical terrain and time.

TABLE 1.1

Sample PMESII-ASCOPE Factors to Characterize an Operational Environment

	Political (P)	Military (M)	Economic (E)	Social (S)	Infrastructure (I)	Information (I)
Areas (A)	Political districts, party support areas	Areas of operations	Market areas, etc.	Religious and tribal lands, parks	Power and water districts, etc.	Media coverage areas
Structures (S)	Polling places, government and party offices, courts	Bases, facilities, etc.	Factories, stores, restaurants, banks, etc.	Churches, mosques, sports venues, tribal buildings, etc.	Roads, rail, power grid, water/sewer system, airports, hospitals, police/fire station	Radio/TV stations, internet infrastructure, schools
Capabilities (C)	Elections and governance, legal system	Military capabilities and weapons	Agriculture, manufacturing, commerce, banking	Religion, sports, tribal support	Transportation, power and water systems, public safety, healthcare	Communications media, education
Organizations (O)	Political parties, government orgs	Military units	Farms, banks, companies	Civic and religious groups, sports teams, tribes/clans, NGOs	Public works orgs, police/fire department, medical personnel	Media orgs, educational orgs
People (P)	Govt. and party leaders, judges and lawyers	Military leadership	Company and bank leaders	Civic, tribal, and religious leaders	Govt. and private sector leaders	Media owners and personalities
Events (E)	Campaigns, elections, coups, diplomacy	Engagements, deployments, exercises	Recessions, collapses, market movements, consumer behavior	Protests, unrest, religious and secular festivals, games	Construction, deployment, failures, repairs	Broadcasts, stories, speeches, social media postings, campaigns

Note: NGOs: nongovernment organizations.

quality of data varies dramatically. Highly integrated and resilient cities such as New York are well charted. The data are not universally well integrated and may well be old. In any operation, the collection of recent data will be required. This can be done using intelligence, surveillance, and reconnaissance (ISR) assets of coalition forces including satellites, aircraft, and unmanned aerial vehicles. Existing assets in the city such as traffic and security cameras and other sensors can be exploited, if available. The ISR needs to encompass the urban terrain, the population, and the infrastructure. Urban terrain features may be obscured by structures, vegetation, weather, or deliberate concealment.

The condition of infrastructure and the level of services provide additional ISR challenges. Stevens University, for example, is using lidar and small robots to chart sewers (Stevens University, 2015). This will help populate the infrastructure part of its virtual Hoboken model, enabling applications such as flood prevention and warning, urban planning and design, and smart city applications for residents and city managers. Additional sensors and data sources, such as social media, can also be used to rapidly assess the availability and the quality of key services in the operational environment.

The softer characteristics of a dense urban area may be more difficult to measure. For urban populations, besides patterns of movement, there are potentially numerous indicators to be exploited. For example, understanding the concerns of populations in a neighborhood is vital to understanding the effects that events and operations might have there. Social scientists in prior decades have relied on survey data obtained through polling. While useful, such data are difficult to obtain on a frequent basis over a large, diverse population. Social media data subjected to sentiment analysis might be one alternative that provide timelier, but perhaps noisier, data. Where that is scarce, proxies for social variables might be achieved from stationary or mobile cameras. For example, the presence of pets and joggers might indicate that people feel safe about a neighborhood (Farry et al., 2014). The condition of roads observed from an overhead imagery might yield clues about the economic conditions in a neighborhood (Owen, 2014).

OSINT including social media can complement ISR and help fill in the gaps. For example, open source information can provide insight on demographics and building purposes/occupants. Open sources include published monographs, reports, websites, commercially available data sets, news, and social media. These include text, images, video, and structured databases. Commercial open sources include financial data (e.g., aggregate credit card spending) and information about local businesses. Open source data providers include city governments (e.g., providing information for citizens), academic institutions, international/nongovernmental organization studies (e.g., the United Nations, World Bank, International Monetary Fund), companies, and individuals (including crowd-sourced data). Establishing the accuracy and the veracity of the crowd-sourced data and other social media can be challenging.

As part of an informal study characterizing the availability of open source data for megacities such as Lagos, we identified a number of data sets from third-party sources, shown in Table 1.2. While we found hundreds of data sets from dozens of sources, most were annual (especially noncommercial sources), related to developmental issues (least common were religious and ethnic data), generally at the province (equivalent to a US state) or the district (equivalent to a US county) level. Disaggregated data within a city were far less common and often self-reported or crowd-sourced.

While open sources can provide economic, demographic, and other indicators, they may be infrequently updated (e.g., annually). Recent social science modeling work—e.g., Kettler and Hoffman (2012)—has exploited news events. Using lightweight natural language

TABLE 1.2

Sample Kinds of PMESII Data for Lagos from Open Sources

Political
- Event data**
- Political appointees by type and gender***
- Courts and court cases***

Economic
- Agriculture*
- Poverty levels*
- Unemployment*
- Exchange rates (country)
- Household expenditure*
- Business information***
- Motor vehicle registrations***
- Number of professionals by sector, gender***
- Number of workers by education, gender***
- Students receiving free meals and/or uniforms***
- Local government revenue, expenditures***
- Number of banks, registered businesses, hotels, restaurants, gas stations, community development centers, market stalls, etc.***

Infrastructure
- Boreholes sunk***
- Existing roadways by surface type***
- Community centers***
- Refuse collection***
- Public toilets constructed***
- Road shape files***

Military
- Event data**
- Number of police stations***

Social
- Population by age and gender**
- Health statistics*
    - Number of doctors, clinics, patients, etc.***
    - Immunization, infectious diseases***
    - Births and deaths***
- Education**
    - Adult literacy***
    - Libraries***
- Crime***
- Corruption reports***
- Youth centers***
- Number of parks, churches, mosques, shrines***
- Marriages and divorces***

Information
- Internet access, cell phone access*

Source: Lagos Bureau of Statistics, *Abstract of Local Government Statistics*, Ministry of Economic Planning and Budget, Lagos State Government, Laos, 2012, retrieved December 9, 2015, from http://www.lagosstate.gov.ng/images/pageimages/downloadfiles/docs/ABSTRACT_OF_LG%202012.pdf.

Note: Asterisks indicate lowest level of availability: *, province; **, local governing area; ***, local council development area.

processing technology, the integrated crisis early warning system (ICEWS)* has extracted over 23 million events from over 35 million news stories spanning 177 countries for the past 20 years (Lockheed Martin Corporation, 2015). An event may describe, for example, a specific instance of a government arresting protesters. The resulting data set provides a rich history of activities, interactions, and sentiment among countries, organizations and groups, and individual actors, as shown in Figure 1.4.

Big data tools may be needed to manage the variety, the volume, and the velocity of relevant data. This includes extract, transform, and load tools that can deal with heterogeneous data formats, including unstructured text. Big data architectures and infrastructure including nonstructured query language repositories and cloud-based computing may be needed. Data can be more useful if they are geolocated, as this provides a way to integrate across different sources. In general, much of the data will be incomplete and of uncertain quality. This will require leveraging best practices and emerging data collection, validation, integration, and analytic techniques.

* ICEWS was a research project funded by the Defense Advanced Research Projects Agency from 2009 to 2012 and later transitioned under subsequent Defense Department funding to operational use as part of a program of record.

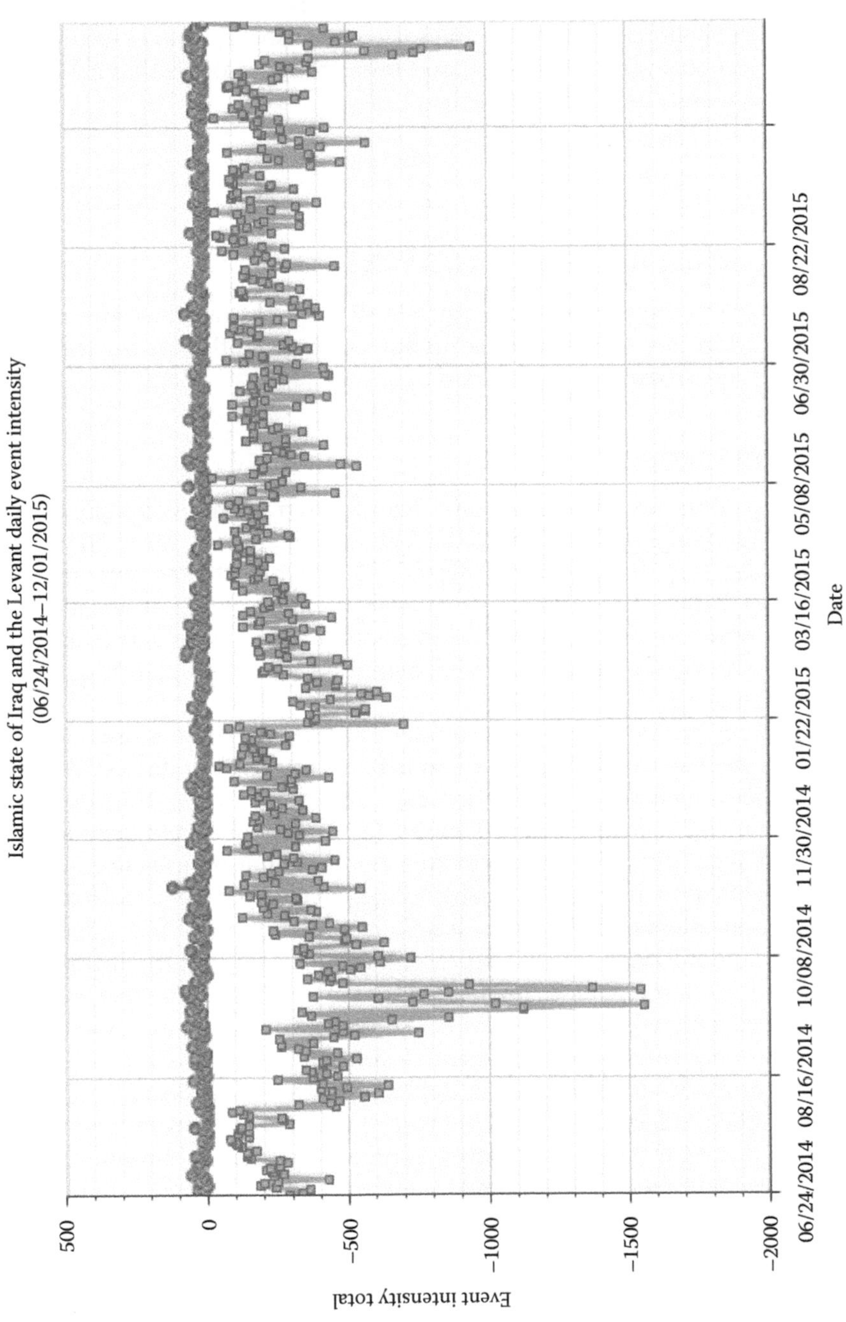

FIGURE 1.4
ICEWS coded news event data shown for events involving the Islamic state in the Levant from mid-2014 through 2015. The *x* axis shows the aggregate hostility level of events (negative = more hostile). The top time series indicates cooperative activity while the bottom time series indicates hostile activity.

KEY CHALLENGES

Computational models of human and urban terrains exist, exploiting different underlying model formalisms. These have been successfully used for understanding and forecasting complex phenomena. The key challenges for ExMo include rapidly identifying, tailoring, composing, provisioning, and interpreting models that may differ in resolution, fidelity, and their inputs/outputs.

Expeditionary Modeling Requirements: Models

Data by themselves are necessary but not sufficient. It requires humans and models to exploit. This section addresses the latter, looking at modeling on the urban, services and population aspects of the urban operations triad.

Modeling Urban Environments

"The context and objectives for urban modeling have grown far more complex over the past two decades" (Waddell and Ulfarsson, 2004). Urban models exist for simulating subsystems at macroscales (e.g., interregional migration) and mesoscales (e.g., intraurban traffic) (Torrens, 2008). Urban design models can examine how people interact with and within urban spaces. Many of these models, however, are not specific to dense urban areas such as megacities. Larger scale models have been applied for urban planning (e.g., land use, regional traffic patterns).

The UrbanSim system has been used as a model of a broad range of areas for urban planning and policy, such as the imposition of an urban growth boundary (Waddell and Ulfarsson, 2004). Policies include transportation capacities (e.g., roads), transport system management (e.g., signals), transportation demand (e.g., carpool incentives), land use/growth management (e.g., urban design), economic development policies (e.g., tax incentives), and environmental policies (e.g., air quality). Agents modeled that create or respond to policies include households, individuals, employers, developers, and governments. These agents operate in an interconnected representation of the land, the infrastructure, the services, and the economy. UrbanSim uses microsimulation and other techniques including discrete choice models, cellular automata, and multiagent simulation to choices among and locations of households, employment, development, and infrastructure.

A key component of many urban modeling applications is traffic. Besides urban planning applications, traffic models are of high interest in dense urban areas where maneuver may be impeded during military operations, civilian evacuations, humanitarian relief supply, etc. Traffic can impact workers getting to their place of employment. In Lagos, over 60% of the economic activity is informal transactions such as roadside selling of all manner of goods (Packer, 2006). Disruption to traffic means the disruption of this economic activity and the influx of goods into the city.

Urban simulation was a focus of US Joint Forces Command's UrbanResolve effort, which has high relevance to modeling megacities (Adamson, 2015). One tool that began under this effort was CultureSim, which represents another type of urban modeling tool and builds on the joint semiautomated forces system (Moyer, 2006). CultureSim generates sociocultural realistic agents moving about a city, i.e., individual people taking trips in vehicles or on foot. As trips are generated, the streets are filled with vehicular and/or pedestrian traffic, according to the road capacities. The patterns of life can become evident over time. Behavior has been added to mimic crowds and responses to events such as natural disasters or bombings.

With adequate hardware, the simulation can support hundreds of thousands of entities. CultureSim consumes and generates a lot of data. Besides demographic data, the input data include terrain data with road information. Road conditions and obstacles are modeled, along with traffic lights. This can be partially generated but requires extensive postprocessing to fix erroneous assumptions by automated ingest tools about road connectivity, directionality, etc. CMSim, a simulation tool derived from CultureSim, was subsequently developed to be more computationally efficient compared to CultureSim by sharing data across computational instances at a more aggregate (road versus entity) level, and it is an order of magnitude faster (Moyer and Macannuco, 2007).

This type of modeling and simulation can still be computationally prohibitive at the megacity scale, so the use of mixed resolution approaches may be desirable. Different parts of the city might be modeled at different resolutions, zooming in to specific neighborhoods on an as-needed basis. This model zoom feature could also be applied by modeling neighborhoods at the population level then zooming down to model specific inter- and intragroup behavior as needed.

Modeling Populations

Computational social science models have shown increasing potential to help understand and predict the behavior of countries, groups, and individuals (Egeth et al., 2014). Equation-based models such as system dynamics models can be useful in capturing quantitative interdependencies in a dynamic system. These models, which can be crafted by subject matter experts or estimated from data, have been used to model a range of relevant phenomena including state stability (Choucri et al., 2007) and army counterinsurgency operations doctrine (Pierson, 2010). Unlike physical system models, social science models cannot generally be tested by comparing their results to physical experiments conducted under laboratory conditions. Thus, model verification and validation remains difficult for these models and a subject of active research.

Statistical models are another type of equation-based model. The ICEWS effort used statistical models estimated from historical data and agent-based models (ABMs) to forecast political instability events of interest in over 167 countries with greater than 90% accuracy and less than 5% false positives (Kettler and Hoffman, 2012). The early model-based event recognition using surrogates (EMBERS) system successfully provided advance warning of protest, epidemics, and other events using a variety of statistical models (Intelligence Advanced Research Projects Activity, 2015). These systems employed mixed methods ensembles of models that exploited diverse data sources including open sources such as news and social media. EMBERS also exploited many weak-signal sources to make its warnings such as using Google Flu Trends, OpenTable (online restaurant reservations), and hospital parking lot imagery to predict the spread of influenza-like illness (Ramakrishnan, 2014).

ABMs can elicit emergent behavior among social agents following simple rules and have been used to model segments of a country's population (Alcorn et al., 2011). Like other simulation models, ABMs can generate data at each time step, and the interactions among agents can illustrate the potential causes of events. ABMs have been combined with equation-based models. For example, ABMs can complement traditional epidemiological (compartmental) models with the former modeling the spread of illness through interactions among individual people (agents) and the latter modeling the population after the disease is more widespread (Bobashev et al., 2007). ABMs are often configured by subject matter experts, which makes them more expensive, although they tend to require less data than empirically derived models.

While some progress has been made in forecasting events, the model-based forecasting of the specific effects of diplomatic, information, military, and economic (DIME) actions proves especially challenging. The effect of DIME actions are highly context dependent, unlike the effects of physical actions (e.g., firing a cannon), which are subject to universal constraints of physics. These effects may be hard to detect as they may be small, delayed, or occur over a long time frame or simultaneously with other actions. Data are often not consistently kept by the military about the efficacy of DIME actions performed. When such data are kept, it may be difficult to isolate the effects of specific actions.

Example: Athena

Several social science models take a hybrid approach to analyzing COAs from a multi-dimensional (PMESII) perspective including the conflict modeling, planning, and outcomes experimentation system (Waltz, 2008) and the national operational environmental model (Salerno et al., 2011). The Athena simulation, developed by the National Aeronautics and Space Administration Jet Propulsion Lab for the US Army Training and Doctrine Command, is a good example of a framework that combines multiple, diverse models in a tightly integrated fashion (Chamberlain and Duquette, 2013). Athena has proven useful in a number of exercises and real-world operations, chiefly by helping analysts compute the effects of complex COAs generally determined through war-gaming. The Athena developers do not claim that Athena is designed for making point predictions (e.g., specific probabilities that an event will occur) but instead is primarily useful for understanding the complex dynamics of a situation over time.

Athena combines over a dozen heterogeneous models that span the PMESII spectrum.* Athena has been generally used by the army and others to quantitatively show how a region or a country may evolve under the influence of multiple, competing political actors, where actors may be regional powers, insurgents, criminals, or other factions trying to gain political control of the neighborhoods (modeling provinces or other areas with the operational environment) by collecting support and influence from civilian groups in each neighborhood. Civilian groups include different demographic (social, ethnic, religious, etc.) groups. Actors influence civilian groups via strategies that comprise several dozens of tactics that span the spectrum of DIME actions such as explicitly supporting (D), conducting influence operations (I), conducting military operations (M) via the actor's forces (force groups), or providing financial aid (E).

Athena includes a six-sector economic model, a combat attrition model, and a belief system model. The last one captures actor and group beliefs on a set of fixed topics and is used to determine the baseline affinity between them based on belief similarity. The belief model also serves as the basis for modeling information operations. Each civilian group has a number of variable satisfaction concerns (in aggregate, their mood) including their perceptions of quality of life, security, autonomy, and respect for their cultural beliefs. Simulation events such as conflict, combat, or infrastructure events such as power outages that will negatively impact the autonomy, the safety, and especially the quality of life.

* While based on models drawn from social science literature, Athena has a heuristic mechanism as well that has not been systematically validated against empirical data.

These effects can spread within and across neighborhoods. Enough negative events in a neighborhood may adversely impact the support for actors and potentially cause a shift of control in a region. Athena does not currently simulate physical terrain, as would be needed for urban modeling.

A key challenge for Athena (and similar simulations) is correctly configuring its constituent models, currently a subject matter expert-intensive process requiring knowledge of how Athena models things, validating those models (typically limited to a face validity check by a subject matter expert familiar with the operational environment, etc.), and extracting meaning from the data generated. Like all models, Athena is an abstraction. Thus, a user must map from their conceptualization of the real world, ideally grounded in empirical data and expertise, into the model's inputs and back from the model's abstract outputs to their world conceptualization.

Composing Models

The value in exploiting multiple models lies in both their ability to address different parts of the problem (e.g., understand different, yet interconnected phenomena) and the ability for different models to present potentially different views of the same phenomena. The latter is found in weather and political forecasting in which multiple, diverse models all produce forecasts of the phenomenon of interest, which are then combined (weighted, etc.) to give a better forecast than any single model would have. We saw this in combining political instability models in ICEWS (O'Brien, 2010). Other values can be found in composing urban and population models, as shown in Figure 1.5, in which human or natural disasters impact the urban environment. This in turn impacts services, quality of life, and group behavior, which may alter the environment.

Putting together multiple models, especially rapidly, is an area of active research. In 2009, the Office of the Secretary of Defense modeling and simulation steering committee produced a report led by the US Naval Research Lab with participation from other

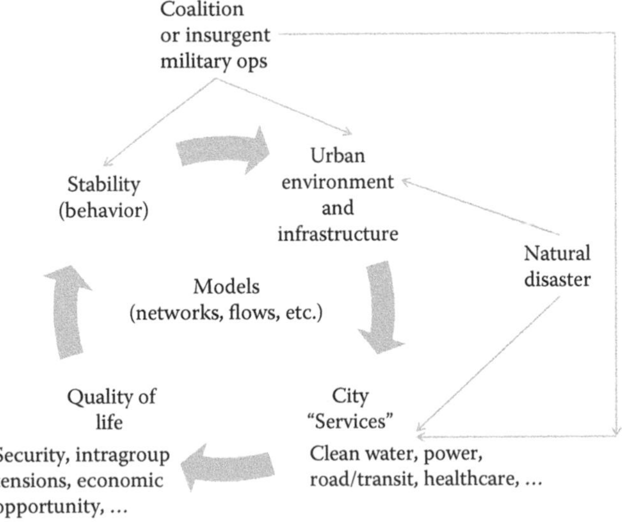

FIGURE 1.5
Notional multimodel composition to understand the impacts of events and actions in a dense urban environment.

government agencies and academia (Modeling and Simulation Steering Committee, 2009). Although the information was collected in 2007–2009, the survey and gap analysis done remains relevant today. Key products included descriptive requirements for characterizing the DIMEL* actions, the PMESII effects, and their interdependencies by mission (via a taxonomy of nearly 150 actions, effects, and overarching factors); a gap analysis that looked at 14 models compared to these requirements; a catalog of over 100 relevant databases (although most were deemed narrow in scope, shallow in depth, or limited to regional focus); a hierarchy of mission-specific measures (metrics and indicators including, from bottom to top, dimensional parameters and measures of performance, effectiveness, force effectiveness, and policy effectiveness); and over 600 model framework requirements identified in five categories (operator interface, system control, model interoperability, system integration and maintenance, system architecture). Key recommendations identified from the study included closing framework deficiencies (e.g., just-in-time models/data to trace assumptions), closing gaps in social theory (e.g., synergistic effects of multiple actors), and analytic explorations (e.g., cyberwarfare effects on society, effects of changes in food supply).

Hard challenges remain in rapidly composing heterogeneous models, especially models not designed to work with one another and where the collection of models might be evolving to meet changing operational needs and data source availability. Several model composition frameworks have been developed. For example, Argonne National Lab's dynamic information architecture system is a flexible, extensible framework for integrating models through a shared object model, providing a level of abstraction and avoiding pairwise model integration (Simunich et al., 2002). Metamodels—models that organize or interconnect other models—are another glue technology that could be developed, leveraging shared ontologies. Some preliminary experimental results for integrated ExMo models are presented in (Kettler and Lautenschlager, 2015).

Conclusion

While the challenges of expeditionary operations in dense, dark, dynamic, and dangerous urban areas are daunting, new automation to exploit diverse data sources to build and drive flexible ensembles of multidisciplinary models would aid commanders in gaining a predictive, broad, quantitative understanding of the changing operational environment. Armed with this knowledge, commanders would be able to better plan to achieve their objectives and avoid actions that cause unanticipated adverse effects and make things worse in fragile environments under crisis. Recent advances in social science and urban modeling, combined with improved access to data and increasing computational capacity, show that this ExMo vision is feasible. More work is needed to put diverse data sources and diverse models together rapidly, validate their combination, and integrate the resulting capabilities with end user tools in a way that fosters understanding and trust. A side effect of this multidisciplinary research will be increased understanding of the core factors that drive complex, population-centric operational environments in general. This knowledge will be useful outside military planning for applications such as smart city management and resilient urban design.

* Diplomatic, information, military, economic, and legal/law enforcement.

DISCUSSION QUESTIONS

1. What are the military challenges of dense urban areas such as megacities?
2. What are the modeling challenges of dense urban areas such as megacities? What emerging techniques are there from other modeling disciplines?
3. What social science advancements would facilitate this vision?
4. What are the data challenges of dense urban areas such as megacities? What are some novel sources of relevant data for megacities?
5. How would military users (commanders and their staff) get feedback from ExMo tools that is understandable, actionable, and trustworthy?
6. How would the Internet of Things, smart cities, and other emerging technologies support this vision?

References

Adamson, W. G. (2015, Spring) Toward better readiness. *The US Army War College Quarterly Parameters*, *45*(1), 45–54.

Alcorn, B., Garces, M., and Hicken, A. (2011) VirThai: A PS-I implemented agent-based model of Thailand as a predictive and analytic tool. *International Studies Association: Annual Convention.*

Benedictus, L. (2015, July 26) How megacities compare: China plans a city 25 times the size of London. *The Guardian*. Retrieved October 25, 2015, from http://www.theguardian.com/cities/shortcuts/2015/jul/26/jing-jin-ji-chinese-megacity-25-times-size-of-london.

Bobashev, G. V., Goedecke, D. M., Yu, F., and Epstein, J. M. (2007) A hybrid epidemic model: Combining the advantages of agent-based and equation-based approaches. *Winter Simulation Conference*, pp. 1532–1537.

Boiney, J., and Foster, D. (2013) *Progress and Promise: Research and Engineering for Human Sociocultural Behavior*. Mclean, VA: MITRE Corporation.

Chamberlain, R. G., and Duquette, W. H. (2013) *Athena in 2013 and Beyond*. Pasadena, CA: Jet Propulsion Lab Publication, pp. 13–9.

Choucri, N., Goldsmith, D., Madnick, S. E., Mistree, D., Morrison, J. B., and Siegel, M. D. (2007) Using System Dynamics to Model and Better Understand State Stability. MIT Sloan Research Paper No. 4661-07.

Egeth, J. D., Klein, G. L., and Schmorrow, D. (2014) *Sociocultural Behavior Sensemaking: State of the Art in Understanding the Operational Environment*. McLean, VA: MITRE Corporation.

Ellis, D. C., and Sisco, J. (2014, April) Understanding megacities RSI: Dhaka's design as an expression of culture and politics. *Understanding Megacities with the RSI Paradigm*, 24–32.

Evans, M. (2015) The megacity myth. *The US Army War College Parameters*, *45*(1), 33–45.

Farry, M., Bullock, B., and Pfautz, J. (2014) Urban socio-cultural modeling with passive sensing. In *Understanding Megacities with the Reconnaissance, Surveillance, and Intelligence Paradigm*. Vicksburg, MS: Topical Strategic Multi-Layer Assessment (SMA) and US Army Engineer Research Development Center (ERDC), pp. 48–57.

Felix, K. M., and Wong, F. D. (2015, Spring) The megacity challenge. *The US Army War College Quarterly Parameters*, *45*(1), 19–31.

Intelligence Advanced Research Projects Activity (2015) *Open Source Indicators (OSI)*. Retrieved November 16, 2015, from http://www.iarpa.gov/index.php/research-programs/osi.

Kettler, B., and Hoffman, M. (2012) Lessons learned in instability modeling, forecasting, and mitigation from the DARPA integrated crisis early warning system (ICEWS) program. *2nd International Conference on Cross-Cultural Decision Making. San Francisco, CA.*

Kettler, B., and Lautenschlager, J. (2015) Expeditionary modeling for population-centric operations in Megacities: Some initial experiments. *Proceedings of the 7th International Conference on Applied Human Factors and Ergonomics (5th International Conference on Cross-Cultural Decision Making)*, Orlando, FL, in press.

Kilcullen, D. (2013) *Out of the Mountains: The Coming Age of the Urban Guerilla.* Oxford: Oxford University Press.

Krulak, C. C. (1999) The strategic corporal: Leadership in the three block war. *Marines Magazine.*

Lagos Bureau of Statistics (2012) *Abstract of Local Government Statistics.* Ministry of Economic Planning and Budget, Lagos State Government, Laos. Retrieved December 9, 2015, from http://www.lagosstate.gov.ng/images/pageimages/downloadfiles/docs/ABSTRACT_OF_LG%202012.pdf.

Lockheed Martin Corporation (2015, November 16) *Integrated Crisis Early Warning System (ICEWS) Dataverse.* Retrieved from Harvard Dataverse: https://dataverse.harvard.edu/dataverse/icews.

Modeling and Simulation Steering Committee (2009, July) *Requirements for a Government-Owned DIME/PMESII Model Suite.* Washington, DC: Office of the Secretary of Defense.

Moyer, D. (2006) Obstacles to producing realistic traffic in a distributed environment. In *Spring Simulation Interoperability Workshop.* Huntsville, AL.

Moyer, D., and Macannuco, D. (2007) A mixed resolution simulation for highly scalable urban and regional populace modeling. *Fall Simulation Interoperability Workshop.*

O'Brien, S. (2010, March) Crisis early warning and decision support. *International Studies Review, 12*(1), 87–104.

Owen, K. (2014, April) Evaluating slum severity from remote sensing imagery. *Understanding Megacities with the RSI Paradigm*, 67–74.

Packer, G. (2006, November 13) The Megacity. *The New Yorker,* p. 64.

Pierson, B. (2010) *A System Dynamics Model of the FM 3-24 COIN Manual.* Warfighting Analysis Division, J8/WAD. US Department of Defense.

Ramakrishnan, N. (2014) Beating the News with EMBERS: Forecasting Civil Unrest Using Open Source Indicators. *2014 Knowledge Discovery and Data Mining Conference.* Association for Computing Machinery, New York.

RAND Corporation (2011) *Standing Up a More Capability Joint Task Force Headquarters: Research Brief.* RAND Arroyo Center.

Salerno, J. J., Romano, B., and Geiler, W. (2011) The national operational environment model (NOEM). In *Modeling and Simulation for Defense Systems and Applications.* Orlando, FL.

Sanborn, J. K. (2015, September 23) Marines on the hunt for gear to combat tomorrow's threat. *Marine Corps Times.* Retrieved from http://www.marinecorpstimes.com/story/military/tech/2015/09/22/marines-hunt-gear-combat-tomorrows-threat/72367754/.

Silver, N. (2012) *The Signal and the Noise: Why So Many Predictions Fail but Some Don't.* London: Penguin Books.

Simunich, K. L., Sydelko, P., Dolph, J., and Christiansen, J. (2002) *Dynamic Information Architecture System (DIAS): Multiple Model Simulation Management.* Lemont, IL: Argonne National Laboratory.

Stevens University (2015) *Weather, Beaten: Stevens Steps up as a Global Leader in Resilience Research.* Retrieved November 16, 2015, from http://www.stevens.edu/news/content/weather-beaten-stevens-steps-global-leader-resilience-research.

Tetlock, P. E. (2006) *Expert Political Judgment: How Good Is It? How Can We Know?* Princeton, NJ: Princeton University Press.

Torrens, P. M. (2008) Modeling megacity futures. In *Symposium on Megacities.* Los Angeles: University of Southern California.

US Army (2012) *ADRP 5-0: The Operations Process.* Washington, DC: US Army.

US Army Capabilities Integration Center (2015, November 16) *The Megacity: Operational Challenges from Force 2025 and Beyond.* Retrieved from ARCIC Unified Quest: http://www.arcic.army.mil/Initiatives/unified-quest.aspx.

US Army SSG Megacities Concept Team (2014) *Megacities and the United States Army: Preparing for a Complex and Uncertain Future*. Washington, DC: Chief of Staff of the Army, Strategic Studies Group, US Army.

US Department of Defense (2007) *Joint Urban Operations: Joint Integrating Concept Version 1.0*. Washington, DC: US Department of Defense.

US Department of Defense (2013) *Joint Urban Operations (Joint Publication 3-06)*. Washington, DC: US Department of Defense.

Waddell, P., and Ulfarsson, G. F. (2004) Introduction to Urban Simulation: Design and Development of Operational Models.

Waltz, E. (2008) Situation analysis and collaborative planning for complex operations. *13th International Command and Control Research Symposium*.

2

More than Just a Story: Narrative Insights into Comprehension, Ideology, and Decision Making

Scott W. Ruston

CONTENTS

Introduction 28
What a Narrative Is Not 29
What a Narrative Is 30
Sociocultural Artifact 30
Narrative as Mode of Comprehension 31
Narrative Comprehension: Structures as Cognitive Schema 32
Narrative Comprehension: Bottom-Up Processing and Data/Schema Mismatch 33
Narrative Rationality 34
Narrative and Ideology 35
Master Narratives 35
Ideological Functions and Narrative 36
Values, Ideology, and Decision Making 37
Actions: Shared Component of Narrative and Decision Making 39
Conclusion 40
Acknowledgments 41
References 41

ABSTRACT As the conflicts that the United States finds itself waging become more and more complex with an increasing civil–military dimension, and as the forces of globalization intertwine the economic, foreign policy, and security interests of more and more states, nonstate organizations and individual communities, the US government and other interested interlocutors require methods and tools for understanding the decision making behavior of these organizations and communities. Narratives are value-shaping, ideology-expressing, and comprehension-guiding phenomena that operate at macro- (cultural), meso- (local, community, subcultural), and micro- (personal) levels and are thus powerful tools for gaining insight into individual, organizational, community, and societal decision making. This chapter explores the phenomenon of narrative in three key dimensions relevant to decision making: as a sociocultural artifact expressing values, influencing culture, and producing ideology; as a cognitive process of understanding and providing frameworks that shape the meaning of actions, policies, events, news, and all manner of information; as a form of rationality distinct from cost-benefit logic, influencing decisions with criteria of familiarity, expectation, and emotion.

Introduction

In this volume's fictional Megacity Xanadu vignette, social divides, economic struggles, and cultural tensions contribute to a sense of volatility. Xanadu might experience any number of forms of social unrest, spontaneous or organized violence, political upheaval, and similar outcomes. Understanding the situations of the Razinis, the Bursuks, and the Çawailans, and how these three ethnic groups may respond to any changes in the status quo, requires a thorough understanding of their economic situation, social variables, resources, and similar quantifiable elements driving social change. It also requires the knowledge of the narratives circulating within Xanadu and among these populations; it will require knowledge of the narratives that help define Razini, Bursuk, and Çawaila cultural or national identities. Why? Because a narrative is more than a synonym for story, and stories are more than simply tales told to children, or a simple recounting of events in the newspaper. Rather, narratives are a rich cultural resource, which illuminates the aspects of how individuals and communities think, feel, and act. Furthermore, a narrative is a cultural phenomenon that simultaneously operates at multiple levels.

Understanding the narratives of the Razini and the Bursuk can aid Xanadan decision makers in forecasting how new policies or initiatives will be received by these communities. Understanding the said narratives can give insight into where Razini or Bursuk allegiances might lie and how they might be influenced. Narratives are born in conflict, and thus narrative analysis can reveal grievances and sensitivities at play and reveal potential narrative landmines affecting how policies and outreach activities are received and understood (Bernardi et al. 2012). The narratives circulating within these communities will also influence the decisions made by Razinis and Bursuks—whether to rise up in protest and whether to do so peacefully or violently; whether a Razini is likely to pursue entrepreneurial efforts or if a Bursuk might seek to return to Bursuka. Increasingly, research from the neuroscience community is demonstrating the neurobiological mechanism of what cognitive narratology and communication theory has established, that narratives directly influence choices, behaviors, and decisions (see, for example, studies by Zak [2015], Barraza et al. [2015], and Green and Brock [2000]). This section aims to explain why narrative is so powerful a tool.

Narrative is critically important for three reasons: (1) while a narrative refers to systems of stories, and thus sociocultural objects, a narrative is also a mode of comprehension, how characters and events/actions are organized in the mind to make meaning and significance; understanding this dual logic of narrative is essential to understand the interplay between cultural frameworks and individual comprehension; (2) leveraging the structure, operation, and meaning of the system components, narratives simultaneously serve as vehicles and creators of values and ideology; (3) the outcomes of decisions are actions, and a narrative structures consequence, meaning, and significance through the relationships of the actions/events, actors/agents, and referents operating within the system.

Before we start examining these features of narrative, particularly relevant to cross-cultural decision making, we should spend a little time establishing an understanding of what a narrative is and what it is not. Fundamentally, *narrative* is simultaneously two things: a sociocultural object that is a system of stories structured for a rhetorical purpose and a cognitive process of understanding that orders data into a specific structure of meaning. *Narrative* is frequently misunderstood as synonymous with story, brand, and message, so let us first look at these areas of distraction and inappropriate synonymity.

What a Narrative Is Not

Narrative is a buzzword frequently heard today in many circles, from marketing professionals to political pundits to government leaders. Marketers use the term when talking about brand identity. "What General Motors (GM) needs out of the Chevy Volt is a new narrative." Marketers are using *narrative* as a shorthand here, representing both an accumulation of new stories that create desirable features and characteristics ascribed to GM and ultimately favorable attitudes by consumers, which ultimately themselves translate into buying behavior. These ascribed characteristics and attitudes followed by actions yield [ideally] a resolution—public opinion of GM as a forward-looking, technologically sophisticated automobile manufacturer—and ultimately financial success for GM. But, for our purposes, it is important to understand that the Volt does not provide or create a new narrative for GM, but rather *participates in a narrative system.*

Political analysts invoke *narrative* to refer to a collective understanding of political trends or of a political candidate. Presidential candidate Barack Obama had a better narrative than John McCain during the 2008 US presidential campaign, analysts would say. "Hope and Change" was a better narrative. "Hope and Change," however, was a theme (again operating within a narrative system), but "Hope and Change" was not the narrative. Both candidates did exploit narrative systems to communicate aspects of their identity. John McCain embraced the moniker *maverick* tying his political unpredictability to a long-valued American ideal of individualism. From old west gunfighters and hotshot Hollywood fighter pilots bearing the name *Maverick* to black turtleneck wearing computer industry nonconformists and numerous other examples, convention-bucking individualists have been consistently valorized in stories circulating within American culture for decades. The stories of these many mavericks participate in a narrative system of individualism that expresses a hallmark trait of the American culture. Similarly, Barack Obama played on his modest family beginnings, challenges growing up, and relative inexperience on the national stage to position himself as the underdog hero, another trope supported by a narrative system born in the American revolution and consisting of historical and fictional stories circulating within the American culture ever since (think of Ben Franklin, Fredrick Douglass, Thomas Edison, and the whole genre of Horatio Alger stories).

Within government circles, policymakers and pundits also frequently misuse the term *narrative.* Defense analysts begrudgingly acknowledge the Islamic State of Iraq and al-Sham (ISIS), Al Qaeda, and similar Islamic extremist groups' adroit use of *narrative* and simultaneously bemoan the United States' inability to gain traction with our own narrative. Part of the problem is that the US governmental bureaucracy and its representatives tend to think and speak singularly about narrative instead of in plural terms (this point will be elaborated on in the following). Thus, they talk of *the* ISIS narrative or *the* US narrative when there are always many circulating. Another part of the problem is the lack of specificity used when invoking the term: frequently, government representatives use *narrative* as the equivalent of *message* or *appeal* and, in this generality, overlook the powerful functions of narrative.

In each of these cases, the speakers are invoking this buzzword expecting an intuitive understanding by their audiences. The word *narrative* certainly invokes connotations of story and of meaning, which are key elements of a narrative but not the entirety of the phenomenon.

What a Narrative Is

If *narrative* is frequently misused as a synonym for brand, theme, or message, how should we understand *narrative* in a manner that allows us to understand its powerful social and cognitive significance? Narrative has many facets that simultaneously function, and thus understanding narrative as a system helps reveal the interplay between these facets. Thus, a *narrative is a system of stories structured in such a way as to achieve a rhetorical purpose or vision*. The multiple facets of a narrative typically manifest in pairs and coalesce into the *dual logics of narrative*. The dual logics of narrative posit that narrative is simultaneously an artifact and a process; form and content; a top-down process and a bottom-up process. Understanding narrative as fundamentally a system allows us to conceptualize how these dual logics function. For our purposes here, we will focus on narrative as a sociocultural artifact and a cognitive process of understanding.

Sociocultural Artifact

First, *narrative* refers to a sociocultural artifact, whether a tangible object such as a novel or a film or a more ephemeral element of culture such as an epic poem recited from memory over generations, or a group of such objects such as a television series, group of movie sequels, and the like. In the same vein, narrative as sociocultural object also encompasses larger groupings of story material, such as collections of newspaper articles or snippets of news broadcasts. In other words, the source material for a narrative, when considered as a sociocultural object, is varied and extends beyond single units of the traditional storytelling mediums (such as book, television, or film). In this form, narrative is a *system of stories*. Thus, in the John McCain example earlier, the stories about his nonconformist, aisle-crossing legislative actions, coupled with his biographical stories of his rule-bending Navy career, coupled with the pop culture stories of characters named Maverick, all coalesce into a narrative system of independence, nonconformity, and individualism.

Definition: *Narrative* is a system of stories structured in such a way as to achieve a rhetorical purpose or vision. This definition borrows from Halverson et al.'s (2011) *Master Narratives of Islamist Extremism*, which defines *narrative* in strategic communication contexts (pp. 13–15 and Chapter 14). This definition is simultaneously congruous with definitions from the fields of literary studies, film and media studies, narratology, and other disciplines, but at the same time, it offers distinct advantages in cross-cultural decision making contexts. First, this definition separates a story (a sequence of actions/events by an actor/agent that contains a resolution or implied resolution) from a narrative (a system of such stories). This distinction, and the emphasis of narratives as systems, facilitates analysis at various levels such as culture-wide, community-based, and personal. This distinction also facilitates understanding how seemingly innocuous or anachronistic stories can be indicators of power cultural influences through their integration with other stories into a narrative system.

Narrative as Mode of Comprehension

In addition to its cultural function as an artifact, narrative also functions as a way of ordering data about the world to make sense and significance and is widely recognized across a variety of fields as a mode of comprehension. From communication studies to psychology to film and literary studies, increasing scholarly attention has focused on how narrative provides a framework with which individuals come to understand the world around them (Branigan 1992, p. xi).

In his *Narrative Comprehension and Film,* film theorist Edward Branigan combines cognitive science and the long history of narratology to offer a clear and detailed analysis of the comprehension process and its relationship to cultural objects. Comprehension involves the acquisition of data by an individual and the organization of this data by way of cognitive schema (mental patterns or structures for organizing pieces of information and their relationship). The schema, Branigan argues, are culturally provided, stemming from the templates of the story forms and story systems prevalent in a culture (Branigan 1992, pp. 13–17). Branigan suggests that the body of classical Hollywood film shares similar narrative structural traits to include abstract/orientation, goals, resolution, climax, and the like. As data are acquired, the individual assigns these nuggets of information a role based on the schema. This process continues until the relationship of the data is revealed and meaning is constructed. If the data arrive and contradict the order, the individual must reorder the existing data to achieve comprehension.

While Branigan's primary example of a narrative schema is based on the narrative structure of a body of Hollywood films, the principle applies to other narrative forms, and an individual may have any number of familiar narrative schemas from which to draw upon in the process of ordering a set of data about the world. The data could come in the form of scenes of a film, as in Branigan's primary analysis, or could be the events from a news broadcast or activities observed by an individual. A given narrative schema might be constructed more broadly with fewer constituent parts or more narrowly and therefore more applicable to certain situations. The narrative arc, in which a conflict gives rise to a desire that is pursued through a trajectory of action ultimately leading to a resolution and some satisfaction of that desire, is one such broad narrative schema that provides a framework for understanding information (for more on narrative arc, see the study by Burke [1989]). With a structure rooted in desire and leading to satisfaction, this particular schema is highly suited to aspirational, future-oriented contexts. Islamic extremist discourse exhibits both this aspirational dimension and also the phenomenon of simultaneous top-down schema application (culturally provided) with bottom-up data processing (individually constructed). In a study conducted on behalf of the US State Department's Digital Outreach Team, the Center for Strategic Communication at Arizona State University analyzed a group of extremist texts invoking the trope *war on Islam* (Corman et al. 2012). These texts contained story clusters featuring actions of steadfast *mujahideen* (holy warriors), actions of victorious battles, a unified *ummah* (a term referring to the Islamic community, not limited by national geographic borders), and other similar features. These stories participate in a narrative system with a war on Islam as the conflict, which generates different desires suitable to the different story clusters: defeat the West, defend the ummah, serve as a hero, establish a caliphate—all of which have potential satisfactions of the desire. When a young person decides to take up arms and wage jihad to achieve this goal or to realize the satisfaction of another narrative system (realize Allah's grace in paradise), they are deciding their path based on their narrative interpretation of the world around them.

Narrative Comprehension: Structures as Cognitive Schema

Scholars since the time of Aristotle have identified narrative structural forms. These forms serve as templates, provided by culture, with which humans organize data into meaningful arrangements of time, causation, and consequence. Aristotle was the first to suggest that narrative has a temporal dimension, composed of a beginning, a middle, and an end, which is now common to most conceptions of narrative in Western culture (Aristotle 1932 [trans.]). Early narrative theorist Gustav Freytag built upon Aristotle's framework and described a structural phenomenon of rising and falling actions, which imposes a further set of relationships between observed or experienced data (Freytag 1900). While the schema Branigan describes is more complex (with eight categorical components), he makes clear an important dimension that allows us to see that the Aristotelian structure and Freytag's triangle and other forms are culturally provided templates. Branigan explains that the temporal order, the rising/falling relationship, or other categorical relationships are imposed during the comprehension process, and that the concept of temporal arrangement (beginning, middle, end) is part of a cognitive schema, but the ordering of which pieces of data (events) into which temporal category is the cognitive activity (decisions) of the individual.

The Aristotelian three-act structure and Freytag's triangle are two broad models of narrative structure common in Western culture. The cognitive schema may be broad such as these, or may be more specific, drawing from patterns common within a culture or a subculture. The police procedural genre is an example of a more specific narrative structure. Typically, this narrative structure follows a chronological sequence, opening with the discovery of a crime and then following the steps the protagonist detectives take as they seek out clues leading to the identification of the perpetrator and the methods and motive of the crime. Events in the past, if presented, are stylistically differentiated (such as a black-and-white footage in television shows or italicized texts in novels), part of the culturally provided template for proper comprehension.

Films, novels, and television shows are not the only source for culturally provided narrative comprehension schema. Take, for example, the victorious battle stories commonly found on Islamic extremist websites. Lundry et al. (2012) studied 271 stories released in November 2010, sourced from extremist websites from Afghanistan and Southeast Asia. Within the subculture of violent Islamist extremism, there exists a common, simple, story form. In it, mujahideen engage in battle with infidel forces and emerge victorious. The battle is engaged either head-on or by way of an ambush. Victory is assured through either

Review: Gustav Freytag was a nineteenth-century German novelist and playwright who provided the European academy a detailed analysis of a narrative structure that is still influential today. While frequently (and erroneously) assumed to be a universal primer for narratives from all cultures, Freytag's analysis of rising and falling actions has influenced literary critics, as well as novelists, playwrights, and screenwriters to this day. When visually depicted as a line graph with time on the x axis and intensity of action related to the central plot conflict on the y axis, the result is a triangle giving rise to the term *Freytag's triangle* or *Freytag's pyramid*. Freytag's model remains a useful tool for analyzing narrative structure, particularly key points of consequence and resolution, and especially when applied to narratives from European, British, and American cultural contexts.

the cleverness or the bravery of the protagonists enhanced by Allah's favor or through the inattention or the unprofessionalism of the oppressing infidel, in this case the Western forces. The specific content varies from story to story, but the form repeats. This pattern provides a cognitive template for understanding the recruitment pitches, for example. When ideologues exhort disaffected young people to strike against an oppressor, the template offers a means to identify the oppressor and a means to visualize an expected victory.

In a converse example, US policy makers assured the US public that the citizens of Iraq and the Iraqi soldiers would greet US forces as liberators, in the face of Saddam Hussein's tyranny. However, this perspective overlooked the dominant narratives of invasion and oppression in the region. A widely known master narrative, one frequently invoked by Islamic extremists in part because of its widespread reach across the Middle East, is the crusader narrative (a *master narrative* is a long-lasting narrative system deeply embedded in a culture maintaining special explanatory power and investing the culture with values, traits, and beliefs; more on master narratives later on). The origin story is the invasion of Muslim lands in the Middle Ages by Christian knights from Europe, and this model story has been repeatedly used over time to explain Western arrivals in the Middle East (Halverson et al. 2011, p. 24). Thus, just as the victorious battle stories provide a cultural template that sets up understandings of conflict between the Western forces and the mujahideen resulting in mujahideen victory, so too does the narrative landscape of the Middle East offering a schema that positions Western forces as the invader and Muslims as the oppressed.

Narrative Comprehension: Bottom-Up Processing and Data/Schema Mismatch

Branigan notes that the process of narrative comprehension is an ongoing process of trial and error fitting the acquired data into schema categories. If, however, the acquired data cannot be organized into the schema categories, one of two outcomes must result: either a new schema must be employed (and the trial-and-error procedure repeated) or confusion (Branigan 1992, pp. 17–20). This explains the common reactions to a film like *Pulp Fiction*, a film in which the storytelling mode contests the expected structure of Hollywood films. The film plays with time and characters, omits connecting scenes, and uses few stylistic cues to signal temporal shifts. This induces confusion while the viewer tries to identify which scene establishes the main plot, which scene provides the climax and the resolution, and which scenes represent intermediate goals. By the end of the film, however, the disjointed scenes have been slotted into suitable schema categories and comprehension is achieved.

This phenomenon of struggling to apply culturally provided cognitive schema was observed during focus group tests as part of a large, cross-cultural narrative research project conducted at Arizona State University. While intended solely to be informative in the production process of experimental stimuli (videos), the focus group observations offer anecdotal evidence of the confusion resulting from a disconnect between top-down schema and bottom-up processing. Three different cultural subgroups (Christian, Muslim, Hindu) were shown videos depicting contemporary stories modeled on well-known narrative structures drawn from two of the subcultures (Christianity and Islam). During the focus group discussion, the Hindu subjects expressed confusion and lack of comprehension of the key points of the videos based on Christian and Islamic narrative structure sources. Most notable to experiment observers was the Hindu subjects' reference to a contemporary American television genre (police procedure docudramas) when discussing one video that involved a missing body. The Hindu subjects discussed the types of clues that the video lacked that would be required to solve the mystery of the missing body.

Christian subjects overlooked missing clues and focused on resolution elements recognizable from the Christian master narrative. The Hindu subjects, lacking this cultural referent to assist in comprehension, employed the different template of a popular television genre's narrative structure to try and make sense of the video.

Narrative Rationality

Why does narrative comprehension matter? It is more than simply because narrative is one of the ways we make sense of the world, a cognitive process of ordering information and actions into a framework of meaning and significance. It is also more than the role of culture providing templates used in this sense-making process. It is because, as the storytelling beings, narrative is another form of rationality that competes with and often supplants logical rationality as the basis of decision making. For example, communication theorist Walter Fisher, in his seminal article "Narration as Human Communication Paradigm" and subsequent work, lays out the role of narrative in communication, understanding, and decision making (Fisher 1984). Writing over a decade before the advent of cognitive narratology that draws on the advancements of psychology and cognitive neuroscience (cognitive narratology is the field that established the simultaneous bottom-up/top-down process of understanding influenced by culturally provided templates described earlier), Fisher's rhetoric-based argument is congruent with these later advances in narrative studies. Fisher argued that logical rationality (cost-benefit analyses and similar weightings of evidence), presumed to be the primary basis of persuasion and argumentation, is an incomplete formulation. In his thinking, the logical rationality approach does not account for emotion, roles of beliefs and values, and many cases in which individuals and groups support positions for which there is no logical explanation.

Fisher argued that since individuals understand their lives as narratives (interlocking systems of stories making sense of characters, actions, events, and resolutions), narrative rationality governs their understanding of arguments as well as decision choices (Fisher 1984). People are persuaded by good narratives, which are narratives that comport with an individual's values and beliefs, and they do so by exhibiting the combination of narrative coherence and narrative fidelity. Narrative coherence is the degree to which a story hangs together and remains internally consistent with the conditions established as defining the narrative world. Deus ex machina and unexplainable actions are violations of the narrative coherence. Working in concert with this internal consistency is narrative fidelity, which is the degree to which a story rings true in accordance with other stories the individual already believes to be true. Narrative coherence, then, is analogous to the bottom-up processing of identifying roles for each story component. Narrative fidelity is analogous to the provision of templates by culture for cognitive understanding.

In *Narrative Landmines: Rumors, Islamist Extremism and the Struggle for Strategic Influence*, Bernardi et al. (2012) recount a situation in the early stages of the Iraq War, when the Iraqi dairy cattle were susceptible to disease made worse by drought conditions (pp. 73–75). Seeing an opportunity to support local Iraqis and demonstrate goodwill, US Army civil affairs teams set out to provide inoculations for the cattle. Rumors spread, however, that the US Army veterinarians were poisoning the cattle. While US forces may have dismissed the rumor as a silly falsehood, it contains both narrative coherence and narrative fidelity. Iraqi dairy farmers saw the veterinarians administer the medication and also saw their cattle die (obviously, some cattle would be beyond saving). The inoculations actually poisoning the cattle does not violate any sort of internal story logic, because the observed events support this outcome. Secondly, the narrative landscape of Iraq is filled with stories of invasion, destruction, and exploitation, especially by Christian Westerners. Stories of poisoned cattle bear much more in common with this rich cultural heritage of invading Christians than do stories of beneficence and support, and Iraqi farmers resisted the US Army program with some actively supporting the insurgent forces. At the core of fidelity is not only a consistency in actions, but also in the conveyed meaning and shared beliefs between those stories already accepted as true and any new stories encountered. The bovine poisoning rumor case illustrates how the understanding of the narrative landscape and how it influences decision making by individuals, groups, and subgroups is crucial. Thus, the key takeaways from the narrative paradigm are that good narratives persuade and inform decisions because they instantiate and make manifest values, beliefs, and ideology and do so by linking present events and actions to historical and cultural referents.

Narrative and Ideology

The centrality of values to Fisher's notion of fidelity highlights an essential function of narratives within culture: they both communicate the values and the beliefs of a culture as well as participate in the creation of these values and the expression of their significance. The history of the American Revolution includes stories of an unresponsive and distant government, stories of taxes and other laws imposed without consent of the governed, and ultimately stories of resistance. These stories coalesce into a system exemplified by the Boston Tea Party, in which rebels took a symbolic action to protest the actions of the British government. The events leading up to the American Revolution and how Americans made sense of them wove into the fabric of the American culture the values of representative government, the rightness of nonviolent protest, and the valorization of underdogs striking against the powerful. The stories of this history, and in particular the exemplar story, convey these values to new generations of Americans.

What sort of values the narrative system supports depends on the relationship of the various components. In the Boston Tea Party example, the British government occupies the role of the antagonist and the Sons of Liberty as the protagonist. A key element cementing the values of individual liberty and representation is that the action taken by the actor/agent (Sons of Liberty) advanced toward the resolution of the Americans earning independence from British rule. Had this story system resolved differently (Americans defeated by the British, for example), the narrative arc resolving in representation, liberty for the American colonies, would not be realized. The Boston Tea Party would not be an exemplar story of a narrative system expressing such values (and might, in fact, be a small story in a narrative system emphasizing loyalty to the crown).

Master Narratives

Owing to its transhistorical nature and its widespread knowledge across a culture, the narrative system exemplified by the Boston Tea Party constitutes a master narrative. In their study *Master Narratives of Islamist Extremism*, Halverson et al. define a master narrative as a system of stories deeply embedded within a culture that communicates specific values, beliefs, and traits over time (Halverson et al. 2011, p. 24). Master narratives contribute to identity formation, as they combine individual conceptualizations of belonging to a culture with those values and ideology expressed by the master narratives. The Boston Tea Party and its associated stories have served as an exemplar story of the American Revolution. The valorization of Samuel Adams, John Hancock, and the other Sons of Liberty connects contemporary members of the American culture to the values of individualism, liberty, and representation. In Muslim culture, for example, the pharaoh master narrative communicates the villainy of tyrants, their blasphemous usurpation of reverence due to Allah, and the righteousness of serving as an agent of Allah against a tyrant. This particular master narrative and its values were referenced by the assassin of Anwar Sadat, Lieutenant Khalid Al-Islambouli. Anwar Sadat earned the enmity of hard line groups after a crackdown on them following the Camp David Accords. The Islamists branded him a pharaoh for his tyrannical rule. Demonstrating the combination of individual belonging with the values of cultural narrative, Lieutenant Al-Islambouli shouted "I have killed the pharaoh" after shooting Sadat during a military parade on October 6, 1981, and repeated the claim at his trial (Halverson et al. 2011, pp. 33, 192).

Ideological Functions and Narrative

Along with producing and expressing values and beliefs, a narrative is also highly intertwined with ideologies circulating within a culture. Simply defined, an ideology is a system of ideas about how the world works or should work. Most approaches to the study of ideology seek to identify the relationship between the expression of ideology—public discourse, visual art, fiction—and the source of ideology—the unconscious, culture or cultural groups, social structures (Herman and Vervaeck 2013). However, this position presumes that ideology is something fixed, a fait accompli that once absorbed by an individual is unchanging. Trethewey et al. (2009) argue that ideology is mutable and constantly subject to change because the system of beliefs and the expression within the social discourse cannot be separated. And narrative is a key venue where this interplay occurs. "Ideology is, fundamentally, what makes power and control possible" from a communication perspective and is thus closely connected to both group and individual decision making (Trethewey et al. 2009, p. 4).

Review: *Ideology* is an often contentious term in mainstream public discourse, one frequently invoked when accusing another person or group of holding unsavory or inappropriate or even morally reprehensible ideas. Frequently in American culture, ideology is positioned as a negative—something the adversary has, such as Soviet communism or Islamist extremism. However, all cultures have ideologies. They are central, just like narrative, to how we make sense of the world. The Oxford English Dictionary defines *ideology* as "a system of ideas or ideals especially one that forms the basis of economic or political theory and policy" (http://www.oxforddictionaries.com). French philosopher Louis Althusser, one of the leading academics to deeply investigate ideology and its social operation, extends this definition to include aspects of material and social practice; that is, the rituals of everyday life are unrecognized aspects of a culture's ideologies (Storey 2006, pp. 3–4). To illustrate these principles and how they apply to cross-cultural decision making, this chapter suggests ideological components of the narratives of the American Revolution but makes no normative nor moral judgment.

Drawing on their analysis of communication, media, sociology, and critical theory, these authors further argue that narrative is one communication venue through which ideology executes its four political functions: naturalizing, obscuring, universalizing, and structuring. Naturalizing turns socially constructed situations into taken-for-granted states of affairs. Obscuring elides details revealing contradictions and power imbalances within the ideological system. Universalizing positions the values and the benefits of the system (as defined by those in power) as useful and applicable to all members of the group. And, an ideological system (leveraging the other three functions) creates rules and arrangements designed to perpetuate the system—this function is structuring (Trethewey et al. 2009, pp. 4–8).

The ideology of America as the icon of liberty and justice is a powerful one, supported by a wide range of stories and narrative systems, including that of the Boston Tea Party mentioned earlier. The stories in this system position the British as the oppressors and the Sons of Liberty as simply fed up colonists as the natural state of affairs. This taken-for-granted status obscures the details of Samuel Adams' and John Hancock's (among others) highly profitable tea smuggling enterprises, which would have been negatively affected by the Tea Act of 1773. Patriotism and liberty are of course values we assign to the Sons of Liberty and believe motivated Adams' and Hancock's leadership, but illegal economic gain does not fit the ideology of liberty and patriotism and thus is obscured by omission. Similarly, by casting the tea situation as a matter of freedom in commerce and taxation, the Boston Tea Party becomes a story universalized for all Americans, and one that allows existing economic power structures (notably the colonists' tea businesses) to remain. The conceptual parallels and ideological functions remain today (with slightly different names), which explains the frequent invocation of the Boston Tea Party in twenty-first century American public discourse.

Likewise, from ISIS to Al-Qaeda an arguable case can be made that these groups' method of leadership and governance is more akin to tyranny than that of Anwar Sadat or Barack Obama or any number of leaders branded pharaoh by Islamist extremists. However, by invoking the widely known master narrative of the pharaoh, the extremists universalize their opposition to a particular leader, obscuring their own brutal and autocratic governance and naturalize the state of affairs that the ummah is oppressed or under attack by these impious tyrants. By connecting the opposition to tyranny as a duty of *sharia*, and by further positioning themselves between the agents of Allah (the Moses role, the Lieutenant Khalid Al-Islambouli role, etc.), extremist leaders create the structure that perpetuates their positions of power, while the foot soldiers of the movement (like Al-Islambouli) choose to take action and are killed (*martyred* in another ideologically laden and complementary narrative system).

Values, Ideology, and Decision Making

Understanding that narrative plays an essential cultural role in creating and expressing ideology, and also understanding that narrative contributes to the functions of an ideological system, can provide significant insight into a culture. This insight can help elucidate what might otherwise seem contradictory or inexplicable dimensions of a culture, and can guide cross-cultural interaction at the organizational or the individual level. Furthermore, numerous studies have explored the link between the values held by an individual and the effect of these values on choices and decision making. Verplanken and Holland (2002) provide an exemplar study of the connection between values and choice behaviors (decisions). After conducting six different studies manipulating personal values and choices, Verplanken and Holland conclude that there exists a profound connection between choices and values (i.e., values guide decisions), but qualify this connection with a caveat: the effects are most significant when the value is cognitively activated and also linked to the individual's self-concept. In Verplanken and Holland's study, the values are activated by word lists, surveys, and asking subjects to imagine persons holding certain values; the self-concept is activated through an examination of a travel story told in the first person.

Verplanken and Holland's use of imagination of persons (characters, which are components of narrative systems) and travel stories, along with the preceding discussion of the role of narratives expressing cultural values, reinforces an otherwise self-evident connection: narratives provide the cognitive activation of values. Master narratives, such as the pharaoh example from Muslim traditions or the Boston Tea Party example from American history, express values widely held by a culture and do so by ascribing these values to the characters represented or associate values with the consequences of events. Narrative systems also play an important role in individual self-concept, another factor cited by Verplank and Holland in the connection of values to decision making. The Lieutenant Al-Islambouli episode described earlier demonstrates a personal dimension and integration with a larger, value-laden narrative system. The pharaoh master narrative contains values of piety, subservience to Allah, and agents of Allah facilitating a punishment for transgressions. Islamist extremists operating in Egypt in the early 1980s branded Anwar Sadat a tyrant and pharaoh. Lieutenant Al-Islambouli projected himself into this narrative system by accepting Anwar Sadat as a pharaoh (as evidenced by Al-Islambouli's claim after shooting Sadat) and taking action as an agent of Allah.

Understanding that narratives operate as systems, simultaneously at the cultural, local, and personal levels, presents the connection between humans' understandings of their sense of self, their sense within a community/culture, and their understanding of the events of the world around them. These narrative systems are also imbricated with the formulation and the reification of the expressed values and cementing these connections between the culture and the self (Verplanken and Holland 2002).

Thus, if we return to the question of liberation or invasion that was made famous by Vice President Cheney during his interview with Tim Russert on *NBC Meet the Press* on March 16, 2003, we can see how different narrative systems from different cultures inform an individual's reactions to the same events. Combining stories of wars to liberate holy cities (origin of the Western connotation of crusade as a righteous fight), stories of American forces aiding overseas allies (World War I), and stories of American forces vanquishing the forces of tyranny (World War II) positions the decision to go to war in Iraq as congruent with American values that are supported by the narratives of American history: the legacy of the American Revolution and both World Wars creates and reinforces the American values of opposition to tyranny and a moral obligation to defeat tyranny on a global stage. Comparisons of Saddam Hussein to Hitler by both Presidents George H. W. Bush and George W. Bush further contribute to this narrative system and associated values. In light of this comprehension framework, the concept of American forces entering Iraq as liberators in 2003 is plausible—there is a narrative logic to the assertion by Vice President Cheney that US troops "will, in fact, be greeted as liberators." However, the narrative landscape of Iraq includes a different set of stories. The crusades is a narrative system of invasion by Christian armies. Stories of the colonial era and the resource exploitation by European companies further fill this narrative landscape with distrust of non-Muslim, Anglo-European strangers. The Western support of Israel, long a reviled antagonist in Arab and Muslim minds, further illustrates an entirely different narrative landscape, one espousing the values of Iraqi nationalism, opposition to Christian invaders and the like.

Actions: Shared Component of Narrative and Decision Making

So far, we have discussed the role narrative plays in shaping the comprehension about the world and also how narratives are connected to beliefs and values. The combination of these two elements informs the narrative paradigm, which asserts that humans make decisions based on a narrative rationale instead of or in addition to a logical rationale. Both of these elements, the role of narrative in comprehension and the role of narrative in belief/value creation and expression, offer insights into cultural differences and their foundations. Greater understanding of these differences and foundations leads to a more adroit operation within a cross-cultural environment. A deeper examination of the narrative structure also uncovers a close connection between narrative and decision making. At its core, a narrative structures meaning through the cause-and-effect relationship of its systemic components, especially actions. And, as the results of decisions are actions, the narrative provides a framework for understanding how decisions might participate in a narrative system, including how a decision might interact with existing actions within a narrative landscape and either advance or obstruct a trajectory toward a desired resolution.

To more fully understand this connection, let us return to a basic form of narrative structure. Narratives are systems of stories that coalesce around a trajectory from a desire (usually born of conflict or lack) and toward a resolution of that desire. In this formulation, a narrative system's key components include conflict/lack (there may be multiple), desire, resolution, events/actions, actors/agents, stories, and linkages to related narrative systems (commonly in the form of archetypal characters and repeating story forms). Forward-looking, aspirational narratives forecast the resolution; historical narratives explain the resolution. Spanning the trajectory are the events that advance or obstruct the progress toward a resolution. Resulting from the combination of the advancement/obstruction actions, the roles played by the agents affecting the actions, and the relationships to other stories, actions, and characters within the narrative milieu, is the structure of meaning created by the narrative. So, for example, the film *Star Wars* is a narrative consisting of a system of interrelated stories: the story of Leia's capture, the story of the droids escape to Tatooine, the stories of Luke's father (a story *told* within the film rather than *enacted,* a difference that matters for artistic analysis but not for our purposes here) and Luke's life on the desert planet, the story of the Rebel Alliance, and so on. This story system reveals Luke's desire to leave Tatooine, which lacks adventure, and soon this desire is further motivated by the main conflict with Darth Vader. Upon finding his uncle's farm destroyed and his family killed, Luke has multiple options to pursue. But, he makes the decision to join the Rebel Alliance. The structure of the *Star Wars* narrative leading up to this point illuminates the narrative rationality of Luke's decision (in the face of the more logical decision to surrender the droids and safely live out a farmer's life in obscurity).

This connection of the narrative components to decision making is evident not only in fictional narratives but also in the frameworks of people's lives. In the example of Lieutenant Al-Islambouli, his exhortation that he had killed the pharaoh provides strong evidence that his decision to join the plot to assassinate Anwar Sadat was influenced by his understanding of contemporary Egyptian politics through the narrative framework invoked by extremist rhetoric, as well as influenced by the desire to connect his life to the values communicated by the pharaoh master narrative.

Conclusion

How then does all of this become relevant to modeling the influences on decision making in a cross-cultural context? How does the knowledge that humans are storytelling beings that comprehend the world around them and their place in it using narrative templates help understand decision making? If logical rationality can be assessed through a cost-benefit analysis or some other objective and empirical method, how can narrative rationality be evaluated to forecast decisions that exhibit narrative fidelity and narrative coherence? What can Xanadu leaders do to forecast potential actions by Bursuk groups and how can Xanadu leaders think about Razini reactions to potential policy proposals?

Obviously, cultural narratives provide a rich repository of values, beliefs, and perspectives that offer in-roads for cross-cultural understanding. And, with the knowledge that narratives are value-shaping, ideology-expressing, and comprehension-guiding phenomena that operate at macro- (cultural), meso- (local, community, subcultural), and micro- (personal) levels, a closer analysis of how desires and conflicts are understood and how actions advance or obstruct progress toward a resolution can provide powerful insights into the decisions of individuals and groups. The war on Islam analysis conducted by the Center for Strategic Communication at Arizona State University includes a cluster of stories imbuing the United States and its allies with traits of treachery, duplicitousness, and betrayal. Within that cluster and related clusters, stories abound that celebrate the steadfastness of the mujahideen in their defense of the ummah in the face of this treachery. There is also a story of the mujahideen coming armed to the negotiating table, in violation of the negotiation terms. However, from a narrative perspective, the decision of the mujahideen to arm themselves is consistent with their understanding of their opponent (treacherous), in accordance with their desire (defend the ummah), and in line with an aspirational resolution of their desire (remain steadfast in defense) (see Corman et al. 2012, pp. 21–25). While an objective analysis of the mujahideen decision to remain armed might center on the violation of an agreement, a narrative analysis reveals a different, but understandable, logic. Similarly, an accumulation of stories from the Razini or Bursuk groups could reveal how these communities or their leaders might approach negotiations, policy changes, or outreach efforts by Xanadu.

A narrative is not a panacea, nor is it always precise, but it is ignored or dismissed at significant peril. While neuroscientists have yet to fully map the narrative comprehension network in the human brain in order to fully explicate all the connections between narrative, memory, belief, attitude, and behavior, there is no doubt that the connection is powerful. Studies showing the activation of motor neurons in response to stories of the same physical motion and studies showing greater response to narrative-based public service announcements provide ample evidence (Tettamanti et al. 2005; Baezconde-Garbanati et al. 2014). The power of narrative as an influence on decision making and cross-cultural environments resides in three aspects of narrative. First, and most importantly, narrative is a mode of comprehension. As a familiar template for processing incoming data about the world, narratives provide a mechanism for humans to quickly decide whether they believe or discard information (narrative fidelity) and how they interpret the relationships of actor/agents to event/actions and what significance to draw. Secondly, a narrative both shapes and expresses ideology and executes the political functions of ideology. Understanding how Razini narratives obscure inherent contradictions within the Razini society and universalize conditions faced by some Xanadu residents can lead Xanadu leaders to more effectively engage with obstructive Razini discourse, for example. Lastly, since narratives structure actions into patterns of significance, a thorough understanding of narrative patterns (and patterns that consistently exhibit narrative coherence, which is one such structuring of actions) will lead to understanding what sorts of decision outcomes are congruent with the narratives that influence a culture, a community, or an individual.

DISCUSSION QUESTIONS

1. What influences your personal decision making, such as consumer choices or voting choices? Do you analyze specific criteria, assign relative value, and calculate cost-benefit? Are you persuaded by stories relating to the choice, such as the life story of a political candidate or stories characterizing a consumer brand? How do these stories intersect with other stories that are meaningful to you (i.e., form a system)?
2. Have you ever observed a group or an organization make decisions that seem illogical to you? What stories circulate within or around that group or organization, and how does the decision relate to outcomes depicted, values expressed, or themes embedded?
3. Examine some key stories from your cultural heritage, whether national or community history, ethnic group, religion, or similar element. How are these stories working together to create meaning and express beliefs? Can you identify contradictions that are obscured or assumptions about the world that are naturalized?
4. Think about the stories and the narrative systems that help define the culture of which you are a part of. Now think about a subset of that culture, a community of which you are a part of. How do the stories intersect? Do they integrate into a system (i.e., a narrative)? What influence do these stories have on the story you tell about your life and your future?

Acknowledgments

This study was supported in part by grants from the Department of Defense (DOD) Human Social Culture Behavior Modeling Program (Office of Naval Research [ONR] Grant #N00014-09-1-0872) and the Defense Advanced Research Projects Agency (Grant #D12AP00074). The author acknowledges the helpful contributions of the Center for Strategic Communication (CSC) grant teams, as well as the guidance of these DOD research communities.

References

Aristotle. 1932 (trans.). W. H. Fyfe. *Poetics*. Available online: http://www.perseus.tufts.edu/hopper/text?doc=Perseus%3atext%3a1999.01.0056.

Baezconde-Garbanati, Lourdes A., Joyee S. Chatterjee, Lauren B. Frank, Sheila T. Murphy, Meghan B. Moran, Lisa N. Werth, Nan Zhao et al. 2014. Tamale lesson: A case study of a narrative health communication intervention. *Journal of Communication in Healthcare* 7/2: 82–92.

Barraza, Jorge A., Veronika Alexander, Laura E. Beavin, Elizabeth T. Terris, and Paul J. Zak. 2015. The heart of the story: Peripheral physiology during narrative exposure predicts charitable giving. *Biological Psychology* 105: 138–143.

Bernardi, Daniel L., Pauline Hope Cheong, Chris Lundry, and Scott W. Ruston. 2012. *Narrative Landmines: Rumors, Islamist Extremism and the Struggle for Strategic Influence*. New Brunswick, NJ: Rutgers University Press.

Branigan, Edward. 1992. *Narrative Comprehension and Film*. New York: Routledge.

Burke, Kenneth. 1989. *On Symbols and Society*. Chicago: University of Chicago Press.

Corman, Steven R., Scott W. Ruston, and Megan Fisk. 2012. A pragmatic framework for studying extremists' use of cultural narrative. In *Advances in Human Factors and Ergonomics 2012-14 Volume Set: Proceedings of the 4th AHFE Conference 21–25 July 2012*. Salvendi, Gavriel, and Waldemar Karwowski, eds. Boca Raton, FL: CRC Press.

Fisher, Walter. 1984. Narration as human communication paradigm: The case for moral public argument. *Communication Monographs* 51: 1–21.

Freytag, Gustav. 1900. *Freytag's Technique of the Drama: An Exposition of Dramatic Composition and Art*. MacEwan, Elias J., trans. Chicago: Scott, Foresman & Company. Available online: https://archive.org/details/freytagstechniqu00freyuoft.

Green, Melanie C. and Timothy C. Brock. 2000. The role of transportation in the persuasiveness of public narratives. *Journal of Personality and Social Psychology* 79/5: 701–721.

Halverson, Jeffry R., Steven R. Corman, and H. L. Goodall, Jr. 2011. *Master Narratives of Islamist Extremism*. Basingstoke: Palgrave Macmillan.

Herman, Luc and Bart Vervaeck. 2013. Ideology and narrative fiction. In *The Living Handbook of Narratology*. Hühn, Peter et al., eds. Hamburg: Hamburg University. Retrieved September 30, 2013. Available online: http://www.lhn.uni-hamburg.de/article/ideology-and-narrative-fiction.

Lundry, Chris, Steven R. Corman, R. Bennett Furlow, and Kirk W. Errickson. 2012. Cooking the books: Strategic inflation of casualty reports by extremists in the Afghanistan conflict. *Studies in Conflict & Terrorism* 35/5: 369–381.

Storey, John. 2006. *Cultural Theory and Popular Culture: An Introduction*. Athens, GA: University of Georgia Press.

Tettamanti, Marco, Giovanni Buccino, Maria Cristina Saccuman, Vittorio Gallese, Massimo Danna, Paola Scifo, Ferruccio Fazio, Giacomo Rizzolatti, Stefano F. Cappa, and Daniela Perani. 2005. Listening to action-related sentences activates fronto-parietal motor circuits. *Journal of Cognitive Neuroscience* 17/2: 273–281.

Trethewey, Angela, Steve Corman, and Bud Goodall. 2009. *Out of Their Heads and Into Their Conversation: Countering Extremist Ideology*. Consortium for Strategic Communication White Paper. Report #0902. Retrieved September 14, 2009. Available online: http://csc.asu.edu/wp-content/uploads/pdf/123.pdf.

Verplanken, Bas and Rob W. Holland. 2002. Motivated decision making: Effects of activation and self-centrality of values on choices and behavior. *Journal of Personality and Social Psychology* 82/3: 434–447.

Zak, Paul. 2015. Why inspiring stories make us react: The neuroscience of narrative. *Cerebrum* February, 1–13.

3

The Spread of Information via Social Media

Brian M. Fairlie

CONTENTS

The Growing Use of Social Media in Global Events 43
Social Media Use in Social Movements 44
Social Media for Misinformation and Propaganda Purposes 45
Dissent: Suppression and Augmentation via Social Media 46
Steps to Help Improve the Spread of Messaging 47
Understanding the Media Environment 47
Audience Attributes: Behavioral Habits, Consumption Habits, etc. 48
Message Length 48
The Value of Hashtags 49
Timing for Optimum Exposure 51
Volume 51
Social Network Analysis 51
User Roles within a Network 52
Narrative Analysis 52
In Summation 53
Case Study: Megacity Xanadu 54
References 56

ABSTRACT Given the growing popularity of digital media, it is important to understand both how information spreads and how such knowledge can be exploited to further spread one's own messaging. Furthermore, with the growing use of social media platforms in events taking place in Ukraine, Egypt, and Turkey, as well as the rise of the Islamic State, it is clear that the role of social media is only likely to grow in the future. With that in mind, it is important to understand the different uses of social media for information dissemination as well as some of the methods available for communicating with target audiences. What follows is a brief guide on social media's growing role in the dissemination of information and how practitioners can better conduct information operations in this relatively new media format.

The Growing Use of Social Media in Global Events

Beyond the relatively mundane application of social media for sharing photos and communicating with friends and family, social media allows the general population a level of social influence potential that has never before been available to the masses. Social media allows completely unknown individuals to express ideas that have the potential to have a global impact. Yet, while social media allows individuals a tremendous ability to express ideas and dissent, it also, simultaneously, provides an avenue for not only the spread of misinformation, but also the very suppression of the dissent intended. This section will examine the role of social media in protests, in misinformation campaigns, and for the suppression of dissent.

Social Media Use in Social Movements

While social media had existed for a number of years before the Arab Spring, its role in coalescing people during the events of 2010 laid bare its potential for changing the course of history. Platforms such as Facebook and Twitter allowed protestors not only to coordinate with one another, but also to rapidly broadcast their messages to international audiences. For instance, during the Egyptian revolution in 2011, protestors commonly used Facebook to organize gatherings and entice recruits (Gustin 2011).

In late 2013, Ukrainian pro-European Union (EU) protestors took to the streets in order to protest the administration of President Viktor Yanukovych. This movement, known as the Euromaidan movement, called for Yanukovych's ousting and closer relations to the EU. Like the Egyptians, Euromaidan protestors used Facebook for organization purposes, such as delivering supplies, organizing meetings, and spreading awareness. However, realizing that local social media platforms (such as Vkontakte [VK]) were not being monitored by Western news outlets, the group took to Twitter to disseminate their message globally (Hilleary 2014). The increased exposure quickly elevated the story to the world's stage and helped with bringing international pressure on the Yanukovych administration.

Social media users have developed even further uses for the medium, such as communications during natural disasters and broadcasting information for public consumption during epidemics. During the Ebola outbreak of 2014–2015, organizations such as the Center for Disease Control and the World Health Organization utilized platforms like Twitter to attempt to quickly provide the public with accurate information. On the other hand, fear and confusion surrounding the disease also led to a great deal of misinformation and fear mongering to run rampant, such as false reports of outbreaks and fake cures (Blair 2014). With that said, social media has also seen itself used for the public's benefit. In April 2015, Nepal suffered a massive earthquake and caught the Nepali government unprepared. People from across the country and beyond worked together and often used platforms like Facebook to arrange rescues, coordinate the delivery of supplies, and help families determine whether loved ones survived the event (Sarmah 2014; *The Christian Science Monitor* 2015).

Another prominent example, one that has been widely studied, was the bombings during the 2013 Boston Marathon. Following the attack, many citizens took to social media to try to determine what had happened, who was responsible, and tell the stories of those who were injured (Ma 2014). What happened instead was a series of wrongful accusations and false stories of victimhood, such as one involving an 8-year-old girl, who was claimed to have died in the attack (Madrigal 2013).

Social Media for Misinformation and Propaganda Purposes

Since the beginning of the conflict in Ukraine in early 2014, social media has seen a rise in the use for misinformation and anti-West propaganda via platforms like Facebook and Twitter. Beyond simple organizing, Euromaidan and pro-Russian protestors both engaged in misinformation campaigns, each portraying the other as responsible for various atrocities (Dettmer 2014). Pro-Ukrainian and pro-Russian users have since been battling one another with posts intended to misinform the general public and memes meant to ridicule the opposition. For example, pro-Russian protestors began disseminating information that suggested that the pro-EU movement was being orchestrated by Ukraine's Right Sector, a far right wing political group. These proregime protestors exploited old fears of Nazi atrocities carried out against the Ukrainian and Russian people. Using social media, users connected the Right Sector and the Euromaidan movement with propaganda via Twitter, Facebook, VK, and various blogs and forums. One of the more notable instances of such propaganda followed a shootout between pro-Ukrainian and pro-Russian forces. Pro-Russian social media accounts began disseminating images of business cards that supposedly belonged to Right Sector's leader, Dmytro Yarosh and claiming that the cards were found at the scene of the fight (Jeltsen 2014). In response to what many viewed as an attempt to portray the right wing group as instigating violence, counterpropaganda began to emerge, designed to mock the original intention of the propaganda piece, involving the business card being visible at various events throughout history (Know Your Meme 2014; Radio Free Europe/Radio Liberty 2014). Another prime example is the resulting disinformation efforts following the downing of Malaysian Airlines flight MH-17 in July 2014. Following the incident, both Ukrainian and Russian supporters began accusing one another of being responsible for bringing down the plane. On VK, Igor Girkin, a pro-Russian separatist leader discussed downing a Ukrainian aircraft, even saying that "We warned them—don't fly in our sky" (*The Interpreter* 2014). After realizing that it was a civilian aircraft, Girkin quickly deleted that post (Miller 2014). In addition, an automated Twitter account that tracks edits made to Wikipedia pages tracked numerous edits to the page were made from Internet protocol addresses belonging to the Russian government (Devichand, Sampat, and Lee 2014).

One popular form of countering disinformation and stoking sentiment against the opposition is the so-called meme. Internet memes are a subcategory of the general meme, a term first coined by Richard Dawkins in his 1976 book, *The Selfish Gene* (Solon 2013). A *meme* is defined as "an idea, behavior, style, or usage that spreads from person to person within a culture (Merriam Webster 2015). Internet memes are similar, except that they spread via the Internet and often mimic some other idea (Solon 2013). In addition, Internet memes leave a trail across social media platform and websites, allowing analysts the ability to potentially trace these ideas back to their origins. Internet memes often, although not always, come in the form of an image with a short amount of text to accompany it. The previously mentioned counterpropaganda surrounding Dmytro Yarosh's business cards is a prime example of a meme. Other popular memes include the "Ridiculously Photogenic Freedom Fighter" or the various image macros discussing Russia's invasion of the Crimean region of Ukraine (Ward 2012; Reilly 2015).

While the pro-Ukrainian/pro-Russian contest continues to serve as a prominent example of misinformation operations, a new and more sinister example has emerged in the form of the extremist organization, the Islamic State. While the group had existed for several years before, it gained the world's attention in June 2014, as it conquered large portions of Iraq in a matter of weeks. In addition to its battlefield successes, the Islamic State (also known as ISIS) announced its presence to the world through its attention-grabbing social media campaigns. With polished videos featuring battlefield footage and violent executions, ISIS took to social media channels like Facebook, YouTube, and Twitter to spread its message and to warn against opposing its goals. In its attempts to establish dominance with images and videos of battlefield victories and brutal executions, ISIS also sought to lure more recruits to its proclaimed caliphate (Masi 2015). Beyond ISIS' impressive visual social media campaigns, ISIS members and its supporters have taken to pushing misinformation and misdirection when suffering setbacks at the hands of the international coalition. As the group loses territory to Iraqi or Kurdish forces, ISIS often tries to change the narrative with gruesome executions; such was the case following the loss of the city of Kobani in Syria (Danin 2015). Within days of fleeing the city, ISIS released the video of the execution of a captured Jordanian pilot, involving him being set on fire while inside a cage. Instead of celebrating the Kurds' battlefield victory, most of the world was concentrating on the horrific immolation.

Dissent: Suppression and Augmentation via Social Media

Social media is also used to find and suppress social media-based dissent. For example, the Russian Federation maintains an army of operators, whose sole job is to propagate pro-Russian narratives and harass those who criticize the Russian government. These individuals monitor chat forums, blogs, and social media platforms, keeping an eye out for anyone speaking ill of the Kremlin (Sindelar 2014). When they do discover such talk, they begin to swarm the opposing user. The term *swarm* refers to a tactic that calls for "striking at a target using dispersed, flexible and seemingly independent social media users. Swarming units operate in a coordinated manner to attack from several directions simultaneously" (Howard 2013). On social media platforms, dissenting users are often bombarded with wild accusations, such as being CIA operatives or fascists, looking to overthrow the Russian government and harm the Russian people. These Russian-backed accusers, commonly referred to as *trolls*, are often paid for each post they make. Naturally the purpose of such activity is to discredit the dissenting users or push the discussion to such a level of absurdity, so that no one else cares to continue following. The Chinese government engages in very similar behavior using the so-called 50 Cent Party (whose name is derived from the amount of money they are paid for each post) (Sterbenz 2014).

Russia has taken such actions a step further by trying to manipulate traditional media. Essentially, the Russian government suspected that traditional media decision makers pay close attention to how they are perceived by social media users. As such, the Russian government has been accused of attempting to alter the content of social media with the intent of convincing traditional media sources to change their reporting on Russia's activities. Vasily Gatov, a Russian media analyst, has previously noted, "Someone sold the thought that Western media, which specifically have to align their interests with their audience, won't be able to ignore saturated pro-Russian campaigns and will have to change the tone of their Russia coverage to placate their angry readers" (Seddon 2014).

On the other hand, ISIS has also seen ordinary citizens take up against them. The hacktivist collective Anonymous has created the #OPISIS, #GhostSec, and #CtrlSec campaigns, which involve hacking ISIS-friendly websites and identifying ISIS supporters on Twitter, for the purpose of reporting them to Twitter for suspension (Gilbert 2015). In December 2014, a supporter of the extremist organization began posting a number of illustrations of anime-like young woman, warning of ISIS' presence across the globe (Know Your Meme 2015). In response, the Japanese ISIS-chan movement emerged. Its aim has been to mock

Social media has also taken on an ugly side, providing ordinary citizens with a new venue to air grievances and seek out revenge against other private citizens in the public forum. Public misinformation (intended or unintended) has led to unfortunate consequences and the persecution of citizens through a system of what could be described as mob justice, allowing these ordinary citizens the means to lash out at others with little fear of retribution, often via social media or social networking sites in what has been coined *outrage culture* (Kain 2015). Punishments have previously included people losing their jobs, death threats, and one's reputation being ruined (Fisher 2015).

ISIS' propaganda efforts with images of Japanese anime, often featuring a young girl dressed in black fatigues, similar to those commonly worn by members of the extremist group. ISIS-chan has gone so far as to establish rules, such as forbidding followers from insulting Islam and banning the use of pornographic images (ISIS Chan 2015; ISIS-Chan2Chan 2015).

Of course, organizations like Anonymous have also worked to support dissent, such as the Occupy Wall Street movement. In September 2011, members of the Occupy movement sought for America to undergo its own revolution, much like the Egyptians did earlier that same year (Bacallao-Pino 2014). Within weeks of the start of the demonstrations on September 17, there were hundreds of Occupy-oriented Facebook pages and other hashtag campaigns calling for similar movements across the United States, such as #OccupyBoston and #OccupyDenver (Bacallao-Pino 2014). However, lacking a clear goal and leadership, the movement has since floundered for the most part (Zara 2014).

Steps to Help Improve the Spread of Messaging

The first section of this chapter examined how social media is commonly used for organization, information dissemination, and suppression of ideas. Implicit in each of these sections is the notion that the individuals attempting to organize, disseminate, and suppress all have the same goal—spread their message as far and wide as possible. Yet, all too often, so-called social media experts engage in messaging endeavors that rely on little more than personal instincts or anecdote for message design. Often, people fail to take the various characteristics and preferences of their audiences into account before designing and sending a message. This section will examine the elements of social media messaging that help accelerate the spread of information. Specifically, this section will explore how to understand the media environment, audience attributes, message length, hashtags, message timing, message volume, social network analysis, and narrative analysis, which can help the social media practitioners more effectively craft messages.

Understanding the Media Environment

Before attempting to craft a message, it is critical for message developers to thoroughly examine the media infrastructure and the social media platform usage of the target audience. Specifically, it is critical for message designers to know the limitations of a particular country or region's telecom infrastructure. What kind of Internet bandwidth does it support? Do they have broadband, digital subscriber line (DSL), or dial-up? What kind of mobile Internet connectivity is supported? Is it 2G, 3G, or 4G? Knowing the particulars of a target audience's media infrastructure will help determine what types of content should be used (market research firms can be valuable sources of information for answering such questions). For areas that have less advanced Internet networks, messaging efforts should consist of content that is less media-oriented or at least restricted to only images or short videos, so as to avoid overloading the network, taxing the end user's device, or costing the user money due to the increase in necessary data to produce the message on their device. Failure to have a solid understanding of the social media infrastructure can be costly. On the other hand, if done right, the rewards can be great, such as with Coca-Cola's "Fanta for the Funny" campaign on Vine (a short-form video sharing platform) (Alford 2014).

Next, practitioners should determine what social media platforms are being used by their target audience. While many platforms, such as Twitter, Facebook, and YouTube, may be popular in Western-based social media markets, these platforms commonly have less penetration in other regions of the world. For example, in Eastern Europe and Russia, VK (a social networking website, in the same style as Facebook) is extremely popular, while in China, Qzone (a blogging website, similar to MySpace) and WeChat (a messaging program) are two of the more dominant platforms (Cosenza 2014). Failing to have a presence on such platforms means that a large portion of the target audiences in these countries will likely be lost.

Audience Attributes: Behavioral Habits, Consumption Habits, etc.

The attributes of the target audience are critical. For example, knowing the literacy rate among the targeted users will determine whether a post should involve text. In some regions of the world where literacy is very low, it might be wise for a post to avoid text altogether and instead rely on an image, a sound file, or perhaps a video. In other contexts, literacy might not be a consideration, but rather, a message designer must be aware of a growing trend of societies' preference for visual media. There certainly appears to be a progression from a text-heavy form of communication, such as e-mail and blogs, to a more visual-based or text-light format, such as tweets. The fast growth in the popularity of platforms like Vine, Instagram, Pinterest, and Tumblr only serves to demonstrate the public's preference for visual social media (Uganec 2013; McInnes 2015). This new form of social media, or social multimedia, involves the use of images, video, and audio content to communicate. Recent studies have shown that the human brain can process images in a little as 13 milliseconds showing that image-based messages are often easier to consume than text (Trafton 2014). YouTube, Instagram, and Internet memes are examples of popular social multimedia. In the case of YouTube, entire conversations can take place via a video, where people can quickly post in reply to one another without needing to know how to read or write.

Message Length

Once it has been decided what type of format would be used (e.g., text, still image, or video), one must then consider the length of the post. Message length can play a large role in whether a user chooses to engage with a post. For example, several studies by Track Social, Buddy Media, and ReelSEO showed that in many cases, tweets should not be more than 100 characters in length (for text) or that a YouTube video should not go over

3 minutes in length (Jarboe 2012; Track Social 2012; Salesforce Marketing Cloud 2013). These studies often examined data on the top social media accounts (often of US origin) and examined the performance of their content over a selected period. The video length can also depend on whether one is engaging the existing fan base or reaching out to new users. Interestingly, another study showed that videos aimed at gaining new audience members work better when they are shorter, while videos aimed at existing audiences can afford to be a bit longer (Oswego Creative 2014). Different audiences often have varying preferences, and content creators must experiment with their content to determine those preferences.

Another important aspect of effective messaging is the use of proper terminology that will resonate with the target audience. Beyond the obvious need to clearly communicate in the language of the people one is attempting to engage, a message designer must also communicate in an engaging way. In most cases, it helps to speak with an audience in a way with which they can more easily associate. Using tools such as Google AdWords, practitioners can research words that are often searched for and more often associated with particular topics within a particular country or language (Google AdWords 2015). Normally used for search engine optimization purposes, this tool can be easily adapted for social media messaging purposes. In addition to using Google AdWords, it helps to research any terms that are unique to the specific topic that a message designer is generating. For example, when messaging on the issue of lesbians, gays, bisexuals, and transgenders to a Russian audience, one might simply look up the direct Russian translation for homosexual, but perhaps a more effective approach would be to use the Russian word *голубой* (pronounced *goluboi*) or *blue*, which has come to serve as a native substitute for *gay* (Moss 2015). Simple text data mining can also be used to identify such unique terms relevant to specific audiences, helping analysts and content creators further improve their ability to connect with their target audiences, by speaking to them on their own terms.

Analysts can also conduct a more detailed level of exploratory data analysis, using advanced social media monitoring tools (such as Sysomos, Synthesio, etc.), analytics supplied by the social media platform itself, or data acquired through an application program interface (API) call to the respective platform (Synthesio 2015; Sysomos 2015). For instance, by using data from Twitter analytics, analysts can use various statistical approaches to better determine what words, among other variables (such as days of the week, hashtags, etc.), are likely effecting the popularity of one's posts. Once a practitioner has determined the variables that show up the most often within a body of tweets, this information can be used as a series of guidelines that will assist with crafting messaging that will likely resonate with the target audiences.

The Value of Hashtags

Hashtags, within most social media platforms, are a way of tagging content for a specific topic, making it easier for others to search for said topic. Hashtags are words or phrases preceded by a hash or a pound sign (#) that allow users to quickly search a particular topic on a variety of social media platforms, most notably Twitter. They also offer a way to create and monitor one's visibility on platforms like Twitter and Facebook. By creating a unique hashtag, practitioners can track the conversation related directly to them or a topic of their choosing and provide what amounts to a rallying point on a particular issue. On Twitter, tweets with hashtags also see a dramatic increase in the number of retweets or shares than those without (Zarrella 2015). It should also be noted, however, that studies have shown no more than two hashtags should be used in a post (Lucie 2012). Spam often contains a large number of hashtags, and one's messaging risks becoming confused with these inauthentic posts.

Practitioners should be aware that hashtags will not always be controlled by the creator. Once a hashtag is created and launched, it belongs to the public, and if one is not careful, a friendly hashtag can be turned against them. In 2012, singer Susan Boyle's record company launched the hashtag #susanalbumparty to coincide with the launch of her new album. Twitter users leapt at the chance to give the hashtag a new, more promiscuous meaning (Waldram 2012). That said, with smart and adaptive strategies, one can keep up with the public and try to head off attempts to hijack or co-opt their hashtag. On the other hand, one may wish to take advantage of other, preexisting hashtags and co-opt them for their own purposes. Taking over a preexisting hashtag or co-opting it can be valuable, although a challenging endeavor. By co-opting a hashtag, one tries to change the meaning of someone else's hashtag to suit one's own purposes. The value behind such a move is that this hashtag already has an existing following, which increases its exposure as one pushes the tag in a new direction. In order to do so, one needs to not only remain persistent in using the hashtag repeatedly, but also get a large number of other users to also begin using the hashtag as quickly as possible, or else fail in the hijacking attempt. In the spring of 2014, the US State Department's @UKRProgress Twitter account had launched #UnitedforUkraine, in advance of the Ukrainian elections. Twitter users, most notably the Russian Ministry of Foreign Affairs, opposed the attempt to hold elections and tried to co-opt the hashtag, although they were ultimately unsuccessful, as the pro-Ukrainian usage overwhelmed the attempted appropriation (Groll 2014; Johnson 2014).

Sometimes, hashtag styles vary between different cultures or languages, and it is beneficial to understand these differences when addressing a particular audience. For example, Arabic script does not lend itself to the traditional hashtag format because all the words are joined together with no spaces in between. Arabic letters merge together, rendering the hashtag indecipherable. Instead, Arabic-speaking users join the words together using underscores. Arabic-speaking users also tend not to use acronyms (Nassr 2014). One of the few exceptions worth noting is the use of داعش or *daesh*, an acronym of the English transliteration for ISIS (ad-Dawlah al-Islāmiyah fīl-'Irāq wash-Shām). There are several reasons as to why this particular hashtag has actually seen wide usage among Arabic-speaking audiences. First, its use has resulted in a furious backlash from ISIS members, which has only encouraged the group's opponents to continue using it. This includes world leaders like US Secretary of State John Kerry and Laurent Fabius, France's foreign minister (Nassr 2014). ISIS likely views this as a challenge to its legitimacy as a sovereign state. Second, it also does not help that the term *daesh* sounds similar to Arabic words like *Daes* ("one who crushes something underfoot") and *dahes* ("one who sows discord") (Harris 2014). Social media users have also been quick to turn *daesh* into *jaish*, which means "donkey," disseminating images of ISIS logos with donkeys worked into the image (Guthrie 2015).

Another important factor that should be considered is the use of regional dialects. When addressing a region- or a country-specific audience, using the local dialect could serve as a way to engender some degree of acceptance. A dialect can offer another important piece of information: a user's location. This is important, as few pieces of information can be more difficult for a user to fake on a social media platform. Other identifiers, like the user's provided location and time zone setting, can be altered by the user, making it difficult to pinpoint where they are located. Even worse is the geotagging option, which provides the exact longitudinal and latitudinal coordinates for every post. This can be turned off by users, and only a fraction of all social media users actually leaves the option active. More importantly, determining the dialect of the target audience also offers the opportunity to better determine the origins of particular users, as it can be quite difficult to actively ensure that one's native dialect is not evident in one's regular communications.

Timing for Optimum Exposure

Timing plays an important role when it comes to improving the reach of one's messaging. By making sure to time the release of the messages when the audience is most often sharing content/active, one can likely increase the number of impressions (the number of times that a platform chooses to show your content) that a post will achieve (Brnrd.me 2015). Peak activity times can sometimes be explained by cultural attributes. In the United States, for example, peak times for Twitter users is generally between 12 and 1 p.m., which coincides with lunch time (Lee 2015). Furthermore, due to the constant production of new content by other users, posts have a short window of opportunity to be seen by target audiences. In fact, according to Wisemetrics, tweets generally have a shelf life of about 24 minutes (Rey 2014). Facebook, on the other hand, is much more difficult to judge. With the recently updated newsfeed algorithm, the staying power of a post really depends on a combination of its relevance and how engaging it is to viewers (Leiter 2015). The ideal move is to publish posts so that the post's shelf life overlaps during these windows of opportunity. For instances when that ideal time in the middle of the night, one can utilize social media property management tools, such as TweetDeck and Hootsuite, to issue posts when it is not convenient for the practitioner (Hootsuite 2015; TweetDeck 2015). However, there are a few potential drawbacks, especially with platforms like Twitter, in that when users reply to these automated posts, there is no one waiting to reply. This can give users the impression that the posts are less genuine and can sometimes conflict with events that occur while the practitioner is away. For example, the morning after the 2012 Aurora, Colorado shooting, a Twitter account for the National Rifle Association landed in hot water when a prescheduled tweet went out that made no mention of the incident. Many criticized the organization for being insensitive to the previous day's event (Woods 2014).

Volume

The volume of posts from a message designer can impact the overall message resonance. In many cases, there may be an ideal number of posts that should be made per day, which applies to every audience. Post too infrequently and one risks missing out on getting content in front of a large portion of a target audience. On the other hand, post too often and one risks annoying the target audience with a deluge of posts. Such annoyance can lead an audience to unfollow or block a message designer's account entirely. One way to explore the appropriate volume of posts is to conduct message testing. For example, while gradually increasing the number of posts made on a daily basis, a researcher could keep an eye on the number of fans/followers that the social media property maintains. Once the number begins to consistently decrease, it is plausible that that the ideal daily post count may have been surpassed.

Social Network Analysis

When it comes to revealing valuable insights into one's target audiences, one of the most valuable tactics available to social media practitioners is social network analysis or social network mapping. It is a method used for visualizing and measuring how people, groups, and organizations are connected when sharing information (KSToolKit 2015). It can be critical in revealing a great deal about a particular conversation and those who are participating in it. More than any other approach discussed in this chapter, social network analysis can often reveal the most actionable information for practitioners. Community analysis, identification of influencers, and narrative analysis are just a few examples of the valuable information that can be gathered through mapping and analyzing social networks. Analysts can use this data to better understand their potential target audiences and craft their messaging accordingly.

Social network analysis involves analyzing the users and the communities or the clusters they form in order to understand how they interact with one another. In the case of social media network analysis, analysts examine these communities in order to understand not only the users involved, but also the conversations involved in a particular set of data. Using tools like Gephi or NodeXL to map out the entire network, practitioners can begin analyzing the various communities that exist within an overall network (Gephi 2015; NodeXL 2015). These communities are often formed around a unique variable, such as an influential user (also referred to as nodes) or a particular topic that users within the overall conversation are discussing among themselves. For instance, when analyzing a broad conversation relating to the Islamic State, one is likely to find communities that are either supportive of the organization or opposed to it. Depending on the practitioner's intentions, being able to distinguish between the two can be incredibly valuable for messaging purposes and intelligence collection. Once these communities are identified, practitioners can begin determining unique characteristics that might be worth knowing, such as derogatory terms used when referring to opposing forces. For example, in the case of pro-ISIS communities, the term *Safavids* is often used in a derogatory fashion when discussing Shiites, as the term refers to an empire that is based in what is now Iran, which is a Shi'a majority state (BBC-Religions 2009). Some Sunnis, such as those belonging to ISIS, view Shiites as heretics (Conant 2014).

User Roles within a Network

By using social network analysis, practitioners can also determine who the most influential users are within a conversation. Knowing the most influential figures within a social network is extremely valuable when conducting information operations, as they are the drivers of the conversation and can serve as indicators of the overall sentiment of their particular community or perhaps even the entire network.

When targeting like-minded communities, it is worth building relationships with influencers and then utilizing them to help with spreading your message to users that you would not normally reach on your own. Practitioners should also profile these influencers, identifying what communities they are most influential in, what issues they tend to discuss most often, and who their followers are. Knowing these points will help when determining which influencers to leverage when one is looking to push a particular narrative.

Analysts can also begin examining network maps to determine the other roles that are normally played within a social network, such as a bridge, which connects separate communities via users within each community. Users in this role are often necessary for a message to go viral, as virality requires that an idea first jump from one community to another. A hub is another role that is featured within social networks. Hubs are users found at or near the center of a specific community or an entire conversation, as their messaging is often shared by others.

Narrative Analysis

Narrative analysis is yet another valuable capability that is made possible through network analysis. Once a conversation has been mapped out and the communities have been profiled to determine who makes up each cluster, practitioners can begin data mining the posts issued from within each community to see not only what issues are being discussed by communities of interest, but also how they are being discussed. Once the narratives have been determined, depending on one's goals, practitioners can work out how to adapt their goals to fit in with the community's interests or perhaps develop effective counternarratives. It is important to understand how an issue is viewed by a community's members before attempting to join the conversation, as one risks coming across as aloof or insincere to their concerns. If this should happen, one can quickly find themselves outside of the community, unable to engage with anyone in any meaningful way. However, once these narratives have been determined, content creators and analysts can begin crafting messaging to counter the points being made by opposing forces or to support and amplify the points made by friendly forces.

In order to properly determine these unique narratives, analysts need to analyze hundreds, if not thousands of posts or documents in a timely fashion. In order to analyze such large amounts of data, the field of text analytics has developed, leading to the creation of programs like SAS Text Miner (SAS 2015). These programs can import thousands of tweets or Facebook posts and analyze these documents in order to pinpoint terms and entities that appear to have value and could possibly be used to help speak to target audiences on their own terms and on issues that they care about.

In Summation

Social media usage only stands to increase in the coming years, as it offers users the ability to instantly communicate with large numbers of people for almost no cost (aside from the cost of Internet access and a device with which to access it). It also offers the means to convey messages through a variety of ways, such as text, images, audio, and video. It can serve practitioners with ways to spread critical information in times of crisis, assisting with organizing rescue efforts and sharing accurate information to people who might not otherwise have access. On the other hand, social media can have destructive effects as well, as it is sometimes used to spread disinformation and misinformation (the difference being that disinformation is done intentionally).

Practitioners and analysts should take advantage of the vast amount of data that is available through social media monitoring tools, as well as data mining and statistical software in order to better craft messaging for specific target audiences. Practitioners should also examine the conversations taking place within communities of interest, as well as the overall observed topics of interest. These tools can also be used to identify influential terminology, hashtags, and users, all of which can be used to increase the likelihood of messaging reaching to the right users. Beyond user data and performance metrics pulled from the digital platforms themselves, practitioners and analysts should also look to tailor their content in order to meet cultural features, behavioral preferences, and technical restrictions of their target audiences. Among other things, this should include their audiences' preferred platforms, telecom infrastructure, and most active times, in order to ensure that their content even has a chance of connecting with the right audiences in the first place.

Practitioners should also keep an eye on continuing evolutions in communication technology and seek to compensate for these changes. Currently, there are two trends that might be worth keeping an eye on. The first of which is what appears to be a progression

ISIS has even officially stated that it uses Telegram as one of its main communication platforms. Despite Telegram's assurances that it will police its own platform, shutting down any ISIS-affiliated accounts that it finds, for every account that is shut down, a new one often takes its place. Furthermore, experts have recently warned that ISIS may even be building its own encrypted messaging app, so that it will not have to deal with interference from outside companies (Groll 2015).

to visual social media. Still images and videos are often understood much faster than text, and audiences' content consumption habits appear to be reflecting this. There might be other possible factors that are contributing to this transition, such as the growing Internet penetration and the spread of Internet-accessible mobile devices across regions where literacy rates are lower and text-heavy content would be less well received. Another trend is the shift from social media websites, like Facebook, to social messaging applications, most notably WhatsApp, WeChat, Telegram, and SnapChat (SnapChat 2015; Telegram 2015; WeChat 2015; WhatsApp 2015). These messaging apps offer many of the same features that social media platforms do but often serve as a direct means of sharing messages between users, as opposed to more of a broadcast method, where many other users can see one's content. Many of these apps offer additional privacy measures, such as end-to-end encryption and ephemeral messaging, meaning that the messages delete themselves once they have been read (Olson 2013). All this ensures that users can avoid attempts to monitor their conversations. This poses a particular problem for those attempting to monitor communications between violent extremist organizations, such as the Islamic State (Engel 2015; Slack 2015). This is already proving a concern for governments across the globe, most recently by the United Kingdom, which commissioned a report on the issue in 2015 (Anderson 2015). The report calls for cooperation between the UK intelligence and law enforcement agencies and the firms that run these messaging apps.

Case Study: Megacity Xanadu

Social media offers the unique ability to monitor the ongoing conversations among the citizens of the megacity Xanadu and Çawaila as a whole. Once a proper search capability is established, government monitors can begin watching for conversations that indicate serious issues, such as general discontent, potential health hazards, rights violations, and so on. They can also begin looking for persons of interest. This could include those who are stirring anger among the population, which could lead to violence, or those who could serve as influential figures who could help with communicating with the target audience, in order to better address the issues facing Xanadu.

Let us assume that government officials decide that they wish to develop a messaging campaign to communicate with teens and young adults of Bursuka and Razini descents. To begin, practitioners should spend some time conducting research of their target audience. This should include determining what social media platforms are commonly used by the population. Are there particular platforms that are used by specific demographic groups or for specific purposes? What about the quality of mobile and landline internet connectivity?

Given that money is an issue with these refugee populations, it is likely that they use cheaper, less advanced mobile devices and that their Internet connection type is on the lower end of the spectrum (something along the lines of dial-up or DSL). Neither of these instances lends themselves to large data and/or graphics-heavy messaging, as the refugees' Internet-enabled devices are not likely to handle the content very well. With that in mind, social media practitioners and content producers need to scale back on any videos or large images, as they will take too long to download and eat up the users' data limits. The reports acquired from market research firms confirm that the refugees tend to use lesser devices and that a microblogging platform, called Pip, is most popular among teens and young adults among the refugee population. Pip is a platform that is very similar to Twitter, and it too offers access to general user data and metadata through API calls. These reports also indicate that these particular users tend to speak in their own native languages when using Pip.

With all this in mind, social media analysts should establish an initial search capability on Pip, using API calls to search for the overall conversation taking place within Xanadu in the native tongues of the Bursuka and the Razini. The analysts should start with a broad search that includes terms that were vetted using basic tools like Google AdWords' related terms search function, as well as data mining blogs belonging to known Bursuka and Razini bloggers, looking for common terms related to workers' rights, societal ills, health issues, and general dissent. These terms should all be included in the initial search.

The analysts should also set geographic search parameters in their search string. Luckily for them, Xanadu is positioned within the region that sits just inside a separate time zone from the rest of Çawaila, Bursuka, and Razinia. Furthermore, given that privacy is not a chief concern among the refugee populations, it is rare that Pip users take the time to change their time zone settings or bother to lie about their location in the user profile. Some users have even elected to turn on the geotagging function on their mobile phones. This all makes ruling out Buruski- and Razini-speaking users that are not among the refugees living in Xanadu far easier. Once these geographic parameters have been included in the search string, the analysts should begin collecting data from the API.

Once some initial revisions have been made to the search string, ruling out spam and other unrelated content, the analysts can begin conducting exploratory data analysis of the data set. First and foremost should be to construct a social network map of the entire data set, as the resulting information will provide the analysts with the most actionable intelligence. For example, this will allow the analysts to profile the various communities that exist within the overall observed conversation. Once the communities have been defined, the analysts can begin determining the various roles that users play within their respective community. Once these have been determined, the analysts should also search to find which among these users are the most influential, so that these users might be leveraged to possibly assist with improving communications with the members of that particular community. If these users are commonly creating dissent, perhaps they need to be monitored or perhaps contacted to see what can be done to help alleviate their situation.

They should also conduct a cluster analysis of text documents of all the posts from each community, so as to discover the unique narratives that are currently taking place within each community. By using cluster analysis, the analysts can also pick out influential hashtags that might be worth including in future messaging, for the purpose of reaching new users and possibly influencing them or perhaps co-opting that hashtag to give it a new meaning, one that benefits Çawaila. With this knowledge in hand, government officials can begin crafting messaging campaigns and taking steps to either address the concerns of Çawaila's refugee populations and perhaps begin monitoring those who might wish to cause harm to the state or to the citizens of Çawaila.

DISCUSSION QUESTIONS

1. As encrypted messaging apps grow in popularity, what are the implications for law enforcement agencies and their abilities to prevent criminals from utilizing said apps to coordinate criminal activities, such as the shooters from the San Bernardino incident were reported to have done?
2. Given the fast and ever-changing field of communications technology, specifically social media and social networking platforms, how are governments looking to respond to the increasing need to revisit previous legislation governing tech-related issues, such as social media users' expectations of privacy?
3. Outside of censorship, how can governments improve their abilities to counter widespread misinformation efforts conducted by opposing countries or organizations?
4. What can governments do to discourage the general public from engaging in attempts to discern and punish citizens in the public forum, as opposed to allowing for the judicial system to determine guilt and dole out punishment?
5. How can governments and law enforcement agencies better counter misinformation during legitimate attempts to spread public awareness on issues such as during natural disasters or outbreaks of disease (i.e., the Boston Marathon bombing or the Ebola outbreak)?

References

Alford, Emily. Video marketing must be tailored to devices [Ad Week]. *ClickZ*. 2014, October 2014. Accessed April 16, 2015. http://www.clickz.com/clickz/news/2373411/video-campaigns-must-be-tailored-across-platforms-ad-week.

Anderson, David. *A Question of Trust Report of the Investigatory Powers Review*. London: Independent Reviewer of Terrorism Legislation. 2015. Accessed July 27, 2015. https://terrorismlegislationreviewer.independent.gov.uk/wp-content/uploads/2015/06/IPR-Report-Web-Accessible1.pdf.

Bacallao-Pino, Lazaro. *Social Media Mobilisations: Articulating Participatory Processes or Visibilizing Dissent?* 2014. Accessed April 13, 2015. http://www.cyberpsychology.eu/view.php?cisloclanku=2014101401&article=3.

BBC-Religions. Islam: Safavid Empire. 2009, September 7. Accessed April 14, 2015. http://www.bbc.co.uk/religion/religions/islam/history/safavidempire_1.shtml.

Blair, Evan. #Ebola lessons: How social media gets infected. *InformationWeek*. 2014, September 9. Accessed April 21, 2015. http://www.informationweek.com/software/social/-ebola-lessons-how-social-media-gets-infected/a/d-id/1307061.

Brnrd.me. Twitter impressions ≠ Reach. 2015, March 22. Accessed April 9, 2015. http://brnrd.me/twitter-impressions-reach/.

Conant, Eve. Iraq crisis: Ancient hatreds turning into modern realities. *National Geographic*. 2014, June 19. Accessed May 2, 2015. http://news.nationalgeographic.com/news/2014/06/140618-iraq-shiite-sunni-isis-militants-maliki-borders/.

Cosenza, Vincenzo. World map of social networks. *Vincos Blog*. 2014, December 1. Accessed April 4, 2015. http://vincos.it/world-map-of-social-networks/.

Danin, Robert. What ISIS hoped to gain from killing the Jordanian pilot. *Newsweek*. 2015, February 4. Accessed March 29, 2015. http://www.newsweek.com/what-isis-hoped-gain-killing-jordanian-pilot-304336.

Dettmer, Jamie. EuroMaidan protesters: We want U.S. protection. *The Daily Beast*. 2014, March 14. Accessed April 5, 2015. http://www.thedailybeast.com/articles/2014/03/04/euromaidan-protesters-we-want-u-s-protection.html.

Devichand, Mukul, Ravin Sampat, and Tse Yin Lee. #BBCtrending: How it trended—The MH17 Blame Game. *BBC News*. 2014, July 18. Accessed March 27, 2015. http://www.bbc.com/news/blogs-trending-28371461.

Engel, Pamela. Inside the app that's become ISIS' biggest propaganda machine. *Business Insider*. 2015, November 21. Accessed December 21, 2015. http://www.businessinsider.com/telegram-isis-app-encrypted-propagandar-2015-11.

Fisher, Max. From gamergate to Cecil the Lion: Internet mob justice is out of control. *Vox*. 2015, July 30. Accessed December 21, 2015. http://www.vox.com/2015/7/30/9074865/cecil-lion-palmer-mob-justice.

Gephi. The open graph viz platform. 2015. Accessed April 5, 2015. http://gephi.github.io/.

Gilbert, David. Anonymous #OpIsis: Hacktivists publish how-to guide for identifying Islamic State twitter accounts. *International Business Times*. 2015, April 14. Accessed March 16, 2015. http://www.ibtimes.co.uk/anonymous-opisis-hacktivists-publish-how-guide-identifying-islamic-state-twitter-accounts-1496378.

Google AdWords. AdWords. 2015. Accessed April 2, 2015. https://www.google.com/adwords/.

Groll, Elias. U.S. and Russia agree: OMG we are all so #UnitedForUkraine. *Passport: Big Stories from a Small World*. 2014, April 25. Accessed March 15, 2015. http://foreignpolicy.com/2014/04/25/u-s-and-russia-agree-omg-we-are-all-so-unitedforukraine/.

Groll, Elias. Welcome to the future of war: ISIS has a smartphone app. *Foreign Policy: Passport*. 2015, December 8. Accessed December 21, 2015. http://foreignpolicy.com/2015/12/08/welcome-to-the-future-of-war-isis-has-a-smartphone-app/.

Gustin, Sam. Social media sparked, accelerated Egypt's revolutionary fire. Wired.com. Conde Nast Digital. 2011, February 11. Accessed March 4, 2015. http://www.wired.com/2011/02/egypts-revolutionary-fire/.

Guthrie, Alice. Decoding Daesh: Why is the new name for ISIS so hard to understand? *Free Word*. 2015, February 19. Accessed April 27, 2015. https://freewordcentre.com/blog/2015/02/daesh-isis-media-alice-guthrie/.

Harris, Kathleen. "Daesh" adopted as new name for ISIS by U.S., France. *CBCnews*. 2014, December 5. Accessed April 7, 2015. http://www.cbc.ca/news/politics/daesh-adopted-as-new-name-for-isis-by-u-s-france-1.2861108.

Hilleary, Cecily. Ukraine's social media revolution years in the making. *Voice of America*. 2014, March 14. Accessed March 5, 2015. http://www.voanews.com/content/ukraines-protest-movement-fueled-by-social-media/1871457.html.

Hootsuite. Hootsuite: Social Media Management Dashboard. 2015. Accessed April 10, 2015. https://hootsuite.com/.

Howard, Russell. Opinion: Digital revolutions gives new life to swarm tactics. *USNI News*. 2013, May 28. Accessed April 3, 2015. http://news.usni.org/2013/05/28/opinion-digital-revolutions-gives-new-life-to-swarm-tactics.

ISIS Chan. ISIS chan in English 英語説明テンプレなど. *ISIS Chan ISIS* ISIL対抗！萌で埋め尽くせISISちゃん. 2015, July 26. Accessed July 31, 2015. http://isis.wiki.fc2.com/wiki/ISIS chan in English 英語説明テンプレなど.

ISIS-Chan2Chan. ISISchan ! ISIS ISIL対抗！萌で埋め尽くせISISちゃん. *ISISchan* ! ISIS ISIL対抗！萌で埋め尽くせISISちゃん. 2015. Accessed June 9, 2015. http://isis-chan2chan.tumblr.com/.

Jarboe, Greg. The best length for a YouTube marketing video? *ReelSEO*. 2012, November 5. Accessed April 3, 2015. http://www.reelseo.com/length-youtube-video/.

Jeltsen, Melissa. Shootout in eastern Ukraine leaves 3 dead. *The Huffington Post*. 2014, April 20. Accessed March 24, 2015. http://www.huffingtonpost.com/2014/04/20/shootout-in-eastern-ukraine_n_5181467.html.

Johnson, Luke. Russia trolls U.S. twitter campaign in Ukraine crisis. *RadioFreeEurope/RadioLiberty*. 2014, April 25. Accessed April 2, 2015. http://www.rferl.org/content/ukraine-us-russia-twitter-trolling/25362157.html.

Kain, Eric. Internet mob justice isn't justice at all. *Forbes*. 2015, May 14. Accessed December 21, 2015. http://www.forbes.com/sites/erikkain/2015/05/14/internet-mob-justice-isnt-justice-at-all/.

Know Your Meme. Yarosh's business card. 2014. Accessed April 2, 2015. http://knowyourmeme.com/memes/events/yaroshs-business-card.

Know Your Meme. ISIS-chan. 2015, February 1. Accessed March 23, 2015. http://knowyourmeme.com/memes/isis-chan#fn2.

KSToolKit. Social network analysis. 2015. Accessed April 8, 2015. http://www.kstoolkit.org/Social%20Network%20Analysis.

Lee, Kevan. New research: The best time to tweet for clicks, retweets, and replies. *Buffer Social*. 2015, June 25. Accessed June 30, 2015. https://blog.bufferapp.com/best-time-to-tweet-research.

Leiter, Melissa. The lifespan of social media posts. *Melissa Leiter Marketing*. 2015. Accessed July 19, 2015. http://www.melissaleiter.com/lifespan-of-social-media-posts/.

Lucie, Mark. Best practices for journalists. *Twitter Blogs*. 2012, September 20. Accessed April 11, 2015. https://blog.twitter.com/2012/best-practices-for-journalists.

Ma, Michelle. Hold that RT: Much misinformation tweeted after 2013 Boston Marathon bombing. *UW Today*. 2014, March 17. Accessed December 21, 2015. http://www.washington.edu/news/2014/03/17/hold-that-rt-much-misinformation-tweeted-after-2013-boston-marathon-bombing/.

Madrigal, Alexis. #BostonBombing: The anatomy of a misinformation disaster. *The Atlantic*. 2013, April 19. Accessed December 21, 2015. http://www.theatlantic.com/technology/archive/2013/04/-bostonbombing-the-anatomy-of-a-misinformation-disaster/275155/.

Masi, Alessandria. ISIS foreign fighter recruitment, social media undeterred by new security crackdowns. *International Business Times*. 2015, February 28. Accessed April 15, 2015. http://www.ibtimes.com/isis-foreign-fighter-recruitment-social-media-undeterred-new-security-crackdowns-1831764.

McInnes, Will. Visual listening: Social media's next frontier. *Forbes*. 2015, May 1. Accessed May 27, 2015. http://www.forbes.com/sites/groupthink/2015/05/01/visual-listening-social-medias-next-frontier/.

Merriam-Webster. Meme. 2015. Accessed March 25, 2015. http://www.merriam-webster.com/dictionary/meme.

Miller, Christopher. Rebel commander backpedals after admitting fighters had missile system in Ukraine. *Mashable*. 2014, July 24. Accessed March 23, 2015. http://mashable.com/2014/07/24/rebel-commander-ukraine-buk-missile-reuters/#KvzHQ_PZrOks.

Moss, Kevin. Why are these pages blue? 2015. Accessed March 24, 2015. http://community.middlebury.edu/~moss/goluboy.html.

Nassr, Wassim. Middle East—French govt to use arabic "Daesh" for Islamic State group. *France 24*. 2014, September 17. Accessed March 9, 2015. http://www.france24.com/en/20140917-france-switches-arabic-daesh-acronym-islamic-state.

NodeXL. NodeXL: Network overview, discovery and exploration for Excel. *NodeXL Network Graphs*. 2015. Accessed May 19, 2015. http://nodexl.codeplex.com/.

Olson, Parmy. Delete by default: Why more Snapchat-like messaging is on its way. *Forbes*. 2013, November 22. Accessed April 16, 2015. http://www.forbes.com/sites/parmyolson/2013/11/22/delete-by-default-why-more-snapchat-like-messaging-is-on-its-way/.

Oswego Creative. *Best Practices for Video Length*. 2014, October 8. Accessed April 8, 2015. http://oswegocreative.com/video-length-recommendations-best-practices/.

RadioFreeEurope/RadioLiberty. Twitter meme mocks Russian claims over Slovyansk shoot-out. 2014, April 20. Accessed March 15, 2015. http://www.rferl.org/content/ukraine-slovyansk-shoot-out-yarosh-meme/25356107.html.

Reilly, Nicholas. Russia has banned memes, so here's the best ones of Vladimir Putin. *Metro: News... but Not as You Know It*. 2015, April 13. Accessed April 5, 2015. http://metro.co.uk/2015/04/13/russia-has-banned-memes-so-heres-the-best-ones-of-vladimir-putin-5147681/.

Rey, Benjamin. Your tweet half-life is 1 billion times shorter than carbon-14's. *Wiselytics*. 2014, March 5. Accessed March 26, 2015. http://blog.wisemetrics.com/tweet-isbillion-time-shorter-than-carbon14/.

Salesforce Marketing Cloud. Strategies for effective tweeting: A statistical review. 2013, March 1. Accessed April 6, 2015. https://www.salesforcemarketingcloud.com/wp-content/uploads/2013/03/Strategies-for-effective-tweeting.pdf.

Sarmah, Sattah. Fighting the endless spread of Ebola misinformation on social media. *Fast Company*. 2014, August 19. Accessed March 15, 2015. http://www.fastcompany.com/3034380/fighting-the-endless-spread-of-ebola-misinformation-on-social-media.

SAS. Text mining software. 2015. Accessed April 6, 2015. http://www.sas.com/en_us/software/analytics/text-miner.html.

Seddon, Max. Documents show how Russia's troll army hit America. *BuzzFeed*. 2014, June 2. Accessed April 1, 2015. http://www.buzzfeed.com/maxseddon/documents-show-how-russias-troll-army-hit-america#.jm9dlErerd.

Sindelar, Daisy. The Kremlin's troll army. *The Atlantic*. 2014, August 12. Accessed March 18, 2015. http://www.theatlantic.com/international/archive/2014/08/the-kremlins-troll-army/375932/.

Slack, James. Fanatics are using secret message apps says anti-terror Tsar: Government set to do battle with web giants including WhatsApp and Facebook after warning in landmark report. *Mail Online*. 2015, June 10. Accessed June 24, 2015. http://www.dailymail.co.uk/news/article-3119167/Fanatics-using-secret-message-apps-says-anti-terror-tsar-Government-set-battle-web-giants-including-WhatsApp-Facebook-warning-landmark-report.html.

SnapChat. 2015. Accessed April 7, 2015. https://www.snapchat.com/.

Solon, Olivia. Richard Dawkins on the Internet's hijacking of the word "meme." *Wired UK*. 2013, June 20. Accessed March 23, 2015. http://www.wired.co.uk/news/archive/2013-06/20/richard-dawkins-memes.

Sterbenz, Christina. China banned the term "50 Cents" to stop discussion of an Orwellian propaganda program. *Business Insider*. 2014, October 17. Accessed April 2, 2015. http://www.businessinsider.com/chinas-50-cent-party-2014-10.

Synthesio. Social media monitoring analysis tools. 2015. Accessed March 22, 2015. http://www.synthesio.com/.

Sysomos. Sysomos: Social media monitoring tools. 2015. Accessed March 13, 2015. http://www.sysomos.com/.

Telegram. Telegram—A new era of messaging. 2015. Accessed December 19, 2015. https://telegram.org/.

The Christian Science Monitor. Social media's big lift after Nepal earthquake. 2015, April 27. Accessed April 10, 2015. http://www.csmonitor.com/Commentary/the-monitors-view/2015/0427/Social-media-s-big-lift-after-Nepal-earthquake.

The Interpreter. Ukraine liveblog day 150: Malaysian passenger jet shot down over Ukraine. 2014, July 17. Accessed April 1, 2015. http://www.interpretermag.com/ukraine-liveblog-day-150-ukrainian-troops-describe-grad-rocket-attack-from-russia/#3391.

Track Social. Optimizing Twitter engagement—Part 3: Tweet length. 2012, October 29. Accessed March 9, 2015. http://tracksocial.com/blog/2012/10/optimizing-twitter-engagement-part-3-tweet-length/.

Trafton, Anne. In the blink of an eye. *MIT News*. 2014, January 16. Accessed March 17, 2015. http://news.mit.edu/2014/in-the-blink-of-an-eye-0116.

TweetDeck. TweetDeck. 2015. Accessed June 4, 2015. https://tweetdeck.twitter.com/.

Uganec, Cameron. Social media and storytelling 4: The growth of visual storytelling. *Hootsuite Social Media Management*. 2013, July 25. Accessed March 25, 2015. http://blog.hootsuite.com/social-media-storytelling-4/.

Waldram, Hannah. #Susanalbumparty: Top five Twitter hashtag PR disasters. *The Guardian*. 2012, November 22. Accessed April 3, 2015. http://www.theguardian.com/technology/shortcuts/2012/nov/22/twitter-susan-boyle-susanalbumparty.

Ward, Alex. Even in a war zone! Ridiculously photogenic Syrian rebel starts a wave of appreciative virals. *Mail Online*. 2012, August 16. Accessed March 15, 2015. http://www.dailymail.co.uk/news/article-2189327/Ridiculously-photogenic-Syrian-rebel-goes-viral.html.

WeChat. WeChat—The new way to connect. 2015. Accessed April 3, 2015. http://www.wechat.com/.

WhatsApp. WhatsApp: Home. 2015. Accessed April 6, 2015. https://www.whatsapp.com/.

Woods, Laura. Pros and cons of scheduling your social media posts in advance. *Social Media Strategies Summit*. 2014, March 27. Accessed April 15, 2015. http://socialmediastrategiessummit.com/blog/pros-cons-scheduling-social-media-posts-advance/.

Zara, Christopher. Occupy Wall Street a failure? OWS 2014 marks survival of brand with blurred purpose. *International Business Times*. 2014, September 17. Accessed March 27, 2015. http://www.ibtimes.com/occupy-wall-street-failure-ows-2014-marks-survival-brand-blurred-purpose-1690198.

Zarrella, Dan. [New Data] use "quotes" and #hashtags to get more retweets. 2013, October 15. Accessed April 16, 2015. http://danzarrella.com/new-data-use-quotes-and-hashtags-to-get-more-retweets/.

4

The Spread of Opinions in Societies

Boleslaw K. Szymanski, Omar Lizardo, Casey Doyle, Panagiotis D. Karampourniotis, Pramesh Singh, Gyorgy Korniss, and Jonathan Z. Bakdash

CONTENTS

Introduction 61
Naming Game-Based Models 62
Threshold Models 66
Tipping Points in Threshold Models 66
Dynamics with Individualized Thresholds 68
Impact of Culture and Gender on Spread of Opinion in Empirical Networks 70
Megacity and Notre Dame NetSense Data 70
Impact of Agent Sociodemographic Characteristics on Activity and Assortative Mixing 71
Race and Ethnicity 74
Class Background 75
Network/Opinion Coevolution: Differential Tie Decay by Dyadic Mix 76
Acknowledgments 82
References 82

ABSTRACT We demonstrate the role of sociocultural factors in opinion dynamics via modeling dynamic social networks using the naming game and threshold models and via analysis of behavioral network data on social interactions collected at the University of Notre Dame. Our analysis shows that the ties among people likely to share opinions tend to persist longer than ties among persons unlikely to share them. Moreover, the partner selection for a tie is correlated with sharing an (political) opinion. Results show that sociocultural factors and social network dynamics together influence the evolution of opinions in multicultural social networks and may have implications for megacities.

Introduction

The nature of opinions and their dynamics has a wide range of impacts ranging from voting to policy decisions to political stability to even military operations (Burstein 2003; Aldrich et al. 2006). Here, we seek to better understand sociocultural factors for opinion dynamics by applying them to two quantitative models of opinion spreading (Castellano et al. 2009) to empirical data: the naming game (NG) model (Steels 1995; Baronchelli et al. 2006; 2008; Dall'Asta et al. 2006) and the threshold model (TM) (Granovetter 1978; Watts 2002). In the NG model, each individual in the network possesses a list of opinions, which comprise its opinion state. Each interaction starts with randomly selecting the speaker first and then randomly choosing the listener among the speaker's neighbors. The speaker then shares a randomly chosen opinion from its opinion state with the listener. In TM, an individual adopts a new opinion only when the fraction of its nearest neighbors possessing that opinion is larger than an assigned threshold. The threshold represents the resistance of the individual to peer pressure.

We structured the study as follows. The NG and TM models are described next, and both are used to simulate the evolution of opinion spread observed in the empirical data. Then the empirical data are analyzed to ascertain the impact of cultural, gender, and structural factors for opinion dynamics. Homophily is observed for multiple factors. People who are more likely to share the same opinion are also more likely to have matched traits (e.g., race, gender, or socioeconomic status) and have a greater probability for maintaining a relationship over time than people whose opinions differ. Results are interpreted for their potential implications to megacities.

Naming Game-Based Models

One of the simplest and most popular models for describing large-scale patterns in opinion spread is the NG model (Steels 1995; Baronchelli et al. 2006, 2008; Dall'Asta et al. 2006). The rules governing the local interactions are as follows. Every node (or individual) in the system contains a list of words. At the beginning of each time step, a speaker and a listener connected to it is randomly chosen, and the speaker shares one word from its list. If the listener's list contains that word, both the speaker and the listener will delete all other words from their lists. Otherwise, the listener will simply add it with no deletion of words on either side.

Here, we use an even simpler variant of the NG model termed the binary agreement model (BAM). In BAM, only two words are present (*A* and *B*), and nodes transition between one of three possible states (*A*, *B*, and *AB*) (Castelló et al. 2009; Xie et al. 2011). The reduction in the number of possible states leads to a drastically reduced time to consensus, now on the order of ln (N) (Baronchelli et al. 2008; Castelló et al. 2009; Zhang et al. 2011). Here, by construction, the states are entirely symmetric with no bias toward either opinion. This game can be altered slightly and made more complex with the addition of committed agents who are set to a singular opinion (for simplicity assumed to be *A*) that will never change (Lu et al. 2009; Xie et al. 2011; Zhang et al. 2011). Committed agents are still able to influence the opinions of the other nodes around them. In the simplest and most direct case, the system starts with some predetermined number of committed agents p, while the rest of the population are all in the opposite opinion *B*. For a finite network given infinite time, a system such as this will always reach consensus on the opinion favored by the committed agents. If the initial population of committed agents is small, a large random deviation is required to achieve consensus, so the time to consensus for such systems is on the order of e^N. However, there is some critical population fraction (*tipping point*) denoted p_c that causes a sharp phase transition in the system (Xie et al. 2011). Above p_c, the time to consensus is quite fast, on the order of ln (N).

The critical population in a fully connected graph is $p_c = 0.09789$ (Xie et al. 2011). Above this population, there exists only one absorbing state at the consensus for *A*. Below, however, there is the consensus-absorbing state in addition to an active steady state and a saddle point.

The model can be extended further by including two competing committed groups, one in state *A* and one in state *B*. Here the populations of the committed agents will be termed p_A and

TIPPING POINTS

The phenomenon of tipping points (Gladwell 2000) in social dynamics can be seen in many real world events throughout history. The explosion of some ideal into the public consciousness causing rapid change after some period of slow growth can be seen in events such as the Arab Spring of 2011, the suffragette movement, and the American civil rights movement. The civil rights movement is of particular interest as it happened right after the African American population in the United States crossed 10%, making it fall right in line with the theory of committed minorities.

p_B, respectively (Xie et al. 2012). The system can enter into different phases depending on the initial setup of the critical population sizes. With these populations initially equal, the system has two active steady states and one saddle point. When the system is in either state, a large fluctuation of opinions by noncommitted nodes is necessary to push the minority population past the saddle point and into the second steady state. When these populations are unequal and below critical values, the saddle point will move to favor the population with the higher number of committed agents. In these cases, the average time to switch from one state to the other is on the order of e^N since a large random fluctuation is needed for such a switch. Finally, when the populations fall outside of their critical values, then the system maintains only a single stable state where the population with the higher number of committed agents dominates. In these cases, the system consensus is reached quickly on the timescale of ln (N).

Next, we consider *waning commitment* a case where committed agents are allowed to move into the mixed state after w consecutive interactions with a node of the opposite opinion. With the ability of the committed nodes to now become normal, the system regains the possible absorbing state for the minority opinion. With a single group of weakly (or partially) committed agents present, the critical population of weakly committed agents required to force a consensus varies with w, the extent of which is seen in Figure 4.1a (Galehouse et al. 2014) and can be numerically approximated by

$$p_c(w) \cong p_c(\infty) + \frac{a}{(b+w)^{\alpha}} \quad \text{with } \alpha \cong 1.73. \tag{4.1}$$

Next, we consider competing weakly committed agents. As seen in Figure 4.1b, the region in a shape of beak emerges where the system will undergo a phase transition when p_A and p_B cross its threshold. The value of w will change the width of the beak, with higher w imposing a narrower beak. In addition, the stable points previously present inside the beak are now absorbing states since the committed population can eventually be swayed.

The idea of waning commitment introduces stubbornness into the system, where the nodes can resist the change without becoming implacable, as is the case with the fully committed nodes. This behavior can be extended to make commitment not inherent to a set of nodes but instead a part of the opinion, enabling it to spread with the opinion. This extension requires eliminating the intermediate state *AB* (Doyle et al. 2016). Instead, each state is assigned an idea inertia denoted by w_A and w_B, respectively, corresponding to how many consecutive times a node in that state must hear the opposite opinion before it switches. If a node listens to an opinion that confirms its own, its count toward switching is reset to zero. Alternatively, if it hears an opinion that is contrary to its own,

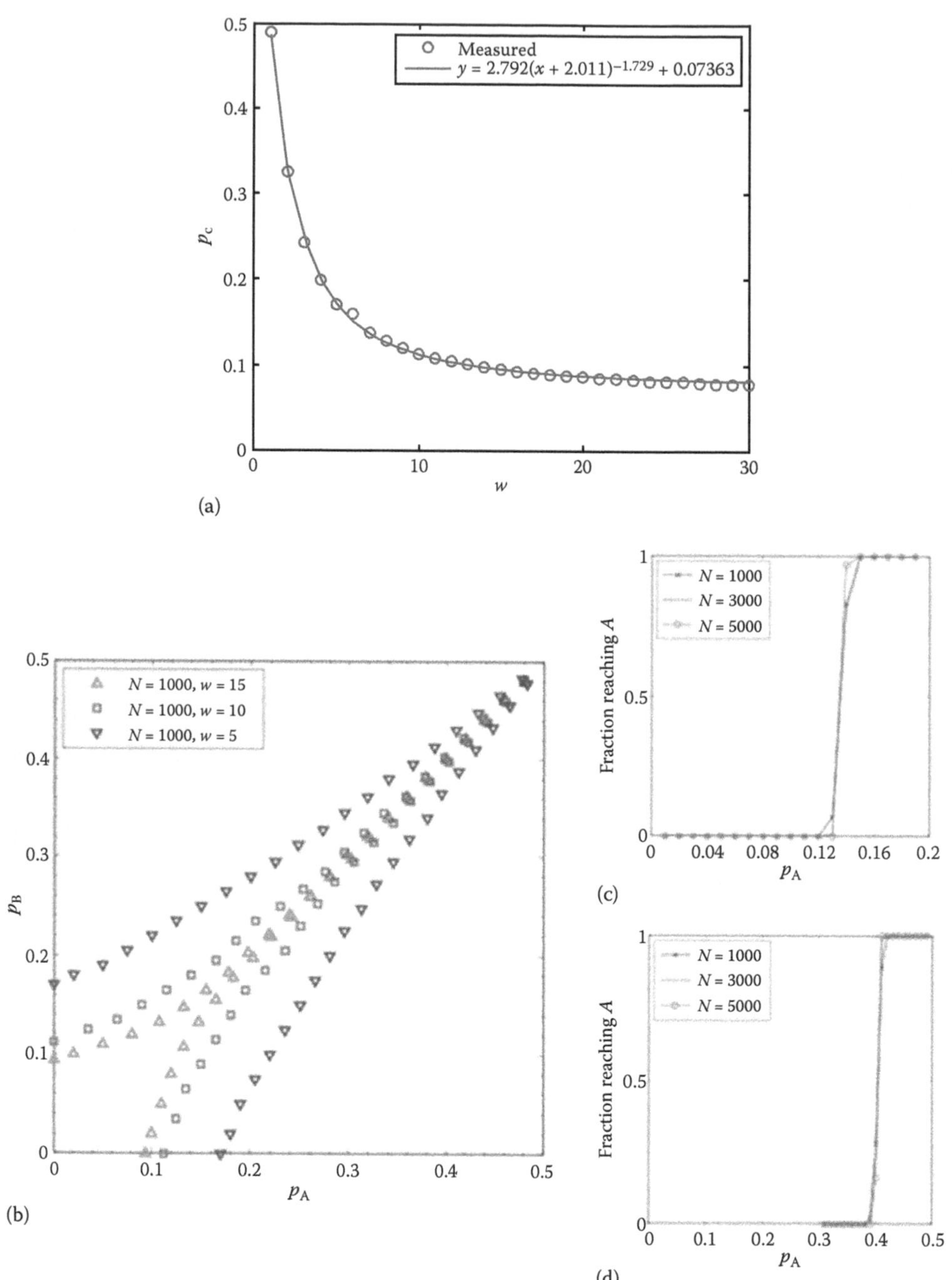

FIGURE 4.1
(a) The critical committed fraction $p_c(w)$, as a function of commitment strength w for Barabási-Albert networks of size $N = 1000$ with average degree $<k> = 10$ (Galehouse et al. 2014). (b) Phase diagram of the system in the presence of competing committed groups holding opinion A and B with initial sizes p_A and p_B, respectively, for Barabási-Albert networks with an average degree $<k> = 10$. (c) The fraction of runs reaching the all-A consensus state along a trajectory $p_A + p_B = 0.2$ in the parameter space for $w = 10$ and $<k> = 10$. (d) The fraction of runs reaching the all-A consensus state along a trajectory $p_A + p_B = 0.8$ in the parameter space for $w = 10$ and $<k> = 10$. (From Galehouse, D., T. Nguyen, S. Sreenivasan, O. Lizardo, G. Korniss, and B. K. Szymanski, *AHFE 2014*, 5:2318–2329, 2014. With permission.)

the node increases its count. Once the count reaches the node's inertia, it switches states and resets its count to zero. When the system is set up similarly to the committed agent systems (with a minority population of opinion A with varying w_A and a majority of opinion B with constant w_B), the critical population of opinion A initiators behaves similarly to that in the waning commitment model. This can be seen in Figure 4.2a and b

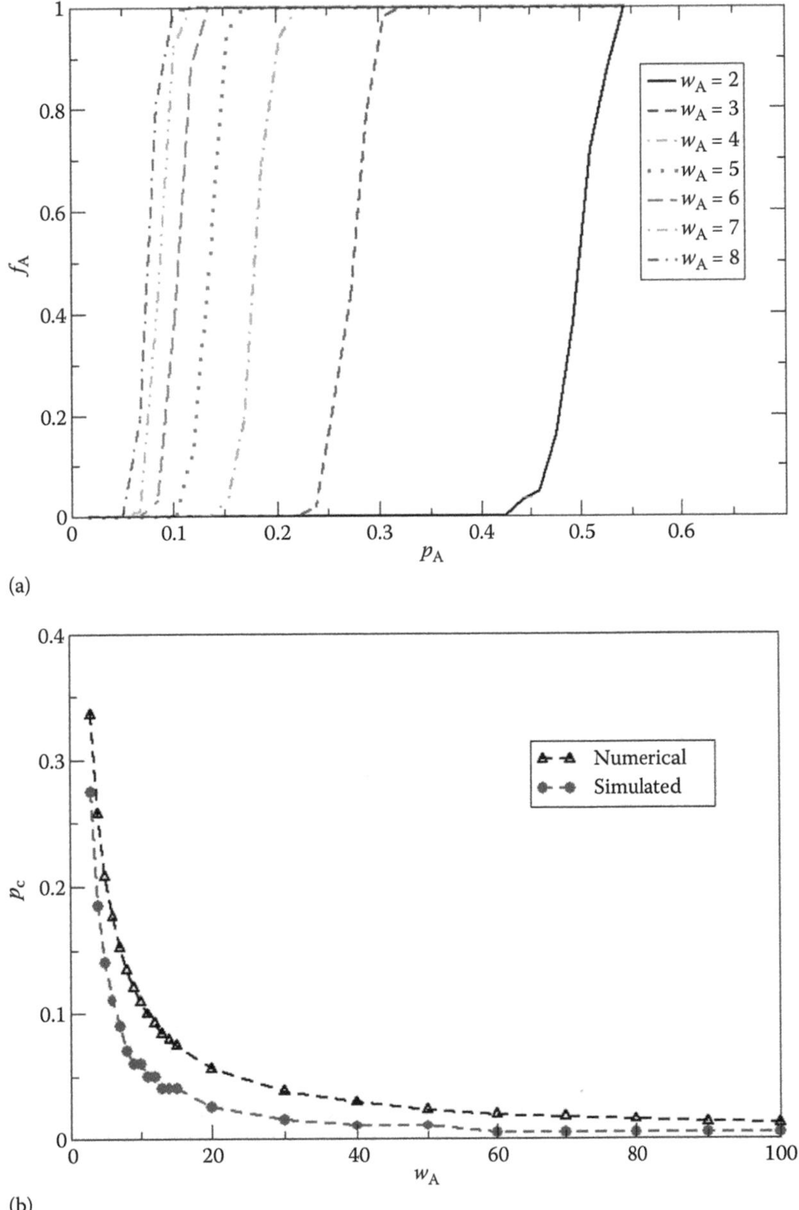

FIGURE 4.2
(a) The fraction of simulation runs that reach consensus on opinion A versus the initial density of agents with opinion A on a complete graph with $N = 1000$ and $w_B = 2$ (Doyle et al. 2015). (b) Individual-based simulation results (circles) and numerical solutions (black squares) for the critical fraction of committed minority p_c as a function of waning commitment w_A.

(circles), where $w_B = 2$ for every simulation. Calculating the rate equations for waning commitment models is difficult because they are non-Markovian (i.e., several prior states may impact the future state). Figure 4.2b shows the dynamics, where the numerical solutions capture the qualitative behavior of the system quite accurately yet consistently overestimates the value of p_c. Nevertheless, such an upper bound is useful as it is a value at which the system will always reach consensus for the minority state given that inertial value.

Threshold Models

One of the simplest models that capture the adoption dynamics of a new opinion, behavior, trait, product, etc., is the TM (Granovetter 1978; Watts 2002; Centola et al. 2007; Singh et al. 2013; Karampourniotis et al. 2015; Ruan et al. 2015; Karsai et al. 2016). In this model, a node's opinion is represented by a binary variable (active or inactive). An inactive node becomes active (i.e., it adopts the new opinion) if at least a threshold fraction (ϕ) of its neighbors is already active. The system continues to evolve until a steady state is reached in which no activations are possible. The adoption threshold ϕ can be the same for every node or it can be randomly drawn from a distribution. The dynamics is also asymmetric since an active node stays active indefinitely. The relative size of the active population in the eventual steady state is also called the cascade size (S_{eq}) (Watts 2002; Singh et al. 2013).

Tipping Points in Threshold Models

First, let us assume that all nodes have an identical threshold ϕ. A small set of nodes holding a new opinion (initiators) can cause activation cascades and even convert the population to the new opinion (global cascades). The structure of the underlying social network plays an important role in determining whether or not such cascades can be triggered by a single initiator (or small single-clique initiators).

For Erdős-Rényi (ER) networks, global cascades cannot be triggered if the average degree of the network is too low (because then the network consists of small isolated clusters) or too high (because it is difficult for the high degree nodes to satisfy the threshold condition). The critical threshold ϕ_c, above which cascades cannot be triggered, is inversely related to the average degree $<k>$ (Watts 2002). Therefore, global cascades can arise only in the intermediate range (cascade window) of $<k>$. Figure 4.3a shows the simulation results of the average cascade size (S_{eq}) as a function of the average degree $<k>$ for a fixed initiator fraction (initiators are selected at random).

In real social systems, adoption thresholds can be very high (Latane and L'Herrou 1996) since individuals are only likely to change their opinion if at least half of their interacting neighbors hold the opposite opinion.

A careful selection (e.g., influence maximization algorithms, heuristic strategies [Kempe et al. 2003; Kitsak et al. 2010; Shakarian et al. 2013; Singh et al. 2013; Morone and Makse 2015]) of the initiators can trigger global cascades, when randomly selected initiators cannot, but they are ineffective when ϕ is very high. Yet global cascades can be triggered by increasing the initiator fraction p so it reaches a critical value p_c. At that point, a discontinuous transition occurs and large cascades appear immediately; see Figure 4.3b. The relevant

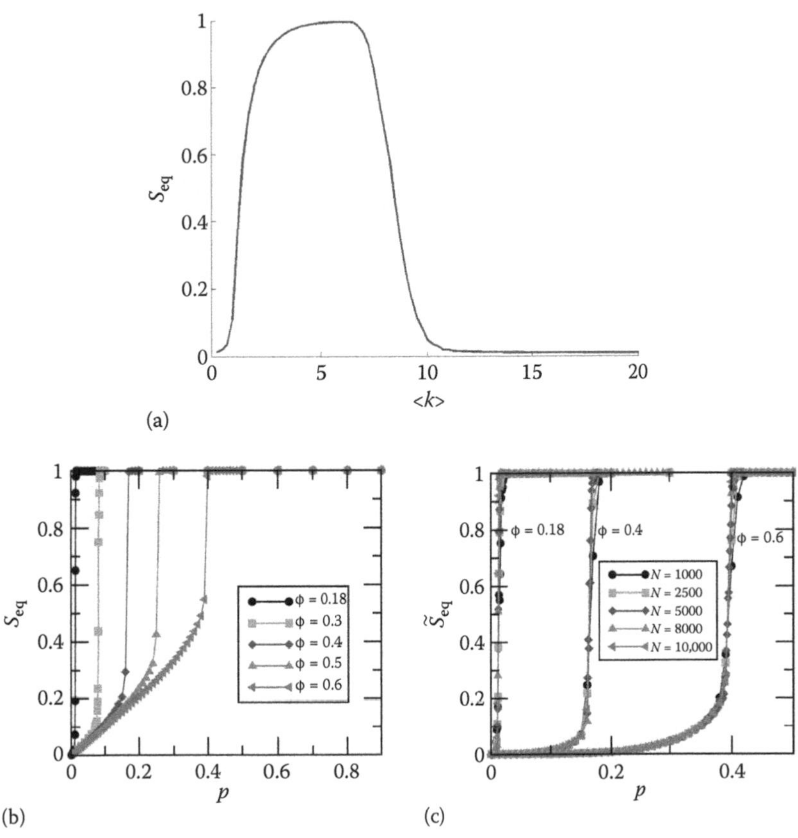

FIGURE 4.3
(a) The cascade window for ER networks, $N = 1000$, $\phi = 0.18$, and initiator fraction $p = 0.01$. Cascade size and scaled cascade size as a function of initiators on ER networks with $\langle k \rangle = 10$. (b) Cascade size S_{eq} as a function of initiators p for ER networks with $N = 10{,}000$ for different values of ϕ. (c) Scaled cascade size $\tilde{S}_{eq}$ versus p for ER networks with different network sizes N and ϕ values. (Reprinted by permission from Macmillan Publishers Ltd. *Sci Rep* (Singh et al. 2013), copyright (2013).)

THRESHOLD MODELS AND TIPPING POINTS

The presence of a tipping point in the TM depends on the distribution of thresholds in the network, which can vary for different communities. For example, communities of very religious people may not be receptive to new opinions. Moreover, some new ideas may be harder to accept than others. For example, surgical anesthesia and antiseptics were both discovered in the nineteenth century, but the former spread much faster than the latter. Furthermore, the speed of spread depends as well on the threshold distribution; as the diversity of thresholds increases, the spread slows down, and at some critical diversity, the tipping point disappears.

quantity, the rescaled cascade size $\tilde{S}_{eq} = (S_{eq} - p)/(1-p)$, discounts initially active nodes and represents the fraction of nodes initially inactive that eventually adopted the active state. The transitions in $\tilde{S}_{eq}$ are shown in Figure 4.3c for different ϕ values and several network sizes; clearly the transition depends only on ϕ and is independent of the system size N. This transition (the emergence of the tipping point) is quite generic in the TM and can be observed in networks with different sizes and average degrees, as well as for different selection methods for initiators (Singh et al. 2013).

Dynamics with Individualized Thresholds

The empirical studies (Latane and L'Herrou 1996; Centola 2010) showed that the threshold varies from node to node necessitating different thresholds at different nodes to capture the complex nature of social influencing.

A natural choice for the distribution of the adoption thresholds is the normal distribution, but then a node has a nonzero probability having a threshold higher than one, making it too stubborn to adopt a new opinion, or a negative threshold, making it a self-activating instigator (Granovetter 1978). An interesting case arises when instigators but not stubborn nodes are present; with a large enough threshold distribution, a cascade can occur without any initiators, mimicking spontaneous riots started by instigators for which an accurate analytical approximation was introduced (Gleeson and Cahalane 2007).

Excluding the presented case, we consider cascades that are initiated only with initiators (Karampourniotis et al. 2015). We assume that every node can potentially be an initiator, and that thresholds are limited to a range (0,1). We consider the impact of the diversity of the threshold drawn from a truncated normal distribution on the cascade size in the presence of multiple initiators. The mean threshold ϕ_0 is kept constant at 0.50, while the standard deviation σ varies from 0 to 0.288. For the simulations, we randomly assign initiators one by one and measure the cascade size. We repeat this process by drawing thresholds from the same distribution. The cascade size S_{eq} for each threshold distribution is obtained by averaging 1000 times on different threshold distribution draws.

Figure 4.4a and b shows the effect of σ on the cascade size S_{eq} for a constant initiator fraction p and a constant mean threshold ϕ_0. As σ increases so does a fraction of nodes whose threshold is far from the average, causing a twofold effect. Those with thresholds far below the average are easily activated, while those with thresholds far above the average are increasingly difficult to activate. Thus, when the initiator fraction is small, S_{eq} is monotonically increasing following the increase of fraction of low threshold nodes. However, when the initiator fractions are large, this increase has a minimal influence because those nodes are likely to be already activated, which leads to the nonmonotonic behavior seen in Figure 4.4a and b.

Following the study by Karampourniotis et al. (2015), we study how the cascade size S_{eq} varies with the initiator fraction p. Figure 4.4c and d shows that increasing the initiator fraction, for a small enough σ, transitions small local cascades to large global cascades which, for synthetic networks, is a discontinuous phase transition; however, averaging over different network ensembles smoothens the transition point (see Figure 4.4c). For a larger σ, the initiator fraction for which the transition occurs is reduced, while for the synthetic networks the spread size still exhibits a discontinuous phase transition. With largely diverse thresholds, a critical initiator size, beyond which cascades become global, ceases to exist, and the tipping point behavior of the social influencing process

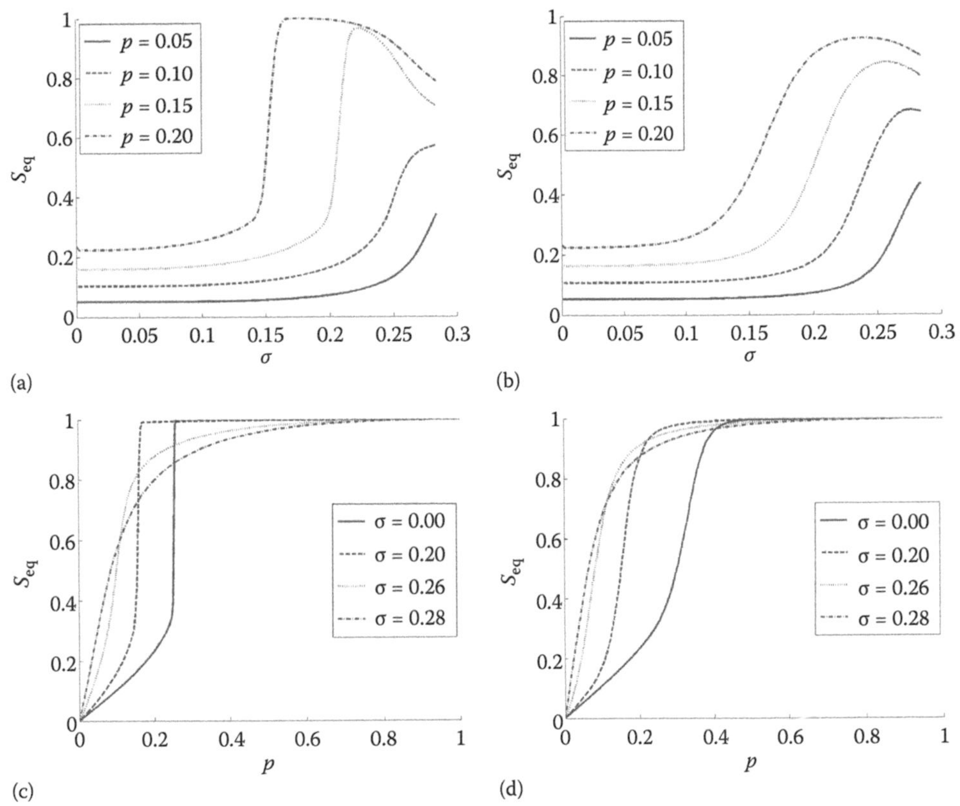

FIGURE 4.4
Behavior of the cascade size at equilibrium S_{eq} for ER networks for the varying standard deviation σ for (a) ER networks and (b) the Facebook network, and for varying initiator fractions p for (c) ER networks and (d) the Facebook network. The ER networks structure is $z = 10$ and $N = 10{,}000$. The Facebook network structure is $z = 43$ and $N = 4039$. The mean threshold is $\phi_0 = 0.50$. The simulations (solid lines) are averaged over 1000 repetitions.
(Continued)

disappears. It is replaced by a smooth crossover governed by the size of the initiators. This property can be important, for example, for a company's marketing strategy of a new product. If the threshold distribution is narrow enough, unless a critical initiator fraction is reached, there is a marginal local spread on a few of the first or second neighbors of the initiators. On the other hand, if the threshold distribution is wide, there is a significant spread.

For ER networks, the analytic estimates based on the treelike approximation model (Gleeson and Cahalane 2007) are in good agreement with the simulations (not shown here). To show this behavior for real networks, we use the TM for a Facebook ego-network (SNAP 2015) in Figure 4.4b and d. In this case, the cascade size change is smoother than that for the ER networks because the communities in the Facebook network have different critical fractions of initiators. A visualization of the spread evolution in the Facebook network (see Figure 4.4f) shows the existence of three communities that are not activated through spread. Finally, a comparison of Figure 4.4e, where σ = 0, and Figure 4.4f, where σ = 0.20, clearly shows that the spread is greater for larger σ (Karampourniotis et al. 2015).

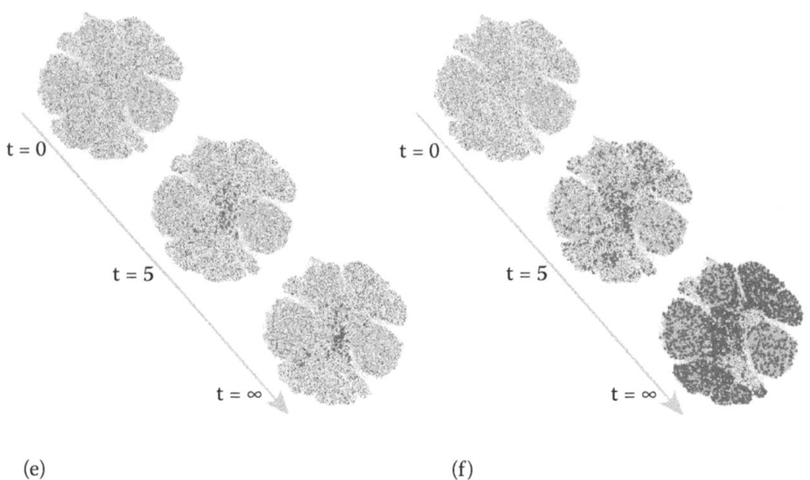

FIGURE 4.4 (CONTINUED)
(e and f) Visualization of the spread of opinion in the TM model on the Facebook egonetwork with the fraction of the randomly selected initiators at $p = 0.20$. The standard deviation of the threshold is (e) $\sigma = 0$, (f) $\sigma = 0.20$. In the color version, the inactive nodes, initiators, and active nodes (through spreading) are marked with green, orange, and red, respectively. In the grayscale version the inactive nodes and initiators are both a light color and indistinguishable between them. (From P. D. Karampourniotis, S. Sreenivasan, B. K. Szymanski, and G. Korniss, *PLoS One*, 10, 143020, 2015. With permission.)

Impact of Culture and Gender on Spread of Opinion in Empirical Networks

Megacity and Notre Dame NetSense Data

Megacity Xanadu is a very large-scale social system whose underlying interactions are structured by cultural, racial, and class divisions. As such, this example, while fictional, resembles the way that most real-life social systems are organized. As has been well known for decades in sociology and anthropology, persons organize their interactions along any distinguishable physically, culturally, biologically, and socially constructed characteristic, with those related to age, gender, and ethnicity (especially when combined to class) having the most influence on who interacts with whom. For instance, Xanadu, first being open to migration and then closing its border, created a correlation between the place of origin and the class location for North Razini immigrants. These immigrants are more likely to live in segregated areas, more likely to interact with coethnics, and less likely to form connections with members of the majority. These are dynamics that unfortunately mirror those of ethnic minorities struggling to assimilate throughout the developed and developing world. As such, studying the link between interaction and exogenous sociocultural markers (such as gender and race) becomes important for understanding these large-scale patterns. This matters even more if it is likely that the link between social characteristics and dynamics of social interaction affects the way in which we understand opinion convergence in large-scale social systems and the persistence of minority subcultures unlikely to be swayed by the majority culture.

We set out to test an overall theory linking homophily and the dynamics of tie formation and decay in real social systems. The study was a smaller scale than a megacity but used an ecologically valid, natural setting: adolescents making the transition to college in a residential campus. Data are obtained from the ongoing NetSense study at the University of Notre Dame (Striegel et al. 2010; Wang et al. 2013). The study equipped a cohort of roughly 200 incoming first year students with smartphones and tracked, among other things, the calls and texts made and received (but not the content of their communications) over a two-year period using a monitoring app employed on each phone. This app logged and then transmitted to a secure database a call data record (CDR) for each communication event. Each CDR contains the phone numbers of the sender and the receiver along with a timestamp indicating when the event occurred. While we have thousands of CDRs for texts and calls involving a NetSense subject and people outside the study, we use only the CDRs in which both parties are in the NetSense study, because for both parties involved in the communication, we have data derived from surveys on their sociodemographic characteristics, tastes, and opinions as well as other variables of interest derived from a survey administered to all NetSense subjects prior to their arrival on campus. However, the activity analysis is based on the entire set of outgoing communications for each participant (complete egonetwork degree) whether the recipient was within the study or not.

Impact of Agent Sociodemographic Characteristics on Activity and Assortative Mixing

In this section, we first summarize the findings concerning the sociodemographic differences in both activity and sociability and assortative mixing (Newman 2003). Sociability effects refer to the higher likelihood of the members of certain categories to form more connections to other persons regardless of the sociodemographic traits of those persons. Studying assortative mixing in the context of modeling opinion spread is important for the following reasons. It is well known that opinions, beliefs, and other cultural commitments tend to be different for members of different sociodemographic groups. This means that when persons interact with those who are similar to them, their current beliefs and values are likely to be reinforced. For instance, in megacity Xanadu, the North Razinis struggle to mesh with the society seemingly both because they are rejected by others and because they have higher levels of in-group homophily. This makes them different from the Bursuks who have weaker homophily patterns and are thus more likely to consider interacting with outsiders. Opinion change in real social networks is more likely to happen for those individuals who are more likely to venture outside of their own group for sociable interactions (Aral et al. 2009). A social system characterized by high levels of assortative mixing is also one that is likely to exhibit opinion polarization rather than homogeneity.

We identify both sociability and assortative mixing effects using exponential random graph models (ERGMs) fitted to the network (Robins et al. 2007). These models allow us to ascertain the statistical dependencies on the network while adjusting for lower order network properties, such as the overall tendency to form edges (density) and the tendency for agents who share neighbors to also be connected to one another. The purpose of ERGMs is to provide a summary description of the local relationship formation mechanisms that generate the global structure of the network. For instance, the NetSense within-study network (depicted in Figure 4.5) may be considered similar to the response or the outcome variable in a regression or a classification model. Here the predictors of whether we observed an edge (or not) are such things as *the propensity of individuals of the same gender/race/religion/class background to form connections* or *the propensity of individuals who share friends to be connected to one another*. In Figure 4.5, for instance, it is clear that individuals who share the same racial/ethnic status tend to cluster together. The ERGM can help us quantify the

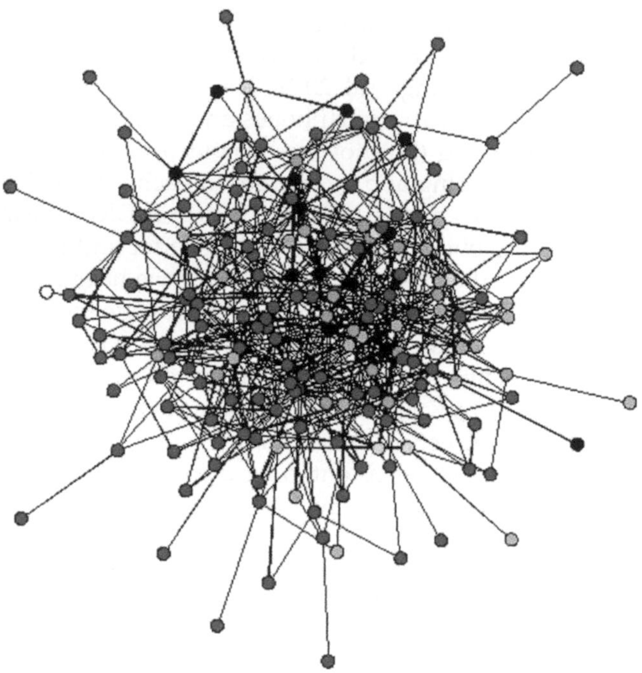

FIGURE 4.5
NetSense within-study network. Vertex color indicates race: dark gray = Caucasian, light gray = African American, gray = Latino, black = Asian, and white = other.

strength of this assortative mixing effect while holding constant other confounding relationship formation mechanisms.

The ERGM specifies the conditional log odds of an edge existing versus not existing between two nodes, conditional on values of covariates and a network configuration:

$$\frac{Pr(X_{ij} = 1 \mid Y = y, X_{ij} = x_{ij})}{Pr(X_{ij} = 0 \mid Y = y, X_{ij} = x_{ij})} = e^{\theta^T \delta_{ij}(x) + \theta_a(y_i + y_j) + \theta_b y_i y_j}, \tag{4.2}$$

where θ_a denotes the (additive) activity parameter and θ_b is the (multiplicative) assortative mixing parameter; y_i and y_j are the values of each vertex on a given sociodemographic trait of interest (in this example, gender, which is a categorical trait with two values: male and female). The other term is for the configuration (endogenous) part of the model. In our case, this involves a parameter for the overall number of edges in the graph (density) and a parameter that counts the number of shared neighbors for each node (transitivity). For the purpose of interpreting the activity and mixing parameter, we can ignore that part of the model because when taking odds ratios, it drops out (assuming that the configuration model is the same for all edges).

What we can compute from this model is the conditional odds ratio. The reference category for these odds is $y_i = y_j = 0$, i.e., the (0,0) cell in the $n \times m$ table cross classifying edges by whether they have or do not have the trait $Y = y$, which in our case is a 2×2 table cross-classifying edge by the gender mix of the vertices attached to them. Thus, for the example used here, Y is the gender, and $Y = 1$ indicates that the node is female.

GENDER DISTINCTIONS IN NETWORK TIES?

Males and females may display gender-specific socializing and communication styles, hence the different numbers of ties observed for ties connecting males to males, males to females, and females to females may arise because of different dispositions toward same-gender and cross-gender interaction.

The conditional odds ratio for an edge having one node with $Y = 1$ versus having both vertices with $Y = 0$ (e.g., odds ratio of male–female tie versus male–male tie) is e^a, the activity parameter. Its value defines the odds that, conditional on one node being male, the other will be female. If the activity parameter is 0, then the odds ratio is 1, making the odds of male–female and male–male ties the same. This means that the counts of male–male and male–female ties should be the same. If the activity parameter is greater than 0, then the male–female ties are more prevalent than the male–male ties; if less than 0, the male–male ties are more prevalent than the male–female ties. The activity parameter reflects the difference in the degree of male versus female, because the only way male–female ties can be more prevalent than male–male ties (assuming random matching, i.e., no homophily) is if females have more ties than males.

If there is no homophily (i.e., random matching), then the odds of a female–female tie versus a male–male tie are e^{2a}. That is, we expect more female–female ties solely because there are females having more edges (i.e., they are more likely to be at the end of a random edge), so that the odds of a female being at both ends of an edge are the exponential of two times the activity parameter. If there is homophily (positive b parameter), then the odds of a female–female tie versus a male–male tie are e^{2a+b}. The $2a$ captures the greater odds of a female–female tie due to the higher activity of females, while the b term adds in increased odds of female–female ties resulting from assortative mixing. Note that the odds of the female–female tie versus the male–female tie are now e^{a+b}. The female–female edge gets a boost over the male–female edge because females have higher activity and because of the activity differences of female–female ties are more prevalent.

For instance, from our parameter estimates (Table 4.1, model 4), we can see that $a = 0.14$ and $b = 0.46$. We coded female as 1 so the reference like above is male–male ties. The odds of a male–female tie versus a male-male tie are $e^{0.14} - 1 = 15\%$, i.e., the former is 15% more likely than the later, reflecting the higher activity of women. Without the

TABLE 4.1

Gender-Based Activity and Assortative Mixing

Parameters	Model 1	Model 2	Model 3	Model 4
Density	–3.24	–3.76	–4.15	–4.51
	(0.04)	(0.08)	(0.08)	(0.10)
Activity (Women)		0.25		0.14
		(0.05)		(0.04)
Mixing		0.51		0.46
		(0.08)		(0.07)
Transitivity			0.99	0.97
			(0.07)	(0.07)

homophily effect, this would imply that female–female ties would be about 32% more prevalent than male–male ties (e.g., $e^{2\times0.14} - 1$). But with the assortative mixing effect included, the odds of a female–female tie are $e^{2(0.14+0.46)} = 2.09$, so more than twice as likely as male–male ties. Finally, we see that female–female ties are $e^{(0.14+0.46)} = 1.82$ times more likely than male–female ties.

In our estimates (Table 4.1), female–female ties are more prevalent than male–female ties, which are more prevalent than male–male ties. This is indicative of the strong female same-gender preference effect coupled with females having a high degree, which implies a tendency for males to not associate with other males because they randomly associate with other people who happen to be more likely females because of their higher activity.

Race and Ethnicity

Table 4.2 shows the result of fitting an ERGM model with separate activity parameters for each racial and ethnic group and a common assortative mixing parameter for all race groups. We find that, in terms of activity, minority groups are more active than Caucasians in their first semester on campus. Recall that the activity coefficients compare the odds of observing an edge featuring a non-Caucasian respondent versus the odds of observing an edge containing two Caucasian respondents. Across all racial

TABLE 4.2

Race-Based Activity and Assortative Socializing

Parameters	Model 1	Model 2	Model 3	Model 4
Density	−3.24	−4.53	−4.15	−5.13
	(0.04)	(0.10)	(0.08)	(0.10)
Activity (African American)		0.99		0.75
		(0.10)		(0.08)
Activity (Latinos)		0.84		0.64
		(0.08)		(0.07)
Activity (Asians)		0.79		0.59
		(0.08)		(0.06)
Activity (Other)		0.75		0.64
		(0.19)		(0.13)
Mixing (All Races)		1.33		1.12
		(0.10)		(0.08)
Transitivity			0.99	0.95
			(0.07)	(0.07)

CULTURAL TIES AND MEASUREMENT

In a majority/minority context, the differences in dynamic stability between intracultural ties and intercultural ties can be used to measure the extent to which minority individuals are accepting the mainstream culture prevailing in the society.

groups (as indicated by the positive activity effects), we find that mixed race edges are more prevalent than the all-Caucasian edges.

Class Background

We find that (Table 4.3, model 3) there are no activity differences based on class background, once we adjust for the statistical propensity to find class-matched edges over class-mismatched ones. In addition, we find that same race dyads are statistically more prevalent than mixed-race dyads (= 1.33), even after adjusting for the propensity of friends of friends to also be connected to one another (= 0.95). Thus, even though Caucasians are less active than non-Caucasians, when they do form ties, they are disproportionately likely to be directed at other Caucasians. Non-Caucasians, on the other hand, have similar same-race preferences, but due to their higher activity, they end up forming more mixed-race connections than we would expect by chance. In our data, we coded class background as follows: first-generation college students were coded as working class (b = 1.06). Neither middle class nor upper class students display any tendency to form ties disproportionately with members of the same class group. Note that the working class affinity effects persist even after adjusting for the tendency toward transitivity.

Also of interest is the time evolution of network activities for different classes of agents, classified according to their personality, gender, and race. The resulting diagrams, shown in Figure 4.6, also provide the range of differences between the respective classes of agents as well as the trends over time. The strong personality, race, and gender effects on the active network size of students after joining the university evolve differently. As shown in Figure 4.6, the race effects persist through the observation period, but the gender effects decline.

TABLE 4.3

Class-Based Activity and Assortative Socializing

Parameters	Model 1	Model 2	Model 3	Model 4	Model 5
Density	–3.24	–3.71	–3.56	–4.15	–4.17
	(0.04)	(0.14)	(0.21)	(0.08)	(0.08)
Activity (Working Class)		0.50	0.21		
		(0.10)	(0.12)		
Activity (Upper Middle Class)		0.11	0.13		
		(0.09)	(0.19)		
Mixing (All Classes)		0.31			
		(0.10)			
Mixing (Working Class)			1.06		0.90
			(0.27)		(0.14)
Mixing (Middle Class)			–0.44		
			(0.38)		
Mixing (Upper Middle Class)			0.11		
			(0.22)		
Transitivity				0.99	0.98
				(0.07)	(0.07)

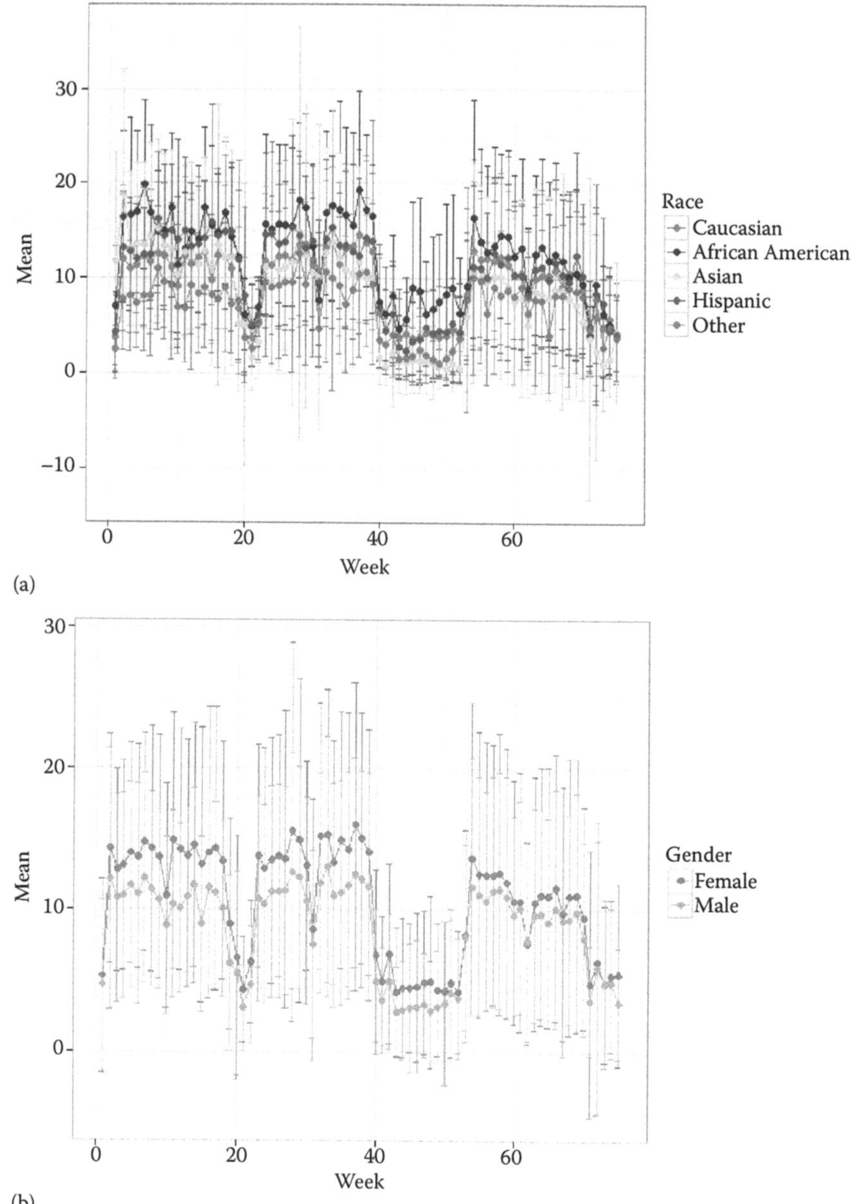

FIGURE 4.6
Race and gender effects on active network size. Error bars represent one standard deviation error. (a) Mean number of contacts per week by race and (b) by gender.

Network/Opinion Coevolution: Differential Tie Decay by Dyadic Mix

A key implication of the dynamic NG model is that the opinion evolution process is qualitatively different when modeled from a static network perspective from when opinions and social ties are considered as coupled dynamic systems. Yet we still have very limited empirical evidence as to whether social ties are subject to differential decay depending on whether persons share a given set of opinions (Noel and Nyhan 2011). After all, if the preexisting sharing of opinions does not impact the dynamic stability of social ties, researchers may be correct in sticking with more tractable models that conceive of the network structure as static (and thus exogenous). In addition, if persons select friends based on the fact that they share an opinion, rather than changing their opinions after they select their friends, then the conditions under which an entire network may be captured by a single opinion (under the static model) change dramatically.

As we noted earlier, each CDR in the NetSense data contains the phone numbers of the sender and the receiver along with a timestamp indicating when the event occurred. While we have thousands of CDRs for texts and calls involving a NetSense subject and people outside the study, we use only the CDRs in which both parties are in the NetSense study, because for both parties involved in the communication, we have data derived from surveys on their sociodemographic characteristics, tastes, and opinions as well as other variables of interest derived from a survey administered to all NetSense subjects prior to their arrival on campus. From these CDRs, we identify 505 ties formed in the first two semesters (40 weeks) among 175 students who remained in the study throughout the first year and for which we have complete data on gender, race, and parental education. We selected these three sociodemographic characteristics because sociological research shows that opinions tend to cluster within same race, gender, and class groupings (Carley 1991; Noel and Nyhan 2011). Therefore, persons of the same race, gender, or class are more likely to share the same opinions, while cross-gender, cross-racial, and cross-class dyads are more likely to have different opinions. Using proportional hazard modeling, the differential decay of the dyad type over time is based on race, class, and gender mixes thus indicating that the network structure is not static but that it coevolves with the opinion structure.

The results are shown in Figure 4.7. We find positive evidence of network/opinion coevolution. As shown in Figure 4.7a, cross-race dyads decay at a faster rate than same-race dyads, with Latino/Latino pairings exhibiting remarkable dynamic durability. Within the cross-race group, some pairings are more durable than others. In particular Asian/Caucasian pairings and Latino/African American pairings decay at a slower rate than other pairs. This is consistent with sociological research that reveals higher levels of agreement among members of these racial categories (Bonilla-Silva 2004). In addition, as shown by Figure 4.7b, cross-class dyads, especially those that feature a lower class person paired with an upper class person, are less durable than other types of dyads, especially dyads that pair two persons of lower class origin.

Sociological research shows that there is more opinion agreement within classes (especially the working class) than there is between the upper and lower classes (Weeden and Grusky 2005). This result is thus consistent with the opinion/network coevolution thesis: the ties between groups with less opinion consensus are less durable than ties within groups with high opinion consensus. This mirrors the fictional social structure of megacity Xanadu, which is roughly divided into a three-part class structure, separating the long-standing residents at the top, from the first wave of Bursuka immigrants in the model, and the more recent wave of North Razini immigrants at the bottom. Applying the results of our statistical model from the NetSense data to this fictional example, our results would thus predict that ties are more likely to be persistent (conditional on formation) between native Çawailans and Bursuks than they are between Razinis and natives. Finally, Figure 4.7c shows that cross-gender pairings are less dynamically less stable than same-gender pairings. Given the large amount of evidence for systematic opinion differences between men and women (Bolzendahl and Myers 2004), this result is consistent with a model in which persons terminate social connections featuring opinion disagreement

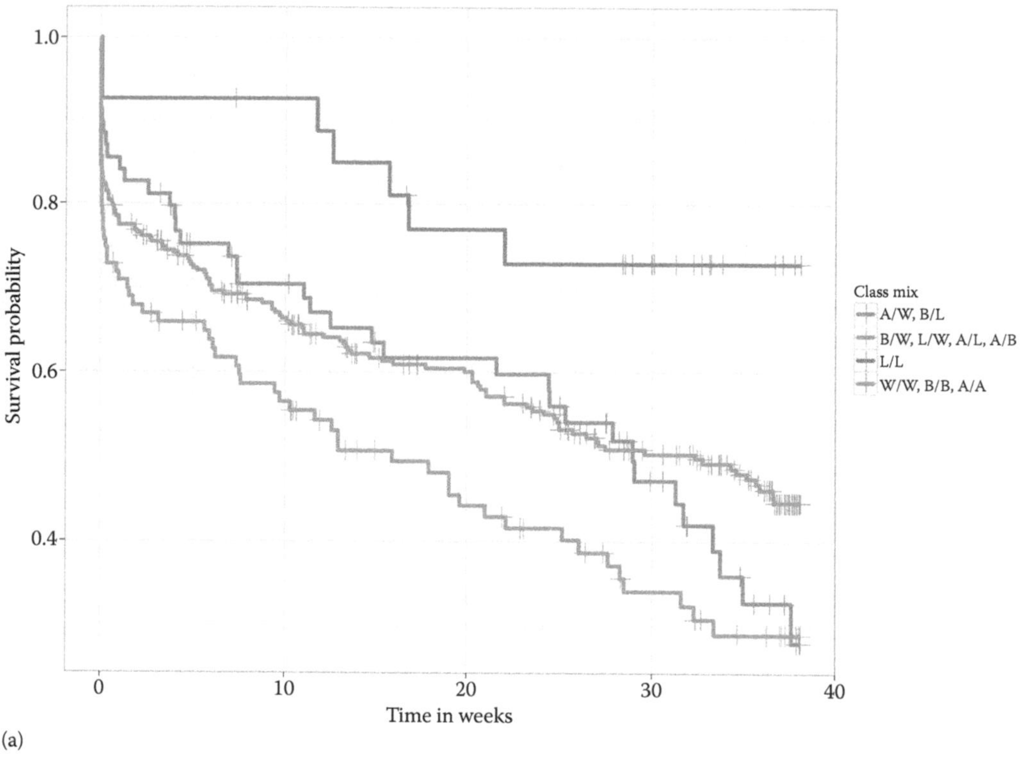

FIGURE 4.7
Survival curves for four sets of dyads classified according to (a) ethnoracial pairing. Estimates of hazard rates (the probability of dissolution at a given time for each dyad type) and survivor probabilities (the probability that a tie will still be active at a given time) are computed using the Kaplan–Meier estimator in which at each time point t, a risk set is specified as those objects that have survived up until t and an event set as those objects that experience an event in the interval from t to Δt. Vertical hash marks indicate right-censored cases (ties active at the end of the observation period). (a) Ethnoracial categories are Caucasians (W), Asian (A), African American (B), and Latino (L). As shown, Latino/Latino (LL) pairs have the slowest decay rate, followed by Caucasian/Caucasian (W/W), African American/African American (B/B), Asian/Asian (AA), Asian/Caucasian (AW), and African American/Latino (B/L) pairs. African American/Caucasian (B/W), Latino/Caucasian (LW), Asian/African American (A/B), and Asian/Latino (A/L) pairs have the fastest decay rates. *(Continued)*

at a faster rate than social network connections featuring opinion agreement (Noel and Nyhan 2011).

Modern social systems are not only beset by divisions based on external or physical features (such as gender and race). People also sort themselves into groups and condition their social interactions based on ideas (such as religious or political ideologies). For instance, in Xanadu, Bursuks have become a religious and cultural minority. This may affect their capacity to integrate and interact with others who share a different belief. Alternatively, if beliefs change depending on who you interact with, then the situation does not seem so hopeless. So the question to answer is do people select ties based on ideological similarity or does ideological similarity emerge after people form ties? In the first case, we should expect the durable existence of ideological enclaves. In the latter case, we could conceivably break these enclaves by inducing interaction across ideological divisions.

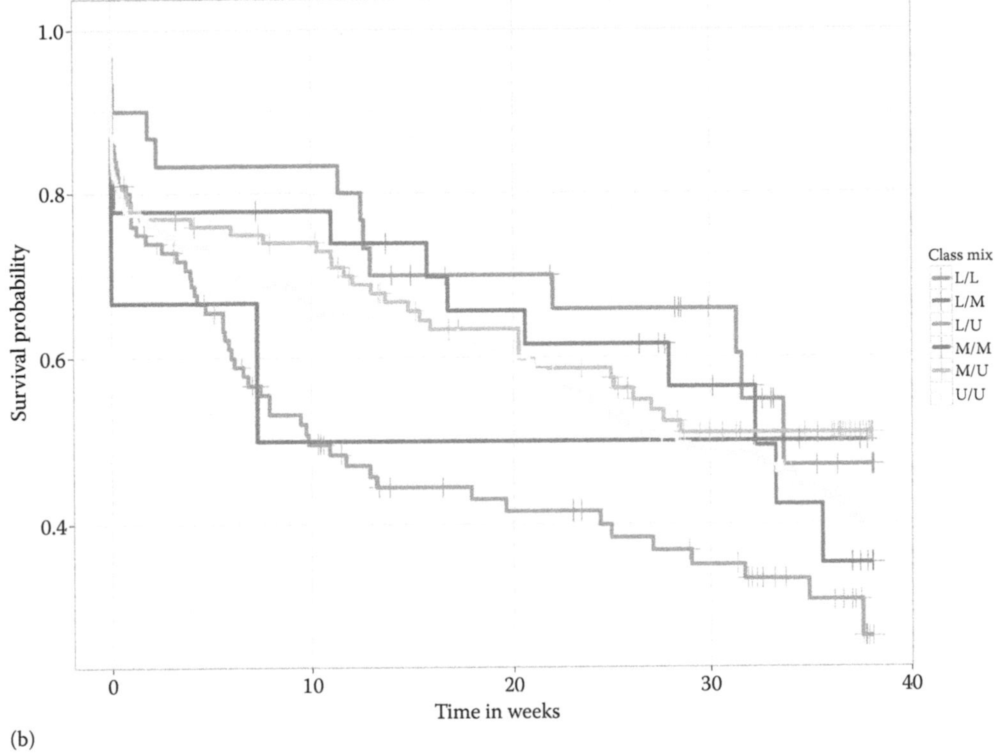

FIGURE 4.7 (CONTINUED)
Survival curves for four sets of dyads classified according to (b) social class pairing. Estimates of hazard rates (the probability of dissolution at a given time for each dyad type) and survivor probabilities (the probability that a tie will still be active at a given time) are computed using the Kaplan–Meier estimator in which at each time point t, a risk set is specified as those objects that have survived up until t and an event set as those objects that experience an event in the interval from t to Δt. Vertical hash marks indicate right-censored cases (ties active at the end of the observation period). (b) Social class categories are upper class (both parents have college degrees) (U), middle class (only one parent has a college degree) (M), and lower class (neither parent has a college degree) (L). The results show that dyads that match upper and lower class persons (L/U) decay faster than other types of dyads. *(Continued)*

In the NetSense project, we set out to test these hypotheses using the case of liberal/conservative placement. Figure 4.8 summarizes the results reported in the study by Wang et al. (2013) based on a study of the coevolution of liberal/conservative opinions among NetSense study participants. We set out to test the hypothesis of whether the overtime correlation between personal opinions and network ties can be accounted for by self-selection of persons into dyads with like-minded others (opinion homophily) or opinion change via influence after people with different opinions became tied. We used stochastic actor-based models as this simulation-based technique allows us to separate the effect of self-selection from opinion change via influence (Snijders et al. 2010). We found that for the most part, persons remained fairly stable in their opinions and that they selected friends based on opinion similarity rather than changing their political opinion based on the influence of people with different opinions. This provides further evidence of another mechanism (self-selection) that may help to account for the coevolution of the opinion distribution and the social network configuration.

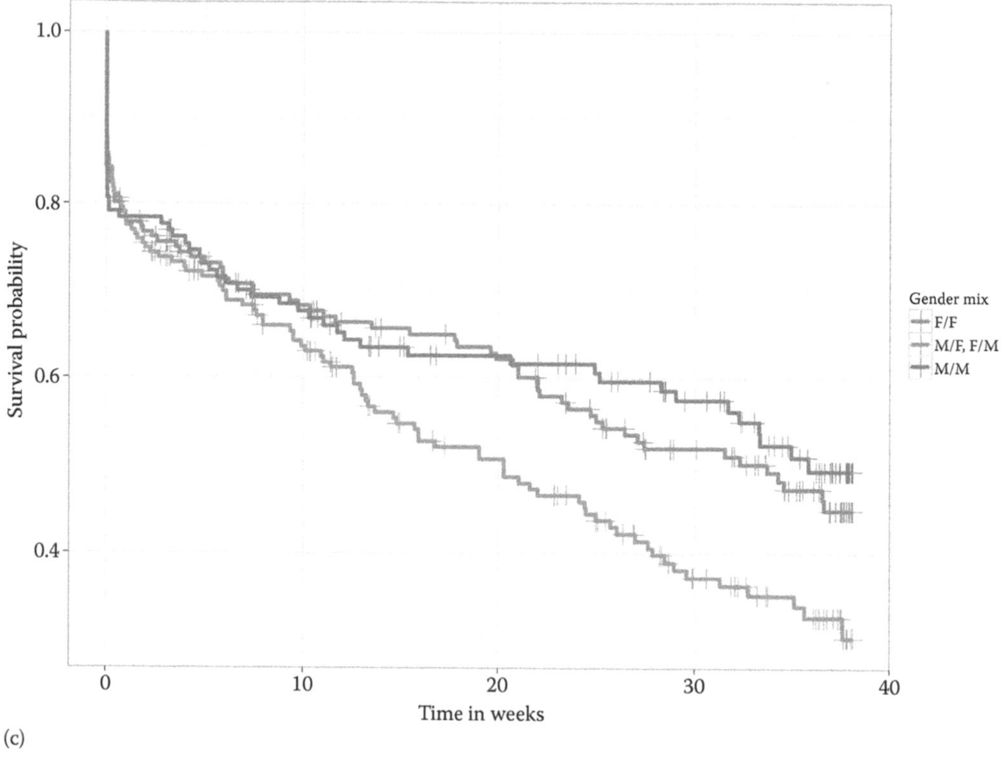

FIGURE 4.7 (CONTINUED)
Survival curves for four sets of dyads classified according to (c) gender pairing. Estimates of hazard rates (the probability of dissolution at a given time for each dyad type) and survivor probabilities (the probability that a tie will still be active at a given time) are computed using the Kaplan–Meier estimator in which at each time point t, a risk set is specified as those objects that have survived up until t and an event set as those objects that experience an event in the interval from t to Δt. Vertical hash marks indicate right-censored cases (ties active at the end of the observation period). (c) Gender categories are males (M) and females (F). The results show that same-gender dyads (F/F and M/M) decay at a slower rate than cross-gender dyads (M/F or F/M).

What lessons do these results have for our understanding of the way that social, cultural, racial, and other divisions structure social systems? First, we should not be overly sanguine or optimistic about opinion change campaigns. Just as in the megacity of Xanadu, we can see that persons structure their interactions based on durable memberships in social groups. These groups could be elective (e.g., based on ideological or membership choices) or they could be nonelective (based on traits such as gender and race). Once the network dynamics are coupled to group membership, a variety of self-reinforcing mechanisms serving to segregate the interaction with like-minded and socioculturally similar others are set off. These mechanisms reify social boundaries and create strong correlations between statuses (e.g., North Razinian descent and lower social class or Bursuka heritage and religious ideology) where initially they were only weak ones. These shape not only the formation of social networks but also their evolution overtime. Thus, even if social systems are expected to increase in terms of size and scale into the future, the underlying dynamics governing the process of tie formation, tie decay, opinion adoption, and opinion change in microlevel social systems will continue to be of primary importance.

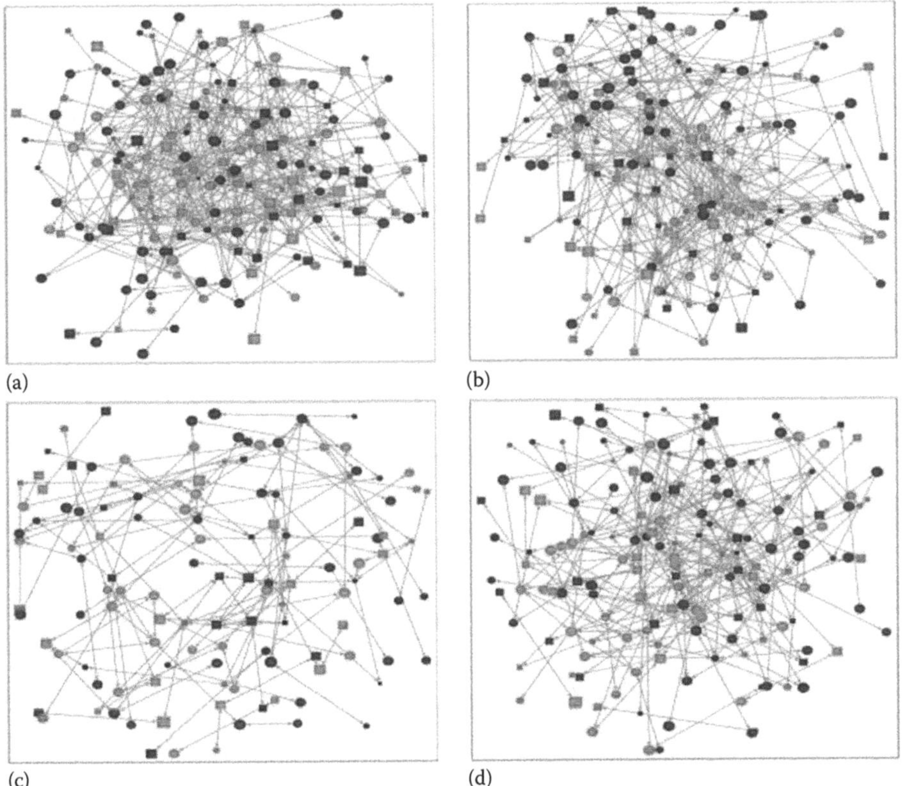

FIGURE 4.8
Evolution of the political opinion network among NetSense study participants (left). Gray nodes indicate conservative respondents and black nodes indicate liberal respondents based on their self-placement on a seven-point scale. Two participants are connected if one sends a short message service or a voice call to another who reciprocates during that time. Participants were surveyed four times in each time. We used stochastic actor-based models (Snijders et al. 2010) to model the effect of opinion similarity and network connectivity on the probability of keeping a tie. We find that for the most part, self-selection of like-minded participants and exogenous opinion accounted for the majority of overtime correlation (77.4%) across different snapshots, with opinion change via influence accounting for only a small portion of the correlation (3.9%). (a) Network in Fall 2011, (b) Network in Spring 2012, (c) Network in Summer 2012, and (d) Network in Fall 2012.

DISCUSSION QUESTIONS

1. Why do the models chosen for complex processes tend to be simple?
2. How would a change in either of the two models discussed in this chapter affect the dynamics of modeled processes?
3. For which trends/ideas/products is the TM a more suitable model than the BAM?
4. What strategies could be efficient for detecting the most influential nodes?
5. What are the unique relationships that are common for male–female ties but not frequent for female–female ties?
6. What relationship could be expected between the fraction of ties connecting minority members with majority members to ties between two members of this minority and the level of acceptance of this minority by the majority?

Acknowledgments

This research was supported in part by the Army Research Office grant W911NF-12-1-0546, by the Army Research Laboratory under Cooperative Agreement Number W911NF-09-2-0053 (the ARL Network Science CTA), and by the Office of Naval Research Grant Nos. N00014-09-1-0607 and N00014-15-1-2640.

References

Aldrich, J. H., C. Gelpi, P. Feaver, J. Reifler, and K. T. Sharp. 2006. Foreign policy and the electoral connection. *Ann Rev of Political Sci* 9(1): 477–502.

Aral, S., L. Muchnik, and A. Sundararajan. 2009. Distinguishing influence-based contagion from homophily-driven diffusion in dynamic networks. *Proc Natl Acad Sci USA* 106(51): 21544–21549.

Baronchelli, A., M. Felici, E. Caglioti, V. Loreto, and L. Steels. 2006. Sharp transition towards shared vocabularies in multi-agent systems. *J Stat Mech: Theory Exp* 2006: P06014.

Baronchelli, A., V. Loreto, and L. Steels. 2008. In-depth analysis of the naming game dynamics: The homogeneous mixing case *Int J Mod Phys C* 19: 785–812.

Bolzendahl, C. I., and D. J. Myers. 2004. Feminist attitudes and support for gender equality: Opinion change in women and men, 1974–1998. *Social Forces* 83(2): 759–789.

Bonilla-Silva, E. 2004. From bi-racial to tri-racial: Towards a new system of racial stratification in the USA. *Ethn Racial Stud* 27(6): 931–950.

Burstein, P. (2003). The impact of public opinion on public policy: A review and an agenda. *Political Res Q* 56(1): 29–40.

Carley, K. 1991. A theory of group stability. *Am Sociol Rev* 56: 331–354.

Castellano, C., S. Fortunato, and V. Loreto. 2009. Statistical physics of social dynamics. *Rev Mod Phys* 81: 591–646.

Castelló, X., A. Baronchelli, and V. Loreto. 2009. Consensus and ordering in language dynamics. *Eur Phys J B* 71: 557–564.

Centola, D. 2010. The spread of behavior in an online social network experiment. *Science* 329(5996): 1194–1197.

Centola, D., V. M. Eguíluz, and M. W. Macy. 2007. Cascade dynamics of complex propagation. *Physica A* 374: 449–456.

Dall'Asta, L. A., Baronchelli, A. Barrat, and V. Loreto. 2006. Non-equilibrium dynamics of language games on complex networks. *Phys Rev E* 74: 036105.

Doyle, C., S. Sreenivasan, B. K. Szymanski, and G. Korniss. 2016. Social consensus and tipping points with opinion inertia. *Physica A* 443: 316–323.

Galehouse, D., T. Nguyen, S. Sreenivasan, O. Lizardo, G. Korniss, and B. K. Szymanski. 2014. Impact of network connectivity and agent commitment on spread of opinions in social networks. *AHFE 2014* 5: 2318–2329.

Gladwell, M. 2000. *The tipping point: How little things can make a big difference*. Little Brown, Boston.

Gleeson, J. P., and D. J. Cahalane. 2007. Seed size strongly affects cascades on random networks. *Phys Rev E* 75: 056103.

Granovetter, M. 1978. Threshold models of collective behavior. *Am J Sociol* 83: 1420.

Karampourniotis, P. D., S. Sreenivasan, B. K. Szymanski, and G. Korniss. 2015. The impact of heterogeneous thresholds on social contagion with multiple initiators. *PLoS One* 10(11): 143020.

Kempe, D., J. Kleinberg, and E. Tardos. 2015. Maximizing the spread of influence through a social network. *Proc Theory of Computing* 11(4): 105–147.

Karsai, M., G. Iñiguez, R. Kikas, K. Kaski, and J. Kertész. 2016. Local cascades induced global contagion: How heterogeneous thresholds, exogenous effects, and unconcerned behaviour govern online adoption spreading. e-print arXiv:1601.07995.

Kitsak, M., L. K. Gallos, S. Havlin, S. Liljeros, L. Muchnik, H. E. Stanley, and H. A. Makse. 2010. Identification of influential spreaders in complex networks. *Nat Phys* 6: 888–893.

Latane, B., and T. L'Herrou. 1996. Spatial clustering in the conformity game: Dynamic social impact electronic groups. *J Pers Soc Psychol* 70: 1218.

Lu, Q., G. Korniss, and B. Szymanski. 2009. The naming game in social networks: Community formation and consensus engineering. *J Econ Interact Coord* 4: 221–235.

Morone, F., and H. A. Makse. 2015. Influence maximization in complex networks through optimal percolation. *Nature* 524: 65–68.

Newman, M. E. J. 2003. Mixing patterns in networks. *Phys Rev E* 67(2): 026126.

Noel, H., and B. Nyhan. 2011. The "unfriending" problem: The consequences of homophily in friendship retention for causal estimates of social influence. *Soc Networks* 33(3): 211–218.

Robins, G., P. Pattison, Y. Kalish, and D. Lusher. 2007. An introduction to exponential random graph (p*) models for social networks. *Soc Networks* 29(2): 173–191.

Ruan, Z., G. Iñiguez, M. Karsai, and J. Kertész. 2015. Kinetics of social Contagion. *Phys Rev Lett* 115: 218702.

Shakarian, P., S. Eyre, and D. Paulo. 2013. A scalable heuristic for viral marketing under the tipping model. *Soc Netw Anal Min* 3: 1225–1248.

Singh, P., S. Sreenivasan, B. K. Szymanski, and G. Korniss. 2013. Threshold-limited spreading in social networks with multiple initiators. *Sci Rep* 3: 2330.

SNAP. 2015. *Stanford Network Analysis Project*. http://snap.stanford.edu/data (accessed 04/23/2015).

Snijders, T. A. B., G. G. Van de Bunt, and C. E. G. Steglich. 2010. Introduction to stochastic actor-based models for network dynamics. *Soc Networks* 32(1): 44–60.

Steels, L. 1995. A self-organizing spatial vocabulary. *Artif Life* 2: 319–332.

Striegel, A., S. Liu, L. Meng, C. Poellabauer, D. Hachen, and O. Lizardo. 2010. Lessons learned from the NetSense smartphone study. *ACM SIGCOMM Comp Comm Rev.*

Wang, C., D. S. Hachen, and O. Lizardo. 2013. The co-evolution of communication networks and drinking behaviors. *Proc AAAI Fall Symposium Series.*

Watts, D. J. 2002. A simple model of global cascades on random networks. *Proc Natl Acad Sci USA* 99: 5766.

Weeden, K. A., and D. B. Grusky. 2005. The case for a new class map1. *Am J Sociol* 111(1): 141–212.

Xie, J., J. Emenheiser, M. Kirby, S. Sreenivasan, B. K. Szymanski, and G. Korniss. 2012. Evolution of opinions on social networks in the presence of competing committed groups. *PloS One* 7(3): 33215.

Xie, J., S. Sreenivasan, G. Korniss, W. Zhang, C. Lim, and B. K. Szymanski. 2011. Social consensus through the influence of committed minorities. *Phys Rev E* 84: 011130.

Zhang, W., C. Lim, S. Sreenivasan, J. Xie, B. K. Szymanski, and G. Korniss. 2011. Social influencing and associated random walk models: Asymptotic consensus times on the complete graph. *Chaos* 21: 025115.

5

Culture's Influences on Cognitive Reflection

Vladimíra Čavojová and Róbert Hanák

CONTENTS

Introduction ... 85
Characterizing Type 1 and Type 2 Processes ... 86
  Measure of Cognitive Reflection ... 88
Cognitive Reflection and Intelligence ... 89
  Cognitive Reflection as Cognitive Style ... 90
  Cognitive Reflection and Resistance to Irrational Beliefs ... 90
  Cognitive Reflection and Biases ... 91
Cultural Differences in Cognitive Reflection ... 92
Concluding Comments ... 97
Acknowledgments ... 98
References ... 99

ABSTRACT This chapter discusses possible cultural influences on cognitive styles. We review the research connected with the use of the cognitive reflection test (CRT), which is sometimes used as a measure of cognitive style (reflective versus intuitive). The CRT is based on performance not self-report and is hypothesized to reflect an individual's ability to suppress intuitive (and wrong) answers and to reflect upon his or her answer. We were not able to find any notable cultural differences between the samples—our review highlighted the need for better education and teaching people to reflect upon their beliefs and intuitions.

Introduction

The distinction between two kinds of thinking and decision making is not new—the recognition that some people prefer to decide by their head while others by their heart is ancient and widespread among lay population and researchers, although the latter usually uses more refined terminology, such as preferred cognitive style, types of processing, etc. Based on a considerable amount of research in the last two decades, a majority of researchers agree that such decision making is characterized by some sort of *dual process theory*. Dual process theories are based on presumptions that human thinking is driven by two different systems or types of processes. Type 1 processes are evolutionary older, autonomous, fast, and effortless. They often include fast and frugal heuristics, so-called rules of thumb that provide sufficiently precise answers for the most cases. Type 2 processes are deliberative, analytical, effortful, and cognitively demanding (thus usually slower). Although there is some consensus about the typical features of these two types of processes, some dual theories argue that the two systems work in parallel and compete for dominance, while others, so-called default interventionist approaches, posit that type 1 processes are always automatically activated and need to be overridden by type 2 processes, when the need arises (Frankish and Evans 2009). Before we move to exploring the possibility that culture influences people's thinking, we outline the recent distinction of these two ways of thinking and explain the terminology used in the rest of the chapter.

Characterizing Type 1 and Type 2 Processes

In dual process theories, rationality has often been (wrongly) associated with type 2 processes, which need to override the default automatic response of type 1 processes. Evans (2010) argues that such overriding depends on cognitions as well as personality. In a similar vein, Stanovich (2011) writes about the individual differences in thinking dispositions, especially the motivation toward rational integration (for discussion, see also the study by Saunders and Over [2009]). However, in a recent article written together by Evans and Stanovich (2013), they both emphasize the fact that rationality is the feature of human beings, not subpersonal processes (i.e., a person can behave rationally, but neither type 1 nor type 2 processes are rational per se).

The view that type 2 processes are behind all rational behavior has been reflected in naming the type 2 processes as rational (e.g., the study by Epstein [2003]) or by arguing that type 1 processes must be overcome in order to avoid biases and errors (heuristic and biases program, [Tversky and Kahneman 1974]). However, we need to consider three issues. First, cultural studies reviewed by Buchtel and Norenzayan (2009) suggest that holistic processing can also be learned, effortful, and normatively correct for solving some problems and that it is likely to reflect the cultural differences in type 2 processing. In other words, in cultures that value holism and relationships more highly, type 2 processes have different attributes than is typically ascribed to them (analytic, decontextualizing, etc.) in the often Western-based literature. Second, Stanovich (2011) recently proposed a taxonomy of errors, which reflects upon a

SOME KEY TERMS

Dual processes—Dual processes are the assumption in many areas of psychology that people engage in two types of processing that contribute to the observed behavior.

Dual types of thinking—Dual types of thinking are the distinct ways of processing information; type 1 is usually referred as more intuitive and type 2 as more deliberate and reflective.

Dual systems—Dual systems are sometimes used as a synonym for types of processing but more associated with an evolutionary distinction.

Modes of processing—Modes of processing are the cognitive styles, which are manifested within the domain of what is regarded as type 2 thinking, usually measured with thinking disposition measures.

newly formed understanding that errors can be caused by not using type 1 processes when it is adequate, not recognizing the need to engage type 2 processing (error of comprehension [Kahneman and Tversky 1982]) or lacking the skills or the knowledge necessary for successfully solving the task (errors of application [Kahneman and Tversky 1982]). In this notion, rationality is identified more with a so-called reflective mind (recognizing the need to intervene in default automatic processing) rather than an algorithmic mind (cognitive capacity). Third, Evans and Stanovich (2013) distinguish between *types* (types 1 and 2 as qualitatively distinct ways of processing information) and *modes* of processing, which are cognitive styles and manifest within the domain of what is regarded as type 2 thinking. Taking cognitive styles as only a variation in the domain of type 2 processes largely explains the many cultural differences reviewed by Buchtel and Norenzayan (2009). In this view, the cognitive styles can be regarded as the manifestation of modes of thinking within type 2 processes.

In his new tripartite model, Stanovich (2011) elaborates on dual process theories but differentiates between the reflective and algorithmic minds at the level of type 2 processes. It is based on extensive research of individual differences and differential association of intelligence with the two minds. The reflective mind is a source of individual differences in rational thinking dispositions and is responsible for detecting the need to override automatic type 1 responses. The algorithmic mind is a source of individual differences in fluid intelligence and is responsible for coming up with the correct response, when the reflective mind detects the need for an override.

The relationship between thinking disposition (cognitive styles) and rational decision making is not straightforward. Some studies have found a positive relationship (Witteman et al. 2009; Sirota et al. 2013); others have found no such relationship (Newstead et al. 2004; Čavojová et al. 2013). One possible explanation for these results could lie in different cultural backgrounds. For example, Buchtel and Norenzayan (2009) reviewed several studies that suggest that not only do Eastern and Western cultures differ in dominant cognitive style (holistic versus analytic), but also these styles are learned in the process of socialization. They assert that cultural differences are best conceptualized as the differences in habits of thought, rather than differences in the actual availability of information-processing strategies in the cognitive repertoire, and that holistic and analytic ways of thinking can be differently encouraged in their development and use by different cultural and situational constraints (Buchtel and Norenzayan 2009, p. 219).

However, another way in which culture can affect how we think about problems (and thus our rationality) is by what we learn during our development (i.e., our mindware). Some of these things relate to the processes that can facilitate rational thinking (e.g., knowing that we have to take into consideration base rates or that anecdotal evidence is not proof). Others can inhibit our thinking—Stanovich (2011) even calls this kind of mindware contaminated. Cultures also affect what is considered as necessary skills to be learned by young ones in a given society (e.g., probabilistic reasoning may not be equally important in all societies).

QUESTIONS TO CONSIDER

Can you think of any belief held in your culture that promotes or inhibits the way you think about problems?

What is your own attitude toward the use of intuition in everyday problem-solving? Is it a way of circumventing effortful deliberation about difficult problems or does it arise after careful consideration of many aspects of a difficult situation?

Thus, culture influences not only what we acquire to our mindware—either deliberately or subconsciously—but also the recommended way to handle complex problems.

Nisbett (2003) describes the homeostatic principle, in which members of cultures acquire different ways of reasoning. He suggests that the differences between thought processes arise not only due to different fundamental beliefs about the world, but rather as a consequence of using different cognitive tools to understand the world. These different cognitive processes then reinforce the beliefs the members of the society hold.

However, as we already mentioned, it seems necessary to distinguish between the modes of processing and the types of processes—while the former mainly refers to cognitive styles and can reflect cultural variations in using the type 2 processes, the latter reflects deeper differences between the autonomous, largely intuitive type 1 processes and the more deliberate type 2 processes that should be culturally more universal.

In this chapter, we focus on a review of research connected with the distinction between reflective and intuitive thinking from various cultural contexts. We specifically chose only one test to review—i.e., the cognitive reflection test (CRT) (Frederick 2005) because (1) as it is sometimes used as a measure of cognitive style (reflective versus intuitive), (2) it is based on performance and not self-report, and (3) it is hypothesized to reflect the individual's ability to suppress intuitive (and wrong) answers and to reflect upon one's answer. Therefore, regardless of the cultural variations in preferred cognitive styles (as modes of thinking), it is supposed to measure an individual's ability to engage in type 2 processing, and this makes it especially interesting for cross-cultural studies.

The aim of the chapter is not to present an exhaustive review but to focus on the specific aspects of cognitive reflection. We primarily focus on three broad areas of research: cognitive reflection and its relation to cognitive ability (cognitive reflection and thinking processes), cognitive reflection and its relation to modes of thinking (thinking dispositions), and cognitive reflection and its relation to various beliefs (content of thinking). We do not expect differences in cognitive abilities in various studies, but we expect different levels of cognitive reflection and different relationships with cognitive styles related to dual process approach, biases, and beliefs. In the next sections, we will review our data (published and unpublished) as well as other studies, focusing preferably on those from non-American samples following the recent critique of the predominance of U.S. college student samples in psychological research by Henrich et al. (2010).

Measure of Cognitive Reflection

The purpose of the CRT is to assess the cognitive ability (Frederick 2005) to suppress an intuitive and spontaneous (system 1 or type 1 processes) wrong answer in favor of a reflective and deliberative (system 2 or type 2 processes) right answer. It is short—it consists only of three questions, and it is usually presented in a free-response format. In contrast with classic insight problems (e.g., the nine-dot problem), it primes an attractive intuitive response—the (incorrect) answer comes to mind easily and the individual has to recognize that it is incorrect and generate the right answer.

The CRT consists of three simple problems, as follows (taken from the study by Frederick [2005]):

1. A bat and a ball cost $1.10 in total. The bat costs a dollar more than the ball. How much does the ball cost? (Intuitive answer: 10 cents; correct answer: 5 cents)
2. If it takes 5 machines 5 minutes to make 5 widgets, how long would it take 100 machines to make 100 widgets? (Intuitive answer: 100 minutes; correct answer: 5 minutes)

3. In a lake, there is a patch of lily pads. Every day, the patch doubles in size. If it takes 48 days for the patch to cover the entire lake, how long would it take for the patch to cover half of the lake? (Intuitive answer: 24 days; correct answer: 47 days)

Recently, there have been some attempts to expand (Toplak et al. 2013; Oppenheimer and Thomson 2015) or modify (Kostovičová et al. 2013) this test, but the majority of the studies reviewed in this chapter used classical the three-item free-response format.

Cognitive Reflection and Intelligence

Cognitive reflection is moderately associated with cognitive abilities (Liberali et al. 2012; Toplak et al. 2011, 2013; Shenhav et al. 2012) and academic achievement (Frederick 2005; Pennycook et al. 2012; Welsh et al. 2011). However, some researchers object to CRT because it is possible that it reflects only the better numeracy skills of participants, not really the cognitive reflection (Cokely and Kelley 2009; Kostovičová et al. 2013; Obrecht et al. 2009).

Reported correlations between CRT and various intelligence tests used (Advanced Progressive Matrices [APM], Vienna Matrix Test [VMT]: Raven Matrices, Wechsler Abbreviated Scale of Intelligence [WASI]: Wechsler's intelligence tests, vocabulary tests, etc.) show typical correlations of about 0.3–0.5, and reported correlations with academic achievement tests (Scholastic Aptitude Test [SAT], American College Test [ACT], Wonderlic Personnel Test [WPT]) are similar—about 0.4. In other words, many people score high in intelligence but miss the correct answers in this test (i.e., they heavily rely on intuitive type 1 processes—no matter their preferred cognitive style), and many people, who score low in intelligence, are able to get the answers in this test right (i.e., they engage effortful type 2 processes), because once a person realizes that the first intuitive answer is wrong, it is not that difficult to compute the right answer.

Most reported results in this section come from U.S. samples. In our study (Čavojová and Hanák, unpublished), using a Slovak sample (Eastern Europe), we found somewhat lower correlations than reported in previous studies ($r_{(406)} = 0.244$, $p < 0.001$), but it could be caused by factors other than culture—our participants were from a less prominent university and mostly women, who generally score lower in CRT (Frederick 2005; Toplak et al. 2013).

A relatively moderate association between cognitive reflection and intelligence can be caused by the differences between optimal and typical performance circumstances (Stanovich 2009). Intelligence test are usually administered under optimal performance circumstances—i.e., people know that they are being tested for their intelligence, they usually have high stakes in performing their best; thus, they are inclined to engage in type 2 processing. However, typical tests of rational or reflective thinking are administered as simple numerical tasks, because we are interested in the typical performance of people—i.e., if they recognize that the task is a tricky one. When the participants suspect the trick beforehand or they are told that a high proportion of even highly intelligent people (such as Massachusetts Institute of Technology [MIT] students) gives at least one wrong answer in 80% of the cases, they are more inclined to engage in type 2 processing and increase their score in the CRT. Therefore, the results strongly depend on the circumstances, in which they were tested. Even small changes in the instruction can change the performance in tasks requiring creative and reflective thinking (Alter et al. 2007; Čavojová 2004).

Cognitive Reflection as Cognitive Style

CRT is generally thought to be correlated with measures of cognitive styles—modes of thinking that prefer deliberative, analytic, and thoughtful processing of information, in contrast with intuitive modes of thinking that rely on automatic processing. Studies using the need for cognition (NFC) scale or the rational–experiential inventory (REI), in which the rationality subscale is based on the NFC, generally support this assumption (Frederick 2005; Liberali et al. 2012; Toplak et al. 2011, 2013; Iyer et al. 2012; Čavojová and Hanák 2014), although the correlations are typically weak, ranging from 0.14 to 0.33. Furthermore, although we usually find correlations with questionnaires measuring preference for the rational mode of thinking, we do not always find expected negative correlations with measures of intuitive mode of thinking, such as faith in intuition or experiential subscale of REI (based on faith in intuition questionnaire). In our study (Čavojová and Hanák 2014), we found no correlations between CRT and any measures of intuitive thinking (we used REI, Myers–Briggs type indicator [MBTI], and cognitive style indicator)*—only MBTI sensing correlated negatively, although weakly with CRT.

In an extensive study, Toplak et al. (2011) found that CRT is a more potent predictor of a wide sample of tasks measuring the rational behavior than other measures of cognitive abilities, thinking dispositions, and executive functioning. Based on these results, they argue that CRT is such a powerful measure because it is performance-based rather than a self-report measure, and that it assesses the resistance to miserly processing in a way that is not addressed by any other test of cognitive abilities or executive functioning.

In our recent study (Čavojová and Hanák, unpublished), we found only weak correlations between CRT and cognitive styles related to rationality (REI-R: $r_{(432)} = 0.155, p = 0.001$; Master Rationality Motive Scale (MRMS): $r_{(432)} = 0.117, p = 0.015$) and no inverse relationship with preference for intuitive processing (REI-E: $r_{(432)} = -0.019, p = 0.694$).

Cognitive Reflection and Resistance to Irrational Beliefs

Recently, several studies examined the relationship between cognitive reflection and inclination toward irrational beliefs—reflecting the assumption that the higher the cognitive reflection, the lower the incidence of irrational beliefs, as suggested by Toplak et al. (2011). Toplak et al. (2011, 2013) in two studies used a classical version of CRT, as well as an extended seven-item version and found close to zero correlations with superstitious beliefs ($r = 0.04$ in the first study and $r = 0.15$ in the second study). In our study (Čavojová and Hanák, unpublished), we examined paranoid ideation, supernatural beliefs, and conspiracy beliefs and found no correlation with cognitive reflection (measured by CRT)—the correlations were virtually zero and similar to the results of the study by Toplak et al. (2011) with their larger sample. It seems that any small correlation that is found between cognitive reflection and paranormal beliefs tends to disappear with larger samples.

Some other researchers examined the relationship between cognitive reflection and religious beliefs. Pennycook et al. (2012, 2013) generally found that the number of correct responses on the three-item CRT was a function of the type of theistic belief. They found significantly more atheists among high scorers in CRT. In three studies with Canadian

* REI (Pacini and Epstein 1999), MBTI (Myers–Briggs et al. 1998), and cognitive style indicator (Cools and van den Broeck 2007) are various questionnaires of cognitive styles, which measure preference for intuition. REI consists of rational and experiential subscale; one of the dimensions of MBTI measures preference for sensing versus intuition, and cognitive style indicator consists of three subscales: knowing, planning, and creating.

samples, they found moderate negative correlations of religious beliefs with CRT (ranging from 0.3 to 0.37). Similar results were found by Gervais and Norenzayan (2012). In studies, where incorrect (intuitive) responses in CRT were scored and put into analyses, participants who gave more intuitive CRT responses reported more confident belief in God on the atheist-believer scale (Shenhav 2012). They found (relatively small) correlations of CRT-intuitive responding with belief in God (r = 0.176), belief in immortal souls (r = 0.141), belief change since childhood (r = 0.192), and conviction of God's existence (r = 0.145). An additional measure of familial religiosity was not correlated with CRT. Similar results were found in a study that examined cognitive reflection and political preferences with libertarians giving the correct answers to CRT at a slightly higher rate than liberals and a moderately higher rate compared to conservatives (Iyer et al. 2012). On the other hand, Kahan (2013) pointed out that participants in many studies are skewed in a liberal direction and when he used a representative sample of U.S. citizens (N = 1750), he found that participants highest in CRT were most likely to display ideologically motivated reasoning (irrespective of their political attitude).

Cognitive Reflection and Biases

Frederick (2005) first tested how CRT predicts time preferences and risk preferences in comparison with standard measures of cognitive abilities. He gave participants several scenarios and they had to choose one of the two options—e.g., whether they would prefer a smaller sum of money now ($100) or a larger sum of money in the future ($1100 in 10 years) or whether they are willing to pay overnight shipping for a chosen book or prefer to wait for a standard delivery without any extra fee, etc. In another set examining risk choices, participants had to choose between certain gains and riskier options with higher expected values, e.g., will you prefer to get $1000 for sure or 90% chance of winning $5000? Participants who scored higher in CRT displayed more patience than low-scoring participants, and they were also more willing to engage in a gamble where they could win a larger amount of money. Moreover, CRT predicted these choices better than other cognitive measures. These results have been generally replicated by others (Oechssler et al. 2009; Čavojová and Hanák 2014; Nofsinger and Varma 2007; Albaity and Rahman 2014; Hardisty and Weber 2009).

Moreover, Hardisty and Weber (2009) studied the preferences of participants not only in a financial domain (as earlier mentioned studies), but also in health and environmental domains. In their study, the participants had to choose between hypothetical financial, environmental, and health gains and losses that took effect either immediately or with a delay of 1 or 10 years. They found out that similar processes drive the discounting of money and air quality (but not health), and that the CRT predicted the discounting of monetary and air quality gains but not the discounting of health gains or losses in any domain. Based on these results, they suggested that cognitive impatience plays a role in the discounting of (relatively abstract) gains but not in time preference for losses or health outcomes (which are generally more affectively charged).

Many researchers have found a relationship between CRT and other cognitive biases. For example, people scoring low in CRT are more overconfident (Hoppe and Kusterer 2011) and are more susceptible to conservativism bias (Albaity and Rahman 2014; Hoppe and Kusterer 2011), base-rate fallacy, and conjunction fallacy (Welsh et al. 2011), although some of these relationships are weak and were not replicated in other studies (e.g., Albaity and Rahman [2014] did not find a relationship between CRT and conjunction fallacy). Cognitive reflection was also found to be related to the anchoring heuristic—low CRT groups anchored on the mean more often than high CRT groups (Moritz et al. 2009). Lastly, compared with low scorers, high scorers on the CRT were more likely to maximize (Koehler and James 2010).

Cognitive reflection was also studied in connection with moral reasoning. For example, Feltz and Cokely (2008) found that the differences in cognitive styles associated with reflection and elaborative information processing in moral judgments. Relatively surprising results were found by Fosgaard et al. (2013), which indicated that the more intelligent (and cognitively reflective) a subject is, the more likely he/she is to cheat in simple gambling task to receive more money.

In another study, Altman and Falk (2009) examined how the arrangement of options (so-called defaults)* influences people depending on their cognitive reflection. In their experiment, people could participate in a simple game (public goods game),† and the defaults were manipulated to be either cooperative or uncooperative. They found out that in the high CRT group, the amount of monetary contribution did not significantly differ between cooperative and uncooperative default options, while in low CRT group, people contributed less when facing an uncooperative default and (slightly) a more undercooperative default option. Less than half (45%) of the subjects with high CRT scores stuck to the default option (either cooperative or uncooperative), whereas roughly two-thirds of low CRT subjects (64%) went along with the default option (either cooperative or uncooperative). Thus, the defaults had a greater impact on the behavior of subjects with lower levels of cognitive reflection, while they leave decisions of more rational subjects unaffected.

Cultural Differences in Cognitive Reflection

Altogether, we reviewed 50 studies using CRT in their methods. For 39 samples, we were able to find or calculate the mean, the standard deviation, and the number and nationality of the participants in each sample (see Table 5.1 for the descriptive statistics). The majority of the samples came from the United States and Canada, although there were several samples from Western Europe (Germany, Belgium, Spain, and Denmark), Eastern Europe (Hungary, Slovakia), South America (Brazil, Argentina), and Asia (Japan, China, and Malaysia).

Because the studies differed in their focus and aims, we decided to look only at the differences between the mean scores in CRT between the samples using participants with various nationalities. We performed a meta-analysis and used a continuous random effects model. The results of our meta-analysis are shown in the Forrest plot (Figure 5.1). For a better presentation, we organized our data first according to the continents and the countries (North America, South America, Asia, and Europe) and then according to the year in ascending order.

* By default, here we mean how the option is phrased—e.g., when the option "to receive newsletters" is checked already, we have to make a conscious decision that we do not wish to receive newsletters and thus uncheck the option. People do not usually read very carefully, and it was shown that they tend to go with this kind of default option also in more serious decision making than receiving newsletters, e.g., when considering organ donations or deciding about their retirement funds (Thaler and Sunstein 2008).

† Public goods game is a social dilemma, where participants are faced with a decision on how much money they will contribute to the common good. Treatments differed in the default option implemented in case a subject did not send back his/her decision form on how much money they will contribute from the reward from participating in the experiment (7 EUR). In the cooperative default option, participants were told that the whole sum of money will be contributed, while in the uncooperative default option participants were told that nothing will be contributed, unless they chose a different contribution level on their decision form.

TABLE 5.1

CRT Descriptive Statistics for Several Studied Samples

Studies	Nationality of the Sample	*N*	Mean CRT Score	SD	Percentage Scoring 0	1	2	3
Frederick (2005)—MIT	United States	61	2.18	0.92	7%	16%	30%	48%
Frederick (2005)—Princeton University	United States	121	1.63	1.06	18%	27%	28%	26%
Frederick (2005)—Web-based studies	United States	525	1.10	1.08	39%	25%	22%	13%
Frederick (2005)—Bowling Green University	United States	52	0.87	1.05	50%	25%	13%	12%
Frederick (2005)—University of Toledo	United States	138	0.57	0.89	64%	21%	10%	5%
Moritz et al. (2009)	United States	319	1.50	1.11	26%	21%	30%	23%
Nofsinger and Varma (2007)	United States	108	1.76	1.13				
Fernbach et al. (2013), experiment 1	United States	167	1.04	1.04	42%	23%	25%	10%
Fernbach et al. (2013), experiment 2	United States	201	1.18	1.14	39%	23%	19%	19%
Liberali et al. (2012), study 2	United States	190	1.5	1.12	24%	27%	23%	26%
Weller et al. (2013), study 1	United States	1970	0.83	0.99	–	–	–	–
Alter et al. (2007), experiment 1: disfluent condition	United States	20	2.49	0.64	–	–	–	–
Alter et al. (2007), experiment 1: fluent condition	United States	20	1.9	0.89	–	–	–	–
Kahan (2013)	United States	1750	0.65	0.95	–	–	–	–
Razmyar and Reeve (2013)	United States	150	0.59	0.91	–	–	–	–
Carpenter et al. (2013)	United States	422	1.78	1.05	14%	27%	26%	33%
Iyer et al. (2012)	United States	9721	1.79	–	–	–	–	–
Corgnet et al. (2014), house money condition	United States	45	1.33	1.17	33%	22%	22%	22%
Corgnet et al. (2014), earned money condition	United States	45	1.47	1.18	27%	29%	16%	29%
Koehler and James (2010), experiment 1	Canada	84	1.67	1.10	21%	18%	33%	27%
Koehler and James (2010), experiment 2	Canada	80	1.65	1.18	23%	25%	18%	35%
Pennycook et al. (2013), experiment 2	Canada	200	0.36	0.37	–	–	–	–
Toplak et al. (2013)	Canada	160	0.49	0.85	–	–	–	–
Toplak et al. (2011)	Canada	346	0.7	0.93	55.8%	–	–	6.6%
Oechssler et al. (2009)	German	564	2.05	1.01	10.1%	17.7%	30.7%	41.5%
De Neys et al. (2011)	Belgium	403	1.12	1.00	–	–	–	–
Bosch-Domènech et al. (2014)	Spain	623	0.77	0.97	–	–	–	–

(Continued)

TABLE 5.1 (CONTINUED)

CRT Descriptive Statistics for Several Studied Samples

Studies	Nationality of the Sample	*N*	Mean CRT Score	SD	Percentage Scoring 0	1	2	3
Brañas-Garza et al. (2012)	Spain	191	0.41	0.66	67%	23%	9%	0%
Liberali et al. (2012), study 1	Brazil	259	1.1	1.17	44%	20%	17%	19%
Campitelli and Labollita (2010)	Argentina	157	0.66	0.91	59%	25%	11%	6%
Akiyama (2013)	Japan	173	2.01	0.99	9%	21%	30%	40%
An et al. (2012)	China	200	1.82	1.22	12.2%	28%	25%	34.5%
Albaity and Rahman (2014)	Malaysia	880	0.88	–	–	–	–	–
Čavojová and Hanák (2014)—College of Public Service	Slovakia	531	0.95	1.13	51.4%	17.9%	15.4%	15.3%
Čavojová and Hanák (unpublished)—Pedagogical Faculty	Slovakia	432	0.77	1.05	56.0%	21.5%	13.7%	8.8%
Kostovičová (2011)—EU students	Slovakia	62	1.21	0.99	–	–	–	–
Kostovičová et al. (2013)	Slovakia	626	1.64	–	–	–	–	–
Sirota and Juanchich (2011)[a]—Comenius University	Slovakia	94	1.3	1.1	58.5%	–	–	32.9%
Ballová Mikušková (unpublished)—Pedagogical Faculty	Slovakia	109	0.82		–	–	–	–
Kiss et al. (2015)	Hungary	60	0.57	0.98	–	–	–	–
Szaszi et al. (2015)	Hungary	210	0.84	1.00	49.5%	26.2%	14.8%	9.5%

[a] Percentages were calculated from the provided data; low CRT group and high CRT groups were created by a median split, where the mean score for low CRT group was 0.5 and the high CRT was 2.5.

In Figure 5.1, the size of the black squares represents the sample size of individual studies. Larger squares mean higher samples, affecting the final result of our meta-analysis. The horizontal lines crossing the squares represent the confidence intervals (CIs) for the mean. The smaller CI, the more homogenous the study sample. For each study, we report the country of respondents and the study reference. The diamond on the bottom right of our graph represents the power-weighted result of our meta-analysis. The width of the diamond depicts the CI for the statistically merged results of the studies. The single vertical line represents the CRT mean score after the meta-analysis of the listed studies. The axis on the bottom of the graph describes the range of average scores reached by the participants—from 0.5 to 2.5. The absolute range of scores in CRT is from 0 to 3.

We can see from the Figure 5.1 that the overall mean score (N = 21,827 participants) was only 1.21 (CI 1.05, 1.37),* which means that despite its short length (only three tasks), CRT is a rather difficult task for the majority of the participants. Across the reviewed studies, there do not appear to be any obvious cultural differences in results. Although, there is a great deal of variance even among the same nationality samples.

* P value < 0.01, heterogeneity: $\tau^2 = 0.25$, $Q(df = 39) = 3100.25$, het. p value < 0.01, $I^2 = 98.74$.

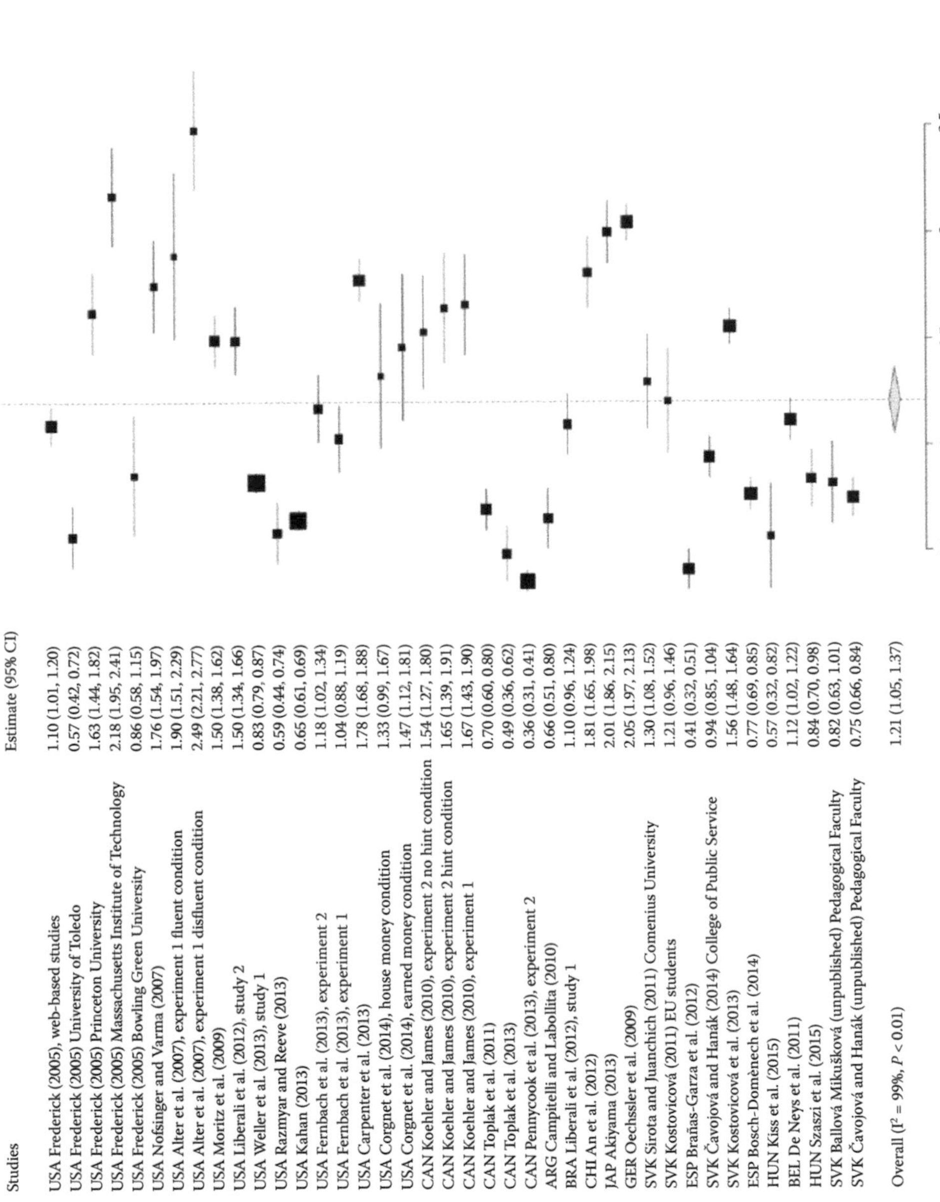

FIGURE 5.1
Mean and CI 95% for CRT scores across the samples. USA—United States, CAN—Canada, ARG—Argentina, BRA—Brazil, CHI—China, JAP—Japan, GER—Germany, SVK—Slovakia, ESP—Spain, HUN—Hungary, and BEL—Belgium.

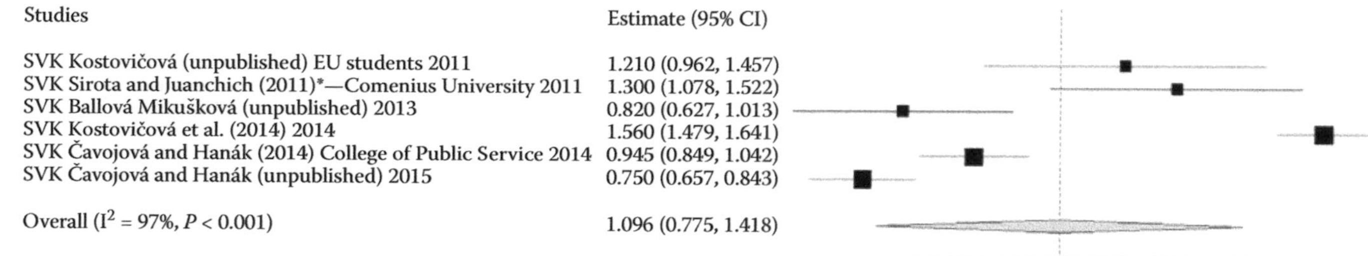

FIGURE 5.2
Mean and CI 95% for CRT scores of the Slovak samples.

Two outliers (MIT sample [Frederick 2005] and disfluent condition in Alter et al.'s [2007] experiment) highlight the fact that cognitive reflection can be enhanced in conditions where the participants realize the need to overcome intuitive responses or when the participants are already very high in numeracy abilities. MIT is among the highest-ranking universities in the world, so we might expect to find there the most reflective individuals; however, even among MIT students, only about half of them solved all three questions correctly. On the other hand, there were only 7% of the participants who did not answer any question correctly, in comparison with one-quarter to two-thirds of the participants in the majority of the other studies. These numbers highlight the fact that people are, indeed, cognitive misers (Stanovich 2011), and to reflect upon one's intuitive response is quite effortful but not impossible as Alter et al.'s (2007) experiment shows.

As we pointed already in the review of the many studies, cognitive reflection tends to be correlated with cognitive abilities and education. The differences between the scores are caused probably more by the participants' education, cognitive abilities, and gender rather than culture. We support this claim by further analysis of the Slovak participants from various studies (Figure 5.2).

Figure 5.2 shows how even the participants in one cultural sample differ in their mean scores between students of cognitive science (Kostovičová et al. 2013) and students of pedagogical faculty (Ballová Mikušková, unpublished; Čavojová and Hanák, unpublished), with students of Comenius University and Economic University somewhere in between (Sirota and Juanchich 2011; Kostovičová 2011). This difference in scores could be simultaneously and independently caused by all three factors mentioned earlier. Cognitive ability and education tend to correlate together—more intellectually gifted students choose better universities. However, gender was also shown to be related with scores of cognitive reflection, with women typically scoring lower and more trusting their intuitive responses. The sample from pedagogical faculty consists mainly of women (usually more than 70% of the sample), and this could affect the scores.

Concluding Comments

The aim of the chapter was to discuss the possible cultural influences on cognitive styles. First, we mentioned some conceptual difficulties as cultures may differ in what they consider as an example of good thinking—essentially what may affect the preferred cognitive styles. It was the main reason why, in this chapter, we focused on reviewing cognitive reflection—regardless of whether we feel that analytic or holistic approach is better for dealing with complex problems in the real world. Moreover, because it is a short, neutral, and simple test, not a self-report, it enables a more objective comparison between cultures, because people are not typically very good at evaluating their own motives, feelings, and abilities (Wilson 2002; Wilson et al. 1993) and the extent that the culture they live in affects them.

We were not able to find any notable cultural differences between the samples—our review rather 10 highlighted the need for better education and teaching people to reflect upon their beliefs and intuitions. The low rate of cognitive reflection even among university samples (most typically used in research) and even among future teachers (some of the samples we reviewed in more detail) is alarming and can pose many challenges in today's multicultural societies, in which people with different cultural and religious backgrounds mix together, such as in the fictional case study of megacity Xanadu.

When we are unable to reflect upon our inherent beliefs and mostly rely on our intuitive responses, our adaptive unconscious (Wilson 2002) can draw on our fear from the unknown leading to unnecessary prejudices and discrimination. People do not want to admit their prejudices, and they rationalize them by quoting arguments seen on the Internet without giving them deeper thoughts. However, it is not the difficulty of the task per se; it is the recognition of the need to reflect on the information, which is the most difficult, and therefore, we consider cognitive reflection such a crucial concept.

Furthermore, as was shown in several studies (Gervais and Norenzayan 2012; Pennycook et al. 2012), analytic thinking and cognitive reflection can promote religious disbelief, which may cause many problems in societies like fictional Bursukas and even more tension with the more secular Çawailan culture. However, the inability to reflect upon one's beliefs leads to strengthening already strong confirmation and belief biases, and this can prevent people from open discussions and exchange of ideas, which could lead to promoting the welfare of both societies. Even worse, it can lead to endorsing beliefs that are actively harmful (i.e., contaminated mindware in Stanovich's terms), and this problem is probably the case for any culture, despite the differences in the content of these beliefs, which can be culturally specific.

We believe that it is sticking to one's already held beliefs and inability to reflect upon them that is more damaging than the differing beliefs and values (either cultural, religious, or others). Encouraging people to reflect on their beliefs and thinking is not desirable for many political systems, but it seems that even Western democratic societies are not very cognitively reflective, as our review has shown. What our results also showed was that there is probably more individual than cultural variation in the level of cognitive reflection, which is both the good and bad news. The good news is that it is our personal choice to reflect upon our thinking and that we can get better at it—and probably use it to be more empathic and just with different cultures, fictional or real. The bad news is that we (and not politicians or someone else) have to do it and it is, indeed, difficult.

DISCUSSION QUESTIONS

1. To what extent does culture influence our ability to reflect upon our dearly held beliefs?
2. How can we teach people to engage in more effortful reflective thinking?
3. If a culture prefers beliefs and certainty more over insecurity of doubt and reflection how can it be made possible to encourage people to reflect upon their own thinking and thinking of others?
4. Is cognitive reflection more of a habit of the mind, which can be taught, or a matter of abilities and motivation?

Acknowledgments

The authors would like to thank Martin Majerník for the help with conducting the meta-analysis. This work was supported by the Slovak grant agency VEGA 2/0064/13 *Expert Decision Making: Using Intuition to Solve Strategic Problems.*

References

Akiyama, E. 2013. How do experienced traders respond to inflows of inexperienced traders? An experimental analysis. Working paper AMSE WP 2013-Nr. 59.

Albaity, M., and M. Rahman. 2014. Cognitive reflection test and behavioral biases in Malaysia. *Judgment and Decision Making* 9 (2): 149–151.

Alter, A. L., D. M. Oppenheimer, N. Epley, and R. N. Eyre. 2007. Overcoming intuition: Metacognitive difficulty activates analytic reasoning. *Journal of Experimental Psychology General* 136 (4): 569–576.

Altmann, S., and A. Falk. 2009. The impact of cooperation defaults on voluntary contributions to public goods. 1–18. Retrieved from http://citeseerx.ist.psu.edu/viewdoc/download?doi=10.1.1.319.7302&rep=rep1&type=pdf.

An, N., X. Shi, and A. Nordvall. 2012. Cognitive ability and psychological biases—Perspectives from Chinese stock indiviodual investors. Master thesis. Umeå: Umeå School of Business. Retrieved from http://umu.diva-portal.org/smash/get/diva2:513883/FULLTEXT01.pdf.

Bosch-Domènech, A., P. Brañas-Garza, and A. M. Espín. 2014. Can exposure to prenatal sex hormones (2D:4D) predict cognitive reflection? *Psychoneuroendocrinology* 43: 1–10.

Brañas-Garza, P., T. García-Muñoz, and R. H. González. 2012. Cognitive effort in the beauty contest game. *Journal of Economic Behavior & Organization* 83 (2): 254–260.

Buchtel, E. E., and A. Norenzayan. 2009. Thinking across cultures: Implications for dual processes. In *In Two Minds: Dual Processes and Beyond*, ed. Evans, J., and K. Frankish, 217–238. Oxford: Oxford University Press.

Campitelli, G., and M. Labollita. 2010. Correlations of cognitive reflection with judgments and choices. *Judgment and Decision Making* 5 (3): 182–191.

Carpenter, J., M. Graham, and J. Wolf. 2013. Cognitive ability and strategic sophistication. *Games and Economic Behavior* 80: 115–130.

Čavojová, V. 2004. Vybrané Aspekty Tvorivosti v Mladšom Školskom Veku a Ich Vzťah K Verbálnej Inteligencii. *Psychológia a Patopsychológia Dieťaťa* 40 (1): 69–76.

Čavojová, V., and R. Hanák. 2014. Assessing fluid rationality and its relation to cognitive styles. In *Proceedings of the 5th International Conference on Applied Human Factors and Ergonomics AHFE 2014*, ed. Ahram, T., W. Karwowski, and T. Marek. Krakow: AHFE Conference.

Čavojová, V., E. Ballová Mikušková, and R. Hanák. 2013. Preferencia K Deliberácii: (ne)istota Úspechu v Kognitívnych Úlohách. In *Kognitivní Věda a Umělý Život*, 53–58. Opava: Slezká univerzita v Opavě.

Cokely, E. T., and C. M. Kelley. 2009. Cognitive abilities and superior decision making under risk: A protocol analysis and process model evaluation. *Judgment and Decision Making* 4 (1): 20–33.

Cools, E., and H. Van den Broeck. 2007. Development and validation of the cognitive style indicator. *The Journal of Psychology* 141: 359–387.

Corgnet, B., R. H. Gonzalez, P. Kujal, and D. Porter. 2014. The effect of earned vs. house money on price bubble formation in experimental asset markets. *Review of Finance*, 1–34. Retrieved from http://rof.oxfordjournals.org/content/early/2014/08/18/rof.rfu031.full.pdf+html.

De Neys, W., N. Novitskiy, L. Geeraerts, J. Ramautar, and J. Wagemans. 2011. Cognitive control and individual differences in economic ultimatum decision making. *PLoS ONE* 6 (11). http://journals.plos.org/plosone/article?id=10.1371/journal.pone.0027107.

Epstein, S. 2003. Cognitive-experiential self theory of personality. In *Comprehensive Handbook of Psychology, Volume 5: Personality & Social Psychology*, 159–184. Hoboken, NJ: John Wiley & Sons.

Evans, J. S. 2010. Intuition and reasoning: A dual process perspective. *Psychological Inquiry* 21 (4): 313–326.

Evans, J. S., and K. E. Stanovich. 2013. Dual-process theories of higher cognition: Advancing the debate. *Perspectives on Psychological Science* 8 (3): 223–241.

Feltz, A., and E. Cokely. 2008. The fragmented folk: More evidence of stable individual differences in moral judgments and folk intuitions. *Proceedings of the 30th Annual Conference of the Cognitive Science Society*, 1771–1776. Austin, TX: Cognitive Science Society.

Fernbach, P. M., S. A. Sloman, R. St. Louis, and J. N. Shube. 2013. Explanation friends and foes: How mechanistic detail determines understanding and preference. *Journal of Consumer Research* 39 (5): 1115–1167.

Fosgaard, T. R., L. G. Hansen, and M. Piovesan. 2013. Separating Will from Grace: An experiment on conformity and awareness in cheating. *Journal of Economic Behavior and Organization* 93: 279–284.

Frankish, K., and J. S. Evans. 2009. The duality of mind: An historical perspective. In *In Two Minds: Dual Processes and Beyond*, ed. Evans, J. S., and K. Frankish, 1–29. Oxford: Oxford University Press.

Frederick, S. 2005. Cognitive reflection and decision making. *Journal of Economic Perspectives* 19 (4): 25–42.

Gervais, W. M., and A. Norenzayan. 2012. Analytic thinking promotes religious disbelief. *Science* 336: 493–496.

Hardisty, D. J., and E. U. Weber. 2009. Discounting future green: Money versus the environment. *Journal of Experimental Psychology General* 138 (3): 329–340.

Henrich, J., S. J. Heine, and A. Norenzayan. 2010. The weirdest people in the world? *The Behavioral and Brain Sciences* 33: 61–83; discussion 83–135.

Hoppe, E. I., and D. J. Kusterer. 2011. Behavioral biases and cognitive reflection. *Economics Letters* 110: 97–100.

Iyer, R., S. Koleva, J. Graham, P. Ditto, and J. Haidt. 2012. Understanding libertarian morality: The psychological dispositions of self-identified libertarians. *PLoS ONE* 7 (8). http://journals.plos.org/plosone/article?id=10.1371/journal.pone.0042366.

Kahan, D. M. 2013. Ideology, motivated reasoning, and cognitive reflection. *Judgment and Decision Making* 8: 407–424.

Kahneman, D., and A. Tversky. 1982. On the study of statistical intuitions. In *Judgment under Uncertainty: Heuristics and Biases*, ed. Kahneman, D., P. Slovic, and A. Tversky, 493–508. Cambridge: Cambridge University Press. http://journals.plos.org/plosone/article?id=10.1371/journal.pone.0042366.

Kiss, H. J., I. Rodriguez-Lara, and A. Rosa-García. 2015. Think twice before running! Bank runs and cognitive abilities. *Journal of Behavioral and Experimental Economics* 02 (2015): 1–8.

Koehler, D. J., and G. James. 2010. Probability matching and strategy availability. *Memory & Cognition* 38 (6): 667–676.

Kostovičová, L. 2011. Why natural frequencies facilitate performance on Bayesian reasoning? Master's thesis. Bratislava: Comenius University. Retrieved from http://alis.uniba.sk/storage/ddp/dostupne/FS/2011/2011-FS-15518/.

Kostovičová, L., M. Konečný, and K. Dudeková. 2013. Čo to Vlastne Meriame? Reflexia Testu Kognitívnej Reflexie. In *Otázniky rozhodovania: Teória, empíria, život, 31. Psychologické dni: Zborník príspevkov*, ed. Sarmány-Schuller, I., 166–171. Bratislava: Stimul.

Liberali, J. M., V. F. Reyna, S. Furlan, L. M. Stein, and S. T. Pardo. 2012. Individual differences in numeracy and cognitive reflection, with implications for biases and fallacies in probability judgment. *Journal of Behavioral Decision Making* 25: 361–381.

Moritz, B. B., A. V. Hill, and K. Donohue. 2009. Cognition and individual differences in the newsvendor problem: Behavior under dual process theory. Unpublished manuscript. Retrieved from http://citeseerx.ist.psu.edu/viewdoc/download?doi=10.1.1.462.9505&rep=rep1&type=pdf.

Myers–Briggs, I., M. H. McCaulley, N. L. Quenk, and A. L Hammer. 1998. *MBTI Manual: A Guide to the Development and Use of the Myers-Briggs Type Indicator* (Third Edit.). Palo Alto: Consulting Psychologist Press.

Newstead, S. E., S. J. Handley, C. Harley, H. Wright, and D. Farrelly. 2004. Individual differences in deductive reasoning. *The Quarterly Journal of Experimental Psychology*, 57 (1), 33–60. doi:10.1080/02724980343000116.

Nisbett, R. E. 2003. *The Geography of Thought*. Boston: Nicholas Brealey Publishing.

Nofsinger, J. R., and A. Varma. 2007. How Analytical Is Your Financial Advisor? *Financial Services Review* 16: 245–260.

Obrecht, N. A., G. B. Chapman, and R. Gelman. 2009. An encounter frequency account of how experience affects likelihood estimation. *Memory & Cognition* 37 (5): 632–43.

Oechssler, J. J., A. Roider, and P. W. Schmitz. 2009. Cognitive abilities and behavioral biases. *Journal of Economic Behavior & Organization* 72 (1): 147–152.

Oppenheimer, D. M., and K. S. Thomson. (2015). Investigation of an alternate form of the cognitive reflection test. In *Final Program & Book of Abstracts*, 172. Budapest: Subjective Probability, Utility, and Decision Making (SPUDM) Conference.

Pacini, R., and S. Epstein. 1999. The relation of rational and experiential information processing styles to personality, basic beliefs, and the ratio-bias phenomenon. *Journal of Personality and Social Psychology* 76 (6): 972–987.

Pennycook, G., J. A. Cheyne, N. Barr, D. J. Koehler, and J. Fugelsang. 2013. Cognitive style and religiosity: The role of conflict detection. *Memory & Cognition* 42: 1–10.

Pennycook, G., J. A. Cheyne, P. Seli, D. J. Koehler, and J. Fugelsang. 2012. Analytic cognitive style predicts religious and paranormal belief. *Cognition* 123 (3): 335–46.

Razmyar, S., and C. L. Reeve. 2013. Individual differences in religiosity as a function of cognitive ability and cognitive style. *Intelligence* 41 (5): 667–673.

Saunders, C., and D. E. Over. 2009. In two minds about rationality? In *In Two Minds: Dual Processes and Beyond*, ed. Evans, J. S., and K. Frankish, 317–334. Oxford: Oxford University Press.

Shenhav, A., D. G. Rand, and J. D. Greene. 2012. Divine intuition: Cognitive style influences belief in God. *Journal of Experimental Psychology* 141 (3): 423–428.

Sirota, M., and M. Juanchich. 2011. Role of numeracy and cognitive reflection in Bayesian reasoning with natural frequencies. *Studia Psychologica* 53 (2): 151–161.

Sirota, M., M. Juanchich, and Y. Hagmayer. 2013. Ecological rationality or nested sets? Individual differences in cognitive processing predict Bayesian reasoning. *Psychonomic Bulletin & Review* 21 (1): 198–204.

Stanovich, K. E. 2009. *What Intelligence Tests Miss: The Psychology of Rational Thought*. New Haven, CT: Yale University Press.

Stanovich, K. E. 2011. *Rationality and the Reflective Mind*. New York, NY: Oxford University Press.

Szaszi, B., A. Szollosi, B. Palfi, and B. Aczél. 2015. The cognitive reflection test revisited: Exploring the ways individuals solve the test. *Cognition*, submitted.

Thaler, R. H., and C. R. Sunstein. 2008. *Nudge: Improving Decisions about Health, Wealth, and Happiness*. New Haven, CT: Yale University Press.

Toplak, M. E., R. F. West, and K. E. Stanovich. 2011. The cognitive reflection test as a predictor of performance on heuristics-and-biases tasks. *Memory & Cognition* 39 (7): 1275–89.

Toplak, M. E., R. F. West, and K. E. Stanovich. 2013. Assessing miserly information processing: An expansion of the cognitive reflection test. *Thinking & Reasoning* 20 (2): 1–22.

Tversky, A., and D. Kahneman. 1974. Judgment under uncertainty: Heuristics and biases. *Science* 185 (4157): 1124–31.

Weller, J. A., N. F. Dieckmann, M. Tusler, C. K. Mertz, W. J. Burns, and E. Peters. 2013. Development and testing of an abbreviated numeracy scale: A Rasch analysis approach. *Journal of Behavioral Decision Making* 26 (2): 198–212.

Welsh, M., N. Burns, and P. Delfabbro. 2011. The cognitive reflection test: How much more than numerical ability? *Csjarchive.Cogsci.Rpi.Edu*, 1587–1592. http://csjarchive.cogsci.rpi.edu/Proceedings/2013/papers/0296/paper0296.pdf.

Wilson, T. D. 2002. *Strangers to Ourselves*. London: The Belknap Press of Harvard University Press.

Wilson, T. D., D. J. Lisle, J. W. Schooler, S. D. Hodges, K. J. Klaaren, and S. J. LaFleur. 1993. Introspecting about reasons can reduce post-choice satisfaction. *Personality & Social Psychology Bulletin* 19 (3): 331–339.

Witteman, C., J. van den Bercken, L. Claes, and A. Godoy. 2009. Assessing rational and intuitive thinking styles. *European Journal of Psychological Assessment* 25 (1): 39–47.

6

Cultural Influences on Cognitive Biases in Judgment and Decision Making: On the Need for New Theory and Models for Accidents and Safety

Atsuo Murata

CONTENTS

Introduction 103
Cognitive Biases and Accidents 104
Cognitive Biases 104
Example of How Cognitive Biases Are Related to Accidents 105
Case Studies of Crashes and a Disaster in Regions Other than Western Countries 106
Aviation 106
Union Carbide's Bhopal Chemical Plant Accident 107
Summary and Call for Additional Theory Development on Cross-Cultural Concept of Safety Culture 108
Summary 108
References 109

ABSTRACT We believe that cognitive biases are ubiquitous in distorted decision making and subsequent accidents. In addition, we contend that cultural differences may potentially compound the effects of cognitive biases in this domain. In this chapter, we briefly examine two case studies of how cross-cultural nuances contribute to crashes or disasters. Based on these analyses, an emphasis is placed on the concept of incorporating cultural differences into a safety culture or a preventive model of crashes or disasters. We suggest that a new theory needs to be built in the domain of safety, which account for cross-cultural nuances as part of the causal processes leading to accidents.

Introduction

Different from traditional economics, a concept known as *bounded rationality* is commonly assumed in behavioral economics (Kahneman 2011). Bounded rationality perspectives generally suggest that people often do not make decisions rationally; rather, people suffer from cognitive biases such as those pointed out by Kahneman (2011) as well as Tversky and Kahneman (1974). They suggest that our cognitive information processing is conducted by one of two systems, which they refer to as System 1 or System 2. System 1 operates quickly, automatically, and intuitively with little or no deliberate effort. System 1 approaches are very simple and intuitive, yet such approaches often suffer from cognitive biases. System 2, on the other hand, is a far more effortful approach to decision making and is generally reserved for deliberation.

While both System 1 and System 2 processes are important to human decision making, this chapter will focus exclusively on System 1 processing because, as noted, these process are most commonly susceptible to cognitive biases. In addition, beyond the role System 1 processing plays regarding cognitive biases, Murata and Nakamura (2014) and Murata et al. (2015) discussed how such System 1 cognitive biases probably play a major role in downstream decision making as it is related to accidents. However, the importance of understanding decision making across different cultures and how this leads to distorted decision making was not discussed in these studies. Essentially, we are suggesting that accidents are often the result of System 1 cognitive biases and related errors. And that such errors can be compounded by cross-cultural differences in cognitive styles. Therefore, we attempt to discuss how cultural differences distort judgment, induce cognitive biases, and eventually lead to mistaken behavior. We begin this discussion with several very brief examples of how cultural differences lead to biased and irrational decision making and induce accidents.

Cognitive Biases and Accidents

Cognitive Biases

Before discussing how cultural differences distort judgment and induce irrational behavior, we will briefly discuss how cognitive biases distort decision making absent cultural issues. As shown in Figure 6.1, it is hypothesized that cognitive biases distort decision making, which subsequently leads to human errors in judgment, decision making, behavior, and might eventually trigger accidents if the commitment to the biased judgment, decision making, and behavior is escalated.

Bazerman and Moore (2001) provide several examples of System 1 heuristics such as the availability heuristic, representativeness heuristic, confirmation bias, affect bias, and anchoring and adjustment. Many of these biases are familiar to most readers, but we provide two quick examples for the uninitiated. The availability heuristic is a function of the vividness of imaginable events. Such vividness can bias our perception of the frequency of such events. Likewise, the confirmation bias is a tendency for people to seek information that confirms their expectations and hypotheses even when information disconfirming their expectations and hypotheses is actually more useful.

Bounded awareness prevents one from focusing on useful, observable, and relevant information. It is assumed that our bounded awareness, as described earlier, and uncertain (risk) situations form the basis of heuristics, overconfidence, and framing. Due to such bounded awareness, we sometimes behave irrationally. We frequently tend to behave irrationally and are, in most cases, unaware of how and to what extent these irrational behaviors influence us. Such irrational tendencies are sure to distort our decisions and, in the worst, cases lead to accidents. Without consideration of our bounded rationality, we cannot properly approach the prevention of accidents and analyze the root cause of accidents.

We should explore how cognitive biases distort decision making, induce preconception, and become a trigger of accidents. To do this, we must further clarify the mechanism on

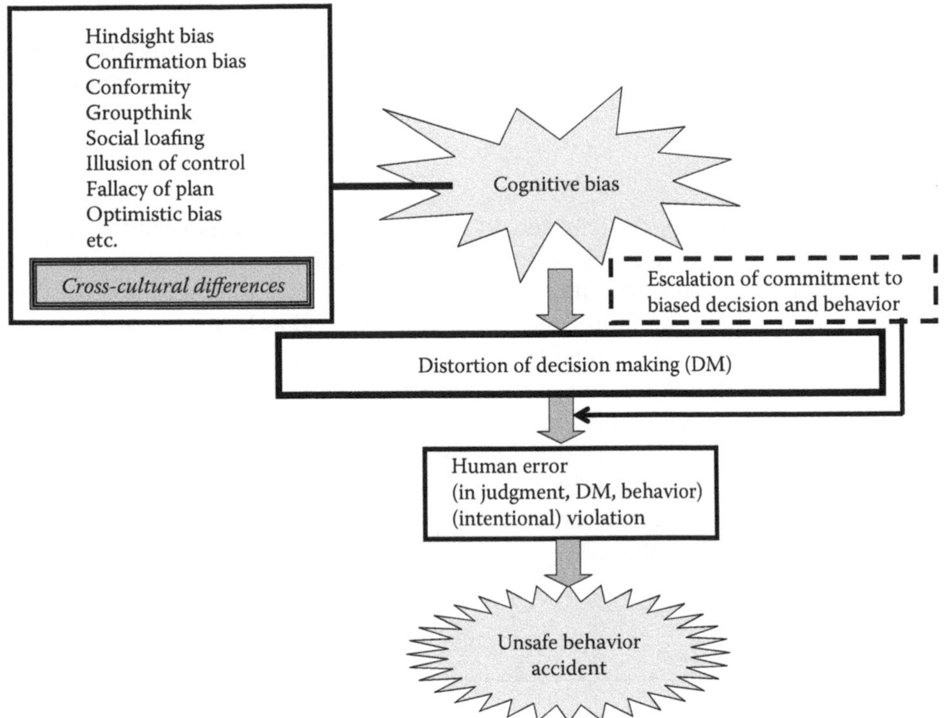

FIGURE 6.1
Relational model between cognitive biases and unsafe behaviors or accidents. (From Murata, A., and Nakamura, T. 2014. Basic study on prevention of human error—How cognitive biases distort decision making and lead to crucial accidents. *Proceedings of AHFE 2014*, 136–141; Murata, A., Nakamura, T., and Karwowski, W. 2015. *Safety* 1: 44–58.)

why we suffer from cognitive biases, what type of cognitive bias is potentially dangerous, and when or how cognitive biases distort decision making and become a trigger of error, violation, and accident. Moreover, we need to identify the causal paths by which cognitive biases, compounded by cultural nuance, induce errors or violations of regulations or safety rules and how this leads to unsafe behaviors or critical accidents.

Example of How Cognitive Biases Are Related to Accidents

One example of how cognitive biases can relate to accidents is provided by Brafman and Brafman (2008) who pointed out that a concept known as the *loss aversion* (an urge to avoid loss) probably contributed to the famous Royal Dutch Airlines (KLM) Flight 4805 disaster. On KLM Flight 4805, a Boeing 747 was leaving Amsterdam and bound for Las Palmas Airport in the Canary Islands. However, a terrorist bomb exploded at the airport flower shop in Las Palmas Airport, and the flight was diverted to another airport. Subsequently, the KLM flight found itself waiting for air traffic controller (ATC) clearance at a foggy and unexpected airport. After waiting for some time, the captain of the KLM flight attempted to take off without the permission from ATC. Unfortunately, a Pan American 747 was parked across the runway, and KLM Flight 4805 crashed into the Pan American aircraft. All crew and passengers lost their lives.

Review: Loss aversion is a human tendency to strongly prefer avoiding loss over and above acquiring gains. As would be expected, the more meaningful a potential loss, the more people are motivated to avoid the loss.

Naturally, multiple factors contributed to the KLM crash, but as noted, one plausible contributing factor is known as loss aversion. This concept is such that the more meaningful a potential loss is, the more likely people are to try and avoid the loss. In the KLM example, the pilot probably had multiple loss factors in mind as he made his decisions. For example, the pilot was probably considering factors such as the mandated rest period for crew, the cost of accommodating the passengers for making them stay at a hotel until the weather improved, and the blot on the captain's reputation for not being punctual in his flight. With all this in mind, the captain may have been preoccupied with the urge to avoid this cascade of losses. With hindsight, it is easy to believe that the captain was foolish in his decision making. Yet, the loss aversion concept is a powerful one in human psychology and may well have been a contributing factor in the KLM disaster.

Case Studies of Crashes and a Disaster in Regions Other than Western Countries

The KLM Flight 4805 example described earlier provides an example of how cognitive biases, such as loss aversion, are probably involved in accidents. This section will expand upon this concept by demonstrating how cultural nuances are probably involved in accidents as well. In this section, we lay out examples of how cultural nuances can generate accidents across the aviation and chemical industries.

Aviation

One example of how cultural differences probably play a role in accidents comes from the analysis of the Korean Air Flight 801 disaster (Boeing 747-3B5; August 1997). In this crash, a possible root cause is a cultural concept known as *power distance.* Power distance (defined by Hofstede [1980] as the extent to which the less powerful members of organizations accept and expect that power is distributed unequally) suggests that high power distance cultures (such as that observed in many East Asian nations such as Korea) tend to display strong obedience to people in high positional rank and considerably less obedience to people in low positional rank. In the case of the Korean Air crash, power distance probably contributed to the crash.

Generally speaking, the landing approach to Guam is straightforward. Guam airport has a glide scope (a technology tool that emits a beam stretching upward to the sky, which helps guide the landing). Unfortunately, the glide scope at the Guam airport was under repair in this instance. What was worse is that the pilots did not recognize that the guide scope was broken in spite of such information being presented to some crew members in advance.

Review: Power distance is the degree to which the less powerful members of organizations accept and expect that power is distributed unequally. This is one of Geert Hofstede's (1980) original cultural dimensions such as individualism–collectivism, masculinity–femininity, and uncertainty avoidance. Later research has discovered additional dimensions as well.

After the analysis of the flight recorder transcript, it was found that power distance might well have been a notable factor in the crash. In the case of Korean culture, the captain of the aircraft is a rather high positional rank. All other crew members are subordinate to the captain and must generally follow the captain's orders without question. According to analysis of the flight recorder, the flight engineer noticed that the captain misunderstood several instructions, yet the overriding cultural framework he was operating within made it such that directly informing the captain of his error would have been nearly impossible in this cultural context. As pointed out by Sohn (1993), in Western cultures, the responsibility of the speaker is to communicate ideas clearly and unambiguously. In Korea, not the speaker but the receiver is responsible for understanding what the speaker communicates. In the analysis of the flight recorder transcript, it seems clear that the flight engineer indirectly attempted to tell the captain that he was wrong about his interpretation of his approach information but never directly warned the captain.

Union Carbide's Bhopal Chemical Plant Accident

Beyond psychological cultural differences such as the power distance concept mentioned earlier, other types of cultural distinctions, such as distinctions between the industrial safety culture between developing countries and advanced countries may be a key in understanding accidents. One possible organizational cultural distinction between developed versus developing nations might be illustrated by the infamous Union Carbide chemical plant accident in Bhopal, India (Reason 1990). In December 1984, the plant released a deadly gas known as methyl isocyanate. At least 4000 people died and over 20,000 people were injured. At the time of the disaster, Union Carbide's profit was declining, especially in this plant. As a result, the plant laid off key personnel who were accustomed to the details of Union Carbide's Bhopal chemical plant. The plant decreased the shift size from 11 to 5 and reduced the maintenance crew by one-half. They further cut the maintenance and other costs (e.g., shut down the refrigeration unit for saving costs and left safety flares and washing towers unrepaired).

We suggest that, in this tragedy, it might be that the safety culture in developing nations may be less of a focus than in developed nations. In this case, the emphasis on economy instead of safety might well stem from the previously mentioned loss aversion (loss of profit due to the safety measures such as enhancing maintenance expenses to enhance safety). Although, to our knowledge, there is no systematic quantitative evidence that distinctions in safety culture between developing and developed nations exist, we believe that this concept is worth serious considerations and future theory development.

Summary and Call for Additional Theory Development on Cross-Cultural Concept of Safety Culture

Although it is recognized that nurturing safety culture is important for safety management (Dessai et al. 2006; Flin et al. 2000), cross-cultural differences are rarely taken into account in the framework of safety culture or man–machine interface. As stated in the Case Studies of Crashes and a Disaster in Regions Other than Western Countries section, cultural differences (Helmreich 1994; Helmreich and Merritt 2000) seem to affect decision making in the safety domain. Throughout our brief case examples, we speculated that cultural differences, as well as cognitive biases, may distort judgment and eventually lead to mistaken behaviors. Therefore, we believe it is critical to build additional safety models and theories that directly incorporate cultural distinctions (whether based in psychology, cultural anthropology, sociology, etc.) into causal chains.

Summary

We have briefly argued that cross-cultural differences might be a potential causal factor in distorted decision making, which could lead to critical crashes or disasters through the examples of accidents, that is, the Korean air crash accident and Union Carbide's Bhopal chemical plant accident.

As well as a variety of biases, the consideration of cross-cultural differences in thinking, behavior, or decision making is important for understanding the root causes of accidents. It is predictable that cultural nuances can lead to unanticipated and irrational decision making and behavior and can eventually be a trigger or a risk leading to a crash or a disaster. Therefore, an emphasis must be placed on considering cultural differences that distort judgment, induce cognitive biases, and eventually lead to a mistaken behavior. Rather, we had better classify the irrational behavior triggered by the cross-cultural nuances (Helmreich 1994; Helmreich and Merritt 2000) as one of the cognitive biases. In conclusion, we must recognize that the consideration of cross-cultural difference, as well as cognitive biases, is important in the area of accidental prevention and analysis.

As seen and recognized from the two examples, many critical accidents originate from cognitive biases (including an irrational behavior that stems from cross-cultural nuances) as one of the main causes of accidents. The correction or the modification of bias in decision making must be one of the promising measures for preventing critical accidents. When the designers, the engineers, and the managers of modern technologies such as transportation systems, nuclear power plants, and social inflation systems do not understand humans' fallibility (error-prone properties) and the cross-cultural difference related to our irrational mind, we tend to design new and improved systems that do not take our limitation (irrationality) into account, that is, man–machine incompatible systems.

Consequently, we may distort our decisions and make serious errors. These distortions or errors lead to critical accidents such ones as analyzed in the Case Studies of Crashes and a Disaster in Regions Other than Western Countries section. Without such understanding of our irrationality together with the cross-cultural nuances, we unwillingly repeat critical accidents and cannot get out of the vicious circles of similar accidents. The understanding of how cognitive biases (including ignorance of the cross-cultural difference) distort decision making and lead to accidents is essential in order to avoid such vicious circles as pointed out by Dekker (2006).

DISCUSSION QUESTIONS

1. How often do you believe you behave rationally?
2. What would be the most difficult element of building new theories of accident generation that incorporate cross-cultural nuances?
3. Do you ever underestimate the effect of cross-cultural differences on the behavior of others (and yourself)?
4. What cultural nuances do you believe should be part of future causal theories of accidents?

References

Bazerman, M. H., and Moore, D. A. 2001. *Judgment in Managerial Decision Making*. Cambridge, MA: Harvard University Press.

Brafman, O., and Brafman, R. 2008. Anatomy of accident. In *Sway: The Irresistible Pull of Irrational Behavior*, eds. Brafman, O., and Brafman, R., 9–24. New York: Crown Business.

Dekker, S. 2006. *The Field Guide to Understanding Human Error*. Farnham: Ashgate Publishing.

Dessai, V. M., Roberts, K. H., and Ciavarelli, A. P. 2006. The relationship between safety climate and recent accidents: Behavioral learning and cognitive attributions. *Human Factors* 48(4): 639–650.

Flin, R., Mearns, K., O'Connor, P., and Bryden, R. 2000. Measuring safety climate: Identifying the common features. *Safety Science* 34: 177–192.

Helmreich, R. L. 1994. Anatomy of a system accident: The crash of Avianca Flight 052. *International Journal of Aviation Psychology* 4(3): 265–284.

Helmreich, R. L., and Merritt, A. 2000. Culture in the cockpit: Do Hofstede's dimensions replicate? *Journal of Cross-Cultural Psychology* 31(3): 283–301.

Hofstede, G. 1980. *Culture's Consequences: International Differences in Work-Related Values*. Beverly Hills, CA: Sage.

Kahneman, D. 2011. *Thinking, Fast and Slow*. London: Penguin Books.

Kahneman, D., and Tversky, A. 1984. Choices, values, and frames. *American Psychologist* 39(4): 341–350.

Murata, A., and Nakamura, T. 2014. Basic study on prevention of human error—How cognitive biases distort decision making and lead to crucial accidents. *Proceedings of AHFE 2014* 136–141.

Murata, A., Nakamura, T., and Karwowski, W. 2015. Influence of cognitive biases in distorting decision making and leading to critical unfavorable incidents. *Safety* 1: 44–58.

Reason, J. 1990. *Human Error*. Cambridge: Cambridge University Press.

Sohn, H. 1993. Intercultural communication in cognitive values: Americans and Koreans. *Language and Linguistic* 9: 93–136.

Tversky, A., and Kahneman, D. 1974. Judgment under uncertainty: Heuristics and biases. *Science* 185(4157): 1124–1131.

Section II

Collecting and Analyzing Data

The first section of this volume discussed the importance of theories in sociocultural research, which provide a conceptual basis for all aspects of the sociocultural modeling process. But if a theory creates the framework, then data are what fill and formalize that structure. Without rich data and validated analyses, no other meaningful steps could be taken in practice—as those who attempt to build models based solely upon theory often discover!

Data Collection

The first three chapters in this section discuss data collection, which presents idiosyncratic challenges for cross-cultural social scientists. For this reason, researchers have devised unique methods for handling the collection of data, ranging from classical ethnography and large-scale population surveys to emerging applications of social media and smartphone data-collection apps. Despite this toolkit of methods, data collection remains one of the most challenging (if not *the* most challenging) steps in the sociocultural modeling process.

Data collection challenges include the following (to name just a few):

- Limited access: It is frequently difficult or dangerous to contact those populations of greatest interest, such as those in conflict zones, disconnected locations, or insular communities.
- Communication barriers: Even when researchers can reach a community, their access may be limited by communication barriers, ranging from simple linguistic differences or literacy limitations to underlying conceptual disparities and fear or mistrust of the researchers.
- Population vulnerability: Simply contacting some populations creates risk for those individuals in a vulnerable community; if others learn of their contact with the researchers—or if members of a vulnerable population seemingly divulge sensitive or unflattering information—their safety may be threatened.

- Ethical ambiguity: Researchers must make numerous ethical decisions, sometimes moment to moment, such as whether the results of their work outweigh the inherent risks to themselves and others. Some researchers, when faced with this challenge, turn to more easily accessed data (e.g., from Western, educated, industrialized, rich, and democratic [WEIRD] countries), choosing safety and convenience over authenticity—and as a result, sometimes inadvertently introduce bias into their work.
- Noisy data: Even the most authentic data, because it describes human behavior, includes redundancies, inaccuracies, and problems with the sensitivity of measures. The very act of measurement may affect the observed phenomena (i.e., the observer effect), and because researchers can never fully rid themselves of bias, they may unintentionally introduce error, too.

Generally, social scientists show acute awareness of these challenges and take great pains to mitigate them, as several chapters in this section reveal.

For instance, Chapter 7 by St. Benoit, Graffeo, Carter, and Swisher discusses high-risk ethnography, that is, the application of ethnography within distressed environments involving conflict, societal upheaval, natural disasters, or other serious dangers. High-risk ethnography is an interdisciplinary scientific approach for conducting time-sensitive research, emphasizing direct collection and analysis of complex social data unique to distressed populations. In other words, this approach addresses many of the data collection challenges outlined in the list earlier, with particular attention paid to the safety of researchers and the population they study as well as ethical concerns and debiasing of the data. In addition to addressing these general concerns, this chapter includes a high-risk ethnography case study, referred to as the *sixty percent mission*. Using this Afghanistan-based case study, the authors describe how high-risk ethnographic techniques led to a 60%–70% drop in aggressive combat activity.

In Chapter 8, Christina Kang expands upon the theme of data collection from difficult-to-access communities; specifically, she outlines five categorical challenges as well as various ethical questions associated with data collection from disconnected communities, such as people in poverty, victims of human trafficking, or those affected by natural disasters and conflict. These societies often lack access to key resources, have distinct sociocultural norms, and have particular vulnerabilities. Researchers must take care to accurately and safely connect with these isolated groups.

Chapter 9 by Combs, Blincoe, Garris, and Vorm discusses a WEIRD problem: A great deal of published social science data, and subsequently the theories and the models derived from those data, are collected from WEIRD populations. In addition to highlighting this issue, the authors review two well-known cross-cultural research frameworks: Markus and Conner's culture cycle and Hofstede's dimensions of culture. The authors encourage fellow researchers to consider an integration of these two models as a framework for exploring hypothesis generation, data collection, and theory building that compares and contrasts WEIRD societies to their non-WEIRD counterparts.

Data Analysis

Raw data, alone, offer little insight—particularly when it involves human behavior. Only by considering data in aggregate, with contextual information and shrewd analyses, can insight be uncovered. Unfortunately, social science data sets are noisy, inconsistent, and distinctly susceptible to observer bias. Even the most accurate observations change (sometimes rapidly and substantially) over time, and social science analysts must have a high tolerance for uncertainty, ambiguity, and logical inconsistencies.

Data analysis challenges include the following (to name just a few):

- Latent versus observable variables: Often the variables of greatest interest in the social sciences, such as collective population attitudes, cannot be measured directly; instead, inferences must be made using careful data collection techniques and shrewd analysis methods that help reveal the underlying latent variables creating the observable and measurable effects.
- Context and implicit relationships: Like latent variables, context and interactions influence observable variables. A given effect may vary widely depending upon a litany of (sometimes nonobvious) contextual factors and implicit relationships among the (observable and latent) variables.
- Data handling: Data handling universally presents challenges, such as labeling, securing, and standardizing data formats. Cross-cultural social science research also faces additional challenges, such as creating and defining common ontologies of codes for organizing phenomena, cross-system sharing, integrating diverse data from different sources, and ethical questions of privacy, protection, and ownership of data.
- Noisy data: As mentioned earlier, almost by definition, data collected in the social sciences suffer from quality issues or noise. Social scientists have developed (and continue to refine) methods for cleaning and transforming noisy data, accounting for natural variability in it, and tolerating—or even capitalizing upon—ambiguities and inconsistencies in data sets.
- Inadvertent bias: Researches can inadvertently introduce error at nearly every stage of cross-cultural research. In the context of data analysis, for instance, inadvertent bias may be created by the nature of analyses used (such as approaching analyses solely from etic perspectives), how variables are selected for inclusion, where analysts draw the thresholds in their data, or the methods they use to address the inherent noise in the data set.
- Interpretability: Although not a unique challenge, necessarily, effectively conveying the results of the analyses in a way that decision makers and other observers can understand remains a critical challenge for sociocultural data analyses.

The final four chapters in this section discuss emerging methodologies, algorithms, and/or computational tools for addressing these challenges.

Chapter 10 by Fenstermacher and Kuznar discusses a novel approach to discourse analysis that helps to more accurately forecast future violent events and inform the development of appropriate courses of action. The chapter begins by reviewing *emic* (or first-person) and *etic* (or third-person, outside observer) perspectives and then shows how traditional etic discourse analyses can be enhanced by the consideration of five discourse features (i.e., integrative cognitive complexity, social identity-based sentiment, idea density, vocabulary diversity, and keyness), which collectively provide greater insight on the emic perspective. By applying these techniques to threat narratives, analysts can identify who, what, and why violence occurs and can possibly predict violent events before they take place.

Chapter 11 by Caglayan and Cassani describes a unique and powerful analysis technique involving semantic knowledge management and entropy calculations. These authors have helped to develop a system that capitalizes on the ubiquity of mobile phones and their extensive—although typically poorly structured—data-capture capabilities. First, they define automated processes that enable semantic encoding of the mobile-collected data. Once the data are stored in a semantic knowledge representation, then anomalies in the data can be uncovered using entropy calculations. *Entropy* is a measure of the uncertainty in a random variable, and detecting it can, for instance, indicate suspicious individuals or activities, which tend to exhibit anomalous behavior even if traditional analysis tools cannot detect those slight deviations from the baseline.

Chapter 12 by Maddux and Appleget offers a case study from the authors' own experience analyzing population surveys for military decision makers. The large-scale, longitudinal survey effort described in this chapter involved data collection throughout Afghanistan's Helmand province. While the results of the project offered useful insights, this chapter primarily explores the analytical methodology used to draw out those insights. In particular, Maddux and Appleget provide detailed discussions on how—and why—to use factor analysis, imputation methods, and the Mann–Whitney test for analyzing such data. They also discuss the importance of providing a user-friendly means of communicating the results to decision makers (who typically lack specialized expertise in social science statistics).

Finally, in Chapter 13, Çakıt and Karwowski discuss the utility of GIS approaches for understanding historical events and predicting future developments. They situate their discussion in the context of a case study involving adverse incidents in Afghanistan. After reviewing the recent history of Afghanistan, the authors present data on terrorist events that occurred in the country from 2004 to 2010. They then demonstrate how the application of spatial and temporal analyses helped uncover patterns in these data. Through these examples, the authors show how these geographic-based analysis techniques may generalize to similar conflict regions.

7

The Sixty Percent Mission: An Introduction to High-Risk Ethnography Process and Protocol in Support of the US Army's Civil Affairs Humanitarian Mission

Tracy Saint Benoit, Clarissa Graffeo, Mark A. Carter, and Col. Richard Swisher (Ret)

CONTENTS

Introduction 116
High-Risk Ethnography 117
High-Risk Environments and the Need for HRE 118
The Role of the HRE Team 118
Concepts and Methods 119
Security and Integration of HRE Team: The Parallel Planning Process 120
Data Analysis 121
Overwatch Activities 122
Social Complexity Data 123
Information and Data 123
Data Management 125
Case Study: The Sixty Percent Mission 128
Conclusion 130
Megacity Use Case 130
References 132

ABSTRACT Many government and nongovernment organizations operate in distressed environments, from natural disaster sites to areas undergoing civil wars or insurgencies. Current, detailed understanding of the population and assessment of the effectiveness of humanitarian services are vital. However, traditional ethnographic methods are often unsuited to such environments. In this chapter, the authors introduce high-risk ethnography, a specialized research protocol for distressed populations and areas with a high risk of violence. The requirements, the concepts, the methods, and the team composition are summarized. Finally, a case study from a field research in Afghanistan during 2007 is provided to illustrate the execution and the value of the methodology.

KEY TOPICS

- Traditional social science methods are often insufficient for distressed environments.
- High-risk ethnography (HRE) is a specialized ethnographic methodology.
- HRE assesses the local populace/environment and the aid organization.
- HRE requires a full research team.

- Physical security for researchers and locals is the primary concern.
- Supporting data analysts provide supplemental information and security.
- HRE teams gather social complexity data (SCD).
- Primary, secondary, and tertiary data are needed for validation/planning.

Introduction

On April 24, 2008, Col. Martin P. Schweitzer, commander of the 82nd Airborne Division's 4th Combat Brigade of the US Army, testified before the House Armed Services Committee on the effectiveness of a new proof of concept project. Referred to in the public press as the *Human Terrain System* (HTS), it proposed to integrate anthropological expertise into military operations. Col. Schweitzer's testimony was based on the first research group or Human Terrain Team (HTT), designated AF1, that operated in Afghanistan's Regional Command East during 2007. AF1 was assigned to the provincial reconstruction team under Col. Schweitzer's command with the primary mission of assessing and evaluating US Army humanitarian projects' effectiveness at meeting the needs of local Afghans. Col. Schweitzer stated the following during the testimony:

> The [HTT] is not an intelligence-gathering tool which is used to "target" individuals. My staff is uniquely organized to run the targeting process and link intelligence systems to time sensitive targeting. The HTT, sourced with anthropologists and social scientists, is not qualified or trained to provide targeting support.
>
> Last year, the Army fielded an HTT to my unit as a proof of concept. The HTT was immediately value added and became mission critical. The team's impacts were exponentially powerful: reduced our kinetic operations, assisted in developing more effective non-kinetic courses of action, improved the unit's overall situational awareness, improved consequence management, increased host nation government support, improved the brigade's humanitarian assistance efforts, improved village assessments, improved information operations capabilities, decreased enemy forces attacks, and decreased ordinary crime in our area of operations. Without the HTT filter on courses of action and the alternative maneuver tools they identified to create the exact same effect, we would have lost double the lives. Using HTT capabilities, we reduced kinetic operations by 60–70%. (US House of Representatives, 2008)

Col. Schweitzer's testimony was the foundation for the subsequent rapid growth of HTS and the increased deployment of additional HTTs sourced and trained by HTS to Afghanistan and Iraq. While much has been written about HTS, the purpose of this chapter is to provide a specific context for Col. Schweitzer's assertions and discuss the use of HRE processes specific to AF1's mission in Afghanistan. It is critical to distinguish between AF1's methodologies and those used by subsequent HTTs.

In practice, AF1 was *not* an HTT like the ones later deployed by HTS, but rather a specialized HRE team. This distinction is critical in order to understand how and why the proof of concept successfully integrated social science within the military domain, ". . . in developing more effective non-kinetic courses of action" (US House of Representatives, 2008).

BIRTH OF THE HUMAN TERRAIN SYSTEM

The first social science team AF1 was originally run under the Foreign Military Studies Office (FMSO) at Fort Leavenworth, Kansas. It was only after the team was well underway in training that the control of the pilot project moved to the Training and Doctrine Command and was renamed *Human Terrain System (HTS).* The original FMSO plan was to field one team and then perform an in-depth study to understand the possibilities of fielding further teams. Unfortunately, AF1's success and following publicity was exploited by the HTS management, and there was a rush to field as many teams as possible. This premature push of an unexamined capability sowed the seeds of HTS' eventual demise.

AF1 implemented specific anthropological field methodologies, team structure, ethical and human subject protection protocols, data, and interdisciplinary social science methodologies established and successfully practiced prior to the advent of HTS. The social science training, planning, and implementation were a direct use of the HRE process and were never replicated by the other HTTs deployed after AF1. The triad of competencies unique to AF1 was:

- The social science-based HRE methodology;
- The Special Forces-based parallel planning process; and
- The critical role of the data manager/analyst in interpreting often complex qualitative and quantitative data, which supported the flow of relevant information in and out of the research site ensuring the security of researchers in the field.

Col. Schweitzer's assertion of "reduced kinetic operations by 60–70%" was the result of these competencies. In this chapter, we provide a high-level summary of the HRE concept and process, illustrating the particular needs and requirements for HRE in comparison to traditional ethnographic practices. We also outline the second generation of data management processes and tools that have evolved to meet the present requirements for appropriate social science methodologies within the HRE context. Finally, we present a case study of a specific civil affairs medical operation to illustrate how these protocols were implemented in the field.

High-Risk Ethnography

The practice of HRE evolved from traditional ethnographic fieldwork practiced by trained anthropologists for decades. Traditionally, ethnographers collect field data by living among research subjects for extended periods. While the populations studied by ethnographers often live in remote and physically challenging conditions, current research needs have emerged in settings characterized by societal upheaval, which introduce additional dangers to researchers and local participants—that is to say, within environments considered high risk.

High-Risk Environments and the Need for HRE

International governmental and nongovernmental organizations involved in rebuilding efforts, disaster relief, medical care, and other services within high-risk environments recognize the need for improved knowledge and understanding about the socially complex environment, particularly as it relates to the dynamic nature of distressed populations during and following major social upheavals (e.g., the study by King [2015]). These high-risk or distressed environments range from catastrophic disaster sites to active war zones and are often characterized by latent or overt social instability and medium- to high-intensity conflict (e.g., civil or ethnic wars, insurgencies).

Due to the extreme physical dangers and arduous environmental conditions posed by these contexts, distressed environments are generally considered no-go zones for academic or humanitarian agency researchers; traditional social science research must frequently rely on at-a-distance analysis based on the aggregation of prior fieldwork and a corpus of potentially dated theoretical knowledge. This circumferential research can be useful but is often insufficient to address the complex and rapidly changing situation characteristics of the exact environments where humanitarian support is most needed. In conjunction with the inherent instability of high-risk environments, globalization, rapid shifting of sociocultural institutions, and almost instantaneous flow of information around the globe via social media have exacerbated the need for temporally sensitive research and analysis. Up-to-date knowledge derived from field research within the distressed environment is *key* to planning and delivering effective, sustainable humanitarian interventions. This begs the question: how can researchers provide political leaders, policy makers, and humanitarian/development experts with an accurate *and actionable* understanding of the distressed human environment?

Conducting fieldwork under these very difficult circumstances is at best challenging and frequently dangerous. The immense social pressures caused by internal social and political instability, natural or human-made disasters, and influence of adversarial foreign actors all negatively impact the research environment. Social structures within indigenous populations invariably become disrupted, and they undergo both temporary and permanent changes, sometimes to the point where they become unrecognizable. Furthermore, these situations often introduce a variety of negative influencers who may impede research and affect local and regional populations. These influencers are often linked to regional or global networks and supported by both formal and informal supply chains within and outside of the indigenous population, as in the case of terrorist or criminal organizations.

Given the risks and the stressors imposed on the researcher, there is a need for specific training in structured, methodologically and ethically sound approaches supported by proactive protective measures for conducting research in support of humanitarian and development interventions. In short, ethnographic research in high-risk or distressed environments requires different skill sets and character traits from those required in traditional fieldwork, and research in high-risk contexts demands a revised methodology and customized training.

In response to this need, the University of Central Florida developed a program for HRE to meet the academic, ethical, and practical requirements of conducting field research under such dangerous and demanding conditions.

The Role of the HRE Team

The HRE team's core function is to develop an understanding of indigenous cultures and values, population concerns, opinions, and attitudes as they stand within the distressed environment, and to track how they change due to the evolving situation. Perhaps more importantly, the HRE team helps identify changing centers of gravity within key population groups, from individuals with important social or political influence to pivotal environmental issues or features. The HRE team provides crucial information needed to effectively calibrate research and analytical parameters along the full spectrum of interventions including political, diplomatic, developmental, educational outreach, and engagement. For organizations operating in distressed areas, the paramount objective is to ensure that humanitarian interventions are effectively identifying and meeting the central needs of the local populace.

A large part of the HRE team's value is vetting potential partners for community engagement. Who are the brokers of trust and influence who can educate us on the nuances of the local problem set, disseminate information within local communities, or ensure that goods and services reach those who need them? Providers of humanitarian services must be able to differentiate between opportunists who seek only financial gain and true local enablers who seek to positively influence the situation. Answering complex issues requires rigorous collection of open-source HRE data to identify partnership candidates and demonstrate their record of measurable effectiveness prior to engagement. Furthermore, HRE research can help identify project templates as part of the vetting that includes assessing whether the activism of our partners is working. Of key importance is also the HRE team's ability to assess and vet the organization's own capacity to provide relevant services and the success of implemented projects. HRE research should never focus solely outward toward the local population but should be holistic.

Concepts and Methods

The specific methodology developed for HRE is radically different from what has been, and is still, taught in today's anthropological institutes. HRE is an interdisciplinary scientific concept and methodology for conducting time-sensitive research, emphasizing the direct collection and the analysis of complex social data unique to distressed populations. It consists of a set of philosophical assumptions, substantive propositions, methodological framings, and methods ranging from ethnography to sensemaking (for sensemaking concepts and methodology, see e.g., the studies by Weick [1995], Kurtz and Snowden [2003], Mills, Thurlow, and Mills [2010], Snowden [2011]).

HRE requires an anthropologist (or a high-risk ethnographer) with specialized training and expertise working under extreme and dangerous conditions. In contrast to traditional ethnographic fieldwork, however, the HRE is not a solo researcher; rather, he or she is one component of an extensive research team including subject matter experts in security, psychology, field medicine, data management, analysis, and logistics. Team members are located both on-site and in the researchers' home country or countries. The combination of field research with robust reachback* for contextual research, information vetting, and security support enables HREs to recognize and monitor the changing environment and provide timely input to decision makers. In keeping with recognized research protocols and humanitarian concerns, the HRE team utilizes strict planning protocols, human subject protections,

* *Reachback* refers to the process of obtaining information, services, and other support from organizations that are not forward deployed.

TABLE 7.1

Comparison of HRE and Traditional Social Science

	HRE	Traditional Social Science
Research Parameters		
Research scope	Focused research relevant to specific outcomes and structured by practical requirements—who to ask, how to ask, what security factors affect answers	Often broadly focused on the general population or a subject of general academic interest; research objectives and human subjects protections defined by traditional institutional review board parameters
Research environment	Distressed populations in unsafe and often nonpermissive environments	Permissive environments
Research context	Populations experiencing major upheavals in relation to traditional lifestyles, norms, and values	Usually benign, peaceful environments where populations engage in stable/traditional lifestyles, norms, and values
Security issues and requirements	Emphasis is on the physical security of the fieldworker and the respondents alike due to the danger of disclosure of private, identifiable information (e.g., data gathered in interviews, filmed on video, or recorded on tape or field notes) or evidence of participation	Traditional social science researchers are bound by an institutional research board human subject protections but presently do not require research restrictions to address special physical security issues for the researcher or research subjects
Experience		
Education level	PhD and/or vetted and trained field researchers	Graduate students to PhD graduates
Prior experience	Minimum of two supervised high-risk environments before HRE, and team is qualified for independent research; experienced in conducting and managing international teams under dangerous conditions	Human subject protections and research methodologies not adapted nor sufficiently comprehensive for high-risk environments
External support	Highly qualified and trained support team providing reachback and logistical support	Research reachback and logistical support limited or nonexistent

robust field methodologies, and rigorous data collection and management processes to facilitate a cyclical process of gathering and continually reassessing research findings.

Table 7.1 provides a broad overview of the differences in training and methods between traditional ethnography and HRE.

Security and Integration of HRE Team: The Parallel Planning Process

The HRE parallel planning process is central to the methodology. It is the foundation of any high-risk research effort and critical to its success. All HRE teams are supervised by a team leader, whose precise qualifications are dependent upon the specific constraints and conditions of the research site. The nature of HRE research in nonpermissive environments requires coordination with the government (in our case, this means the US government), for which the team leader is primarily responsible. Unlike traditional academic research, which is conducted solo or with a detached research unit operating at a distance from other actors, the HRE team operates as a fully functioning component of the humanitarian operations in progress. This requires additional steps during the planning and execution process to align objectives, communicate between the HRE team and other personnel, and regularly perform debriefs and revise plans in response to emerging data.

Due to the nature of high-risk environments, the security for team members and local national participants is the priority during the parallel planning process and overrides all other data collection and research plans. Generally, a two-person personnel security detachment provides security for the researcher in the field. In some cases, cultural restrictions may prohibit the two-person security protocol. In these cases, the team must review the advantages and the danger areas or concerns and how to mitigate them. The team leader has the final authority on security issues and can withdraw the team and cancel the research if he or she believes security protocols were or could be compromised.

As an example, the HRE team leader for AF1 was a former Special Operations officer. In addition to his military experience, he was uniquely trained to understand, support, and integrate research requirements into the security constraints found in civil affairs humanitarian missions in Afghanistan, in 2007. Working with the HRE researcher, the team data manager, the analysts, and the team leader would collaborate and plan the integration of the HRE team with the civil affairs unit; the HRE team became fully functioning members of the civil affairs humanitarian mission. The sixty percent mission (mentioned in the quote by Col. Schweitzer earlier and discussed further in the following) required the HRE researcher to support the civil affairs medic as part of the medical outreach to Afghan villagers.*

The parallel planning process incorporates a unique blend of military expertise, research planning, and actual implementation of supporting humanitarian services in the field. It is a dynamic and continuous process of assessment, refinement, and dialogue among team members, reachback, and supported agency or unit. The ultimate goal is to ensure that research can be conducted as safely as possible in disaster or conflict environments and can maintain direction and actionability for the mission at hand.

Data Analysis

The analytical component of the HRE team is responsible for providing detailed data and security analysis based on the deployed team's needs. The analysis component introduces an additional level of personnel and time overhead compared to traditional ethnography, which must be incorporated into the research plan. The operations tempo and the availability of secondary information on the region will dictate secondary research time requirements. As a rough example, the AF1 deployment required a minimum of 6 hours of research and analysis by the stateside team for every hour of field operations.

Premission planning requires a great deal of analytical preparation for the specific area of interest in each HRE deployment. Operating alongside the military necessitates the integration of analytical tradecraft with social science research methodology to answer the HRE team's information needs. Regardless of the partner organization, however, the analysis team must develop an evolving data store including primary, secondary, and tertiary data sets to meet potential support requirements. Fusing current field-collected data with publicly sourced news information, historical data, and existing research forms the HRE team's holistic understanding of the environment as well as response or mitigation strategies for issues arising during the research period.

* HRE team members are trained in basic field medical skills and combat lifesaving. Medical training is a required core skill for all team members.

DATA SETS

Primary data: Primary data are collected by an ethnographer from surveys, interviews, and direct observations. Format: text, image, video, audio, or physical artifacts. They are cataloged using the date/time of collection, responsible ethnographer, informant(s), background situation, relevant atmospherics, and an identification element.

Secondary data: Secondary data are information from outside sources such as statistical data, peer-reviewed documents and articles, related observations, and reporting. They are used to validate, confirm, explore, and expand the content of primary data. They must be cataloged using the date interest, provenance, and an identification element.

Tertiary data: Tertiary data are the resulting information from the evaluation of primary and secondary data. They are used to present analytical results. They must be cataloged using date interest, analyst, provenance, and an identification element.

The HRE process also requires the generation and the management of additional data for validation purposes. The scope of these data will necessarily vary based on the research environment and the aims of each project, but AF1's data may provide an illustrative example. For roughly every 10 kB of primary data ingested by the AF1 analysis team, approximately 10 MB of secondary data and 20 MB of tertiary data were generated to validate and corroborate the collected primary data. In essence, the analysis to support primary data collection operations logarithmically increases the validated secondary and tertiary holdings for each mission. The process used to validate the data within the HRE team is an analytical methodology named recombinant data processing and analysis (RDPA). RDPA is a continuous evaluation and analysis cycle that consumes pre-, trans-, and postmission data to evolve information and understanding based on historic and emerging data to accurately reflect the state of the observed population throughout the mission cycle.

The security issues present in HRE research environments require the analysis team to perform a secondary function in providing overwatch or security monitoring.

Overwatch Activities

Concurrent with data management and analytical activities, a critical function of the data management component within the HRE team is providing additional mission overwatch,

RULE OF THUMB

Practical experience in HRE environments has demonstrated a correlation between the volume of primary data and the volume of secondary data necessary for validation. It is essential that the provisions be made to store collected data and relevant metadata. Data structures must make provisions for structured and unstructured data. The lack of available data storage will result in data loss and affect the analysis of primary and secondary analysis results.

OVERWATCH REQUIREMENT

HRE requires external monitoring to ensure the safety and the security of the HRE team. There are three main facets to overwatch activities: threat monitoring (threats such as hostile activity, epidemiological, and weather threats), logistics monitoring (logistical issues such as communications, travel, and sustainment), and health monitoring (specifically the health of deployed team members). HRE teams do not often have the resources in the field to maintain situational awareness.

i.e., supplementary security monitoring and planning. While the team leader and local partners track and ensure the security on the ground, the data management/analysis staff observes and tracks external threat data streams to provide additional warning and indication support to field personnel. These threat streams include information regarding weather; epidemiological, ground, and air movement threats; and direct or indirect hostile threats to the team. Potential threats are relayed to the ground team as conditions warrant, allowing the HRE to focus on the mission without direct regard to potential threats.

The security concerns and rapid review of data required for directed research in distressed environments necessitates a larger research team, tighter information sharing and integration with partners, and additional security procedures in comparison to traditional ethnographic fieldwork or sociological research. HRE protocols must be rehearsed, reviewed, and conducted with all relevant stakeholders to be successful. When the activities of the full HRE research team are combined, it results in the ability to perform rapid sensemaking within the operational environment that is valid, cross-referenced, and subject to review.

Social Complexity Data

The end products of HRE fieldwork are data and information that support the development of conclusions or recommendations for decision makers. SCD are first-hand information relayed by indigenous population sources on their perceptions, concerns, opinions, needs, hopes, prejudices, and attitudes. SCD require that data and information are validated and aggregated with existing data sets to enhance and augment humanitarian and development interventions. The data inform our knowledge about the society under distress.

Information and Data

The question of what SCD are, contrasted against traditional ethnographic data, is valid. HRE data are nominally the same as traditional ethnographic data with several caveats:

- HRE data collection is conducted in environments and during situations of high human stress with significant risks to the observer and the observed.
- Collection often occurs in time-compressed, limited-information, broad research-objective environments where lives are at stake.
- The research objective is normally in response to a stakeholder's need for immediate answers to complex, potentially intractable questions.

TABLE 7.2

HRE Data Categories

Data Category	Data Forms	Purpose	Creator
Primary	Field notes, planning documents, interviews, surveys, multimedia, and statistical annotations	The ethnographic mission record; forms the set of information developed by the ethnographer during the course of the mission	Ethnographer
Secondary	Documents, graphics, multimedia, database records, statistical results	Data and data elements used to describe, validate, and inform the ethnographer on the local population throughout the HRE mission cycle	Support team
Tertiary	Context renderings, geospatial data and rendering, relational graphs and a–b data, temporal graphs and rendering data, statistical analysis results	Formed from the integration of primary and secondary data as a result of analysis and data validation process	Ethnographer and support team

The pressures involved in an HRE mission may result in the introduction of conscious or unconscious data errors that can affect research outcomes, even beyond standard issues of subjectivity and bias. The majority of cognitive errors are the result of stressors involved during hostile actions directed at or surrounding the HRE, when ". . . temporary object representation can become chaotic, and information can be mixed up within and between object. . . . known as illusory conjunction" (Keebler et al., 2015*). Furthermore, the potential for cognitive bias is an ever-present risk regardless of the experience or the maturity of the ethnographer (Heuer, 1999). Therefore, the criticality and traceability of ethnographic data is a primary concern of the ethnographer and the support team.

The information and data produced by HRE include qualitative and quantitative data that are both structured (e.g., a database) and unstructured (e.g., a set of documents). The forms of data range from documents and delimited files to video, audio, and in some cases forensic data. There are three basic categories of information in HRE operations: primary data, secondary data, and tertiary data. Table 7.2 provides a general overview of these categories.

SCD provide the relevant foundational data and information for contextualization within the HRE operational environment. Historical information supports contextualization by allowing longitudinal understanding and permitting organic and dynamic analyses to continually inform the HRE and research team. SCD, when seen temporally, are defined by the historical context elements generated through research performed prior to the HRE deployment. Historical information is refined by research into the current situation's context.

The concept of a *near past occurrence window* aggregates the historical and current contexts to establish the basis for current observations. When the historical and current contexts are combined, they form the substantive components for determining potential near-future projections, estimations of what may happen in the short term based on historical and observed actions. The development of predictions from social science information is based on well-established methods commonly found in marketing and clinical science research

* While the reference addresses the identification of friend or foe, the concept has applicability to the HRE context in that complex observations and evaluations are performed during stressful situations.

DATA SENSITIVITY

The management of HRE and associated data incorporates numerous sources of information. There are a number of protocols that must be observed to protect the information. Some sources are classified and must be protected under the applicable security rules and procedures. It is critical that all social science informant information be protected as established by title 45 of the Code of Federal Regulations (CFR) part 46, the Common Rule. Each US government agency has a specific guidance.

See http://www.hhs.gov/ohrp/humansubjects/commonrule for more information.

that determine the potential for action when a target population experiences external stimuli. During the course of data analysis, "the process of machines learning from data … uncovers what drives people and the actions they take. … With the new knowledge gained, prediction is possible" (Siegel and Davenport, 2013).

In practice, SCD form an evolving information chain supporting decision- and policymaking to understand the local populations, the core of existing and emerging issues, the potential intervention strategies, and how local populations may respond to the intervention. The importance and the sensitivity of SCD require that the collected data and information must meet the highest academic and professional standards. Utilizing the National Science Foundation data standards (Office of Management and Budget, 2002) for data reliability, quality, validity, and integrity ensures that the HRE data meet or exceed scientifically acceptable data conventions.

Central to the development of SCD is the operations required to form raw data into usable data to support comprehensive relational, temporal, geospatial, and statistical data visualizations for use in the rapid sensemaking process. Data operations are a critical function within HRE operations, as they support the long-term viability of the data as well as the immediate safety of the HRE team. Data used for evaluations and conclusions must be auditable and available for peer review. A supplementary provision is the ability to evaluate HRE methods and techniques for efficacy and ethical conduct to ensure human research subject protection.

Data Management

The management of SCD is essential to the overall HRE effort. The volume of data generated can become overwhelming depending on the mission objectives and the level of effort. Primary data require processing, validation, and protection. Secondary data require processing, validation, and metadata instantiation. Tertiary data require generation, validation, and cross-referencing. Figure 7.1 provides a notional timeline for contextual data generation points that are inclusive within the data collection process. Of note, the ethnographer's field notes require collection, transcription, storage, and protection. The protection of field notes is paramount as the Common Rule applies to all government agencies and activities funded by the government funds* (Office of the Federal Register, 2015; US Department of Health and Human Services, 2009). This is a fundamental concern for data

* The Common Rule (Human Research Subject Protection) is codified under title 45 of CFR part 46 and specifically applies to each government agency, including the Department of Defense (DoD). The DoD Common Rule specifications are addressed as statute in title 32 of CFR 219 and as regulation in Department of Defense Directive (DODD) 3216.02. Violations can result in civil and criminal liabilities.

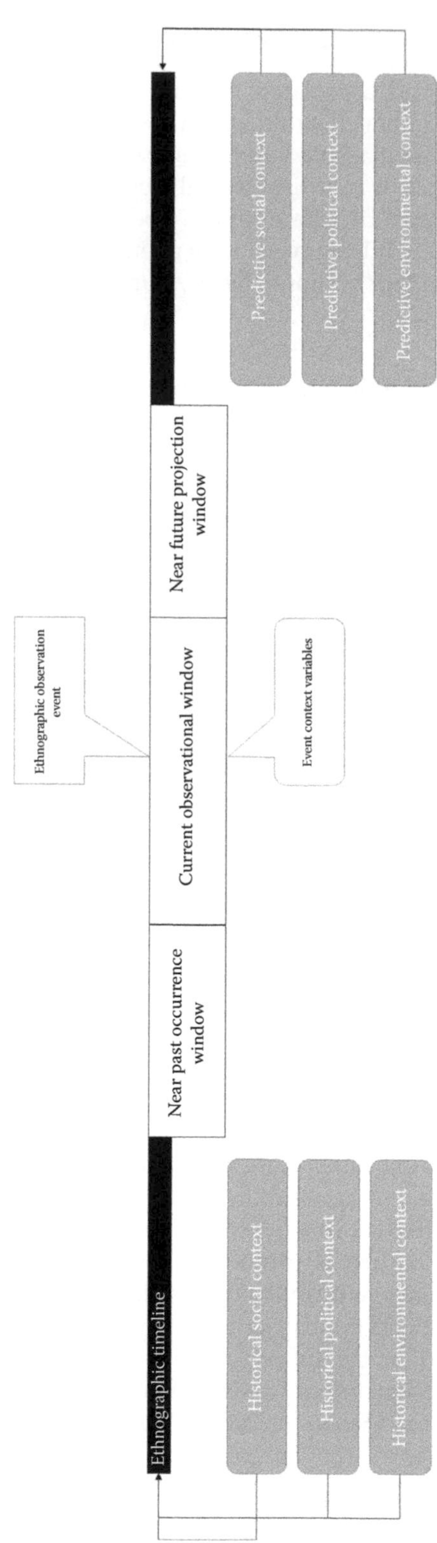

FIGURE 7.1
Research context timeline concept. (From Carter, M. A. 2015a. *High Risk Ethnography Operational and Technical Implementation*. Unpublished Work.)

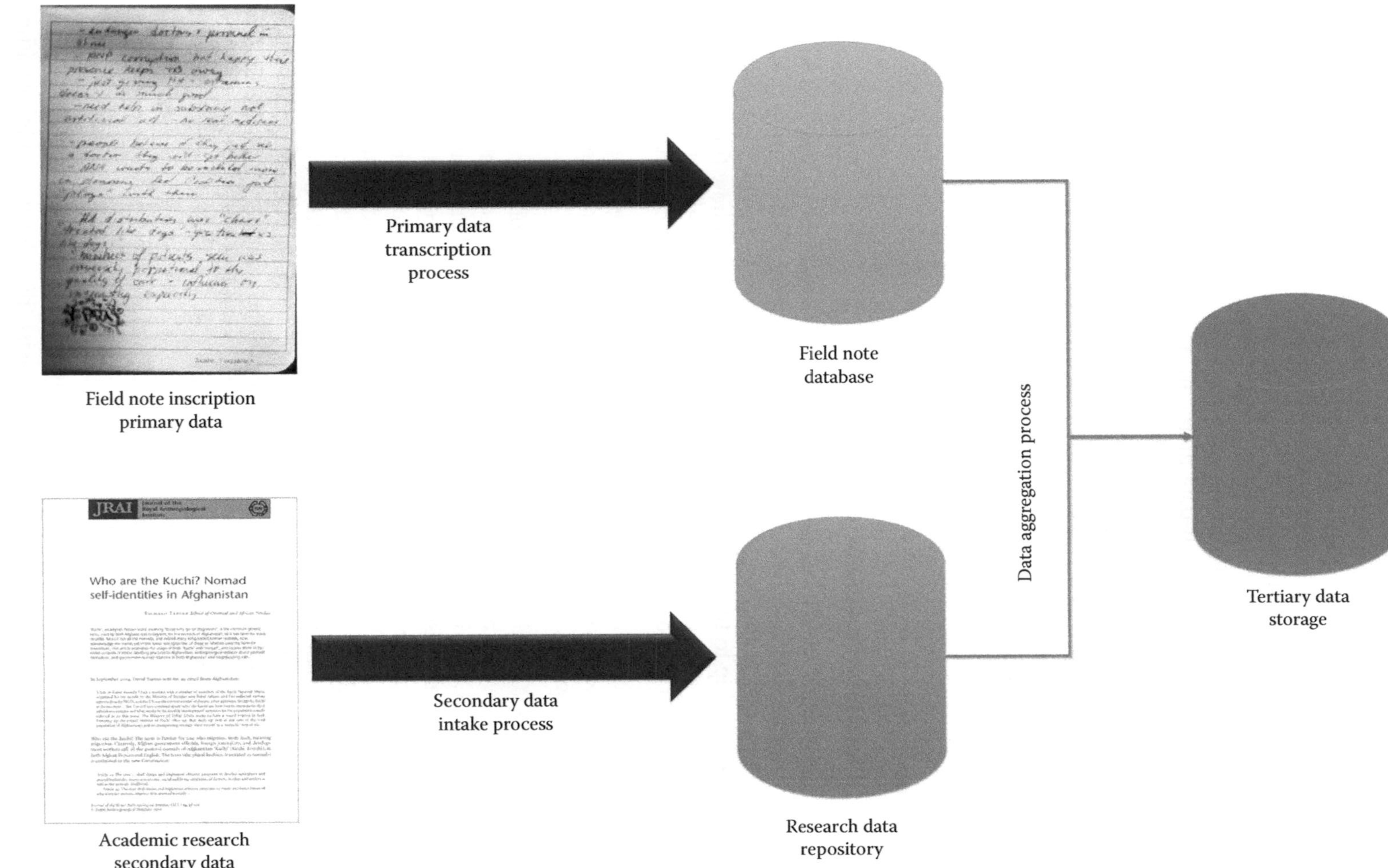

FIGURE 7.2
SCD storage framework. (From Carter, M. A. 2015b. *Visualizing Ethnographic Data Interactively*. Unpublished Work.)

management personnel and necessitates the use of data structures that can secure primary data, store secondary data and supporting metadata, and store aggregated tertiary data as part of larger frameworks for data storage and interaction. A general structure that supports this framework is depicted in Figure 7.2.

The implementation of an environment to store HRE data can be as simple as building a file structure or as complex as implementing a semantic context-capable data framework. There are three practical considerations related to data storage: scalability, replication, and hardware/software upgradability. Furthermore, a common set of operational procedures must be used and enforced so that all data maintain uniform structural elements such as metadata. The management of HRE data cannot be performed by a general user, as the ability to represent relevant context could be severely jeopardized. Furthermore, content reliability, validity, quality, and integrity can suffer enormous problems in accuracy, thereby affecting data utility as a whole. One current structure compatible with these requirements is the Civil Information Management Data Processing System (CIM-DPS) or the Distributed Common Ground System (DCGS).

Case Study: The Sixty Percent Mission

The unnamed mission in Col. Schweitzer's testimony, referred to as the *sixty percent mission* and described there for the first time publicly, was responsible for a 60–70% drop in kinetic activity (i.e., military action involving lethal force). This case study will highlight the HRE protocols, the relationship to the unit and the local population, and how this augmented the unit's understanding not only of the Afghan village's needs but also of the their own role and responsibilities in providing more effective medical and veterinary services.

NOTE: Due to security concerns and strict adherence to human subject research protections requirements, we will not divulge the location of the village or the identities of individual Afghan/US personnel.

In the late summer of 2007, AF1 was requested to participate in a major operation in eastern Afghanistan. The team was to embed with a unit providing medical and veterinary services to the local population. Our mission was to act as contributing members of the medical and veterinary teams in order to assess the unit's effectiveness in providing culturally relevant services and to develop long-term strategies for continuing these services with full participation of the Afghan partners. The village had critical medical needs, and the unit focused on providing sound medical care and medical supplies to avoid a humanitarian crisis.

During this mission, the composition of the core AF1 team included one anthropologist trained as a high-risk ethnographer, one team leader (former Special Forces officer with army medical training), one researcher in training, and two class 2 interpreters. The composition and size of the team is important since the team is fully integrated into the maneuver unit. AF1 was expected to work alongside the US and Afghan medical team as active participants in the operation while observing as ethnographers. It was critical to the HRE mission that the team brought years of experience working in similar environments and was fully qualified in combat environments.

AF1 was split into two components: the HRE and the female interpreter were assigned to assist the medical team in conducting a mobile medical clinic; the team leader, the interpreter, and the researcher in training were tasked to assist the veterinarian in providing basic care to livestock. The split teams addressed the gender division found in the Pashtun village. The assignment was clear for both teams: assess the medical and veterinary services and their effectiveness to meet the village's needs, and if the services were deficient, identify why the services were not successful.

The split team's methodology is the foundation of the basic HRE protocol. The protocol is iterative and most importantly not done in isolation of the field site. This distinction is critical. We acknowledge that no ethnographer can thoroughly understand all factors and variables affecting they events as they occur in a dynamic and changing conflict environment. For over 21 days, AF1 interviewed the members of the medical team and their Afghan counterparts, documenting the processes and procedures of the medical services during implementation while also interviewing male and female Afghan patients. At this point, the role of the team's data manager and analyst becomes essential in order to augment and contextualize the ethnographer's view where necessary.

The medical team found itself in intense negotiations with Pashtun's village doctor. He had originally accepted the presence of the unit's medical clinic, but due to increasing threats from the Taliban to himself and his village, he later refused to allow the medical operation to take place and wanted the unit to immediately leave the area. The AF1 team, through extensive interviews, observations, and analysis, found several key issues not previously known:

- Fact: The geography of the village isolated it during the winter from the main trading road, so people and goods could only move during certain months of the year.
- Fact: Because of this physical constraint, the village doctor was also a central purveyor of wood; this was critical to the economic well-being of the village and the outlying areas that were dependent upon wood as fuel during the winter months.
- Fact: The US medical unit's presence placed political and economic pressure on the doctor since Taliban retaliation had been threatened if the village did not continue trade with them.

Despite the best intentions of the medical team and the fact that the village desperately needed the services the medical operation could provide, it was perceived by the Afghan doctor and the villagers as a negative and dangerous event to the safety of the village. After discovering these complicating factors, the AF1 team worked with the medical team to negotiate an assessment of acceptable services. Working closely with the village doctor's wife and other women in the village, the female members of the AF1 team, along with the female medic were able to devise a culturally appropriate alternative: midwives' pre- and postnatal education.

Through the iterative HRE interview and analysis protocols, we were able to identify the key role of midwives in women and children's health. In keeping strict human subject protections in place and using sound ethnographic methodologies supported by full data analysis, we were able to understand the Afghans' needs on the Afghans' terms. By teaming with the Afghan women, the AF1 HRE and the unit medic established services that would not require the intervention and long-term presence of US military personnel. The village doctor in turn agreed to introduce the unit commander to the other village elders as a gesture of good will. During the time the unit remained in the village area, kinetic activity dropped. We do not claim that the medical unit and AF1 team's efforts directly caused the drop in kinetic activity, but the postmission analysis with Col. Schweitzer's staff did indicate a correlation between the unit's presence and the amount of kinetic activity during this time. Furthermore, the longitudinal review of the AF1 deployment using both HRE-collected and open source information has established pre- and postmission increases and decreases in violent activity surrounding HRE activities, although causality is difficult to determine.

It is important to understand that this observation was a team effort. The HRE team did not have the available resources on-site to evaluate the geospatial and intelligence situation as the event unfolded. Unit intelligence assets did not have the requisite information to inform the team. Furthermore, due to environmental constraints, it was difficult to communicate through traditional military channels. The data management/analytical staff maintained a 24/7 overwatch communicating information requirements and response through any means available, which included satellite phones, short message service text messaging through satellite uplink, locally purchased cell phones, and plain old telephone system when necessary. Where possible, data files and reports were transmitted, or if not possible, communicated verbally as necessary. The objective was to provide clear, concise, relevant, and critical information as needed and when requested by any means possible. The stateside staff had to assume the operations tempo of the HRE team and be available when and where necessary. Ultimately, these communication channels, combined with sound methodology and integrated teamwork, resulted in a successful operation.

Conclusion

The sixty percent mission and Col. Martin Schweitzer's subsequent testimonial regarding a drop of "60–70% kinetic activity" has been controversial and until now has not been publicly discussed. In this chapter, we have introduced the HRE protocol used during the preliminary AF1 deployment and presented an overview of the special protocols, field methodologies, data management, and analysis procedures that distinguish HRE from traditional social science. In the case study of the AF1 humanitarian mission, we demonstrated how and why Col. Schweitzer could make this assertion to the House Armed Services Committee. This methodology supports incorporating research into an organization's standard operations in order to address highly distressed environments while also addressing the common difficulties in assessing project and organizational effectiveness. HRE is a specialized interdisciplinary field science, and with a properly trained experienced and skilled team, HRE continues to provide sound research and analysis to ensure that humanitarian aid is logistically, culturally, and ethically effective for the populations in need.

Megacity Use Case

Minah, a 17-year-old Razinia refugee cries out in pain as the field nurse tries to dress the infected wound on her foot. Minah, her 7-month old baby boy, her elderly mother, and her 12-year-old brother made the dangerous journey across the border to neighboring Xanadu. Like many Razinia women, Minah and her mother both lost their husbands in the 10-year civil war. As widows in their country, they were shunned and constantly lived in threat of starvation. Minah's brother, Himal, supported the family by selling metals he recovered in the massive garbage heap next to their village. After their village was overrun by violent combatants and their meager home burned, Minah and her family along with the surviving villagers fled for what they hoped would be a better life in Xanadu. An HRE prepares the antibiotics and the bandages as the nurse cleans and dresses the wound. The other HRE team member who is fluent in Minah's dialect quietly explains to Minah how often to change the bandage.

The HRE team NW1 (North Ward Refugee Camp 1 [NWRC1] Xanadu) has placed a four-person team with a world-renowned medical nongovernmental organization (NGO). The medical NGO has established a small field clinic in NWRC1. The camp is run by the United Nations High Commission for Refugees (UNHCR), and despite overcrowding, limited resources, and simmering social tensions, the field clinic still manages to provide basic medical services to the refugees.

Prior to arriving in the North Ward Refugee Camp, the HRE team spent several months reviewing all available information to understand the volatile situation in the megacity and in Xanadu. The HRE analytical team researched available information on the history of the area, ethnographic studies, geographic data, economic data, demographic information, security studies, political information, and religious practices, as well as weather and health studies gleaned from open, closed, and social media sources. The information was aggregated into a set of basic reports that provided a background for the field teams to study during premission preparation. The data and information were stored in an internet-accessible data storage structure so that the field team could access them when needed. Supporting a long-term analysis, a list of information gaps is prepared for the field team to provide as the need becomes available.

As part of the parallel planning process, the team outlines specific protocols, human subject protections, logistics, and security measures to meet the unique requirements for extensive on-site fieldwork in a dangerous and high-risk environment. The analytical team provides operational overwatch, evaluating news and social media streams, assessing if the field team has drawn the attention of hostile actors. If a risk is observed, the analytical team advises the field team of the threat and the potential for aggressive activity, so that the field team can identify contingencies and possible COAs. When the analysis and study is completed, the HRE team has developed a working knowledge of the security situation in the operating area.

Working closely with the medical NGO, the HRE team leader coordinates and establishes the roles and mission of the HRE team. As in any high-risk field scenario, the highest priority in equal measure is given to the safety and the security of the team, supported organization, and local population served. The HRE team leader ensures that HRE team NW1 is medically trained and fully prepared to support the NGO and the UNHCR. The research protocols will be secondary to providing a fully functioning HRE team to the NGO's field clinic and medical mission.

As part of their field duties, NW1 will interview incoming patients on their life narratives in conjunction with their medical needs. Information about the interviewed patients will be stored in accordance with the Common Rule guidance supplemented by more stringent HRE human subjects protection protocols. No identifying or sensitive information will ever be stored and in some cases is not documented. This is to assure that the first principle of ethical social science, "Do No Harm," is abided by as humanely as possible.

The primary purpose of the research is to enhance the NGO's understanding of the conditions and issues that affect their ability to provide meaningful medical services. The

HRE team, working with specific patients/refugees such as Minah and her family, will gain a more in-depth understanding of the social forces at play in the camp and what small-scale humanitarian interventions can be done to better the lives of refugees such as Minah and her family. The secondary purpose is to add to the body of ethnographic knowledge to inform the scientific community as well as governmental and quasigovernmental decision makers.

DISCUSSION QUESTIONS

1. If you were going to deploy an HRE team from your organization, how might you integrate the HRE process with your organization's existing procedures? How could you adopt your organizational processes and requirements to the rigorous human protection protocols required?
2. Is it permissible to store identifying information concerning social science informants interviewed or observed during an HRE mission on an enterprise or standalone storage architecture? If yes, why? If no, why not? What risks could the data be exposed to?
3. What potential ethical concerns should an investigator be aware of prior to, during, and after an HRE mission with regards to information derived from observations or interactions with the population being studied?
4. Why is the overwatch function of a reachback team important? What sources of information can support the overwatch activities? Do these sources of information have potential reliability problems and why?
5. Is it important for reachback analysts to validate primary ethnographer inscriptions? Why is it important? Are there measures that could be instituted to prevent problems from occurring during the data validation process?
6. What is your understanding of contextual analysis? What are the opinions of others in your research group concerning contextual analysis?

References

Carter, M. A. (2015a). *High Risk Ethnography Operational and Technical Implementation*. Unpublished Work.

Carter, M. A. (2015b). *Visualizing Ethnographic Data Interactively*. Unpublished Work.

Heuer, R. J. (1999). *Psychology of Intelligence Analysis*. Washington, DC: Central Intelligence Agency. Retrieved from https://www.cia.gov/library/center-for-the-study-of-intelligence/csi-publications/books-and-monographs/psychology-of-intelligence-analysis/PsychofIntelNew.pdf.

Keebler, J. R., Sciarini, L. W., Jentsch, F., and Nicholson, D. (2015). *A Cognitive Basis for Friend-Foe Misidentification in Combat*. Orlando, FL: University of Central Florida. Retrieved from http://www.researchgate.net/publication/239612283_A_Cognitive_Basis_for_Friend-Foe_Misidentification_in_Combat.

King, C. (2015, July 6). The decline of international studies. Retrieved July 6, 2015, from https://www.foreignaffairs.com/articles/united-states/2015-06-16/decline-international-studies.

Kurtz, C. F., and Snowden, D. J. (2003). The new dynamics of strategy: Sense-making in a complex and complicated world. *IBM Systems Journal*, *42*(3), 462–483.

Mills, J. H., Thurlow, A., and Mills, A. J. (2010). Making sense of sensemaking: The critical sensemaking approach. *Qualitative Research in Organizations and Management: An International Journal, 5*(2), 182–195.

Office of Management and Budget. (2002). Guidelines for ensuring and maximizing the quality, objectivity, utility, and integrity of information disseminated by federal agencies. *67*(36), 8452-8460. Retrieved from http://www.gpo.gov/fdsys/granule/FR-2002-02-22/R2-59.

Office of the Federal Register. (2015). Protection of human subjects. *Code of Federal Regulations, 32*(219). Retrieved from http://www.ecfr.gov/cgi-bin/text-idx?tpl=/ecfrbrowse/Title32/32cfr219_main_02.tpl.

Siegel, E., and Davenport, T. H. (2013). *Predictive Analytics: The Power to Predict Who Will Click, Buy, Lie, or Die*. New York: Wiley.

Snowden, D. (2011). Naturalizing sensemaking. In Mosier, K. L., and Fischer, U. M. (Eds.), *Informed by Knowledge: Expert Performance in Complex Situations* (pp. 223–234). New York: Taylor & Francis.

US Department of Health and Human Services. (2009). Protection of human subjects. *Code of Federal Regulations 45 CFR 46*. Retrieved from http://www.hhs.gov/ohrp/humansubjects/guidance/45cfr46.html.

US House of Representatives. (2008). *Statement Of: Colonel Martin P. Schweitzer Commander, 4/82 Airborne Brigade Combat Team United States Army Before The House Armed Services Committee, Terrorism & Unconventional Threats Sub-Committee and the Research & Education Sub-Committee*. Washington, DC: 110th Congress, 2nd Session.

Weick, K. E. (1995). *Sensemaking in Organizations*. Thousand Oaks, CA: Sage.

8

Challenges in Connecting with the Disconnected: An Introduction to Connecting with Communities Disconnected from the Rest of the World

Christina S. Kang

CONTENTS

Introduction 136
The Need to Connect 137
Implementation Challenges 138
Challenge 1: Limited Resources 138
Case Study: Victims of Forced Labor Face Challenge of Limited Resources 138
Discussion on Limited Resources 140
Challenge 2: Poor Infrastructure 140
Case Study: Nepalis Face Poor Infrastructure after 2015 Earthquakes 140
Discussion on Poor Infrastructure 143
Challenge 3: Harsh Physical Environments 144
Case Study: Lord's Resistance Army-Affected Regions Face Harsh Physical Environments 144
Discussion of Harsh Environments 145
Challenge 4: Unstable Sociopolitical Environment 146
Case Study: Afghans Face Unstable Sociopolitical Environments 146
Discussion of Unstable Sociopolitical Environment Factors 147
Challenge 5: Other Sociocultural Differences 148
Case Study: One Earth Designs Faces Sociocultural Differences from Their Nomadic Community Users in the Himalayan Plateau 148
Discussion of Other Sociocultural Differences 149
Summary 149
Ethical Concerns 150
Concern 1: Poor Security and Safety 151
Concern 2: Unintended or Negative Influence 151
Concern 3: Bystander Responsibility 152
Concern 4: Exploitation 152
Concern 5: Poor Data Quality and Validation 153
Concern 6: Data Ownership and Access 154
Concern 7: Privacy 154
Summary 155
Road Forward 157
Megacity Use Case 157
Limited Resources 157

Poor Infrastructure 158
Harsh Physical Environment 158
Unstable Sociopolitical Environment 158
Other Sociocultural Differences 159
References 160
Suggested Reading 162

ABSTRACT Disconnected communities hold valuable information that can benefit a range of sectors, from counterinsurgency to profit making. Despite a plethora of communication modalities at our fingertips, it remains difficult for the connected world to effectively and efficiently communicate with these disconnected communities due to both the implementation challenges and the ethical concerns. This chapter introduces a framework for understanding the implementation challenges faced across sectors, domains, and geographies, and it offers examples of how different organizations have overcome these challenges. A framework around the potential ethical concerns is also introduced and discussed in brief.

KEY TOPICS

- The desire to connect with disconnected communities spans sectors and motivations.
- Challenges include resources, infrastructure, physical/sociopolitical environments, and sociocultural differences.
- Various ethical concerns are considered, including bystander responsibility, exploitation, and privacy.

Introduction

Type a message.

Press send.

Receive a message.

Respond accordingly.

In today's global communication environment, you are probably one of many that take this two-way communication pathway for granted (and may pound your head against the wall when there is just a minute of lag). Our options for reliable communication pathways range from basic face-to-face chats to text message flood warnings and video teleconferencing. Even the headlines from programs working with vulnerable communities boast that everyone now has a cell phone, but as the following statistics will show, we are far from being a fully connected world.

The disconnected comprise a broad range of communities cut off from the rest of the world for various reasons. This population includes an estimated 700 million people in poverty (Cruz et al. 2015), 21 million victims of human trafficking (International Labor Organization [ILO] 2012), and 148.2 million people are affected by natural disasters and conflict (Barajas and Design 2015). While the disconnected span diverse geographies and cultures, many share common obstacles involving limited access to resources (including, but not limited to, information) and differing sociocultural norms and values, which in turn create implementation challenges for connecting with them. In addition, limited access makes many of these communities particularly vulnerable, bringing rise to ethical concerns when trying to communicate with these communities for a purpose that may benefit the initiators.

Viable modalities for connecting with populations may include in-person discussions, paper/posters notices, radio, video, short message service (SMS), voice call, interactive voice response, smartphone apps, and web-based forms, but even when they are available, applying these tools to disconnected communities becomes an art as well as a science. More formally said, it requires an interdisciplinary, multimodal, experimental approach.

This chapter studies different implementation challenges to consider when connecting with the disconnected. It explores organizations on the cutting edge of this challenge, denoting the implementation challenges they faced and how they overcame them. Finally, the chapter closes with a brief discussion of ethical concerns to keep in mind.

The Need to Connect

The need to connect with the disconnected has become a problem across sectors, countries, and cultures. From information operations for counterinsurgency to market research for business, and from data collection for an academic study to monitoring and evaluation of developmental programs, every sector has a need to connect with these vulnerable communities who are often disconnected from the rest of the world by technological or social constraints. A short list of reasons for connecting is as follows:

- Data collection for the following:
    - Citizen journalism (e.g., Twitter)
    - Academic research (e.g., Center for Civic Media)
    - Monitoring and evaluating a development program (e.g., United States Agency for International Development [USAID])
    - Market research (e.g., Nielsen)
    - Needs assessments (e.g., USAID)
    - Reporting warnings/alerts (e.g., Invisible Children, LaborVoices)
    - Identifying and prioritizing areas and people (e.g., disaster relief)
    - Government intelligence
- Influencing for the following:
    - Military information operations
    - Adjusted behavior for self-improvement, such as health programs
    - Education of both academic and life skills
    - Marketing of both commercial and social products and services
    - Postdisaster actions

Despite the breadth of reasons for connecting, the implementation challenges and the ethical concerns remain universal. Let us delve deeper—first into the implementation challenges.

Implementation Challenges

Limited resources, poor infrastructure, harsh physical environments, unstable sociopolitical environments, and sociocultural differences represent common challenges. The following table (Table 8.1) summarizes these challenges.

The following are a few real-world examples of organizations trying to connect with disconnected communities. As the examples show, the solutions they devised not only require technology but also a deep understanding of the communities themselves, creative incentive mechanisms and appropriate methodological designs. After they are presented, we will explore the lessons learned from these examples and discuss how these organizations overcame the challenges. The more generalizable points from these examples will inform the model presented later in this chapter.

Challenge 1: Limited Resources

Case Study: Victims of Forced Labor Face Challenge of Limited Resources

Meager wages, 18-hour workdays, lack of healthcare, hazardous conditions, and debt bondage—almost 21 million people are victims of forced labor, often facing these kinds of conditions and suffering violations against their basic human rights. Often by the mere misfortune of being born into extreme poverty, these individuals are suppressed and exploited for the profit of others—an estimated US$150 billion in profits every year (ILO 2012). Some of the more fortunate victims have access to cell phones with SMS and voice call capabilities,

TABLE 8.1

Framework of Challenges for Connecting with the Disconnected

Challenge	How It Affects Connection	Example Facing This Challenge
Limited resources	Resources enable connection (e.g., finances, technology, education)	Garment factory workers unable to afford smartphones or credits on feature phones
Poor infrastructure	Infrastructure allows the technology to function	Nepal 2015 earthquake collapsed the existing power and IT infrastructure
Harsh environment	The environment erodes or inhibits the viability of the technology and the infrastructure	Areas affected by the LRA are covered in dense forests and vast savannahs across a wide geography
Unstable sociopolitical environment	The sociopolitical environment affects the technology, the infrastructure, the community trust, and the security in intense and unpredictable ways	War in Afghanistan over many years, involving penetrating conflict and violence
Other sociocultural differences	Sociocultural differences between the connected and the disconnected create barriers to understanding	International NGO connecting with a nomadic Himalayan community to improve air quality and health

but they are blocked from seeking help or reporting labor violations due to digital or language illiteracy, lack of awareness about the options or the reasons for reporting a violation, sociocultural aversion to reporting to an unknown source, potential repercussions from the violators, limited accessibility to reporting options, and lack of incentives to speak up about their poor conditions. This leads not only to labor trafficking and violations but also to massive casualties such as those documented in the April 24, 2013 collapse of the Rana Plaza factory complex in Bangladesh. While factory audits are intended to catch many of these problems before they become a catastrophe, they occur sporadically and rarely incorporate worker input (Leiber 2013; Desai 2014; Kohl Gill, personal communication).

LaborVoices, a tech startup based in Silicon Valley, is overcoming many of these challenges through their approach to data collection from workers, particularly workers from garment factories and agricultural industries who have access to a phone. While founder Kohl Gill's initial motivation for this effort was empowering workers and addressing human rights violations, LaborVoices is not philanthropy; rather, they market their products as an early warning system for large corporations to avoid catastrophes leading to bad press. Their clients receive access to an online dashboard with information about their factories, which alerts them of workplace issues before they become emergencies. Simultaneously, this system gives workers a voice. Through extensive and iterative user research with the workers, LaborVoices has identified the challenges that the workers face in reporting issues, and in turn, has designed data-collection campaigns appropriately. They use an agile approach, adjusting their efforts based on real-time data (Leiber 2013; Desai 2014; Kohl Gill, personal communication).

Through the user research, LaborVoices found that the main obstacles to collecting data from workers on labor conditions and violations (traditionally through hotlines) were the following:

- Unreliability of existing resources (e.g., busy tone when multiple people called in)
- Lack of awareness on what to report on and how (or to whom) to report
- Unavailability during hours when the workers were free
- Cost (e.g., lack of smartphones and finances for voice calls and SMS on feature phones)
- Limited or no connection with the companies using the factories or the auditors who often post their contact information on websites that the workers cannot access or are unaware of

To overcome these challenges, LaborVoices developed their SmartLine platform with the following features that enable workers to report labor violations and other issues such as building damage, fire safety issues, and minor but important modifications that would greatly improve working conditions:

- Inbound and outbound calls with persistent user profiles
- Ability to handle simultaneous calls (i.e., no busy tone)
- Survey, information bulletin, and grievance tools
- Anonymous reporting
- 24/7 availability
- Free calls for the caller

- Iterative approach with adjustments in the connection strategy based on real-time data
- Expansion of the platform to other use cases, including those that make the data transparent and directly beneficial to the workers themselves

In the first two weeks after its launch in Bangladesh, LaborVoices received over 600 calls from about 200 unique users, a majority of whom had heard of the system by word of mouth. By the end of 2015, LaborVoices had reached nine countries and several tens of thousands of workers. Gill says that the data are "anonymized, aggregated across workers, and vetted for accuracy" both to protect the workers from negative repercussions and to provide higher-quality data to the companies. Since their initial launch, LaborVoices has expanded their platform to other use cases, such as offering crowdsourced market information to choose the best employers. As their slogan explains, LaborVoices helps solve two global problems with one elegant solution: (1) transparent supply chains for corporate brands and (2) a voice for global workers (Leiber 2013; Desai 2014; Kohl Gill, personal communication).

Discussion on Limited Resources

The challenges LaborVoices faced in connecting with garment factory workers are not unique. Technology and (lack of) resources needed to obtain and use it directly affects the way a community can communicate. Although technology penetration is advancing at a rapid rate, 4.4 billion global citizens remain disconnected from the Internet, and of these, 3.2 billion (or 75% of those offline) are concentrated in just 20 countries, as seen in Figure 8.1 (McKinsey & Company 2014). Even mobile subscriptions are limited to only half the world's population. This represents a vast improvement from the 1990s (during which only one in five persons had mobile access) but still far from 100% (Groupe Speciale Mobile Association 2015). Furthermore, between 1.1 and 2.8 billion individuals "cannot get online via the mobile network because they do not live within sufficient mobile network coverage" (McKinsey & Company 2014). A segment of this population are too young to be connected, are choosing to remain disconnected, or connected via others around them, but this figure still poses a problem to those that want to connect with the disconnected population.

Limited resources mean more limited technology. According to a 2014 study by McKinsey & Company, "most of the world's population, some 64 percent, live in rural settings, where poor infrastructure, healthcare, education, and employment, impede Internet adoption, the study says. In India, for instance, roughly 45 percent of the population lives without electricity, making Internet access all the more unthinkable. [This affects education as well.] Roughly 900 million people who aren't connected to the Internet are illiterate." This study goes on to discuss the barriers that limited resources, incentives, and user capability cause for internet access.

These challenges affect a large portion of the disconnected. As the LaborVoices case study exemplified, careful user research, user-appropriate design, and agile and local partnerships are key to overcoming such challenges.

Challenge 2: Poor Infrastructure

Case Study: Nepalis Face Poor Infrastructure after 2015 Earthquakes

The 7.8 magnitude earthquake on April 25, 2015, in Nepal followed by a 7.3 quake on May 12, 2015, is a recent, painstaking example of how poor infrastructure can pose major

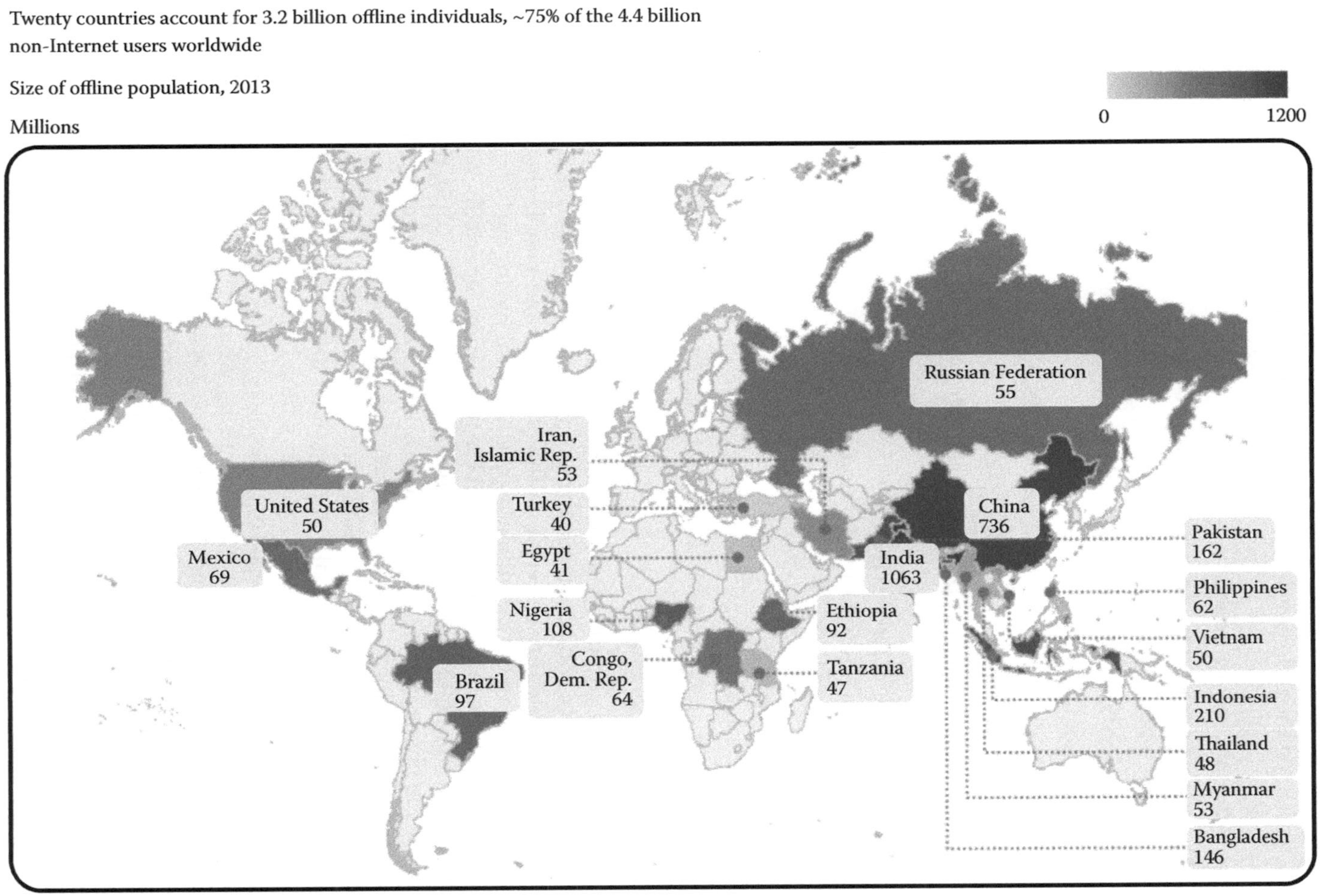

FIGURE 8.1
Map of non-Internet users worldwide in 2013 (McKinsey & Company 2014).

LIMITED RESOURCES: KEY PRINCIPLES

In addition to the barriers described by McKinsey's report, other challenges—many of which affected the factory workers in the example earlier—include the following:

- Low income and affordability
    - Low income or consumer purchasing power
    - Total cost of ownership for device
    - Cost of SMS, voice, and/or data plan (many rely on prepaid credits)
    - Consumer taxes and fees
    - Cost of infrastructure access such as electricity
- Low awareness and user-centric design
    - Low awareness of appropriate technology or relevant use cases to be connected
    - Lack of relevant content and services (e.g., language, culture, examples, race)
    - Sociocultural obstacles to adoption
- Limited education
    - Low digital literacy
    - Low language literacy
    - Low awareness on reasons to connect (e.g., obtaining resources and safety)

challenges to collecting and disseminating information—in this case, on the breadth and depth of damages, areas, and people requiring immediate and long-term attention, scope of the problem, and matching of people to the resources and people they need. Despite an estimated 23 million of the 27 million people in Nepal having mobile phone subscriptions (Ncell et al. 2015) and a reported 86% of the population reportedly relying on mobile phones to communicate (Nepal Telecommunications Authority 2015), the earthquakes caused many cell phone towers and power lines to collapse, disconnecting the Nepalis and their information technology (IT) to various degrees.

In response to the 2015 earthquakes, Kathmandu Living Labs (KLL), a Nepali nonprofit technology company, launched QuakeMap.org. KLL took a multimodal approach with QuakeMap, allowing data submission through a broad array of technologies including SMS, e-mail, smartphone app, web form, OpenStreetMap, Twitter, and Ushahidi. This multimodal approach enabled users to report using whatever modality and connection they had available to them at the time. Other local NGOs and community members set up stations with ham radio (also called amateur radio) to connect and transfer vital information where other connections could not be achieved (Geuss 2015). KLL worked closely with third-party data collectors, such as foreign aid workers or local citizens who had maintained or obtained network access (e.g., via satellite, portable backpacks, generators). These third-party collectors could submit reports on damage caused by the earthquake, people trapped, blocked roads, help wanted, shelter areas, and distribution areas, and their inputs would update the QuakeMap (Forrest 2015).

KLL also collaborated with international organizations, including Digital Globe. Digital Globe collected and analyzed geospatial data to supplement local data collection. Through crowdsourcing (identifying differences via before/after geospatial imagery), people around the world contributed to the data collection effort, documenting the impact of the disaster and where to prioritize response efforts. KLL and Digital Globe recognized that connecting with these communities was urgent but difficult through more traditional means. Thus, they used unique strategies to connecting through environmental imagery and crowdsourcing (DigitalGlobe 2015).

In addition to the major challenges of chaos and casualties, rubble, and overall destruction postearthquakes, organizations trying to connect with Nepalis faced a myriad of poor infrastructure challenges, including the following:

- Loss of data connectivity
- Many collapsed cell phone towers
- Limited Internet
- Power loss
- Information chaos with limited organization of how to rebuild/reconnect

KLL and Digital Globe overcame these challenges through creativity and collaboration, connecting with the affected populations and their environments to quickly and accurately prioritize efforts and save lives. Key features in their approach included the following:

- Multimodal options for users with different levels of connectivity and access to submit information (i.e., technology does not equal connection—poor infrastructure can restrict connection even when the technology itself is accessible)
- Third-party mechanisms like Twitter and Ushahidi that have established connection platforms and presence in the communities
- Collaborations with local and international volunteers, organizations, and communities
- Crowdsourced geospatial analysis to reach the areas human voices could not

KLL reported that the Nepal Army "used 300 QuakeMap reports marked as 'critical' for prioritizing and planning the rescue operations and helped many other partners prioritize disaster relief" (KLL 2015).

Discussion on Poor Infrastructure

Nepal postearthquake provides an obvious example of poor infrastructure, but many communities face limited technology infrastructure every day, either due to acute incidents or pervasive lack of development. Many communication modalities require electric and IT infrastructure (e.g., cell towers, satellites, radio towers) or portable/temporary alternatives to properly function. Poor infrastructure development, or acute conflict and natural disasters like the Nepal 2015 earthquakes, can cause infrastructure to collapse or lose functionality, crippling the technology and the connectivity. With over 148.2 million people affected by natural disasters or conflict (often causing existing infrastructure collapses [Barajas and Design 2015]) and over 1 billion people without access to electricity (World Bank 2012), poor infrastructure remains a critical challenge.

Infrastructure development such as building new cell towers or even more innovative solutions like Google's Project Loon ("a network of balloons traveling on the edge of space,

POOR INFRASTRUCTURE: KEY PRINCIPLES

Built upon the 2014 McKinsey study, the following infrastructure barriers can be generalized from the Nepal earthquake example:

- Limited network coverage
- Limited adjacent infrastructure (e.g., electricity)

designed to connect people in rural and remote areas, help fill coverage gaps, and bring people back online after disasters" [Google 2015]) are potential long-term solutions, but relying on the modes of communication that the community already uses, combined with multimodal options that can be affordably and quickly launched and maintained, can be the critical component to effective connection despite poor infrastructure.

Challenge 3: Harsh Physical Environments

Case Study: Lord's Resistance Army-Affected Regions Face Harsh Physical Environments

Vast mountains, deserts, thick forests—geographic challenges are deeply embedded in the landscape and often difficult to change, inhibiting not only the movements of people but also the movement of information and ideas. The Lord's Resistance Army (LRA) was responsible for over an estimated 100,000 deaths and 2.5 million civilian displacements between 1987 and 2012 across Central Africa (United Nations [UN] Security Council 2013). The LRA became renowned for its ability to effectively navigate rough terrain, vast savannas, and dense forests—areas that inhibit observation and also block network signals, creating a physical challenge to connect or track its members. Some of these areas were "almost entirely inaccessible with no high frequency radios or mobile phone service, making it difficult for locals to report attacks, and enabling the guerilla-style LRA forces to operate with little interference" (Agger 2012).

Invisible Children and the US Military collaborated on their efforts to connect with the disconnected individuals in the LRA and facilitate safe defections. These organizations used creative multimodal approaches including aerial leaflet drops, aerial loudspeakers, shortwave radio broadcasts, mobile cinema displays, establishment of "safety zones in which LRA combatants [could] seek amnesty and support" (Maclin 2015), and even leaflets nailed onto trees in the dense forests where LRA members were thought to traverse (Lombardo 2014). Messaging centered on how to safely surrender (e.g., with directions to safe reporting sites) or promised to help individuals look for missing family members and children. Fliers often contained photos of recently escaped LRA defectors with their families to encourage others who wanted to leave but either did not know how to surrender securely or worried if their families were still alive and safe (Vandiver 2013).

In parallel, Invisible Children collaborated with The Resolve LRA Crisis Initiative and with the US Military to collect timely data from the affected communities about the situation on the ground, including LRA activities, attacks, recent abductions (in hopes of quickly rescuing victims), LRA capacities, and supply routes. To collect daily data from communities across these harsh and diverse physical environments, Invisible Children set up an early warning radio network with both existing and donated high-frequency (HF) radios in affected communities. They trained local community members as radio operators, educated local members to report LRA activity to the HF radio operators, and scheduled time twice daily for the radio operators to report LRA activity to the Dungu and Obo hubs in the Democratic Republic of the Congo and the Central African Republic, respectively. The hubs captured this information in Excel spreadsheets for coding and dissemination (Ross 2013).

These organizations faced many challenges connecting with LRA members and local communities, including the following challenges related to the physical environment:

- Dense forests
- Rough terrain
- Vast savannahs
- Mountains
- Diverse geographies and climates spanning across four countries

While the fight against LRA continues, several lessons emerged from this example. The Invisible Children, The Resolve LRA Crisis Initiative, and the US Military used intersectoral collaboration, environmentally appropriate design, close engagement with the local communities, and careful user design to improve the quality and the quantity of their connections as demonstrated by the increased number of defections and reports of LRA activity. Key features of their solution include the following:

- Multimodal information distribution
- Iterative approach
- Expansion of the connection capabilities to multiple use cases
- Collaborations with local and international volunteers, organizations and communities
- Basic technology (e.g., HF radios, paper postings)
- Education, empowerment, and close collaboration with members of the communities they were trying to connect with

According to Invisible Children, the LRA crisis tracker was used to report 3182 LRA-related incidents, a reported 80% increase from before its existence (Invisible Children 2014).

Discussion of Harsh Environments

Harsh and changing physical environments make it difficult to develop and access both power and IT infrastructure. Populations are also generally smaller in these environments, making them a lower priority for many development efforts. Even when infrastructure does exist, extreme temperatures, climates, and topographies prove to be hostile environments for the technology system, like those in the areas affected by the LRA.

As exemplified by Invisible Children and the LRA crisis tracker, solutions using basic technology, close collaborations with the local communities who know how to navigate the rough and harsh terrain, and intersectoral collaborations to increase the resources can enable effective communication to both directly and indirectly collect data in harsh environments where more traditional data collection means may not be possible.

HARSH ENVIRONMENTS: KEY PRINCIPLES

Harsh and changing physical environment barriers include:

- Rough terrain (e.g., vast mountains, deserts)
- Extreme weather conditions (e.g., temperature, precipitation, dust storms)
- Environmental factors harmful to connection technology and infrastructure (e.g., excessive dust/sand, high humidity)
- High variability (e.g., drastic and unexpected changes in the barriers)

Challenge 4: Unstable Sociopolitical Environment

Case Study: Afghans Face Unstable Sociopolitical Environments

Apps (2015) reported that 92,000 Afghans were killed; 3495 coalition soldiers were killed; and $1 trillion in US taxpayer funds were spent during the war in Afghanistan. There is no question that the Afghanistan war, the longest overseas conflict in American history, as of 2015, has affected the lives of many (Dyer and Sorvino 2014).

Data and information have played a crucial part on the efforts in Afghanistan. Militaries, other government organizations, NGOs, academics, corporations, and journalists have strived to influence behavior through information and to collect data from the local communities about construction plans, clinic locations, terrorists, farmers, and more. As the UN Population Fund (2012) put it, "For nearly three decades, the Afghan population has been struggling to survive a myriad of social, political, economic and health problems. Getting reliable data has been a serious challenge due to inaccessibility and insecurity of certain regions. The lack of reliable and current population data has added to the impediments to development planning in Afghanistan."

Beer for Data was the de facto tagline for the effort Dave Warner undertook in Afghanistan to collect information about anything and everything—in exchange for beer (through his own bar, the Taj) to those who provided it. The information was then made open, accessible by anyone who asked for it. Collected information ranged from construction plans to election polling sites and health clinic locations. Warner also experimented with a variety of methods to crowdsource valuable information from the public, focusing on modalities they already used and trusted, such as local radio stations and starting with innocuous information requests (e.g., "Who is your favorite cricket player?") before delving into more valuable requests (e.g., "What is the price of milk today?"). Warner met face to face to develop personal relationships with the local community members throughout Afghanistan, and his methodology was replicated and expanded upon by other companies after Warner left Afghanistan (Calvert 2013). A 2012 study by the RAND Corporation found that "the biggest [psychological operations (PSYOPS)] successes have been in the area of face-to-face communication, . . . key leader engagements, and establishing individual relationships with members of the Afghan media, supporting Warner's efforts" (Munoz 2012).

Coalition forces and NGOs in Afghanistan faced a myriad of challenges, including those specific to the unstable sociopolitical environment, including the following:

- Mistrust of coalition forces and international NGOs
- Fear of retaliation from insurgency groups
- Mistrust of safety of digital communications (mainly oral communication tradition)

Warner addressed these issues through a creative model with features including the following:

- Multiple modalities
- User-appropriate incentives
- Safe, comfortable environment
- Community immersion
- Collaboration with local radio stations
- Anonymized reports
- Open access to anonymized information
- Intersectoral collaboration and data sharing

RAND also recommended the following features for improving the effectiveness of information and psychological operations, in particular, that can be applied to other efforts connecting with communities that face unstable sociopolitical environments (Munoz 2012):

- Subject matter expert input to define best practices from people who have experience with the communities
- Local focus groups and public opinion surveys
- Piloting of messages to pretest the effect before the full launch
- Collaboration with an established leader or communicator
- Harmonization between disparate efforts

The general principles that Warner used can be applied to data collection across cultures and geographies: understand the local communication mechanisms, interests, and challenges and develop data collection efforts around them. Warner collected a wide breadth and depth of data through multiple modalities from the disconnected, sharing the data with various efforts, including the Pentagon and the UN (Calvert 2013).

Discussion of Unstable Sociopolitical Environment Factors

In 2013 alone, there were 251 security incidents reported, 45 highly violent political conflicts, and 51.2 million people "forcibly displaced because of conflict" (Barajas and Design 2015). Areas struck with extremism, war, or other conflict pose violence and other physical threats to the members of the community—often exacerbated by communicating with Westerners or anyone that is seen to have different values and motivations from those of the local violent powers.

UNSTABLE SOCIOPOLITICAL ENVIRONMENTS: KEY PRINCIPLES

The following challenges from unstable sociopolitical environment factors emerge from the Afghanistan example that can be generalized to other similar communities:

- Fear and mistrust of outsiders
- Mistrust of technology, security, and privacy
- High variability

Challenge 5: Other Sociocultural Differences

Case Study: One Earth Designs Faces Sociocultural Differences from Their Nomadic Community Users in the Himalayan Plateau

Around 3 billion people—the majority of whom live in developing countries—rely on biomass fuels for cooking every day, while over 4 million people die "prematurely from illness attributable to the household air pollution from cooking with solid fuels" (World Health Organization 2014). In 2012, China alone reported that "more than 500,000 deaths from acute lower respiratory infections related to indoor air pollution" (Seow et al. 2014). In 2012, nomadic communities in the eastern edge of the Himalayan plateau in China were also victims of this; complaints about respiratory problems were widespread, and roughly one child for every 100 families would die every other year. To help solve this problem, Scot Frank, cofounder of One Earth Designs (OED), began working with one of those communities (comprising about 200 families). This nomadic community, like many others, was mostly disconnected; living in extremely harsh physical environments (e.g., high altitude, extreme temperatures), possessing poor literacy, and including only few affluent members (mostly men) who owned cell phones or satellite television. Even those with connectivity mainly utilized it for voice calls to individuals they already knew and trusted.

When OED started its work, the community was running out of fuel and needed an alternative option, but available alternatives were too expensive, inaccessible, or not appropriate for their needs. OED set out to understand those needs and determine why the available alternatives are not meeting them. Through observation, interviews, surveys, and over two years of living and traveling with the community, OED conducted a thorough needs assessment and market research. They developed personal relationships with the members of the community, gained trust, hired local community members to not only translate but also help understand the purpose and goals of the effort. OED ultimately designed and launched SolSource, a parabolic solar–thermal cookstove that is affordable, lightweight and portable, resilient in extreme climates and terrains, accessible, effective at grilling a steak in just 5–10 minutes, and (most importantly) desirable to the local community.

In addition to many of the other obstacles described earlier, OED faced the following key challenges related to the differences between the OED founders' and the local community's sociocultural differences:

- Relationship-based society and fairly closed network
- Chinese versus English languages
- Participant fatigue leading to not getting their many questions answered
- Preconceived notions that OED was like most other groups—coming in for one to two surveys, building a school, and departing, never to be seen again
- Patriarchal society where despite women being the cooks, the men insisted on answering most questions (often with different interpretations from the women)
- Difficult to get permission for the foreign men to speak with the local female users

OED slowly overcame these challenges over the two years. Key features they used to connect include the following:

- Full immersion living with the community the way they live and becoming part of the community
- Learning the local language

OTHER SOCIOCULTURAL DIFFERENCES: KEY PRINCIPLES

Key factors of these challenges include the following:

- Different languages
- Varying incentives
- Unaligned value systems
- Incongruous neural networks and logic chains
- Preconceived notions and stereotypes
- Wariness of the unknown and the unproven

The list continues. Connecting with the disconnected is hard and as much an art as it is a science.

- Skilled (and also) multilingual employees
- Question prioritization to avoid fatigue and collecting data through observation
- Close collaboration with community members (personal and professional)
- Decoy use distracting men while women talked to women—the main users of interest
- Appropriate timing to connect when men were not around
- Trust building

Through this user-focused, agile approach, OED worked closely with about 20 community members. Over 200 families purchased the SolSource cookstoves—almost 100% penetration—and they are still being used in 2016, 3 years after purchase. According to Frank (personal interview), on average, the products have paid for themselves in under a year because their owners do not have to purchase fuel for the SolSources, resulting in approximately 70% reduction in both fuel usage and emissions (the remaining 30% occurs when SolSource is not an option).

Discussion of Other Sociocultural Differences

The first four challenges described earlier are types of sociocultural differences, but this section refers to other challenges that occur due to sociocultural differences between the connected and the disconnected that were not captured earlier.

Living with the disconnected community of interest is not the best (nor even a realistic) option for every project, but every effort can look for creative opportunities and embedded collaborators to help learn from and positively influence the lives, minds, and hearts of the local people.

Summary

The preceding sections described five—*daunting!*—challenges that inhibit data collection from disconnected communities. Fortunately, many organizations have made progress

TABLE 8.2

Emerging Framework of Implementation Challenges

Challenge	Key Factors
Limited resources	• Low income and affordability • Low awareness and user-centric design • Limited education
Poor infrastructure	• Limited network coverage • Limited adjacent infrastructure (e.g., electricity)
Harsh environment	• Rough terrain (e.g., vast mountains, deserts) • Extreme weather conditions (e.g., temperature, precipitation, dust storms) • Environmental factors harmful to connection technology and infrastructure (e.g., excessive dust/sand, high humidity) • High variability (e.g., drastic and unexpected changes in the factors)
Unstable sociopolitical environment	• Fear and mistrust of outsiders • Mistrust of technology, security, and privacy • High variability
Other sociocultural differences	• Different languages • Varying incentives • Unaligned value systems • Incongruous neural networks and logic chains • Preconceived notions and stereotypes • Wariness of the unknown and unproven

in overcoming these barriers. Table 8.2 summarizes the emerging framework of these challenges.

The key is to take the time to understand the way the communities already communicate—what is their default method for receiving and sending the types of information you want to relay and collect? Which modalities do they use? How often do they communicate? In many situations, the way that one receives information may differ from the way they relay it, and what seems like minute details (such as the gender of the voice that asks a question) may be the difference between a successful and a failed data collection effort. This is all assuming, of course, that you have already overcome the first challenge of understanding what data to collect and from whom. This question becomes not just an implementation and technical one, but also an ethical one.

Ethical Concerns

Data! The new currency for everything from market research to monitoring and evaluation, from triaging medical needs to triaging resources after a natural disaster, and from identifying terrorist activity to human trafficking networks. It is easy to forget the ethical concerns and the less obvious implications of collecting data from individuals. Reliable, accurate, structured, and real-time information in these environments is rare and valuable. The average citizen is a key resource for gaining access to this information, but using people in an already vulnerable community to fulfill our own missions comes with many ethical concerns.

Perhaps you are thinking, "this doesn't apply to me; I'm not collecting data, just trying to influence behavior positively, such as like increasing awareness about washing hands."

Unfortunately, communicating with disconnected populations is never that simple. Ethical concerns include poor security and safety, unintended influence, bystander responsibility, exploitation, poor data quality and validation, data access and ownership, and privacy concerns. There are no easy answers, but hopefully asking these questions will help. While each concern could be discussed in a book of its own, below is a brief introduction of some of the most notable ethical considerations when contacting disconnected populations.

Concern 1: Poor Security and Safety

Setting up an SMS system for local community members to report on terrorist and criminal activities in the areas of the Democratic Republic of Congo, the Central African Republic, South Sudan, and Uganda may initially seem like a win-win situation, but this creates danger for those who submit information. They may become targets of the terrorists they are reporting on. Over 20% of the communities in the LRA-affected regions were internally displaced and too afraid to return home. In this harsh reality where death may be a consequence of reporting on activity, or even the appearance of cooperating with foreigners, researchers must take extreme care. Key questions to consider include the following:

- What is the risk of an individual or a community being identified through the mechanisms the data are being collected?
- To what degree, and in what way, is the information being collected valuable?
- Are the data providers fully informed and aware of both the risk and the potential reward for their cooperation and choosing to participate on their own free will?
- Are the technologies and the mechanisms being used appropriate and secure for protecting informants as much as possible?
- Is it possible to anonymize the data collected?

Concern 2: Unintended or Negative Influence

Subtle influences can be harder to detect or take seriously than imminent danger, but these can create just as much harm if not carefully considered. Every action and each word can subtly influence a community or an individual for better or worse. An effort researching the influence of different narratives on a population's sentiment toward terrorism may seem innocuous, but what if the narrative proves extremely successful, causing such strong emotions of fear and hatred that hate crimes increase? An intersectoral study of 330 randomly selected households in rural western Kenya found evidence that "being surveyed can change later behavior and related parameter estimates" because they inform and influence the "non-conscious, low-effort cognition" of the surveyed. The study proposed that surveyors aim for less exposure to more individuals ("infrequent survey visits to larger samples") versus higher exposure to a smaller subset. While this study focused on surveys and a social effort, the lessons are transposable to all that want to connect with the disconnected (Zwane et al. 2010).

Along a different vein of unintended influence, initiators may offer financial (e.g., free minutes) or other (e.g., beer) incentives in exchange for data, but these incentives may encourage alcoholism or false reporting, reliance on providing data for minutes, or encouraging use of phones when it cannot be afforded. While it may be impossible to consider all the unintended consequences of the ways and reasons for connecting, it is important to be introspective of the possibilities and agile enough to adjust accordingly.

Finally, connecting with the disconnected is a priority for not only those intending to do good for that community or the world, but also those intending to do harm or further their own agendas without regard to what happens to the disconnected themselves. The same best practices for how to influence the disconnected to wash their hands or report on terrorist activity can also be used to influence them to suppress women or commit terrorist acts.

Key questions to ask include the following:

- What unintended consequences are most likely to occur?
- How important is it to connect and to connect in this way?
- How can I minimize my influence? How can I minimize the level of contact that I have with them?
- Are there any possible life-threatening or otherwise dangerous consequences?
- Does the data already exist in some form or with another organization?

Concern 3: Bystander Responsibility

Researchers collecting data from vulnerable communities will inevitably face situations where people are suffering in ways the researchers could (technically) help ameliorate. What if you are collecting data for a healthcare project to prevent polio and you meet hundreds of children still suffering from the disease? What if you are collecting data to identify postearthquake needs and see a child trapped under a building in front of your eyes? What if you see 1000 children trapped? What if you are collecting data on human trafficking and see a woman being raped? An entire brothel of women? It is easy enough to say "that's not my job" or "how will I achieve my data collection goals if I'm stopping every minute to help another person" or "I need to focus on the bigger picture," but is it that simple? Of course not everyone can or should intervene, but remembering that you are benefiting from the connection and being introspective is the first step. Key questions to consider include the following:

- What are we, as connectors for whatever purpose, obligated to do as human beings?
- Where is the line? What if the people in need are not right in front of you—what if they are behind a closed door? What if you only hear their cry for help but do not see it; what if you do not hear or see it but know it is there?
- How can I design my effort to empower the disconnected?
- Are the individuals and communities I am engaging with in need of something or some service I can provide or fulfill?

Concern 4: Exploitation

With data as a new currency, there is the potential of unintentional (or intentional) exploitation of vulnerable communities through data collection from these communities. Even in cases where data are being collected in order to ultimately better serve these communities (e.g., collecting data to inform the monitoring and evaluation of development programs), often the data are used to help an organization further their own priorities and goals whether it be promoting their own military or diplomacy agendas or making more money. Collectors can profit from gaining better marketing, influencing behavior of the disconnected to act toward agendas they are not aware of and could even potentially harm them, winning more grants, or even directly monetizing the data. In order to avoid exploiting these already vulnerable, often exploited communities, we must ask the following questions:

- What are the local community members receiving in return for providing data?
- How is the connection influencing them and their lives? If data are being collected, how are they being used? If information is being sent, what is being communicated and influenced?
- How can we both achieve our own goals and avoid exploiting vulnerable communities purely because they do not have the experience, the knowledge, or the power to defend themselves?

In many cases, exploiters may not realize they are being exploitative or be in denial of the fact. For example, one segment of anti–human trafficking involves educating buyers of sex to not stop buying, but rather to ask the right questions before they purchase to ensure they are not exploiting victims of human trafficking. (The debate continues on whether prostitution can be voluntary, but that is a subject for a book of its own.) It is not enough to respond to obvious or glaring exploitations—it is our responsibility to investigate potential forms of exploitation, understanding that many times the data collectors or information disseminators automatically hold an unjust power over the disconnected purely by having more information, resources, and options.

Concern 5: Poor Data Quality and Validation

The implications of misusing or analyzing data can be huge, never mind the issue of collecting biased, skewed, or otherwise inaccurate data. Statistics can be extremely powerful but also manipulated, for instance, by selectively including parts of the data or using particular analysis methods. Consider also unintentional errors; for instance, an aid project may conduct a survey on local community development wants, an admirable goal, but the outcomes of incorrect data collection could create more harm than good. Imagine that an astounding 70% of the community responds that they want a way of monitoring community crimes, then the project successfully designs and launches a crime-monitoring program only to realize years later that a majority of the respondents were male, and the system is mainly used to monitor women's behavior, criminalizing them for walking alone or for speaking with other men.

Poor data quality and validation extends beyond ethical concerns, but key ethical data quality questions include the following:

- What assumptions am I making about the community I am connecting with and the information that I am obtaining?
- Am I connecting with an accurate sample based on demographics, geography, sociocultural factors, and other attributes that are pertinent to my effort?
- Is there a way to validate my data?
- What are the caveats associated with the data I have collected?

Concern 6: Data Ownership and Access

David Loshin (2002) and the Office of Research Integrity (n.d.) describe data ownership as "both the possession of and responsibility for information [where] ownership implies power as well as control. The control of information includes not just the ability to access, create, modify, package, derive benefit from, sell, or remove data, but also the right to assign these access privileges to others." While Loshin is mainly referring to data ownership in the context of academic research, his research applies to ethical concerns around data ownership and access when data are collected from the disconnected. In many situations where the disconnected are being connected for data collection purposes, the collectors (often the economic and educational elite) are the data owners and thus hold not only data, but also a social, power dynamic. Too often, the individuals providing the data remain unaware of what the data are being collected for and how those data are being used. Often these contributors often lack ownership, access, or awareness of the data. LaborVoices addressed this issue in 2015 by creating a platform specifically for the workers from whom they collected data, making those data accessible to workers for job hunting (Kohl Gill, personal communication). Making the data open source and accessible to the disconnected can be one solution but may not work for all efforts.

Key questions to consider include the following:

- Who has control over the collected data? Who has access?
- How am I benefiting from this data and how are the contributors benefiting?
- How can I enable data access, ownership, and derived benefit for the disconnected?
- Am I using this data in a way the individuals contributing the data would approve of?

Concern 7: Privacy

Edward Snowden became known worldwide, and closely linked to the issue of data privacy, after he leaked details of Internet and phone surveillance by the US National Security Agency (BBC 2014). This scandal on the violation of citizens' privacy spurred public outrage and debates that still persist years later and have impacted the US 2016 presidential campaigns. This debate includes government data privacy concerns, as well as how other institutions handle data, including technology (e.g., Google and Facebook) and consumer goods sectors. For example, in 2012, Target used data to figure out a teenage girl was pregnant and sent her coupons (a form of targeted marketing) before her father was aware of her pregnancy (Hill 2012). This spiraled into a large debate on data privacy. Despite the surge of interest in the issue of data privacy for the connected world, too often this issue does not surface when discussing data for the disconnected. As more and more data collection initiatives arise in disconnected communities and as they become more connected, it is critical that we consider data privacy for these communities as well. Best practices and principles of more seasoned fields requiring data collection such as medical research, technology, and marketing may be useful to investigate.

Information collected for the right reasons can be used for bad intentions, if it falls into the wrong hands. For example, the International Organization of Migrants maintains a database of human trafficking victims, and Polaris similarly has a database of details about victims, traffickers, and buyers from the hotline calls that they receive. While they use these data to help the victims and fight human trafficking, such data in the wrong hands (e.g., through getting hacked or poor data security) could be used by the traffickers to further exploit vulnerable individuals.

Key questions to consider include the following:

- Is the privacy of the contributors protected?
- Are the contributors aware that data are being collected and how they are being used?
- Did the contributors give their permission for the data to be collected and for them to be used in the way they are used?
- Is personally identifiable information (PII) being used, and if yes, does it need to be? If yes, is it appropriately protected and restricted?

Summary

Connecting with the disconnected is still a nascent concept, with increasing but still insufficient studies and best practices. With the myriad of practical challenges outlined in the first section of this chapter, it has been difficult to effectively engage these communities, as well as to consider and address the associated ethical concerns.

Table 8.3 summarizes this section with a framework of the types of ethical concerns that may require consideration.

TABLE 8.3

Emerging Framework of Ethical Concerns

Concern	Description	Key Questions to Consider
Poor security and safety	Disconnected individuals in unstable or restricted communities can face threats and dangers from real or perceived connection with foreign contacts as well as for reporting certain types of information.	• What is the risk of an individual or a community being identified through the mechanisms the data are being collected? • To what degree is the information being collected valuable and how is it valuable? • Are the data providers fully informed and aware of both the risk and the potential reward for their cooperation and choosing to participate on their own free will? • Are the technologies and the mechanisms being used appropriate and secure for protecting informants as much as possible? • Is it possible to anonymize the data collected?
Negative or unintended influence	Communication is proven to directly influence behavior. There will almost always be unintended influences—some of which may be harmful or not apparent for long periods. Best practices for connecting can also be used for intentional harm.	• What unintended consequences are most likely to occur? • How important is it to connect and to connect in this way? • How can I minimize my influence? How can I minimize the level of contact that I have with them? • Are there any possible life-threatening or otherwise dangerous consequences?

(*Continued*)

TABLE 8.3 (CONTINUED)

Emerging Framework of Ethical Concerns

Concern	Description	Key Questions to Consider
Bystander responsibility	Many disconnected communities are also vulnerable and face poverty and social injustices. As initiators and beneficiaries of connecting, we may have a certain degree of bystander responsibility to act on our observations.	• What are we, as researchers, market researchers, and data collectors for whatever purpose it may be, obligated to do as good Samaritans and human beings? • Where is the line? What if the people in need are not right in front of you—what if they are behind a closed door—what if you only hear their cry for help but do not see it—what if you do not hear or see it but know it is there. • How can I design my effort to empower the disconnected? • Are the individuals and communities I am engaging with in need of something or some service I can provide or fulfill?
Exploitation	Possibility of intentional or unintentional exploitation is real as we hold a social and data ownership power dynamic over the disconnected—often vulnerable communities.	• What are the local community members receiving in return for providing data? • How is the connection influencing them and their lives? If data are being collected, how are they being used? If information is being sent, what is being communicated and influenced? • How can we both achieve our own goals and avoid exploiting vulnerable communities purely because they do not have the experience, the knowledge, or the power to defend themselves?
Poor data quality/ hard to validate	Reliable control groups are often difficult, if not impossible, to have when connecting with disconnected communities for practical reasons.	• What assumptions am I making about the community I am connecting with and the information that I am obtaining? • Am I connecting with an accurate sample based on demographics, geography, sociocultural factors, and other attributes that are pertinent to my effort? • Is there a way to validate my data? • What are the caveats associated with the data I have collected?
Data access/ ownership	The connected often have the majority of ownership and access to collected data, which can create an unbalanced power dynamic.	• Who has control over the collected data? Who has access? • How am I benefiting from this data and how are the contributors benefiting? • How can I enable data access, ownership, and derived benefit for the disconnected? • Am I using these data in a way the individuals contributing the data would approve of?
Privacy	Collecting data on private information can have both intentional and unintentional negative consequences.	• Is the privacy of the contributors protected? • Are the contributors aware that data are being collected and how they are being used? • Did the contributors give their permission for the data to be collected and for them to be used in the way they are used? • Is PII being used, and if yes, does it need to be? If yes, is it appropriately protected and restricted?

These ethical concerns should be considered at all steps of connecting with a community: before, during, and after. The influences of a campaign can sometimes be seen immediately, but often take years and even decades to emerge. Only by asking these questions at every stage can we progress our understanding of how to connect with the disconnected more effectively and ethically.

Road Forward

Connecting with the disconnected is an art and a science that every sector faces, but none have yet mastered. As such, each effort to connect with disconnected communities is treated like a microstartup or an experiment—having to understand the end population, develop an effective capability to connect, and reach the desired set of people (analogous to clients for a startup). Lean startup principles *Validate Learning* and *Build–Measure–Learn* apply here. The key to an effective and efficient connection is an agile, iterative, and appropriate design that puts the community members at the center of establishing an open two-way communication (Ries 2011). The challenges, the concerns, and the examples introduced in this chapter are just the start; many challenges remain in effectively, efficiently, and ethically connecting with the disconnected. I challenge you to not only consider the information from this chapter, but also really use it as a motivator. Discuss the challenges. Identify more. Dig deep to ask those ethical questions you would rather run away from. Connect.

Megacity Use Case

The Razini migrant and Bursuka refugee communities in Xanadu, as described in the use case from the preface of this book, are good examples of disconnected communities, and they demonstrate the challenges that organizations may face when trying to connect to such societies. Let us explore these challenges in the hypothetical context of Xanadu.

Limited Resources

A local private hospital offers free public health services in the Razini wards, direly needed, but discovers that there are few to no patients visiting their clinics or partaking in their services, despite the informational flyers the hospital posted throughout the wards and a phone number to call or send an SMS for more information. A local NGO supported by Razini staff goes into the community to understand why no one is going to the clinic; the organization discovers that no one is aware of it. They had seen the flyers, but being illiterate, the Razinis had passed by them without a second thought. The organization also discovers that the unemployment rate is extremely high among the Razinis, and most cannot even afford feature phones, but they do have radios and listen to them intently for all of their news and information. Based on these insights, the hospital advertises their services through a local radio station popular among the Razinis and creates visual flyers that show the location of the clinic and a depiction of Razinis getting healthcare. They hire the few Razinis in the community that have feature phones to be communication hubs—allowing other members of the community to use their phones free of charge to send SMS to the clinic. In the following weeks, the hospital is bustling with Razinis!

Poor Infrastructure

A disastrous typhoon hits Xanadu, flooding the city. Citizens congregate on building tops and hills, stranded on these islands without clean water, food, or medical support. An international relief organization collaborates with the local military to provide disaster relief but finds pure chaos—cell phone and power towers are down, hundreds of organizations and thousands of volunteers are arriving to help but require coordination and management, and they do not know where their help is urgently needed the most or even how to reach those in need. The relief organizers decide to take a multimodal approach to finding and prioritizing victims of the flood: (1) airdrop multipurpose cards that contain information on how to get to the closest relief center, how to set up a help signal using locally available materials that can be seen from the air, and how to turn the cards into water purification filters; (2) collaboration with a geospatial imagery technology company to crowdsource before and after images of Xanadu to identify areas that appear most impacted or where people have constructed help signals; and (3) training of local and international volunteers to triage aid needs on the ground through a smartphone app that functions without a connection and transmits data through SMS, data, bluetooth, and WiFi whenever one modality becomes available.

Harsh Physical Environment

As the Razinis were making the harsh journey from their home country to Xanadu, an international aid organization tried to raise awareness about safe options for migration, to help the Razini refugees avoid the dangers they faced during the journey and the conflict upon arrival. They tried donating smartphones, but the Razinis were moving through dense and humid jungles where moisture quickly damaged the devices. They tried donating HF radios and training the migrants, but the radios were too heavy for the migrants to transport during the trek through humid 115°F climates. They tried SMS campaigns through features phones, but the density of the jungles blocked the signals. Finally, through collaboration with an international technology company offering mobile, robust, waterproof, and heat-resilient smartphones and temporary balloon hotspots; the aid organization set up communication hotspots along the migration path, advertising it as not only a source of information but also a way for the migrants to communicate with their loved ones and receive news.

Unstable Sociopolitical Environment

A regional law enforcement group is trying to address the severe organized crime issues by the mafia in the northern wards, where the Razinis have settled, but they have little insight into real-time events in the ward. By the time the law enforcement hears about a crime, it is too late to intervene and the evidence has disappeared. To crowdsource mafia activity from the people experiencing it firsthand, the law enforcement group sets up radio ads, visual flyers, and SMS reporting through the communication power users previously established by the hospital in the example earlier. They receive few reports. Through a discussion over tea with the power users, with whom the police have developed close personal relationships, the law enforcement organization realizes that the Razinis fear the mafia too greatly to report on their activities, mistrust the reporting mechanism to keep the informants safe, and doubt the police's ability to improve the situation. Over the next year, the law enforcement group, working with the power users, develops a relationship with Razini community leaders, helping them understand the harm the mafia causes to their community, the safety precautions that will help protect and anonymize reporters, the support services available to the community, and the immediate responses criminal activity reports will receive. Through collaboration with local NGOs, a network of volunteers agrees to provide support and protection services for those who report, anonymizing the reporting at all levels of the communication process so that it cannot be traced, immediately and reliably responding to the reports, and closely working with the community leaders to educate their community on these features. The reports surge and the law enforcement group is able to identify and prosecute key mafia leaders.

Other Sociocultural Differences

A foreign atheist academic researcher seeks to study the negative impacts of migration on Bursuka families. She travels to Xanadu to visit one of the Bursuka refugee camps, armed with both paper and digital surveys translated to the local language of Xanadu, ready to meet Bursukas, interview them, and ask them to fill out the surveys for her study. The researcher knocks on many doors, her T-shirt drenched in sweat from the humidity, but every door closes on her, and the few women and children who give her time only speak broken Xanadu and seem nervous interacting with her. One of the local hospital workers she befriends advises her over coffee based on his 15 years of experience working with the Bursukas: (1) the male religious clerics form the centers of gravity for the Bursukas, (2) there is poor integration between the culture, the activities, and the language of the local Xanaduans and Bursukans, (3) there is great skepticism about foreigners that come in and out of the refugee camps with no consistency or true understanding because the Bursukans have been exploited and ignored, and (4) it is considered disrespectful for women to show their bare arms. The researcher covers her arms, translates her surveys into Bursukan, and works with her friend over the next two years to build relationships with the religious clerics. She helps them understand her effort and the potential for more positive impacts on the community based on her research. The clerics connect her with a few literate men and women in the community who have mobile phones and have learned the researcher's language through working with other foreign workers. Together, they collect honest responses from the community.

The Xanadu case is just one hypothetical example of disconnected communities and the challenges that organizations wanting to connect may face. It is critical to note that every community and connection effort is different and requires agility to leverage best practices and lessons learned from other efforts while maintaining enough flexibility to adjust throughout the process. Hopefully, over time and with collaboration and information sharing, the numbers of the disconnected will decrease, and the knowledge around how to most effectively and efficiently connect with the disconnected will become more of a science.

DISCUSSION QUESTIONS

1. In what situations and for what purposes would you want to connect with the disconnected? What are your motivations and what would both parties benefit or lose from the connection?
2. What challenges does the community you are trying to engage with face? What implementation challenges and ethical concerns are likely to arise during an effort to connect?
3. How does the community you are trying to engage with currently communicate the types of information you are trying to send and/or collect? Does it make sense to use these same modalities for your effort? What do they value and what do they fear?
4. How can you maintain agility throughout your effort while maintaining consistency? What needs to remain constant and what can remain variable? How can you quickly test theories before launching the main effort and iterate throughout?

References

Agger, K. (2012, November). *Field Dispatch: Chasing the Lord's Resistance Army: Challenges Faced by Ugandan Soldiers Pursuing the LRA*. Enough Project. Retrieved from http://www.enoughproject.org/files/ChasingTheLRA.pdf.

Apps, P. (2015, October 1). *The War in Afghanistan Has So Far Cost $33,000 Per Citizen. And Will Not End Well*. London: Reuters. Retrieved from http://blogs.reuters.com/great-debate/2015/10/01/the-war-in-afghanistan-cost-of-33000-per-citizen-and-will-not-end-well/.

Barajas, L., Card, B., and Design, B. (2015). *World Humanitarian Data and Trends 2014*. Jerusalem: UN Office for the Coordination of Humanitarian Affairs. Retrieved from http://www.unocha.org/data-and-trends-2014/downloads/World%20Humanitarian%20Data%20and%20Trends%202014.pdf.

BBC (2014, January 17). *Edward Snowden: Leaks that Exposed US Spy Programme*. London: BBC. Retrieved from http://www.bbc.com/news/world-us-canada-23123964.

Calvert, B. (2013, July 1). The merry pranksters who hacked the Afghan War. *Pacific Standard Magazine*. Retrieved from http://www.psmag.com/books-and-culture/the-merry-pranksters-who-hacked-the-afghan-war-60873.

Cruz, M., Foster, J., Quillin, B., and Schellekens, P. (2015, October). Ending extreme poverty and sharing prosperity: Progress and policies. *Policy Research Note #3*. Washington, DC: World Bank Group. Retrieved from http://pubdocs.worldbank.org/pubdocs/publicdoc/2015/10/109701443800596288/PRN03-Oct2015-TwinGoals.pdf.

Desai, K. (2014, April–July). *Supplier Analytics: Improving Workplace Safety and Employee Satisfaction through Anonymous Worker Feedback: Results from the Bangladesh Garment Factory Pilot, April–July 2014*. Millbrae, CA: LaborVoices. Retrieved from https://s3.amazonaws.com/lv-documents/LaborVoices+Case+Study+-+Bangladesh+Garment+Factory+Pilot+2.pdf.

DigitalGlobe (2015, April 29). *DigitalGlobe Releases Preliminary Nepal Earthquake Crowdsourcing Results*. Westminster, CO: DigitalGlobe. Retrieved from http://www.digitalglobeblog.com/2015/04/29/digitalglobe-releases-preliminary-nepal-earthquake-crowdsourcing-results/.

Dyer, G., and Sorvino, C. (2014, December 14). *$1tn Cost of Longest US War Hastens Retreat from Military Intervention*. Financial Times. Retrieved from http://www.ft.com/cms/s/2/14be0e0c-8255-11e4-ace7-00144feabdc0.html#slide0.

Forrest, A. (2015, June 15). *Crisis Mapping Technology Boosts Relief Efforts in Nepal*. New York: Forbes. Retrieved from http://www.forbes.com/sites/adamforrest/2015/06/15/crisis-mapping-technology-boosts-relief-efforts-in-nepal/.

Geuss, M. (2015, April 28). *After Nepal Earthquake, People Turn to Ham Radio.* ARS Technica. Retrieved from http://arstechnica.com/gadgets/2015/04/after-nepal-earthquake-people-turn-to-ham-radio/.

Google Project Loon (2015). *Loon for All.* Mountain View, CA: Google. Retrieved from https://www.google.com/loon/.

Groupe Speciale Mobile Association (2015). *The Mobile Economy, 2015.* London: Groupe Speciale Mobile Association. Retrieved from http://www.gsmamobileeconomy.com/GSMA_Global_Mobile_Economy_Report_2015.pdf.

Hill, K. (2012, February 6). *How Target Figured Out a Teen Girl was Pregnant Before Her Father Did.* New York: Forbes. Retrieved from http://www.forbes.com/sites/kashmirhill/2012/02/16/how-target-figured-out-a-teen-girl-was-pregnant-before-her-father-did/2/.

International Labour Office (ILO) (2012). Special action programme to combat forced labour (SAP-FL). *ILO Global Estimate of Forced Labour Results and Methodology.* Geneva: ILO. Retrieved from http://www.ilo.org/wcmsp5/groups/public/—ed_norm/—declaration/documents/publication/wcms_182004.pdf.

Invisible Children (2014). *LRA Crisis Tracker.* San Diego, CA: Invisible Children. Retrieved from http://invisiblechildren.com/program/lra-crisis-tracker/.

Kathmandu Living Labs (KLL) (2015). *Quakemap.org.* Chundevi: Kathmandu Living Labs. Retrieved from http://kathmandulivinglabs.org/project/details/quakemap-org.

Leiber, N. (2013, May 16). *LaborVoices, Labor Link Help Workers Report Sweatshop Factory Conditions.* New York: Bloomberg Business. Retrieved from http://www.bloomberg.com/bw/articles/2013-05-16/laborvoices-labor-link-help-workers-report-sweatshop-factory-conditions.

Lombardo, T. (2014, October 13). *Army: Messaging Leads to Mass Defections from Kony.* Springfield, VA: Army Times. Retrieved from http://archive.armytimes.com/article/20141013/NEWS/310130052/Army-Messaging-leads-mass-defections-from-Kony.

Loshin, D. (2002, June 8). *Knowledge Integrity: Data Ownership.* Retrieved from http://www.datawarehouse.com/article/?articleid=3052.

Maclin, B., Kelly, J., Auletta-Young, C., and Cragin, W. (2015, November). *We Mobilized Ourselves.* Cambridge, MA: Harvard Humanitarian Initiative. Retrieved from http://www.hhi.harvard.edu/sites/default/files/publications/hhi_lra_report_we_mobilized_ourselves.pdf.

McKinsey & Company Technology, Media, and Telecom Practice (McKinsey & Company) (2014, October). *Offline and Falling Behind: Barriers to Internet Adoption.* McKinsey & Company. San Francisco, Singapore, Johannesburg, Silicon Valley, and Brussels. Retrieved from http://www.mckinsey.com/~/media/mckinsey/dotcom/client_service/media%20and%20entertainment/pdfs/offline%20and%20falling%20behind%20%20full%20report1118final.ashx.

Munoz, A. (2012). *US Military Information Operations in Afghanistan: Effectiveness of Psychological Operations 2001–2010.* Santa Monica, CA: RAND National Defense Research Institute. Retrieved from http://www.rand.org/content/dam/rand/pubs/monographs/2012/RAND_MG1060.pdf.

Ncell, Flowminder, and Worldpop (2015, May 8). *Nepal Population Estimates as of May 1, 2015.* Nepal: Ncell, Flowminder, and Worldpop. Retrieved from https://data.hdx.rwlabs.org/dataset/population-movements-after-the-nepal-earthquake-v-1-up-to-1-may-2015/resource_download/fc242f46-1929-4850-9e79-262e84314d88.

Nepal Telecommunications Authority (2015, March). *MIS Report.* Nepal Telecommuncations Authority. Retrieved from http://cdn.arstechnica.net/wp-content/uploads/2015/04/NTA-MIS-94.pdf.

Office of Research Integrity (n.d.). *Data Ownership.* Washington, DC: US Government Department of Health and Human Services. Retrieved from https://ori.hhs.gov/education/products/n_illinois_u/datamanagement/dotopic.html.

Ries, E. (2011, September 13). *The Lean Startup: How Today's Entrepreneurs Use Continuous Innovation to Create Radically Successful Businesses.* New York: Crown Business.

Ross, S. (2013, October 10). *Can Radios Stop the Lord's Resistance Army in Congo?* London: African Arguments. Retrieved from http://africanarguments.org/2013/10/10/can-radios-stop-the-lords-resistance-army-in-congo-by-scott-ross/.

Seow, W. J., Hu, W., Vermeulen, R., Hosgood, H. D., Downward, G. S. et al. (2014, October). Household air pollution and lung cancer in China: A review of studies in Xuanwei. *Chinese Journal of Cancer*. Retrieved from http://www.ncbi.nlm.nih.gov/pmc/articles/PMC4198749/#b3.

UN Population Fund (2012, June 29). *Conducting a Social, Demographic and Economic Survey of Afghanistan*. United Nations Population Fund (UNFPA). Retrieved from http://www.unfpa.org/news/conducting-social-demographic-and-economic-survey-afghanistan.

UN Security Council (2013, May 29). Demanding that Lord's resistance army end all attacks, security council calls for full implementation of regional strategy in Central Africa. *United Nations Meetings Coverage of 6971st Meeting*. Geneva: UN. Retrieved from http://www.un.org/press/en/2013/sc11018.doc.htm.

Vandiver, J. (2013, February 28). *US Special Ops, Activists Working Together Against LRA*. Washington, DC: Stars and Stripes. Retrieved from http://www.stripes.com/news/us-special-ops-activists-working-together-against-lra-1.210040.

World Bank (2012). *Data*. Washington, DC: World Bank. Retrieved from http://data.worldbank.org/.

World Health Organization (2014, March). *Household Air Pollution and Health*. Geneva: World Health Organization. Retrieved from http://www.who.int/mediacentre/factsheets/fs292/en/.

Zwane, A. P., Zinman, J., Dusen, E. V., Pariente, W., Null, C. et al. (2010, November 29). Being surveyed can change later behavior and related parameter estimates. *PNAS Early Edition*. Washington, DC: National Academy of Sciences. Retrieved from http://economics.mit.edu/files/6765.

Suggested Reading

Chessell, M. (2014). *Ethics for big data and analytics*. New York: IBM. Retrieved from http://www.ibmbigdatahub.com/sites/default/files/whitepapers_reports_file/TCG%20Study%20Report%20-%20Ethics%20for%20BD%26A.pdf.

DiCicco, M. (2012, September 18). MCIOC exercise gives marines relationship practice. *Quantico Sentry*. Retrieved from http://www.quanticosentryonline.com/news/article_fd28f026-ae8b-548a-be1e-c44716a104f1.html.

Enough Project (n.d.). *The LRA in Congo, CAR, and South Sudan*. Enough Project. Retrieved from http://www.enoughproject.org/conflicts/lra/congo-car-south-sudan.

Finley, K. (2015, October 6). *Thank (Or Blame) Snowden for Europe's Big Privacy Ruling*. Boone, IA: Wired. Retrieved from http://www.wired.com/2015/10/tech-companies-can-blame-snowden-data-privacy-decision/.

Greene, L., and Mamic, I. (2015, February). The future of work: Increasing reach through mobile technology. *ILO Asia-Pacific Working Paper Series*. Geneva: ILO. Retrieved from http://www.ilo.org/wcmsp5/groups/public/—asia/—ro-bangkok/—sro-bangkok/documents/publication/wcms_342162.pdf.

Laframboise, N., and Loko, B. (2012, October). Natural disasters: Mitigating impact, managing risks. *IMF Working Paper*. Washington, DC: International Monetary Fund. Retrieved from http://www.imf.org/external/pubs/ft/wp/2012/wp12245.pdf.

Ronan, P., Transier, K., Poole, S., Rotshuizen, S., Marie-Regnault, C., Elam, J., and Dougan, L. (2014, October). *Tracking Joseph Kony: A Rebel Leader's Nine-Year Odyssey*. The Resolve LRA Crisis Initiative, LRA Crisis Tracker, and Invisible Children. Retrieved from http://reports.lracrisistracker.com/en/tracking-kony/.

Timm, T. (2015, October 8). *The Snowden Effect: New Privacy Wins Await After Data Transfer Ruling*. London: The Guardian. Retrieved from http://www.theguardian.com/commentisfree/2015/oct/08/snowden-effect-new-privacy-wins-data-transfer-ruling.

9

*They Are Beyond WEIRD: Helpful Frameworks for Conducting Non-WEIRD Research**

David J. Y. Combs, Sarai Blincoe, Christopher P. Garris, and Eric S. Vorm

CONTENTS

Introduction 164
Some Perspectives on Cross-Cultural Research in the Social/Behavior Sciences 164
Addressing the Research Gap: Helpful Frameworks 166
Markus and Conner's Culture Cycle Framework 166
Individual Level 166
Interaction and Institution Levels 166
Conceptual Ideas Level 167
Hofstede Dimensions 168
Power Distance 168
Uncertainty Avoidance 168
Masculinity–Femininity 169
Individualism–Collectivism 169
The Newer Dimensions 170
Integrating the Culture Cycle and the Hofstede Dimensions 171
Individualism–Collectivism within the Culture Cycle 171
Implications for Future Research 175
Conclusions 175
Megacity Use Case 176
Xanadu 176
Information Processing within the Culture Cycle 177
References 178

ABSTRACT The majority of research and theory in the social sciences, particularly psychology, comes from so-called Western, educated, industrialized, rich, democratic (WEIRD) societies. The emphasis on WEIRD societies has created a significant knowledge gap. This chapter begins by describing why cross-cultural research is so important, as evidenced by some of the many differences between WEIRD and non-WEIRD societies. Next, the chapter explains two prominent frameworks for culture: the Markus and Connor's culture cycle and the Hofstede cultural dimensions. We then propose an integration of these frameworks than can organize existing research and generate new research questions and hypotheses, thus closing the gap in our understanding of WEIRD versus non-WEIRD worlds. The chapter closes with a case study application of the integrated framework.

* Portions of this chapter are drawn from Chapter 2 of Combs, Blincoe, and Vorm (2015). Trust, attitudes, and social influence: The cross cultural social psychology of counterinsurgency. *Department of Defense Manual.* Available at http://pksoi.army.mil/ or from the first author of this chapter.

KEY TOPICS

- WEIRD societies are Western, educated, industrialized, rich, and democratic.
- Data from WEIRD samples problematically dominate social science.
- Essential, sometimes foundational, differences exist between WEIRD and non-WEIRD.
- Markus and Conner's culture cycle and Hofstede's culture dimensions are useful frameworks for non-WEIRD research.

Introduction

For nearly 15 years, the United States engaged in conflicts related to the global war on terror. The major engagements have occurred in Iraq and Afghanistan, with additional activities in places such as Somalia, Yemen, Pakistan, and Libya. Within this overarching conflict, the missions of the United States and its coalition allies became less about conventional state-on-state warfare and instead focused almost entirely on less conventional approaches, such as irregular warfare, counterinsurgency, counterterrorism, and stabilization operations (Ucko, 2009). As others (e.g., the study by Kilcullen [2010]) have noted, such missions generally involve substantial interactions with local populations and therefore require a tremendous amount of social–cultural/social science-based knowledge. While some US DoD research programs have made progress on this front,* much work remains to be done.

We hope to achieve three goals in writing this chapter. First, we point out that whereas the last several decades have seen considerable progress and interest in cross-cultural social science research, that literature problematically relies on samples that are WEIRD (Henrich, Heine, and Norenzayan, 2010). Second, we briefly describe two well-known theoretical frameworks in the cross-cultural social sciences (i.e., Markus and Conner's culture cycle [2013] and Hofstede, Hofstede, and Minkov's culture dimensions [2010]) and suggest that the integration of those frameworks might help researchers generate new hypotheses and projects that, if tested in multiple cultural settings, might help further resolve the field's relative lack of understanding of non-WEIRD cultures. Finally, we conclude with an application of the integrated culture cycle and cultural dimension frameworks to the Xanadu megacity case example woven throughout this volume.

Some Perspectives on Cross-Cultural Research in the Social/Behavior Sciences

It should come as no surprise to behavioral scientists† that much of their theory and research occurs within the context of WEIRD societies. As far back as 1946, commentators lamented

* E.g., The MINERVA Initiative, the Human Social Culture Behavior Modeling Program.
† E.g., psychologists, economists, sociologists, political scientists.

that "the existing science of human behavior is largely the science of [college] sophomores" (McNemar, 1946), and as Sears (1986) described, college sophomores are often undesirable research participants due to their "less crystallized attitudes, less formulated senses of self, stronger cognitive skills, stronger tendencies to comply with authority, and more unstable peer group relationships." A more recent comprehensive, and critical, assessment of the behavioral sciences' reliance on a narrow segment of humanity for research comes from Henrich, Heine, and Norenzayan (2010). They found that in the psychological literature, 96% of test participants were from Western societies (e.g., North America, Europe), with 68% from the United States alone; ultimately, ". . . 96% of the psychological samples come from countries with only 12% of the world's population." Despite these patterns in hypothesis testing, Henrich, Heine, and Norenzayan pointed out that many researchers make claims about global generalizability without "even a cautionary footnote about these inferential extensions."

In their review, Henrich, Heine, and Norenzayan summarized major distinctions among three telescoping levels of culture (i.e., industrialized societies and small-scale societies, WEIRD and non-WEIRD societies, and Americans versus other citizens of the WEIRD world), which influence factors such as perspectives on economic fairness, spatial cognition, and categorization of biological entities. Perhaps one of the more striking distinctions between WEIRD societies and non-WEIRD societies (what Henrich, Heine, and Norenzayan call *small-scale societies* in their paper) is a distinction in visual perception processes. For instance, Henrich, Heine, and Norenzayan noted that some research has even demonstrated how certain nonindustrial societies are less susceptible to optical illusions than people from some industrial societies. Similarly, some non-WEIRD societies are more able to see (and remember seeing) background information from visual scenes than people from some Westernized societies (Chua, Boland, and Nisbett, 2005).

Henrich, Heine, and Norenzayan suggested that while some distinctions between WEIRD and non-WEIRD societies might well be anticipated, the fact that such cultures can differ in processes as foundational as visual perception may surprise even seasoned researchers. Our cultural background can literally change "what we see and what we don't see" (Heine, 2008). Cross-cultural differences in such basic processes are a stark reminder that research done exclusively with WEIRD participants should rarely be assumed to generalize beyond those contexts without strong scientific evidence to the contrary.

Of course, consistencies do exist. In the field of anthropology, Brown (1991) outlined a comprehensive list of human universals based on field research. These universals ranged from color perception and labeling (e.g., black) to taboos and norms (e.g., the incest taboo versus the norm of reciprocity). In the field of psychology, Williams and Best (1990) collected data on gender stereotypes from 30 nations and revealed a striking consistency across these nations in the designation of specific qualities as being typically male or female. In the area of personality psychology, the five-factor taxonomy of personality traits has successfully fit personality data from more than 36 different countries (McCrae, 2002).

In summary, whereas the existence of human universals has been empirically demonstrated, research also reveals that human processes can substantially vary across cultures. The risk of inaccurate generalization increases when comparing Western samples (WEIRD) to non-Western (non-WEIRD) samples. Although generalizing across cultures should typically be avoided *in the absence of supporting data*, it is unfortunately a fairly common practice in social science.

Addressing the Research Gap: Helpful Frameworks

Given the historical reliance on WEIRD samples in the social sciences, it will take decades to fill the research gap. A strong theoretical framework can help ground subsequent hypotheses, direct research, and inform data collection (of course, this is true whether the subject of the research is cross-cultural or not). This section describes two well-known theoretical frameworks, Markus and Conner's culture cycle and Hofstede's famous culture dimensions; it then describes a simple integration of the two and how this may prove useful for those generating new cross-cultural hypotheses and research.

Markus and Conner's Culture Cycle Framework

Markus and Conner define culture as "the ideas, institutions, and interactions that tell a group of people how to think, feel, and act" (Markus and Conner, 2013). This definition emphasizes the multiple levels at which culture occurs, specifically, the individual, interaction, institution, and conceptual idea levels—*each of which influences and interacts with the other levels* (see Figure 9.1 for a visualization of Markus and Conner's idea). In the following, we briefly explain each level of their model.

*Individual Level**

At the lowermost level of Markus and Conner's framework, the *individual level* describes the part of a person that pays attention, feels emotions, learns and remembers information, makes decisions, and acts, tying together his or her days and years. At this level, culture is how people make meaning out of their experiences and decide what to do with their lives. Most WEIRD people believe that individuals have a single cohesive self, but a closer look reveals multiple elements (e.g., the employee, the little league coach, the aunt, or the uncle) that can switch on or off depending on the situation (Markus and Conner, 2013). This partially explains how some individuals behave differently according to their current situation. As individuals connect with other people and things in the environment, culture occurs at the interaction and institution levels.

Interaction and Institution Levels

We experience the interaction level of the culture cycle most often. Our interactions, how we think, feel, and act in association with people and objects, are guided by implicit and explicit norms or expectations. Mundane cultural products, like stories, tools, or advertisements, make certain ways of interacting easier than others. In turn, every interaction occurs within the institutional level of the culture cycle. Institutions, including legal, religious, economic, and scientific, dictate rules for individuals and their innumerable interactions. Even when we are unaware of the full set of laws and policies and stories that these institutions dictate, they "exert a formidable force, silently allowing certain practices and products while forbidding others" (Markus and Conner, 2013).

* Each level description in the following is based on Markus and Conner (2013).

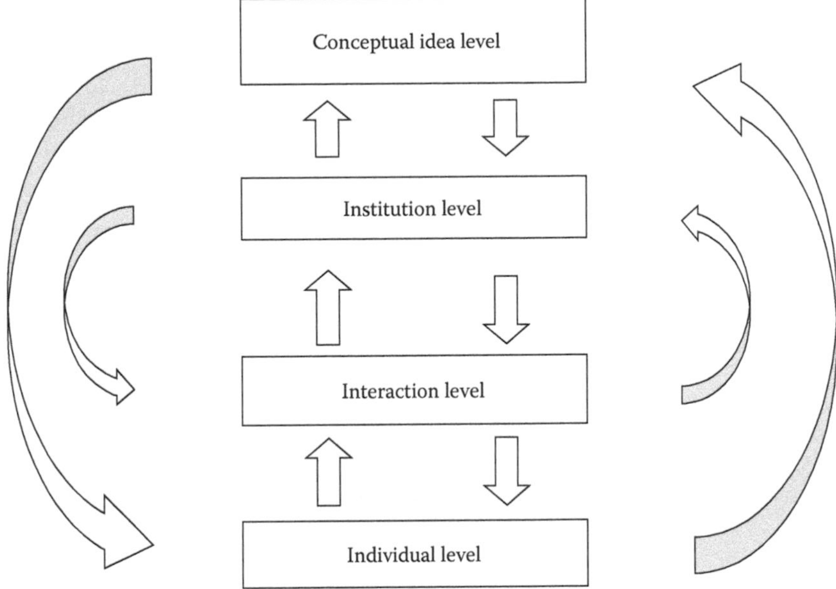

FIGURE 9.1
Overall model by Markus and Conner (2010) depicting their culture cycle. (Adapted from Markus, H., and Kitayama, S. [2010]. *Perspectives on Psychological Science*, 5(4), 420–430.)

Conceptual Ideas Level

The uppermost (and most abstract) level of this cultural model "is made up [of] the central, usually invisible [conceptual] ideas that inform our institutions, interactions, and ultimately, our individual self. Like the unseen forces that hold our planet together, these background ideas hold our cultures together. Because of them, cultures have an overarching pattern" (Markus and Conner, 2013). They offer a particular culture's answers to big questions like "how did we, and the world, get here?" Just as research in physics has identified the invisible but powerful gravitational, electromagnetic, and nuclear forces that shape the physical world, research on cultures has identified powerful ideas that shape human behavior (Combs, Blincoe, and Vorm, 2015). These big ideas cannot be directly altered by individuals; they are too deeply ingrained. But as individuals exert pressure to shift patterns of interaction and adapt their institutions, big ideas follow. Markus and Conner use the American civil rights movement to exemplify these relations. To alter the big idea that some groups are inherently superior to others, intervention and change need to take place at lower levels in the culture cycle. Individual attitudes could be altered through education and contact with others of different races or ethnicities. Abolishing Jim Crow laws would be an important step at the institutional level. With substantive, pervasive change at the individual, interaction, and institutional levels, changes can occur at the level of cultural ideas.

The next section of this chapter describes Hofstede's famous culture dimensions. We contend that these dimensions easily fit within Markus and Conner's culture cycle, at the conceptual idea level. In our thinking, these dimensions belong at the conceptual idea level because Hofstede and others have often written about these dimensions using nations as the unit of analysis. Therefore, it seems these concepts would not obviously belong in the individual, interaction, or institution levels of the culture cycle—rather, they probably impact these levels from their position in the cultural ideas level.

THE CULTURE CYCLE

Remember, Markus and Conner's model has four levels: the individual level, the interaction level, the institution level, and the conceptual idea level. The levels are interrelated; each can affect all of the others.

Hofstede Dimensions

In the 1970s, Geert Hofstede analyzed survey data from 100,000 IBM workers in more than 50 countries. His survey results showed that these countries prioritized different values, and those values could be described along four dimensions: power distance, uncertainty avoidance, masculinity–femininity, and individualism–collectivism (Hofstede, 1980; also see Hofstede, 2011). Later research has identified two additional dimensions: long-term–short-term orientation and indulgence–restraint (Minkov and Hofstede, 2010).* These six dimensions are briefly reviewed briefly in the following.

Power Distance

Power distance refers to "the extent to which the less powerful members of institutions and organizations within a country expect and accept that power is distributed unequally" (Hofstede, Hofstede, and Minkov, 2010). Nations that score higher on this concept tend to expect and accept that some people are powerful and others are not. Nations that score lower on this measure tend to expect and accept that power will be more evenly distributed among people. In countries high on the power distance dimension, people generally see hierarchy as reflective of inherent and acceptable inequalities; thus, individuals tend to accept hierarchical structures with no further justification needed for their positions within them. One empirical example of the impact of power distance comes from a project that compared test participants from a high power distance (Hong Kong) culture to a low power distance (the United States) society on how they perceive insults. A research team theorized that in a business context, people from high power distance nations would perceive insults directed by superiors at subordinate employees as more acceptable than would people from low power distance nations. Using a realistic vignette, the researchers indeed found that the Hong Kong citizens felt insults from a superior to a subordinate were more acceptable than did participants from the United States (Bond et al., 1985).

Uncertainty Avoidance

Because the future is unknown, a degree of ambiguity surrounds existence. Ambiguity, naturally, can produce anxiety. Uncertainty avoidance refers to how (and if) members of a society feel threatened by such ambiguity. As Hofstede and his team noted, feeling threatened by ambiguity is not the same as feeling threatened by risk; societies who dislike ambiguity may behave in risky ways in order to reduce uncertainty. Hofstede and his team reported that societies high in uncertainty avoidance often have more concern about truth and less room for relativism or doubt (e.g., in religious belief or moral stances). For example, students in high uncertainty avoidance societies sometimes report

* For ease of referencing, our descriptions of the Hofstede dimensions are from Hofstede, Hofstede, and Minkov (2010).

disliking teachers who admit they that do not know the answer to a question. On the other hand, students from low uncertainty avoidance societies report more respect for teachers who admit to not knowing something. Advertising campaigns, as cultural products that readily embody cultural ideas, have reflected the location of different countries on the spectrum of uncertainty avoidance. Highly avoidant nations make more extensive use of experts (such as doctors) in advertisements; less avoidant nations are more likely to rely on humor in advertising (Hofstede, Hofstede, and Minkov, 2010).

Masculinity–Femininity

The masculinity–femininity distinction is the most controversial of Hofstede's cultural conceptual ideas. As a result, some researchers use alternative names (e.g., quality of life–quantity of life or "performance oriented–cooperation oriented").* Hofstede and his team suggested that the masculinity end of this cultural conceptual idea represents "a preference in society for achievement, heroism, assertiveness, and material rewards for success. Society at large is more competitive." In comparison, societies more representative of the femininity dimension will preference "cooperation, modesty, caring for the weak, and quality of life. Society at large is more consensus-oriented." Additionally, low masculinity (i.e., high femininity) countries have more fluid sex roles. For example, equal shares of responsibility are given at home and at work, and both men and women are permitted to show tenderness and concern within their interactions. This is the only one of Hofstede's dimensions for which a systematic gender difference occurs. In other words, within a given society, such as the United States, researchers see a systematic difference on this dimension, with men scoring more toward the masculinity end and women more toward the femininity end. But cultures still differ from one another on the dimension, such that across all members of the society (men and women averaged together), certain cultures show more masculine characteristics than others (Hofstede, Hofstede, and Minkov, 2010).

As an example, the United States scores quite high on masculinity, whereas the Netherlands scores as a strongly feminine society. Illustrative of this difference, Hofstede notes that Americans freely talk of their achievements and are motivated by the possibility of showing their successes to others. In comparison, the Dutch often define success quite differently, emphasizing a work–life balance. When Hofstede and his team compared the games of children in the United States to those of children in the Netherlands, they found distinct patterns. In the United States, male school children were more likely to choose games that allowed for competition and success, whereas female children were more likely to choose games that allowed them to enjoy each other's company. These game preferences did not emerge along male–female lines in the Netherlands (Hofstede, Hofstede, and Minkov, 2010).

Individualism–Collectivism

Whereas power distance, uncertainty avoidance, and masculinity-femininity conceptual ideas are critical elements of Hofstede's cultural dimensions, none has generated nearly the level of research attention as the individualism–collectivism dimension.†

* For purposes of consistency in the literature, we use the original masculinity–femininity language.

† In many ways, the individualism–collectivism concept has completely dominated the other cultural conceptual ideas proposed by Hofstede. Ultimately, the research community should take steps to balance the overall field of research.

At its most basic, the distinction between individualism and collectivism can be reduced to the differences in societies between the role of individual persons and the role of groups. In individualist societies, the individual person and his or her interests tend to take precedence over the interests of the group. For example, a person's choice about which college to attend to might be a personal decision on which his or her family has little impact. On the other hand, in collectivist societies, the individual person and his or her interests tend to take a back seat to the interests of groups, particularly the family. When picking a college, an individual is more likely to subordinate his or her own desires and choices to the desires of the family. Hofstede suggests that one (of many) interesting way that this cultural idea manifests is in a language nuance. People from individualist societies grow up learning to say the word "I" and think of themselves as "I." They are a unique person, an individual different from others. On the other hand, people from collectivist cultures often grow up thinking of themselves in terms of "we"—"the group is distinct from other groups." Along these lines, one research team pointed out that "the English language, spoken in the most individualist countries . . . is the only one we know that writes 'I' with a capital letter." Or as one medieval Arabic saying states, "The satanic 'I' be damned" (Figure 9.2) (Hofstede, Hofstede, and Minkov, 2010).

The Newer Dimensions

As noted, in the decades since Hofstede identified the four dimensions described earlier, additional research has identified the dimension of long-term–short-term orientation and the dimension of indulgence–restraint. For brevity, we summarize those dimensions here. Interested readers should see the study by Hofstede, Hofstede, and Minkov (2010) for additional details.

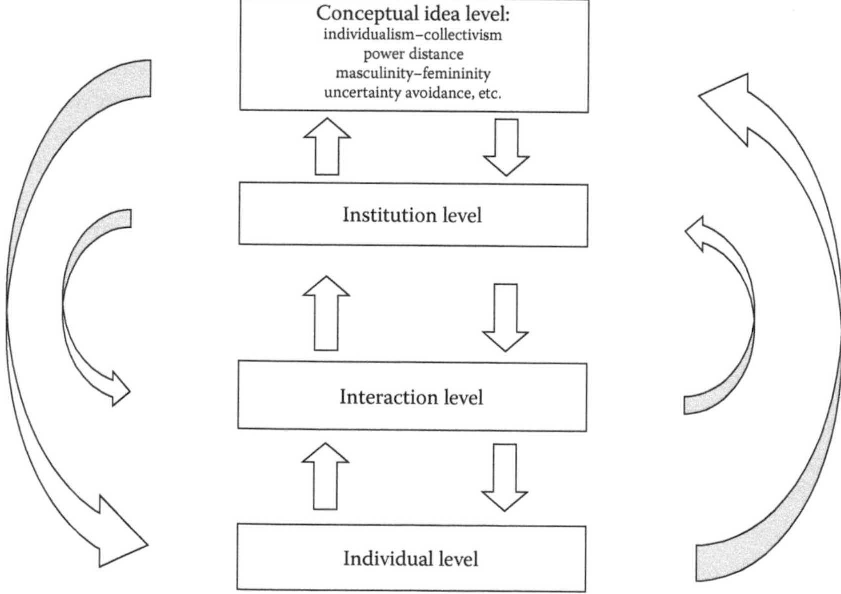

FIGURE 9.2
Illustration by Markus and Conner (2010) depicting their culture cycle. (Adapted from Markus, H., and Kitayama, S. [2010]. *Perspectives on Psychological Science*, 5(4), 420–430.)

The long-term versus short-term orientation dimension reflects how cultures consider their past, present, and future. Low-scoring societies emphasize the maintenance of tradition and look suspiciously upon change. Cultures that score high encourage the preparation for the future, through means such as education; they emphasize a pragmatic approach. In the indulgence–restraint dimension, indulgence represents a relatively permissible attitude toward the gratification of basic and natural human drives related to enjoying life and having fun. Alternatively, in societies that emphasize restraint, there are strict social norms regulating and suppressing the gratification of such needs.

Integrating the Culture Cycle and the Hofstede Dimensions

As described earlier, Markus and Conner's culture cycle has four interrelated levels—the individual, interaction, institution, and conceptual idea levels. They suggest that each of these connects with and influences the others; however, they do not propose a particular set of major conceptual ideas, nor do they discuss Hofstede's work in the context of the culture cycle. They instead focus on two styles of self, the independent and the interdependent, that cooccur for each person at the individual level. The ratio of independence and interdependence is in turn linked to hemispheric, gender, racial, and class cultures, among others (Markus and Conner, 2013). For example, Western, white, Global North, and middle-class cultures (think WEIRD) are classified as independent, whereas Eastern, non-white, Global South, and working-class cultures are classified as interdependent. In our thinking, if a researcher has a strong sense of Hofstede's dimensions and inserts them into the conceptual idea level of the culture cycle, then he or she will be in a position to make predictions about other levels of the model.

In what remains of this chapter, we propose a purposeful integration of Hofstede's value dimensions of culture into the highest level, i.e., the conceptual idea level, of the Markus and Conner's culture cycle. This integration offers theorists and researchers a fertile ground for organizing existing research findings and generating new hypotheses. To illustrate this potential integration we use the individualism–collectivism dimension and trace some of its possible effects on other levels of the culture cycle.

Individualism–Collectivism within the Culture Cycle

According to Markus and Kitayama (1991), individualism–collectivism (what we consider a cultural conceptual idea) influences the individual level process of how individuals define the self. Individuals who develop in individualist cultural settings are more likely to develop what Markus and Kitayama call a predominantly independent self, meaning a self-concept that exists independent of others. On the other hand, people who develop in collectivist cultural settings tend to be predominantly interdependent, meaning that the self-concept includes, or is interdependent with, other people.

QUICK SUMMARY: HOFSTEDE DIMENSIONS

Hofstede originally discovered four cultural dimensions in his famous IBM study—individualism–collectivism, masculinity–femininity, power distance, and uncertainty avoidance. Later research added two more dimensions: long-term–short-term orientation as well as indulgence–restraint.

The independent self is a unique person. In American culture, which embraces the idea of individualism, people tend to be more independent. They see themselves as special and stable over time and situations. For example, when asked to describe themselves, Americans tend to mention their special, stable qualities. An American might describe himself as a Libertarian, an atheist, a bookworm, and shy. These are parts of the person that will not easily change. They are things that make this person unique. By contrast, the interdependent self is "relational, similar to others, adjusting to their situations, rooted in traditions and obligations, and ranked in pecking orders" (Markus and Conner, 2013). Japanese culture, which embraces the idea of collectivism, tends to produce interdependent people. For interdependent people, the sense of self might change from situation to situation. The self is often flexible, more willing to adapt over time and for different social roles. When asked to describe themselves, Japanese respondents might mention flexibility and interconnections with others. For example, a person might describe himself as an obedient son, generous brother, and quiet employee who tries to help coworkers. These descriptions emphasize roles and flexibility (Figure 9.3).

As would be expected, this distinction between independent and interdependent people at the individual level (which is driven by differences at the conceptual idea level) affects the interaction level. According to the Globe Study (House and Hanges, 2004), people from individualist or collectivist societies (i.e., independent/interdependent people, at the individual level) tend to order their interaction level priorities somewhat differently. In individualist cultures, typically, a person's thoughts, feelings, motivations, and interaction level behaviors often emphasize individual advancement. For example, individualists usually embark upon a career, make money, and use that money as they see fit. Perhaps they purchase a new car or set aside money for vacations.

On the other hand, in collectivist cultures people are, from their birth onward, "integrated into strong, cohesive in-groups, which throughout people's lifetimes continue to

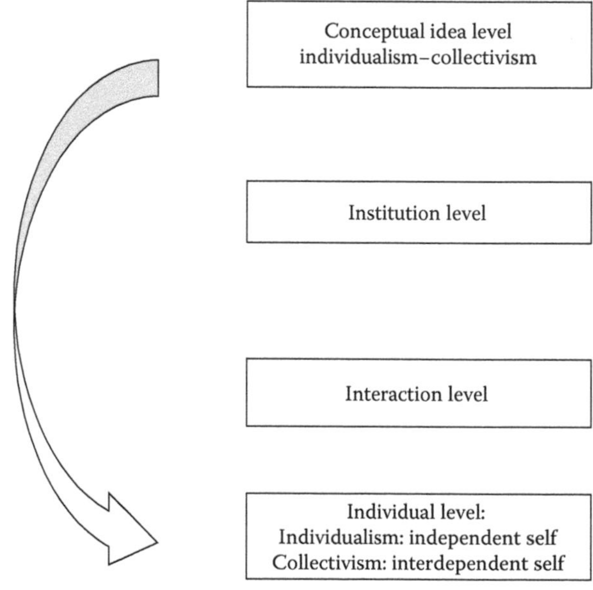

FIGURE 9.3
Illustration by Markus and Conner (2010) depicting their culture cycle. (Adapted from Markus, H., and Kitayama, S. [2010]. *Perspectives on Psychological Science*, 5(4), 420–430.)

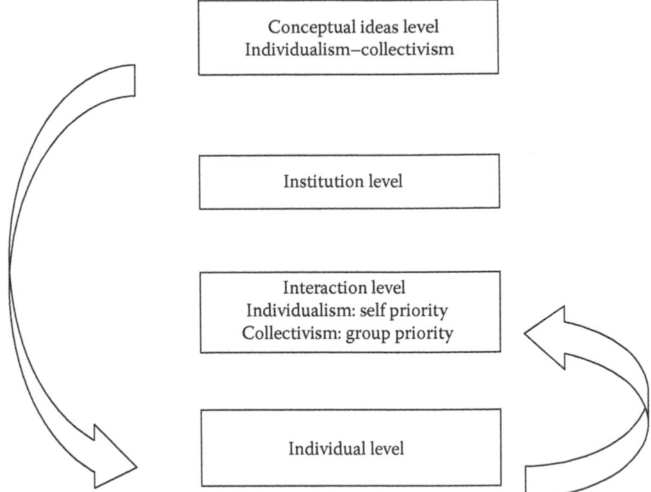

FIGURE 9.4
Illustration by Markus and Conner (2010) depicting their culture cycle. (Adapted from Markus, H., and Kitayama, S. [2010]. *Perspectives on Psychological Science*, 5(4), 420–430.)

protect them in exchange for unquestioning loyalty" (Hofstede, Hofstede, and Minkov, 2010). Correspondingly, a person's thoughts, feelings, motivations, and behaviors often focus on group advancement. Collectivists, compared to individualists, will be far more likely, at the interaction level, to share career wealth with family. As Hofstede, Hofstede, and Minkov (2010) put it, "the loyalty to the group that is an essential element of the collectivist family also means that resources are shared. If one member of an extended family of 20 persons has a paid job and the others do not, the earning member is supposed to share his or her income in order to help feed the entire family. On the basis of this principle, a family may collectively cover the expenses for sending one member to get a higher education, expecting that when this member subsequently gets a well-paid job, the income will also be shared" (Figure 9.4) (Hofstede, Hofstede, and Minkov, 2010).

So far, we have tried to point out ways in which the individualism–collectivism distinction, if considered at the conceptual idea level of the culture cycle, affects the individual level of the cycle (the independent/interdependent concept of self), and how the individual level of the cycle in turn affects the interaction level. How might the interaction level of the cycle impact the institution level? Although the empirical research on this matter is mixed,* some perspectives have suggested that businesses in collectivist nations (such as Japan, a notably collectivist society according to Hofstede) are more likely to emphasize values such as collective responsibility for decisions, group success (rather than individual success), and collective decision making. On the other hand, individualism may be manifested at the institution level in the value managers place on individual achievement, drive, and ambition. Some perspectives suggest that similar patterns exist in sports. For example, although anecdotally, one report (Nippon Professional Baseball: Major League Baseball 2003) points out that Japanese baseball players change teams far less frequently than American players because of a mentality that a team is a family in Japan compared to a job in the United States. In addition, one comparative analysis (Ikei, 2000) of Japanese and American baseball culture indicates that Japanese baseball managers are often former star

* See the study by Beechler et al. (1996) for a review and somewhat contrarian empirical findings.

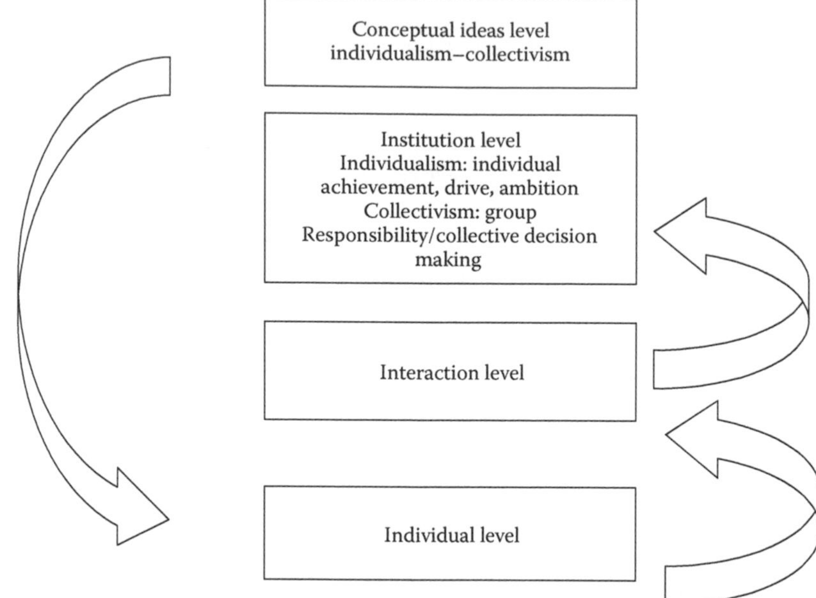

FIGURE 9.5
Illustration by Markus and Conner (2010) depicting their culture cycle. (Adapted from Markus, H., and Kitayama, S. [2010]. Cultures and selves: A cycle of mutual constitution. *Perspectives on Psychological Science*, 5(4), 420–430.)

players of a team; they may not be especially qualified to manage but are considered part of the team family. American managers, on the other hand, may not be former players at all; instead, they are generally selected for their managerial acumen. In another sports-related example of how individualism-collectivism probably affects the institution level of the culture cycle, a group of American and Japanese bilingual observers were trained to code media coverage of the 2000 and 2002 Olympics. The researchers were interested in how American and Japanese coverage differed. They found that American journalists talked more about the individual characteristics of athletes and attributed failure and success to individual characteristics, of either the competitor or the American athlete. The Japanese media, however, discussed failure as the result of not trying hard enough to do what was expected. The Japanese, both the press commentators and the athletes themselves, made twice as many negative comments as the Americans (Figure 9.5) (Markus et al., 2006).

INDIVIDUALISM–COLLECTIVISM

One hypothesis that could be derived from perspectives on individualism–collectivism as it relates to baseball might be how this cultural dimension impacts the tactics used by players and managers in places like Japan and the United States. Might a quantitative comparison between Japanese baseball and American baseball reveal differences in the volume of sacrifice bunts, sacrifice flies, etc.? One reasonable hypothesis might be that players in the Japanese leagues might engage in sacrifice for their teams more often than players in places like the United States might. As discussed, in collectivist societies, there is a far greater emphasis on group success than individual success.

Implications for Future Research

For researchers and analysts, we believe that a strong understanding of Hofstede's major cultural conceptual ideas can generate policy recommendations (as noted in the Xanadu case example) and research ideas. These major cultural conceptual ideas (e.g., Hofstede's dimensions) have been tested globally. This means that if practitioners or researchers find themselves in a situation in which they have little to no understanding of the individual, interaction, or institution levels of a nation or a region, knowing how a society is characterized via Hofstede's dimensions will place them in a strong position to generate initial policy ideas and/or research hypotheses.

For example, if an intelligence officer needs to brief a team of humanitarian aid workers on how best to interact with locals in Indonesia after a terrible storm, the officer, simply knowing that Indonesia is a strongly collectivist society, can immediately begin formulating hypotheses about the kinds of individual, interaction, and institution level cultural nuances he or she might encounter. Such experience would similarly provide an intelligence officer, for instance, additional insights on Ethiopia—another nation that is strongly collectivist. That is, regardless of the particulars, people in both Indonesia and Ethiopia share ways of thinking in terms of "we" rather than "I," prioritize harmonious friendships and family relationships of individual self-advancement, adjust themselves to the expectations of others, and root themselves in tradition. Clearly, the specific nuances of how these dimensions manifest themselves will differ by nation. But the broad themes remain similar. (Of course, it is still important to examine how Indonesia and Ethiopia are similar or different on the other Hofstede conceptual ideas as well. Just because they are similar on the individualism–collectivism spectrum does not mean that they will be similar on dimensions such as power distance.)

Conclusions

This chapter briefly outlined the currently WEIRD state of cross-cultural social science research, described two prominent models or frameworks for culture, recommended an integration of those models, and provided examples of how their integration can help researchers and analysts better understand and predict behavior—ultimately helping to close the gap in the research knowledge between WEIRD and non-WEIRD cultures.

As noted earlier, the social and behavioral sciences are plagued by an overreliance on WEIRD models, WEIRD research, and WEIRD researchers. WEIRD researchers often assume, implicitly or explicitly, that their WEIRD theories, tested with WEIRD participants, are general to all of humanity. Such assumptions can (and do) frustrate practitioners who want to leverage social science research but often find that results obtained in WEIRD contexts fail to replicate in places like Afghanistan. The first author of this chapter has had many discussions with frustrated military personnel and diplomats who have followed the guidance of WEIRD research when working in non-WEIRD contexts only to see the guidance fall completely flat. As a result, some practitioners simply give up on the social sciences as a source of insight altogether.

We believe that this is not the time to become skeptical of the social sciences. The field as a whole has a wealth of knowledge about WEIRD societies. Cooperatively, governments, militaries, diplomatic communities, and academics can lead the charge for greater cross-cultural understanding through rigorous, systematic research. Furthermore, we believe that a simple integration of Markus and Conner's culture cycle with Hofstede's cultural dimensions will assist with framing and forming new hypotheses. Perhaps, decades from now, when these research gaps are closed—no one will seem *weird* at all.

Megacity Use Case

Earlier, we suggested a way in which Markus and Conner's culture cycle and Hofstede, Hofstede, and Minkov 's major culture dimensions (i.e., conceptual ideas), when integrated, could provide a useful framework for understanding existing sociocultural research and encouraging additional non-WEIRD exploration. Specifically, Markus and Conner's culture cycle provides a helpful taxonomy of the levels of analysis ranging from major culture conceptual ideas all the way to individual level thoughts and perceptions, and Hofstede's cultural dimensions provide a helpful way to make predictions within that cycle. In the section that follows, we use the Xanadu megacity case example to illustrate how the integration of the culture cycle and the culture dimensions can aid practitioners. Specifically we attempt to show how a research analyst can make solid recommendations to superiors about an information campaign if he or she is grounded in the integration of the culture cycle and Hofstede's individualism–collectivism dimension. Since the case example is about an information campaign (the kind of activity we believe might be especially suited to leveraging this kind of perspective), we constrain our example to how the conceptual idea level impacts the individual level of the model.

Xanadu

As Xanadu grows and the influx of refugees continues, the government of Çawaila and the broader western world grow concerned for regional stability and devote increased attention to the city. Although Çawaila has seen coups, civil unrest, and insurgency for decades, a new government has just been elected, and the subsequent calm permits enough breathing room to try and establish a functioning and inclusive democracy. Officials in the new government, although internationally supported, fear future unrest. In order to get ahead of any possible crisis, the newly elected (but cash-strapped) government of Çawaila reaches out to a global research and marketing organization for assistance. Julie, a freshly minted PhD graduate, is part a team of analysts tasked with providing perspectives to Çawaila on the people groups within the city; this information will be used to design information campaigns. Unfortunately, the turmoil that has characterized Xanadu for the last several years means that accurate demographic information, let alone data from standardized opinion polling, is unavailable or, in the case of data from the UN or the World Bank, out of date. Time and funding are insufficient to commission a reliable census. The possibility of using social media data to understand the city's population has been rendered moot by the Çawaila government's legal team in order to protect citizen privacy.

Julie is in the difficult position of trying to provide actionable information for decision makers—while having effectively no data to work with. Can Julie provide more than anecdotal information from singular contacts and sources she may know in the city? As it turns out, with a thorough understanding of Hofstede's dimensions of culture and their place in the Markus and Conner culture cycle, Julie has a wealth of information at her disposal. Although the cultural dimensions have not been formally measured in Xanadu, Julie is aware that prior research categorizes most of the nations from which the refugees are coming as strongly collectivist.* At the conceptual idea level, a common collectivist theme motivates and structures the institutions, interactions, and individual selves she is tasked to understand. Acknowledging the differences that nevertheless exist, Julie can anticipate that most people in the city tend to think in terms of "we" rather than "I," prioritize harmonious friendships and family relationships, adjust themselves to the expectations of others, and root themselves in tradition. She also remembers two pieces of critical information regarding information processing that might be especially interesting to people trying to generate information campaigns in Xanadu.

Information Processing within the Culture Cycle

Equipped with the knowledge that Xanadu's residents are influenced by the conceptual idea of individualism–collectivism, she can anticipate certain behaviors at the individual level. Specifically, she knows that research has linked individualism with a more analytic individual level information processing and collectivism with holistic processing. Individualist/analytic thinkers tend to "see objects as discrete and separate from their environments; they see events as moving in linear fashion when they move at all; and they feel themselves to be personally in control of events even when they are not" (Nisbett, 2003). On the other hand, non-Westernized collectivists tend to engage in more holistic thinking. They will "see a great deal of the field, especially background events; they are skilled in observing relationships between events; they regard the world as complex and highly changeable and its components as interrelated; they see events as moving in cycles between extremes; and they feel that control over events requires coordination with others" (Nisbett, 2003).

Furthermore, the individualism versus collectivism distinction affects how people process the information they see. For example, Julie once read a research paper that described collectivist and individualist participants watching two brief film clips of almost identical scenes. Their task was to identify how the clips differed. In reality, the film clips differed in terms of the background (e.g., the shape of an airport control tower), the foreground (e.g., the landing gear of an aircraft was up or down), and in the relationships between objects (e.g., a helicopter moved closer to a single-engine plane). Individualist test participants noticed more differences in the foreground of the video while collectivist test participants noticed more differences in the background of the video (Masuda and Nisbett, 2002). To better understand these findings, other researchers used eye-tracking equipment. They found that individualist people rarely looked away from the focal points of the scene. On the other hand, collectivists looked at the focal point only briefly and then quickly shifted their gaze to gather information from the rest of the background (Chua, Boland, and Nisbett, 2005).

With this information about information processing distinctions between individualist and collectivist cultures in hand, Julie can approach her leadership and recommend that any information campaign they propose to the Xanadu government should (1) avoid a campaign that relies on simple logic and linear thinking in favor of a campaign that is more holistic and emphasizes relationships between people and events and (2) carefully construct background information. Knowing that the collectivists within Xanadu are

* She, and you, can obtain such data from http://geert-hofstede.com/countries.html.

more likely to see and process background information than people from more individualist societies, messaging could be inserted into the background of scenes.

By integrating the Hofstede cultural dimensions into the conceptual idea level of the Markus and Conner culture cycle, Julie is able to anticipate possible effects of the conceptual ideas of individualism and collectivism on the individual level information processing. Her perspectives, while not a silver bullet for campaign design, helps her team provide culturally relevant information that could have easily gone overlooked. She is aware that the individualism and collectivism distinction, at the conceptual idea level, provides even more predictive power at the individual level of the culture cycle. However, her team informs her that due to space constraints, she needs to simply footnote a reference for such additional information, which she does.*

DISCUSSION QUESTIONS

1. Markus and Conner's culture cycle suggests that there are four interrelated levels of culture. Is this framework compelling? Should there be more (or fewer) levels of the cycle? If so, why?
2. Beyond Hofstede's culture dimensions, what else could be considered cultural conceptual ideas?
3. We referenced research and perspective from sports, which indicated that individualism–collectivism probably has an impact on values and approaches in various sporting institutions. How might the other Hofstede dimensions impact sports?
4. The Xanadu megacity case example laid out ways in which individualism–collectivism could affect other levels of the culture cycle. Can you think of ways in which the other Hofstede dimensions affect the other levels of the cycle? For example, how might power distance have an impact at the institution level?
5. Where would religions (such as Islam, Christianity) sit within the culture cycle? Ought they be considered conceptual ideas (like Hofstede's dimensions) or institutions? Does religion impact Hofstede's dimensions—do the dimensions impact religions?

References

Beechler, S., Najjar, M., Stucker, K., and Bird, A. 1996. Japaneses-style versus American-style human resource management overseas: Examining whether the data support the "facts." Working Paper Series Center on Japanese Economy and Business Graduate School of Business Columbia University.

Bond, M., Wan, K., Leung, K., and Giacalone, R. 1985. How are responses to verbal insult related to cultural collectivism and power distance? *Journal of Cross-Cultural Psychology*, 16: 111–127.

Brown, D. E. 1991. *Human Universals*. Philadelphia, PA: Temple University Press.

Chua, H. F., Boland, J. E., and Nisbett, R. E. 2005. Cultural variation in eye movements during scene perception. *Proceedings of the National Academy of Sciences USA*, 102: 12629–12633.

* Julie then refers her colleagues to the study Combs, Blincoe, and Vorm (2015), for a much deeper discussion.

Combs, D., Blincoe, S., and Vorm E. 2015. Trust, attitudes, and social influence: The cross-cultural social psychology of counterinsurgency. *Department of Defense Manual*. Available at http://pksoi.army.mil/.

Heine, S. J. 2008. *Cultural Psychology*. New York: W. W. Norton.

Henrich, J., Heine, S., and Norenzayan, A. 2010. The weirdest people in the world? *Behavioral and Brain Sciences*, 33: 61–135.

Hofstede, G. 1980. *Culture's Consequences: International Differences in Work-Related Values*. Beverly Hills, CA: Sage.

Hofstede, G. 2011. Dimensionalizing cultures: The Hofstede model in context. *Online Readings in Psychology and Culture*, 2.

Hofstede, G., Hofstede, G. J., and Minkov, M. 2010. *Cultures and Organizations: Software of the Mind* (Rev. third ed.). New York: McGraw-Hill.

House, R., and Hanges, P. 2004. *Culture, Leadership, and Organizations: The GLOBE study of 62 societies*. Washington, DC: Sage Publications Inc.

Ikei, M. 2000. Baseball, Besuboru, Yakyu: Comparing the American and Japanese games. *Indiana Journal of Global Legal Studies*, 8: 73–79.

Kilcullen, D. 2010. *Counterinsurgency*. Oxford: Oxford University Press.

Markus, H., and Conner, A. 2013. *Clash! 8 Cultural Conflicts that Make us Who We Are*. New York: Hudson Street Press.

Markus, H., and Kitayama, S. 1991. Culture and the self: Implications for cognition, emotion, and motivation. *Psychological Review*, 98: 224–253.

Markus, H., and Kitayama, S. 2010. Cultures and selves: A cycle of mutual constitution. *Perspectives on Psychological Science*, 5(4): 420–430.

Markus, H. R., Uchida, Y., Omoregie, H., Townsend, S. S. M., and Kitayama, S. 2006. Going for the gold: Models of agency in Japanese and American culture. *Psychological Science*, 17: 103–112.

Masuda, T., and Nisbett, R. E. 2002. *Change Blindness in Japanese and Americans*. Unpublished manuscript. Ann Arbor, MI: University of Michigan.

Minkov, M., and Hofstede, G. 2010. Hofstede's fifth dimension: New evidence from the world values survey. *Journal of Cross-Cultural Psychology*, 43: 3–14.

McCrae, R. R. 2002. NEO-PI-R data from 36 cultures: Further intercultural comparisons. In McCrae, R. R., and Allik, J. (Eds.), *The Five-Factor Model of Personality across Cultures* (pp. 105–126). New York: Kluwer Academic/Plenum Publishers.

McNemar, Q. 1946. Opinion attitude methodology. *Psychological Bulletin*, 43(4): 289–374.

Nippon Professional Baseball: Major League Baseball. 2003. Differences between Nippon professional baseball and major league baseball. Available at http://www.umich.edu/~wewantas/brooke/differences.html.

Nisbett, R. E. 2003. *The Geography of Thought: How Asians and Westerners Think Differently—And Why*. New York: Free Press.

Sears, D. O. 1986. College sophomores in the laboratory: Influences of a narrow data base on social psychology's view of human nature. *Journal of Personality and Social Psychology*, 51(3): 515–530.

Ucko, D. H. 2009. *The New Counterinsurgency Era: Transforming the U.S. Military for Modern War*. Washington, DC: Georgetown University Press.

Williams, J. E., and Best, D. L. 1990. *Sex and Psyche: Gender and Self Viewed Cross-Culturally*. Newbury Park, CA: Sage.

10

Deciphering the Emic Perspective in Data in Order to Assess Threat

Laurie Fenstermacher and Lawrence A. Kuznar

CONTENTS

Introduction 182
The Relationship of the Researcher/Analyst to Data 182
What Are the Data? 183
The Problem of Inference 183
Whose Interpretation of What? 184
What Is a Threat Narrative? 185
How Are Threat Narratives Assessed? 186
The Emic Perspective, Threat Narrative and Forecasting Human Behavior 187
In Search of an Emic Perspective 188
The Lens of Social Identity: Forecasting Methodologies 190
The Lens of Affect: Dynamics of Conflict 192
Text Analytic Tools for Forecasting Violence Based on Multiple Discourse Markers 195
How Important Is the Emic Perspective for Forecasting? 198
Conclusions and Looking Forward 199
Megacity Use Case 200
References 201

ABSTRACT Methodologies and text analytic algorithms that consider the emic (or first-person) perspective can help analysts interpret threat narratives, enabling them to better forecast violent events and prepare appropriate COAs. This chapter discusses the rationale, the approach, and some results for these methodologies and algorithms. It includes several real-world examples of their application, such as from the Islamic State and the Boko Haram, political and military actions between India and Pakistan, and competing actors in Philippines and Egypt. Empirical testing found that the emic discourse markers correlated with actual violence 69% of the time (much higher than the standard etic approaches); this increased to an 86% correlation when the emic and etic models were integrated.

KEY TOPICS

- Discourse analysis to help forecast violent events and inform courses of action
- Brief overview of emic and etic perspectives and methodological considerations
- Interpretative frameworks (such as social identity theory) applied to discourse analysis

- Discourse features: integrative complexity, sentiment, idea density, vocabulary diversity, keyness
- Development of a discourse processing tool to identify intent and forecast behaviors

Introduction

"Hindsight is 20/20." The post hoc discovery of information or a behavior pattern that, if discovered prior to the event would have enabled the prediction or the forecast of that event, often provides a forensic diagnosis for many violent events. And, the diagnostic information or behavior pattern is often in the form of verbal or physical cues provided by violent offenders (Defense Science Board Report 2012, p. 19).

For example, prior to the attack at Columbine High School, shooter Eric Harris had openly threatened classmates and described pipe bombs on his website, seemingly broadcasting his intent to engage in violence. In the year preceding the attack, Harris' website became increasingly filled with suicidal anger and rage expressing hatred in general as well as toward specific individuals, including the desire to kill certain individuals—essentially a target list: ". . .I am coming for EVERYONE. . .and I WILL shoot to kill. . . ." (Owen 2001). A classmate, Brooks Brown, stated (about Harris' website) that "Eric was saying how he was gonna blow people up. 'Hey, I'm making pipe bombs. I've got the designs for them on my Web site. I'm gonna kill these people. Here's why.' That's a level beyond making a joke" (Kohn 2001).

Even apparently obvious signals, such as these, are frequently missed altogether, disregarded, or buried in noise. And explicit verbal threats are typically not communicated publically, but rather to friends or family members (Davis et al. 2013). Often, however, there are implicit verbal (or textual) patterns that can be used to forecast violent events. These patterns frequently occur well in advance of the event, but they are nuanced, requiring the development of a different lens to find and interpret discursive patterns and behaviors associated with social identity, affect, and integrative cognitive complexity. The US Air Force Research Laboratory, in partnership with analysts at the US National Air and Space Intelligence Center, developed methodologies and text analytic algorithms exploiting these patterns in order to enable analysts to make meaning about threats and threat narratives based on *emic*, or first-person, perspectives in addition to forecasting violent events.

Prior to discussing the rationale, the approach and some results for these methodologies and algorithms, we address the epistemological difficulties with assessing threat narratives that originate from viewpoints or cultures foreign to our own, and we provide guidelines concerning the issues analysts must bear in mind to create valid understandings of others' threat narratives.

The Relationship of the Researcher/Analyst to Data

Data never speak for themselves. A host of intermediary variables modifies, distorts, and reveals what the data can represent. Therefore, capturing the intended meaning of a threat

In order to assess one's ability to capture a meaningful reality from data, various issues, described in the subsections below, must be addressed.

narrative and its interpretation by its intended audience require a careful consideration of how biases transform data into an analyst's perceptions.

What is the data or the information from which an analyst assesses or makes meaning about a threat? The analyst(s) will almost never have any direct contact with the individual(s) involved in a violent event; he or she only knows the threat/adversary through an examination of documents and intelligence derived from other human and technical means of collection. To understand the meaning of a threat—not only the *what* and *who* but the *why*—requires the application of the analyst's body of knowledge to provide a contextual foundation from which to reason about the available evidence, all while avoiding various analytic pitfalls (e.g., biases, evaluation of evidence).

What Are the Data?

In the analysis of discourse, the data consist of unstructured text (words, phrases, sentences, paragraphs) from written sources and utterances, including extemporaneous or scripted speeches.

Human beings possess an incredible ability to extract meaning from discursive data. However, the meanings humans extract are notoriously subjective, idiosyncratic, unstable, and influenced by context. In order to overcome these distortions, the analyst must realize that no meanings in data are given, but are the result of inferences the analyst makes from data.

The Problem of Inference

The analyst is burdened with having to clearly convey and justify what he or she thinks the data represent. This alone demonstrates that the data do not speak for themselves; an interpretation of the data must be backed up with arguments for the logical connection from the analyst's interpretation to the original data themselves, to demonstrate why alternative interpretations are not better. The data do not speak for themselves; they are spoken for. A broadly scientific approach is recommended to establish the validity of analysts' interpretations in order to insure that the logical connections between the data and the analyst's inferences are clear and both empirically and theoretically supported.

The scientific approaches to interpretation of data have nothing at all to do with quantification, use of technology, or proving any final truths about the world. They are characterized by transparency, rigor, logical connections to data, reliability (which is analogous to trainability), and vulnerability to better interpretations (Kuznar 2009). Scientific approaches to data interpretation/analysis result in an interpretation that is not idiosyncratic to the analyst, requiring the analyst to be explicit about how he or she interprets the data, to be rigorous and consistent in the application of his or her criteria for interpreting the data, and to be open to the fact that his or her interpretations are only ever partial, flawed, and provisional, and therefore open to reinterpretation. Challenging interpretations (competing hypotheses) must adhere to the same criteria and be reliable; that is, other analysts, applying the same methods to the same data, should generate the same interpretation. These are the hallmarks not only of the scientific method, but also of the scientific attitude that transcends both the sciences and the humanities (Bronowski 1951).

Archaeologists, anthropologists, and others deal with this very problem in that they must infer what data represent while limiting distortions based on their cultural and theoretical biases. The meaning of the data depends on context (e.g., other artifacts found in the same location) and the theoretical question a cultural anthropologist or an archaeologist is exploring. For example, consider a seashell is found with a chipped edge 10,000 ft in altitude in the Andes in an ancient wood structure with projectile points and bones of animals. It is a knife. This is very meaningful to an anthropologist who wants to reconstruct the daily economic life of ancient Andean people.

The same seashell, found at 10,000 ft on a pile of stones with charred bones and broken, highly decorated pots, is more likely to be a sacrificial offering representing an ocean or a water deity (there are bridging arguments based on what is known about Andean cultures to support this). This is of less interest to the economic anthropologist, but highly significant to the anthropologist who wants to understand ancient Andean religion. Finally, if a broken seashell is found on the Pacific beach with nothing else around, it is a broken seashell. No anthropologist cares about this. In each case, the possibility exists that the shell does not have the meaning assigned to it; however, the context, combined with other knowledge, makes each interpretation more likely correct. Furthermore, the theoretical question increases the relevance of the interpretation. It is important to be explicit about the assumptions and the bridging arguments regarding the context and how that gives meaning to the artifact/data.

For instance, one contributing author had the experience of encountering a church in a remote valley in the treeless highlands of Peru with the remains of three bonfires in front of it. He confidently turned to a villager and remarked, "This must be a representation of the Trinity." His informant looked at him blankly and responded, "No, we found enough wood for three fires this year, usually we only have enough for one." The anthropologist prematurely used his own understanding of Christianity to interpret the data, but the informant's emic perspective was entirely different and immanently pragmatic in this case. Interpretations are tricky things, sometimes meaning is deeply embedded in layers of meaning only an insider can understand.

Whose Interpretation of What?

But what meanings are analysts trying to discover? Anaïs Nin was quoted as saying, "We don't see things as they are, we see them as we are." Human cognition and behavior are shaped or influenced by beliefs and assumptions about life and reality, the various dimensions of worldview. Worldview is the lens through which a person views and interprets reality. To interpret or make meaning regarding the behaviors of another, an analyst should understand the worldview of that individual or group. Richards Heuer (1999, p. 33) stated that to ". . . see the options faced by foreign leaders as these leaders see them, one must understand their values and assumptions and even their misperceptions and misunderstandings. Without such insight, interpreting foreign leaders' decisions or forecasting future decisions is often little more than partially informed speculation. Too frequently, foreign behavior appears 'irrational' or 'not in their own best interest.' Such conclusions often indicate analysts have projected American values and conceptual frameworks onto the foreign leaders and societies . . ." Instead, if we are interested in knowing what others' texts mean, then the key is to know what those texts mean to them.

TABLE 10.1

Islamic State Emic and Etic Example

	Emic	Etic
Emic	The apocalypse is near.	The Tigris and Euphrates Rivers are drying up as a sign of the apocalypse.
Etic	People believe that there is a prophecy that the Tigris and the Euphrates will dry up.	The Tigris and the Euphrates are drying up as a consequence of global climate change

This native participant viewpoint or first-person perspective is *emic*, a term originally derived from *phonemic* by linguist Kenneth Pike. This is contrasted with the *etic* (derived from *phonetic*) perspective, the detached *observer* or third-person viewpoint. The goal of the emic analysis of data was to understand which set of sounds conveyed specific meaning to the native speakers of a language (Pike 1998). When applied to the study of human behavior, the emic viewpoint results from studying behavior as from inside the system (Pike 1967), attempting to "capture participant's indigenous meanings of real-world events" (Yin 2010) and look at "things through the eyes of member of the culture being studied" (Willis 2007). This contrasts with the etic perspective, encompassing the external view of a culture, a language, meaning associations, and real-world events (Lett 1990, p. 130). Anthropologist Marvin Harris pointed out that the etic perspective amounts to an outsider observer's perspective as constrained by the scientific criteria outlined earlier (Harris 1979, pp. 34–41; 1999, pp. 32–33, 35). However, since etic perspectives are constrained by their own epistemological biases, if only the etic perspective is considered or included, there is the possibility of overlooking hidden nuances, meanings, and concepts (Morris et al. 1999).

Table 10.1 shows an example of the emic and etic interpretations from the Islamic State.

In the context of analysis for understanding threat narratives and forecasting violent events, the *etic* and *emic* terms contrast the subjective/worldview-filtered view of a situation/event/issue with that of a more objective/scientific/measurement-based view. While both etic and emic perspectives are an important "part of any understanding" (Agar 2011, p. 39; Harris 1999, p. 33), the current balance of sensor data collection/processing/analysis overwhelmingly supports the etic perspective. An "outsider's (etic) perspective can never fully capture what it really means to be part of the culture" (Olive 2014). Emic information provides unique insights and important early indicators or signals of impending action/violence as well as important context to enable meaning-making beyond the target detection or identification (Fenstermacher 2015). One way to integrate these perspectives in order to make meaning about threats is to use threat narratives as a construct, pieced together from the physical and verbal data available in order to assess and forecast threats.

What Is a Threat Narrative?

A threat narrative, as discussed in this chapter, is not simply an utterance (or set of utterances) that is threatening. A threat narrative is the externalized behavioral storyline in which knowledge and subjective reality regarding threat is expressed (Moon 2009). It includes the answer to questions as perceived by the object of study: Is there a threat? What is motivating the threat? What is the intent? What are the capabilities of the threat? How can the threat be deterred or defended? Where did the threat come from? Why does

THREAT NARRATIVE

A threat narrative is the externalized behavioral storyline in which knowledge and subjective reality regarding threat is expressed. It is not simply a threatening utterance (or set of utterances).

it exist? What are the *right* actions regarding the threat? What are the preferred future outcomes? The focus on the threat narrative implicitly involves an understanding of the what, who, and how related to the threats and also enables the development of an understanding of why—providing a context for forecasting events as well as key clues regarding potential COAs.

A threat narrative is a construct, a framework that enables meaning-making about a threat and forecasting of events. It integrates multiple perspectives, the emic or first-person/participant perspective and the etic or third-person/observer perspective. It creates a tomographic view (one that elucidates and integrates the aspects of a worldview as well as penetrates multiple layers of meaning) of how the other (an individual or group) views themselves, others, and the world. Through an understanding of the threat narrative, one can begin to make inferences about the motivation(s) and the intent as well as the potential COAs in order to influence or shape behaviors. Equally important, a threat narrative provides a guide for the collection, the analysis, and the integration of data and information (Fenstermacher 2014).

How Are Threat Narratives Assessed?

The behavioral storyline of the other is found in what they say, write, and do. Underpinning this behavioral storyline is a worldview, the lens through which a person or a group views reality. Thus, a systematic assessment of the aspects of worldview, as expressed in physical and dialectical manifestations, provides an approximation of the threat narrative. The worldview-based constituents of a threat narrative include the emic perspective of the past, the present, and the future.

To do so, an analyst seeks the answers to the following questions:

- Is there a threat? If so, what is the threat? What is motivating the threat? What is the intent of those who pose the threat? What are the capabilities of those who pose the threat? What are the vulnerabilities of the threat? How is it structured? How does it function? What is its nature? What are our (the subject of study) capabilities, both tangible and intangible, to deter, defend against, or defeat the threat?
- Where did this threat come from? Why does this threat exist? Why is the threat the way it is and not different? How did it originate? What kind of explanations can be put forward? How did we (the subject of study) get to where we are regarding our ability to deter, defend against, or defeat the threat?
- Where does the subject of the study perceive he or she is or should be going (what is the desired future)? How does the threat influence this perceived future?

In addition, the worldview dimensions regarding beliefs (existential, evaluative, proscriptive, and prescriptive) are assessed. A good starting point is to assess the beliefs regarding: human nature (moral orientation, mutability, complexity—are humans good or evil; is this temporary or permanent?), will (free will, determinism and rational/irrational roots of behavior—are humans impacted more by biological or environmental/social factors?), cognition (human thoughts and minds—what is the most valid source of knowledge: science, tradition, an authority, or intuition?), behavior (focus of behavior, behavior guidelines—is the past, the present, or the future more important in decision making? Are moral guidelines absolute or relative?), interpersonal (characteristics of interpersonal and intergroup relationships—is the authority structure hierarchical or lateral with shared power?), truth (is truth relative or universal? is what they believe true inclusive or exclusive for their identity group?), and world/life (purpose of life and place of humanity in the cosmos—what is the nature of the divine?) (Koltko-Rivera 2004). An understanding of these beliefs informs the answers to the following questions:

- What does the subject regard as true or false?
- How does the subject evaluate reality, especially the reality of threat? What does the subject believe he/she should strive for?
- How does the subject believe he/she should act in response to a perceived threat? What are the subject's general principles governing the organization of his/her actions to deter, defend against, or defeat the threat? What is the subject willing to do or not to do to deter, defend against, or deter the threat? What is the subject's preferred means to an end?

The process of assessing threat narratives helps to develop a context to make meaning from information collected; for example, understanding the language used by an in-group and the encoded meaning based on values, referents. For example, in a recent discourse analysis study on the Islamic State based on translated documents, one of the documents contained the word *mischievous*. This did not attract any particular attention, since the word is not typically associated with threat, particularly imminent threat. However, contextualization based on deciphering aspects of worldview and threat narrative could sensitize an analyst to coded language being used. This includes understanding prescriptive beliefs, such as the recognition that *mischievous* is actually a reference to a specific interpretation of a Quran passage which rationalizes violence (Quran 5:33)—essentially a veiled alert, a signal of impending violence.

The Emic Perspective, Threat Narrative and Forecasting Human Behavior

Forecasting is difficult for several reasons, including the fundamental unpredictability of certain classes of events and of human behavior itself, given the facts that the past is an imperfect indicator of the future and the difficulty in discerning a signal from a noise. Some event patterns can be detected post hoc but cannot be used to forecast events, while other patterns are not causal and cannot be perceived or predicted (Kurtz and Snowden 2003). In addition, patterns change, as humans adapt to different situations or vary from a certain model (e.g., rational actor model in which it is assumed that behaviors/decisions always serve to maximize utility).

Often, the strategy employed to anticipate future threats is to extrapolate from past events; however, this is often not effective, akin to "drivers looking through the rear view mirror while convinced they are looking ahead" (Taleb 2007). Overreliance on past observations ignores the less tangible/directly visible aspects of past cases, leading to the production of false trend models and erroneous expectations. A focus on the emic perspective and the investigation of the elements of a worldview as part of the assessment of a threat narrative move the understanding of a situation/threat beyond descriptive (identifying who and what) to the evaluative (why), enabling predictive assessment or forecasting.

In Search of an Emic Perspective

There has long been an interest in linking what people say to what they do, especially in identifying cues in language that help to forecast or anticipate violent events. Much of the focus in text analytics has been on developing algorithms for entity extraction or sentiment analysis. These are both useful; however, they are not effective for assessing threat narratives or enabling forecasts of violent events. The gold standard of techniques for forecasting instability/violence has traditionally been a technique called *event coding* (O'Brien 2010)—essentially forecasting future events based on the type and the frequency of events that occurred in the past. Subsequent research efforts demonstrated that the inclusion of sentiment variables markedly improved the ability to forecast the advent and the cessation of violence (Shellman, Levey, and Young 2013).

Thus began a sequence of US Air Force Research Lab research efforts focused on the identification and the interpretation of discourse and discursive behaviors that enable an anticipatory analysis (early indicators or signals), which precedes a future event or behavior (e.g., violent event). The research focused on developing both methodologies and text analytics to find and interpret discursive patterns and practices related to social identity, affect, integrative cognitive complexity, trustworthiness, and worldview.

These behaviors have been described and studied in previous research and are based on social identity theory, insights from neuroscience, political psychology, critical discourse analysis, and studies of worldview. The social identity theory (Tajfel and Turner 1979, 1986) posits that people identify themselves with various social categories or groups in order to improve their self-esteem, categorize the groups with which they identify (self-categorization), and tend to accentuate differences or similarities (beliefs, norms, style of speech, attitudes, etc.) relative to other groups, comparing their groups (in-group) to others (out-group[s]) along dimensions in which the in-group is judged positively and the out-group is judged negatively (Hogg and Abrams 1988). When groups (are perceived to) compete for resources or status, they are viewed as threats to the in-group, and discrimination, prejudice, (Stephan and Renfro 2002) and thus potentially violence can result.

Social identity (in-group/out-group) has been identified as a factor in motivating violence. As threats from out-groups increase, in-group identification increases and, as

COMMON TEXT ANALYSIS METHODS

Some common text analysis methods include entity extraction, sentiment analysis, and event coding. Entity extraction involves automatically extracting knowledge and themes (document metadata) from unstructured text documents. Sentiment analysis (or opinion mining) involves identifying and categorizing opinions to determine whether the attitude toward a particular topic is positive, negative, or neutral. Finally, event coding is an approach to labeling events (typically drawn from news reports) using common, comparable metadata; today, nearly all event coding involves automated, machine-coding versus human-coded event data.

in-group identification increases, out-group derogation increases, as does the likelihood of violence against that group (Pyszczynski 2013). Social identity, along with the diffusion of responsibility, deindividuation, and obedience, is the mechanism for reducing the psychological costs of engaging in violence (Borum 2010). Richard Bolt (2003) wrote that "language is inescapably a badge of identity" and Peter Burke (1993) wrote that "speaking the same language, or variety of language, is an equally effective way of distinguishing oneself from other individuals or groups" (Burke 1993).

Words or linguistic choices are often not neutral. For example, the choice between *stubborn* and *steadfast* or the choice between *hits* and *is abusive* betrays a value judgment. Because discourse is not neutral and people identify themselves as part of an in-group that is differentiated from out-group(s), the discourse tends to reflect the assumptions, the beliefs, etc., of the in-group. The boundary maintenance between groups that are good or like us and those that are unlike us or bad and perspective taking or perspective giving forms a significant—albeit often subconscious—part of discourse.

Matsumoto, Hwang, and Frank (2014) postulated that, in the case of terrorism or extremist violence, there is a trajectory in affect from anger to contempt to disgust. All three of those emotions would be coded by sentiment analysis as negative, but they have strikingly different impacts on the perceptions of threats and resultant behaviors. Disgust, in particular, is a strong signal of impending violence, ". . .disgust transforms aggression (sometimes constructive) into hostility (usually not) and anger into hatred." Bandura (1990) outlined a number of mechanisms used to disengage the moral sanctions to violence, including dehumanization (e.g., the Nazi cartoons depicting Jewish people as rats during the lead up to the Second World War).

Integrative complexity is defined "in terms of degrees of differentiation and integration." *Differentiation* refers to the ability to distinguish different viewpoints on an issue and to acknowledge them, and *integration* is defined as the "conceptual connection[s] among differentiated dimensions" (Baker-Brown et al. 1992, p. 393). An integrative view acknowledges the mutual influence and the interdependence of different dimensions. Very low integrative complexity is binary, black-or-white thinking, all-or-nothing judgments, general inability or unwillingness to accept uncertainty and divergent viewpoints, and a desire for rapid closure. Conversely, very high integrative complexity is characterized by a high acceptance of uncertainty as well as the ability to synthesize opposing viewpoints or multidimensional integration of opinions. Numerous studies on the relationship between integrative complexity and subsequent behaviors demonstrated a link between a drop in integrative complexity and subsequent hostility/violence (Suedfeld and Tetlock 1977; Suedfeld 2010).

Worldview encompasses the sets of beliefs and assumptions that people use to filter or describe reality. A given worldview includes assumptions on a variety of topics, including the human nature, the meaning and nature of life, and the composition of the universe itself. Worldview defines "what can be known or done in the world, and how it can be known or done. In addition to defining what goals can be sought in life, a worldview defines what goals should be pursued" (Koltko-Rivera 2000, p. 2). The assumptions regarding physical and social realities that "may have powerful effects on cognition and behavior" (Koltko-Rivera 2004). Worldview is a socioculturally, historically, influenced conceptual framework used to make sense or meaning of what is, explain where it all came from, and what ought to be. This framework encompasses beliefs about what is possible and what is not, what is good or evil, what behaviors are desirable or undesirable.

The Lens of Social Identity: Forecasting Methodologies

The initial US Air Force Research Laboratory discourse analysis research focused on the development of multilingual methodologies (Arabic and Pashto), documented in primers transitioned to operational customers, including the National Air and Space Intelligence Center. The methodologies enable the detection and the interpretation of the discourse patterns related to social identity (in-group/out-group) (Fenstermacher, Kuznar, and Yager 2012). The initial methodology was critical discourse analysis-based, focusing on language form, particularly lexicalization, quotations, references and allusions. In addition, discourse mechanisms were identified that strengthen the language of social identity, including nominalization, evidentiality, authority, intensifiers, and attention-direction.

The discussion of one's in-group and out-group tends to correlate with positive terminology and negative terminology, respectively. This is a result of *lexicalization*, the process by which words are chosen to describe a particular event or entity. Both the ways in which *quotations* are introduced (using speech act verbs with power-related requirements like *decree* or *require*, or playing down power by writing that a leader says or agrees) and the types of quotations are important (authors are more likely to include quotes from someone they agree with or consider part of their in-group than from someone they are merely quoting to get all sides of an issue). The references that people make can indicate the way they conceptualize their world. Overt references to *us* and *them* and *ours* and *theirs* are used along with more subtle indicators, all of which demarcate sides. The use of allusions is an example of discourse being used to evoke other discourses. Understanding the reference, the context, and the connotations can elucidate the purpose (e.g., demonstrating that the author is part of the in-group). Understanding the cultural context in which discourse takes place is essential for interpreting the significance of quotations, the ways of referencing others, and the allusions to other events and texts.

Focusing on these in-group/out-group-related discursive mechanisms, a case study was conducted with documents provided by the US National Air and Space Intelligence Center to identify the ways in which these discursive mechanisms manifest in Arabic discourse. The result of this was a methodology for Arabic in-group/out-group discourse, based on Van Dijk's (2003, 2006, 2011) critical discourse analysis framework, documented in a methodological primer delivered to the US National Air and Space Intelligence Center (Toman et al. 2010). This primer has been used to successfully forecast several real-world events.

In order to assess the extensibility and the robustness of this methodology, a subsequent study was conducted. This second study integrated insights, through the application of

LANGUAGE FORM ELEMENTS

Nominalization, evidentiality, authority, intensifiers, and attention-direction are constructs from linguistic studies. Nominalization is the use of a nonnoun word (e.g., a verb or an adjective) as a noun (e.g., "He wanted a change."). Evidentiality refers to the existence and nature of evidence provided for a given statement. Authority refers to the explicit or the implied authority of the writer. Intensifiers are rhetorical devices designed to increase the salience, the magnitude, or the impact of a statement (e.g., "very, very important"). Finally, attention-direction devices, as their name indicates, service to direct the reader/listeners' focus to a particular topic or object (e.g., gestures such as pointing or expressions such as "this one").

grounded theory (Strauss and Corbin 1990), from both linguistically trained Arabic speakers as well as Arabic speakers with a more colloquial understanding of how in-group/out-group sentiments are expressed, creating a natural methodology that did not require formal training or critical discourse analysis expertise.

A refined methodology (Table 10.2) was developed based on relevant discursive practices and patterns (themes and rhetorical devices), including intensifiers (used to strengthen arguments) used to express social identity (Toman et al. 2010). The refined methodology differed from the methodology documented in the methodology primer (Toman et al. 2010) in that it identified language effects (e.g., dehumanization or victimization) and rhetorical devices (e.g., attention, attribution of virtue) important for identifying in-group/out-group discourse as opposed to language forms (e.g., lexicalization, allusions). Many of the language effects represented by the factors relate to phenomena and mechanisms from social psychology and critical discourse analysis (e.g., mechanisms for moral disengagement [Bandura 1990] and in-group euphemization and out-group derogation [Van Dijk 2003]). The same grounded theory approach resulted in a similar methodology (to identify and interpret the language associated with an in-group/out-group) for Pashto discourse (Kuznar and Yager 2012). Key themes expressed in both Arabic and Pashto included intimacy, power, virtue, and honor/shame. *Shaheed*, or martyr, was also important for the expression of in-group identity in Pashto (Fenstermacher, Kuznar, and Yager 2012).

The social identity-based methodology developed by the US Air Force Research Laboratory was subsequently used to identify leading indicators for political and military actions between India and Pakistan and by North Korea's Kim Jong Un and Syria's Bashar al-Assad. A time series analysis of the occurrence of themes in 206 speeches by Indian and Pakistani leaders between 1998 and 2008 demonstrated the following statistical patterns (Kuznar et al. 2012):

- Themes concerning military, war, terrorism, courage, and contested regions were more common before the conflicts; military themes and courage are especially salient among Pakistan's ethnic groups, and reference to contested regions (especially Kashmir) invoke nationalistic themes.

TABLE 10.2

Key In-Group/Out-Group Factors in Arabic Discourse

In-Group Treatment	Attribute	Out-Group Treatment
Much attention	Amount of attention	Not represented
Fully represented	Opinions represented	Not represented
Respectful, human terminology	Reference terminology	Disrespectful, inhuman terminology
With good entities; against bad entities	Groupings	With bad entities; against good entities
Close to *us*/the world	Intimacy	Distant from *us*
Powerful/involved	Attributed power	Weak/useless
Glorified/canonized	Attributed virtue	Immoral/irresponsible
Neutral/cooperative	Attributed motivations	Nonneutral/has negative motivations
Bad attributes diminished; has fundamentally good nature	Attributed nature	Good attributes diminished; has fundamentally bad nature
Victimized/sufferer	Victimization	Victimizer/aggressor

Source: Toman, P., Kuznar, L., Baker, T., and A. Hartman, *Analysis of Discursive Accent and Discursive Practices I&W*, AFRL Technical Report AFRL-RH-WP-TR-2010-0128, Wright-Patterson Air Force Base, Dayton, OH, 2010. With permission.

- Themes concerning pride, courage, shaheed, and duty were more common before the nuclear and missile testing events while nuclear themes were notably absent before the tests but present at other times. The invocation of themes such as pride, courage, duty, and even martyrdom represents the use of deep-seated cultural themes within these societies to justify a decidedly novel and modern military action. The absence of references to nuclear and missile testing before a test, when such references occur otherwise, may indicate implicit or explicit deception.
- Peace events were typically preceded by prosocial themes, *including trade, cooperation, progress, trust, and dignity and respect.*

A study of Kim Jon Un's speeches during his first year in power replicated the same pattern with regard to missile testing; the North Korean leadership would speak about nuclear and missile tests when not intending to carry these actions out and would be silent regarding nuclear and missile capabilities a month before executing these actions (Kuznar 2013). As with the Indian and Pakistan leaders, Kim Jong Un also stressed themes of national pride before testing missiles.

A recent study of Bashar al-Assad's public discourse (124 speeches from 2000 to 2013) demonstrated that Assad's use of figurative language (metaphor, allusions, symbolism), intensifiers, and conspiracy themes statistically increased before his regime took violent action (siege of Daraa and Homs, Ghouta gas attack). Interestingly, Assad's use of more culturally based rhetoric did not increase or decrease before the military action, perhaps reflecting his Western education and analytical personality (Kuznar 2014), indicating that while broad patterns can hold in diverse cultural settings (southern Asia, North Korea), outliers (i.e., black swans) do occur and must be considered.

This methodology provides useful insights on how a group sees themselves and others and, more importantly, on their attitudes and likely behaviors. It can be used to forecast events, to rapidly characterize the tenor of the documents and to provide a more nuanced understanding of a group's language use and discursive practices for use in analysis or information operations.

The Lens of Affect: Dynamics of Conflict

Many standard approaches to forecast behaviors have focused largely on static factors (e.g., poverty, grievances) at the expense of dynamic factors (e.g., who is doing what to whom, attitudes, and/or sentiment of key actors/groups). Subsequent research explored the link/influence between the affects expressed in discourse and behaviors (O'Brien, Shellman, and Covington 2013), focusing on the dynamics between the government, the dissidents (a nonstate actor who uses nontraditional means [protest, violence] to meet their political needs, goals), and the population (Shellman 2006a,b, 2008). Conflict is conceptualized (see Figure 10.1) as a process in which competing actors (individuals, groups, states) make interdependent decisions regarding the escalation or the deescalation of conflict. The population is considered a key actor because both dissidents and (most) governments vie for their support.

Text analytic algorithms extracted events as well as affect from data (events data from the Factiva corpus and sentiment and emotions data from blogs) gathered in the Philippines and Egypt from 2001 to 2012. Emotions data were derived from extracted sentiment expressions by mapping the expressions to eight affect categories, such as anger,

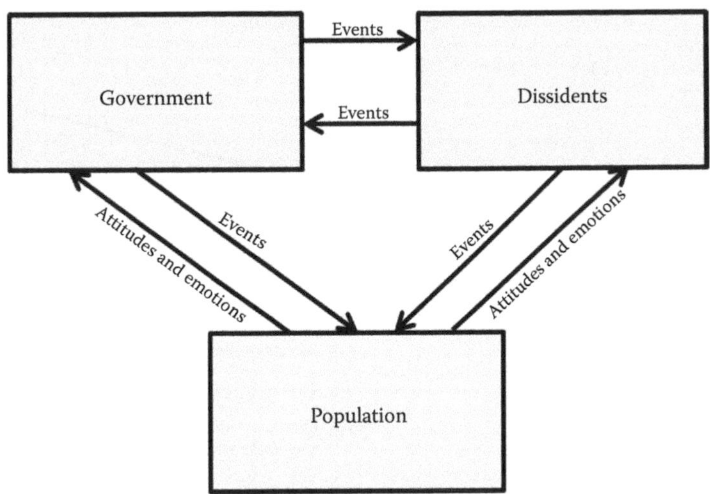

FIGURE 10.1
Conceptualization of conflict as strategic interactions. (From Shellman, S. M., *Conflict Management & Peace Science*, 23, 73–90, 2006; *Security Studies*, 15, 563–599, 2006. With permission.)

fear, or disgust (Plutchik 1980). For example, *The Swiss government lauded Suu Kyi for her courage* would be coded as a positive sentiment and under the joy affect class; whereas for *The King is appalled*, the sentiment would be coded as negative and the affect class as disgust. The relationship models of the government, the dissidents, and the population were estimated using autoregressive conditional heteroskedasticity due to the nonconstant variance in the time series data. The focus was on understanding the relationship of the impact of both the societal emotions on dissident and government behaviors and the government and dissident behaviors on societal emotions. The quantitative models of the effects of emotions on behaviors of competing actors in Egypt and the Philippines illuminated similarities and differences in the influence of one group/organization on the other; for example, how affect (e.g., fear) expressed by the population influenced dissident behavior and government behaviors.

For example, in Egypt and in the Philippines, intense societal fear of dissidents and societal disgust toward the government was associated with increases in dissident hostility (see Figure 10.2), and, conversely, societal anger toward dissidents was associated with a reduction in dissident hostility. However, differences arose in response to disgust expressed by the population toward the government. In the democratic Philippines, the government appeared to view negative social attitudes toward dissidents, principally anger, as a pretext to justify cracking down on dissidents through the escalation of repressive behaviors; however, these behaviors eased in cases where the society began to fear the government or display disgust toward it. This change in behavior makes sense to the extent that democratic leaders require the support of the people to attain or retain political office. Alternatively, the authoritarian Egyptian government reacted to similar forms of disgust by intensifying repression (see Figure 10.3). In a time series analysis of the emotion variables, only societal anger could be explained by government and dissident behavior indicators across both cases. Dissident hostility tended to reduce societal anger while government hostility tended to inflame it.

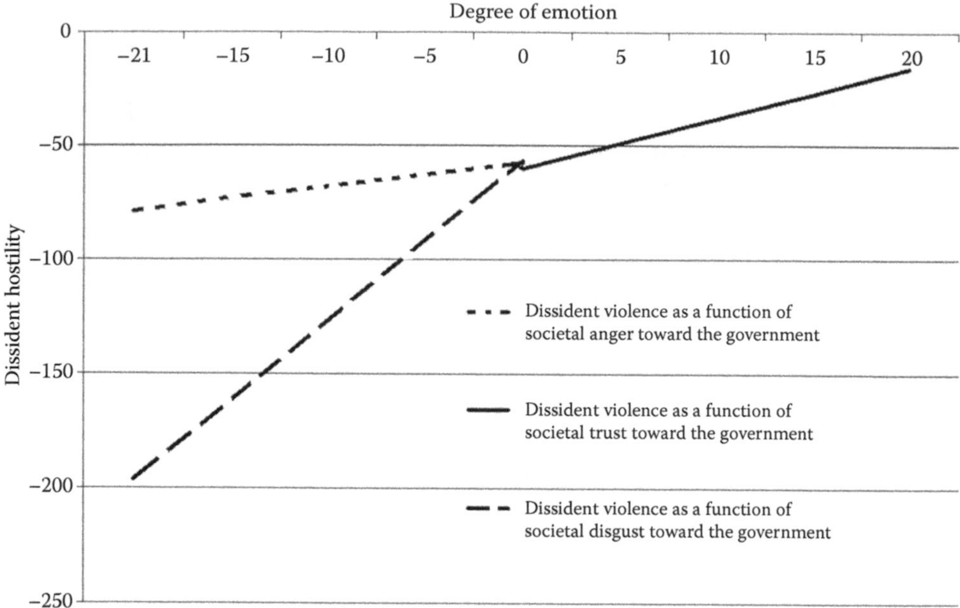

FIGURE 10.2
Impacts of societal emotions on Egyptian dissident hostility, 2001–2012. (From Fenstermacher, L., Kuznar, L., Yager, M., and Shellman, S., *Advances in Cross-Cultural Decision Making*, 61–71, 2014. With permission.)

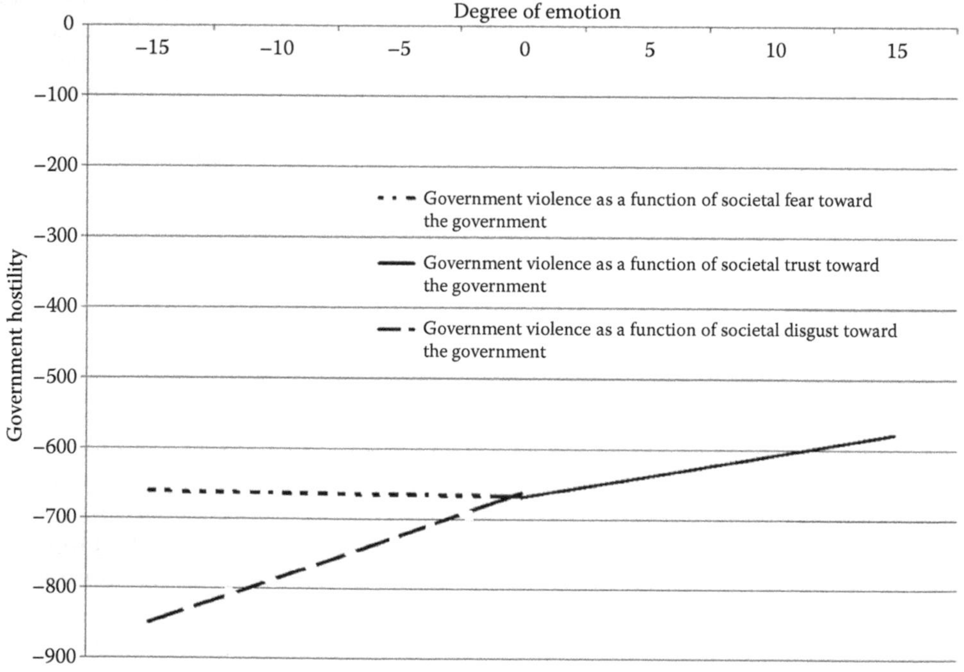

FIGURE 10.3
Impacts of societal emotions on Egyptian government hostility, 2001–2012. (From Fenstermacher, L., Kuznar, L., Yager, M., and Shellman, S., *Advances in Cross-Cultural Decision Making*, 61–71, 2014. With permission.)

Text Analytic Tools for Forecasting Violence Based on Multiple Discourse Markers

To increase the analytic throughput, it is desirable to have semiautomatic tools that enable an analyst to detect, monitor, and focus on emergent and existing threats. Research was conducted to investigate ways to automate the processing of discourse to extract discourse features in order to identify intent and/or forecast behaviors. The initial text analytic tool extracted, measured, and modeled five discourse features: integrative cognitive complexity, sentiment/affect expressed by in-groups toward out groups (social identity), idea density, vocabulary diversity, and keyness (O'Brien, Shellman, and Covington 2013).

Integrative complexity (or integrative cognitive complexity), as explained previously, is the extent to which a person is open to new ideas/perspectives and is able to integrate them. Integrative cognitive complexity is a measure intended to approximate the one used by Baker-Brown et al. (1992) in which the sum of the density of high (>2) integrative cognitive complexity (language) indicators is divided by the number of words in the document (and then multiplied by 100 for easier visualization).

Idea density is a psycholinguistic measurement for determining the propositional idea density (PD) of a corpus based on part-of-speech tags. It can be approximated by the number of verbs, adjectives, adverbs, prepositions, and conjunctions divided by the total number of words (Snowdon et al. 1996) and is perhaps the opposite of integrative cognitive complexity in that the greater use of adjectives may be used to reinforce a point, not necessarily to make a more subtle point using grammatical connectives which, because they are not separate pieces of information, do not count as ideas. Vocabulary diversity is the range of vocabulary used in a text. It is congruent with integrative cognitive complexity in that a wider range of vocabulary usage is one indicator of integrative complexity and vocabulary diversity is higher when many topics are discussed.

Keyness is the frequency of the use of certain words in a text relative to comparative texts (a way to track the salience of topics over time). Keyness, expressed in terms of statistical significance, is a measure of the extent to which words or phrases are used more frequently in one text than in a comparison set of texts. To compute keyness, a G-test is computed, which approximates a χ^2 test.

$$G = 2\,(O1L1 + O2L2), \text{ with } L1 = \ln\,(O1/E1) \text{ and } L2 = \ln\,(O2/E2),$$

with

$$O1 = \text{observed number of occurrences of the word in text 1, and}$$

$$O2 = \text{observed number of occurrences of the word in text 2.}$$

$$\begin{aligned} E1 &= \text{expected number of occurrences of the word in text 1 if the texts were alike} \\ &= (\text{occurrences of word in both texts}) \times (\text{length of text } 1/\Sigma\,(\text{length of texts } 1,2)) \end{aligned}$$

$$\begin{aligned} E2 &= \text{expected number of occurrences of the word in text 2 if the texts were alike} \\ &= (\text{occurrences of word in both texts}) \times (\text{length of text } 2/(\text{length of texts } 1,2)) \end{aligned}$$

G is interpreted as χ^2; the higher the χ^2 is, the more statistically significant the difference. The utility of keyness was explored for tracking the salience of topics discussed by groups over time and their relationship to the intensity levels of violence.

The sentiment/affect of the in-group (Naxalites) toward various out-groups (Muslim dissident organizations, prodemocracy groups, government, police, military, and other state officials) was measured by extracting language related to anger, fear, and disgust (negative sentiment). The expectation was that the expression of anger or especially disgust toward an out-group would be associated with increased violence (Matsumoto, Hwang, and Frank 2014) and the expression of fear would be associated with decreased violence.

These features (integrative cognitive complexity, sentiment/affect of in-group toward out-group, idea density, vocabulary diversity, and keyness) were extracted from 35–50 page online newsletters regularly produced by the People's War Group (PWG) from 1998 to present. The PWG is a violent organization associated with the Naxalite communist movement in India, and the newsletter serves to justify opinions, euphemize the group, and derogate their opponents, among other objectives.

It is no surprise that there was clear statistical evidence showing that a significant decline in cognitive complexity on the part of the Naxalites was associated with increased levels of violence, consistent with previous findings (Toman et al. 2010; Satterfield 1998). Idea density and vocabulary diversity also correlated with Naxalite violence, and they may serve as orthogonal measures of cognitive complexity (see Figure 10.4). Keyness is similar to topic tracking, and it is useful for tracking the use of discourse over time and potentially

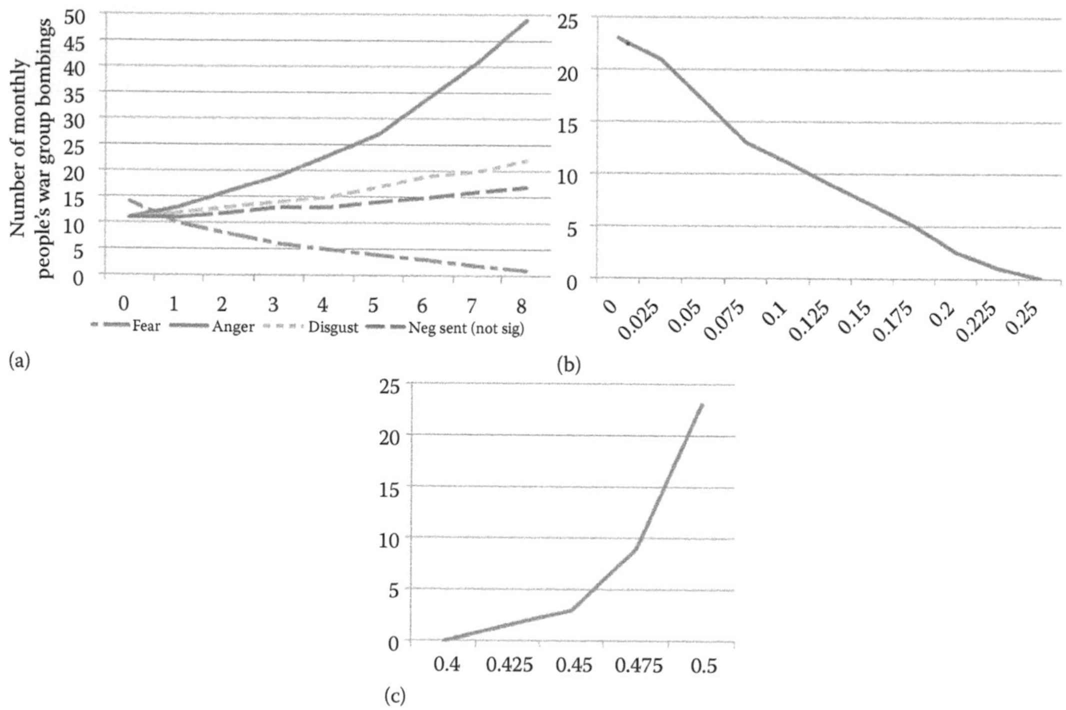

FIGURE 10.4
Effects of various discourse indicators on PWG bombings in India: (a) In- to out-group expressed emotion, (b) cognitive complexity, and (c) idea density. (From O'Brien, S., Shellman, S. M., and Covington, M. *Automated Discourse Analysis*, AFRL Technical Report AFRL-RH-WP-TR-2013-0036, Wright-Patterson Air Force Base, Dayton, OH, 2013. With permission.)

useful for identifying early indicators of violence including certain uses of imagery and propaganda.

Figure 10.5 shows the power of using these discourse markers together to predict violence in the case of the PWG. The solid line shows a monthly frequency count of PWG bombings throughout India. The dotted line shows the predicted level of bombings using only measures of vocabulary diversity, idea density, cognitive complexity, and in to out-group expressions of fear, anger, and disgust. Each variable is statistically significant, and there is a 91% correlation between the model-fitted values and the actual number of monthly bombings.

The model only included automated discourse indicators (cognitive complexity, sentiment/affect toward out-group, idea density, and vocabulary diversity). There were no lagged dependent variables, structural indicators, and event-based or repression-based measures included in the model. And yet, the discourse markers alone yielded an incredibly, perhaps surprisingly, powerful model of group-level violence. A second case study on al-Qaeda extracted discourse variables ([integrative] cognitive complexity, idea density, vocabulary diversity, use of interrogatives/subordinate clauses, and in-group affect [anger, disgust] toward out-group) from 290 al-Qaeda statements from the Haverford corpus and the discourse from 10 issues of al-Qaeda's *Inspire Magazine*. These variables were used to forecast al-Qaeda attacks between 1998 and 2011 (National Consortium for the Study of Terrorism and Responses to Terrorism events data), resulting in a 0.73 correlation between actual and predicted attacks.

Subsequent research explored enhancements to the original discourse analysis algorithms with new metrics, independent complexity indicators, such as differentiation and integration, and content analysis (Pennebaker, Booth, and Francis 2007; Pennebaker and Chung 2008) that identified characteristic words and phrases used before, during, and after attacks (e.g., *us* versus *them*, loaded language such as *imperialist*, *injustice*, *atrocity*, loyalty rhetoric such as *truth* and *allegiance*, and hedging rhetoric such as *perhaps* and

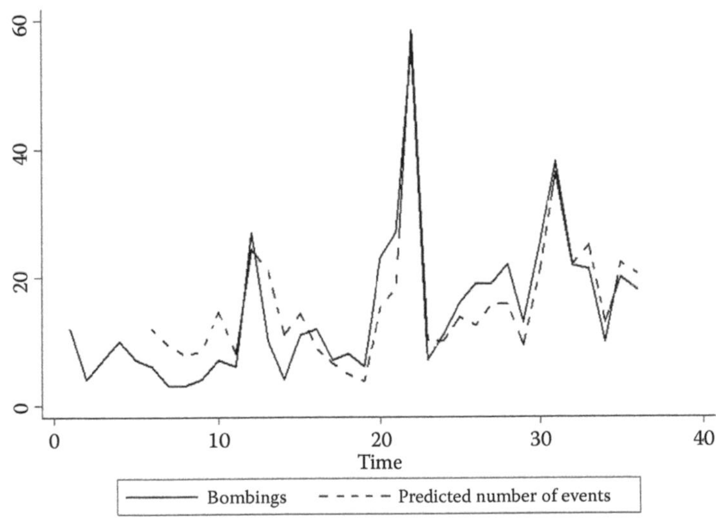

FIGURE 10.5
Actual versus predicted PWG bombings using autogenerated discourse variables. (From O'Brien, S., Shellman, S. M., and Covington, M. *Automated Discourse Analysis*, AFRL Technical Report AFRL-RH-WP-TR-2013-0036, Wright-Patterson Air Force Base, Dayton, OH, 2013. With permission.)

somewhat). The content analysis essentially identifies the perspectives or the alternatives that are being differentiated and integrated in the integrative cognitive complexity measures.

How Important Is the Emic Perspective for Forecasting?

A recent case study investigated the use of three different text analytic algorithms for forecasting violence by the Boko Haram, a violent extremist group in Nigeria (Fenstermacher 2015). The data used were from news feeds, social media, and Boko Haram YouTube videos (transcripts). Separate statistical forecasting models were developed for the features/markers extracted from the three text analytic methods. The first was based on event analysis (O'Brien 2010); that is, coding events (kidnapping, coup, bombing, etc.). The second was based on sentiment/affect analysis (e.g., society to Boko Haram disgust), and the third was based on the discourse analysis variables previously discussed including those based on content analysis (e.g., Boko Haram to government hostility), social identity, integrative complexity, idea density, and vocabulary diversity. The events analysis-based forecasting resulted in the predicted events correlating with actual events 36% of the time. The result based on sentiment analysis was slightly higher, 49%. The result based on the discourse markers was 69% (see Figure 10.6). Figure 10.7 shows the results based on an integrated or fused model (all three sets of variables in a single forecasting model) was 86%.

Clearly, the incorporation of the participant or emic perspective (sentiment and discourse analysis) is useful for forecasting; although, the best forecasting performance overall resulted from combining the events analysis (etic) and the emic perspectives, as one would expect. This is a small step toward information fusion incorporating both *etic* and

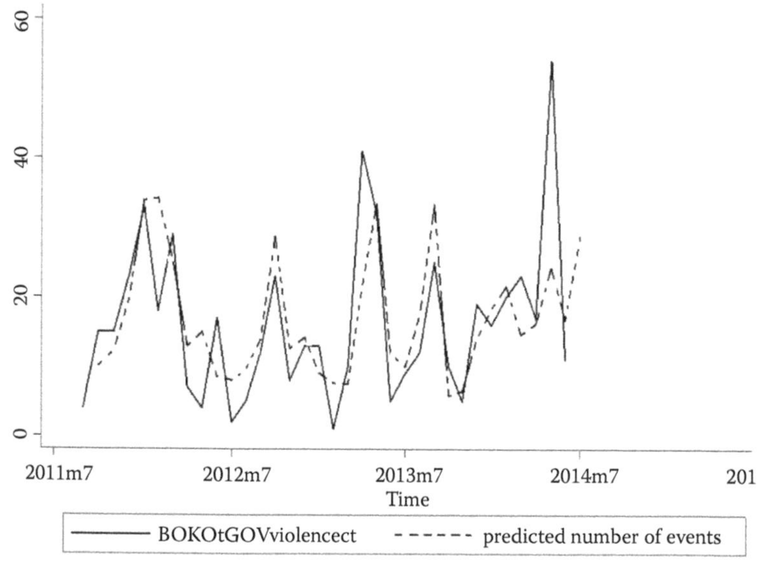

FIGURE 10.6
Forecasting Boko Haram violence based on discourse variables ($r = 0.69$). (From Fenstermacher, L., Kuznar, L., Yager, M., and Shellman, S., *Advances in Cross-Cultural Decision Making*, 61–71, 2014. With permission.)

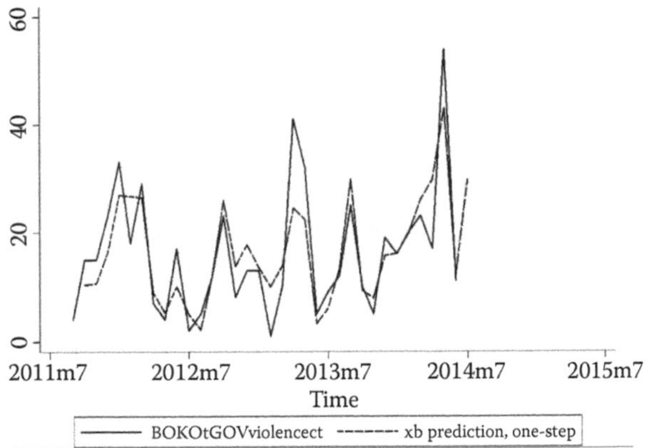

FIGURE 10.7
Results of integrated forecasting model for Boko Haram violence. (From Fenstermacher, L., The importance of the "emic" perspective in information fusion. In *Proceedings SPIE Signal Processing, Sensor/Information Fusion, and Target Recognition XXIV*, 20–22 April 2015, Baltimore, 9474, xcv–xcviii, 2015.)

emic perspectives, and much more research is needed to develop methods that incorporate the sources of information other than text; however, the prospects are promising.

Conclusions and Looking Forward

There are verbal cues contained in discourse (i.e., what people say or write) that telegraph intentions to engage in violent behaviors in nuanced ways. These cues are found in language associated with the expression of affect (e.g., fear, anger, and disgust), the ability to differentiate and integrate different perspectives (integrative complexity) and social identity. The emic or first-person perspective cues are pieces of an overall threat narrative. A threat narrative, interpreted broadly, is a useful construct that enables meaning-making about threats through a systematic consideration of the various aspects of a worldview based on physical and discursive-based data. Before an analyst can capture the aspects of the emic perspective, he/she needs to ensure that the meanings given to the data are based on scientific approaches to inference that is based on rigorous and consistent interpretations and reinterpretations. In addition, the analyst must avoid mirroring by understanding the values, the assumptions, the beliefs and misperceptions, or the misunderstandings of the subject of their analysis.

Advanced methodologies and text analytic algorithms developed by the US Air Force Research Laboratory enable the identification, the extraction, and the interpretation of discourse markers associated with the emic perspective and enable analysts to make meaning about a threat narrative. The methodologies and the algorithms are a big step forward in enabling the anticipatory analysis of violent events. Importantly, the verbal cues on which methodologies and algorithms are typically based occur well in advance of the actual event; thus, they provide leading indicators of violence, enabling either the collection of other information to confirm or action to accelerate, mitigate, or prevent the event.

Due to the number of gray zone operations the United States is currently engaged in, operations that are characterized by "ambiguity on the nature of the conflict, the parties involved, and the validity of the legal and political claims at stake" (Oldham 2015), analysts need a powerful set of methods and tools to forecast and make meaning about threats (not only the what, but the how, the why, and the what should I do). Forecasting approaches based solely on the emic perspective fall short of 100% accuracy. To form a more complete picture or understanding of the situation/threat requires the integration of emic and etic perspectives. As Lt. Gen. Robert P. Otto (US Air Force) expressed in the Air Force Intelligence Surveillance and Reconnaissance (AF ISR) 2023 vision document, "our nation needs AF ISR to once again anticipate strategic surprise, baseline the adversary, and conduct ISR operations in challenging, cross-cultural environments across the range of military operations" (AF ISR 2023, p. 2).

Megacity Use Case

The methods and the tools described in this chapter enable an analyst to decipher the discourse about and between salient individuals/groups in a way that elucidates signals of impending conflict or cooperation. The key actors in the megacity Xanadu use case include the government, the Bursukans, the North Razinians, the potential swing actors (e.g., criminal groups), and the population. All these actors have different perspectives on the past, the present, and the future, different ideas about right and wrong, justice, etc. Understanding this, to the extent possible, allows for a more nuanced understanding of the situation, including the dynamics and likely behaviors/futures in megacity Xanadu.

At a top level, the methods and the tools discussed in this chapter enable the capabilities to both monitor and detect changes in fractionalization, affect, and integrative complexity in the human environment, at various levels: individuals, groups/organizations, and the system of systems (government, population, groups). Markers extracted from discourse (e.g., speech, text, video transcripts, blogs, websites, speeches, social media) can be integrated with events data in order to forecast ethnic/religious violent events. The affect analysis can be used to identify patterns of influence in the system of systems (e.g., how something expressed by the Bursukans or the North Razinians will influence the behaviors/actions of the government or vice versa).

One can peel the onion with these methods and tools and explore some interesting questions regarding the impact of environment/structural conditions (e.g., economic conditions, food prices, unemployment) or government policies/behaviors. For example, they could be used to examine likely effects, predictions of economic activities/relations, and relative deprivation across subpopulations/regions within the megacity. In a way, this would be assessing not only the threat narrative but also the economic narrative. Additionally, the interactions between various groups (Bursukans, North Razinians, criminal groups, etc.) and the population could be examined by assessing how the integrative complexity and the idea density ebb and flow with behavioral indicators of either violence or cooperation and/or economic behaviors (e.g., black market transactions, economic growth, relative deprivation, grievances).

In addition, the methods and the tools discussed could be used to understand and assess appropriate COAs to stabilize behavior/social relations or mitigate conflict. The discourse analysis methods/tools could be used to monitor in-group/out-group derogation and in-group glorification. As these change, the COAs could be assessed based on counterfactual analysis (akin to placebo testing where the difference between the outcomes stemming from modeling with and without an intervention or a COA is assessed). Based on this assessment, messages/communications could be changed to shift and stabilize behavior/social relations. The impacts of government repression on various groups within Xanadu could be examined to assess what types of policies/actions quell and which ones exacerbate tensions/violence. Furthermore, how do these repressive policies affect the discourse indicators above which, in turn, impact observed behavior?

Messaging/influence strategies can also be informed. The methods and the tools can be used to monitor the content themes within social media; for example, recruitment of rebel/terrorist groups, recruitment of refugees by violent groups, use of religious rhetoric, *us/them* themes, extremist lingo. This understanding could inform the development of various COAs to influence recruiting and mitigate potential violence.

Finally, beyond forecasting violence or developing COAs to mitigate violence, the methods and the tools discussed could inform a more general understanding of megacity Xanadu. For example, how does discourse impact economic behaviors such as black market activities, economic/product innovation, taxes paid, government regulation of markets? An analysis could be conducted analyzing the discourse of the government using the methods and tools. This could inform an understanding of how consistent or even truthful the government is; for example, how does its cognitive complexity, idea density, content themes, in-group/out-group rhetoric jive with their behavior? What is the relationship between the government discourse and economic growth/decisions, relative deprivation over time, government repression, or government cooperation?

DISCUSSION QUESTIONS

1. How do emic and etic interpretations differ, and how does integrating the two perspectives provide a more robust analysis?
2. How does the assessment of threat narratives (as a construct described in this chapter) differ from approaches that focus on identifying utterances that happen to be threatening?
3. What is social identity theory, and how can it influence the interpretation of discourse?
4. What do these five discourse features represent: integrative cognitive complexity, sentiment/affect expressed by in-group toward out groups (social identity), idea density, vocabulary diversity, and keyness? How does the consideration of these discourse markers help predict future violence?

References

Air Force Intelligence Surveillance and Reconnaissance (AF ISR). AF ISR 2023: Delivering decision advantage: A strategic vision for the Air Force ISR enterprise. 2013. http://www.defenseinnovationmarketplace.mil/resources/AF-ISR_2023.pdf. (accessed November 12, 2015).

Agar, M. 2011. Making sense of one other for another: Ethnography as translation. *Language & Communication* 31(1), 38–47.

Baker-Brown, G., Ballard, E. J., Bluck, S., de Vries, B., Suedfeld, P., and Tetlock, P. E. 1992. The conceptual/integrative complexity scoring manual. In *Motivation and Personality: Handbook of Thematic Content Analysis*, ed. Smith, C. P., Atkinson, J. W., McClelland, D. C., and Veroff, J., 401–418. New York: Cambridge University Press.

Bandura, A. 1990. Mechanisms of moral disengagement. In *Origins of Terrorism: Psychologies, Ideologies, Theologies, States of Mind*, ed. Reich, W. Cambridge: Cambridge University Press.

Bolt, R. K. 2003. *Language and Social Identity*, ed. Bolt, R. K. Westport, CT: Praeger.

Borum, R. 2010. Understanding Terrorist Psychology. In *The Psychology of Counter-Terrorism*, ed. Silke, A., 19–33. Oxon: Routledge.

Bronowski, J. 1951. *The Common Sense of Science*. London: Faber and Faber.

Burke, P. 1993. *The Art of Conversation*. Ithaca, NY: Cornell University Press.

Davis, P. K., Perry, W. L., Brown, R. A., Yeung, D., Roshan, P., and Voorhies, P. 2013. Using Behavioral Indicators to Help Prevent Violent Acts: A Review of the Science Base. http://www.rand.org/content/dam/rand/pubs/research_reports/RR200/RR215/RAND_RR215.pdf (accessed October 21, 2015).

Defense Science Board. 2012. Task Force Report: Predicting Violent Behavior: Misperceptions about Targeted Violence Figure 3. http://www.acq.osd.mil/dsb/reports/PredictingViolentBehavior.pdf (accessed November 12, 2015).

Fenstermacher, L. 2014. Information fusion: Telling the story (or threat narrative). In *Proceedings SPIE Signal Processing, Sensor/Information Fusion, and Target Recognition XXIII*, 5–8 May 2014, Baltimore, 9091(9091OQ) 1–11.

Fenstermacher, L. 2015. The importance of the "emic" perspective in information fusion. In *Proceedings SPIE Signal Processing, Sensor/Information Fusion, and Target Recognition XXIV*, 20–22 April 2015, Baltimore, 9474, xcv–xcviii.

Fenstermacher, L., Kuznar, L., and Yager, M. 2012. Analysis of discourse for indications and warnings. In *Advances in Design for Cross-Cultural Activities Part II*, ed. Schmorrow, D. D., 230–240. Boca Raton, FL: CRC Press.

Fenstermacher, L., Kuznar, L., Yager, M., and Shellman, S. 2014. Meaning making regarding threat narrative based on discourse analysis. In *Advances in Cross-Cultural Decision Making*, ed. Schatz, S., Cohn, J., and Nicholson, D., 61–71. Krakow: AHFE Conference.

Harris, M. 1979. *Cultural Materialism: The Struggle for a Science of Culture*. New York: Random House.

Harris, M. 1999. *Theories of Culture in Postmodern Times*, Walnut Creek, CA: Altamira Press.

Heuer, R. J. 1999. *Psychology of Intelligence Analysis*. Fairfax, VA: Center for the Study of Intelligence, Central Intelligence Agency. https://www.cia.gov/library/center-for-the-study-of-intelligence/csi-publications/books-and-monographs/psychology-of-intelligence-analysis/ (accessed October 21, 2015).

Hogg, M. A., and Abrams, D. 1988. *Social Identifications: A Social Psychology of Intergroup Relations and Group Processes*. London: Routledge.

Kohn, D. 2001. Columbine: Were there warning signs? Should school officials have known? *60 Minutes*. http://www.cbsnews.com/news/columbine-were-there-warning-signs/ (accessed October 20, 2015).

Koltko-Rivera, M. E. 2000. The worldview assessment instrument (WAI): The development and preliminary validation of an instrument to assess world view components relevant to counseling and psychotherapy. *Dissertation Abstracts International*, 61(04), 2266B. (UMI Microform No. 9968433).

Koltko-Rivera, M. E. 2004. The psychology of worldviews. *Review of Psychology* 8(1), 3–58.

Kurtz, C. F., and Snowden, D. J. 2003. The new dynamics of strategy: Sense-making in a complex and complicated world. *IBM Systems Journal* 42(3), 462–482.

Kuznar, L. A. 2009. *Reclaiming a Scientific Anthropology*, Second Edition. Walnut Creek, CA: AltaMira Press. Original edition, 1997.

Kuznar, L. A. 2013. *Discourse and Cognitive Complexity Analysis of Kim Jong Un's Speeches April 2012–March 2013*. Strategic Multilayer Assessment (SMA), OSD/ASD (R&E)/RSD/RRTO, Briefing, 17 April 2013, Arlington, VA.

Kuznar, L. A. 2014. Thematic analysis of Bashar al-Assad's speeches. In *A Multi-Disciplinary, Multi-Method Approach to Leader Assessment at a Distance: The Case of Bashar al-Assad Part II: Analytical Approaches*, ed. Cabayan, H., and Wright, N., 27–61. Arlington, VA: Strategic Multilayer Assessment (SMA) Periodic Publication, OSD/ASD (R&E)/RSD/RRTO.

Kuznar, L. A., and Yager, M. 2012. *Analysis of Pashtun Narratives: Report on Results. What Cues Do Pashto Speakers Use in Understanding How to Draw In- and Out-Group Distinctions?* Air Force Research Lab (AFRL) Report, Contract FA8650-10-C-6106. Dayton, OH: Wright-Patterson Air Force Base.

Kuznar, L. A., Yager, M., St. Clair, C., and Stephenson, A. 2012. *Cognitive Complexity Assessment of Pakistan–India Regional Narratives*. Report prepared for Air Force Research Laboratory SAMOA Project (FA8650-10-C-6106). Dayton, OH: Strategic Multilayer Assessment (SMA), OSD/ASD (R&E)/RSD/RRTO.

Lett, J. 1990. Emics and etics: Notes on the epistemology of anthropology. In *Emics and Etics: The Insider/Outsider Debate*, ed. Headland, T. N., and Pike, K. L., 127–142. Newbury Park, CA: Sage.

Matsumoto, D., Hwang, H., and Frank, M. G. 2014. Emotions expressed in speeches by leaders of ideologically motivated groups predict aggression. In *Behavioral Sciences of Terrorism and Political Aggression* 6(1), 1–18.

Moon, W. H. 2009. Worldview—Subjective Reality—Threat Narrative: The Macro View. Working Paper.

Morris, M. W., Leung, K., Ames, D., and Lickel, B. 1999. Views from inside and outside: Integrating emic and etic insights about culture and justice judgment. *Academy of Management Review* 24(4), 781–796.

O'Brien, S. 2010. Crisis early warning and decision support: Contemporary approaches and thoughts on future research. *International Studies Review* 12(1), 87–104.

O'Brien, S., Shellman, S. M., and Covington, M. 2013. *Automated Discourse Analysis*. AFRL Technical Report AFRL-RH-WP-TR-2013-0036. Dayton, OH: Wright-Patterson Air Force Base.

Oldham, C. 2015. SOCOM: Navigating the Gray Zone. http://www.defensemedianetwork.com/stories/socom-navigating-the-gray-zone/2/ (accessed October 21, 2015).

Olive, J. L. 2014. Reflecting on the tensions between emic and etic perspectives in life history research: Lessons learned. *Forum: Qualitative Social Research* 15(2), 1–9. http://www.qualitative-research.net/index.php/fqs/article/view/2072/3656.

Owen, B. 2001. Governor's Review Columbine Commission: The Report of Governor Bill Owen's Columbine Commission, May 2001. http://www.colorado.gov/columbine/Columbine_20_WEBFULL.pdf.

Pennebaker, J. W., and Chung, C. K. 2008. Computerized text analysis of al-Qaeda transcripts. In *A Content Analysis Reader*, ed. Krippendorff, K., and Bock, M. A., 453–65. Thousand Oaks, CA: Sage.

Pennebaker, J. W., Booth, R. J., and Francis, M. E. 2007. *Linguistic Inquiry and Word Count. LIWC (2007)*. Austin, TX: LIWC. https://www.liwc.net.

Pike, K. L. 1967. *Language in Relation to a Unified Theory of the Structure of Human Behavior*, Second Edition. The Hague: Mouton.

Pike, K. L. 1998. A linguistic pilgrimage. In *First Person Singular III: Autobiographies by Scholars in the Language Science: Studies in the History of the Language Sciences 88*, ed. Koerner, F. K., 144–159. Amsterdam/Philadelphia, PA: John Benjamins Publishing Company.

Plutchik, R. 1980. *Emotion: Theory, Research, and Experience: Vol. 1. Theories of Emotion 1*. New York: Academic Press.

Pyszczynski, T. 2013. Appendix B: Group identity, intergroup conflict and support for violent political action. In *Mixed Initiative Discourse Analysis System*, ed. Patten, T., Romero, V., Koelle, D., Carlson, E., Pfautz, J., and Pyszczynski, T. AFRL Technical Report AFRL-RH-WP-TR-2013-0029. Dayton, OH: Wright-Patterson Air Force Base.

Satterfield, J. M. 1998. Cognitive-affective states predict military and political aggression and risk taking: A content analysis of Churchill, Hitler, Roosevelt and Stalin. *Journal of Conflict Resolution* 42(6), 667–690.

Shellman, S. M. 2006a. Leaders and their motivations: Explaining government-dissident conflict-cooperation processes. *Conflict Management & Peace Science* 23(1), 73–90.

Shellman, S. M. 2006b. Process matters: Conflict and cooperation in sequential government-dissident interactions. *Security Studies* 15(4), 563–599.

Shellman, S. M. 2008. Coding disaggregated intrastate conflict: Machine processing the behavior of substate actors over time and space. *Political Analysis* 16(4), 464–477.

Shellman, S. M., Levey, B., and Young, J. K. 2013. Shifting sands: Explaining and predicting phase shifts by dissident organizations. *Journal of Peace Research* 50(3), 319–336.

Snowdon, D. A., Kemper, S. J., Mortimer, J. A., Greiner, L. H., Wekstein, D., R., and Markesbery, W. R. 1996. Linguistic ability in early life and cognitive function and Alzheimer's disease in late life: Findings from the nun study. *JAMA* 275, 528–32.

Stephan, W. G., and Renfro, C. L. 2002. The role of threat in intergroup relations. In *From Prejudice to Intergroup Emotions: Differentiated Reactions to Social Groups*, ed. Mackie, D. M., and Smith, E. R. 191–206. New York: Psychology Press.

Strauss, A., and Corbin, J. M. 1990. *Basics of Qualitative Research: Grounded Theory Procedures and Techniques*. Thousand Oaks, CA: Sage Publications.

Suedfeld, P. 2010. *The Scoring of Integrative Complexity as a Tool for Forecasting Adversary Intentions*. DRDC Report DRDC Toronto CR 2010-039. Ottawa, ON: Defence Research and Development Canada.

Suedfeld, P., and Tetlock, P. E. 1977. Integrative complexity of communications in international crises. *The Journal of Conflict Resolution* 21(1), 169–184.

Tajfel, H., and Turner, J. C. 1979. An integrative theory of intergroup conflict. In *The Social Psychology of Intergroup Relations*, ed. Austin, W. G., and Worchel, S. Monterey, CA: Brooks/Cole.

Tajfel, H., and Turner, J. C. 1986. The social identity theory of inter-group behavior. In *Psychology of Intergroup Relations*, ed. Worchel, S., and Austin, L. W. Chicago: Nelson-Hall.

Taleb, N. N. 2007. *The Black Swan: The Impact of the Highly Improbable*. New York: Random House Publishers.

Toman, P., Kuznar, L., Baker, T., and Hartman, A. 2010. *Analysis of Discursive Accent and Discursive Practices I&W*. AFRL Technical Report AFRL-RH-WP-TR-2010-0128. Dayton, OH: Wright-Patterson Air Force Base.

Van Dijk, T. A. 2003. Critical discourse analysis. In *The Handbook of Discourse Analysis*, ed. Schiffrin, D., Tannen, D., and Hamilton, H. E., 352–371. Malden, MA: Blackwell Publishers.

Van Dijk, T. A. 2006. Politics, ideology and discourse. In *Politics & Language, Encyclopedia of Language and Linguistics* (Second Edition, Vol. 9). Boston: Elsevier.

Van Dijk, T. A. 2011. Introduction: The study of discourse. In *Discourse Studies: A Multidisciplinary Introduction*, ed. Van Dijk, T. A. Los Angeles: Sage.

Willis, J. W. 2007. *Foundations of Qualitative Research: Interpretive and Critical Approaches*. Thousand Oaks, CA: Sage.

Yin, R. K. 2010. *Qualitative Research from Start to Finish*. New York: The Guilford Press.

11

Collecting Data and Semantic Content via Mobile Devices

Alper Caglayan and Laura Cassani

CONTENTS

Introduction 206
Semantic Knowledge Representation 207
Mobile Data Collection to Support Semantic Knowledge Management 209
Data Analysis and Fusion 211
Conflict Detection 211
Inconsistency Detection 212
DL Query for Inconsistencies 212
Paraconsistency Reasoners 213
Probabilistic Semantics for Anomaly Detection 214
Entropy-Based Anomaly Detection 214
Detection of Information Measures in Time Using KL Divergence 215
Drill Down of Anomalous/Suspicious Behavior 216
Future Research 217
Conclusion 218
Megacity Use Case 219
References 220

ABSTRACT Although the security community has made large strides toward information dominance, the problem of harnessing big data for national security purposes is still an evolving science. Part of this trend involves the increasing use of mobile phones, particularly in the developing world. To leverage this advancement, we have researched and developed a knowledge management cloud service and a mobile data collection and situational awareness capability. We apply an innovative approach that incorporates both logical and probabilistic methodologies to automatically detect anomalies within the data for actionable insight.

KEY TOPICS

- Semantic knowledge management represents data in a knowledge graph.
- Information is represented as nodes and connected by properties that denote relationships and links.
- Semantic queries are performed by traversing the knowledge graph.

- Mobile data can be collected and automatically represented in the knowledge graph, thereby semantically enriching it by connecting it to other linked entities.
- Anomalies can be detected within the data through logical inconsistencies represented in the graph and through entropy-based measures.

Introduction

Although the security community has made large strides toward information dominance, the problem of harnessing big data for national security purposes is still an evolving science. There are now more than five billion mobile phone subscribers in the world, and with the mobile trend increasing in relevance, particularly in the developing world, the volume and velocity of the data being collected is exponentially increasing. With this trend, mitigating the undesirable *garbage out* consequences of *garbage in* for intelligence data (content) and metadata (annotations) becomes of paramount importance.

Traditionally, when a relational database store is used, an extract–transfer–load (ETL) process is employed for cleaning the quality of data from structured sources. The major disadvantages of relational solutions are (1) a lack of relationship representation between multiple different domain's metadata; (2) an inability to define operational tactics, techniques and procedures, and environmental constraints; and (3) static intelligence product templates that are hard to leverage given the operational tempo. In contrast, semantic store-based solutions offer several advantages by tracking where each piece of information came from, capturing the metadata about the knowledge, thus enabling operators and analysts to conduct sophisticated semantic queries for analyzing information, enabling quicker responses to new information for an automated planning of intelligence assets. Most importantly, this approach enables semantic reasoners to generate new knowledge specific to a domain covered by the underlying ontology.

When the collected data are stored in a graph database, then semantic ETL offers data cleansing beyond removing the syntactic errors in collected data. In semantic ETL, the data are transformed into a knowledge graph and stored in a triple store.

We have developed a semantic knowledge management cloud service and a mobile data collection and situational awareness capability that supports the automated semantic enrichment of mobile collected data while employing our semantic extension to ETL for improved data quality. Our innovative approach incorporates both logical and probabilistic methodologies to automatically detect anomalies within the data for analyst insight. The semantic representation and querying drives the detection of hard inconsistencies, while the entropy drives the detection of soft inconsistencies. Since we embed the entropy into the semantic formulation, this solution provides a unified coherent system for the management of semantically enriched mobile collected data.

EXTRACT–TRANSFER–LOAD

ETL stands for extraction–transformation–loading and is the process of populating a data warehouse from federated sources by extracting (i.e., pulling the data), transforming (i.e., cleansing, enhancing, and transforming), and loading the data (i.e., storing the transformed data in the data warehouse using the warehouse database schema).

ENTROPY

The use of entropy-based methods for anomaly detection and classification is well represented in the cyber security domain. However, applying entropy to understand population dynamics and to detect suspicious information in patterns of behavior is a novel application of this methodology.

Entropy is the measure of the uncertainty in a random variable. Using entropy calculations, we can detect anomalies that cause abrupt changes in the distribution of values, as well as those that slowly change the values. If an anomaly occurs, no matter how slowly it changes the value, it can be detected once the relative entropy increases to a certain level. We incorporate probabilistic information into our ontology by extending our semantic formalism. Using the concept of entropy, this enables us to abstract a set of related assertions into a single assertion. We will demonstrate the applicability of this technology for deriving insights into population dynamics and through the identification of illicit patterns of behavior relevant to the nascent underground power structures and economies that have taken hold in the fictional megacity of Xanadu, especially in the northern wards.

Semantic Knowledge Representation

The sheer quantity of data in the information environment often prohibits effective and efficient manual processing of data into information products and knowledge. Machines must be employed at all levels for filtering, cleaning, and transforming the data and also at higher levels to mimic human judgment for the purposes of flagging, triaging, linking, and interpreting the information. To address this challenge, it is necessary to enable a large body of interconnected data and to use intelligent systems (both human and machine) to extract actionable knowledge. For a machine to be able to aid a human in such a high-level task, it must hold a representation of background knowledge and be able to infer new information based on separate but related pieces of data. In summary, knowledge management is central to information dominance, and the representation of the context is the basis for discerning meaning of its subjects, which brings the formulation into the field of semantic knowledge management.

SEMANTIC KNOWLEDGE MANAGEMENT

Semantic knowledge management involves representing data in the form of a graph, where information is represented as nodes and connected by properties that denote relationships and links. The knowledge base can then be queried by traversing the graph, which contains rich information about the data through the links between entities. Graph data technology is increasingly being used in the commercial sector as well, with Google's Knowledge Graph and Facebook's Graph Search capabilities at the cutting edge of applied research.

To leverage machine automation for generating semantic annotations, the underlying logical foundation for our research is description logics (DL), which are subsets of first-order logic. In our work, we map the mobile collected data and fuse them with additional knowledge sources onto assertions using resource description framework (RDF) triples, represented in the *subject, property, object* format. Together with web ontology language (OWL), RDF provides an explicit and intuitive representation of common knowledge representation syntax for data transfer in the form of RDF triples. Knowledge representation is separated by DL into terminological knowledge (TBox) for representing ontology classes and axioms about classes (e.g., class/subclass hierarchy, equivalent classes) and assertional knowledge (ABox) for describing specific objects and facts (e.g., type, property value) about individual objects. The TBox represents ontology classes and attributes, hierarchy of concepts, and relationships between concepts. The ABox describes specific entities and facts about individual entities. The combination of TBox and ABox defines the full ontology, which is commonly referred to as the knowledge base. This representation, and semantic knowledge management principles in general, makes it possible to leverage semantic reasoners, which enables us to detect and reason about suspicious and conflicting information automatically (Horridge et al. 2008).

In our research, we utilize SPARQL-DL, an RDF query language, in addition to OWL-DL for formulation. SPARQL-DL provides OWL-DL-like semantics for SPARQL basic graph patterns, which involves both ABox queries and mixed TBox and ABox queries over DL ontologies. As a starting point to formulate the semantics of information, we define two assertions to conflict with each other when the subject instance S_1, has the two different values for the same property P. For instance, <subject S_1, property P, object V_1> and <subject S_1, property P, object V_2> conflict with each other. We define a set of instance assertions to be inconsistent when the assertions violate the properties of the underlying classes. For instance, <S, P, V_1> holds in the subject domain for all instances of the class S. In that case, the instance assertion <S_1, P, V_2> is inconsistent with the class property. For suspicious information, we define a set of assertions to be suspicious when the entropy of the value distribution differs from that encountered before. For instance, we analyze the suspicious nature {<S_1, P, V_1>, . . . , <S_n, P, V_n>} by computing the entropy of the stream {V_1, . . . , V_n>} for the instances {S_1, . . . , S_n} of the same class S.

Based on our research, we identified three primary benefits of semantically representing information, allowing semantic databases to be much more expressive than traditional databases. First, unlike in a traditional relational database that forces its data into a fixed schema that was designed before the first piece of data was recorded, a semantic database has a dynamic schema. As the world changes, the database schema can be easily updated to represent the data in an evolving way while maintaining usability. Secondly, a relational database's schema can represent only a small set of relationships through the particular ways its tables are joined (e.g., one to one, one to many); a semantic database can express arbitrary relationships among its fields through the use of a graph structure. Additionally, these relationships end up defining what each field means in such a way that other semantically enabled systems would be able to dynamically interoperate or even collaborate with

WHAT ARE ONTOLOGIES?

An ontology is a formal naming and definition of the types, the properties, and the interrelationships of the entities within a particular domain.

it, removing the danger of stove-piped systems that are so often referenced as a key issue. Thirdly, the information in semantic databases is formatted in such a way that a semantic reasoner can use DL to infer new information from existing records and their relationships (Nardi and Brachman 2007). In a semantic knowledge base, class-level relationships are represented in the database itself, enabling the automated use of logic to conclude that there is a conflict between records.

Mobile Data Collection to Support Semantic Knowledge Management

Given an understanding of semantic knowledge management constructs, we have developed a mobile device application (app) to automatically semantically enrich and fuse mobile-collected data for analysis and visualization. Our research transforms data collection from an unstructured process to a mobile device-enabled approach that leverages the advantages of structured forms. We use semantic technology to author, publish, and distribute forms to mobile devices. When the structured form is submitted to the mobile device, it is mapped to a semantic form to automatically generate semantic annotations from the mobile collected data for semantic enrichment. The semantic representation of the data enables users to organize, analyze, visualize, and create custom information products from the aggregated data. Mobile users can receive geospatially enabled notifications and alerts to improve their situational awareness and refine data-collection plans from the field. Using the mobile device's global positioning system locator, we automatically capture the coordinates where the assessment took place to enable geospatial analysis and visualization. Figure 11.1 denotes an example of our mobile app being used to collect information from a vulnerable population.

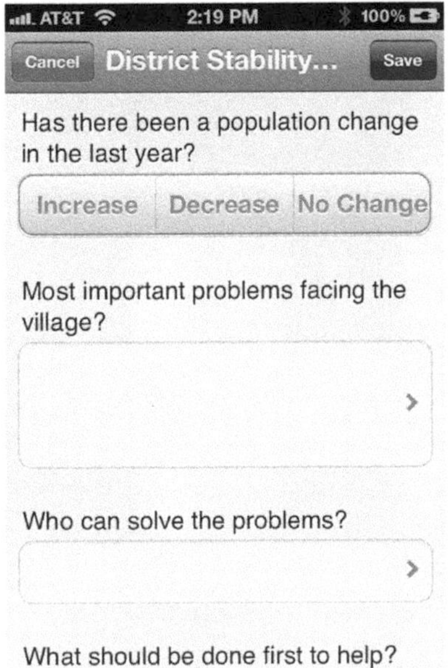

FIGURE 11.1
Structured form for data collection within a mobile app.

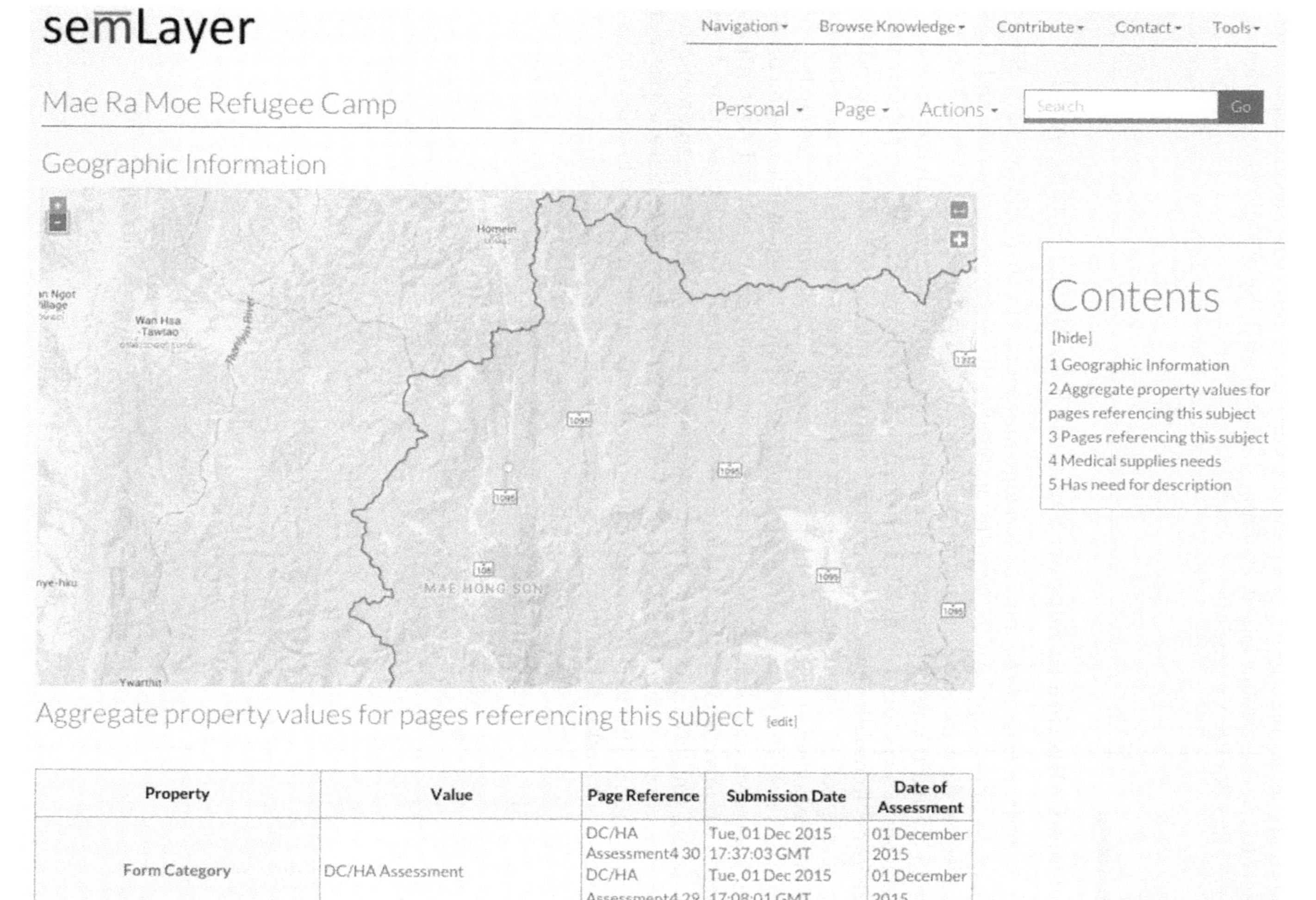

Property	Value	Page Reference	Submission Date	Date of Assessment
Form Category	DC/HA Assessment	DC/HA Assessment4 30	Tue, 01 Dec 2015 17:37:03 GMT	01 December 2015
		DC/HA Assessment4 29	Tue, 01 Dec 2015 17:08:01 GMT	01 December 2015

FIGURE 11.2
Form submitted from the mobile app-created page in knowledge base.

To semantically enrich and fuse mobile-collected data, our app allows users to download forms to their mobile device from a central knowledge base and populate the form with submissions that automatically create pages in the semantic knowledge base for each submission. As the forms are brought in to the semantic knowledge base, the form fields map onto the semantic forms so that the data are automatically annotated in the subject, property, object format. By tagging data in this format, we are able to generate semantic queries to analyze and visualize the data. For example, mobile data collectors interacting with the public can record the sentiment of an interviewee using positive, negative, or neutral annotations. This technology can enable semantic queries to aggregate data across dimensions such as understanding demographic data for those individuals that reported a negative view in a particular location. These types of structured questionnaires can be utilized to gain more robust situational assessments of areas of interest. For example, understanding infrastructure requirements and conducting engagements with key leaders and civilians can inform stakeholders about population sentiments on various key issues for more data-driven and informed policies on both neighborhood and subdistrict levels. Figure 11.2 shows a page that has been automatically created in the knowledge base from mobile collected data.

Data Analysis and Fusion

With the mobile-collected data in a structured format in our semantic knowledge base, we can then begin the task of data analysis and fusion. Our solution uses the Protege reasoner to interpret query results and present them to the user. The semantic processing workflow is as follows:

- An ontology editor enables the addition of an assertion from new data.
- The semantic query module finds all related assertions as a subgraph.
- The semantic reasoner checks ABox/TBox to find inconsistencies and conflicts.
- The semantic reasoner is extended with rules beyond general ontology-class reasoner to detect application specific knowledge.
- A suspicious, conflicting, inconsistent (SCI) data interpreter analyzes results for presentation to the analyst in a coherent manner.

Our approach is a combination of several methods, query types, and applications of rules. We will describe our techniques for detecting conflicting and inconsistent information separately and also explain the diversity in methodologies within those categories.

Conflict Detection

We define conflicting information as logical anomalies in the ABox, or instance-level, data (versus the TBox or class-level data). First, there may be inference steps involved in discovering the conflict. Furthermore, a conflict may be uncovered via interpretations or calculations derived from the semantic data. For example, to determine anomalies related to reports of an individual's location, we can calculate the distance between two locations that were reported in intelligence data from the field. The query therefore aims to determine whether two reports of the location of a person are in conflict with one another. If report 1 says that Ibrahim is in Burj Abi Haidar at 10:00 a.m., and report 2 says that Ibrahim is in Kafr Batna at 10:30 a.m. on the same day, it is not immediately obvious whether these two reports are in conflict; it is necessary to have information about how far apart the two locations are and how fast Ibrahim would have to travel in order to go from one location to the other within the allocated time. If the two locations are in very close proximity, it may be possible for him to be spotted at both locations within a short interval of time. In our research, we then use a query that automatically calculates the distance between locations as identified in reports.

A SPARQL query first returns the two reports, then with SPARQL built-in functions, determines the distance between the two cities, 85.5 miles. Since the two reports are 30 minutes apart, the query can also do the division and determine that Ibrahim would have to be traveling 171 mph in order for both report 1 and report 2 to be accurate. Assuming that Ibrahim did not have access to high-speed transit, such as a private plane, it is impossible for him to have travelled that fast; therefore, the two reports are in conflict and can be automatically flagged as such to the analyst.

As this example illustrates, our solution detects conflicts using a two-step process. First is the use of a SPARQL query to receive relevant results (e.g., all recent reports about Ibrahim's location). SPARQL is a standard semantic query language based on the structured query language syntax but applicable to semantically encoded data. We then interpret the query results using a semantic interpreter on behalf of the analyst. Our interpreter makes use of the OWL API standard, a very large Java library that allows programmatic control of the reasoner and modifications to the ontology and query.

Inconsistency Detection

In our approach, there are several different ways to represent inconsistent information.

For example, a semantic reasoner can automatically find an inconsistency by applying a general TBox-level rule to the ontology. However, stringent ontological rules cannot be applied to real-world data with any useful consequences because as soon as an ontology is deemed inconsistent, no classical reasoning can be performed on the ontology. This is because of (what is termed in logic) the principle of explosion. Once there is an inconsistency in an ontology (e.g., both a statement and its inverse are in the ontology), no logical distinctions can be made between true and false statements. There are a few ways to deal with this problem:

1. Once an inconsistency is found in the ontology, resolve or remove it.
2. Use a paraconsistency reasoner, one that works on consistent subsets of ontologies.
3. Do not impose rules on the ontology that can trigger inconsistencies, but instead query for conflicting information and tag results as conflicting.

Given that the real world is often inconsistent and illogical, resolving/removing inconsistencies would seriously reduce the expressiveness of the ontology. The two other alternatives—DL query and paraconsistency reasoner—are described in the following.

DL Query for Inconsistencies

Another way to qualify how entities are related is through logical rules on the properties that join them. Properties can be tagged as being transitive, functional, symmetric, etc.

Some of these rules primarily serve to restrict the usage of the property. For example, if *birthmother* is tagged to be functional, then the reasoner will not allow an entity to have more than one birthmother. In other cases, the rules help the reasoner derive new information. For example, if *hasSibling* is tagged to be transitive and the ontology states that *personA hasSibling personB* and *personB hasSibling personC*, then a reasoner will derive that *personA hasSibling personC*. In our ontology regarding Ibrahim, we set the hasSibling relationship to be transitive. In the ontology, we also asserted the following:

Abdullah hasSibling Ibrahim
Ibrahim hasSibling Tevfik

Therefore, the reasoner can deduce from the assertions and detail information relating to the individual's social network the following conclusion that "Abdullah hasSibling Tevfik" with the explanation citing the transitivity property.

These logical rules help in discovering inconsistencies when the inconsistencies are between inferred pieces of information and not information explicitly stated in the ontology. For the example earlier, the ontology also included an assertion "Abdullah notRelatedTo Tevfik." This assertion is inconsistent with the assertion that they are siblings. To detect this inconsistency, we performed the following query that looks for all pairs of persons that are joined by these conflicting relations, *hasFamilialRelation* and *notRelatedTo.*

This produces the following results: Abdullah Al-Asiri and Tevfik al-Asiri. This application is useful as it can enable the automatic flagging of inconsistent relations in a social network to analysts. As data are collected in the field through the mobile app with civilian engagements and interviews, a robust social network can be built up in the knowledge base. Utilizing these principles, analysts can be assisted by a machine to automatically determine inconsistent and anomalous entities and relationships for further review in the network defined through the ontology and semantic rules.

Paraconsistency Reasoners

Besides the approach described earlier of detecting inconsistencies, one can use a paraconsistency reasoner to perform reasoning and detect anomalies even in an inconsistent ontology, or one that contains syntactic errors. The goal of reasoning on inconsistent ontologies has been illustrated through multiple new approaches that have been implemented in the Large Knowledge Collider (2014) platform. One such approach (Fang and Huang 2010) is through the use of minimal inconsistent sets to determine consistent subsets that can be reasoned on. In this way, even though the ontology as a whole is not consistent, it may be that the subset of the ontology that is relevant to a particular query is consistent, and therefore classical reasoning can be applied to that subset. Another approach involves contrastive reasoning (Fang, Huang, and van Harmelen 2012) or reasoning with qualifications.

```
SELECT ?z ?x
WHERE
   { ?x pre2:hasFamilialRelation ?z ;
   pre2:notRelatedTo ?z .}
```

A third approach includes what is called four-valued logic (Maier, Ma, and Hitzler 2013). In classical logic, all assertions can have two values: true or false. In four-valued logic, an assertion can have four values: (*n*) nothing is known about it, (0) false, (1) true, (b) both true and false. Using this value set essentially allows for inconsistencies to exist and for reasoning to consider all values of an assertion, returning possibly multiple results but not an explosive number of results.

Probabilistic Semantics for Anomaly Detection

Entropy-Based Anomaly Detection

Let us consider the utility of these probabilistic semantics for anomaly detection methods in the context of an example. To return to the fictional use case, we can use this methodology to detect suspicious information about high-value targets in an effort to combat the growing mafia presence in Xanadu. Over a baseline period, a high-value target exhibits a certain routine, which is not completely rigid but which has clear statistical characteristics. Intuitively, suspicious behavior is simply a behavior that is outside of our expectation, based on prior knowledge of a situation. When the high-value target suddenly changes his daily routine, it is in a way that might be subtle and not noticed by a human, but the entropy calculations indicate a clear anomaly. We can define suspicious behavior as anomalous changes in the typical behavior of an entity or conflicting dissemination of information by distinct entities. Using this definition, we can use entropy—basic entropy, Kullback–Leibler (KL) divergence, and conditional entropy—to uncover suspicious information. Also, for understanding high-value targets, a conflict may be uncovered through interpretations or calculations derived from the semantic data.

In our approach, to detect suspicious and deceptive information, we use a measure of the change in the entropy of a stream of information using KL divergence. For suspicious information, we define a set of assertions as suspicious when the entropy of the value distribution differs from that encountered before. As mentioned earlier, entropy is the measure of the uncertainty in a random variable. Using entropy calculations, we can detect anomalies that cause abrupt changes in the distribution of values, as well as those that slowly change the values. If an anomaly occurs, no matter how slowly it changes the value, it can be detected once the relative entropy increases to a certain level. For instance, we can analyze the suspicious nature $\{<S_1, P, V_1>, \ldots, <S_n, P, V_n>\}$ by computing the entropy of the stream $\{V_1, \ldots, V_n>\}$ for the instances $\{S_1, \ldots, S_n\}$ of the same class S.

We incorporate probabilistic information into our ontology by extending our subject, property, value formalism to subject, property, random variable value. This enables us to abstract a set of related RDF assertions into a single assertion using the concept of entropy. Suppose our sources inform the following:

(KeyLeaderOfXanaduDistrictA, hasName, personW) → 2 times
(KeyLeaderOfXanaduDistrictA, hasName, personX) → 3 times
(KeyLeaderOfXanaduDistrictA, hasName, personY) → 3 times
(KeyLeaderOfXanaduDistrictA, hasName, personZ) → 2 times

We can then calculate the entropy of this information and generalize these 10 RDF assertions with one. In this case, the entropy = $-0.2 \times \log_2 (0.2) - 0.3 \times \log_2 (0.3) - 0.3 \times \log_2 (0.3) - 0.2 \times \log_2 (0.2) = 1.97$.

Then we add a new statement to our ontology summarizing the current entropy state of this information by

(KeyLeaderOfXanaduDistrictA, hasNameEntropy, 1.97)

In the case earlier, the entropy is high because the three possible outcomes for the key leader's name appear almost equally, i.e., there is no one name that is overwhelmingly stated as the one. In this way, the entropy gives a measure of the conflict/inconsistency in the data. Suppose, conversely, that our intel said

(KeyLeaderOfXanaduDistrictA, hasName, personW) → 5%
(KeyLeaderOfXanaduDistrictA, hasName, personX) → 90%
(KeyLeaderOfXanaduDistrictA, hasName, personY) → 1%
(KeyLeaderOfXanaduDistrictA, hasName, personZ) → 4%

In this case, the entropy is only 0.81, as one name, PersonX, is represented a large majority of the time. In other words, in this case, there is very little conflict over the reported key leader's name.

The maximal entropy (strongest conflict) occurs when each outcome is observed in equal proportions. The entropy for N number of outcomes where the probability of each outcome is the same (i.e., $p = 1/N$) is

$$H(X) = -\text{Sum}_i[p(x_i) \log (p(x_i))] = -Np \log (p) = -N(1/N) \log (p) = -\log (p) = -\log (1/N).$$

So for a piece of information with N possible outcomes, the entropy is in the range $[0, -\log (1/N)]$.

Detection of Information Measures in Time Using KL Divergence

For certain kinds of information, such as the location of a person of interest over time, or the sentiment toward a leader in a country, the presence of a multiplicity of outcomes is normal. In these cases, an anomaly exists when the distribution of outcomes significantly changes over time. Such anomalies occur, for example, when a person of interest changes his routine or when the distribution of sentiment toward a leader increases or drops significantly. For detection of changes like these, it is useful to consider previous distributions of outcomes for a piece of information, X, as the baseline probability distribution, p. Then the current distribution of outcomes to evaluate, call it m, can be compared to the baseline to detect a significant change. The cross entropy is defined as

$$H(p, m) = -\text{Sum}_i[p(x_i) \log (m(x_i))]$$

and indicates how well the new distribution estimates the baseline. If $p = m$, then the cross entropy is at its minimum and is simply the entropy of X. The more distinct m and p are, the higher the cross entropy is, and the less likely it is that the sample error can account for their differences. The KL divergence, defined as

$$D_{KL}(P|M) = H(p,m) - H(p),$$

is a measure of the change in the entropy of the system. Therefore, we can use the KL divergence to measure the significance of the temporal change of the entropy. For example, perhaps there is a stream of information of the location of a person of interest, Ibrahim. Over a nine-day period, three reports of his whereabouts were logged, once in the morning, the afternoon, and the evening. Over the first four days, Ibrahim's locations were distributed according to the pie chart on the left in Figure 11.3. The pie chart on the right includes his locations over all nine days of analysis.

In the first case, the entropy of the person's location is 0.98, and we consider this the entropy of the baseline distribution. It is clear that he has changed his routine and is now visiting a new location, An Nadhim, over the course of his day. The new entropy is 1.48, indicative that there is more diversity in his whereabouts, especially considering that the maximum entropy for a variable with three outcomes is 1.58. The KL divergence of the new with the old distribution is 0.30, which is about a 30% change from the baseline distribution's entropy. In this case, our solution would compute the entropy divergence and alert the analyst, while presenting the baseline and new distributions for analysis.

Drill Down of Anomalous/Suspicious Behavior

Some pieces of information are substantially interrelated with other concepts such that a deeper investigation may reveal deeper insights into the anomalous or suspicious behavior. In our example earlier, Ibrahim's location is dependent on the time of day, and so it may be that his overall distribution of locations does not significantly change, but his reported locations at a particular time of day do significantly change. In this case, the piece of information—Ibrahim's location—can be conditional on another variable, such as the time of day.

Conditional entropy is defined for joint entropy conditioned over the value of one of the random variables:

$$H(Y|X) = -\text{Sum}_i(\text{Sum}_j)[p(y_1, x_j) \log p(y_1|x_j)],$$

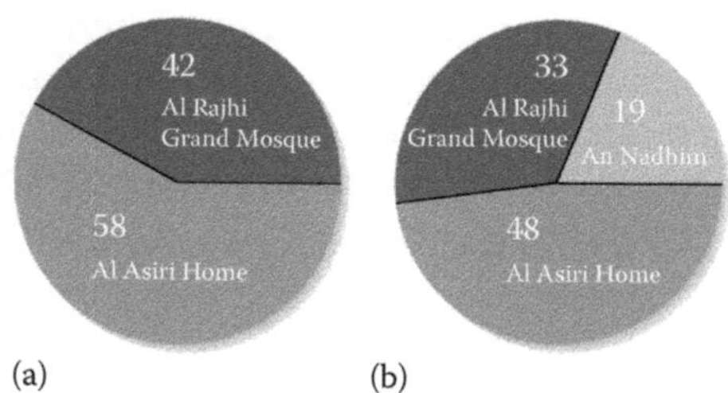

FIGURE 11.3
Distribution of Ibrahim's locations over a four-day (a) and nine-day (b) period. In (a) two locations, a mosque and the home location of the target are identified. In (b) the mosque, home, and the neighborhood An Nadhim are identified as locations.

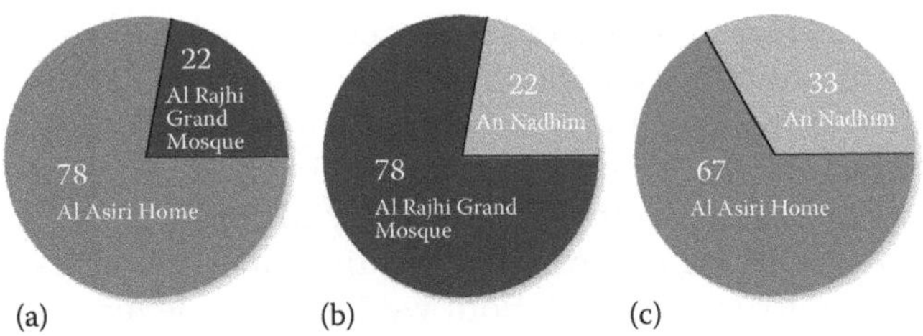

FIGURE 11.4
Distribution of Ibrahim's locations at (a) morning, (b) afternoon, and (c) night.

where $p(y, x)$ is the joint probability distribution of x and y. Here we can see the distribution of Ibrahim's locations in Figure 11.4 when conditioned on the time of day.

The distribution is significantly different in the evening from the other times of day, and this may be tagged as suspicious and brought to the analyst's attention via an alert.

We can define suspicious behavior as anomalous changes in the typical behavior of an entity or the conflicting dissemination of information by distinct entities. Using this definition, we have shown that by using entropy—basic entropy, KL divergence, and conditional entropy—we can uncover suspicious information.

Future Research

Based on our research and development in this domain, we envision integrating our semantic and entropy techniques into a full-scale anomaly detection engine. As new intelligence data comes in, it is inserted into the semantic data store and enriched with contextual knowledge. The anomaly detection engine runs through several default queries, such as spatiotemporal inconsistency and performs those relevant to the incoming data to find any logical anomalies. Similarly, entropy calculations on data distributions modified by the incoming data are performed in order to discover any probabilistic anomalies because incoming intelligence effectively changes the distribution of values for the monitored properties. For logical anomalies, the related data and any logical inference steps are then displayed; for probabilistic anomalies, visualizations demonstrating the entropy anomalies are displayed. Additionally, the user will be able to accept or reject the anomaly. If the anomaly is accepted, its description as an anomaly will be reinserted back into the semantic store and will become a part of the machine and the human knowledge of the intelligence domain. If rejected, the system will adjust internal parameters to better reflect the user's concept of an anomaly. In this way, we incorporate machine learning to improve and converge the system's knowledge and judgment with that of the user.

P[Location | Time = morning] = .76
P[Location | Time = afternoon] = .76
P[Location | Time = evening] = .91

We also aim to investigate the use of stream entropy algorithms for the scalability of our approach to high-velocity data streams. For instance, consider the computation of conditional entropy that indicates the extent to which one value in a subject, property, value format can be predicted given the value of another value. With pairs of distributions for conditional entropy dictated by different interests of multiple analysts, the amount of processing and memory required increases quadratically. In order to develop a scalable entropy computation, we intend to leverage our own research on stream entropy computation (Arackaparambil et al. 2012), developed for monitoring network activity in cybersecurity applications (Arackaparambil et al. 2010). Our stream entropy computation uses sampling theory-based on the hierarchical sample sketch algorithm of Bhuvanigiri and Ganguly (2006) and favorably compares to the implementation of the Lall et al. (2006). We will also investigate Jensen–Shannon divergence, a measure of distance (i.e., difference) between two distributions that behaves as a metric (e.g., for three distributions, pairwise distances obey the triangle inequality). Unlike KL divergence, which yields infinity for any two distributions one of which contains a category not contained in another, the Jensen–Shannon divergence does not have this issue.

Conclusion

To understand population dynamics, as information is collected on the mobile application, operators ask questions related to the critical needs of various segments of the society to gain new insight from the identification of anomalies. For example, they might ask questions when civilians report specific problems with key services or particular distrust of local officials and governmental institutions. Entropy semantics for the data collected by mobile devices will automatically determine the important changes in the information product results, such as changes in population sentiment over time in a particular area or spatiotemporal inconsistencies for determining patterns of behavior for high-value targets. Our semantic knowledge management solution eliminates the challenge of form management for mobile data collection campaigns, providing a cloud service for collection, aggregation, analysis, and collaboration. We use semantic annotations as a semantic enrichment representation for knowledge management, leveraging machine automation to capture semantic enrichment through form-based data collection, linking categories, properties, and values in the data. The data are collected with contextual knowledge captured in the knowledge base ontology ingested from other data repositories. Semantic inferencing can be used to detect new relationships in the data for greater data aggregation and analysis and anomaly detection.

While there are systems that attempt to semantically represent information, they are often unable to reasonably deal with information anomalies that arise. Similarly, there are systems that calculate statistics on streams of information but in inflexible and inadaptable ways that do not result in powerful information anomaly detection. Our research uses the principles and tools of the semantic web to encode information in expressive and adaptable ways so that the knowledge base can truly be a representation of known information. Using logical inferencing and information entropy, we can detect SCI information and then flag them for the attention of an analyst.

Megacity Use Case

For practical purposes, let us consider how semantic knowledge management and anomaly detection in intelligence data can be applied to the fictional use case of the city of Xanadu presented in this book. Our research proposes to use semantically enriched data in an innovative way that incorporates both logical and probabilistic methodologies to automatically detect anomalies within the data for actionable insight. Semantic representation and querying drives the detection of hard inconsistencies, such as specific values that conflict, while entropy drives the detection of soft inconsistencies that are more probabilistic in nature. Since we embed the entropy into the semantic formulation, this solution provides a unified coherent system for the management of semantically enriched mobile-collected data.

In our approach, we use entropy to measure the level of conflict and inconsistency, specifically detecting suspicious information by assessing the entropy of a stream of information and using KL divergence to measure change in the entropy. For suspicious information, we define a set of assertions as suspicious when the entropy of the value distribution differs from the value encountered before. We incorporate probabilistic information into our ontology by extending our semantic formalism. This enables us to abstract a set of related assertions into a single assertion through use of the concept of entropy. Using these methods, an analyst can derive insights into population dynamics, for instance, to identify illicit patterns of behavior relevant to the nascent underground power structures and economies in the megacity of Xanadu.

Consider a situation where various operators throughout Xanadu are collecting information via their mobile applications. The operators ask Xanadu residents questions relating to the critical needs of various segments of the society, such as issues with key services or particular distrust of local officials and governmental institutions. In these cases, the semantic encoding plus anomaly detection approach can provide new insights based upon the identification of anomalies, highlighting when an anomaly exists if the distribution of outcomes significantly changes over time, such as if the distribution of sentiment toward a leader increases or drops significantly.

Additionally, in an effort to combat the growing mafia presence in Xanadu, we can use this methodology to detect suspicious information for high-value targets. Over a baseline period, a high-value target exhibits a certain routine that is not completely rigid but has clear statistical characteristics. Intuitively, suspicious behavior is simply a behavior that is outside of our expectation, based on prior knowledge of a situation. When the high-value target suddenly changes his daily routine (in a way that might be subtle and not noticed by a human), the entropy calculations can indicate a clear anomaly. We can define suspicious behavior as anomalous changes in the typical behavior of an entity or the conflicting dissemination of information by distinct entities. Using this definition, we have shown that by using entropy—basic entropy, KL divergence, and conditional entropy—we can uncover suspicious information.

Also, for understanding high-value targets, a conflict may be uncovered through interpretations or calculations derived from the semantic data. As we discussed the main body of the chapter, reports may place a high-value target at different locations on the same day. Drawing upon the semantic encoding, the system can calculate the distance between two locations and determine whether two reports of the two sightings conflict with one another. Our calculations can automatically determine if the information is in conflict and flag it to the analyst using the KL divergence to measure the significance of the temporal change of the entropy. Using these techniques, concerned parties could monitor the social landscape of Xanadu, gaining insight on illicit underground behaviors and providing early detection of anomalies that foreshadow potentially dangerous future events.

DISCUSSION QUESTIONS

1. Why is it necessary to consider the collection of data from mobile devices as an important data collection tool?
2. What does semantic knowledge management mean, and why is it useful to represent information in a knowledge graph?
3. How does the semantic representation of data support automated querying for detecting anomalies within data?
4. In broad terms, how can entropy be used to detect anomalies within data?

References

Arackaparambil, C., S. Bratus, J. Brody, and A. Shubina. 2010. Distributed monitoring of conditional entropy for anomaly detection in streams. Paper presented at the *IEEE Workshop on Scalable Stream Processing Systems (SSPS), Atlanta, GA.*

Arackaparambil, C., G. Yan, S. Bratus, and A. Caglayan. 2012. On tuning the knobs of distribution-based methods for detecting VoIP covert channels. Paper presented at the annual *Hawaii International Conference on System Sciences (HICSS-45) Insider Threat Modeling, Detection, and Mitigation Minitrack, Grand Wailea, Maui, Hawaii.*

Bhuvanagiri, L., and S. Ganguly. 2006. Estimating entropy over data streams. In *Proceedings of ESA.*

Fang, J., and Z. Huang. 2010. A new approach of reasoning with inconsistent ontologies. In *Proceedings of the 4th Chinese Semantic Web Symposium.*

Fang, J., Z. Huang, and F. van Harmelen. 2012. A method of contrastive reasoning with inconsistent ontologies. *The Semantic Web: Lecture Notes in Computer Science* 7185: 1–16.

Horridge, M., J. Bauer, B. Parsia, and U. Sattler. 2008. Understanding entailments in OWL. In *Proceedings of the Fifth OWLED Workshop on OWL: Experiences and Directions, 7th International Semantic Web Conference (ISWC-2008), Karlsruhe, Germany* 432.

Maier, F., Y. Ma, and P. Hitzler. 2013. Paraconsistent OWL and related logics. *Semantic Web Journal* 4(4): 395–427.

Nardi, D., and R. J. Brachman. 2007. An introduction to description logics. In *The description Logic Handbook*, ed. Baader, F., D. Calvanese, D. L. McGuinness, D. Nardi, and P. F. Patel-Schneider, 1–40. New York: Cambridge University Press.

Lall, A., V. Sekar, M. Ogihara, J. Xu, and H. Zhang. 2006. Data streaming algorithms for estimating entropy of network traffic. In *Proceedings of ACM SIGMETRICS* 34(1): 145–156.

The Large Knowledge Collider. 2014. http://www.larkc.org/.

12

Measuring Changes in Attitudes: Using Factor Analysis to Track Population Attitudes Spatially and Temporally

Joseph Maddux and Jeffrey Appleget

CONTENTS

Introduction 222
Background 222
Statistical Primer 223
Methodology 225
Case Study: Afghanistan Survey 230
  History 230
  Results of Factor Analysis 231
  Deriving Indicators from Factors 232
Analytics Suite for Displaying Measured Change 233
  Displaying Spatiotemporal Indicator Change 233
  Mean Respondent Display 235
  Proportionality of Indicator Support Display 236
Conclusion 237
Megacity Use Case 238
References 240
Appendix 240

ABSTRACT This chapter discusses a way to analyze population surveys. Concepts and techniques such as factor analysis, exploratory data analysis, nonparametric statistics, data cleansing, and hypothesis testing are covered. A case study of Afghanistan's Helmand province is used to demonstrate the application of the methodology and its usefulness to an analyst. The objective of this chapter is to show a way of uncovering hidden information within survey data and a method to display those results to decision makers.

KEY TOPICS

- Factor analysis used to help uncover hidden information
- Data cleansing and recoding Likert scale responses
- Deriving meaningful indicators from factor analysis
- Measuring change using nonparametric statistics and hypothesis testing
- Visual displays of analytically derived survey metrics

Introduction

Population surveys can provide valuable insight on population beliefs and attitudes on a wide range of topics. Countless surveys are conducted each year for just that purpose. Survey questionnaires can range anywhere from dozens to hundreds of questions. Professionally designed, quality surveys are typically costly and require many human hours to complete. After collecting the data, the experts who designed and administered the survey often clean the data and report cursory results to the sponsoring agency (e.g., a military combatant command or NGO), but detailed survey analysis is typically left to the sponsoring agency. Unfortunately, many survey sponsors lack the personal expertise or other resources necessary to conduct such analyses. As a result, the sponsoring agency may be unable to unlock any value from the study or may inadvertently perform biased analyses. For instance, well-meaning staff may brief leadership on survey results item by item (where an item is a survey's question and response) or the survey results may be culled through to provide insights by a pick-up team trying to salvage value from an expensive investment. The worst case, which we have seen many times, occurs when the sponsoring agency simply parks a survey's results with the hope that someone else will do something useful with the data in the future.

This chapter outlines the authors' experience with such a situation, and we describe, in a plain and accessible way, how we analyzed the given survey data and provided valuable insights to the sponsoring agency. The overall effort, supported by the Naval Postgraduate School, involved a five-year research effort that spawned 16 student theses and over a dozen presentations at conferences in the United States and at international venues.

Background

In June 2010, the United States Government Accountability Office (GAO) released the findings of a study conducted for United States Africa Command (AFRICOM), one of nine US military combatant commands around the world. The study examined AFRICOM's effectiveness in carrying out its mission in five areas with respect to activity planning and implementation. Among the findings, the GAO report stated, "AFRICOM is generally not measuring the long-term effects of activities" and by not doing so they had no way of understanding how their actions impacted their area of responsibility (GAO, 2010).

In response to the GAO report, AFRICOM developed a comprehensive method of measuring long-term effects, known as the strategic measuring and reporting tool (SMART), which was intended to use diverse data to inform measures of effectiveness (mission achievement metrics) and measures of performance (measures of a system's quantifiable performance features) to help AFRICOM decision makers monitor the long-term effects of their actions. In essence, SMART helps decision makers within AFRICOM to conduct their strategic planning. The SMART system is populated with objective, subjective, and perceptive indicators. Objective indicator data are gathered from reputable open sources, such as the World Bank, the CIA Fact Book, or US embassies within Africa. Subjective indicator fields are scored by some of these same embassies, along with other regional or country experts, and are guided by a strict set of scoring criteria designed to mitigate over- or underinflation of scoring. Objective and subjective indicators are plentiful. However, AFRICOM personnel stated the number of high-quality perceptive indicators (i.e., indicators derived from popular perceptions) needed to balance their model was lacking; the few perceptive indicators they did incorporate into SMART were retrieved mainly from arbitrary selection of survey questions.

We developed a process that would use surveys to help solve this problem. AFRICOM possessed perception data from Africa's Sahel region, via surveys from six countries in the Sahel. The surveys had been conducted annually over a four-year period; an iteration of the survey is known as a wave. Each survey instrument contained anywhere from 75 to 100 questions, depending on the country and the wave; for the most part, the survey questions remained quite consistent over time. Together, the consistency of location and questions allowed us to analyze these surveys temporally and spatially to show if, where, and how population attitudes had shifted over time.

Statistical Primer

Before sharing our personal experiences, it is first important to understand some of the basic analytical techniques used to help analyze the data: factor analysis and imputation.

Factor analysis is a statistical technique that creates clusters (or factors) within the data. This serves two purposes: first, to reduce large amounts of observed data into a smaller set of unobserved data and, second, as a method for identifying latent traits from question-level survey data. Factor analysis aids the interpretation of survey results whenever the phenomenon of interest is complex and not directly measurable via a single question. In such situations, it is necessary to ask a series of questions about the phenomenon and then appropriately combine the resulting responses into a single measure or factor. Such factors then become the observed measures of the unobservable or the latent phenomenon (Fricker et al., 2012). Factor analysis is a method to extract valuable insight from the entire data set rather than just hand-selecting a few items to analyze and leaving the remaining data archived.

Figure 12.1 illustrates the idea of factor analysis with six observed variables (i.e., survey question responses) that can be effectively summarized in terms of two latent variables (factors). Note that the survey question responses are observed with error (denoted by ε), and the question responses are weighted linear combinations of the factors (where the weights or the loadings are shown as λ_{ij}). Performing a factor analysis effectively models the observed variables as linear combinations of factors, where the analyst has prespecified the number of factors such that the model covariance matrix closely matches the sample covariance matrix of the observed variables (Fricker et al., 2012).

With few exceptions, we have found that factor analysis usually groups items that have an obvious common theme or topic. The items serve as an observed variable (a question that was asked directly), and the factors serve as unobserved variables (latent traits) that were not specifically identified or purposefully addressed in the survey instrument.

In order to use factor analysis on surveys, the data may require recoding (e.g., Likert scale questions assigned numerical values), and some accommodation for missing survey data must be made. For missing data, casewise deletion (deleting the entire respondents' survey record because she failed to answer a single question) results in the loss of valuable data;

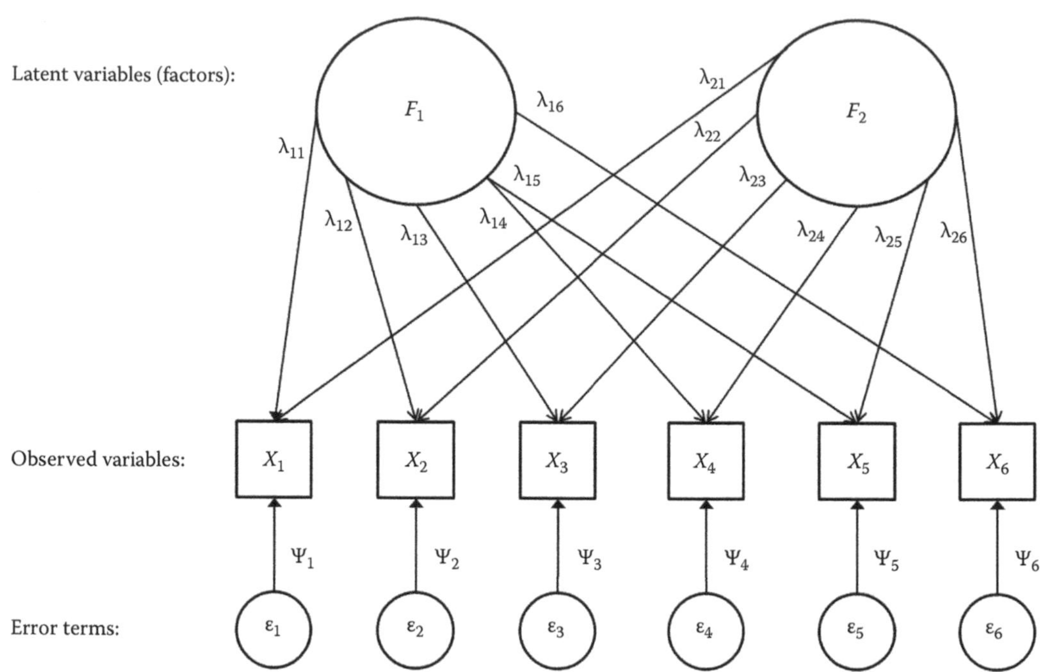

FIGURE 12.1
An illustrative example of factor analysis. (From Fricker, R. D., Jr., Kulzy, W. W., and Appleget, J. A., *Phalanx*, 45, 4, 30–34, 2012. With permission.)

thus, various imputation techniques (techniques to fill in missing responses) are needed. Andrew Gelman and Hill (2006) discuss a variety of imputation techniques in their book *Data Analysis Using Regression and Multilevel/Hierarchical Models*. For this research, we used a process known as hot deck imputation, which, in our experience, yields effective results. This technique finds a replacement for a missing value by comparing the demographic data of the respondent whose value is missing (the receiver) to the demographic data of respondents who answered the particular question. These respondents who gave a valid answer to the question are known as donors. Once a set of donors is compiled, an answer from a randomly selected donor is used in place of the missing value for our receiver. We used this process for all missing values in the survey.

DATA ARE TOO VALUABLE TO THROW AWAY

One way to deal with missing data is to use casewise deletion, meaning if one or more of a respondent's answers are missing, then that entire record is deleted as if the interview of that person never took place. In ecological studies, this is typically a bad idea. In our earlier research, we found that missing as little as 2.4% of the answers would lead to the loss of nearly 68% of all respondents if casewise deletion were employed. Thus, in practice, any of a number of imputation techniques, i.e., methods that substitute a missing value with a valid response, are superior to casewise deletion.

Methodology

Because our goal was to identify the changes in population attitudes from one wave to the next, we had to create a methodology that leveraged factor analysis to provide a consistent way to examine factors over time. The first challenge was the survey instrument itself. The survey slightly changed from wave to wave, so questions that might comprise a factor in one wave may not even be asked in the subsequent wave. In order to avoid that challenge, we created a list of what we called *enduring questions*, which included questions appearing in each wave of the survey instruments. This allowed us to create factors from comparable questions.

The factor analysis process, itself, presented another challenge. If the factor analysis is applied to the survey data from the same region over different waves, the questions that comprised a factor may have different weights (λ_{ij} or loadings) for different waves. In some cases, an identified factor included more or fewer questions from wave to wave. This occurred when a question had a loading greater than 0.4 in one wave but less than 0.4 in a different wave. We determined that, for consistency, we would perform a factor analysis for a base wave, typically the first or the last wave, and fix the questions and loadings for the factors. We called this schema the *indicator* (since it would not truly be a factor in subsequent wave).

For the purposes of illustration, we present results from a factor analysis performed for each of three different countries for the same wave; the survey asked how each country's population perceived its government. We focused on four governance-related factors derived from sample data: trust in policy makers, trust in agencies, democracy, and trust in government and democracy (see Table 12.1). These four factors were derived from 12 survey questions that varied in definition across the three countries. The numbers in the three columns represent the loadings that correspond to each of the questions on the left, and the boxes show how the factors are defined in terms of the questions and associated loadings. From these 12 questions, the data related to country A formed two factors: trust in policy makers and trust in Agencies. The former comprised questions about the country's president and prime minister, while the latter included six questions asking about the respondents' trust in a series of government organizations and agencies. Contrasting country A with country B, we see country B's trust in policy makers factor also included trust in the national assembly, while its trust in agencies factor did not include this question nor the question about trust in political parties. In addition, country B had a democracy factor comprising two questions about democracy, while the same two questions for country A never coalesced into a factor. Finally, for country C, all of these questions formed one overarching factor about both trust in government (policymakers and agencies) and democracy. Note that the blank cells corresponded to questions with small loadings, which were subsequently set to zero. It is common practice when using surveys to have a break point in which small factor loadings are set to zero, that is, completely removed from the factor (Neill, 2012). For this research, variables with loadings less than 0.4 were removed.

These indicators, derived through factor analysis, then became the measures of effectiveness used for trend analysis on various topics of interest. We used nonparametric statistical techniques to detect spatial and temporal changes in these indicators throughout each country from wave to wave. Then, we used exploratory data analysis, displaying spatial and temporal changes that occurred using an analytics suite to build a better knowledge base of popular perceptions. While exploratory data analysis not only gave us a better understanding of if and where positive changes were happening, it also gave us

TABLE 12.1

Comparing Factor Analysis Results for Three Countries

Survey Question		Country A Loadings		Country B Loadings		Country C Loadings
How much trust do you have in President [NAME]?	Trust in Policy Makers	0.80	Trust in Policy Makers	0.91	Trust in Government and Democracy	0.82
To what extent do you approve of [President NAME]'s overall performance?		0.52		0.75		0.68
How much trust do you have in Prime Minister [NAME]?		0.58		0.83		0.80
How much trust do you have in your national assembly?	Trust in Agencies	0.63		0.64		0.82
How much trust do you have in your political parties?		0.67				0.77
How much trust do you have in your armed forces?		0.60	Trust in Agencies	0.54		0.72
How much trust do you have in your courts and judges?		0.57		0.46		0.77
How much trust do you have in your local police?		0.56		0.64		0.76
How much trust do you have in your local government?		0.63		0.45		0.81
How much trust do you have in your religious leaders?						0.66
How satisfied are you with the way democracy works in your country?			Democracy	0.80		0.50
In your opinion, how stable is democracy in your country?				0.83		0.54

Source: Fricker, R. D., Jr., Kulzy, W. W., and Appleget, J. A., *Phalanx*, 45, 4, 30–34, 2012. With permission.

the opportunity to further explore why positive changes did (or did not) occur by digging deeper into a particular area of interest. The mechanics of how one might conduct such analysis is outside the scope of this paper.

An example of how this worked may prove helpful. Table 12.2 shows a notional data set with six questions and six respondents. Each Likert scale response was recoded as −2, −1, 0, 1, or 2. The most negative response was recoded as a −2, and the most positive response as a 2. At first glance, an obvious pattern may not be noticeable. But a closer inspection reveals that each respondent tended to answer questions 1, 4, 5, and 6 the same way. In fact, in this extreme example, respondents answered these questions exactly the

TABLE 12.2

Recoded Questions Used for Factor Analysis

	Questions					
Respondent	**Q1**	**Q2**	**Q3**	**Q4**	**Q5**	**Q6**
A	1	−2	−2	1	1	1
B	2	0	0	2	2	2
C	−2	1	1	−2	−2	−2
D	1	−2	−2	1	1	2
E	0	2	2	0	0	0
F	−1	2	2	−1	−1	−1

same. Similarly, all the respondents answered questions 2 and 3 the same way. Between respondents, the answers to questions 2 and 3 sometimes differ, but the respondents who answered question 2 with a particular value always answered question 3 with that same value. Even if it were possible to notice this pattern within this notional data set, it would be impossible to expect anyone to notice such patterns in data sets with hundreds of questions and thousands of respondents. Table 12.3 shows each of the two factors and the question loadings for each question in the notional data (Table 12.2). Note the 0s listed in each column. This indicates, for example, that questions 2 and 3 did not load on to factor 1, and questions 1, 4, 5, and 6 did not load on factor 2.

After recoding the notional survey responses to numeric values and conducting a factor analysis as described in the last section, we calculated an indicator value, the score for each respondent in each survey wave. To do these calculations, we used two matrices. The first matrix, A, was the recoded survey responses from Table 12.2. It was an $n \times m$ matrix where n was the number of respondents in the survey wave and m was the number of enduring questions from the survey. The second matrix, B, was the survey's factor analysis output from Table 12.3. This matrix was an $m \times p$ matrix where m was (as in matrix A) the number of enduring questions from the survey and p was the number of factors generated from the factor analysis. To calculate each score, we used matrix multiplication on the two matrices as described in Equation 12.1. The output of Equation 12.1 for each survey wave was an $n \times p$ matrix where n and p are as described previously.

Matrix multiplication

$$\mathbf{AB}_{np} = \sum_{i=1}^{m} A_{nm} B_{mp} \tag{12.1}$$

TABLE 12.3

Factor Analysis Output

Question	Factor 1	Factor 2
Q1	0.53	0
Q2	0	0.68
Q3	0	0.72
Q4	0.47	0
Q5	0.49	0
Q6	0.62	0

TABLE 12.4

Respondent Scores

Respondent	Score	
	Factor 1	**Factor 2**
A	2.11	−2.8
B	4.22	0
C	−4.22	1.4
D	2.73	−2.8
E	0	2.8
F	−2.11	2.8

As shown in Table 12.4, each respondent in the survey now has a score (result of Equation 12.1) for each indicator. A sample calculation follows: for respondent A in Table 12.2, we took each of his recoded answers and multiplied them by the corresponding factor loading and the sum those products. So we found the sum of (1 × 0.53) + (−2 × 0) + (−2 × 0) + (1 × 0.47) + (1 × 0.49) + (1 × 0.62) is equal to 2.11. This meant respondent A's score for factor 1 for our notional survey was 2.11. We used this process for each respondent in each wave of each of our actual surveys to calculate all indicator scores for all survey respondents. Throughout this process, we saved the respondents' home district or province to represent geographic changes, which will be discussed later.

For this research, we wanted to know if the change had occurred for a particular indicator from one wave of a survey to the next. In order to do this we used the Mann–Whitney test, as defined by Conover (1999), to compare the scores between waves of the survey to determine if the scores tended to be equal, improving, or declining between survey waves. The Mann–Whitney test is the nonparametric equivalent of a two-sample *t*-test, which is used to compare the means of two data sets and assumes that the means are normally distributed. One key difference between the Mann–Whitney test and the standard two-sample *t*-test is that the *t*-test determines if the means are equivalent while, the Mann–Whitney test determines if one set of data tends to have larger values (scores) than the other, regardless of the means of each data set. As a result, the Mann–Whitney test is robust to the influences of outliers.

To conduct the Mann–Whitney test, we used geographic subsets of each survey wave by parsing the data into district subsets. Our hypotheses are formally stated in the following, where X is the set of scores from a particular district in one wave of the survey and Y is the set of scores from the same district in some later wave of the survey:

- H_0: $P(X > Y) = P(X < Y)$
- H_1: $P(X > Y) \neq P(X < Y)$

A less formal statement of the hypotheses is

- H_0: The probability that an X observation exceeds a Y observation is 0.5
- H_1: The probability that an X observation exceeds a Y observation is not 0.5

To conduct the test, we let n be equal to the number of respondents in the wave we measured from, called the *from wave*, and let m be equal to the number of respondents in the wave we measured to, the *to wave*." Thus, $N = n + m$ (these variables will be used later in this section). We then created a table with N rows and three columns. Column one is the indicator score, column two (Source) indicates which sample the score came from (i.e., from wave or to wave), and the third column indicates the rank of the score in comparison to the other scores in the data set. In the case of ties among scores, of which there were many due to the large sample size in our survey, we averaged the ranks of each of the ties and each tie score received the same ranking. Table 12.5 shows the output of these calculations for a hypothetical example. We then calculated a test statistic, T, which determined any significant statistical difference between the two waves. The calculation of T is quite simple as shown in Equation 12.2. It is a summation of the ranks of each score in the from wave.

Test statistic for the Mann–Whitney test

$$T = \sum_{i=1}^{n} Rank(From\ Wave_i) \tag{12.2}$$

Unfortunately, as mentioned before, there were a high number of tied ranks. Even in the small example earlier, 10 of the 12 scores are tied with at least one other score. Due to this high number of ties among scores we "subtracted the mean number of observations in each data set from T and divided by the standard deviation" (Conover, 1999) to get a new test statistic known as T_1 (Equation 12.3). In this equation, $\sum_{i=1}^{N} R_i^2$ refers to the sum of squares of all N ranks or average ranks actually used in both samples" (Conover, 1999).

Test statistic for Mann–Whitney with many ties among ranks

$$T_1 = \frac{T - n\frac{N+1}{2}}{\sqrt{\frac{nm}{N(N-1)}\sum_{i=1}^{N} R_i^2 - \frac{nm(N+1)^2}{4(N-1)}}} \tag{12.3}$$

TABLE 12.5

Example of Rankings

Score	Source	Rank
2.2	From wave	1.5
2.2	From wave	1.5
2.3	From wave	4
2.3	From wave	4
2.3	From wave	4
2.4	To wave	7
2.4	To wave	7
2.4	To wave	7
2.5	To wave	9.5
2.5	To wave	9.5
3.8	To wave	11
5.8	From wave	12

THE NORMAL DISTRIBUTION

Recall from your basic statistics class the normal curve (i.e., bell curve) where half of the observations fell on either side of the data's mean. Also recall that standard deviation (σ), as it relates to the normal curve, predicts roughly 68% of the observations fall within one σ of the mean and roughly 95% fall within 2σ. The threshold for the T_1 test statistic falls just under 2σ, 1.96σ to be precise. If T_1 is greater than 1.96 or less than –1.96, that indicates a statistically significant difference between two the means. Furthermore, rejecting the null hypothesis implies that there is no statistical difference between the means of the two samples.

Once we derived the T_1 test statistic, we used it to find a p value. This p value indicated, at a given confidence level, whether to accept or reject the null hypothesis; for this research, we used a 95% confidence level ($\alpha = 0.05$) for each hypothesis test. If the p value supported accepting the null hypothesis that indicated there was no statistically significant change of the indicator from one wave to the next. However, if the p value supported rejecting the null hypothesis that indicated there was a statistically significant difference between the two waves—but we did not know whether the change was higher or lower (i.e., better or worse). To resolve this, we created a function using *R*, which, instead of requiring the output of a p value, required the output to be –1 if the change was negative, 1 if the change was positive, or 0 if no change occurred. We used the same T_1 test statistic calculated earlier, and the function returned to –1 if T_1 was in the upper tail of a standard normal curve, 1 if it was in the lower tail, or 0 otherwise.

As an example of this, we calculated Equations 12.2 and 12.3 using the data in Table 12.5. Here $T = 27$ and $T_1 = -1.956$. This T_1 value told us to reject H_0 with $\alpha = 0.05$. Rejecting H_0 is equivalent to saying that there is a statistically significant difference between the two data sets, and the value of T_1 told us that, in this case, the scores in the to wave tended to be larger than those in the from wave (i.e., generally increasing scores over time). Thus, we could state that there had been a positive change for this indicator.

We used this particular hypothetical example to point out the value of using this methodology versus relying on comparing the means of two sets of numbers. The mean of our from wave scores was 2.85, while the mean of our to wave scores was 2.67. This illustrates the virtue of using the Mann–Whitney test to assess trends. While the mean of the populations' scores would falsely lead us to believe the populations' scores trended down from the from wave to the to wave, the Mann–Whitney test concludes that the to wave scores tended to be larger than the from wave scores in our example. It is clear by inspection that even with the presence of an outlier in the from wave, the to wave scores in Table 12.5 tended to be higher than those of the from wave.

Case Study: Afghanistan Survey

History

Once the focus of Operation Enduring Freedom shifted from a traditional kinetic fight to counterinsurgency, so did the emphasis on improving infrastructure, enhancing youth education, increased employment, provision of healthcare for the populace, along with better governance and development. Additional efforts were introduced to gauge the perceptions of the local population through surveys. Afghanistan's administrative borders follow a model of country, province, district, and city similar to the way the United States follows a country, state, county, city model. Afghanistan's Ministry of Interior officially recognizes 34 provinces and 398 districts. The ability to make judgments at various levels of granularity largely depends on the survey's sample size in that geographic level.

For this case study, we used a survey that focused on Afghanistan's Helmand province. The granularity of the survey was down to the district level since the sample size of each district was large enough to allow for statistically significant judgments on each district. The survey was initially conducted in October 2010 and was conducted on a roughly quarterly (two–four months) basis. Each wave of this survey had approximately 4000 respondents that were fairly evenly distributed among nine of Helmand's 12 districts; there were nine such waves for us to conduct our analysis. A *survey wave* is defined as an iteration of conducting the entire survey process from survey design to compiling the final reports.

Key areas of discussion within the survey included (in no particular order) the following:

- Performance of the district governor in governance, infrastructure, and employment opportunities
- General satisfaction with infrastructure, education, and health services
- Performance of the Afghan National Army (ANA) and the Afghan National Police (ANP)
- Ability of various levels of leadership to solve disputes fairly, address corruption, and provide security to their districts
- Expectations of future infrastructure, education, and health services
- Fairness of district, provincial, and national level judicial system

After selecting our questions from our nine waves as per the methods described earlier, we were left with 94 enduring questions. Recall that for our purposes, enduring questions were those perception questions that appear in all waves we chose to analyze. For this research, we used the last wave of the survey for factor analysis, which allowed us to establish a baseline for each survey from which to make comparisons. The factor analysis resulted in the reduction in the data from 94 observed variables (survey questions) to 22 unobserved variables (factors). Only those questions that had a factor loading of 0.4 or higher and −0.4 or lower were kept, and the loadings of all other questions were set to 0 (example in Table 12.4). Each factor consisted of at least two and up to as many as 12 observed variables.

Results of Factor Analysis

The 22 factors produced by the factor analysis on the survey can be viewed in the Appendix. One notable observation is that this survey allowed for a deep understanding of the Helmand province because of the large sample sizes from each district. The fact that there were so many factors was a product of the consistency of the questions from wave to wave, which allowed for a large number of enduring questions. For comparison sake, the research we conducted on other survey instruments in different Afghan provinces produced only about 10 factors. While the general nature of the questions making up each factor was not significantly different from those in the other surveys, these questions included greater detail about each of the topics. For example, several separate factors used questions that specifically discussed the ANP, ANA, International Security Assistance Force (ISAF), the Taliban, the government of the Islamic Republic of Afghanistan (GIRoA), the members of Afghanistan's parliament, provincial government, district governments, and so on. The precision of the questions plus the factor analysis approach enable analysts to gain a much more specific perspective from this survey without becoming overloaded by an overly broad range of topics.

Deriving Indicators from Factors

The factors that we derived were initially labeled factor 1, factor 2, etc. However, since the desired end state was to find meaningful insights from the survey, we gave each factor a more descriptive name. A selection of factor analysis results from a notional survey (Table 12.6), and the Helmand survey (Table 12.7) illustrates the process for naming the indicators.

The results displayed in Tables 12.6 and 12.7 carry a similar theme in that each question asked something about security in a particular area. Due to the persistence of security and safety themes throughout the questions in Table 12.6, we called it *satisfaction with security and safety in local area* and refer to this as the indicator name from here on. Similarly, in Table 12.7, the security theme appears again, this time with a focus on roads. In this case, we name the *satisfaction with security along roads* indicator. All the factors were named in a similar fashion. The exact name of the indicator is not important; rather, it is important that we now have an analytically derived indicator that we can measure against. Each question carries a particular weight, the factor loading, which will be used later to make comparisons.

Perhaps the most important takeaway from this is that no single person chose this grouping of questions. In essence, the group of respondents as a whole decided that the

TABLE 12.6

Factor 4 as Derived by Factor Analysis on Notional Survey Data

Question	Factor Loading
How is the security situation in your *mantaqa* (village)?	0.4778
How safe do you feel using the roads in your district?	0.6055
Is security in your mantaqa better, the same, or worse than it was six months ago?	0.4079
How safe are the children in your village when they go to school and study in school?	0.5034
How safe do you feel traveling outside of your mantaqa during the day?	0.7209

TABLE 12.7

Factor 5 as Derived by Factor Analysis on Actual Survey Data

Question	Factor Loading
Do you agree or disagree that the district government has improved security along roads *in your area* over the last three months?	0.4727
Do you agree or disagree that the district government has improved security along roads *in major urban areas* over the last three months?	0.4149
Do you agree or disagree that the district government has improved security *in your area* over the last three months?	0.4124

topics of these questions were alike enough that, individually, each respondent tended to answer each of the set of questions in a similar manner. (That last sentence often causes confusion. It is not that everyone in the survey answered all questions of the indicator the same way. It is that each person, individually, tended to answer each of the indicators' questions similarly.)

Analytics Suite for Displaying Measured Change

Displaying Spatiotemporal Indicator Change

The statistical significance of the change is an important first step in understanding an area of interest. Relaying that the information to decision makers is perhaps the most challenging aspect. To accomplish this, we used one of the most common, readily available pieces of software: Microsoft Excel. Using Excel, we created an analytics suite called the measure of indicator change (MICh) tool. MICh was designed to be user friendly, even by first-time users. The MICh analytics suite shows three separate metrics of each indicator: the change of the indicator across waves as derived from our Mann–Whitney test, the mean of the indicator in each wave by region, and the proportion of support for each indicator, by district, over the duration of the survey.

MICh uses the indicator scores and the 1, 0, and –1 codes indicating positive, neutral, or negative change as the inputs for analysis. An input form (see Figure 12.2) prompts the user three items: first, the wave he/she would like to measure from, the from wave; second, the wave the he/she would like to measure to, the to wave; and lastly, an option to choose one of the indicators. The user can select different waves or a different indicator from the drop-down box. Upon clicking View Map, the user is presented with a display of the three metrics for the satisfaction with security along roads indicator.

The first metric is a map (Figure 12.3) of an area in Afghanistan that was commanded and controlled by the ISAF Regional Command Southwest. The area actually covered two provinces; however, the survey only polled residents of the Helmand province. Of the Helmand province districts depicted, three districts, along with the Nimroz province, are crossed out, indicating that the surveys did not poll any respondents from these areas, and therefore, they received a rating of *Not Measured*. Also visible are six districts coded white, indicating a positive change for the indicator. The Reg-e Khan Neshin district is light gray;

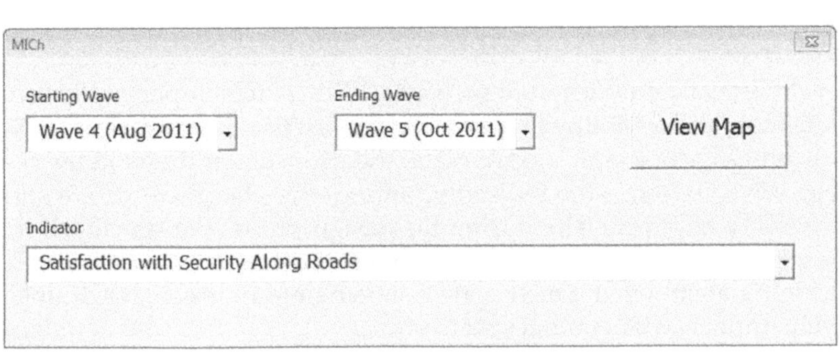

MICh

Starting Wave
Wave 4 (Aug 2011)

Ending Wave
Wave 5 (Oct 2011)

View Map

Indicator
Satisfaction with Security Along Roads

FIGURE 12.2
Wave and indicator choice form.

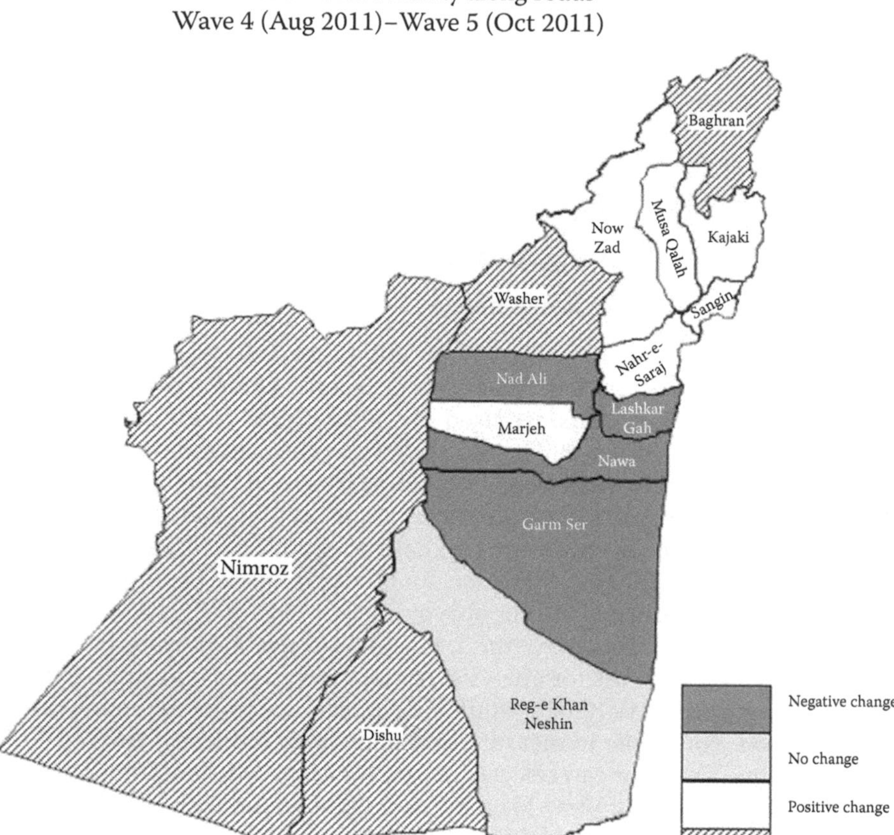

FIGURE 12.3
Map of statistically significant change.

MAKING ANALYTICS USER-FRIENDLY

When developing an analytics suite such as the MICh, it is important to consider the entire audience and your ability to deploy the results. Because Excel is so widely available, it was an easy choice for us. We made the drop-down menu items as intuitive as possible. We considered the indicator naming, the placement of the date ranges, and the visibility of our metrics during the design phase. We specifically chose the odd color scheme of purple, brown, and gray to indicate change to cause the user to pause and think about what it meant. We felt using red or green, for example, would instinctively imply bad or good, respectively.

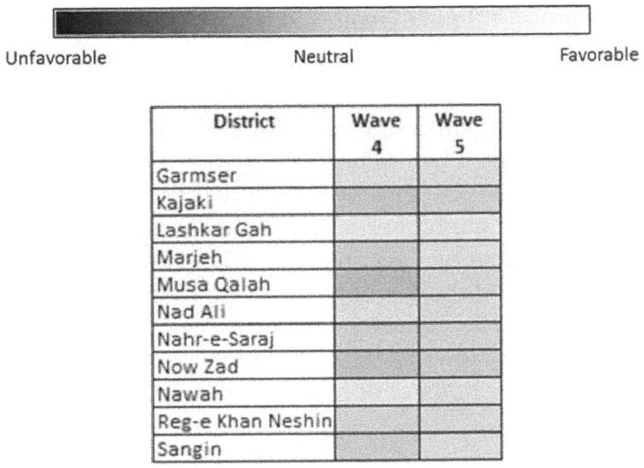

Unfavorable Neutral Favorable

District	Wave 4	Wave 5
Garmser		
Kajaki		
Lashkar Gah		
Marjeh		
Musa Qalah		
Nad Ali		
Nahr-e-Saraj		
Now Zad		
Nawah		
Reg-e Khan Neshin		
Sangin		

FIGURE 12.4
Status of indicator.

this indicates that there was no statistically significant change there. Lastly, four districts are dark gray; this indicates that there was a negative change for the indicator in each of these four districts. This graphic simply shows whether there has been a change or not and its valence (better or worse). This map alone does not necessarily tell the whole story of how the general sentiment relates to this indicator. In other words, we know that the northern districts were more satisfied with the security along the roads in wave 5 than they were in wave 4 but not the magnitude of the change. To help answer that question, the user can look at Figure 12.4, which is simultaneously displayed on the screen.

Mean Respondent Display

Figure 12.4 displays a grayscale chart ranging from black to white for each of the measured districts. A black cell indicates the sentiment of the indicator (during the time of the surveys' wave) to be most unfavorable, white indicates most favorable, and gray indicates neutrality. We used a gradient scale to display all possible color indications from black to gray to white. The color of each cell represents the mean score of the indicator for each district in the province. The extremes of these scores are set by multiplying –2 by each factor loading for the indicator to get the most unfavorable score possible, and multiply the same factor loadings by 2 to get the most positive score possible. Each indicator will have a different value for the most favorable and most unfavorable scores because each factor has different numbers of questions and different loading values. Using this color scheme, along with the statistical significance of change information from Figure 12.3, offers a better understanding of the full picture. For example, we know from Figure 12.3 that there was a negative change for the satisfaction with security along roads" indicator in the Lashkar Gah district. The additional information in Figure 12.4 broadens our understanding. We can see that in Lashkar Gah for both waves 4 and 5, the general sentiment for this indicator is positive. Considering both pieces of information, the takeaway is despite a decline from wave 4 to wave 5 for Lashkar Gah's satisfaction with security along roads indicator, the population's outlook for this topic is still quite favorable.

Proportionality of Indicator Support Display

The third analysis to be displayed is the proportion of support for an indicator over the duration of the survey for each district and the province (examples in Figures 12.5 and 12.6). These charts give a slightly different perspective of the indicator than that shown in Figure 12.4. The values displayed by the line charts were derived from the set of respondent scores by district or province, similar to Figure 12.4. This time, however, they are shown as a percentage where the higher values indicate more support for the indicator, enabling a broader trend perspective. As an example, if the range of unfavorable to favorable were 0–10, respectively, and the average score for a district was 7.5, we would say that the proportion of support for the indicator was 75%. Additionally, we see error bars displayed on the charts, which represent a 95% confidence level of the proportion. The error bars are directly affected by the number of respondents for the survey. Therefore, the error

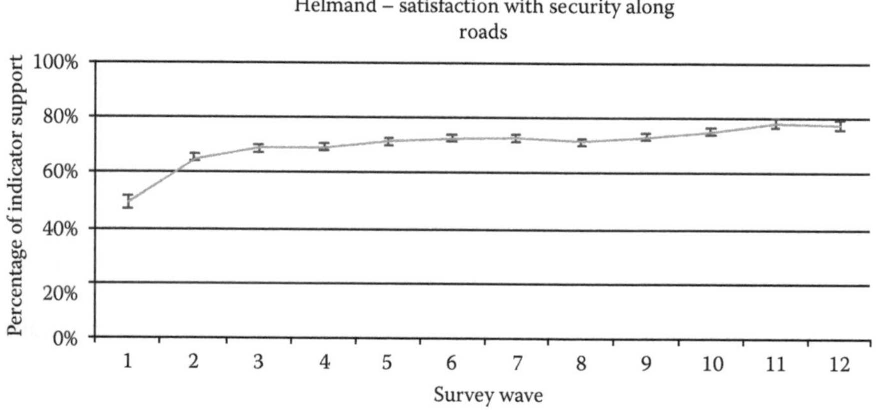

FIGURE 12.5
Proportion of support for the Helmand province.

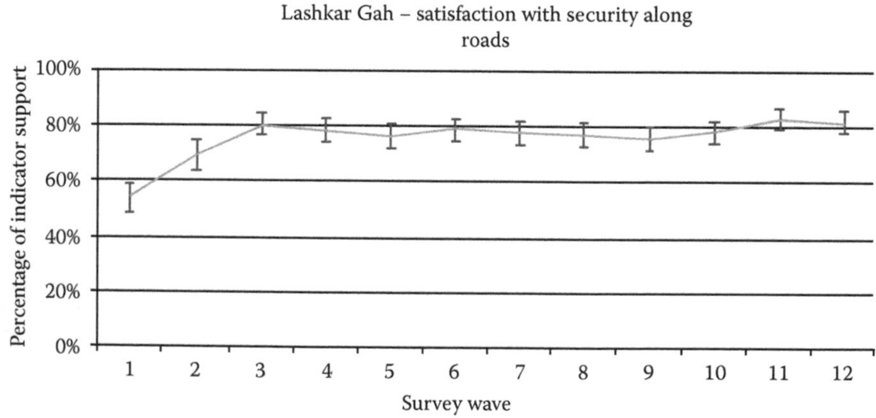

FIGURE 12.6
Proportion of support for Lashkar Gah.

ERROR BAR CALCULATION

Error bars were calculated from the standard error formula, $\sqrt{\frac{p(1-p)}{n}}$, where p is the proportion of support and n is the sample size. All our sample sizes were large enough to assume a normal distribution. In order to display the 95% CI around the point estimate, p, we multiplied the standard error by ±1.96 to get the upper and lower bounds of the CI. Note the different error bar ranges on each chart. As n grows, as in Figure 12.5, the error bars shrink. Conversely, when n is not as large, as in Figure 12.6, the error bars are much wider.

bars are much narrower in the provincial level charts than they are in the district level charts because of the larger sample sizes at the provincial level.

This third feature of the analytics suite shows that the satisfaction with road security has been quite high in Lashkar Gah, and across the Helmand province in general, since about wave 3 of the survey. We can now state the following: According to the survey, the sentiment regarding Lashkar Gah's satisfaction with security along roads showed a statistically significant decline from wave 4 to wave 5; however, the average Lashkar Gah resident remained generally positive about road security, and that the sentiment has continued to be quite steady for the duration of the survey.

The MICh is designed to be able to display these metrics based on surveys for any two periods for every indicator. This equates to thousands of comparisons to be measured at the district and provincial levels of the Helmand province with only a few clicks. The metrics are easy to understand, and the availability of Excel to run the analytics suite is nearly universal.

Conclusion

Population surveys provide valuable insights into a population's beliefs and attitudes. In this chapter, we demonstrated ways to increase the usefulness of surveys by using factor analysis to uncover latent, unobserved variables resident within the survey data. After a discussion and a demonstration of the methodology, we applied our technique to a population surveys from AFRICOM and the Helmand province, Afghanistan. By using our methodology on these surveys, we showed how an analyst could uncover important themes from the respondents' points of view and how they may have changed over time.

We then discussed a tool, MICh, which serves as a visual display for three metrics derived from our analysis. The MICh tool also enables exploratory data analysis, designed to inform and hopefully to incite further areas to study by providing quick and easy to understand metrics of the analytically derived and measured indicators. The MICh can help leaders use survey data to understand popular perceptions of topics and provide them a means to use that information to inform decisions. The MICh is scalable and able to be modified to a useful state for many different levels of responsibility and leadership for a wide range of organizations when developing and analyzing long-term plans involving populations and measuring the effects of past programs and policies for those populations that an organization may have employed.

Megacity Use Case

Nations or local governments concerned with improving the economic, social, and religious unrest in the fictional megacity Xanadu could benefit from the type of more nuanced structured survey analysis described in this chapter. In fact, this analysis is perfectly suited to provide insights for this type of situation. Assuming a valid survey could be periodically conducted in the area to provide multiple sets of survey responses, an analyst could use our methodology to gain a better understanding regarding shifts in perceptions of different sects of the population over the duration of the survey waves.

Certainly, challenges exist in the area of Xanadu in obtaining valid, consistent survey data. The continually shifting demographics of the city could lead to skewed results that would make analysis results tougher to interpret. In a city like this, there would likely be a plethora of transients who work in the city but live elsewhere, migrants, and homeless people who could not be surveyed for various reasons. Additionally, even the identities of neighborhoods of permanent residents may change over time. However, while the spatial trends might be difficult to track, other demographics could be measured such as cultural affiliation, religious background, economic standing, or education. Any of these demographics could still be tracked in order to elicit value from the survey by using our methodology even if the spatial aspect does not exist. Measuring the temporal changes of these alternate demographics can still lead to valuable understanding of the people, their key issues, and how their perceptions of those issues have changed for better or worse. In fact, in many areas that we have seen from our research of African countries and Afghanistan, using arbitrary boundaries such as districts, provinces or states may be the least useful demographic to measure since many respondents identify more with a tribe or a culture than they do with borders.

Societies need to be able to entrust their government in order to thrive. In the case of Xanadu, many of the residents of the city would be likely left fending for themselves, as corruption runs rampant throughout the city. The residents are constantly threatened by the local, unofficial power centers of whichever ward of the city they call home or where they work. Considering the scarce resources the city has to combat these problems, and their seemingly lack of interest in doing so in some areas, they could use help from the international community to assist them in employing their existing resources to make the city a more stable one. Using the MICh analytic suite described, an organization can assess over time where help is needed or to whom (culturally, socially, or otherwise) help should be given.

A spatial analysis of survey indicators would likely show obvious differences between the Razinis in the north and the Bursukans in the west. It could potentially confirm that the Razinis appear to have found themselves in the untenable position of being exploited by streetwise criminal organizations by feeding on their desperate desire to survive. Survey analysis could also elucidate both Razini and Bursukan attitudes regarding faith in the local government and their ability to provide good working conditions and a means to earn an honest living, as well as the extent to which religious leaders have effectively shielded the Bursukans from the tyranny happening in the north.

While Afghanistan presented a different set of problems than Xanadu, survey analysis enabled a much greater understanding of the people's problems and concerns. The major issues in Afghanistan dealt with grievances with local government officials, infrastructure, and security. Security and feelings of safety were major concerns amongst Afghans due to the seemingly constant roadside bombs and artillery fire. Once the ISAF was able to quell most of that activity, we saw economic activity in the district centers begin to thrive. School attendance increased and, in general, the people felt safe traveling on the roads to other districts to work or purchase goods and services. We were able to see, out of the 14 districts in the Helmand province, that all but one of those districts displayed positive changes in each of the perception indicators. The only holdout was the Sangin district. In Sangin, the insurgents still held key terrain and had freedom of movement in the area. While our survey analysis did not discover this activity, it supported our intelligence being collected from other means. We were able to work with the department responsible for spurring economic development within the Helmand province to focus certain resources to the Sangin area. The provincial and district governments in Afghanistan also had limited resources at their disposal similar to the government in Xanadu. The international community and the local governments of Xanadu could be helped in a similar fashion to the people and the lower-level governments in Afghanistan.

In the end, this methodology of analyzing surveys is about furthering the supply chain of information. It is about either taking data that already exists or creating the data through fielding surveys, and turning it into useful, meaningful information that can be acted upon. The many real-life areas that inspire the fictional city of Xanadu are precisely the population types that solid population survey analysis supports the best: people who feel, or do not even realize, that they do not have a voice and people who feel like their current situation could be improved but have no means to relay that information to the outside world. It is important for us, the civilized international community, to assist wherever we can and use our abundant resources (money, education, innovation) to help those oppressed people. Due to the nature of these areas being split among many factions and spread over typically vast areas, it is often difficult to gain a solid understanding of their situation as a whole society. This is one way we can further that understanding and make an improvement, even if only a small one, to the world.

DISCUSSION QUESTIONS

1. What is the purpose of factor analysis?
2. What is the nonparametric equivalent of the two-sample *t*-test and how do they differ?
3. What happened once the focus of Operation Enduring Freedom shifted from a traditional kinetic fight to counterinsurgency?
4. What software system serves as the basis of MICh and why was it chosen?
5. Name and describe the three metrics produced by MICh.

References

Conover, W. (1999). Some methods based on ranks. In W. Conover, *Practical Nonparametric Statistics* (pp. 269–286). New York: John Wiley & Sons.

Fricker, R. D., Jr., Kulzy, W. W., and Appleget, J. A. (2012). From data to information: Using factor analysis with survey data. *Phalanx*, 45:4, 30–34.

Gelman, A., and Hill, J. (2006). Missing-data imputation. In A. Gellman, *Data Analysis Using Regression and Multilevel/Hierarchical Models* (pp. 529–543). New York: Cambridge University Press.

Government Accountability Office (GAO). (2010). *Defense Management: Improved Planning, Training, and Interagency Collaboration Could Strengthen DOD's Efforts in Africa.* Washington, DC: US Government.

Neill, J. (2012). *Exploratory Factor Analysis*, Survey Research and Design in Psychology, Lecture 5, slide 82. Accessed at http://www.slideshare.net/jtneill/exploratory-factor-analysis, May 19, 2012.

Appendix

TABLE A.1

Indicators Derived from Factor Analysis on Afghanistan Survey Discussed in Case Study

Satisfaction with Provincial Government (Provincial and Court Systems)
Confidence in ability to resolve dispute in fair manner—Provincial Governor
Confidence in ability to resolve dispute in fair manner—District Huquq
Confidence in ability to resolve dispute in fair manner—Provincial Huquq
Confidence in ability to resolve dispute in fair manner—District Prosecutor
Confidence in ability to resolve dispute in fair manner—Provincial Prosecutor
Confidence in ability to resolve dispute in fair manner—Lashkar Gah (LKG) Court
Confidence in ability to address corruption—Provincial Governor
Confidence in ability to address corruption—District Huquq
Confidence in ability to address corruption—Provincial Huquq
Confidence in ability to address corruption—District Prosecutor
Confidence in ability to address corruption—Provincial Prosecutor
Confidence in ability to address corruption—LKG court
Confidence in the Legitimacy of the ANP
Satisfaction/Dis—Role in keeping district secure—Police (ANP)
Agree/Dis—National police (ANP) is honest and fair with the local people
Agree/Dis—National police (ANP) is unprofessional and poorly trained
Agree/Dis—National police (ANP) help improve the security in my area
Agree/Dis—National police (ANP) efficient arresting to bring to justice
Agree/Dis—National police (ANP) in my region act in interests of local people
Agree/Dis—National police (ANP) are not sanctioned when they misbehave
Confidence in ability to resolve dispute in fair manner—Police
Confidence in ability to address corruption—Police
Satisfaction/Dis—Role in keeping district secure—District governor/authority
Agree/Dis—(generally speaking) GIRoA are doing a good job

(Continued)

TABLE A.1 (CONTINUED)

Indicators Derived from Factor Analysis on Afghanistan Survey Discussed in Case Study

Satisfaction with Employment Opportunities, Bazaars, and Justice Systems
Agree/Dis (3 months, district) District governor (DG) has improved employment opportunities.
Agree/Dis (3 months, district) DG has increased effectiveness—Justice (State)
Satisfaction/Dis—Services in area—Government Justice Services
Expectations (next 6 months, district)—The central bazaar
Expectations (next 6 months, district)—Availability economic opportunities/jobs
Expectations (next 6 months, district)—Effectiveness state justice system
Confidence in the Legitimacy of ISAF
Satisfaction/Dis—Role in keeping district secure—ISAF/Foreign forces
Confidence in ability to resolve dispute in fair manner—Foreign forces
Confidence in ability to address corruption—Foreign forces
Satisfaction with Security along Roads
Agree/Dis (3 months Dist) DG has improved security—along roads in my area
Agree/Dis (3 months Dist) DG has improved security—roads major urban areas
Agree/Dis (3 months Dist) DG has improved security in my area
Confidence in the Legitimacy of Local Militias
Satisfaction/Dis—Role in keeping district secure—Local warlords militia
Confidence in ability to resolve dispute in fair manner—Local militia
Confidence in ability to address corruption—Local militia
Satisfaction with Utilities (Electricity, Drinking Water, and Irrigation Systems)
Agree/Dis (3 months, district) DG has improved electricity supply
Agree/Dis (3 months, district) DG has improved irrigation services
Agree/Dis (3 months, district) DG has improved household/drinking water
Expectations (next 6 months, district)—The supply of electricity
Expectations (next 6 months, district)—Availability of irrigation water
Expectations (next 6 months, district)—Availability of safe drinking water
Expectation of District Improvements over the Next Six Months
Expectations (next 6 months, district)—Availability of education services
Expectations (next 6 months, district)—Ability to move freely and safely
Expectations (next 6 months, district)—The security in my area
Expectations (next 6 months, district)—Availability of irrigation water
Expectations (next 6 months, district)—Access to information and news
Confidence in the Legitimacy of the ANA
Confidence in ability to resolve dispute in fair manner—Army
Confidence in ability to address corruption—Army
Confidence in the Legitimacy of the Taliban
Satisfaction/Dis—Role in keeping district secure—Taliban
Confidence in ability to resolve dispute in fair manner—Taliban
Confidence in ability to address corruption—Taliban
Satisfaction with Availability and Access to Government Healthcare Services
Agree/Dis (3 months, district) DG has improved healthcare services
Satisfaction/Dis—services in area—Government health services
Expectations (next 6 months, district)—Availability of health services

(Continued)

TABLE A.1 (CONTINUED)

Indicators Derived from Factor Analysis on Afghanistan Survey Discussed in Case Study

Satisfaction with Availability and Access to Government Education Services
Agree/Dis (3 months, district) DG has improved education services
Satisfaction/Dis—services in area—Government Education services
Expectations (next 6 months, district)—Availability of education services
Confidence in the Legitimacy of Members of Parliament
Confidence in ability to resolve dispute in fair manner—Members of Parliament
Confidence in ability to address corruption—Members of Parliament
Acceptance of Western Societal Norms
Acceptable/Not—Listening to music
Acceptable/Not—Watching television
Acceptable/Not—Talking about issues related to money
Confidence in the Legitimacy of the District Governor
Confidence in ability to resolve dispute in fair manner—District Governor
Confidence in ability to address corruption—District Governor
Satisfaction with the Performance of GIRoA
Agree/Dis—GIRoA listen/acts on behalf of people rather than own interests
Agree/Dis—(generally speaking) GIRoA are doing a good job.
Satisfaction with Security in General
Rate—Change in security situation in District (compared to 6 months ago)
Rate—(in general) security situation in area
Satisfaction with District Governments Use of Taxes to Improve Services and Access to News
Agree/Dis (3 months Dist) DG has provided better access to information news
Agree/Dis (3 months Dist) DG has used taxes to improve public services.
Confidence in the Legitimacy of the District Court
Confidence in ability to resolve dispute in fair manner—District Court
Confidence in ability to address corruption—District Court
Confidence in the Legitimacy of the Provincial Governor
Confidence in ability to resolve dispute in fair manner—Provincial Governor
Confidence in ability to address corruption—Provincial Governor
Confidence in the Legitimacy of the District Huquq
Confidence in ability to resolve dispute in fair manner—District Huquq
Confidence in ability to address corruption—District Huquq
Confidence in the Legitimacy of the District Prosecutor
Confidence in ability to resolve dispute in fair manner—District Prosecutor
Confidence in ability to address corruption—District Prosecutor

Note: The indicators (unobserved variables) are in bold; the questions (observed variables) that derive each indicator are directly beneath the indicator.

13

Gaining Insight by Applying Geographical Modeling

Erman Çakıt and Waldemar Karwowski

CONTENTS

Geographical Aspects of Terrorism 244
Background 245
  Potential Use of Geographical Modeling in Terrorism Studies 245
  Nation-Building 245
  Understanding Cultures: Integration of HSCB and GIS Modeling 246
Researching Design and Methodology 247
  Study Area 247
  The Data Set Description 247
  Point Density Analysis 251
Results and Discussion 251
  Descriptive Statistics of Adverse Events by District 251
  GIS-Based Analysis and Visualization of Adverse Events: The Case of Afghanistan ... 252
    Mapping Regional Differences and Getting the Picture 252
    Three-Dimensional Display of Adverse Events 254
    Mapping Adverse Events and Compared with Population Density 254
    Geographical Modeling Using Point Density Analysis 255
Concluding Remarks and Further Work 259
Megacity Use Case 263
Acknowledgments 264
References 265

ABSTRACT This chapter is a review of adverse events throughout the war in Afghanistan by representing the mapping of these events. Three types of adverse events were considered in terms of the number of people killed, wounded, and hijacked and their total number in the active war theater of Afghanistan over the period of 2004–2010. For the purpose of understanding the patterns of adverse events, the results can be summarized by visualizing the occurrence of incidents by region; the emphasis was on analyzing the number of people killed, wounded, and hijacked to determine the risk of different parts of Afghanistan.

KEY TOPICS

- GIS to analyze adverse events (e.g., terrorism)
- Applications of spatial analysis and point density analysis
- Example application using a terrorist event dataset with over 30,000 records
- Several geographically oriented models performed
- Benefits of geographical modeling in decision making highlighted

Geographical Aspects of Terrorism

Terrorism can be described as "the unlawful use or threatened use of force or violence by a person or an organized group against people or property with the intention of intimidating or coercing societies or governments, often for ideological or political reasons" (Denning 2000). As the rates of terrorism on local, regional, and international scales continue to increase, understanding and countering terrorism is more important than ever before. The understanding of terrorism and counterterrorism should begin with a solid grasp of the historical patterns of terrorist incidents. For the purpose of understanding these patterns, geographically oriented modeling should be performed.

In the context of military research, the geographical aspects of terrorist incidents are essential to the understanding of key historical events, which may provide useful information to develop predictive models of future terrorist incidents. The geographical aspects of terrorist incidents are also important in determining key strategies and tactics to be employed for similar scenarios in an active war theater in the future. Thus, spatial and temporal pattern analysis may provide information on, and an understanding of, the geographic activity and the basic patterns of terrorist incidents.

This chapter focuses on terrorist incidents, which are denoted here as adverse events, including the number of people killed, wounded, and hijacked, and the total number of individuals affected, in the active war theater of Afghanistan over the period of 2004–2010. Here, the authors apply a GIS approach to determine and provide knowledge about the patterns in adverse events that are necessary to elucidate and combat terrorism in an active war theater.

The chapter is organized as follows. The section "Background" provides a background, which deals with a literature review, including studies of spatial and temporal patterns using GIS in terrorist incidents, and introducing human social culture behavior (HSCB) modeling. The section "Research Design and Methodology" gives a description of the study area, the data set, and the methods used for pattern analysis and visualization. The section "Results and Discussion" applies the methods for pattern analysis and visualization to analyze the adverse event dataset. Finally, the section "Concluding Remarks and Further Work" concludes the chapter.

ANALYZING TERRORIST INCIDENTS WITH GIS

The geographical aspects of terrorist incidents are essential to the understanding of key historical events, which may provide useful information to develop predictive models of future terrorist incidents. Geographical aspects of terrorist incidents are also important in determining key strategies and tactics to be employed for similar scenarios in an active war theater in the future. Geographical modeling is performed in some areas of conflict research; however, it could be better developed and utilized in countering terrorism. Although geographically based terrorism studies have gained increasing interest in academic literature in the twenty-first century, the geographical aspect of terrorist incidents has traditionally been lacking from most studies on terrorism research, and it remains considerably unfamiliar to many researchers in this field.

Background

Potential Use of Geographical Modeling in Terrorism Studies

It is possible to discover representative patterns in adverse activity or behavior over time and space by analyzing geospatial intelligence on reported incidents, as stated by the director of National Intelligence, Open Source Center (Federation of American Scientists 2009). The applications of geographically oriented modeling, in the domain of terrorism, is a relatively recent trend, and the work of spatial and temporal analysis of terrorist attacks is becoming increasingly important in the literature. Several studies have employed spatial and temporal analysis to terrorism events in Afghanistan (Open Source Center [OSC] 2009; Shroder 2005; Beck 2003), Iraq (Medina et al. 2011; Siebeneck et al. 2009; Johnson and Braithwaite 2009), Israel (Berrebi and Lakdawalla 2007; Brown et al. 2004), the United States (Webb and Cutter 2009), Spain (LaFree et al. 2011), and Turkey (Rusnak et al. 2012; Demirci and Suen 2007).

Techniques applied for detecting, monitoring, and estimating spatial patterns for crime analyses may also be beneficial in conducting studies of terrorist incident data. This is due to the fact that terrorist incidents are crimes, and thus, target selection for both incident and crime data is not random in space and time. In the specific case of crime prediction, semantic data for identifying the incidents is highly acceptable, as it is necessary to support decision making processes and, in general, to prevent and correct policies (Kumar and Chandrasekar 2011). Several studies of crime pattern analysis approaches that are applicable to terrorism research include point pattern analysis (Liu and Brown 2003), cluster detection methods (Corcoran et al. 2003), time series forecasting with the help of GIS (Polat 2007), and spatial–temporal prediction modeling (Kumar and Chandrasekar 2011).

Nation-Building

In April 2002, in an attempt to make Afghanistan a better place, the US Congress approved over $38 billion in humanitarian and reconstruction assistance to Afghanistan from 2001 to 2009 (Çakıt and Karwowski 2015). In early 2002, the Afghan government built their army—the ANA with a target goal of 70,000 troops. With the help of the US military, the ISAF defended the Kabul region with 4000 non-US soldiers. The US government provided aid for Afghan-nation rebuilding, and around 8000 US and allied troops—mostly based north of Kabul—for conducting counterterrorist operations across the country.

In 2002, Afghanistan was considered among the 10 bottom countries socioeconomically, and there was no human capital to build on (The 2013 Human Development Report). The international community, including the United States, promised over $5 billion in aid and started the work of helping to rebuild Afghanistan (Collins 2011). After the attacks on 9/11, the US government signed agreements with the energy-rich countries bordering Afghanistan. The main objectives of these agreements were to increase economic liberalization and attract investments from foreign capital. The total amount of US assistance was categorized into four portions (Tarnoff 2010). The main portion—approximately 56% of the total amount—was given to the Afghan Security Forces Fund. This portion includes the training of Afghanistan security forces and their equipment.

The second largest amount is composed of economic, social, and political development efforts, and it comprises approximately 31% of the total amount. A third portion of assistance, humanitarian aid, was implemented mainly through the United States Agency for

International Development and international organizations; it constitutes about 4% of the total aid provided since 2001. The last portion of the aid program is counternarcotics, and it comprises approximately 9% of the total aid provided.

The impact of these infrastructure developments remains scarcely explored, and the relationship between infrastructure development and terror is not clear cut (Child 2014). Between 2005 and 2009, Böhnke et al. (2010) conducted a survey of the northeastern region of Afghanistan. Researchers tested the effect of aid in conflict zones; they concluded that participants who already felt secure were more likely to feel positively about the aid. However, Afghan-nation rebuilding was not so successful from 2001 to August 2009, when the second presidential election occurred. During this period, there was a negative relationship between the number of military forces and safety in Afghanistan, as the number of terrorism events tripled between 2002 and 2007 and persisted through the summer of 2009 (Kamrany 2009). The economic and reconstruction efforts are part of the irregular warfare missions, which are highlighted by today's military. To support these efforts, the US military has encouraged various programs to understand the effects of social and cultural factors on human behavior, especially to the domain of HSCB modeling.

Understanding Cultures: Integration of HSCB and GIS Modeling

The US DoD defines *irregular warfare* as "a violent struggle among state and non-state actors for legitimacy and influence over the relevant population(s)" (Irregular Warfare Joint Operating Concept 2007). In particular, this type of warfare uses disproportionate force to convince and hassle the enemy in regions where opposing forces are smaller and less effective (Clancy and Crossett 2007). Since 2007, the US military has made adjustments to its force structure in recognition of the challenges posed by irregular warfare. An article by Bhattacharjee (2007) (entitled Pentagon asks academics for help in understanding its enemies) reported on the growing interest of a field of social science called HSCB modeling, which helps to guide organizations like the US military in developing a better understanding of cultures encountered while operating overseas (Drapeau and Mignone 2007). The overarching aim of HSCB modeling is to enable the DoD and the US government to better organize and control the human terrain during nonconventional warfare and other missions (HSCB Modeling Program Newsletter 2009). HSCB models are increasingly being considered during the development of current and future operational requirements. The US military recently recognized the vast importance of the consideration of sociocultural factors, which have been summarized as follows (Pool 2011):

- Being respectful and sensitive to local people
- Understanding local culture, custom, and history deeply
- Being capable of communicating in their language, at least, at the introductory level
- Understanding the tribal nature and their leaders

The application of a GIS to social sciences is an emerging research area (Goodchild et al. 2000). Considering sociopolitical problems, such as terrorism, through a geographic lens leads to a greater understanding not only of the location of terrorists and their activities, but also of spatial processes, behavior of organizational units, social network connections, sociocultural regions, social systemic operations, and changes in space and time (Bahgat and Medina 2013). HSCB models, in comparison to GIS, are designed to characterize the

US DoD HSCB MODELING PROGRAM

The US DoD's HSCB modeling program was established in 2008 in response to a study that found major gaps in the department's ability to model HSCB; that study recommended an increased investment to support the research and development with an emphasis on product maturation and transition. The resulting HSCB modeling program included four phases and multiple research tracks, including collecting sociocultural data, analyzing and visualizing those data, developing models based on the data to forecast future actions, and conducting sociocultural training. More information can be found at http://www.dtic.mil/biosys/hscb-mp.html.

behavior and the structure of organizational units at both the macrolevel (economies, politics, and sociocultural regions) and at the microlevel (terrorist networks, tribes, and military units) (Stanton 2007). Thus, in order to be successful in understanding the effects of social and cultural factors on human behavior, in both current and future operational requirements, one must pay particular attention to and integrate both the HSCB and GIS models.

Researching Design and Methodology

Study Area

Afghanistan lies in central Asia and is split into 34 provinces. Each province has its own capital and a provincial administration; the provinces are subdivided into 400 districts. Figure 13.1 illustrates a map of districts, which are grouped by province. Each district, which is represented by a district governor, has a city or a certain number of villages. Afghanistan boarders Pakistan on the south and east, Iran on the west, and Turkmenistan, Uzbekistan, and Tajikistan on the north; it also boarders China on the far northeast. Afghanistan is 652,000 km^2 in size, and it is somewhat smaller than Texas in the United States. Afghanistan has a population of approximately 32 million people. Based on the UN Human Development Index, which is an index calculated according to the health, education, and economic life of people, Afghanistan has been ranked 175th out of 185 member states of the UN (The 2013 Human Development Report). Afghanistan's particular geography is relatively inaccessible. Therefore, this situation makes it difficult to secure the population and to improve their economic situation.

The Data Set Description

This study uses data describing adverse events in Afghanistan provided by HSCB program management. The adverse event data sets include information collected through news feeds, blogs, reports, and various databases. The collected adverse event data set was processed by text processing techniques and analytics to generate a suitable set of data to be used in the content of an HSCB project.

The adverse event data set includes information regarding the date of the event; the number of people killed, wounded, and hijacked; the province, city and district;

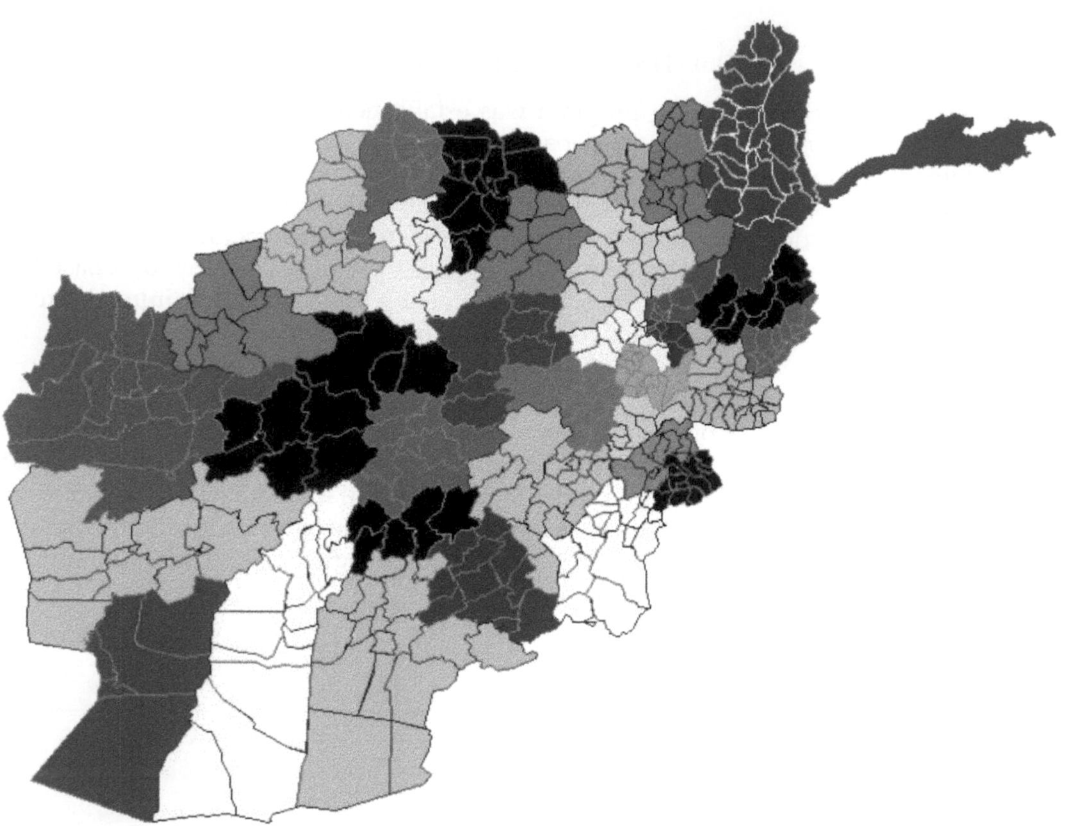

FIGURE 13.1
A map of the districts, which are grouped by province.

a description of the event; the location in terms of latitude and longitude coordinates; and a simple event summary (Cakit and Karwowski 2015; Çakıt et al. 2014). We considered three types of variables in terms of the number of people killed, wounded, and hijacked that took place during the period of 2004–2010. These variables provide new information about adverse events represented by a data set with over 30,000 records, where the information about adverse event patterns is represented on a regional level and yearly.

Mapping the number of adverse events by region and population provides location-based information to elucidate and understand hidden patterns behind the adverse events. The UN identified seven subregions in Afghanistan that assisted in planning; these seven subregions can be classified into three main regions: northern (northeastern, northwestern), central (eastern, central, western), and southern (southeastern, southwestern). Figure 13.2 illustrates the Afghanistan map that was divided into seven subregions and three main regions for pattern analysis. Table 13.1 represents each region according to different numbers of provinces, districts, and number of records. Some data sets among adverse event files have more data points for specific periods. However, the Afghanistan geo-referenced worldwide incidents tracking system (WITS) data include the widest range of data points during the period of 2004–2010 and the most complete data among other datasets. Thus, the WITS data were selected for the purpose of mapping adverse events and variables.

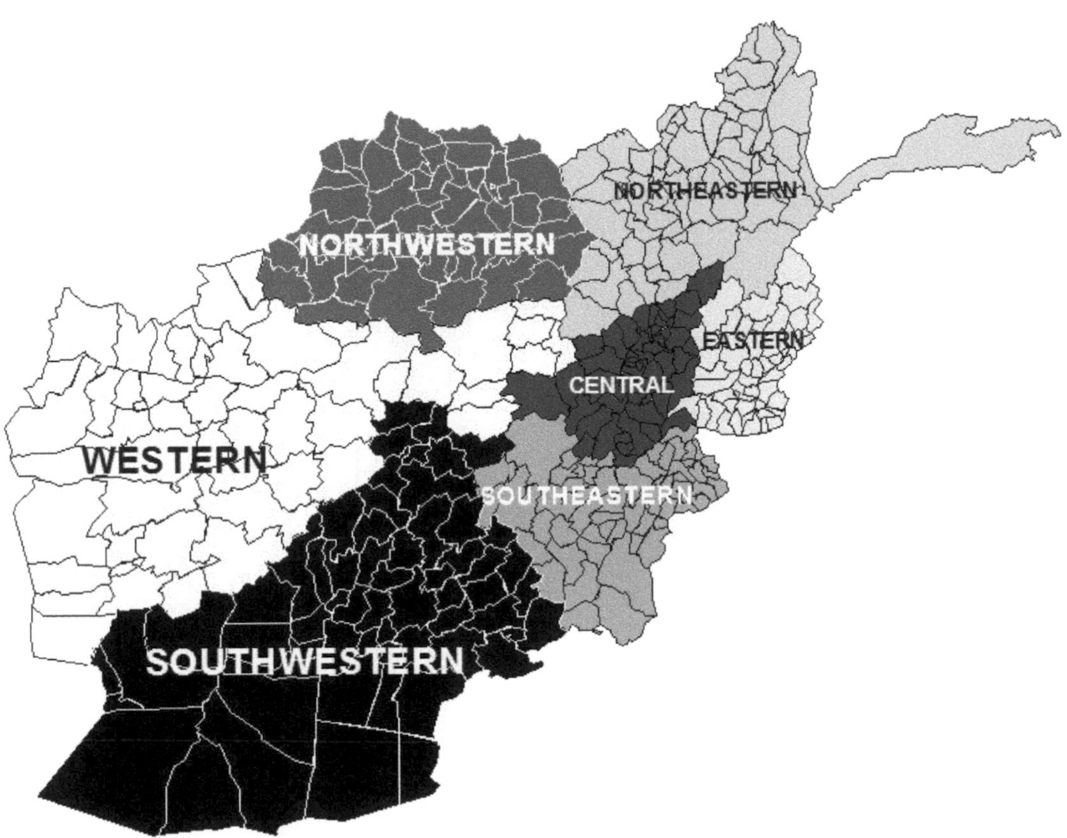

FIGURE 13.2
Regions of Afghanistan.

TABLE 13.1
Province and District Information for Each Region

Region	Number of Provinces	Number of Districts	Total Number of Records
Central	6	55	4620
Eastern	4	50	4200
Northeastern	4	67	5628
Southeastern	4	62	5208
Western	5	51	4284
Northwestern	5	55	4620
Southwestern	6	60	5040
Afghanistan (Total)	34	400	33,600

This data set was acquired monthly at the district level, where it was limited to incidents occurring between 2004 and 2010. The data set was sorted based on the order of year, month, and province information using MS Excel; it was then imported into the ArcMap 10.2 software (Figure 13.3). The address points are represented using the display XY data function. All these variables were mapped into GIS format; the output of the shape file is shown in Figure 13.3. The output provides information about the distribution of all adverse events throughout Afghanistan.

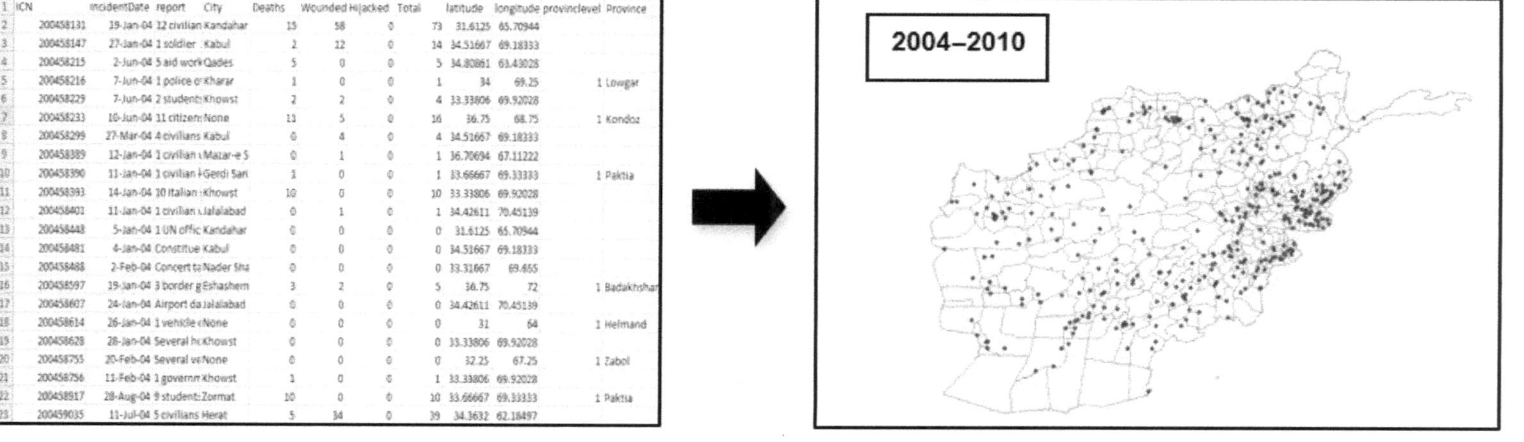

1	ICN	incidentDate	report	City	Deaths	Wounded	Hijacked	Total	latitude	longitude	provinclevel	Province
2	200458131	19-Jan-04	12 civilian	Kandahar	15	58	0	73	31.6125	65.70944		
3	200458147	27-Jan-04	1 soldier	Kabul	2	12	0	14	34.51667	69.18333		
4	200458215	2-Jun-04	5 aid work	Qades	5	0	0	5	34.80861	63.43028		
5	200458216	7-Jun-04	1 police o	Kharar	1	0	0	1	34	69.25	1	Lowgar
6	200458229	7-Jun-04	2 student	Khowst	2	2	0	4	33.33806	69.92028		
7	200458233	10-Jun-04	11 citizen	None	11	5	0	16	36.75	68.75	1	Kondoz
8	200458299	27-Mar-04	4 civilians	Kabul	0	4	0	4	34.51667	69.18333		
9	200458389	12-Jan-04	1 civilian	Mazar-e S	0	1	0	1	36.70694	67.11222		
10	200458390	11-Jan-04	1 civilian	Gerdi Sari	1	0	0	1	33.66667	69.33333	1	Paktia
11	200458393	14-Jan-04	10 Italian	Khowst	10	0	0	10	33.33806	69.92028		
12	200458401	11-Jan-04	1 civilian	Jalalabad	0	1	0	1	34.42611	70.45139		
13	200458448	5-Jan-04	1 UN offic	Kandahar	0	0	0	0	31.6125	65.70944		
14	200458481	4-Jan-04	Constitue	Kabul	0	0	0	0	34.51667	69.18333		
15	200458488	2-Feb-04	Concert ta	Nader Sha	0	0	0	0	33.31667	69.655		
16	200458597	19-Jan-04	3 border g	Eshashem	3	2	0	5	36.75	72	1	Badakhshar
17	200458607	24-Jan-04	Airport da	Jalalabad	0	0	0	0	34.42611	70.45139		
18	200458614	26-Jan-04	1 vehicle	None	0	0	0	0	31	64	1	Helmand
19	200458628	28-Jan-04	Several h	Khowst	0	0	0	0	33.33806	69.92028		
20	200458755	20-Feb-04	Several ve	None	0	0	0	0	32.25	67.25	1	Zabol
21	200458756	11-Feb-04	1 governm	Khowst	1	0	0	1	33.33806	69.92028		
22	200458917	28-Aug-04	9 student	Zormat	10	0	0	10	33.66667	69.33333	1	Paktia
23	200459035	11-Jul-04	5 civilians	Herat	5	34	0	39	34.3632	62.18497		

FIGURE 13.3
The general concept of mapping latitudes and longitudes into GIS format. The points illustrate the locations of adverse events in Afghanistan.

Point Density Analysis

A GIS is a computer-based information system for dealing with spatial data; it was first developed as a tool for the storage, the retrieval, and the representation of geographic information (Fotheringham and Rogerson 2013). Here, the term *geographic* refers to the fact that data in a GIS are grouped in terms of latitude and longitude coordinates; the term *information* refers to the fact that the data set in a GIS is grouped to represent useful knowledge using colorful maps and images and statistical summaries, such as graphs and tables; finally, the term *system* refers to the fact that a GIS consists of various interrelated and linked components with different functions (Bonham-Carter 2014). The density function spreads out a measured quantity of an input data set throughout a landscape to produce a continuous surface (Environmental Systems Research Institute [ESRI] 2013). Density maps are useful for illustrating areas where point locations are concentrated. In the ArcGIS product, from the ESRI of Redlands, California, (website: www.esri.com) ArcMap 10.2 software, there are three potential density mapping tools—kernel density, line density, and point density. For this study, a point density analysis was conducted for the purpose of calculating a magnitude per unit area from point features that fall within a neighborhood around each cell. In other words, points that fall within the area are summed and then divided by the corresponding area size to calculate density value for each cell.

Results and Discussion

Descriptive Statistics of Adverse Events by District

As a starting point, we performed various descriptive statistics, including calculation of the sum, mean, median, standard deviation, minimum, and maximum values of adverse events over the period of 2004–2010. During this period, 9371 adverse events took place in which 12,144 people were killed, 17,527 were wounded, and 2758 were hijacked. As tabulated in Table 13.2, the mean number of adverse events remained at 0.27 per district month, with a mean of 0.36 per district month people killed, 0.52 per district month wounded, and 0.08 per district month hijacked per month. The standard deviation, minimum, and maximum values in Table 13.2 show high dispersion in the distribution of total number of adverse events and the number of people killed, wounded, and hijacked.

Figure 13.4 provides information about regional differences for the number of people killed, wounded, and hijacked and a total number of adverse events that occurred during the time period of 2004–2010. Based on Figure 13.4, the southwestern region of Afghanistan

TABLE 13.2

Descriptive Statistics of Adverse Events per District per Month (2004–2010)

Variable	Sum	Mean	Median	Standard Deviation	Min	Max
Number of people killed	12,144	0.36	0	2.15	0	103
Number of people wounded	17,527	0.52	0	4.08	0	261
Number of people hijacked	2758	0.08	0	1.23	0	156
Total number of adverse events	9371	0.27	0	1.09	0	38

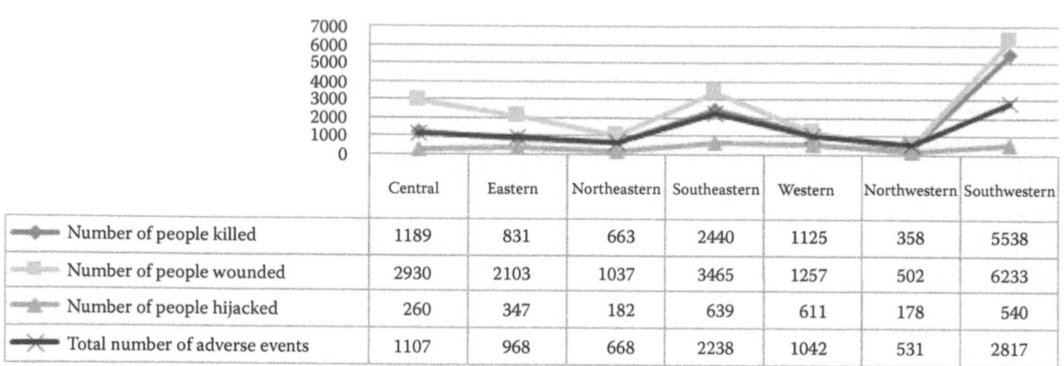

	Central	Eastern	Northeastern	Southeastern	Western	Northwestern	Southwestern
Number of people killed	1189	831	663	2440	1125	358	5538
Number of people wounded	2930	2103	1037	3465	1257	502	6233
Number of people hijacked	260	347	182	639	611	178	540
Total number of adverse events	1107	968	668	2238	1042	531	2817

FIGURE 13.4
Regions with total number of each variable (2004–2010).

experienced the highest number of people killed (5583) and wounded (6233) and the greatest total number of adverse events (2817). The southeastern region of Afghanistan had the second highest number of people killed (2440) and wounded (3465) and a total number of adverse events (2238).

For the number of people hijacked, the southeastern Afghanistan region had the highest event levels, with a total of 639, and the western region of Afghanistan had the second highest event levels, with a total of 611. Table 13.3 provides information on the total number of people killed, wounded, and hijacked; the total number of adverse events; and the mean number of events by district in each region that took place between 2004 and 2010.

Based on the information in Table 13.3, the number of people killed, wounded, and hijacked and the total number of adverse events was the lowest in 2004 and the highest in 2010. Between 2004 and 2010, there was an increasing trend, in general. The corresponding mean by district values varied between 0 and 30. The southwestern region had the highest mean by district values than other regions and the whole of Afghanistan for all variables. On the other hand, the northwestern region had the lowest mean by district values when compared to the other regions and the whole of Afghanistan for all variables. Furthermore, when we compared the variables against each other, the number of people hijacked had the lowest values in total and on average by district than the other variables.

The monthly based pattern analysis for the total number of adverse events suggests interesting trends. Based on Figure 13.5, the number of adverse events varies on a seasonal basis. For instance, the total number of adverse events in the summer months is higher than the total number of adverse events recorded in the winter months. This seasonal change is due largely to the characteristics of the area and the harsh climate, which likely significantly impact terrorist group's abilities and strategy. The poppy harvest season, which is typically in March–April–May timeframe, and the well-known fighting and non-fighting seasons might be another reasons for the seasonal change.

GIS-Based Analysis and Visualization of Adverse Events: The Case of Afghanistan

Mapping Regional Differences and Getting the Picture

We examined how adverse events are distributed across space in different regions of Afghanistan. Figures 13.6 through 13.8 illustrate the density of adverse events during

TABLE 13.3

Yearly Total and Average Number of Events by District in Each Region between 2004 and 2010

		Number of People Killed		Number of People Wounded		Number of People Hijacked		Total Number of Adverse Events	
Region	**Year**	**Total**	**Mean by District**	**Total**	**Mean by District**	**Total**	**Mean by District**	**Total**	**Mean by District**
Central	2004	24	0.43	56	1.02	3	0.05	17	0.31
	2005	27	0.49	62	1.13	1	0.02	49	0.89
	2006	92	1.67	296	5.38	5	0.09	112	2.04
	2007	268	4.87	391	7.11	30	0.55	151	2.75
	2008	213	3.87	489	8.89	65	1.18	186	3.38
	2009	295	5.36	1053	19.15	84	1.53	226	4.11
	2010	270	4.91	583	10.60	72	1.31	366	6.65
Eastern	2004	16	0.32	65	1.30	15	0.30	28	0.56
	2005	56	1.12	75	1.50	10	0.20	65	1.30
	2006	76	1.52	160	3.20	9	0.18	93	1.86
	2007	133	2.66	260	5.20	30	0.60	115	2.30
	2008	120	2.40	482	9.64	48	0.96	86	1.72
	2009	260	5.20	476	9.52	63	1.26	232	4.64
	2010	170	3.40	585	11.70	172	3.44	349	6.98
Northeastern	2004	20	0.30	14	0.21	0	0	7	0.10
	2005	6	0.09	9	0.13	0	0	9	0.13
	2006	18	0.27	62	0.93	0	0	32	0.48
	2007	137	2.04	227	3.39	11	0.16	46	0.69
	2008	41	0.61	61	0.91	12	0.18	51	0.76
	2009	134	2.00	185	2.76	49	0.73	202	3.01
	2010	307	4.58	479	7.15	110	1.64	321	4.79
Southeastern	2004	66	1.06	43	0.69	0	0	30	0.48
	2005	76	1.23	84	1.35	3	0.05	77	1.24
	2006	221	3.56	530	8.55	79	1.27	265	4.27
	2007	422	6.81	491	7.92	117	1.89	342	5.52
	2008	486	7.84	604	9.74	141	2.27	365	5.89
	2009	566	9.13	854	13.77	149	2.40	511	8.24
	2010	603	9.73	859	13.85	150	2.42	648	10.45
Western	2004	25	0.49	58	1.14	2	0.04	14	0.27
	2005	28	0.55	35	0.69	8	0.16	24	0.47
	2006	130	2.55	124	2.43	12	0.24	83	1.63
	2007	194	3.80	187	3.67	16	0.31	97	1.90
	2008	132	2.59	202	3.96	248	4.86	115	2.25
	2009	287	5.63	313	6.14	133	2.61	288	5.65
	2010	329	6.45	338	6.63	192	3.76	421	8.25
Northwestern	2004	0	0	2	0.04	0	0	3	0.05
	2005	17	0.31	53	0.96	0	0	17	0.31
	2006	30	0.55	32	0.58	0	0	34	0.62
	2007	25	0.45	59	1.07	1	0.02	34	0.62
	2008	51	0.93	37	0.67	1	0.02	44	0.80
	2009	92	1.67	66	1.20	35	0.64	115	2.09
	2010	143	2.60	253	4.60	141	2.56	284	5.16

(*Continued*)

TABLE 13.3 (CONTINUED)

Yearly Total and Average Number of Events by District in Each Region between 2004 and 2010

		Number of People Killed		Number of People Wounded		Number of People Hijacked		Total Number of Adverse Events	
Region	Year	Total	Mean by District	Total	Mean by District	Total	Mean by District	Total	Mean by District
Southwestern	2004	138	2.30	111	1.85	5	0.08	51	0.85
	2005	472	7.87	432	7.20	91	1.52	250	4.17
	2006	684	11.40	903	15.05	66	1.10	342	5.70
	2007	766	12.77	781	13.02	85	1.42	333	5.55
	2008	954	15.90	1016	16.93	76	1.27	374	6.23
	2009	1144	19.07	1240	20.67	104	1.73	551	9.18
	2010	1380	23.00	1750	29.17	113	1.88	916	15.27
Afghanistan	2004	289	0.72	349	0.87	25	0.06	150	0.38
	2005	682	1.71	750	1.88	113	0.28	491	1.23
	2006	1251	3.13	2107	5.27	171	0.43	961	2.40
	2007	1945	4.86	2396	5.99	290	0.73	1118	2.80
	2008	1997	4.99	2891	7.23	591	1.48	1221	3.05
	2009	2778	6.95	4187	10.47	617	1.54	2125	5.31
	2010	3202	8.01	4847	12.12	951	2.38	3305	8.26

the period of 2004–2010. A map of the provinces, which are color grouped by region, is also shown in the following. The size of the circle in each region reflects the number of people killed, wounded, and hijacked; the larger the circle, the more individuals affected.

During the period of 2004–2010, most of the adverse events occurred on the Pakistan border along the southwestern, southeastern, and eastern regions. In summary, mapping adverse events makes hotspot analysis more advantageous and contributes worthy information to those responsible for operations in the region. In addition, visualizing the number of people killed, wounded, and hijacked by region reveals useful information about specific patterns of activity.

Three-Dimensional Display of Adverse Events

A three-dimensional view is also useful in comparing the number of events that occurred in different locations. Figure 13.9 provides information about a three-dimensional intensity view of the total number of people killed, wounded, and hijacked. Based on Figure 13.9, the total number of people killed, wounded, and hijacked was greatest in the southeastern region of Afghanistan. This view helps decision makers to compare the actual number of people killed, wounded, and hijacked with other locations.

Mapping Adverse Events and Compared with Population Density

The rural female, rural male, urban female, and urban male population densities were mapped with the total number of people killed, wounded, and hijacked. The corresponding information is provided in Figures 13.10 through 13.13. The eastern, northeastern, and central regions, which have large numbers of people killed, wounded, and

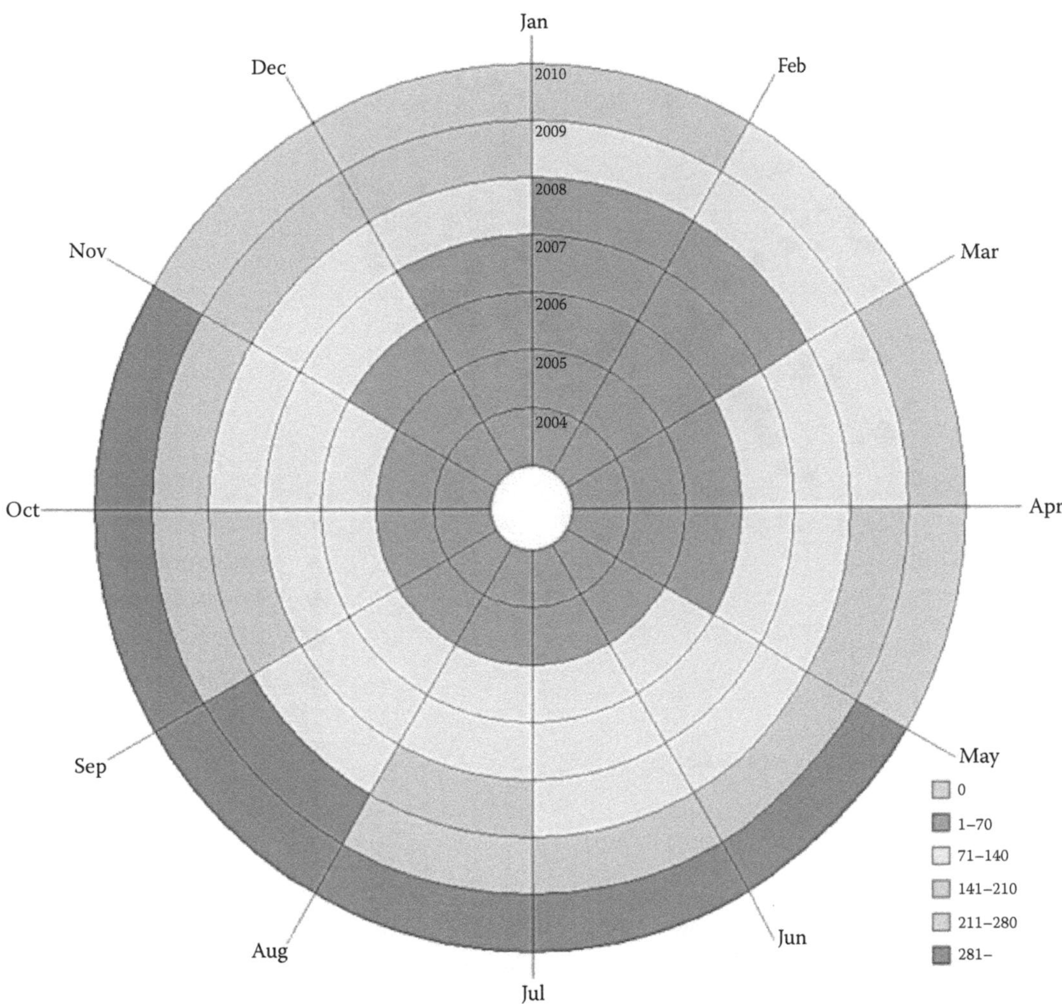

FIGURE 13.5
Monthly based pattern analysis for total number of adverse events (2004–2010).

hijacked, also have high rural female and male population densities. On the other hand, in addition to the northeast region, there are two high urban male and female population density areas—shown in the central and southwestern regions—that have a large number of people killed, wounded, and hijacked. These areas contain major urban population centers located in the Kabul and Kandahar provinces. These centers are two of the most convenient locations for terrorist-related activities in Afghanistan. As expected, the regions with a high number of adverse events where those in which the population density was also high.

Geographical Modeling Using Point Density Analysis

Using ESRI's ArcMap 10.2 software, the variable called *total* (the sum of the *number of people hijacked*, *number of people wounded*, and *number of people killed* variables) in the

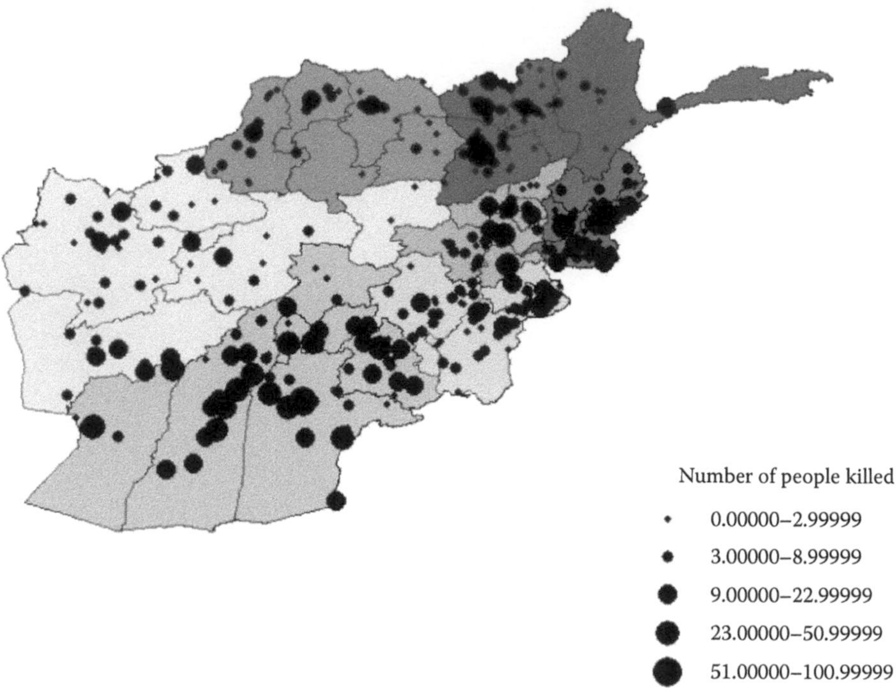

FIGURE 13.6
Number of people killed by region (2004–2010).

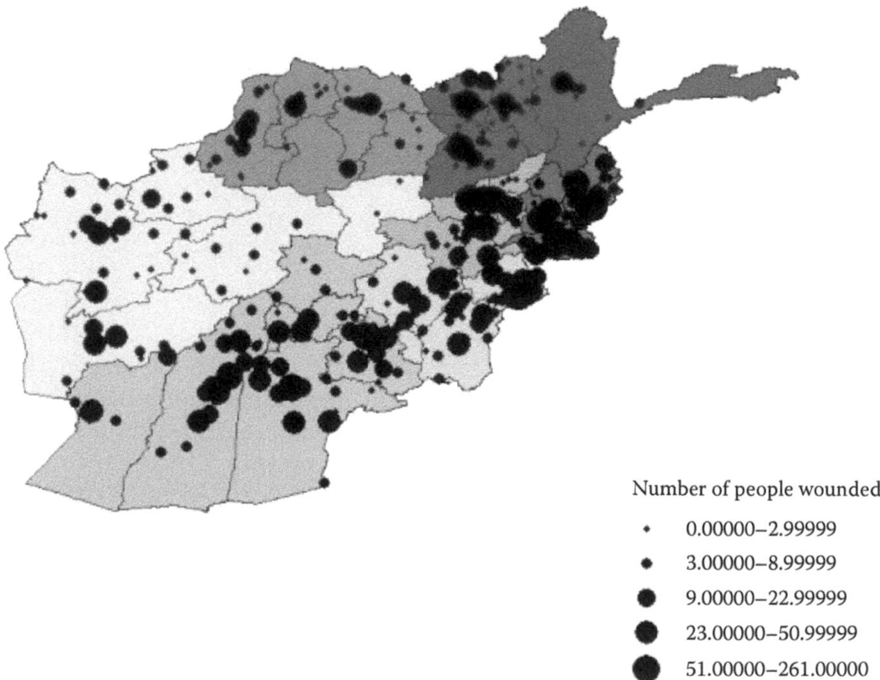

FIGURE 13.7
Number of people wounded by region (2004–2010).

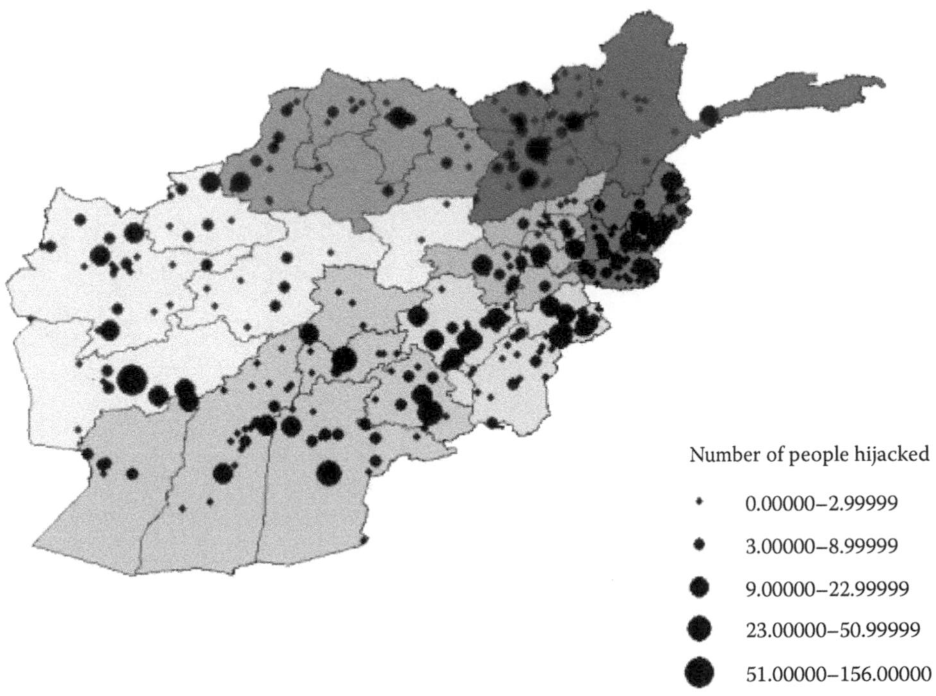

FIGURE 13.8
Number of people hijacked by region (2004–2010).

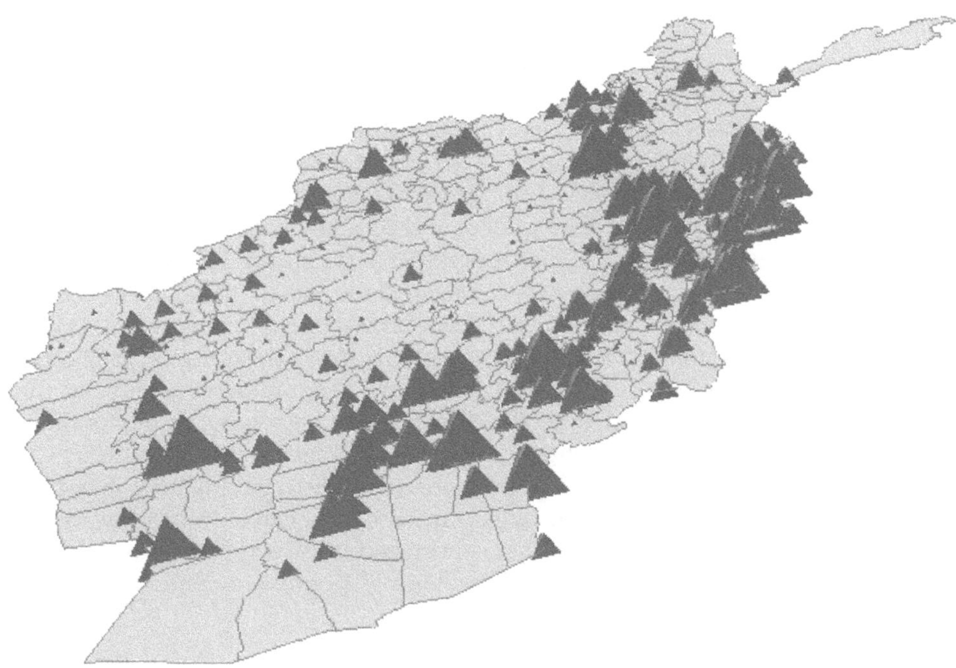

FIGURE 13.9
Three-dimensional view of total number of people killed, wounded, and hijacked density (2004–2010).

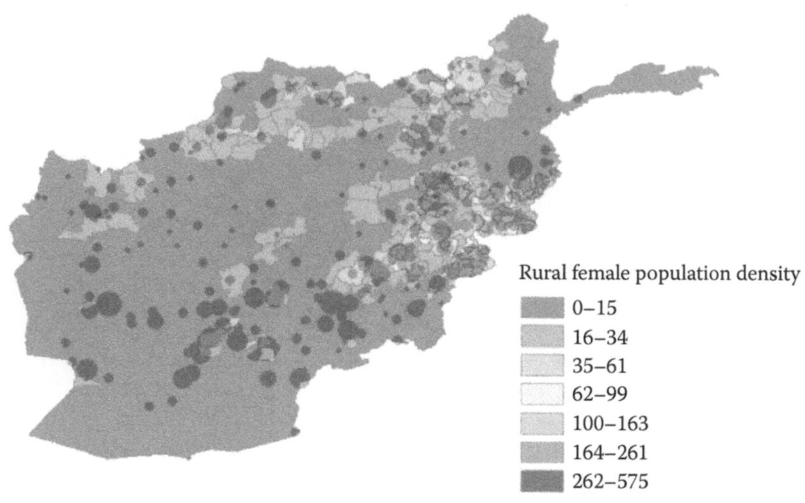

FIGURE 13.10
Mapping rural female population density with total number of people killed, wounded, and hijacked.

WITS event file (from 2004 to 2010) was added to the data frame. After the adverse events were mapped in the ArcGIS environment, a point density analysis was conducted to detect high adverse event density areas. Figure 13.14 illustrates where the number of adverse events is particularly high in consideration of the population density of that district.

A legend is provided in the right-hand corner of each map where high and less activities of the total number of people killed, wounded, and hijacked are illustrated. All data during the period between 2004 and 2010 are represented in Figure 13.15, and the individual years that show differences between years can be observed visually. Mapping the number of adverse events by districts can help decision makers to explore problematic areas in more detail. Most problematic districts accumulate in the eastern region of the country. This high concentration can be explained by its close geographic proximity to the lawless border regions of Pakistan. Most of the districts that have large numbers of adverse events also have major population centers. It is due to the fact that it concentrates around Kabul and Kandahar, which are highly populated areas. Districts that have major population centers on the west side have also large numbers of adverse events. Based on these results, major population centers and higher concentrations of economic development will be more sensitive to adverse events than less densely populated centers and less well-developed areas.

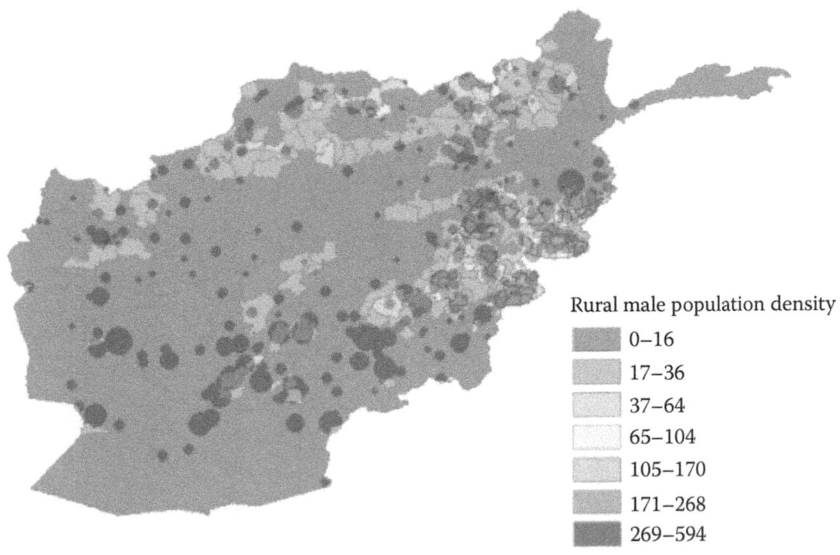

FIGURE 13.11
Mapping rural male population density with total number of people killed, wounded, and hijacked.

Concluding Remarks and Further Work

In daily life, individuals and organizations make spatial decisions (GIS-based decisions) frequently, such as when selecting a grocery store in which to shop, a route for a specific destination, a place to build a factory, etc. Military decision makers are not much different in this respect. They take into account the realities of spatial decision making when allocating resources and managing infrastructures for logistics purposes. This chapter represented adverse event data in terms of the number of people killed, wounded, and hijacked in the active war theater of Afghanistan over the period of 2004–2010. The visualization of each map helps the reader to understand implicit trends; it also provides new information about adverse events represented by a data set with over 30,000 records, where the information about adverse event patterns is represented on a regional level and yearly.

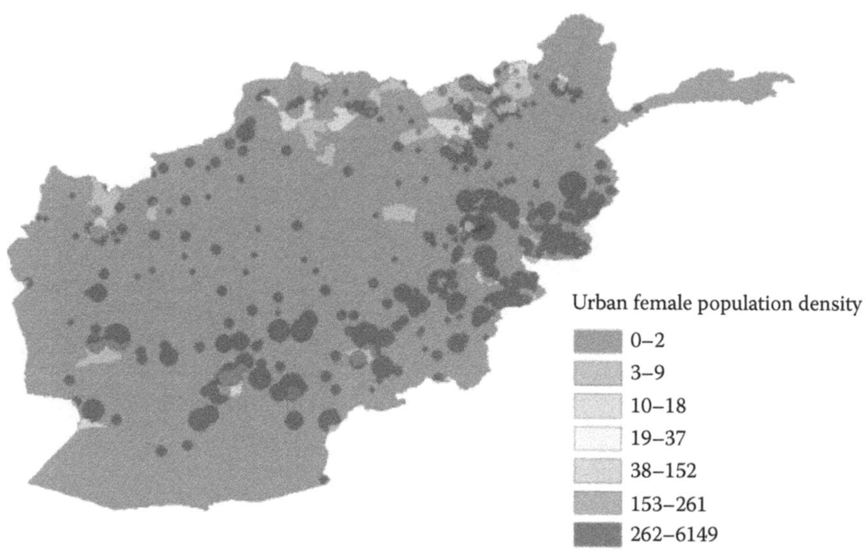

Total number of people killed, wounded and hijacked

0.00000–5.00000

5.00001–21.00000

21.00001–52.00000

52.00001–106.00000

106.00001–261.00000

FIGURE 13.12
Mapping urban female population density with total number of people killed, wounded, and hijacked.

Based on the results obtained, it was concluded that the frequency of adverse events has significantly increased from 2004 to 2010. The number of people killed, wounded, and hijacked and the total number of adverse events were at their lowest point in 2004 and at their highest point in 2010. When the regions were compared to one another, the southwestern region had the highest mean by district values than any other region and the whole of Afghanistan for all variables. On the other hand, the northwestern region had the lowest mean by district values than any other region and the whole of Afghanistan for all variables. When we compared the variables against each other, the number of people hijacked had the lowest values in total and on average by district than any of the other variables. As expected, the regions with the highest number of adverse events were also regions with high population density, at least in Afghanistan. For the purpose of understanding patterns of adverse events in an active war theater, the results can be summarized by visualizing the occurrence of incidents by region; the emphasis was on analyzing the number of people killed, wounded, and hijacked to determine the risk of different parts of Afghanistan.

In summary, this chapter represents the benefits of employing GIS-based modeling in decision making under the study of terrorism. Although it is not possible to determine every detail of adverse events using pattern analysis, analyzing and better understanding

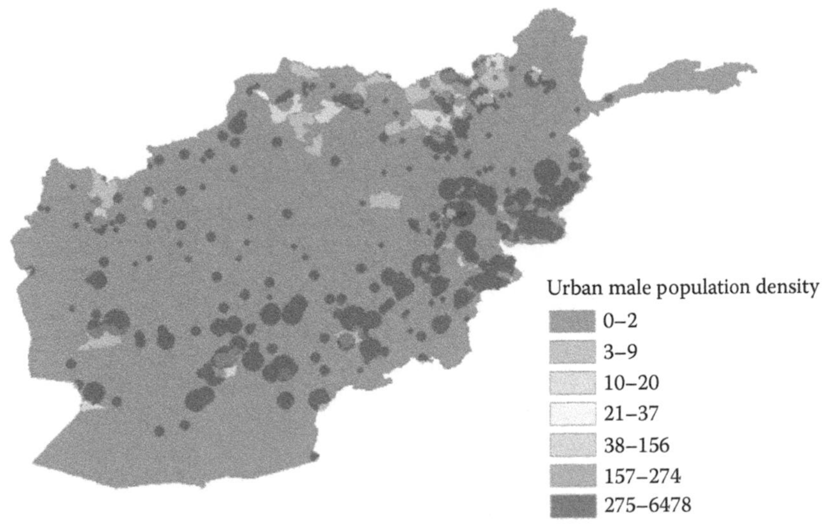

Total number of people killed, wounded and hijacked

- 0.00000–5.00000
- 5.00001–21.00000
- 21.00001–52.00000
- 52.00001–106.00000
- 106.00001–261.00000

FIGURE 13.13
Mapping urban male population density with total number of people killed, wounded, and hijacked.

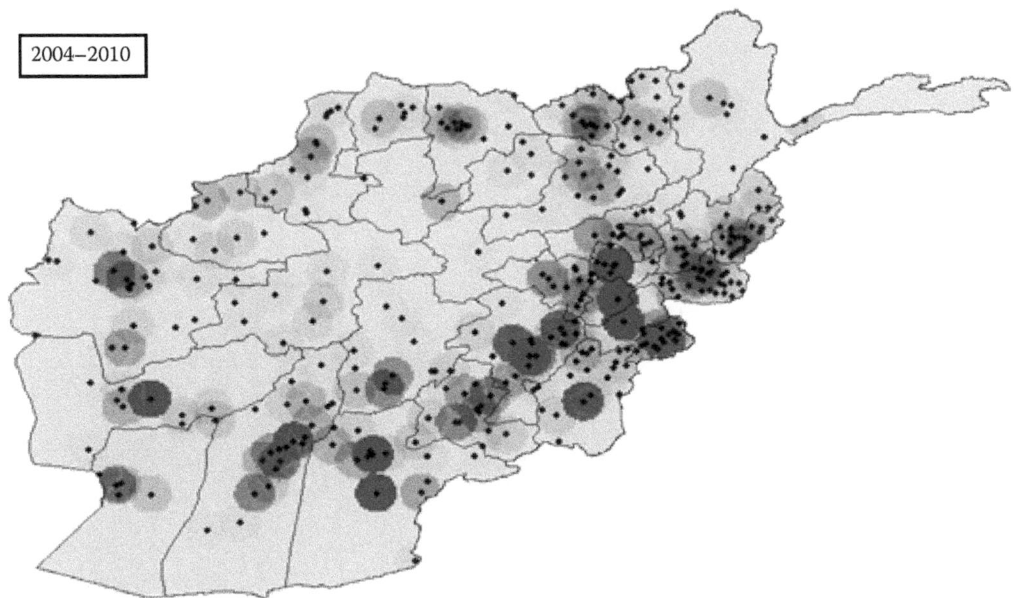

FIGURE 13.14
A point density analysis of total number of people killed, wounded, and hijacked from 2004 to 2010.

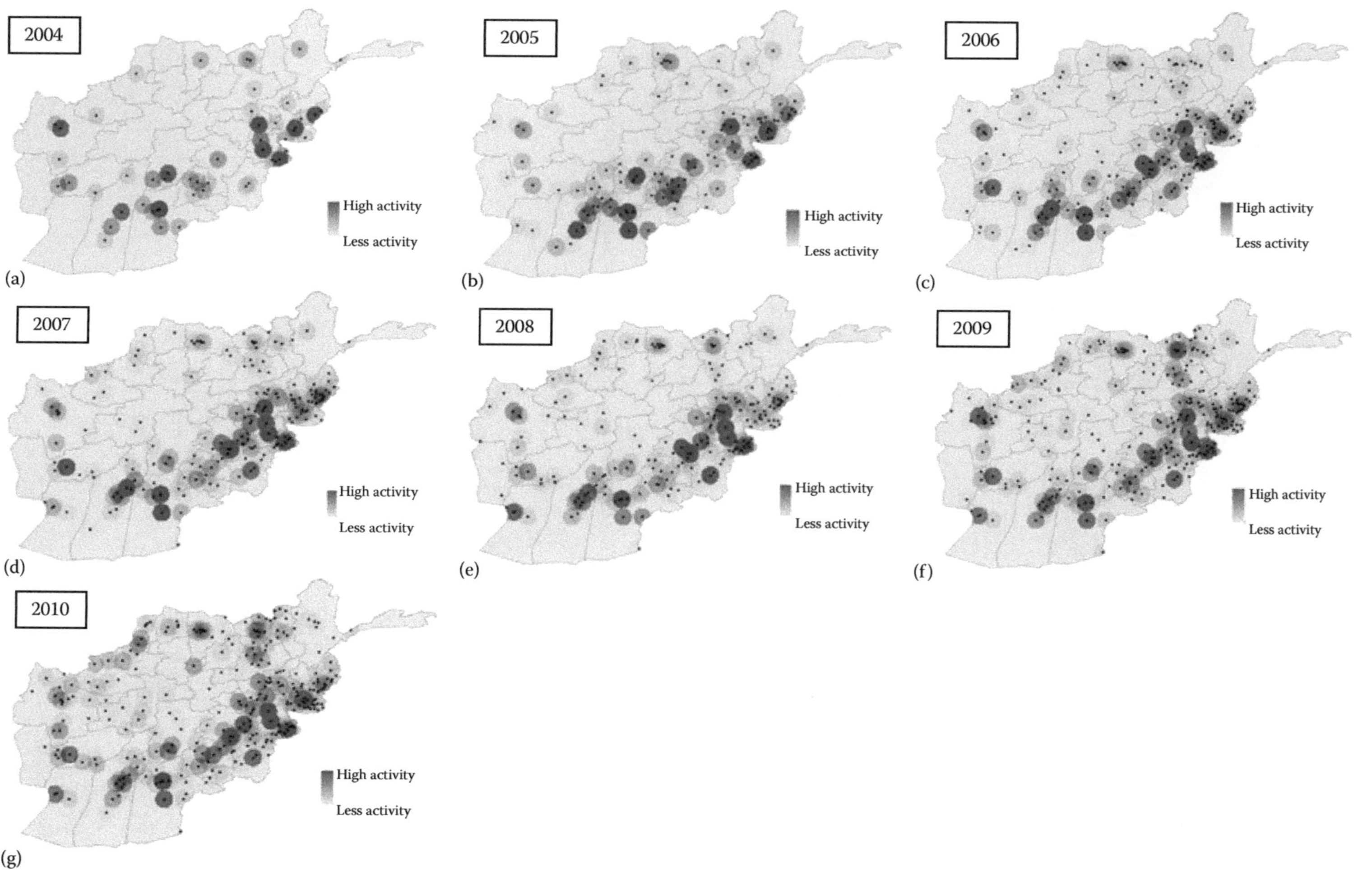

FIGURE 13.15
A point density analysis of total number of people killed, wounded, and hijacked for each year: (a–g) 2004–2010, respectively.

these patterns, both spatial and temporal, may provide useful information about the strategies and the tactics to be employed for similar scenarios in an active war theater in the future. For future studies, it is anticipated that by examining these GIS features, this chapter may increase attention to the geographical studies in literature and their contributions to military decision making.

Megacity Use Case

Megacities are growing rapidly and becoming more connected. The nature of each megacity's complexity differentiates it from the rest. For state governments, the ability to deal with megacity growth and maintain their security has become increasingly difficult. No two megacities exhibit the same stability, which means that the stability level of each megacity requires unique understanding.

Recent trends in the economic development of the megacity Xanadu have meant that its borders teem with migrant workers. Megacity Xanadu's state government needs to adjust its economic structure to recognize the challenges of meeting the growing needs and demands of its population. The incremental migration rate in megacity Xanadu creates some challenges for mapping the sociocultural integration of urban and rural migrants. These challenges mean that megacity Xanadu's stability is a function of several characteristics, including population density, availability of facilities, cost of land, and security. New approaches, methodologies, and policies need to be considered for better managing and handling the megacity's challenges. The management of megacity Xanadu requires GIS analysis for many activities, as it has the potential to be an essential approach that combines the socioeconomic and cultural elements in megacity management.

GIS is a powerful tool for various kinds of analysis. Because GIS is able to represent useful knowledge using maps and related spatial data sets based on physical, political, social, and cultural factors, it is a potent tool for defining regions based on the criteria deemed important to a given research problem. For instance, when a database with the demographic potential of a region is well built, the needs of the population in this region can be predicted and thus allow the government to plan for needed sociocultural projects. Furthermore, a geographical approach is at the merging point of many disciplines. Most research areas can be addressed using a geographical approach. In daily life, individuals and organizations make spatial decisions (GIS-based decisions) frequently, such as when selecting a grocery store in which to shop, a route for a specific destination, a place to build a factory, etc. The state governments of megacities are not much different in this respect. They apply GIS as a basic tool that supports decision making in city management, and it has the potential to play a major role in different decision making phases by including social, economic, and demographic factors while considering a megacity as a closed system. All these phases may be detailed and analyzed with sets of different GIS layers and bringing different types of information together. GIS can help megacity governments monitor the growth and the changes within their city and predict possible areas of risk in real time and in a shorter timeline as compared to other traditional management tools. GIS-based tools can be applied in new management tools for traditional city management activities. Another unique feature of using GIS is that it allows the integration of different sources of information such as infrastructure development data, population data, security, economic activity data, etc.

Megacity Xanadu's state government might take into account the realities of GIS-based decision making when allocating resources and managing infrastructures to identify areas of instability. Furthermore, GIS-based tools may contribute to interoperability and integration within the megacity. Integrated GIS information may help megacity management make unified decisions that are timely and less costly. Also, the decision making process, when applying GIS overlay analysis, allows insufficient resources to be prioritized, particularly in emergency and risky situations. More specifically, GIS overlay mapping may be applied to provide a useful way to evaluate current or future phases and to compare their efficiencies in specific locations regarding to the projected demand arising from growth in populations and economic activities. When this approach is applied to megacity Xanadu, it is possible to identify primary and secondary zones. With this basic information, its state government can study and plan for new development projects appropriate for the available resources.

In summary, we conclude that GIS can perform the role of a processing platform. All information for megacity Xanadu would be integrated, stored in a central database, and used by many institutions. More specifically, the citizens of megacity Xanadu could individually contribute by gathering geospatial information, which would then be processed and integrated. Instability results from both sociocultural and sociopolitical issues. Understanding the basic demands of the population under these issues with the help of a GIS approach will allow better allocation of resources and management of infrastructure development. Furthermore, the information may be useful to the megacity Xanadu community for planning and to overcoming the natural drivers of instability.

DISCUSSION QUESTIONS

1. What are the benefits of employing GIS-based modeling in decision making under the study of terrorism?
2. If the geographical aspects of incidents are essential, do you think that the decision makers should use this information to determine key strategies and tactics to be employed for similar scenarios in the future?
3. How did the regional and seasonal differences affect the occurrence of adverse events?
4. How can the state government of a megacity use GIS tools to identify areas of instability?

Acknowledgments

This study was supported in part by Grant No. 1052339, Complex Systems Engineering for Rapid Computational Socio-Cultural Network Analysis, from the Office of Naval Research (ONR). The authors acknowledge the helpful guidance of ONR program management and the contributions of the technical team.

References

Bahgat, K., and R. M. Medina. 2013. An overview of geographical perspectives and approaches in terrorism research. *Perspectives on Terrorism* 7(1), 38–72.

Beck, R. A. 2003. Remote sensing and GIS as counterterrorism tools in the Afghanistan war: A case study of the Zhawar Kili region. *The Professional Geographer* 55(2), 170–179.

Berrebi, C., and D. Lakdawalla. 2007. How does terrorism risk vary across space and time? An analysis based on the Israeli experience. *Defence and Peace Economics* 18(2), 113–131.

Bhattacharjee, Y. 2007. Pentagon asks academics for help in understanding its enemies. *Science* 316(5824), 534–535.

Bonham-Carter, G. F. 2014. *Geographic Information Systems for Geoscientists: Modelling with GIS*. Elsevier, Philadelphia, PA.

Böhnke, J. R., J. Koehler, and C. Zürcher, 2010. *Assessing the Impact of Development Cooperation in North East Afghanistan 2005–2009: Final Report*, ed. Bundesministerium für wirtschaftliche and Zusammenarbeitund Entwicklung. German Federal Ministry for Economic Cooperation and Development, Berlin and Bonn.

Brown, D., J. Dalton, and H. Hoyle. 2004. Spatial forecast methods for terrorist events in urban environments. *Lecture notes in Computer Science* 3073, 426–435.

Cakit, E., and W. Karwowski. 2015. Assessing the relationship between economic factors and adverse events in an active war theater using fuzzy inference system approach. *International Journal of Machine Learning and Computing* 5(3), 252–257.

Child, T. B. 2014. Reconstruction and insurgency: The importance of sector in Afghanistan. Available at http://www.europeanpeacescientists.org/T.B.Child_Stuart%20Bremer%20Winner%202014.pdf (accessed September 20, 2015).

Clancy, J., and C. Crossett. 2007. Measuring effectiveness in irregular warfare. *Parameters* 37(2), 88.

Collins, J. J. 2011. *Understanding War in Afghanistan*. NDU Press, Washington, DC.

Corcoran, J., I. Wilson, and J. Ware. 2003. Predicting the geo-temporal variation of crime and disorder. *International Journal of Forecasting* 19, 623–634.

Çakıt, E., W. Karwowski, H. Bozkurt, T. Ahram, W. Thompson, P. Mikusinski, and G. Lee. 2014. Investigating the relationship between adverse events and infrastructure development in an active war theater using soft computing techniques. *Applied Soft Computing* 25, 204–214.

Çakıt, E, and W. Karwowski. 2015. Understanding patterns of infrastructure development in the active theater of Afghanistan over the period 2002–2010. *Procedia Manufacturing* 3, 3876–3882.

Demirci, S., and I. S. Suen. 2007. Spatial pattern analysis of PKK-KONGRA GEL terror incidents in Turkey: 2003–2004. In *Understanding and Responding to the Terrorism Phenomenon: A Multi-Dimensional Perspective* 21, 153.

Denning, D. 2000. Cyberterrorism. Testimony before the Special Oversight Panel on Terrorism, Committee on Armed Services US House of Representatives, Georgetown University. Nova Science Publishers, New York.

Drapeau, M. D., and B. K. Mignone. 2007. Culture, conflict, and… climate? *Science* 316(5831), 1564–1564.

Environmental Systems Research Institute (ESRI) . 2013. ESRI. Available at http://www.esri.com (accessed August 1, 2015).

Federation of American Scientists. 2009. A GEOINT analysis of terrorism in Afghanistan. *Secrecy News*. Available at http://www.fas.org/blog/secrecy/2009/05/05 (accessed May 3, 2015).

Fotheringham, S., and P. Rogerson (eds). 2013. *Spatial Analysis and GIS*. CRC Press, Boca Raton, FL.

Goodchild, M. F., L. Anselin, R. P. Appelbaum, and B. H. Harthorn. 2000, April. Toward spatially integrated social science. *International Regional Science Review* 23(2), 139–159.

HSCB Modeling Program. 2009. Human social culture behavior modeling program. Available at http://www.dtic.mil/biosys/docs/HSCB-news-spring-2009.pdf (accessed March 15, 2015).

Irregular Warfare Joint Operating Concept. 2007, September 11. Irregular warfare joint operating concept. Available at http://www.fas.org/irp/doddir/dod/iw-joc.pdf (accessed March 20, 2015).

Johnson, S. D., and E. Braithwaite. 2009. Spatio-temporal modeling of insurgency in Iraq. In Freilich, J. D., and G. R. Newman (eds.), *Reducing Terrorism Through Situational Crime Prevention: Crime Prevention Studies*, Vol. 25. Criminal Justice Press, Monsey, New York, pp. 9–32.

Kamrany, N. M. 2009. After the Taliban: Nation-building in Afghanistan (review). *The Middle East Journal* 63(4), 669–671.

Kumar, M. V., and C. Chandrasekar. 2011. Spatial clustering simulation on analysis of spatial-temporal crime hotspot for predicting crime activities. *International Journal of Computer Science and Information Technologies* 2(6), 2867–2874.

LaFree, G., L. Dugan, L. M. Xie, and P. P. Singh. 2012. Spatial and temporal patterns of terrorist attacks by ETA 1970 to 2007. *Journal of Quantitative Criminology* 28(1), 7–29.

Liu, H., and D. E. Brown. 2003. Criminal incident prediction using a point-pattern based density model. *International Journal of Forecasting*, 19, 603–622.

Medina, R. M., L. K. Siebeneck, and G. F. Hepner. 2011. A geographic information systems (GIS) analysis of spatiotemporal patterns of terrorist incidents in Iraq 2004–2009. *Studies in Conflict & Terrorism*, 34(11), 862–882.

Open Source Center (OSC). 2009. Afghanistan-geospatial analysis reveal patterns in terrorist incidents 2004–2008. Available at http://www.fas.org/irp/dni/osc/afghan-geospat.pdf (accessed February 21, 2015).

Polat, E. 2007. Spatio-temporal crime prediction model based on analysis of crime clusters. Master thesis. Geodetic and Geographic Information Technologies. Department of Geodetic and Geographic Information Technologies. Middle East Technical University, Ankara, Turkey.

Pool, R. 2011. Sociocultural data to accomplish Department of Defense missions: Toward a unified social framework: Workshop summary. United States. Office of Naval Research, National Research Council (US). Planning Committee on Unifying Social Frameworks, & National Academies Press (US). National Academies Press, Washington, DC.

Rusnak, D. M., L. W. Kennedy, I. S. Eldivan, and J. M. Caplan. 2012. Analyzing terrorism using spatial analysis techniques: A case study of Turkish cities. In *Evidence-Based Counterterrorism Policy*. Springer, New York, pp. 167–185.

Shroder, J. 2005. Remote sensing and GIS as counterterrorism tools in the Afghanistan war: Reality, plus the results of media hyperbole. *The Professional Geographer* 57(4), 592–597.

Siebeneck, L. K., R. M. Medina, I. Yamada, and G. F. Hepner. 2009. Spatial and temporal analyses of terrorist incidents in Iraq, 2004–2006. *Studies in Conflict & Terrorism* 32, 591–610.

Stanton, J. 2007. Evolutionary cognitive neuroscience: Dual use discipline for understanding & managing complexity and altering warfare. *International Studies Association Conference, Portugal*. Available at Social Science Research Network, http://ssrn.com/abstract=1946864.

Tarnoff, C. 2010, June. *Afghanistan: US foreign assistance*. Library of Congress Congressional Research Service, Washington DC.

The 2013 Human Development Report. 2013. The Rise of the South: Human Progress in a Diverse World. HDRO (Human Development Report Office). United Nations Development Programme. pp. 144–147. Retrieved May 20, 2015, http://hdr.undp.org/en/2013-report.

Webb, J., and S. L. Cutter. 2009. The geography of US terrorist incidents, 1970–2004. *Terrorism & Political Violence* 21, 428–449.

Section III

Building and Validating Sociocultural Models

In a rapidly changing world, leaders across the public and private sectors are searching for innovative, real-time (or near real-time) aids for decision making. In the age of globalization and big data, we are now inundated with information. The question is how can these data be collected, analyzed, and presented in ways that reduce the cognitive load on the human? The ability to take in massive amounts of data, sift through those data, and allow the human to explore the most critical pieces relevant to their missions would greatly enhance decision making.

Modern recognition of the operational value of understanding sociocultural underpinnings of foreign entities to support military operations stems from the nineteenth-century work of T. E. Lawrence, or "Lawrence of Arabia." The importance of sociocultural understanding has continued to be prioritized in contemporary military campaigns, such as Iraq and Afghanistan, where winning hearts and minds was an official counterinsurgency strategy of the US military. Today, although the landscape of warfare spans from traditional boots-on-the-ground campaigns to cyberspace, the need to understand and engage foreign populations still exists.

The global population has been, and will continue to be, affected by a number of nontraditional warfare campaigns. For instance, cyberattacks not only target governments, e.g. the 2014 US government Office of Personnel Management breach, but also affect financial institutions, corporations, businesses, and even individuals. Tactics utilized by the Islamic State have dissolved all barriers that may have once been upheld by adversaries in distinguishing combatants from noncombatants, keeping no one safe from harm. The events of the Russia–Ukraine–Crimea conflict have fostered international concern with "gray zone" conflicts, in which creative and/or constrained political, economic, and military responses must be considered to avoid all-out war. And the nexus of power vacuums, organized crime, and conflict continue to incite political instability around the globe.

The pace at which decisions must be made in these types of environments has encouraged the development and validation of sociocultural models that can provide a number of capabilities to decision makers, to include indications and warnings, scenario building, and conflict simulations, or war-gaming. Although regional and cultural subject matter experts (SMEs) can provide invaluable insight into their fields of expertise, automated technologies built upon validated sociocultural models can collect, organize, tag, and visualize massive amounts of data in real-time—something that is simply impossible for even the best SMEs and analysts to accomplish.

This section, "Building and Validating Sociocultural Models," highlights current research in the area of sociocultural model development and validation. The chapters in this section explore research questions spanning from sociocultural model requirements and design, to use of specific models within a suite of decision aid technologies, and to identifying current gaps in the sociocultural modeling domain.

Chapter 14, "Sociocultural Capability Requirements across All Phases of Military Operations," by Perry, investigates the current state of sociocultural capabilities in the US military and conducts a gap analysis on which of the identified, required capabilities the military has access to in order to successfully execute missions across a variety of operations.

Chapter 15, "Trust in Models," by Benjamin, Vadakkeveedu, and Ramachandran, focuses on the need for better methods and competencies for developing and maintaining sociocultural models to enhance trust in this domain. The authors identify the aspects of trust that aid the decision making process; in particular, uncertainty, and explore the effects these factors have on trust in sociocultural modeling-enabled decision making.

Chapter 16, "Validating Causal and Predictive Claims in Sociocultural Models," by Sliva, Reilly, Chamberlain, and Casstevens, explores the difficulty of producing successful causal and predictive sociocultural models. The authors suggest an approach to validating such sociocultural models based on the invalidation and refinement of causal and predictive relationships that includes a breadth of techniques, enabling the authors to utilize the strengths of each.

Chapter 17, "Rapid Generation of Political Conflict Simulation for Scenarios around the World," by Silverman, Sun, Weyer, and Bharathy, discusses a prototype system developed by the authors to aid in the rapid development of conflict simulations. The prototype provides models and simulations to support and influence decision making processes.

Chapter 18, "Detailed Model Development Case Study: The Peace Game," by Powers, describes a war-game-based software package developed to support the decision making processes and COAs taken in conflict resolution, peace-building, and stability operations. The Peace Game offers a flexible software package that enables decision makers to tailor the software's simulations to their specific needs.

Bibliography

Kaldor, M., and I. Rangelov. *The Handbook of Global Security Policy*. Chichester: Wiley Blackwell, 2014.

Lawrence, T. E. *Seven Pillars of Wisdom: A Triumph: The Complete 1922 Text*. Blacksburg, VA: Wilder, 2011.

Lawrence, T. E. *Revolt in the Desert*. New York: Barnes & Noble Books, 1993.

Schmorrow, D. *Sociocultural Behavior Research and Engineering in the Department of Defense Context*. http://handle.dtic.mil/100.2/ADA549230, 2011.

United States. 2007. *The U.S. Army/Marine Corps Counterinsurgency Field Manual: U.S. Army Field Manual No. 3-24: Marine Corps Warfighting Publication No. 3-33.5*. Chicago, IL: University of Chicago Press.

14

Sociocultural Capability Requirements across All Phases of Military Operations

Walter L. Perry

CONTENTS

Introduction 270
The Missions 270
  Operational Phases 271
  Missions by Phase Description 272
The Sociocultural Capabilities 273
  What Is a Sociocultural Capability? 273
  Sociocultural Capability Categories 274
Sociocultural Capabilities by Operational Phase 277
Gaps and Mitigation Measures 278
Megacity Use Case 282
  Peacekeeping Operations in Çawaila 282
Conclusion 284
References 286

ABSTRACT In modern military operations, it is critical that commanders be sensitive to the social and cultural norms of populations with whom they must interact. Most military operations have the mission of promoting stability and reconstruction or eventually evolve to such after a major combat as in Iraq. Gaining the cooperation and the active assistance of the local population is critical to ensuring a secure environment. This chapter identifies the sociocultural capabilities needed to execute missions and tasks across the full spectrum of military operations, identifies which of these the military has access to, and proposes mitigating measures to close requirement gaps.

KEY TOPICS

- Missions and phases of military operations
- Required sociocultural capabilities (SCCs) and categories
- SCCs required by military operational phase
- Gaps in SCCs and mitigating measures

ARMY AND MARINE CORPS VIEW

...staffs should identify and analyze the culture of the society as a whole and of each major group within the society. Social structure comprises the relationships among groups, institutions, and individuals within a society; in contrast, culture (ideas, norms, rituals, codes of behavior) provides meaning to individuals within the society (Department of the Army 2006, pp. 3–6).

Introduction

In modern warfare, commanders and their staffs must be sensitive to the social and cultural norms of host nations. Most military operations have the mission of promoting stability and reconstruction or eventually evolve to such after a major combat as in Iraq.

Gaining the cooperation and the active assistance of the local population is critical to ensuring a secure environment. The Army and Marine Corps doctrine manual on counterinsurgency operations emphasizes the importance of understanding the local culture and the social structure.

Although this addresses counterinsurgency operations, the Army and the Marine Corps recognize that understanding a society's culture and social structures is necessary in all phases of military operations.

This chapter identifies the SCCs needed to execute missions and tasks across the full spectrum of military operations, identifies which of these the military has access to, and proposes mitigating measures to close requirement gaps. The analysis draws on the CBA process.

The Missions

The full spectrum of military operations is described in terms of operational phases. Joint Publication 3-0 (Joint Chiefs of Staff 2010), *Joint Operations*, describes the military operations phasing model. This model is designed to organize operations by helping commanders and their staffs to think through an entire campaign from the early shaping phase through the combat phases to reestablishing civil authority in the host nation. There are six phases identified. Although presented as a sequence, there can be a considerable overlap of the phases of an operation, thus precluding a distinct separation between them. This reality highlights the need to integrate sociocultural factors throughout a military operation.

CAPABILITY-BASED ASSESSMENT

A capability-based assessment (CBA) is designed to identify solutions to provide the war fighter with capabilities needed to accomplish missions. A CBA consists of three parts: (1) the identification of needed capabilities, (2) the identification of gaps in the required capabilities, and (3) the identification of means to fill the capabilities gaps.

Operational Phases

We describe the activities associated with each of the operational phases along with the missions drawn from the sources cited earlier. For each phase, we summarize the mission set by citing the major objectives associated with the phase. The full list of missions is included in Table 14.1.

- Phase 0: Shape—The Shape phase consists of benign military operations designed to enhance international legitimacy and gain international support for the US military and the national objectives. The focus is on building local capacity, developing host nation military capabilities, and dissuading potential adversaries. The

TABLE 14.1

Military Missions by Phase

Phase	Missions
Phase 0: Shaping	Build local capacity and support infrastructure projects.
	Develop host nation military capabilities and solidify relationships with friends and allies.
	Dissuade or deter potential adversaries.
	Improve information exchange and intelligence sharing, and provide US forces with peacetime and contingency access.
Phase I: Deter	Employ the joint force to demonstrate capability and resolve.
	Deter criminal and violent extremist activities.
	Mobilize and tailor the force to address the crisis in coordination with other government agencies (OGAs), IGOs, and NGOs.
	Support the command's operational concept by obtaining overflight permissions and planning for force protection and logistics.
Phase II: Seize the initiative	Disable or destroy the enemy through kinetic means.
	Limit collateral damage to infrastructure and partner country populations.
	Provide assistance to relieve conditions that precipitated the crisis.
	Influence and inform the local population.
Phase III: Dominate	Prevent instability from spreading to adjacent areas.
	Influence population to counter anti-US messages.
	Disrupt activities of violent extremist organizations.
	Control the situation.
	Overmatch the enemy by employing the full joint force.
	In irregular warfare, dominate the environment using conventional and unconventional, information and stability operations.
Phase IV: Stabilize	Improve local governance.
	Assist in the provision of basic services and rebuild, improve, and protect infrastructure.
	Build capacity of local law enforcement and justice system and partner with local authorities to increase security and restore order.
	Ensure that conditions leading to the original crisis do not recur.
	Begin redeployment operations as soon as possible.
Phase V: Enable civil authority	Build, support, improve, and secure legitimate civil government and enable civil authority's provision of services.
	Coordinate actions with multinational IGOs and NGOs.
	Persuade locals to regard US and local civil authority favorably.
	Disrupt and deter human/drugs/weapons trafficking and erode support for criminal and violent extremist organizations.

objective is to create a favorable opinion of the United States and to establish or reinforce relations with friendly countries.

- Phase I: Deter—In phase I, the objective is to demonstrate resolve by engaging in preparatory actions that demonstrate that a joint force is both capable and committed to deterring enemy activity. Some of the activities in this phase include demonstrating capability, deterring criminals and violent extremists, and gaining the command's support.
- Phase II: Seize the initiative—If adversaries cannot be deterred, the United States may move quickly to launch combat operations to seize the initiative, regardless of whether it is a hostile nation state or a nonstate group posing the threat. In noncombat operations, the objective is to apply the appropriate force to gain the initiative. Phase II activities can include conducting offensive combat operations, killing or capturing an enemy's senior leaders, reducing an enemy's freedom of movement, dislodging enemy forces from their positions, and degrading an enemy's capability, or undertaking acts of violence.
- Phase III: Dominate—Phase III focuses on breaking the enemy's will for organized resistance or, in noncombat situations, controlling the operational environment. If the United States succeeds in seizing the initiative from an adversary, it can attempt to dominate the environment and prepare to transition from offensive operations to longer-term stability operations. In phase III, the United States might conduct continuous operations aimed at suppressing threat networks and isolating threat groups, including violent extremist organizations, operating within a partner nation.
- Phase IV: Stabilize—Phase IV applies when there is not a functioning government in the host nation or only a weakly functioning government. This scenario may arise from the collapse of a regime. The phase is characterized by the transition from combat operations to stability operations. In this phase, the United States may be compelled to step in and use military forces to provide basic security, rebuild critical infrastructure, and ensure that the essential needs of the host nation population are met.
- Phase V: Enable Civil Authority—In phase IV, the joint force has assisted the reestablishment of governance by a host nation. In phase V, the joint force focuses on providing support to a legitimate, functioning government, one that it may have helped establish. The support to a functioning government is provided in accordance with some agreement. US military forces will typically work in coordination with other US government agencies, foreign government agencies, and NGOs and international government organizations (IGOs).

Missions by Phase Description

We identified 27 missions across the full spectrum of military operations (Table 14.1). These missions were culled from several joint staff, major command, and service documents. We reviewed Joint Publication 3-0 (2010), which describes the activities that commanders might anticipate in various phases of a theater campaign. We consulted the planning guidance and tasks assigned to combatant commanders in the *Joint Strategic Capabilities Plan, 2010* (Joint Chiefs of Staff 2008). We reviewed Joint Warfighting Center materials to identify tasks organized by type of operation. These tasks were developed for mission analysis and published in the *Universal Joint Task List*, 2002, a Chairman of the Joint Chiefs of Staff manual (Joint Chiefs of Staff 2002). Finally, we took future missions into consideration by reviewing scenario data published for each of the selected combatant commands in the multiservice force deployment documents promulgated by the joint staff.

The Sociocultural Capabilities

Across the globe, in commands, theaters, and campaigns, there is an increasing recognition that sociocultural knowledge and expertise are critical for the success of military missions. The US military and its allies are increasingly dealing with internal conflicts within states, responding to complex humanitarian emergencies and other situations far outside the scope of traditional state-on-state warfare. In these new operational environments, the enemy is often not a clearly discernible group and may change rapidly from day to day. Such environments call for new capabilities that can be used to assess and make productive use of operationally relevant information regarding the social and cultural landscapes of the area of operations. As recent campaigns have shown, this is no small feat. Human terrain is inherently local and by its nature rapidly evolving. Our next task then was to develop a list of SCCs necessary to meet this challenge.

What Is a Sociocultural Capability?

The first step in developing a list of required SCCs required developing a working definition of the term. We assembled a team of military experts and social science experts with applied research backgrounds in anthropology, psychology, political science, and public health.

The team conducted a literature search for descriptions of how sociocultural knowledge and research had been used to inform military operations, as well as calls for improvements in the use of sociocultural knowledge for future military operations. This involved consulting doctrine, military documents such as reports from HTT, academic journals and books, and existing RAND Corporation literature on the topic. In addition, we vetted the definition of SCCs with military and nonmilitary experts during one-on-one and group interviews we conducted over the course of the study. This process produced the earlier definition.

We further defined the italicized elements of the earlier definition in Table 14.2.

Our SCC definition includes a *know* element (knowing the social and cultural landscape) and a *do* element (using information to plan, prepare, execute, and assess a mission).

TABLE 14.2

Sociocultural Component Definitions

Component	Definition
Operationally relevant	Pertaining to the requirements of the supported unit and mission
Social landscape	A particular group's political, economic, institutional, kinship, and class structure and organization
Cultural landscape	Meanings, beliefs, attitudes, behavioral norms, values, and perceptions shared by and distributed among a group
Human resources	Resources that support an individual's or a unit's SSC development. Examples include interpreters/linguists, SMEs, tribal leaders, key figures in local governance, NGOs, social scientists
Material resources	Information storage and sharing systems; software for data analysis; local media for dissemination and influence, etc.

SOCIOCULTURAL CAPABILITY

An SCC is the ability to employ *human and material resources* to gather, synthesize, analyze, interpret, and share *operationally relevant* information about a population's *social and cultural landscapes* and use this to inform and assist the planning, preparation, execution, and assessment of a military operation.

SOCIOCULTURAL CATEGORIES

Sociocultural categories consist of planning, communication and coordination, gaining detailed cultural awareness, engaging the population, and assessing the impact of this engagement.

Modern social science views the social and the cultural as causally intertwined; thus, the term *sociocultural* is a reflection of the myriad causal forces and processes that influence patterns of thought and behavior in a society or a group.

Sociocultural Capability Categories

Based on interviews with approximately 30 military and nonmilitary experts and an extensive literature review, we reduced a rather long list of SCCs to a manageable set of 12, which we divided into five categories.

The categories were organized to approximate a chronological progression of how SCCs might be used throughout the planning and the execution of military missions. We describe the categories as follows, and Table 14.3 lists the associated required SCCs.

- Planning: The commander must have the ability to discover the features of the human terrain that are most operationally relevant—in other words, to know what needs to be known. Furthermore, he must have the ability to make use of sociocultural information to inform mission objectives. This requires the ability to identify relevant sociocultural information and use it to plan operations.
- Coordination and communication: One critical SSC is the capacity to communicate information and coordinate productively with other parties operating in the area. This requires the ability to communicate findings and information to both military and nonmilitary organizations and to work well with organizations in the region to plan collaborative activities.
- Gaining detailed cultural awareness: The most important and arguably most difficult capability is the capacity to obtain valid and reliable operationally relevant information regarding the social and cultural landscapes of an area. Some of the requirements in this category include the ability to acquire knowledge of the culture, social structure, political dynamics and economics of the area and how they interconnect, design, and administer efforts aimed at obtaining this information and verify the accuracy and the reliability of the information.

TABLE 14.3

Sociocultural Capabilities by Category

Category	Required SCCs	References
Planning	The ability to identify operationally relevant sociocultural information categories and requirements	Salmoni and Holmes-Eber 2008
	The ability to use sociocultural information to design missions and activities such as possess access to best practices for using sociocultural information to inform or alter mission objectives	Salmoni and Holmes-Eber 2008; Flynn et al. 2009; Medby and Glenn 2002
Coordination and communication	The ability to effectively communicate findings and information to both military and nonmilitary organizations	Shaley 2007
	The ability to store and establish access to information for multiple agencies and organizations	Department of the Army 2006; Kjeldsen 2006; Byman 2001
	The ability to work with organizations, both the United States and foreign, in the region to plan collaborative activities	Medby and Glenn 2002; Byman 2001
Gaining detailed cultural awareness	The ability to acquire knowledge of the culture, social structure, political dynamics, and economics of the area and how they interconnect, such as (1) understand where to go to in order to get information, (2) have access to documents describing the local culture, and (3) possess access to experts (HTTs, SMEs, civil affairs operatives, etc.)	Medby and Glenn 2002
	The ability to design and administer appropriate efforts to gain detailed situational awareness, such as (1) possess the ability to design and administer surveys, interviews, focus groups and conduct ethnographic observations and (2) possess the ability to conduct quantitative and qualitative analyses and make appropriate inferences	Morgan 1993
	The ability to verify the accuracy and the reliability of the information gained, such as (1) have procedures in place to cross-check the accuracy and the reliability of information and (2) update with new information as it becomes available	LeGree 2010; Medby and Glenn 2002
Engaging the population	The ability to communicate with the local population, such as (1) possess survival language skills for limited interactions; (2) possess the ability to communicate fluently for in-depth interactions; (3) have access to translators/interpreters if needed; and (4) if translators/interpreters are needed, understand how to use them appropriately	

(*Continued*)

TABLE 14.3 (CONTINUED)

Sociocultural Capabilities by Category

Category	Required SCCs	References
	The ability to productively interact with the local community, such as (1) understand the meaning and the significance of local culture; (2) possess awareness of appropriate social etiquette, with people and groups across the social hierarchy; (3) possess the ability to build effective relationships with local institutions and persons of influence or high status; and (4) possess locally appropriate influence, negotiation, and conflict resolution skills	Jandora 2006; LeGree 2010; Abbe and Halpin 2010; Jager 2007; Salmoni and Holmes-Eber 2008; Wunderle 2006; Abbe et al. 2008; Cushner and Brislin 1995; Cushner and Landis 1996; Stenmark 2006; Black and Porter 1991; Brett and Okumura 1998; Gelfand et al. 2000; Morris et al. 1998; Ohbuchi and Takahashi 1994; Sheer and Chen 2003; Sullivan et al. 1981; Tyler et al. 2000
Impact assessment	The ability to obtain reliable and relevant information on postengagement conditions to assess the efficacy of employing SCCs, such as (1) possess the ability to develop a credible sociocultural impact assessment plan, (2) possess knowledge of various assessment methods, (3) possess the ability to access trained personnel and resources required to conduct proper assessments, (4) possess the ability to assess the impact of US operations on the local sociocultural environment (may focus on balance of power, traditional practices, etc.), and (5) possess the ability to assess the impact of the sociocultural milieu on US operations	DoD 2011
	The ability to understand the implications of local changes for successive engagements, such as the ability to integrate relevant information and conclusions from impact assessment efforts into planning processes for successive missions	

SMART CARDS

When the United States invaded Afghanistan in 2001 and Iraq in 2003, "culture" was not part of the vocabulary of war. The US had established major military bases in Saudi Arabia, Qatar and, later, Kuwait following the 1990 Iraqi invasion of its neighbor to the south. Veterans of the subsequent Gulf war recall that certain units developed informational and training materials concerning Arab and Muslim societies, including a small pamphlet or "smart card." But this effort was fleeting (Davis 2010).

- Engaging the population: Increasingly, through the use of provincial reconstruction teams, female engagement teams, and the like, the US military is recognizing the need for and difficulty of engaging in sensitive interactions with the local population in a culturally appropriate manner.

This requires an ability to communicate with the local population, build trust, and productively interact with them. A number of initiatives attempting to address these challenges have been undertaken to include smart cards described earlier.

- Impact assessment: After SCCs have been employed in the pursuit of mission success, it is important to assess the impact of these efforts in order to inform future mission planning. This requires an ability to obtain reliable information on postengagement conditions and the ability to understand the implications of local changes for successive engagements. For example, the effectiveness of language training courses is routinely assessed. Within the Army, language tests are mandatory for soldiers who have received foreign language training at government expense (Department of the Army 2009). However, annual retesting of speaking skills is not currently required for all language-dependent military occupational skills. The defense language transformation road map sets proficiency level goals for language professionals (DoD 2005).

Sociocultural Capabilities by Operational Phase

With the missions and the required SCCs identified, the next step was to identify which SCCs were needed to accomplish the missions across all phases of military operations. We proposed to develop a matrix with the 27 missions as rows and the 12 SCCs as columns. The result was to be a matrix with 324 cells containing the manner in which the SCC column can assist in accomplishing the row mission. Figure 14.1 illustrates the matrix structure. We refer to the cell entries as *intersections*.

While useful for validation purposes and to produce a large base of raw data for further work, this level of resolution was found to be excessively specific and therefore not likely to be useful for military planners or other operatives across the diversity of conditions found in different combatant commands and operations. Thus, once we created the set of mission activities

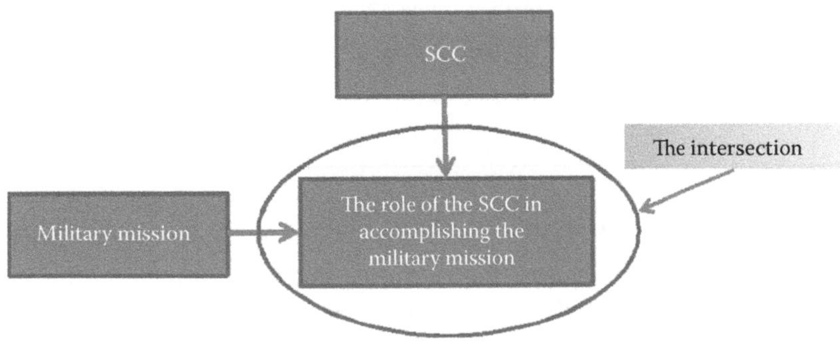

FIGURE 14.1
Intersections of SCCs by mission activities.

TABLE 14.4

Sample Intersections of Phases by SCCs

SCC Categories	Planning	Coordinating and Communicating
Operational Phases		
Phase 0: Shape	Develop compelling messages aimed at both potential adversaries and supporters that decreases negative and increases positive perceptions of US actions in the region, minimizing the use of threats or punitive language if possible.	Gain access to appropriate communications media to target specific audiences in ways that reach them in time to influence their response.
Phase I: Deter	Develop messages that inform the population of (1) how the US military's demonstration of resolve deters enemy activity and (2) how the preparatory actions they are observing contribute to a capable joint force.	Disseminate messages developed using a wide range of media to reach a diverse audience to include the local government, the military, and the local population.
Phase II: Seize the initiative	Develop the commander's critical information requirements aimed at understanding the local population demographics and cultural sites to help avoid unnecessary collateral damage. Develop messages to inform the population of the military operations to the extent possible.	Disseminate messages that inform the local population of the operations underway. Aim at convincing the populace that the military will not deliberately destroy homes and culturally significant sites.
Phase III: Dominate	Advise the commander to avoid collateral damage when conducting combat operations by identifying culturally significant sites as well as major population centers. Develop messages to inform the population of military operations to the extent possible.	Disseminate messages that inform the local population of the operations underway. Aim at convincing the populace that the military will not deliberately destroy homes and culturally significant sites.
Phase IV: Stabilize	Advise the commander on ways to engage the population that account for cultural sensitivities. Provide humanitarian assistance and engage in nation-building.	Develop messages that convince the local population that the US humanitarian assistance and nation-building efforts are in their best interests.
Phase V: Enable civil authority	Partner with and provide support to the legitimate government while continuing to provide humanitarian assistance.	Disseminate messages developed using a wide range of media to reach a diverse audience to include the local government and military and the local population.

and SCCs described earlier, we created a diminished matrix consisting of the six phases as rows and the five categories as columns. This resulted in a more manageable matrix consisting of 30 cells. The SCC categories and phases intersect at a more general level. Table 14.4 illustrates the process by depicting two of the five category columns and all six of the operational phases.

Gaps and Mitigation Measures

Our research consisted of reviewing the pertinent sociocultural literature and military documents in addition to conducting several interviews of military and nonmilitary personnel. In the process, we identified several gaps in the provision of SCCs, organized along our five major categories.

TABLE 14.5

SCC Gaps and Mitigation Measures

	Doctrine		**Organization**					**Training and Education**				**Material**	**Leadership**			**Personnel**				
Gaps	**1**	**2**	**1**	**2**	**3**	**4**	**5**	**1**	**2**	**3**	**4**		**1**	**2**	**3**	**1**	**2**	**3**	**4**	**5**
Planning																				
Lack of appropriate personnel				X		X							X	X		X	X	X		X
Plans not sensitive to local context			X		X	X	X	X			X	X								
Plans based on hunches			X		X	X	X	X			X	X								
Kinetic bias	X	X															X			
Coordination and Communication																				
Problems within the military							X					X								
Ineffective outreach			X		X		X	X		X								X		
Military versus nonmilitary conflict					X	X				X								X		
Gaining Detailed Cultural Awareness																				
Unsure where to access				X		X	X	X				X	X							X
Poor or lack of validity checking						X						X							X	
Unclear roles													X							
Experts not leveraged				X	X	X		X							X					X
Lack of culture-general tools										X										
Poor local area of interest information			X		X		X	X			X	X			X					X
Lack of methodological training and expertise						X					X	X	X	X		X	X	X		
Lack of appropriate personnel				X		X							X			X	X	X		X

(Continued)

TABLE 14.5 (CONTINUED)

SCC Gaps and Mitigation Measures

	Doctrine		Organization					Training and Education				Material	Leadership			Personnel				
Gaps	**1**	**2**	**1**	**2**	**3**	**4**	**5**	**1**	**2**	**3**	**4**		**1**	**2**	**3**	**1**	**2**	**3**	**4**	**5**
Engaging the Population																				
Poor basic language skills			X					X						X	X	X	X	X		
Lack of fluency			X										X	X		X	X	X		
Too few interpreters and translators									X									X		
Poor etiquette			X					X							X			X		
Poor relationship-building methods			X		X		X	X		X								X		
Poor conflict resolution ability			X				X	X		X				X	X	X		X		
Negative perceptions of military			X		X		X	X		X								X		
Impact Assessment																				
Not incorporated in standard operating procedures	X	X										X								
Lack of appropriate personnel				X		X					X		X			X	X	X		X
Poor/biased assessments						X					X	X							X	
No assessments of SCC training						X													X	

TABLE 14.6
Gap Mitigation Measures

DOTMLPF Category	No.	Gap Mitigation Measures
Doctrine	1	Incorporate SCC information into long-term strategic planning.
	2	Develop doctrinal publication that provides guidance on how to do the above, and incorporates lessons learned from employment of SCCs over the last decade.
Organization	1	Repeat deployment to same local area.
	2	Utilize resident SCC expertise.
	3	Expand and systematize cooperation with NGOs and IGOs.
	4	Systematize collaborative links with colleges, universities, research centers/think tanks.
	5	Increase pass-through of sociocultural information during RIP/TOAs.[a]
Training and education	1	Conduct predeployment training in sociocultural aspects of local areas of interest.
	2	Train for the proper use/caveats of using translators and interpreters.
	3	Develop training in culture-general skill sets (nonverbal communication, cultural relativism, empathy, etc.).
	4	Have a training track in data acquisition, verification, analysis, interpretation.
Material		Develop standardized field-based platforms for data acquisition (geospatial, survey, social network, etc.), as well as both field-based and centralized standardized analysis and interpretation tools with broad accessibility.
Leadership	1	Have education tracks for senior military leaders in social science and related fields.
	2	Bolster existing education efforts in SCC areas in academies and civilian schools.
	3	Support commanders' individual initiatives for predeployment training in sociocultural arena.
Personnel	1	Hire tracks to aid recruitment of military personnel with SCC skills.
	2	Have career tracks that incentivize the development of SCCs and balance rewards for kinetic and nonkinetic skills.
	3	Include assessments of SCC-related skills in officers' performance evaluations.
	4	Develop review boards to assess the quality of sociocultural information, education, and training.
	5	Develop a reference list of regional and thematic experts for consultation on sociocultural issues.

[a] RIP/TOA stands for relief in place/transfer of authority. It is the process whereby deployed units are replaced by incoming units in an area of operations such as Afghanistan or Iraq.

DOTMLPF

DOTMLPF means doctrine, organization, training and education, material, leadership, personnel, and facilities.

We noted that not all gaps constituted a complete lack of capability. Some were due to limited access to required capabilities or limited knowledge that such capabilities existed. In addition, competing priorities resulted in precluding access. The following is a summary of our major findings and recommendations. Table 14.5 lists all the gaps we uncovered in this research along with the measures we recommended to mitigate the gaps arranged by DOTMLPF. We have omitted facilities from the list, as it had not been mentioned in any of our interviews. The numbers at the head of each column refers to the coded gap mitigation measure listed by DOTMLPF category in Table 14.6.

Megacity Use Case

Peacekeeping Operations in Çawaila

Violence has erupted in the north wards as the Razinis attempt to gain absolute control of the wards and threaten to expand their aggression throughout the megacity of Xanadu. The government of Çawaila fears that the violence will lead to a full-scale insurgency. Çawailan forces are weak, prompting the government to ask the UN for a peacekeeping force to suppress the Razini militants. A UN resolution was passed authorizing a peacekeeping force, and the United States has agreed to lead it. US forces are now preparing to deploy to megacity Xanadu, and they are planning operations to include using their SCCs to subdue the insurgents and gain the support of the local population in the rebelling wards.

Operation Restore Hope consists of one mechanized brigade with various support troops commanded by Major General Adam Belfer. MG Belfer views the operations in Xanadu to consist of phase III (dominate) and phase V (stabilize) operations in that order. Consequently, he sees his immediate objectives as consisting of (1) gaining control of the situation in the north wards, (2) preventing the instability from spreading to other parts of the city, (3) disrupting Razini violence, and (4) convincing the local population in Xanadu to support US forces in the area. He feels that these tasks need to be accomplished before stability operations can begin.

MG Belfer recognizes that to be successful, he needs the support of the local population. The insurgent Razinis have an advantage in that they are part of Xanadu's polyglot society, whereas the US forces are outsiders, likely to pack up and leave at any time. Consequently, MG Belfer turns to his civil affairs team and his human terrain specialists to advise him during the planning and execution phases of his mission. His sociocultural team sees their major tasks to be the following:

- Assisting the commander plan operations: The military aspects of the operation in the north wards can be very easily hampered if the troops confront a hostile population. The sociocultural team is therefore working hard to gather information about the culture in the area and craft messages aimed at both the insurgent Razinis and the supporters of US intervention that decreases negative and increases positive perceptions of US actions.
- Informing the forces of local cultural norms: Although the commander understands the sociocultural imperative, not all the troops do—after all, they were trained to fight, not act as social workers. The sociocultural team therefore needs

to craft messages that transmit their findings about the local cultural environment to the military and other nonmilitary organizations that may be operating in the area. They must also advise the forces on how to work with other organizations.

- Understanding the local culture: The multiple ethnic groups in Xanadu greatly complicate this task. In addition, dealing with the dozens of languages and dialects makes coordination and communications problematic. The sociocultural team needs to canvass the population to understand their grievances. They do this using surveys and focus groups and by conducting ethnographic observations. What they learn directly informs military operations.
- Working with the local population: This last task can be difficult in that it is not always possible to identify the leaders (religious or secular) in a polyglot population like Xanadu. The language skills alone needed to effectively engage the population can be daunting. Interpreters and translators can be used, but their allegiances can sometimes taint their work. This can hamper the command's ability to build effective relationships with local institutions and persons of influence.

Although the sociocultural aspects of the operations were problematic, MG Belfer's force was able to control the situation in the north wards and prevent the insurgency from spreading. The peace, however, was fragile. Competing ethnic groups in the aging refugee camps could cause another eruption at any time. Consequently, MG Belfer now turned his attention to stabilization. In this phase, he viewed his objectives as (1) working with the local Çawaila and Xanadu governments to improve their governance of the city, (2) provide some relief to the refugee camps by providing basic services and rebuilding needed infrastructure, (3) help the Çawaila and Xanadu governments build a better law enforcement and justice system.

This phase of the operation required even more support from the sociocultural team. Almost all military activities in this phase consist of interactions with the local populations and governments.

- Improving governance requires interactions with identified local leaders and influential groups. The objective is not to install a democratic government, but one that is acceptable to the population. This means one that can effectively and even handedly deal with the large refugee population. The military leaders interacting with these entities must understand the local etiquette as well as the cultural norms.
- Providing basic services is largely a logistics and engineering operation, but to be successful, local contractors and laborers need to be involved in the work. Like improving governance, this requires interactions with the local population—this time local contractors. Because money is involved, rivalries can erupt into serious conflicts rather quickly. The military personnel dealing with these issues need to understand the local views on equity when dealing with contract awards. They also need to understand local sensitivities to employing various ethnic groups to perform the work.
- Building a law enforcement and a justice system again means working with locals—this time with existing law enforcement personnel and local courts. In Xanadu, the law enforcement is in shambles (as witnessed by the unchecked mafia-like operations of the Razinis) with no one in control except for the US military. Civil justice

is equally fractured with each ethnic group parceling out justice according to their customs. Rebuilding will require the identification of trusted leaders in the community able and willing to cooperate with the US military and other organizations to create fair and equitable law enforcement and civil justice system. The sociocultural team will advise the trainers how to deal with these leaders. Creating a disciplined, multiethnic police force requires considerable cultural awareness.

The US peacekeeping force is still in Xanadu. They are making progress, but ethnic tempers continue to flare now and then. Some semblance of order has been restored to the government; new infrastructure projects are underway; and the local police are actually patrolling the wards. Operation Restore Hope, although not a resounding success, has managed to considerably improve conditions in Xanadu. And the mission continues.

Conclusion

On the whole, we found that substantial progress has been made in terms of building SCCs, and that when gaps emerged, there were at least limited attempts to address unmet needs. Personal initiative accounted for a large percentage of these ad-hoc efforts and attests to the growing appreciation among military personnel for the important role of SCCs in achieving mission goals. Although we provide recommendations for how to fill specific gaps, finding ways to encourage and support personal initiatives will not only help meet sociocultural needs as they arise in specific locales, but also nurture the development and testing of original ideas that could be used to address sociocultural needs more broadly. Issues that on the surface appeared unrelated to sociocultural issues were often the main impediments to building greater sociocultural knowledge and skill sets: a position's promotion potential, the length and repetition of deployments to an area, the commissioning of assessments, and the quality of the rotation process. Resolving these overarching issues would likely produce marked improvements over multiple SCC categories and as such might be cost-effective areas for focusing future SCC reform efforts.

Our research team used the DOTMLPF framework to provide recommendations to mitigate the gaps (Table 14.6).

- Doctrine: We recommend a doctrinal or doctrine-like publication that focuses on lessons learned in the provision of SCCs during the past decade, as well as the use of SCCs in long-range strategic planning.
- Organization: We recommend organizing repeated deployments to the same local areas as well as improving processes during relief in place/transfer of authority to transfer sociocultural knowledge and ensure the proper hand off of relationships with key leaders. As well as building greater sociocultural capacities within the armed forces, we recommend expanding and systematizing relationships with NGOs and IGOs, as well as universities and think tanks that provide access to SCCs.
- Training: We recommend expanding the provision of sociocultural knowledge specific to the precise area predeployment (rather than regional or national training), expanding training in the use of interpreters and translators, and also

expanding training in culture general skill sets, including empathy, suspension of judgment, and ability to adapt to new cultural environments.

- Material: We recommend establishing systems for the standardized acquisition, analysis, and interpretation/validation of both existing sources of information and new streams of data from units deployed and collecting data in the field. As a model, we can point to the Serengeti system being developed by the AFRICOM.*
- Personnel: First, we suggest instituting specialized hiring efforts and promotional tracks to both attract personnel with sociocultural skill sets and to reward the expansion of these skills during service members' careers. We also recommend making SCC-related parameters a systematic part of officers' performance evaluations and implementing efforts to monitor and assess sociocultural training and education.

DISCUSSION QUESTIONS

The following are questions and our recommendations for discussion. The questions focus on the identified SCCs needed to support all phases of military operations and the recommendations focus on closing the gaps implicit in those questions.

1. Planning: What are the sociocultural knowledge challenges associated with planning military operations? What SCCs are needed to support strategic planning? How can sociocultural knowledge of the local population improve mission and activity planning?
2. Coordination and communication: What are the cross-cultural communications gaps with regard to interactions between the military and non-military organizations? What cross-cultural communication difficulties exist when coordinating among different military entities? What problems arise when collaborating with IGOs and NGOs?
3. Gaining detailed cultural awareness: How can we better provide access to up-to-date, verified information about specific areas of interest? What systematic approaches and systems for obtaining, analyzing, storing, and sharing sociocultural information should be institutionalized in the military? What additional military and nonmilitary methodological and area expertise is needed to cover the wide range of areas of interest covered by the military?
4. Engaging the population: How can the military increase training in languages and/or use of interpreters and translators? How can the military create more capacity to build productive relationships with local populations, engage local leaders, and resolve conflicts in culturally appropriate ways?
5. How can impact assessments for the provision of SCCs be systematically incorporated in standard operating procedures? How can military sociocultural education and training be systematically assessed?

* Serengeti is an Africa-centric data repository that is populated with open source data according to a specific architecture and taxonomy. It was developed in AFRICOM's knowledge development division. See Lee (n.d.).

References

Abbe, A., and S. M. Halpin. The cultural imperative for professional military education and leader development. *Parameters*, Vol. 39, No. 4, pp. 20–31, Winter 2009–2010.

Abbe, A., L. M. Gulick, and J. L. Herman. Cross-cultural competence in army leaders: A conceptual and empirical foundation. ARI Study Report No. 2008-01. Arlington, VA: US Army Research Institute for the Behavioral and Social Sciences, 2008.

Black, J. S., and L. W. Porter. Managerial behaviors and job performance: A successful manager in Los Angeles may not succeed in Hong Kong. *Journal of International Business Studies*, Vol. 22, pp. 99–113, 1991.

Brett, J. M., and S. Okumura. Inter- and intracultural negotiation: US and Japanese negotiators. *Academy of Management Journal*, Vol. 41, pp. 495–510, 1998.

Byman, D. Uncertain partners: NGOs and the military. *Survival*, Vol. 43, No. 2, pp. 87–114, Summer 2001.

Cushner, K., and R. W. Brislin. *Intercultural Interactions: A Practical Guide*. Second ed. Thousand Oaks, CA: Sage Publications, 1995.

Cushner, K., and D. Landis. The intercultural sensitizer. In *Handbook of Intercultural Training*, Thousand Oaks, CA: Sage Publications, pp. 185–202, 1996.

Davis, R. Culture as a weapon system. *Middle East Report*, Vol. 255. As of July 20, 2010: http://www.merip.org/mer/mer255/davis.html, Summer 2010.

Department of the Army. Counterinsurgency. FM 3-24/MCWP 3-33.5. Washington, DC: Department of the Army, 2006, December.

Department of the Army. Army foreign language program. Army Regulation (AR) 11-6, Washington, DC: Department of the Army, 2009, August 31.

DoD (Department of Defense). *The Defense Language Transformation Roadmap*. Arlington, VA: DoD, 2005, January.

(DoD). DoD regional and cultural capabilities: The way ahead. *Department of Defense June 2007 Summit*. As of February 25, 2011: http://www.culture.af.edu/PDF/CulturalCapabilities.pdf.

DoD. *Strategic Plan for Language Skills, Regional Expertise, and Cultural Capabilities 2011–2016*. Arlington, VA: DoD, 2011, February.

Flynn, M. T., M. Potting, and P. D. Batchelor. *Fixing Intel: A Blueprint for Making Intelligence Relevant in Afghanistan*, Washington, DC: Center for a New American Security, 2009.

Gelfand, M. J., and N. Dyer. A cultural perspective on negotiation: Progress, pitfalls, and prospects. *Applied Psychology: An International Review*, Vol. 49, pp. 62–100, 2000.

Holiday, H. *Improving Cultural Awareness in the US Military*. Carlisle, PA: US Army War College, 2008.

Jager, S. M. On the uses of cultural knowledge. *Monograph*. Carlisle, PA: Strategic Studies Institute of the US Army War College 2007.

Jandora, J. Military cultural awareness: From anthropology to application. Landpower Essay No. 06-3, Arlington, VA: Institute of Land Warfare, 2006, November.

Joint Chiefs of Staff. Universal joint task list. CJCSM 3500.04C. Washington, DC: DoD, 2002.

Joint Chiefs of Staff. *Joint Strategic Capabilities Plan*. Washington, DC: DoD, 2008.

Joint Chiefs of Staff. *Joint Operations*. Joint Publication 3-0. Washington, DC: DoD, 2010, March.

Kjeldsen, S. *Training Impact Assessment of the United Nations Civil-Military Coordination (UN-CMCoord) Course*. Geneva: United Nations Office for the Coordination of Humanitarian Affairs, 2006.

Lee, B. A. US Africa command: Delivering cultural competence. Briefing. IKD Knowledge Development Division, SERENGETI, US Africa Command. Stuttgart, Germany, (n.d.).

LeGree, L. The battle for the minds: IO and COIN in the Pashtun Belt. *Military Review*, pp. 21–31, 2010, September–October.

Medby, J. J., and R. W. Glenn. Street smart: Intelligence preparation of the battlefield for urban operations. MR-1287-A. Santa Monica, CA: RAND Corporation. As of February 25, 2011: http://www.rand.org/pubs/monograph_reports/MR1287.html, 2002.

Morgan, D. L. ed. *Successful Focus Groups: Advancing the State of the Art.* Thousand Oaks, CA: Sage Pubilcations, 1993.

Morris, M. W., K. Y. Williams, K. Leung, R. Larrick, M. T. Mendoza, D. Bhatnagar, J. Li, M. Kondo, J. Luo, and J. Hu. Conflict management style: Accounting for cross-national differences. *Journal of International Business Studies*, Vol. 29, pp. 729–747, 1998.

Ohbuchi, K., and Y. Takahashi. Cultural styles of conflict management in Japanese and Americans: Passivity, covertness, and effectiveness of strategies. *Journal of Applied Social Psychology*, Vol. 24, pp. 1345–1366, 1994.

Salmoni, B. A., and P. Holmes-Eber. *Operational Culture for the Warfighter: Principles and Applications.* Quantico: Marine Corps University Press, 2008.

Shaley, O. Constructing multi-dimensional perspective of multiculturalism in armed forces. In Cees M. Coops & Tibor Szvircsev Tresch, eds., *NATO Defense College: Cultural Challenges in Military Operations*. Rome: NATO Defense College, pp. 130–140, 2007.

Sheer, V. C., and L. Chen. Successful Sino-Western business negotiation: Participant's accounts of national and professional cultures. *Journal of Business Communication*, Vol. 40, pp. 50–85, 2003.

Stenmark, T. *Looking for Gold Nuggets in the Melting Pot: Language, Cultural Awareness, and the Fourth Generation Warrior*. Maxwell Air Force Base, AL: Air Command and Staff College, 2006, April.

Sullivan, J., R. B. Peterson, N. Kameda, and J. Shimada. The relationship between conflict resolution approaches and trust—A cross cultural study. *The Academy of Management Journal*, Vol. 24, pp. 803–815, 1981.

Tyler, T. R., E. A. Lind, and Y. J. Huo. Cultural values and authority relations: The psychology of conflict resolution across cultures. *Psychology, Public Policy, and Law*, Vol. 6, pp. 1138–1163, 2000.

Wunderle, W. D. *Through the Lens of Cultural Awareness: A Primer for US Armed Forces Deploying to Arab and Middle Eastern Countries*. Fort Leavenworth, KS: Army Combined Arms Center: Combat Studies Institute Press, 2006.

15

Methods to Characterize and Manage Uncertainty for Sociocultural Applications

Perakath Benjamin, Kalyan Vadakkeveedu, and Satheesh Ramachandran

CONTENTS

Motivations ... 290
Characterizing Uncertainty ... 291
  Data Uncertainties ... 292
  Uncertainty Representation ... 292
  Uncertainty Caused by Communication Mechanisms ... 295
  Data to Decisions—Characterizing How Uncertainty Propagates ... 295
    Propagation of Uncertainty in the Data-to-Decisions Process ... 297
    Data Preprocessing ... 297
    Data Processing ... 297
    Information Processing ... 297
    Knowledge Refinement ... 297
Reducing and Managing Uncertainty ... 298
  Measuring and Reducing Uncertainty in Sociocultural Models ... 298
    Information Theoretic Methods for Characterizing and Reducing Uncertainty ... 299
    Reducing Uncertainty Using Bayesian Fusion ... 303
  Identifying and Resolving Inconsistencies ... 304
    Source Weighting ... 304
    Reasoning-Based Methods for Resolving Conflicts ... 305
Megacity Use Case ... 308
  Introduction ... 308
  Using Multisource Analytics and Uncertainty Reduction for Crisis Response Management in Xanadu ... 308
    Illustrative Xanadu Crisis Management Scenario ... 308
  Example Crisis Response Management Application Architecture ... 309
  Reducing Uncertainty Using Bayesian Fusion to Support Enhanced Crisis Response Situational Awareness ... 309
Summary and Conclusion ... 312
References ... 313

ABSTRACT Driven by the premise that reducing uncertainty in the sociocultural decision information will increase trust, this chapter advocates the use of uncertainty reduction and uncertainty management methods for sociocultural applications. The chapter first describes the importance of uncertainty as an important determinant of trust for sociocultural model-based decision making. Methods to characterize the uncertainty and the propagation of uncertainty through the data-to-decisions process are described. Next, several uncertainty reduction and resolution methods are outlined. Finally, an example application scenario is presented that shows the utility of uncertainty management methods for sociocultural applications. The central value proposition is that the reduction of uncertainty in sociocultural decision information will likely lead to better and more timely decision making.

KEY TOPICS

- Trust and uncertainty are related concepts, important to sociocultural models.
- Characterizing uncertainty is key to the design and the evaluation of models.
- Uncertainty reduction and management can be accomplished by information fusion methods.
- Example application demonstrates uncertainty modeling and reduction methods in the real world.

Motivations

This chapter is motivated by the need for better methods and capabilities for determining and managing trust in sociocultural models and applications. Trust is an overloaded concept that has been studied by several communities of interest, social science, psychology, economics, information and data sciences, and management science (Hirshleifer and Riley 1992; Jøsang and Presti 2004; Robinson 1996; Watson and Moran 2005). Trust is a complex and multifaceted concept and is related to concepts such as risk, uncertainty, security, reliability, and reputation.

We focus attention on the aspects of trust that help improve human decision making. Furthermore, we focus on the aspects of trust that involve the use of models, data, information, and knowledge used to make decisions.

We contend that trust and uncertainty are related concepts: any reduction in uncertainty associated with a model is directly proportional to the level of trust associated with the model. This chapter focuses attention on the technical problems of modeling and managing uncertainty associated with the data, the information, and the models that are used for sociocultural applications. A central idea that motivates this chapter is this: unquantified uncertainty in models implies untrustworthy models. Therefore, quantifying uncertainty using theoretically sound methods is a necessary step to understanding, reducing, and managing uncertainty and trust in sociocultural models. The remainder of the chapter will give attention to the characterization and the management of uncertainty for sociocultural applications.

UNCERTAIN IS UNTRUSTWORTHY

The primary motivation for this chapter is the idea that unquantified uncertainty in models implies untrustworthy models.

There are many proposed definitions of uncertainty in the literature (Ayyub and Gupta 1994; Henrion and Shachter 1990; Morgan and Henrion 1998). Uncertainty with respect to an attribute refers to the possible deviation from the attribute's true value but without precise knowledge of the magnitude. Uncertainty encompasses several concepts such as error, accuracy, precision, validity, quality, variability, noise, completeness, confidence, and reliability.

Uncertainty exists in models for different reasons. First, the techniques used to build the model may have assumptions that are invalid. For example, mathematical models are often simplified representations of the phenomena being studied. Note that good modeling practices seek to induce the greatest simplifications while providing an adequately accurate representation of the processes affecting the phenomena of interest. However, oversimplifications lead to unanticipated uncertainties in the model.

Second, the model uncertainties are often driven by the inherent uncertainties in the data used to build the model. That is, uncertainties may arise because of the natural variability in the phenomena being measured by these data. Data-intensive application situations such as planning and intelligence analysis are often fraught with significant uncertainty.

Lastly, model uncertainties occur through the process by which data are converted to information and knowledge that is used for decision making (data operations, information processing and interpretation, classification, etc.). In this chapter, we provide a description of how uncertainties build and propagate from the attributes of the raw data, through data processing, information processing, and knowledge refinement.

Uncertainty modeling has a particularly rich history in scientific applications in which error-prone measurements or incomplete sampling often result in uncertain data. In economic and other predictive models, it is common for uncertainty to result from approximate projections about future behavior. Many times, even when precise data are available, uncertainty may be intentionally introduced to conserve system resources while data are being transmitted or processed (data condensation, glossy data compression). Another characteristic of uncertainty is that, because of the resolution of the data set or the analysis, it is not necessarily an absolute.

Why conduct uncertainty analysis? In many complex sociocultural systems, decision theoretic models are the imperative given the large number of factors and decision variables. In addition, many of these factors are either unknown or are characterized by imperfect knowledge. A systematic uncertainty analysis provides insight into the level of confidence in the estimates provided by the decision theoretic models and can aid in assessing how various possible model estimates should be weighed. Furthermore, it can lead to the identification of the key sources of uncertainty (such as data gaps), which merit further investigation, as well as the sources of uncertainty that are not important with respect to a given response. The purpose of the quantitative uncertainty analysis is to use currently available information for quantifying the degree of confidence in the existing data and models.

Characterizing Uncertainty

This chapter focuses attention on sociocultural models and applications that are significantly influenced by data. We will describe how uncertainties arise in data and how these uncertainties propagate to the information, knowledge, and models and applications that are derived from the data.

Data Uncertainties

Data uncertainties occur at what we refer to as the *source level* (for example, an imagery sensor versus a radar sensor). Uncertainties are often described in terms of source credibility and reliability. With human data sources (e.g., a news reporter), the reputation of the person is a significant consideration in quantifying the uncertainty in the gathered information and conveyed by this type of source.

Once we get past the data source, data uncertainties are influenced by factors relative to how the phenomena are sensed and how the data are acquired and stored. For sensor data types, it depends on the type and level of sophistication of the sensor relative to the phenomena being measured (e.g., cameras for visible and infrared imagery). We focus attention in this chapter on unstructured data (text data) that are sensed and recorded by human sensors. There are several distinctive characteristics of human sensors leading to unique types of errors associated with the (human) collection, storage, and conveyance of text data. The causes for the uncertainty with human sensor-derived text data include human error, misinformation, bias, differences in human capabilities for sensing, perceiving, and communicating what is observed.

Data uncertainties may arise from measurement errors; these in turn can involve (1) random errors, (2) systematic biases, or (3) inaccuracies in the assumptions used to infer the actual quantity of interest from a surrogate or a proxy variable. Other potential sources of data uncertainties in estimates of parameters include misclassification, estimation of parameters through small data sets, and estimation of parameters through nonrepresentative data sets. Furthermore, the uncertainty in a model application arises from uncertainties associated with measurement data used for the model evaluation. While dealing with uncertainty (and making decisions under its presence), there is a significance in understanding the nature as well as the typology of uncertainty.

Lastly, we note that there is an important challenge that is currently unaddressed in sociocultural models and applications: the problem of quantifying the uncertainty associated with information that is jointly derived from multiple and disparate sources of data. For example, suppose that there are two sources of data, D1 and D2, with associated uncertainties, U1 and U2, respectively. If D1 and D2 are used to generate information that is used in sociocultural model M, how will the uncertainty associated with M (say U (M)) be computed? Intuitively, U (M) will be dependent on U1 and U2. The section "Reducing and Managing Uncertainty" of this chapter will describe uncertainty reduction methods that provide an approach to address this problem.

Uncertainty Representation

From a modeling perspective, uncertainty is the incompleteness in knowledge (in either information or context) that causes model-based predictions to differ from reality in

UNCERTAINTY CATEGORIES

Uncertainty is prevalent in most decision making environments; the choice of how it is characterized is critical to better decision making. Certain representation methods are more suitable when the uncertainty is more objective (probability theory methods), others are more suitable when the uncertainty is more subjective (evidence theoretic and Bayesian methods), and some are more suitable when the uncertainty is vague (rough set theory, fuzzy set theory).

a manner described by some distribution function. It is clear from this definition that uncertainty implies the possibility of multiple outcomes and an ability to mathematically model these ranges of possibilities. Overall, there are various approaches for representing uncertainty, and the type of method that is suitable for a particular situation depends on the deeper semantics associated with what uncertainty refers to within the domain. Uncertainty within an environment could be classified as (1) epistemic (systematic)—if it results from a limited knowledge of the agent that generates the assertion or (2) aleatory (statistical)—if it is intrinsic in the observed world. Depending on the features of the source that generates the uncertain observations, it could also be classified as (1) objective, if the uncertainty derives from a repeatable observation, and (2) subjective, if the uncertainty in the information is derived from an informal evaluation. Furthermore, uncertainty could also be semantically categorized based on several criteria including ambiguity, inconsistency, vagueness, incompleteness, and empirical evidence. The appropriate representation method will depend upon the type of uncertainty that is to be modeled and how closely the representation methods match the requirements of the domain in terms of data, complexity, prior knowledge, etc. We provide a brief overview of different uncertainty modeling methods:

- Classical set theory: One approach to express uncertainty is through sets that represent mutually exclusive possibilities (possible outcomes, events, measurement, etc.). Here, the uncertainty is represented in terms of the nonspecificity (multiple possible outcomes) that is represented by the set. Larger sets (greater possible outcomes) correspond to less specific predictions (greater nonspecificity) than smaller sets (lesser possible outcomes). Full specificity is achieved only when one outcome is possible.
- Probability theory: Perhaps the most widely used method for representing uncertainty is through probabilistic measures. Here, uncertainty is expressed in terms of a measure on subsets of a universal set of possibilities, also called the sample space. The uncertainty measure is a real-valued function that maps every subset of the universal set to the 0–1 interval. This mapping for any subset, called *probability of the subset*, expresses the likelihood for the subset; in other words, the likelihood that the desired outcome or event belongs to the subset. Here, the uncertainty arises from the conflict among on-intersecting (mutually exclusive) subsets of the universal set that has a positive probability. These sets conflict as they are each possible since they have greater that zero likelihood, yet only one of these subsets can be true—and that is how the uncertainty is characterized or represented. Reasoning systems such as Bayesian networks (BNs) leverage probabilistic measures as the underlying uncertainty representation mechanism.
- Evidence theory: Evidence theory, also called Dempster–Shafer (DS) theory, proposed by Dempster (Dempster 1967) and Shafer (Shafer 1976) provides another framework that allows for representation and reasoning with incomplete information. It is designed to deal with the distinction between uncertainty and ignorance by (1) rather than computing the probability of a proposition, it computes the probability that the evidence supports the proposition, and (2) rather than estimating probabilities, it uses belief intervals to estimate how close the evidence is to determining the truth of a hypothesis. There are three basic functions in DS theory:
    - The belief mass function—The belief mass function specifies the belief mass distribution over all possible subsets of a frame of discernment (set of all

hypotheses). The belief mass function is similar to the probability distribution function except with one important difference—the values are assigned to sets of possibilities rather than single events; in other words, it assigns masses to all the nonempty subsets of the propositions that compose a system.

- The belief function—The belief in a set of elements, say subset A of the overall set of hypotheses, represents the total belief that one is based on the evidence obtained. It is the sum of all the belief masses assigned to elements that are contained in set A and the belief mass assigned to set A itself. Unlike the probability theory, Bel(A) = 0 represents the lack of evidence about A, whereas Prob(A) = 0 represents the impossibility of A.
- The plausibility function—The plausibility in a set, say A of a frame, represents the maximum possibility that set A is true given all the evidence. It is equal to the sum of the belief masses over all the subsets that have nonzero intersection with the set A. Overall, the evidence theory allows for the representation of ignorance and allows the reasoning system to be skeptical.

- Fuzzy set theory: In classical set theory, an element either belongs to the set or it does not belong to the set; in other words, elements have a purely bivalent membership function vis-a-vis the set. By contrast, fuzzy sets [13] permit the graded assessment of the membership of elements in a set, which is prescribed through a membership function valued in the real unit interval [0, 1]. Fuzzy sets are capable of expressing nonspecificity as well as vagueness (ambiguity). Nonspecificity relates to the imprecision connected to set sizes (cardinality). Vagueness on the other hand relates to the imprecision of definitions, particularly imprecisions that are inherent in the definitions of linguistic terms. In fuzzy sets, the membership is not purely in terms of presence (belongs to the set) or absence (does not belong to the set), but rather expressed in terms of the degree of presence/absence, and this degree of presence/absence is the way that uncertainty is characterized or represented.
- Fuzzy measure theory: The fuzzy measure theory (Lin and Lee 1996) is the extensions of classical measure theory (which provides the foundations for advanced probabilistic theories), where the measure in the classical sense is a systematic approach to determining the size of any subset of the overall sample space. At a very high level, the fuzzy measure theory relaxes some of the assumptions of the classical measure. In particular, these more generalized measures involve relaxing the additive property by the weaker property of monotonicity. The fuzzy measure theory introduces several special classes of measures, each of which is characterized by a special property. These measures used include plausibility and belief measures, as well as classical probability measures. The main difference between the fuzzy measure theory and the fuzzy set theory is that within the fuzzy set theory, the conditions for the membership of an element into a set are vague, as opposed to the fuzzy measure theory where the conditions are precise, but the information about an element is insufficient to determine whether it satisfies those conditions.
- Rough set theory: A rough set is an imprecise representation of a crisp set in terms of two subsets, a lower approximation and an upper approximation. Rough sets and fuzzy sets are different strands of the same overall approach to generalize classical set theory to model uncertainty. A rough set can be viewed as a set that is coarsely described, whereas a fuzzy set is a class (collection of sets) with unsharp boundaries. The lower and upper approximations for a rough set could itself be imprecise or fuzzy.

Uncertainty Caused by Communication Mechanisms

The process of conveying data and information between different agents and components within a sociocultural model-based system introduces additional uncertainties. The quality and reliability of the communication has the potential to degrade the reliability and the robustness of the systems that depend on these types of communication.

In sociocultural models, the term *communication* refers to either direct human-to-human communication or through digital mediums such as telephones and computers. When used in the first sense, i.e., human-to-human communication, uncertainty and differences in what the listeners perceive compared to what the speaker intends are caused by a multitude of factors such as errors in comprehension, differences in understanding between the speaker and the listener of semantic word senses, contextual information that is implicitly known to the speaker but not to the listeners, etc. When communication refers to the interchange of information through a technological medium, usually where the speaker and the listeners are separated by longer distances and by the very nature of the technologies, there is uncertainty inserted to the transmitted data called channel distortion and noise, which manifests as static in voice communication, grainy and intermittent pictures and video streams, etc. Continuing breakthroughs in technologies are reducing the error rates in digital communications and keeping them within acceptable limits. In the context of sociocultural models, characterizing and quantifying uncertainty in communication in both the senses are important to improve trust in the models.

Data to Decisions—Characterizing How Uncertainty Propagates

Data has been defined as agent-discernable differences in the physical states of the world (Boisot and Canals 1998). This definition presupposes that such agents would need to expend energy using their sensing apparatus to recognize/register data about the observed real-world phenomena. Information is defined as significant regularities in data that are extracted by an agent. This definition presupposes that there is a relationship between the information and the agent and that conventions have been established by these agents to determine what constitutes significant regularities. We define *knowledge* as action-enabling information. Implicit in this definition is that the agents make decisions to take the actions that are enabled by the knowledge. Agents use contextual information and background knowledge in order to discern knowledge. Also, agents have stored mental models and expectations that may be modified with the accrual of new knowledge (Boisot and Canals 1998).

We now describe a characterization of uncertainty in terms of the knowledge creation process (the data-to-decisions process) described in the previous paragraph (Figure 15.1).

UNDERSTANDING UNCERTAINTY PROPAGATION

Characterizing the data-to-decisions process is crucial to understanding how uncertainty propagates through the different stages of the data processing pipeline. A key benefit of understanding the uncertainty propagation process is the insight provided about the methods to control and manage uncertainty in the decision making process.

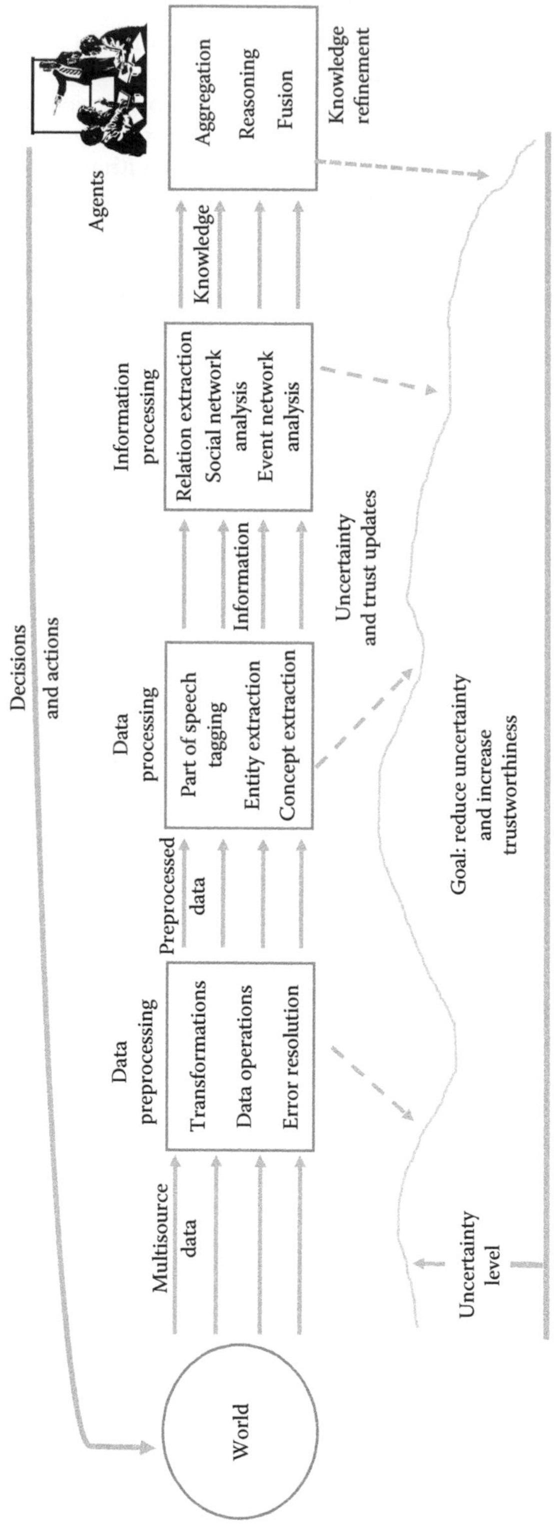

FIGURE 15.1
Characterizing uncertainty propagation as part of the data-to-decisions process.

Propagation of Uncertainty in the Data-to-Decisions Process

There is inherent uncertainty in real-world phenomena and in the observed data about what agents gather and use for sociocultural models and applications.

Data Preprocessing

Data preprocessing refers to the operations used to prepare and transform the raw data into a form that allows for more refined data processing. The examples of specific types of data preprocessing activities include transformations, data operations, and error resolution. For text data preprocessing, the tokenization operation is performed to convert raw text into tokens that include sentence boundaries (e.g., periods), within sentence markers (e.g., commas) and words. An example of a source of uncertainty is the misidentification of sentence boundary markers. These types of errors, if unresolved, will often propagate and become compounded in the downstream data and information processing activities.

Data Processing

Data processing refers to the processing steps that are performed on the preprocessed data and converts data to information. For text data processing, examples of data processing activities include part of speech tagging, entity extraction, and concept extraction. For example, the word *painting* may be used as a noun or a verb. A source of uncertainty for this example activity is misclassification of an occurrence of *painting* (noun versus verb). Again, the ramification of this uncertainty, if unresolved, will cause additional errors in a downstream information processing activity leading to a compounding of uncertainty associated with the model that uses the information associated with the *part of speech* tag. The correct part of speech tagging, on the other hand, will reduce the uncertainty associated with the models that use this information.

Information Processing

Information processing refers to the activities that convert information to knowledge (Figure 15.1). For text information processing, examples of information processing activities include relation extraction, social network analysis, and event network analysis. For these activities, the misclassification of relationships, social network structures, and event network structures would increase the uncertainty that would likely propagate into the sociocultural models that use this type of knowledge. On the other hand, the correct identification of these deep semantic relationships has the potential to significantly reduce the uncertainty of the models that use this knowledge.

Knowledge Refinement

Knowledge refinement includes knowledge validation, aggregation, and uncertainty reduction. Validation involves identifying inconsistencies in knowledge and resolving the inconsistencies. Inconsistency identification may be done manually by inspection or by automated methods that use deductive reasoning methods (Brachman and Levesque 2004). Resolving inconsistencies is more difficult and is usually done by humans. The successful resolution of uncertainties will lead to a reduction in uncertainty in the knowledge used by the model.

Aggregation refers to the activity of composing uncertainty values from leaf/component level uncertainty values (Benjamin et al. 1998). Probability and statistics-based techniques are used to perform this uncertainty aggregation in a scientific manner. Some uncertainty aggregation methods are also used to reduce the overall uncertainty. Uncertainty reduction is accomplished through different methods that are described in the section "Measuring and Reducing Uncertainty in Sociocultural Models." Finally, when the knowledge validation activity is complete, the decision maker is provided with action-enabling information (i.e., knowledge) at different levels of abstraction.

The next section describes methods to reduce and manage uncertainty.

Reducing and Managing Uncertainty

One of the important questions, when using a sociocultural model with multiple data sources and various layers of data and information processing, is how much trust can be placed on the conclusions of the model? In the section "Motivations," we discussed the relationship between trust and uncertainty in the context of sociocultural models. The question of quantifying trust in a model can be partly addressed by determining the methods to measure and reduce the overall uncertainty in the model's output such as inferred facts, actionable knowledge, and decisions. Two approaches to deal with this question are described in the section "Measuring and Reducing Uncertainty in Sociocultural Models."

Another important question in dealing with multisource models is how to resolve data inconsistencies, where data points are in direct conflict with each other. The imperfections at various stages of data collection and processing, and the subjective nature of sociocultural models, can cause inconsistencies in the data collected. Two approaches to resolve inconsistencies in the data are discussed in the section "Identifying and Resolving Inconsistencies."

Measuring and Reducing Uncertainty in Sociocultural Models

In order to reduce uncertainty in sociocultural models, we need to define uncertainty and propose methods to measure it. The field of information theory, based on the contributions of Claude E. Shannon (Shannon 1948) and many other prominent scientists, has developed practically applicable definitions and methods to measure uncertainty. Some of the key concepts from information theory that are relevant to model uncertainty are presented here. A deeper treatment of the subject matter can be found in information theory textbooks (Cover and Thomas 2006).

USING INFORMATION THEORY FOR MANAGING SOCIOCULTURAL MODEL UNCERTAINTY

Uncertainty reduction and management is beneficial to many application areas. An information theoretic approach, used successfully in communication engineering, will likely be useful for managing uncertainty in sociocultural models.

Information Theoretic Methods for Characterizing and Reducing Uncertainty

Here we describe various uncertainty reduction methods that can be used based on information theory. Some of the important uncertainty models are presented here, and their relevance in uncertainty measurement and reduction is discussed with examples.

Entropy

Entropy is the measure of the average surprise or information content in a data source such as a human speaker or an e-mail document. It is assumed that the list of all possible outcomes is known, and the probability of each of those outcomes is also known. For a random variable X with a probability density function $p_X(x)$, entropy $H(X)$ is defined as

$$H(X) = -\sum_{x \in \mathcal{X}} p_X(x) \log_2 [p_X(x)]. \tag{15.1}$$

Entropy has many equivalent definitions depending upon the units of measurement. When the earlier definition is used, with the base of the logarithm as 2, the unit of measurement is bits.

Let us take the example of the uncertainty associated with a model's results. Let us denote the model output by random variable M. One of the ways to characterize the model uncertainty is using the probability of error of the model. If, for instance, the model's output m has a 25% probability of error, then the model's results have a 25% chance of being incorrect and a 75% chance of being correct. The probability density of this model's output is shown below in Table 15.1:

The entropy of this model's output $H(M)$ is given as

$$\begin{aligned} H(M) &= -\sum_{m \in \mathcal{M}} p_M(m) \log_2 [p_M(m)] \\ &= -[0.75 \log_2 (0.75) + 0.25 \log_2 (0.25)] \\ &= 0.8113 \text{ bits.} \end{aligned} \tag{15.2}$$

In the case given earlier, the number of possible output states considered is two (correct and incorrect). In that case, the probability density function gives a sequence of values describing the uncertainty of the output. The entropy of the output can provide a more concise definition of the model's uncertainty instead of the vector of values given by the probability density function.

TABLE 15.1

Probability Density Function for Model M

m	$p_M(m)$
Correct	0.75
Incorrect	0.25

INFORMATION VERSUS UNCERTAINTY

The term *information* is used interchangeably with reference to uncertainty in the information theory literature. The rationale for this usage is that by observing the outcome of a highly uncertain event or data source, we gain more information from it on average. Therefore, the average information content of a highly uncertain event is higher than that of a highly certain event.

In the extremities, when the model is always correct or incorrect, the entropy computation will yield a value of 0 bits, which agrees with the intuitive notion that there is no uncertainty.

Consider a scenario where there are two sociocultural models that represent the same phenomenon. In section "Reducing Uncertainty Using Bayesian Fusion to Support Enhanced Crisis Response Situational Awareness," we present a model to identify a drug gang leader. Suppose that there is another model that is also available that also identifies a gang leader. In order to decide which of the two models to trust more, we can use the entropy of each of the models as a measure of the uncertainty in the models and choose the model with less uncertainty as the more reliable or trustworthy model.

Joint Entropy

If there are two models, their combined uncertainty levels can be represented using joint entropy. Joint entropy is a measure of the total average surprise or information content in two or more data sources such as the outputs of a group of models. If the sources are independent, then the joint entropy is the sum of the entropies of the individual sources. When the sources are not independent, the overall information content is less than the sum of the individual entropies. For a set of two data sources modeled as random variables X, Y, with joint probability density $p_{X,Y}(x, y)$, the joint entropy is $H(X, Y)$ defined as

$$H(X,Y) = -\sum_{x\in\mathcal{X},y\in\mathcal{Y}} p_{X,Y}(x,y)\log_2[p_{X,Y}(x,y)]. \tag{15.3}$$

This definition is easily extensible to more than two variables. The total average uncertainty or the joint entropy of a group of independent random variables is the sum of their individual entropies. Whenever there is dependence between the variables, the total uncertainty and the joint entropy decreases.

Consider the following example of uncertainty in the outputs of two different models M_1 and M_2. Let the joint probability of the outputs of these two models $p_{M_1,M_2}(m_1,m_2)$ be given as

$$p_{M_1,M_2}(m_1,m_2) = \begin{cases} 0.60, \text{ if } m_1 \text{ is correct and } m_2 \text{ is correct} \\ 0.15, \text{ if } m_1 \text{ is incorrect and } m_2 \text{ is correct} \\ 0.15, \text{ if } m_1 \text{ is correct and } m_2 \text{ is incorrect} \\ 0.10, \text{ if } m_1 \text{ is incorrect and } m_2 \text{ is incorrect} \end{cases} \tag{15.4}$$

This joint probability is shown in Table 15.2 as 2 × 2 cells. The corresponding marginal distributions can be computed by adding the two rows for $p_{M_1}(m_1)$ and adding the two columns for $p_{M_2}(m_2)$.

TABLE 15.2

Joint and Marginal Probability Density Functions for Models M_1 and M_2

		m_2 = correct	m_2 = incorrect	
	$p_{M_1,M_2}(m_1,m_2)$			$p_{M_1}(m_1)$
m_1 = correct		0.60	0.15	0.75
m_1 = incorrect		0.15	0.10	0.25
	$p_{M_2}(m_2)$	0.75	0.25	

The joint entropy $H(M_1, M_2)$ is computed as

$$\begin{aligned} H(M_1,M_2) &= -\sum_{m_1,m_2} p_{M_1,M_2}(m_1,m_2)\log_2[p_{M_1,M_2}(m_1,m_2)] \\ &= -\left[0.60\log_2(0.60)+0.15\log_2(0.15)+0.15\log_2(0.15)+0.10\log_2(0.10)\right] \\ &= 1.5955 \text{ bits.} \end{aligned} \tag{15.5}$$

The sum of the entropies of each individual model, M_1 and M_2, $H(M_1) + H(M_2)$ is given as

$$\begin{aligned} H(M_1)+H(M_2) &= 2\left\{-\left[0.75\log_2(0.75)+0.25\log_2(0.25)\right]\right\} \\ &= 1.6226 \text{ bits.} \end{aligned} \tag{15.6}$$

We can see that $(M_1) + H(M_2) > H(M_1, M_2)$. Therefore, the models M_1 and M_2 are not independent and may be combined to reduce the overall uncertainty. Specifically, the models M_1 and M_2 are not independent and knowing the outcome of both the models together can reduce the overall uncertainty.

In order to illustrate the application of the joint entropy concept, suppose we have two models M_1 and M_2 representing some sociocultural phenomenon such as the one described in the example in the section "Using Multisource Analytics and Uncertainty Reduction for Crisis Response Management in Xanadu." Furthermore, suppose that the joint entropy of the two models is less than the sum of the entropies of the individual models. Then we know that by combining the two models M_1 and M_2, we will reduce the overall uncertainty about the phenomenon. On the contrary, if the joint entropy was equal to the sum of the individual entropies, there is no opportunity to reduce the uncertainty.

Conditional Entropy

Conditional entropy can be used to measure the residual uncertainty in a model, given another model. Conditional entropy is a measure of the average surprise or uncertainty in a data source such as a model's output, given the outcome of another data point such as another model's output. If the two data sources compared are independent, then the first sources output does not reduce any uncertainty about the output of the second source's output. When they are dependent, the second source's uncertainty is reduced by knowing the output of the first source. Mathematically, conditional entropy $H(M_2|M_1)$ is defined as

$$H(M_2|M_1) = -\sum_{m_1,m_2} p_{M_1,M_2}(m_1,m_2)\log_2\left[\frac{p_{M_1,M_2}(m_1,m_2)}{p_{M_1}(m_1,m_2)}\right]. \tag{15.7}$$

In the earlier example, the conditional entropy of M_2 given that M_1 is given as

$$\begin{aligned} H(M_2|M_1) &= -\left[0.6\log_2(0.6/0.75) + 0.15\log_2(0.15/0.75)\right. \\ &\quad \left. + 0.15\log_2(0.15/0.25) + 0.1\log_2(0.1/0.25)\right] \\ &= 0.7842 \text{ bits.} \end{aligned} \tag{15.8}$$

Model M_2 has the same probability density as model M discussed in the section "Entropy" (Table 15.1). Therefore, the entropy of model M_2 is $H(M_2) = 0.8113$ bits. We can see that $(M_2) > H(M_2|M_1)$. Therefore, as we have seen in the joint entropy computation, the models M_1 and M_2 are not independent, and observing the outcome of model M_1 can lead to a reduction in the uncertainty about model M_2. These two conclusions, using joint entropy or conditional entropy, are equivalent and are interchangeable.

Conditional entropy may be used to conclude whether we can combine the outputs of two models to reduce the overall uncertainty (increased trust). The two models could be representing the same phenomenon such as the one presented in section "Reducing Uncertainty Using Bayesian Fusion to Support Enhanced Crisis Response Situational Awareness."

Mutual Information

Mutual information may be used to model the uncertainty that is shared by two models. If the mutual information between two models is greater than zero, the two models share uncertainty, and this uncertainty is quantified by the value of this mutual information. The difference between the sum of the individual entropies of models M_1 and M_2 and the joint entropy $H(M_1, M_2)$ is a measure of the uncertainty that is shared by the two models. The mutual information between models M_1 and M_2 $I(M_1, M_2)$ is defined mathematically as

$$I(M_1,M_2) = \sum_{m_1,m_2} p_{M_1,M_2}(m_1,m_2)\log_2\left[\frac{p_{M_1,M_2}(m_1,m_2)}{p_{M_1}(m_1)^* p_{M_2}(m_2)}\right]. \tag{15.9}$$

For the example discussed earlier, the mutual information between models M_1 and M_2 $I(M_1, M_2)$ is given as

$$\begin{aligned} I(M_1,M_2) &= \left\{0.6\log_2\left[0.6/(0.75\cdot 0.75)\right] + 0.15\log_2\left[0.15/(0.75\cdot 0.25)\right]\right. \\ &\quad \left. + 0.15\log_2\left[0.15/(0.25\cdot 0.75)\right] + 0.1\log_2\left[0.1/(0.25\cdot 0.25)\right]\right\} \\ &= 0.0271 \text{ bits.} \end{aligned} \tag{15.10}$$

Note that the mutual information $I(M_1, M_2)$ can also be computed as

$$I(M_1, M_2) = H(M_1) + H(M_2) - H(M_1, M_2) = 1.6226 - 1.5955 = 0.0271 \text{ bits.} \tag{15.11}$$

Similarly, we can also see that the difference between the uncertainty in the outcome of model M_2 and the conditional uncertainty in the outcome of M_2 given that M_1 is $I(M_1, M_2) = H(M_1) + H(M_2) - H(M_2) - H(M_2|M_1) = 0.8113 - 0.7842 = 0.0271$ bits $= I(M_1, M_2)$.

From the above equation, we see that the three alternative ways of computing the mutual information $I(M_1, M_2)$ gives the same value of 0.271 bits. That is, models M_1 and M_2 have a common uncertainty of 0.0271 bits. By knowing the outcome of one of the two models, we can reduce the uncertainty in the knowledge of the other outcome by 0.0271 bits.

In general, if the mutual information between two models M_1 and M_2 is greater than zero, we can reduce the overall uncertainty of the models by combining the outputs of the two models M_1 and M_2.

The information theoretic methods rely on knowing the ground truth about the model's accuracy using historical data. Ground truth may be determined by (1) learning from training data collected from past experience and (2) from human experts' assessments.

In summary, we described three methods of identifying opportunities to reduce uncertainty: (1) joint entropy, (2) conditional entropy, and (3) mutual information. These three methods are equivalent; any of these may be used to measure and reduce sociocultural model uncertainty.

Reducing Uncertainty Using Bayesian Fusion

Bayesian fusion refers to the use of BNs in order to reduce the uncertainty. BNs may be used to model the knowledge about a domain where various statements are known with differing levels of uncertainty (Murphy 2012). The uncertainty in information is modeled using probabilities. Various data sources and statements are modeled using Bayesian random variables with associated probability (or conditional probability) functions. The cause and effect relationship between the random variables are represented as an arc from the cause variable to the effect variable. BNs form a directed acyclic graph. BNs allow inferences about the probability of various statements relevant to the context. Combining various sources of information to perform conclusions from all the sources combined is called Bayesian fusion.

In this section, we describe an example of Bayesian fusion that is useful in combining information from multiple sources, such as the output of various models with varying levels of uncertainty, to reduce the overall uncertainty. The scenario allows modeling various sources of uncertainties in the data as well as the reliabilities of the sources themselves using probabilities.

Let us consider an example, where three sources S1, S2, and S3 are providing information about the same phenomenon using a statement F with conditional probability functions P1, P2, and P3. P1 represents the probability that statement F is true, if source S1 is reliable. This probability is represented as a conditional probability table. Similarly, P2 and P3 also represent the conditional probability that statement F is true given their respective sources S2 and S3. In addition, let us assume that the reliability of each of the models from historical knowledge about the domain as R1, R2 and R3. R1 is the probability that source S1 is reliable represented as a probability function. The following Bayesian model is a fusion model that combines the various models to perform the overall inference about the phenomenon of interest.

The overall probability is computed based on Bayes rule as

$$P(F = \text{true}) = \sum_{i=1,2,3} P_i\left(F = \text{true} \middle| S_i \text{ is reliable}\right) R_i(S_i \text{ is reliable}). \tag{15.12}$$

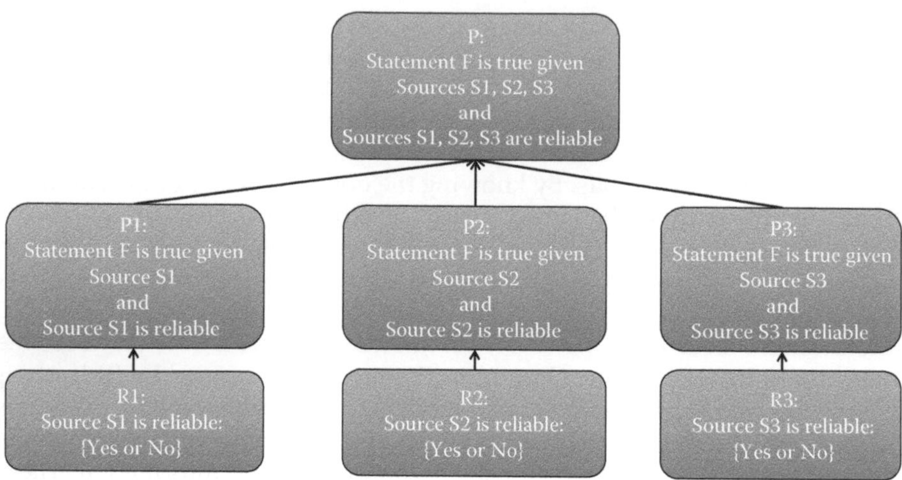

FIGURE 15.2
Example Bayesian fusion model.

The multisource data of different types are processed using a suite of text and data analytical methods as shown in Figure 15.2.

One of the important questions in BN modeling is how to arrive at the probability functions P1, P2, and P3 and R1, R2, and R3, i.e., the model input parameters. The two approaches are (1) learning from training data sets and (2) using human expert knowledge. Detailed discussions of Bayesian learning approaches can be found in *Learning in Graphical Models* edited by Michael Jordan (Jordan 1999).

In this example, we used a binary valued variable F to represent the validity of a statement. It is possible to apply the same technique to model variables that can have more than two possible values. The use of Bayesian fusion to reduce uncertainty is illustrated using a simple application example in the section on Megacity Use Case.

Identifying and Resolving Inconsistencies

Data collected from the real world often have uncertainty and inconsistencies. It should be noted that these two problems are related, and the solution approaches draw from overlapping sets of techniques. Two approaches to reduce uncertainty were discussed in the previous section. Identifying and resolving inconsistencies require modeling the information available and a systematic method to arbitrate among conflicting pieces of information. We present two such approaches in the next sections.

Source Weighting

Consider a scenario where the data collected from the open source media (such as blogs, news sites, and social media sites) provide two pieces of conflicting information. A simple method to resolve the conflicting pieces of information is to evaluate the relative trust levels of the various sources. The source trust-based method is similar to the PageRank algorithm (Brin and Page 1998) in the sense that they utilize information about the source of information in computing the rank. In other words, if one of the conflicting pieces of information is found in a credible source, such as the National Aeronautics and Space

Administration (NASA) website, and the opposing piece of information is found on a personal blog, we would likely place a higher degree of trust on the NASA source.

The trust model can be based on a score of 0.0–1.0 scale, where 1.0 represents full trust. By using this weighting method, we can use probabilistic methods to combine trust values as shown in the BN methods. This method will also work with multiple types of conflicts and multiple levels of trust in the source.

A simpler but realistic scenario involves having multiple sets of unknown social media users (such as Twitter users) that provide conflicting pieces of information. In such a scenario, a voting-based method where each piece of information is weighted according to the number of sources that agree with it can be employed. This method could suffer from spam, where intentional human and computer bot-based pollution of the information pool can introduce errors, resulting in very high uncertainty.

Another relevant aspect of source trust weighting methods is to be aware of the importance of timeliness in the value of data. In other words, even if the data is received from a reliable source, if the data is obsolete, its trust needs to be weighted down compared to other more current data sources.

Reasoning-Based Methods for Resolving Conflicts

Logical reasoning methods are used to identify and resolve conflicts in the knowledge base. For example, a common sense rule would indicate that the same person cannot be at different places at the same time. The execution of such a rule using a deductive reasoning system may be used to diagnose this conflict. By modeling the individual sources' confidence in the data and the reliability of each of those sources, techniques such as the Markov logic network (MLN) are applied to arbitrate among conflicting pieces of information. The various probability function models of the MLN can be learned from historical data sets.

In logical reasoning systems, the known facts about a domain are recorded, and the rules of deductive inference are used to derive new information and facts. The production of information through reasoning methods provides a powerful mechanism to reduce uncertainty. Moreover, reasoning methods are used to reduce and manage uncertainty through conflict detection and conflict resolution.

An important example of a reasoning system is the first-order predicate logic. Logic-based systems are augmented with probabilities in order to model situations that involve uncertainty. An example of a probabilistic logical system is the MLN (Richardson and Domingos 2006). In order to understand MLNs, we need to understand first-order predicate logic systems as well. We present these two systems in the sections "First-Order Logic" and "Markov Logic Networks."

First-Order Logic

First-order logic is a set of sentences constructed using constants, variables, predicates, and functions that capture the knowledge in a particular domain. The constants represent the entities in the domain of interest, such as people or objects, e.g., *Adam, Bryan, Car,* and *Pencil.* Variables are placeholders that can take the value of one of the constants in the domain, denoted by symbols, for instance, *X, Y.* Predicates describe the relationship between objects, such as *Friends, Colleagues,* or attributes of objects such as *plays Basketball.* Functions represent mapping from object tuples to objects, such as *FatherOf, StudentOf,* etc. The knowledge in a domain can be captured using first-order logic for knowledge representation and inferencing. The selection of symbols for constants, variables, functions, and predicates is called an *interpretation,* which is required to perform reasoning on the domain.

An expression of objects in the domain is called a *term*, e.g., *Car, Y, FatherOf(Y)*. A sentence or a formula is a predicate symbol applied to a tuple of terms; for instance, *Colleagues(X)*, called an *atomic* formula. A sentence can take truth values of true or false. Formulas can also be constructed applying connectives and quantifiers to other formulas. In other words, if $S1$ and $S2$ are formulas, then $S1 \vee S2$ represents a disjunction that is true when either $S1$ or $S2$ is true. $\neg S1$ represents a negation that is true when $S1$ is false. $S1 \wedge S2$ represents a conjunction, which is true when both $S1$ and $S2$ are true. The universal quantifier represents *for all x S1 is true*, represented by the formula $\forall x, S1$. The existential quantifier represents *there exists an x for which S1 is true*, represented by the formula $\exists x, S1$.

Positive literal is an atomic formula and negative literal is a negated positive literal. A term containing no variables is called a *ground term.* A possible world applies a truth value to all possible ground atoms, which can be ground terms or atomic formulas constructed from ground terms. A formula $S1$ is satisfiable if there is at least one possible world for which that formula evaluates to true. A formula $S1$ entails a knowledge base KB if $S1$ is true in all worlds where KB is true. The refutation of a formula $S1$ with respect to a knowledge base KB is achieved if $KB \cup \neg S1$ is not satisfiable. Entailment is often proven using refutation. An important inferencing problem in the first-order logic is to decide a knowledge base that entails a formula. In automated reasoning, it is beneficial to represent the formulas into conjunctive normal form (CNF), which is a conjunction of clauses that are a disjunction of literals. Every knowledge base with only first-order logic formulas can be converted to CNF. First-order logic is not decidable in the general case. Therefore, restrictions such as one or less positive clauses per literal, called *Horn clauses*, are used; e.g., in the Prolog programming language.

As noted earlier, first-order logic is used to reduce the uncertainty through automated reasoning by generating new information, through conflict identification and conflict resolution.

Consider the scenario where we have the two models that conclude that (1) person 1 was seen at location A at 4:00 p.m. GMT on January 1, 2010, and (2) person 1 was seen at location B at time 4:00 p.m. GMT on January 1, 2010. We know that the same person cannot be at two different places at the same time. Therefore, the conclusions of the two models are inconsistent with each other and at least one of the two models' conclusions is false. Logical reasoning methods can be used to identify such inconsistencies, and if more information is available that corroborates one of the presented conclusions, we can determine that the other conclusion as false and eliminate it from the knowledge base using logical reasoning. However, when the number of statements in the knowledge base is large, such logical reasoning methods become computationally very time consuming (Domingos and Webb 2012; Russel 2015). Therefore, faster approaches such as MLNs (presented in the section "Markov Logic Networks") are required.

Markov Logic Networks

An approach to reduce uncertainty for multimodal multisource data fusion-based applications is to combine the power of automated reasoning systems, such as with probability-based uncertainty modeling and reasoning methods, such as probabilistic graphical models (e.g., BNs). When specific human expertise in a domain is required to be replicated using automated or semiautomated reasoning, the human knowledge can be captured using MLNs. In the context of the event detection problem and the source term estimation problem, the expert knowledge could be rules or patterns observed by human experts. Such rules can be modeled using MLNs and can be used for optimal estimation and detection for false alarm reduction and uncertainty reduction in the estimates. This section describes the MLN modeling and logical reasoning and model parameter learning approach.

MLNs are a generalization of limited subsets of first-order logic to first-order logic. The first-order logic allows no contradictions within a knowledge base. In MLNs, if a formula contradicts with the rest of the knowledge base, it is still possible to be true, but less probable. The contradicting formula is assigned a lower weight, compared to the higher probable formulas. This modification allows MLNs to represent real-world knowledge that contains contradictions and uncertain information.

An MLN N is a set of pairs (F_i, w_i) where F_i is a first-order logic formula and w_i is a weight assigned to the formula, signifying its likelihood. A Markov network M can be constructed from N, and a finite set of constants $C = \{c_1, c_2, \ldots c_K\}$ and inferencing and learning can be performed on this Markov network M.

$$P(X = x) = \frac{1}{z} exp\left\{\sum_i [w_i \cdot n_i(x)]\right\},$$

where, $n_i(x)$ is the number of grounding of F_i that are true in x and z is a normalization factor.

A typical inferencing problem is to compute the probability that formula F_i is true given that formula F_j is true, which can be computed as

$$P\left(F_i \middle| F_j, C, N\right) = P\left(F_i \middle| F_j, M\right) = \frac{P\left(F_i \wedge F_j \middle| M\right)}{P\left(F_j \middle| M\right)} = \frac{\sum_{x \in \mathcal{X}_{F_i} \cap \mathcal{X}_{F_j}} P\left(X = x \middle| M\right)}{\sum_{x \in \mathcal{X}_{F_j}} P\left(X = x \middle| M\right)}. \quad (15.13)$$

Learning the weights from a database of closed ground atoms is an intractable problem. A more tractable and reasonable solution, used in natural language processing and spatial statistics, is to optimize the pseudolikelihood:

$$P_w^*(X = x) = \prod_{l=1}^{n} P_w\left[X_l = x_l \middle| MB(X_l)\right],$$

where $MB(X_l)$ is the state of the Markov blanket of X_l in the data.

Similar to first-order logic-based methods, MLNs provide a robust approach to reduce uncertainty by combining probability techniques with deductive reasoning techniques.

MLNs can be used to represent facts such as the conclusions of various models and detect and eliminate inconsistencies. An MLN is used to increase the speed of resolving inconsistencies in the knowledge base.

PRACTICAL BENEFITS TO SOCIOCULTURAL MODELING

The ideas and the methods described in this chapter have the potential to benefit a wide variety of practical applications. The megacity use case will provide a flavor for the practical benefits of these ideas and methods.

Megacity Use Case

Introduction

In an evolving crisis situation, such as a category three hurricane or organized criminal activity (looting with killing), emergency response requires quick and effective situational awareness to mobilize and direct resources and perform prevention, mitigation, and recovery actions for the affected people. Uncertainty, time criticality, and fast-changing nature of emergency situations make situational awareness a difficult problem. Uncertainty is introduced by incomplete data, conflicting data, sources with varying reliability, and imperfections in analysis algorithms. Time criticality is introduced by the need for a fast response in crisis situations as well as the changes in the state of entities, such as motions of people and vehicles, spreading of fires or floods, and changes in weather patterns, causing partial or complete obsolescence of information.

Many of the existing analytics-based methods either fall short of exploiting all sources of data or produce a large number of false positives (Wachowicz and Hunter 2005), making effective decision making and crisis response planning difficult. Many of the existing multisensor fusion methods (Hall and Llinas 2001) use data from a single modality such as different types of imagery (or different types of text sources). Currently, large amounts of human capital such as large teams of analysts and crowdsourcing participants are required to connect the dots among multimodal data sources. This approach is not scalable with an ever-increasing number of sensors and social media data sources. Automated or semiautomated methods are required for uncertainty reduction in crisis management situations, such as event detection, anomaly detection, and hot spot detection to greatly improve accuracy, speed, and scalability.

Using Multisource Analytics and Uncertainty Reduction for Crisis Response Management in Xanadu

This widespread adoption of social media provides a means for communication previously unavailable to large segments of the world's population. Disaffected populations that previously lacked the means to shape political debate or affect social change have now been given a significant voice. The use of social media may already have played a critical role in several recent rebellions including the Moldavian revolution, the student uprising in Iran, and the Arab Spring. There have been reports that the State Department asked Twitter to delay a scheduled maintenance event so that Twitter feeds coming out of Iran would not be disrupted during the Iranian student uprising of 2009.

Illustrative Xanadu Crisis Management Scenario

Consider the following hypothetical scenario in the megacity of Xanadu. The Xanadu police chief, Mr. Dawn asks his assistant, Mr. Dusk, to assess the current crime threat level the day before a religious festival of the year. Increased attention has focused on an emerging drug-related gang working in partnership with the human trafficking gang leaders.

- Trigger event: On the day before the religious festival, suppose that an intercepted cell phone message indicates the potential arrival of a suspicious drug gang leader (say X) in a certain suspicious vehicle (say Y) to a specific location in the trouble-prone west side of Xanadu. Suppose that the main Xanadu bank and the largest one-stop shopping complex (the Xanadu Megastore) are also located in this part of the city.

Assume that a social network analysis identifies a network of key members of the drug gang associated with *X*. Fast forward to the next day (festival day) and suppose that a suspicious tweet is detected between two people within *X*'s social network. A geolocation analysis reveals the movement of four individuals in the network to a roadside area close to the festival location. Uncertainty reduction methods are used to reveal the identities of *X* and *Y* with an increased level of confidence.

- Festival day: Suppose that there is an unexpected change in weather at 0900 hours of the festival—increased risk of tornados blowing in from the west of Xanadu. Several tweets are posted about funnel clouds near the neighboring town of Denkali, two Instagram pictures of funnel clouds in Denkali. The center of this system is expected to pass Xanadu at 1000. In the meantime, Chief Dawn draws up an interdiction plan for *X* with *Y* with the optimal time opportunity window being 1030 through 1100 hours. Analyst Dusk appraises Dawn of the imminent weather threat and provides an alternative route for the interdiction operation. Finally, Chief Dawn orders the interdiction at 1015 hours. The operation is successfully completed by 1045 hours, more than seven hours ahead of festival start time (1800 hours). The earlier example scenario, although simple, is illustrative of a typical sociocultural model-based crisis response management application.

Example Crisis Response Management Application Architecture

Multisource data of different types are processed using a suite of text and data analytical methods as shown in Figure 15.2. Social media analytics methods are used to generate topics, sentiments, and social network information that are useful for threat assessments. These analytics methods also produce an estimate of the uncertainty associated with the analytical results, for example, the degree of confidence about the identity of the drug gang leader *X* based on a social media source and the degree of confidence in different types of social network relationships. The results of the analytics processing from the different data sources are input to a multisource information engine that produces more accurate, robust, and actionable information for enhanced situational awareness as shown in Figure 15.3. The information fusion also reduces the uncertainty of the information produced by the analytical methods applied to each source of data.

Reducing Uncertainty Using Bayesian Fusion to Support Enhanced Crisis Response Situational Awareness

In this example application situation, suppose that Bayesian fusion is used to reduce the uncertainties associated with the data and the models for enhanced situational awareness in this example crisis situation. The three sources of data shown in Figure 15.3 contain potential evidence of three people—Person#1, Person#2, and Person#3. In this example, one of these three people is the actual drug gang leader X. Suppose further that Person#1 is 80% more likely to be the suspect (Person X) according to the incident report, which is reliable 75% of the time. Also suppose that Person#2 and Person#3 appear as possible suspects in the social media and signals data once each. The social media and signals data sources

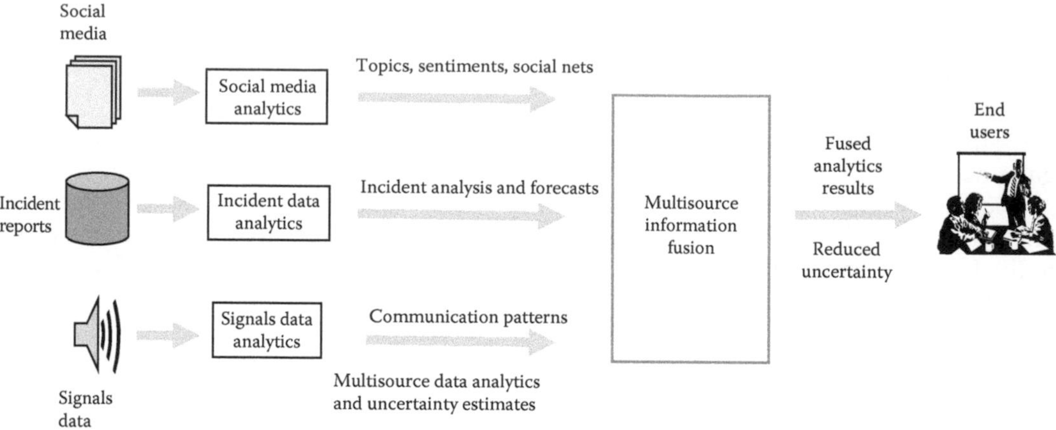

FIGURE 15.3
Multisource analytics and fusion sociocultural application architecture.

both have somewhat equal (60% and 50%) reliabilities and less than the incident report (at 80%); therefore, their overall likelihood being Person *X* is lower than that of Person#1.

These source reliability models are represented in Tables 15.3 through 15.5.

These probabilities are modeled as the bottom row of three random variable nodes in Figure 15.4.

The conditional probability of Person#1, Person#2, or Person#3 being the leader, given the fact that the incident report is reliable, is given by the conditional probability tables (Tables 15.6 through 15.8).

These conditional probabilities are modeled as the middle layer of three nodes in the BN model shown in Figure 15.4. By combining all three sources of information and their

TABLE 15.3
Probability Density Function for Incident Report Reliability

IR Is Reliable	p_{IR} (IR Is Reliable)
Correct	0.75
Incorrect	0.25

TABLE 15.4
Probability Density Function for Social Media Source Reliability

Social Media Source Is Reliable	p_{SM} (m | Social Media Source Is Reliable)
Correct	0.60
Incorrect	0.45

TABLE 15.5
Probability Density Function for Signals Data Reliability

Signals Data Is Reliable	p_{SD} (m | Signals Data Is Reliable)
Correct	0.50
Incorrect	0.50

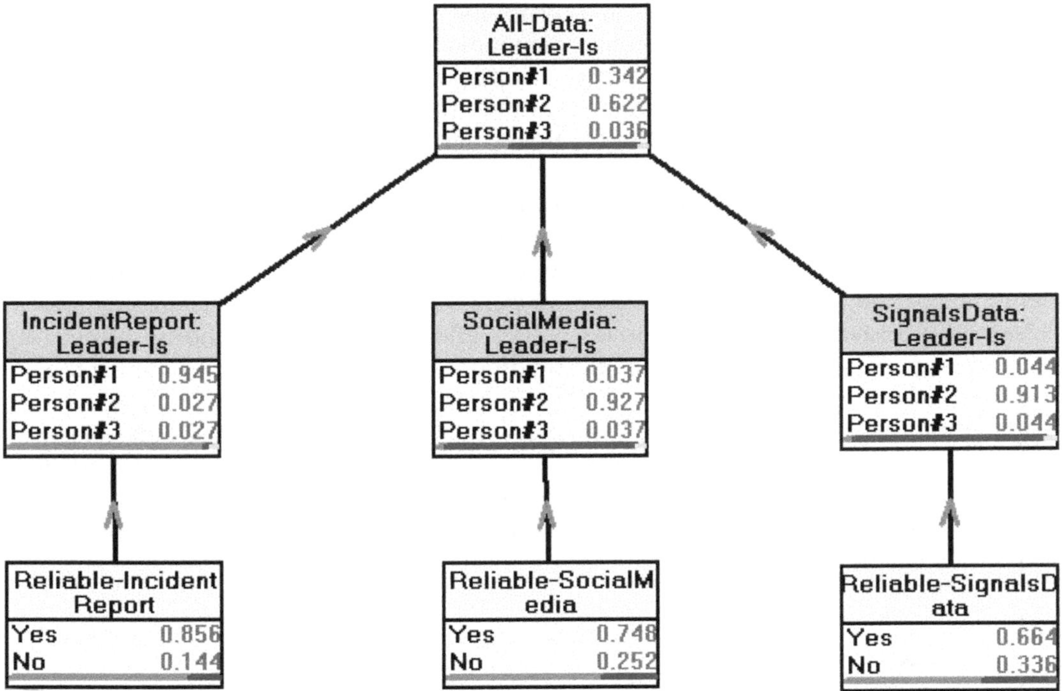

FIGURE 15.4
Using Bayesian fusion for uncertainty reduction: an example.

TABLE 15.6
Conditional Probability Function for Leader Identity Given the Incident Report Source Reliability

	P **(*x* Is the Leader|Incident Report Source Is Reliable)**		
Incident Report Source Is Reliable	***x*** **= Person#1**	***x*** **= Person#2**	***x*** **= Person#3**
Yes	0.80	0.10	0.10
No	0.3333	0.3333	0.3334

TABLE 15.7
Conditional Probability Function for Leader Identity Given the Social Media Source Reliability

	P **(*x* Is the Leader|Social Media Source Is Reliable)**		
Social Media Source Is Reliable	***x*** **= Person#1**	***x*** **= Person#2**	***x*** **= Person#3**
Yes	0.10	0.8	0.10
No	0.3333	0.3333	0.3334

TABLE 15.8
Conditional Probability Function for Leader Identity Given the Signals Data Source Reliability

	P **(*x* Is the Leader|Signals Data Is Reliable)**		
Signals Data Source Is Reliable	***x*** **= Person#1**	***x*** **= Person#2**	***x*** **= Person#3**
Yes	0.10	0.80	0.10
No	0.3333	0.3333	0.3334

reliabilities, we can compute the overall probability function of the ringleader as shown in the top node in Figure 15.4. Based on the computations, we can see that person 2 has about a 62% chance of being the gang leader (person X), person 1, 34% ,and person 3, about 4%.

The earlier example, although simple, illustrates the use of the Bayesian fusion method for reducing the uncertainty associated with the information used for a sociocultural decision support application.

When encountering real-world problems, there could be more or fewer data sources available, and this model is extensible as new sources or modes of crisis management data are added or some existing modes are removed. This model can also be updated as new information or change in some piece of information that contradicts already known information is available. These changes will be effected through (1) adding or deleting new nodes or (2) modifying the prior and conditional probability tables for each of the BN variable nodes.

It is to be noted that in real-world scenarios involving multiple sources and pieces of partial evidence that run in large volumes of data, and uncertain and even contradictory pieces of information, a rigorous information fusion method such as the Bayesian modeling works really well. One of the main strengths of using BN is that it is extensible and its generation can be automated.

Summary and Conclusion

This chapter started by identifying the importance of uncertainty as an important determinant of trust for decision making with sociocultural models and applications. Methods to characterize uncertainty were outlined. The propagation of uncertainty through the data-to-decisions process was described. Next, several uncertainty reduction and resolution methods were outlined. Finally, an example application scenario was presented that shows the utility of uncertainty management methods for sociocultural applications.

Driven by the premise that reducing uncertainty in the sociocultural decision information will increase trust, this chapter advocates the use of uncertainty reduction and uncertainty management methods for sociocultural applications. Characterizing and managing uncertainty is important because of the significant influence of uncertainty on the determination of trust. If the premise of the chapter is valid, then the reduction of uncertainty in sociocultural decision information will likely lead to better and more timely decision making.

DISCUSSION QUESTIONS

1. Does reducing uncertainty increase trust in sociocultural models?
2. What are some of the factors that cause uncertainty in socio-cultural models?
3. Why is the quantification of uncertainty important for sociocultural models?
4. Why is trust important for decision making in applications involving sociocultural models?
5. What are some ways to reduce the uncertainty associated with sociocultural models?

References

Ayyub, B. M., and M. M. Gupta. Editors. 1994. *Uncertainty Modeling and Analysis: Theory and Application*. Philadelphia, PA: Elsevier Science.

Benjamin, P., M. Erraguntla, D. Delen, and R. Mayer. 1998. Simulation modeling at multiple levels of abstraction. In Medeiros, J., E. F. Watson, J. S. Carson, and M. S. Manivannan (eds), *Proceedings of the 1998 Winter Simulation Conference*, pp. 391–398. Washington, DC. Los Alamitos, CA: IEEE Computer Society Press.

Boisot, M., and A. Canals. 2004. Data, information and knowledge: Have we got it right? *Journal of Evolutionary Economics* 14: 43–67.

Brachman, R. J., and H. J. Levesque. 2004. *Knowledge Representation and Reasoning*. San Francisco: Elsevier.

Brin, S., and L. Page. 1998. The anatomy of a large-scale hypertextual web search engine. *Computer Networks* 30(1–7): 107–117.

Cover, T. M., and J. A. Thomas. 2006. *Elements of Information Theory*. Hoboken, NJ: John Wiley & Sons. Second Ed.

Dempster, A. P. 1967. Upper and lower probabilities induced by a multivalued mapping. *The Annals of Mathematical Statistics* 38(2): 325–339.

Domingos, P., and W. A. Webb. 2012. A tractable first-order probabilistic logic. *Proceedings of the 26th AAAI Conference on Artificial Intelligence*. https://www.aaai.org/ocs/index.php/AAAI/AAAI12/rt/captureCite/5028/5350/BibtexCitationPluginhttp://dblp.uni-trier.de/rec/bibtex/conf/aaai/DomingosW12

Hall, D. L., and J. Llinas. 2001. *Handbook of Multisensor Data Fusion*. Boca Raton, FL: CRC Press.

Henrion, M., and R. Shachter (eds.). 1990. *Uncertainty in Artificial Intelligence 5, Machine Intelligence and Pattern Recognition*. Vol. 10, Amsterdam: Elsevier, North Holland.

Hirshleifer, J., and J. G. Riley (eds.). 1992. *The Analytics of Uncertainty and Information*. Cambridge: Cambridge University Press.

Jordan, M. (ed.). 1999. *Learning in Graphical Models*. Cambridge (USA): The MIT Press.

Jøsang, A., and S. L. Presti. 2004, April. Analysing the relationship between risk and trust. In Dimitrakos, T. (ed.), *Proceedings of the Second International Conference on Trust Management (iTrust'04)*. pp. 135–145. Oxford. Berlin: Springer-Verlag.

Lin, C. T., and C. S. G. Lee. 1996. *Neural Fuzzy Systems A Neuro Fuzzy Synergism to Intelligent Systems*. Upper Saddle River, NJ: Prentice Hall.

Morgan, M. G., and M. Henrion. 1990 (reprinted 1998). *Uncertainty: A Guide to Dealing with Uncertainty in Quantitative Risk and Policy Analysis*. New York: Cambridge University Press.

Murphy, K. P. 2012. *Machine Learning: A Probabilistic Perspective*. Cambridge: The MIT Press.

Richardson, M., and P. Domingos. 2006. Markov logic networks. *Machine Learning* 62(1–2): 107–136.

Robinson, S. L. 1996. Trust and breach of psychological contract. *Administrative Science Quarterly* 41(4): 574–599.

Russel, S. 2015. Recent developments in unifying logic and probability. *Communications of the ACM* 58(7): 88–97.

Shafer, G. A. 1976. *Mathematical Theory of Evidence*. Princeton, NJ: Princeton University Press.

Shannon, C. 1948. A mathematical theory of communication. *The Bell System Technical Journal* 27: 379–423, 623–656.

Wachowicz, M., and G. J. Hunter. 2005. Dealing with uncertainty in real-time knowledge discovery process. In van Oosterom, P., S. Zlatanova, and E. Fendel, *Geo-Information for Disaster Management*, pp. 789–797. Berlin, Heidelberg, New York: Springer.

Watson, S., and A. Moran (eds.). 2005. *Trust, Risk and Uncertainty*. New York : Palgrave Macmillan.

16

Validating Causal and Predictive Claims in Sociocultural Models

Amy Sliva, Scott Neil Reilly, John Chamberlain, and Randy Casstevens

CONTENTS

Introduction 316
Challenges of Complex Causal/Predictive Patterns 318
Validation Methodology for Causal Analysis in Sociocultural Models 319
Model (In)Validation and Causal/Predictive Hypothesis Generation 320
Granger Causality for Analyzing Temporal Offsets in Causal/Predictive Relationships 321
Dynamic Time Warping for Uneven Temporal Relationships 321
Convergent Cross Mapping for Bidirectional Causality 322
Qualitative Feature-Based Analysis 323
Case Study: Validating a Model of Poverty and Violent Conflict 325
Discussion and Conclusions 331
Megacities Use Case 332
Acknowledgments 333
References 334

ABSTRACT To better understand and describe the world around them, social scientists need to be able to create, analyze, and validate social, political, and economic models that include causal and predictive elements. Causality is, however, notoriously difficult to analyze. To address this complexity, we describe a suite of causal/predictive analysis techniques adapted from a variety of social, natural, and computational science applications, specifically chosen for their unique applicability to the problems of analyzing complex causes and effects in social science and demonstrate how these methods can be used to understand the causal relationship between poverty and conflict.

KEY TOPICS

- Causal and predictive models of sociocultural processes
- Granger causality
- Dynamic time warping
- Convergent cross mapping
- Qualitative causal features

Introduction

To better understand the world around them and ultimately produce theories that can facilitate better decision making or policy interventions, social scientists need to be able to represent, reason about, and (to the extent possible) validate social, political, and economic models that include causal and predictive relationships. In fact, it is largely the ability to capture these sorts of causal relationships that makes sociocultural models useful for predicting the future and enables policy makers to use them to make informed decisions. To leverage these sociocultural models to their fullest extent, it is crucial that researchers are able to validate that the relationships captured in these models are in fact accurate and a sound basis for decision making or alternatively to recognize invalid models and work to refine theories of sociocultural interaction.

Causal and predictive relationships are notoriously difficult to analyze and validate, especially in sociocultural modeling where there are numerous, complex, and subtle causal/predictive linkages to understand and where rigorously controlled experiments are often difficult to design and implement, requiring researchers to rely on observational data.

Furthermore, existing tools are helpful but not sufficient for addressing the task at hand. For instance, existing quantitative/statistical tools that are often used to analyze data (e.g., Stata, R, SAS, and Excel) lack the analytical depth and social science focus necessary to identify causal relationships and provide a deeper understanding of sociocultural theories. In fact, typical statistical analyses, such as regression analysis, often make it harder to separate the concepts of causation and correlation. For instance, if *A* and *B* are correlated, we have little insight into how they are actually interacting, but if we consider the temporal fact that changes in *A* occur before the changes in *B*, we can at least question the validity of models positing that *B* causes *A*. At the other end of the spectrum, qualitative researchers often view causality as deterministic, rather than probabilistic. This approach finds possible causes/predictors for some phenomenon *A* of interest, such that without *B*, *A* would not have occurred. By focusing on analysis of counterfactuals, potential causal/predictive candidates can be invalidated by a single deviation from the overall pattern [1]. However, these approaches have difficulty selecting or ruling out competing causal/predictive links [2].

While human social science experts often have good insights about the causal/predictive relationships in their field of study, they quickly become overwhelmed by the amount and the complexity of data necessary to validate their models or tease out more subtle

Causality is extremely difficult to analyze in part because there is no agreed upon definition of what makes a relationship causal. Goldthorpe [3] identifies three possible definitions that have been prominent in social science research. First is causality as robust dependence, where causation implies association or correlation with strong predictive power. Another approach views causation as consequential manipulation, i.e., observing different consequences when the causal variable is manipulated or varied. Finally, a growing field of study views causality as a generative process, where the relationship is determined by some underlying social process or mechanism, which itself may be unobservable.

relationships, so we believe that it is important to focus on automated/computational tools to support the human experts to validate or refine their models of sociocultural systems. In this chapter, we focus on two types of analysis that combine the best of both human knowledge and computational power to jointly enhance the validation of causal/predictive models. In the first type of analysis, model invalidation, the human scientist has developed a model of a sociopolitical phenomenon that they want to compare against available data in an attempt to find flaws with the theory (in practice, proving a theory is correct from data is impossible in many cases, especially when using only observational data, so we focus on analysis techniques that attempt to invalidate the theory and find flaws with the expert's theory). In the second type of analysis, we use automated techniques to identify and hypothesize novel causal/predictive relationships that have not been provided by the human expert. This is a form of machine learning or data mining, which is notorious for identifying spurious relationships. In this case, the role of the human expert is to vet the automated hypotheses, refine them, and then decide how/if they belong in their theoretical model of the world.

In exploring automated causal/predictive analysis techniques and weaknesses of many existing tools, we have identified three properties that make causality analysis particularly difficult in social science research: (1) the temporal nature of causation means that the causes and the effects can be separated in time (and by nonuniform amounts), making correlation analyses ineffective; (2) it is challenging to tease apart the causes and the effects in cases of bidirectional or cyclic causality; and (3) qualitative features and effects are difficult to capture in quantitative (e.g., statistical) models (e.g., instability in factor X leads to stability in factor Y).

We have not identified any automated causal/predictive analysis technique that addresses all these problems. Therefore, rather than focusing on any one approach, we introduce a suite of causal/predictive analysis techniques—Granger causality, dynamic time warping (DTW), convergent cross mapping (CCM), and qualitative feature analysis—adapted from a variety of social, natural, and computational science applications. Each has its own strengths and weaknesses, and they have been specifically chosen for their applicability to the challenging causal/predictive patterns in social science research we have identified. Social scientists can use these analytic techniques, either individually or in combination with one another, to validate their causal/predictive hypotheses and produce more robust understandings of the nature of causal/predictive relationships in complex systems.

More interesting, perhaps, is how these causality analyses can be used to invalidate or refine the causal relationships. Researchers can identify cases where the hypothesized

The validation of social science models is an active—and sometimes contentious—field of research commonly referred to as verification and validation (V&V). In general, V&V looks at two model criteria: (1) does the model correctly represent relationships between the variables (verification) and (2) does the model output conform to an external entity, such as a real-world social system. Verification is relatively straightforward to determine—did you build the model you intended to build? However, validity is extremely challenging in social science, where it is difficult and often impossible to run rigorous, controlled experiments to test the correctness of a social theory. More on this discussion can be found in the study by Lustick and Tubin [4].

causal relationships are not empirically supported by the data, indicating the need for refinement of the existing model (through deeper analysis, wider data collection, or adjustment of contextual assumptions) or the development of new models that can incorporate these cases. This step can actually be an ongoing, iterative process of evaluation and refinement of the causal model as new data are acquired or new cases are examined. Continued evaluation can enable the systematic initial testing of causal hypotheses in a big data test bed, helping researchers identify new cases or overlooked variables that might be worth exploring with more in-depth qualitative analysis or validate their theories in terms of general patterns of causality that might suggest new avenues of exploration.

Note that in this chapter, we do not address the interesting and challenging problem of teasing apart causality and prediction as part of the (in)validation process. The sociocultural models in which we are interested, however, include both purely the predictive and causal elements, and the analysis and validation techniques we propose will largely work for both as long as we use them properly. We have attempted to be clear in this chapter about what we believe is actually possible with the approaches described.

The remainder of this chapter is organized as follows. In the section "Challenges of Complex Causal/Predictive Patterns," we highlight some specific issues related to causal/predictive analysis in social/political science. In the section "Validation Methodology for Causal Analysis in Sociocultural Model," we present an overarching methodology for social science modeling that we used to help describe how our tools and methods can be used effectively. In the section "Model (In)Validation and Causal/Predictive-Hypothesis Generation," we describe a suite of analysis methods that can be used to validate (or invalidate) causal/predictive relationships and present an iterative model refinement process that uses these tools to better understand and validate sociocultural models. In the section "Case Study: Validating a Model of Poverty and Violent Conflict," we present the results of empirical evaluations using these various approaches, examining the nature of the causal/predictive relationship between poverty and violent conflict as a case study. Finally, in the section "Discussion and Conclusions," we summarize our approach and results.

Challenges of Complex Causal/Predictive Patterns

The complexity of sociocultural systems poses a challenge for researchers attempting to discover or validate causal/predictive linkages. We identified the following three properties of social systems that we believe are particularly challenging to analyze or validate effectively using traditional techniques. We believe that these properties also provide opportunities to leverage new insights from a variety of causal/predictive analysis techniques to more thoroughly analyze social science theories.

1. Temporal offsets: The temporal nature of causation can be both challenging for correlation-based analysis methods and powerful for teasing apart causes from effects. Intuitively, effects cannot precede causes in time. Current quantitative approaches to validating causation based on statistical analyses often address correlation as opposed to causation [5]. These methods, however, do not account for the temporal gaps between causes and effects, so the correlation values or parameters do not often accurately reflect the strength of a causal/predictive relationship and cannot distinguish a clear sequence to make robust claims about which is the

cause and which is the effect. An additional complexity is that many phenomena do not exhibit fixed temporal offsets, either because of actual variability (e.g., an increase in crime might happen anywhere from 6 to 12 months after an uptick in unemployment), or due to sampling rates. These variable temporal offsets make it difficult to validate causal/predictive relationships from data even using advanced time series models.

2. Bidirectional causality: Some systems have variables that causally interact in both directions. A simple example is a cyclic predator–prey system where changes in either population will cause a reaction in the other. For example, a decrease in the prey population (perhaps through predation) will cause a decrease in the predator population, which will subsequently cause an increase in the number of prey. Existing methods in both qualitative and quantitative approaches have difficulty representing and validating bidirectional causality. When these effects are offset in time, analysis can be even more challenging, because the correlations can be masked in the observed data even though the underlying interactions are very clear and useful for causal inference.
3. Qualitative features and effects: Not all models of social or political systems have simple, quantitatively measurable relationships between variables. For instance, if a period of social stability tends to precede a period of economic growth (or vice versa), there may be little statistical correlation between these variables. In fact, correlation-based analyses may actually make this type of relationship more difficult to discern, because the relevant concepts encompass potentially complex combinations of values for a single variable that must be related to occurrences in another variable. For example, social stability may reflect a period of only minor changes in the amount of dissatisfaction with the government where the definition of minor is determined relative to the known historical, political, and social contexts. However, despite these measurement challenges, these patterns are still associated with qualitatively meaningful concepts (e.g., social stability) that are important to capture and reason about.

Validation Methodology for Causal Analysis in Sociocultural Models

The increasing availability of diverse data sets can provide a rich resource for identifying or validating causal relationships using a combination of the analysis techniques described earlier. In fact, the variety of the data and the complexity of the possible causal/predictive interactions in sociocultural models make it possible—and, we believe, necessary—to apply different types of analysis and validation techniques depending on the research question and characteristics of the data being used. Using this collection of causality analysis techniques, researchers can validate various aspects of their theories and hypotheses and develop more robust causal/predictive theories.

Goldthorpe [3] describes a procedure for causal analysis in social science that serves as a general template for our proposed validation methodology for sociocultural models: (1) establish a phenomenon of interest, (2) hypothesize causal relationships, and (3) test the hypotheses of the causal process. We examine each of these steps in more detail as follows to explore sample workflows that can leverage diverse data sources and combine a variety of analytic techniques to address the challenging causal properties (i.e., temporal offsets, bidirectional causality, and qualitative features and effects) defined in the section "Challenges of Complex Causal/Predictive Patterns." To reiterate, we do not advocate one particular approach to validating causal/predictive relationships in sociocultural models, but rather want to provide a step toward facilitating the design and the implementation of social science research protocols that can iteratively refine existing models and theories and assist in the creation of new hypotheses for exploration.

1. Establish a phenomenon of interest: This first step in the model validation and analysis process may seem trivial, but clearly establishing the questions to be answered and the observations to be explained is actually quite important when constructing a coherent research design and assessing whether the proposed sociocultural hypotheses are supported by the available data sources and observations.
2. Hypothesize causal relationships: To understand the potential phenomena of interest, social scientists must hypothesize the existence of some causal process to construct a sociocultural model that can be tested and validated. These hypotheses may be generated by qualitative case studies, quantitative analyses of correlational evidence, or may be motivated by new theoretical assumptions. During this step, researchers also hypothesize the relevant form of the causal dynamics, such as the possible time lags or the strength of the causation, indicating which of the suite of causality analysis and validation tools may be most appropriate for assessing the proposed model.
3. Test the hypothesis of the causal process: One of the biggest challenges for social science researchers is step 3 of Goldthorpe's causality analysis process: how do they test their findings regarding causal relationships to validate and refine their hypotheses? A part of this challenge lies in the nature of causality and the sociocultural systems under investigation—causal interactions are just very complex, as discussed in the section "Challenges of Complex Causal/Predictive Patterns." However, some of the difficulty is related to the gap in available approaches between the need to combine qualitative, quantitative, and theory-driven analytic techniques and the availability of the proper tools to implement this methodology. The growing availability of digital data sets provides an opportunity for an extensive test bed within which to validate—or more importantly, perhaps, to invalidate and iteratively refine—causal hypotheses to help social scientists produce more robust understandings of complex sociocultural systems. To validate relationships in diverse data, researchers must have a wide range of methods at their disposal that can assess causation, especially with regard to the challenging causal patterns found in social science data.

Model (In)Validation and Causal/Predictive Hypothesis Generation

In this section, we present several methods for analyzing and invalidating causal/predictive relationships in sociocultural models. These techniques, from a variety of social, natural, and computational science applications, were chosen for their unique applicability to the challenges identified in the section "Challenges of Complex Causal/Predictive Patterns" but are not meant to be an exhaustive list of options. Rather, this suite is the starting point for a continuously evolving battery of tools to help facilitate the identification of and/or the invalidation of complex dependencies that might otherwise be missed, assumed away, or taken for granted, and rule out spurious relationships that might not hold under the given conditions.

Granger Causality for Analyzing Temporal Offsets in Causal/Predictive Relationships

We begin with a discussion of the Granger causality, a popular statistical technique familiar to many researchers but worth including for its ability to address the temporal aspects of causality and prediction. Granger causality [6,7] was originally introduced as a method for time series analysis in economic models to help deal with the problem of temporal offsets (described earlier) and the inherent temporal ordering implied by causality. It can, however, be adapted as a kind of invalidation test for causality in sociocultural data. Granger causality makes two assumptions: (1) the effect does not precede the cause and (2) the causal variable provides information about the effect that would otherwise be unavailable. The formal definition of Granger causality is a multivariate autoregressive process [8], where we compare the results of regressing a variable against time-lagged versions of itself and the possible causal condition [9].

- Definition 1 (Granger cause): The temporal variable X *Granger causes* temporal variable Y iff $P\left(Y_t \mid Y_{t-1}^{t-L}\right) \neq P\left(Y_t \mid Y_{t-1}^{t-L}, X_{t-1}^{t-L}\right)$ where L is the maximum time lag, a_i and b_j are parameters in a linear combination, ϵ_1, ϵ_2 are error terms, and

$$P\left(Y_t \mid Y_{t-1}^{t-L}\right) = \sum_{l=1}^{L} a_l Y_{t-l} + \epsilon_1 \tag{16.1}$$

$$P\left(Y_t \mid Y_{t-1}^{t-L}, X_{t-1}^{t-L}\right) = \sum_{l=1}^{L} a_l Y_{t-l} + \sum_{l=1}^{L} b_l X_{t-l} + \epsilon_2. \tag{16.2}$$

A variable X is a Granger cause of Y if Y can be better predicted using the histories of X and Y rather than just of Y alone. The Granger causality focuses on the predictive power of causal relationship and does not attempt to separate causation from prediction [3]. This viewpoint may enable researchers to determine the strength of their causal/predictive assumptions over a large quantity of data. Validating a Granger causal relationship using the Granger causality can be done through hypothesis testing. If Equation 15.2 is statistically more accurate than Equation 15.1 using an F statistic, then a Granger causal relationship between X and Y can be considered valid for prediction. Using this same technique, we can provide evidence for invalidating both predictive and causal claims.

Dynamic Time Warping for Uneven Temporal Relationships

In sociocultural systems, variability in human behavior and infrequent data sampling can produce uneven temporal delays or introduce additional artifacts, making causality difficult to validate as temporal offsets between causes and effects (or predictor and predicted) vary over time. For example, suppose researchers hypothesize that lower employment rates cause an increase in crime anywhere from 6 to 12 months in the future. The causal/predictive link may be verified within some individual cases of interest, but may be difficult to generalize due to the challenge of temporal inconsistencies when applied to additional data.

Causal relationships with uneven temporal offsets cannot be captured by most standard statistical analyses, including the Granger causality, which assume a stationary process. To validate these uneven causal/predictive linkages, we borrowed and extended the DTW algorithm from gait recognition [10,11]. DTW has been traditionally used to identify a person's gait from two motion curves, recognizing the correlated pattern of movement (i.e., the characteristics of someone walking slowly and quickly should be the same, even though the time between the steps may be compressed or extended). DTW compares the two time series to find the optimal alignment by warping one series—that is, stretching or shrinking it along its time axis to align the two curves. From a technical perspective, DTW computes a minimal warp path, which identifies the minimum amount of stretching or shrinking required to produce the highest possible correlation between two time series. A formal definition is given in the following.

- Definition 2 (Warp path): Given two time series X and Y of size n and m, respectively, a warp path W is a sequence $W = w_1, w_2, \ldots, w_K$, where K is the length of the path and each element $w_k = (i, j)$ represents a mapping between point i in X with point j in Y. The optimal warp path minimizes the sum of the distances between the mapped points as

$$\text{argmin Dist}(W) = \sum_{k=1}^{k=K} \text{Dist}(w_{ki}, w_{kj}), \tag{16.3}$$

where Dist(W) is the distance of warp path W and Dist(w_{ki}, w_{kj}) is the distance between point i in series X and point j in series Y.

Because causality can only impact the future, we have enhanced DTW to handle the one-directional case in a new algorithm ForwardDTW. Rather than matching points by warping the time series data in both directions, ForwardDTW only matches the points in X with future values of Y. ForwardDTW allows us to use DTW to analyze causal relationships with uneven time lags—the smaller the warp distance between X and Y, the stronger the causal/predictive link. A user can specify this threshold to determine when a relationship will be considered invalid. Again, DTW may be most valuable from the perspective of invalidating causal/predictive assumptions or indicating circumstances from a large collection of cases when the insights of a causal/predictive hypothesis do not hold, identifying possible candidates for further in-depth case analysis or suggesting refinements to the existing model.

Some advantages of DTW over other time series analyses are that it can account for missing data and compare series with different timescales or sampling frequencies. DTW is also very straightforward to represent visually, making the results easy to interpret by human analysts despite the mathematical nature of the approach.

Convergent Cross Mapping for Bidirectional Causality

Granger causality and DTW can identify or validate causal/predictive relationships and consider complex temporal interactions. However, many social systems contain feedback relationships, where the dependency between variables is bidirectional. For example, declining economic output may increase levels of political violence, which further depresses the economy. These causal relationships may be intuitive for social scientists but are extremely difficult to validate in actual data, where correlation analyses (as well as the previous two analysis techniques described) can completely obscure this cyclic linkage.

To analyze bidirectional causality, we introduce the CCM approach [12], a recent advance in biological studies used to model causal feedback relationships, such as predator/prey systems. This method can be adapted to model behavior in social and political systems, providing a new opportunity to generalize cyclic causal theories in sociocultural models.

To use CCM, we assume the existence of some underlying dynamic, generative process that can be projected on to the variables of interest, X and Y. This projection consists of a set of vectors for variables X and Y called the *shadow manifolds*, essentially estimating how the unobservable process generates the observed values for X and Y.

- Definition 3 (Shadow attractor manifold): For a time series variable X, the *shadow attractor manifold* M_X consists of points $x(t) = (X(t), (X(t - \tau), X(t - 2\tau),\ldots, X(t - E\tau)))$, where τ is a sampling time lag and E is the manifold dimension.

For subsets of time series X and Y of length L, we can construct manifolds M_X and M_Y. CCM will then determine how well local neighborhoods—small regions of M_X—correspond to neighborhoods in M_Y. If X and Y are causally linked, there will be a one-to-one mapping between points in M_X and M_Y. To create this cross mapping, we use a neighborhood in M_X to predict the values of contemporaneous points in M_Y and compute the correlation ρ between the predicted values. If a causal relationship exists, predictions of Y from X (and vice versa) will improve as the amount of data (L) increases, i.e., the mapping of X and Y will converge to perfect predictability $\rho = 1$.

Qualitative Feature-Based Analysis

The previous approaches have focused on how to use quantitative data to analyze human-generated causal/predictive hypotheses or further validate correlational analyses. However, some causal/predictive models require the development of qualitative theories to interpret the complex interactions between variables. In these cases, the interesting properties of a social or a political system are not descriptive of a single value of some variable in the data set, but may comprise a context-sensitive pattern of values.

A simple example is to consider the concept of a spike in a particular variable, such as the number of violent crimes in a city over time. A spike intuitively denotes some kind

> CCM is already having an impact in real-world policy debates. The developers of this approach applied their method to understanding the dynamics of sardine and anchovy populations in California fisheries. Conflicting hypotheses have assumed a competition between the species that leads to fluctuations in their numbers, while another asserts that external environmental factors impact both populations. Using CCM, it was determined that there is actually no direct interaction between sardines and anchovies, but that they are both impacted by a common environmental driver related to water temperature, altering decades of fishery management policies.

of relatively extreme increase in the number of crimes but is otherwise imprecise. What magnitude of change determines a spike? Over what time frames should the increase and subsequent decrease occur? These definitions are inherently related to the specific context under investigation—the size of spikes in crime in Oslo may constitute only the daily variations in criminal activity in Juarez, but may be no less significant in their particular context.

While it is certainly possible to capture some of these qualities statistically (e.g., through controls in multivariate regressions), causal/predictive interactions between these qualitative patterns may be difficult to express quantitatively. Often, the data series themselves will not be related in terms of correlation-based measures, despite obvious relationships in qualitative features. For example, suppose stability in the unemployment rate is known to produce a decrease in crime. The raw data will exhibit little or no correlation—crime decreases as unemployment remains stable—but can be explained as the causal interaction of two qualitative features.

To help social scientists identify and analyze causal/predictive relationships that are more readily described as qualitative features than as strictly quantitative relationships, we borrow a technique from signal processing called featurization. Featurization is the process of identifying interesting segments of a time series and associating them with a qualitative concept that can be used in a causal/predictive theory. For example, a researcher may first search through an economic time series and select periods of economic instability, then search a criminal activity time series to extract periods of increasing crime. These individual feature definitions can be approximated by mathematical formulae. This mathematical estimation facilitates the partial automation of the qualitative feature selection process where computational data mining techniques can either learn by example to find additional instances of patterns initially provided by the researcher or identify potentially interesting subsets of the data set that the user can associate with the relevant qualitative concepts. By featurizing the data into qualitative patterns, researchers can analyze causal/predictive relationships between variables that exhibit little or no quantitative correlations.

To support these two approaches, we developed a structural featurization language using six common morphologies (or shapes) that can comprise qualitatively meaningful patterns in a time series [13]. These morphological features are shown in Figure 16.1. More complex patterns can be created by combining multiple elements of this language into compound concepts. For example, a single cycle of economic activity may be an upward spike followed by a downward spike. An entire time series can be represented as a sequence of these components, which can then be chosen by the researcher for a more in-depth analysis.

Once a time series has been mapped to a sequence of qualitative features, it is possible to validate the causal/predictive links. For instance, if a model for a particular city indicates

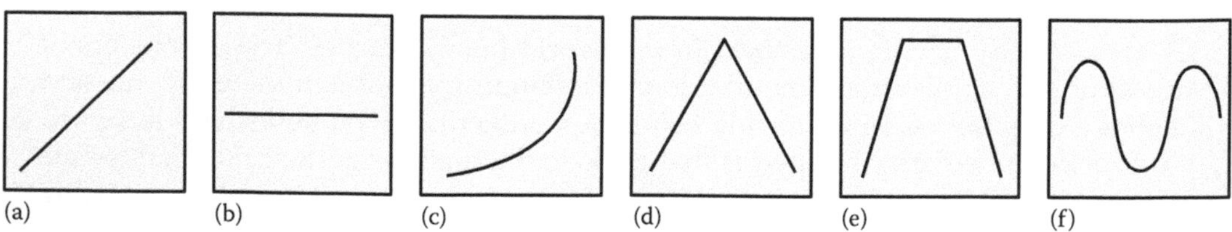

FIGURE 16.1
Morphological components describing qualitative features: (a) slope, (b) constant, (c) exponential, (d) triangle, (e) trapezoidal, and (f) sinusoidal.

that increases in crime (*C*) follow periods of economic instability (*E*) (within some time window), we can acquire additional time series data related to crime and economic activities, convert the series into qualitative features related to increases in crime and economic stability and search for cases where instances of *C* are not preceded by *E*s, where *E*s are not followed by *C*s, and where the specified pattern is actually found in the data.

Case Study: Validating a Model of Poverty and Violent Conflict

The analysis techniques described earlier provide methods that social scientists can use to validate—or, more precisely, invalidate—their hypotheses for sociocultural models. Using a variety of causal/predictive analysis techniques makes it possible to characterize these complex relationships, as well as identify places where models need further refinement or exploration. To illustrate the efficacy of this validation approach, we have implemented prototypes of the Granger causality, the DTW, the CCM, and the qualitative featurization/analysis algorithms as part of our research software framework called the model analyst's toolkit (MAT). MAT [14–17] was developed for social scientists, and it provides tools for exploring and visualizing data as well as for creating, refining, and validating models; a beta version is freely available for the academic and government research communities.

As a case study, we demonstrate a representative exploration of the causal/predictive relationship between poverty and conflict. A large body of literature exists that explores the conflict trap—the process whereby countries get stuck in a repeated pattern of violent conflict and economic underdevelopment [18]. There have been several studies evaluating the causal/predictive link between these two features using standard statistical approaches, with some finding evidence for poverty driving societies into conflict [19,20], while others [21] indicate that civil conflict may be the cause of depressed economic growth. Using the methods described in the previous section, we can better untangle and characterize this relationship and gain insight into the sociocultural processes that lead to the conflict trap.

When validating causal/predictive relationships in such models, the choice of data is itself a challenge, as many phenomena of interest involve complex and abstract concepts—such as poverty and conflict—that are difficult to represent as measurable variables. To model conflict we use the Uppsala conflict data program/Peace Research Institute Oslo data set [22], which tracks the incidence and the intensity of global armed conflict between 1946 and 2013. To capture the notion of poverty, which is not merely a measure of income,

The conflict trap was introduced by Collier et al. [18] in a World Bank research report exploring the possible role of the international community in averting civil wars. In a subsequent study [23], Collier identifies three additional causal development traps that stand as barriers to economic growth in developing countries: (1) the natural resource trap, whereby countries rich in natural resources may have underdeveloped industrial sectors or less accountability to a citizen tax base; (2) the governance trap, where poor policies both prevent economic growth and deter external investment; and (3) the landlocked trap, where landlocked countries with bad neighbors may be unable to tap into the global economy.

but also of relative well-being, we use two variables from the World Bank world development indicators data set [24]—infant mortality rate, measured as the number of infants per thousand live births that die each year, and gross domestic product (GDP), to measure the overall level of development. We consider conflict as both an ordinal variable ranging from 0 to 3 indicating the intensity of a conflict in a given year and as a numerical value with counts of the battle deaths due to conflict within a country. We focused on the timeframe from 1960 to 2013 as both data sets were more complete for this period.

In our first experiment, we analyzed the relationship between poverty and conflict using the Granger causality, varying the amount of history used to make predictions between 1 and 10 years. We considered two possible models of the relationship between poverty and conflict: (1) conflict causes poverty and (2) poverty causes conflict. Figure 16.2 shows the percentage of countries in which each of these poverty and conflict models was supported by a Granger causality test (i.e., the resulting p value from the F-test was below 0.05). Out of the 100 countries under study, we only saw evidence that conflict causes poverty (the solid line in Figure 16.2) in about 30% of the cases when the maximum amount of history included was set to one year. The number of countries exhibiting this causal linkage slightly reduced as the amount of history was increased. While this result may appear to be counterintuitive (we would not expect the inclusion of more historical data to result in less predictive power), this phenomenon might actually be indicative of the complex relationship between conflict and poverty. The Granger causality assumes a linear relationship between the cause and the effect, which may not be accurate in some sociocultural models, particularly in cases where there is a cyclic feedback relationship. When looking at only one year of historical data, the Granger causality performs well because it is only taking a very small, localized view of the data to create a simplistic model. As more historical data are added to the Granger causality computation, this simple linear model is less reflective of the real-world dynamics and may be confounded by the more complex relationships present in the data set.

Interestingly, the hypothesis of a causal relationship from poverty to conflict (the dashed line in Figure 16.2) is statistically significant in an increasing number of countries as the amount of history included in the analysis increases. The increasing support for the model that poverty causes conflict as the amount of history is increased may indicate

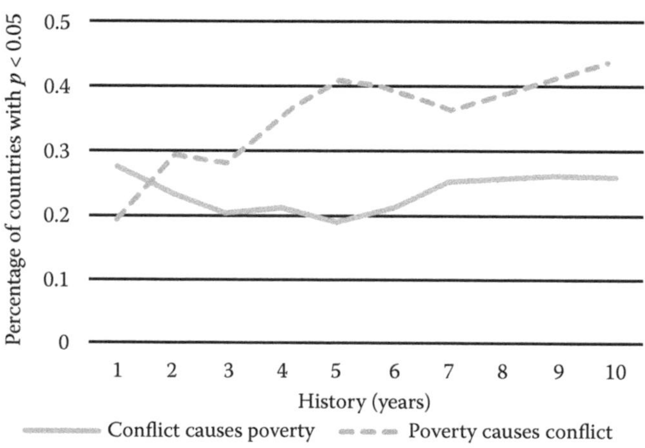

FIGURE 16.2
Results from the Granger causality analysis with increasing time lag.

the persistent nature of poverty and conflict as conditions with longer duration impacts, where longer historical perspectives can provide better predictive models. However, this same result may also be caused by uneven temporal lags between instances of poverty and resulting conflict (e.g., a conflict may occur anywhere from 1 to 10 years after poverty begins to rise), which cannot be precisely captured by the static model assumed by the Granger causality. While the results of the Granger causality tests cannot definitively validate one hypothesis over another, they do indicate that there is some relationship of interest here worth further investigation, using more advanced methods to capture the complicated causal relationships.

To examine the possibility of varying time lags in the relationship between poverty and conflict, our second experiment uses DTW to warp the poverty and conflict time series and compute the improvement in the correlation when we account for possible uneven temporal effects. Figure 16.3 shows an example of using DTW analysis for data from Senegal, where the algorithm attempts to match points in the poverty time series with potential future points in conflict to test for a causal relationship, looking for the mapping that will maximize the overall correlation between the time series. Here, in spite of allowing for uneven temporal delays of up to three years, the original correlation of 0.456 is only slightly improved to the postwarp correlation of 0.468. The cost of the transformation is not excessive in this case, but that is largely because further modifications do not help. That is, by warping the poverty data in the right of the graph further right by a year, we can create a slightly better correlation, but the system has searched for other transformations of up to three years and none of them provide a better correlation score. That should not be too surprising looking at the data as the spikes in conflict are aligned with a relatively stable and high poverty, and the flattening of the conflict curve is associated with a gradual decrease in poverty, but the curves are not the kinds of things that can be aligned and correlated in a standard statistical manner. For the same reason, attempting to use DTW to examine the causal relationship in the other direction, from conflict to poverty, also failed to provide any useful results.

For our third experiment, we used CCM to further characterize the causal relationship between conflict and poverty, which appears to be more complex than the linear model used in the Granger causality or the offset model in DTW. Social processes are often best described by complex dynamical systems, with multiple layers of feedback and interaction, and CCM can help identify these more complex causal interactions, particularly reciprocal or bidirectional causality. Because CCM examines the relationships between projections of the time series, we normalized the data to measure the percentage of change at each time point to account for the vastly different scales of conflict casualties, infant mortality, and GDP. As described in the section "Convergent Cross Mapping for Bidirectional Causality," CCM indicates a causal relationship if the percentage of predictability of one time series from another converges toward 1 (e.g., perfect predictability) as the amount of data increases. We observed convergence in 80% of the countries supporting the hypothesis that conflict causes poverty and in 12% of the countries for poverty leading to conflict. While the Granger causality found a likely predictive link from poverty to conflict, the low predictive power produced by the CCM analysis is an indication that this hypothesis—or this particular version of the hypothesis using infant mortality and conflict deaths—may not be a valid model and may need to be further refined to adequately represent the true dynamics of the underlying social processes. However, the current CCM results may be skewed by the fact that 59% of the countries experienced no conflict in the period under study, making it very easy for the algorithm to achieve perfect predictability of zero conflict in these cases. These results are further skewed a bit by the fact that CCM typically

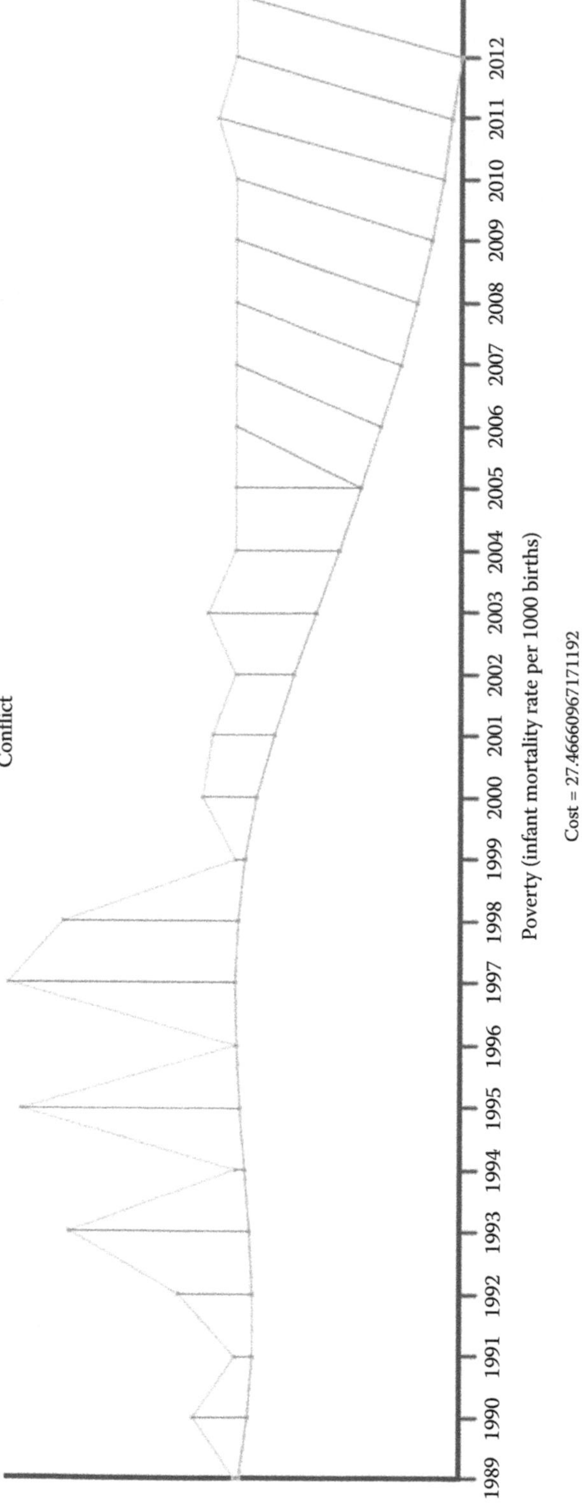

FIGURE 16.3
Results from DTW to examine the causality between conflict and poverty in Senegal.

needs more data to converge than we had available in this data set, so the fact that we reached convergence in as many cases as we did can likely be treated as a lower bound on what CCM would find with additional data.

Figure 16.4 shows an example of this convergence pattern for Comoros, showing how as the amount of data increases, the ability to predict the amount of poverty (as measured by GDP per capita) from the level of conflict converges toward 1. This result is a good indicator that, at least in Comoros, there is a causal link from conflict to poverty, although there is not enough data to see complete convergence (one of the downsides of CCM is the amount of data it requires to reach convergence in many cases). By using CCM, we cannot definitively validate a causal linkage from conflict to poverty given such examples, but this approach does provide additional support for this sociocultural model.

While Granger causality, DTW, and CCM can be useful analytic tools, as we saw in the DTW analysis, the nature of our case study data is not ideal for these types of methods that look explicitly for point-by-point relationships across the time series. However, even though the relationships between the conflict and the poverty data are difficult to quantify through these types of measurements, we found that they can be reasonably described through qualitative featurization analysis. While conflict and poverty are linked to one another, this phenomenon does not manifest as similar patterns of proportional increases or decreases in values offset in time. Instead, across the countries studied, we saw that rapid increases in conflict or periods of recurring conflict are associated not with similar fluctuations in poverty, but by continually decreasing or statically depressed levels of economic activity and by statically high levels of infant mortality. Similarly, we found that

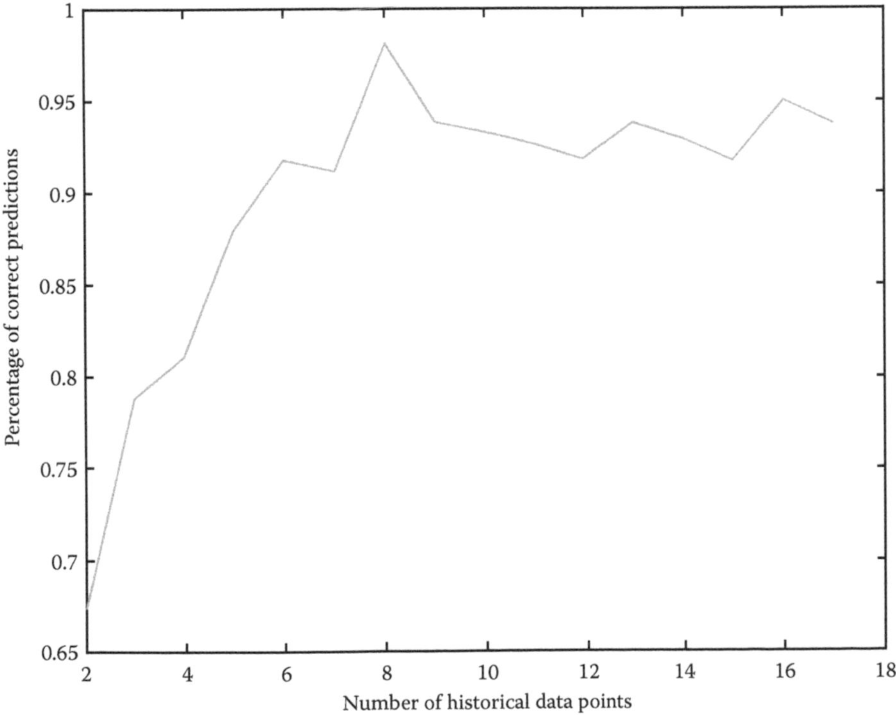

FIGURE 16.4
Results from the CCM analysis illustrating convergence indicative of a causal link from conflict to poverty.

when the conflict ended, we saw decreases in poverty follow. In essence, this illustrates the notion of the conflict and poverty traps, where violence is associated not with rapid declines into poverty, but with sustained levels of minimal development.

Figure 16.5 shows an example using the qualitative feature-based approach to analyze the data from Senegal. The top plot indicates the GDP in current US dollar from 1989 to 2013, the middle plot shows the number of battle-related deaths, and the bottom chart is the infant mortality rate. The human-guided qualitative featurization algorithm has divided these data series into important component pieces representing distinct features, which are indicated by the boxes highlighting different regions of the graphs. In the top graph for the GDP in Senegal, the automated feature search algorithm identified two distinct features—the first, from 1989 to 2001, is a period of relatively constant GDP, and from 2001 to 2008, another feature captures an increase in the GDP. Similarly, in the bottom graph showing infant mortality rates, the automated feature extraction algorithm identified two distinct features—a period of constant (high) infant mortality rates from 1989 to 1999 and a period of decreasing mortality rates between 1999 and 2009. The features in the

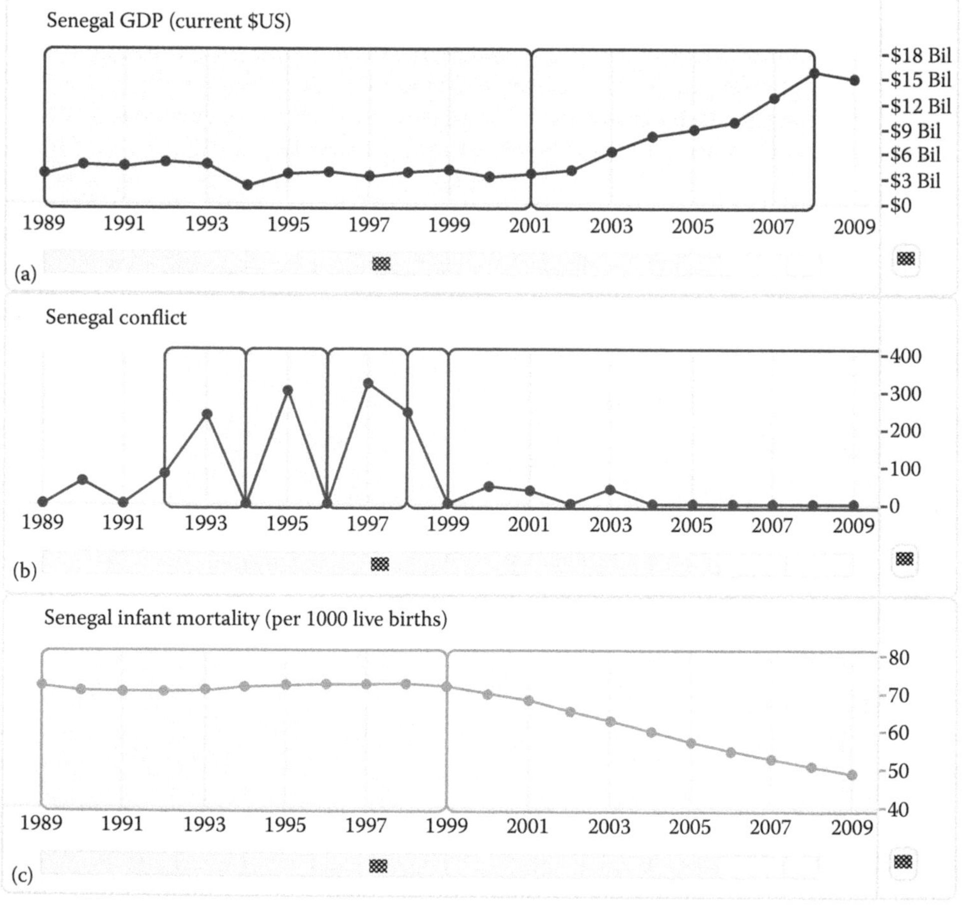

FIGURE 16.5
Qualitative featurization showing the relationships between poverty and conflict: (a) Senegal GDP, (b) Senegal conflict, and (c) Senegal infant mortality.

middle graph of conflict events in Senegal were found through a collaborative human–machine process. First, the human analyst highlighted the region from 1992 to 1994 as a spike in conflict. The feature search algorithm was then able to use this exemplar to find other instances of this feature, also highlighting the periods from 1994 to 1996 and from 1996 to 1998 as spikes in conflict. In addition, the automated featurization also identified a decrease in conflict between 1998 and 1999 as well as a period of relatively constant (low) conflict levels from 1999 to 2010.

From these qualitative features, it is evident that there was a period of violent conflict from 1989 to about 2004, with several spikes in the number of casualties. During this same time, infant mortality was consistently high and the GDP was consistently low. However, after 2004, GDP and infant mortality both began to steadily improve, while the conflict remains very limited. By using these features to represent concepts such as spikes in conflict and high infant mortality, we can identify causal patterns between these more complex features that are not visible when doing a lower-level comparison of individual time points.

Discussion and Conclusions

Sociocultural models provide an opportunity for researchers and policy makers to gain a deeper understanding of the world around them and facilitate decision making based on this knowledge of social dynamics. However, due to the inherent complexities of human behavior, and the inability to conduct carefully controlled experiments in most sociocultural domains, the validation of these models remains extremely challenging. In this chapter, we have presented an approach to the validation of sociocultural models based on the value of invalidation and refinement of causal and predictive relationships, using a variety of methods—Granger causality, DTW, CCM, and qualitative feature-based analysis—to identify and analyze these relationships in observational sociocultural data. Of course, the suite of tools presented here is certainly not complete, and this approach can be further expanded with other quantitative and qualitative causality analysis methods.

The rigor of this approach comes from its breadth of techniques and recognition of their individual strengths and weaknesses, which can be used in combination with one another to explore various dimensions of possible causal/predictive linkages, invalidate spurious hypotheses, and iteratively refine the models. In the given case study analyzing sociocultural models of the processes relating poverty and conflict, we showed how the Granger causality and the CCM both provide some preliminary support for the validation of the causal hypothesis regarding conflict and poverty, but their inconclusive results indicate shortcomings in the existing model (and availability of data) and the need for a deeper analysis. Using qualitative feature-based analysis, we were able to further analyze these relationships between complex patterns that are difficult to describe in terms of purely mathematical correlations. Each of these causality analysis approaches helps to clarify the nature of the causal relationship between poverty and conflict, producing a body of evidence that can provide researchers with a variety of techniques to enhance their understanding of causal patterns and improve the validity of their sociocultural models.

Megacities Use Case

Megacities, such as Xanadu, are becoming an increasingly common phenomenon, as urban areas around the globe continue to see an influx of populations, creating the potential for new economic, social, and environmental challenges. To adequately address these emerging problems and develop mitigating policies, it is critical that we develop a deep understanding of the complex sociocultural dynamics that influence why people migrate to urban areas and how various social processes interact in these extremely dense population centers. However, in any real-world situation, there are limited available resources or feasible policy options, and running a controlled experiment to identify the best approaches is rarely feasible. Instead, we have to rely on available sociocultural models of the world to make informed decisions. Useful models can answer questions like the following: What is the relationship between high unemployment among the refugees from Bursuka and North Razinia and black market criminal activity? Would heightened law enforcement help control organized crime in the city or further inflame ethnic and racial divisions? Can the risk of violent social unrest be reduced through provision of health and social services? Unfortunately, there are many more theories than there are correct models. Knowing when a model is correct (or, at least, consistent with currently available data) and in which contexts it can be used to predict the effects of various policy decisions is critically important in making reliable decisions but poses challenging problems. Using the approach described in Chapter 17, we can (in)validate and refine our sociocultural models to gain a better understanding of the possible impacts of these policy options and how we might achieve the best outcomes in megacities like Xanadu.

The first steps in our validation process are to identify the particular sociocultural phenomena of interest and to generate hypotheses of the possible causal/predictive relationships that might exist. As a starting point, we can base these hypotheses on existing sociocultural theories, which can be validated and refined to improve their applicability to the specific situation in Xanadu. For example, we may have preliminary theories that say that the increases in the population density and the growing percentage of migrants have a causal influence on increased criminal activity or that ethnic heterogeneity combined with high unemployment rates and greater competition for resources (e.g., water, healthcare) can lead to violent social unrest.

Causal analysis methods, such as Granger causality, DTW, CCM, and qualitative feature-based analysis, provide a suite of tools that can be used to analyze and attempt to find problems with available models, allowing researchers to better understand the causal mechanisms at play in Xanadu's strained economic and social situation. The longitudinal data for cities such as Xanadu may be extremely difficult to obtain—often, the cases of greatest interest may be newly emerging or previously understudied, leading to a dearth of historical data from which to generate and validate relevant sociocultural models. Similar to the case study presented in Chapter 17, we can also use cross-sectional data from other megacities with high rates of poverty and crime (e.g., Mumbai or Rio de Janeiro) or analogous situations (e.g., the Kukuma refugee camp in Kenya and the camps for displaced Syrians refugees in Turkey, the social and the economic status of Palestinians in Israel, and European policies in response to migrants from North Africa) to provide evidence for generalizable causal patterns, combining this with specific information available for Xanadu, such as the size of the refugee camps or the recent demographic trends, that can help to identify the shortcomings in our models as they may apply to the Xanadu situation. By applying a variety of causality analysis techniques, researchers can characterize different aspects of the complex sociocultural processes.

Additionally, by using real-world data to validate these sociocultural theories, it is possible to identify patterns that were previously unknown in the Xanadu case. For instance, it might be that the North Razinis' organized crime and social structure has a long history in North Razinia and was instrumental in the collapse of the monarchy, creating new social dynamics imported to Xanadu with the migrants that may be leading to greater instability and growing criminal activity. Conversely, pockets of stability, such as the culturally distinct Bursukan social group, may exhibit unique characteristics—religious hierarchies, physical proximity to water sources—that can help explain their relative success and indicate causal pathways that can perhaps be generalized to the larger population of Xanadu to improve the social conditions. These new hypotheses can also be validated and analyzed using available data sources, helping researchers produce more reliable models that can be used to support decision making that accounts for and leverages these established causal and predictive patterns.

DISCUSSION QUESTIONS

1. Many automated analysis methods cannot reliably separate predictive relationships from causal relationships. What are some of the implications of this? When can predictive models be reasonably used to support decision making? Are there ways to help differentiate predictive and causal relationships?
2. In the section "Challenges of Complex Causal/Predictive Patterns," we review a number of challenges when identifying causal/predictive patterns in data, but this is not intended to be an exhaustive list. What other kinds of challenges might be encountered when extracting causal/predictive models from data?
3. When building data-driven models about Xanadu, it might not be possible to collect all the necessary data to create reliable models. What are some options for building models and operating in the face of models with reliable certainty? Can data from other cities be used? What are the potential challenges of doing so? Are there ways to help human decision makers use these models more effectively?
4. In this chapter, we focused on the invalidation of models (figuring out when a model is wrong). What are some challenges of this approach? An even harder problem is to positively validate models. What are some of the challenges in doing so, especially in a scenario like Xanadu?

Acknowledgments

This chapter is based upon the work supported by the US Navy, under Contract No. N00014-12-C-0653. Any opinions, findings, conclusions, or recommendations in this material are those of the author(s) and do not necessarily reflect the views of the US Navy. The authors would like to thank Dr. Harold Hawkins of the Office of Naval Research for his assistance and guidance throughout this effort.

References

1. Munck, G. L. (1998). Canons of research design in qualitative analysis. *Studies in Comparative International Development*, 33, 18–45.
2. Brady, H. E., Collier, D., and Seawright, J. (2006). Toward a pluralistic vision of methodology. *Political Analysis*, 14, 353–368.
3. Goldthorpe, J. H. (2001). Causation, statistics, and sociology. *European Sociological Review*, 17, 1–20.
4. Lustick, I. S., and Tubin, M. R. (2012). Verification as a form of validation: Deepening theory to broaden application of DoD protocols to the social sciences. *Proceedings of the 4th International Conference on Applied Human Factors and Ergonomics, San Francisco.*
5. Harrell, F. E. (2001). *Regression Modeling Strategies: With Applications to Linear Models, Logistic Regression, and Survival Analysis*. Berlin, Springer.
6. Granger, C. W. J. (1980). Testing for causality: A personal viewpoint. *Journal of Economic Dynamics and Control*, 2, 329–352.
7. Granger, C. W. J. (1969). Investigating causal relations by econometric models and cross-spectral methods. *Econometrica*, 37, 424–438.
8. Box, G. P., and Jenkins, G. M. (1970). *Time Series Analysis: Forecasting and Control*. San Francisco, Holden Day.
9. Freeman, J. R. (1983). Granger causality and the times series analysis of political relationships. *American Journal of Political Science*, 27, 327–358.
10. Myers, C., and Rabiner, L. (1981). A level building dynamic time warping algorithm for connected word recognition. *IEEE Transactions on Acoustics, Speech and Signal Processing*, 29(2), 284–297.
11. Chan, P., and Salvador, S. (2007). Toward accurate dynamic time warping in linear time and space. *Intelligent Data Analysis* 11, 5.
12. Sugihara, G. et al. (2012). Detecting causality in complex ecosystems. *Science*, 338(6106), 496–500.
13. Olszewski, R. T. (2001). *Generalized Feature Extraction for Structural Pattern Recognition in Time Series Data*. Pittsburgh, PA: Carnegie Mellon University.
14. Sliva, A., and Neal Reilly, S. (2014). A big data methodology for bridging quantitative and qualitative political science research. *Proceedings of the American Political Science Association*.
15. Neal Reilly, S. (2010). *Validation Coverage Toolkit for HSCB Models*. R09005-03. Cambridge, MA: Charles River Analytics Inc.
16. Neal Reilly, W. S., Pfeffer, A., and Barnett, J. (2010). A metamodel description language for HSCB modeling. *Advances in Cross-Cultural Decision Making*. Eds. Schmorrow, D., and D. Nicholson. Boca Raton, FL: CRC Press.
17. Neal Reilly, W. S. et al. (2011). *A Computational Toolset for Socio-Cultural Data Exploration, Model Refinement, and Model Validation*. Chantilly, VA: HSCB Focus.
18. Collier, P., Elliott, V. L., Hegre, H., Hoeffler, A., Reynal-Querol, M., and Sambanis, N. (2003). *Breaking the Conflict Trap: Civil War and Development Policy*. Washington, DC: World Bank and Oxford University Press.
19. Collier, P., and Hoeffler, A. (2004). Greed and grievances in civil wars. *Oxford Economic Papers*, 56, 563–595.
20. Braithwaite, A., Dasandi, N., and Hudson, D. (2014). Does poverty cause conflict? Isolating the causal origins of the conflict trap. *Conflict Management and Peace Science*, 33(1), 45–66.
21. Djankov, S., and Reynal-Querol, M. (2008). Poverty and Civil War: Revisiting the Evidence. Washington DC: Center for Economic Policy Research (CEPR) No. 6980.
22. Themnér, L., and Wallensteen, P. (2014). Armed conflict, 1946–2013. *Journal of Peace Research* 51(4).
23. Collier, P. (2007). *The Bottom Billion: Why the Poorest Countries Are Failing and What Can Be Done about It*. Oxford: Oxford University Press, pp. 17–37.
24. The World Bank. (2013). *World Development Indicators*. Washington, DC: World Bank.

17

Rapid Generation of Political Conflict Simulations for Scenarios around the World

Barry G. Silverman, David Q. Sun, Nathan Weyer, and Gnana K. Bharathy

CONTENTS

Introduction and Purpose 336
Architecture and Cycle of the Case-Based Generator 337
  Background on StateSim 339
The Generator Elicitation Screens 340
  Factions 341
  Actors 344
  Institutions, Resources, and Economics 345
CBR and the Repository of Past Cases 349
Conclusions and Next Steps 354
Megacity Use Case 355
  Analysis of the Megacity Use Case to Fill Generator Screens 355
    Introduction 355
    Generator Pull Phase 355
    Possible Variants in Modeling Phase and Use Case Application 358
Acknowledgments 359
References 359

ABSTRACT This chapter describes a prototype of a rapid scenario generator that has been demonstrated in a field setting in 2014. The generator consists of a set of screens that assists a human in quickly specifying political conflict scenarios around the world. The goal is to produce models that can simulate the scenarios and thereby help to assess the impact of the alternative COA the user might take to influence the scenario outcomes. The generator is able to speed up scenario construction since it uses analogical or case-based reasoning (CBR), and it is built upon a repository of hundreds of past StateSim models of political groups, actors, institutions, and other conflict scenario elements. This chapter introduces the generator purpose in the section "Introduction and Purpose." Background on CBR and StateSim are in the section "Architecture and Cycle of the Case-Based Generator." This walks through a fictional state use case to create a scenario and a set of StateSim models in the section "The Generator Elicitation Screens." We then explain the repository of past models and some tools to explore what is in it in the section "CBR and the Repository of Past Cases." Finally, it discusses the results to date and next steps in the section "Conclusions and Next Steps." At the end of the chapter, we discuss the detailed application of our platform to the attached megacity use case.

KEY TOPICS

- Rapid scenario generation
- Political conflict simulation
- Case-based reasoning
- Agent-based modeling
- Sociocognitive models
- Mesolevel modeling of state instability and dynamics

Introduction and Purpose

According to expert practitioners and researchers in the field of human behavior modeling (Silverman et al. 2007; Zacharias et al. 2008; Numrich and Picucci 2012), a common central challenge now confronting designers of human behavior modeling (HBM) applications is to increase the realism of the synthetic agents' behavior and coping abilities. It is well accepted in the HBM community that cognitively detailed, thick models are required to provide realism in models where synthetic agents are endowed with cognition and personality, physiology, and emotive components.* To make these models work, one must find ways to integrate scientific know-how from many disciplines and to integrate concepts and insights from hitherto fragmented and partial models from the social sciences, particularly from psychology, cultural studies, economics, and political science. One consequence from such integration of multiple and heterogeneous concepts and models is that we frequently end up with a large feature space of parameters that then need to be filled in with data.

In recent years, modeling methodologies have been developed to help construct models, integrate heterogeneous models, elicit knowledge from diverse sources, and also test, verify, and validate models (see the study Bharathy and Silverman [2012], or Silverman et al. [2007], for example). However, these methodologies have required extensive use of manual labor to develop each model. In light of this obstacle, we believe that the development of automatic techniques would significantly improve the efficiency of the model generation process. In this chapter, we will explore how model generation can be sped up with the help of analogical or case-based reasoning (CBR) from a repository of past model components. The first approach analogical reasoning is defined as the process of solving new problems or configuring new designs based on the reuse of similar cases that worked in the problem solver's past. We all use this type of thinking every day—Analogical reasoning permits the recall and the reuse of past cases across domains. The second approach CBR refers to previous cases drawn only from the same domain as the problem case.

In this chapter, we use a modified version of CBR described in the section "Architecture and Cycle of the Case-Based Generator." The section "The Generator Elicitation Screen" then takes the reader through a walkthrough of the main steps and the screens of the generator culminating in a working StateSim model (a fictional political dilemma is provided in Megacity Use Case that is referred to in order to illustrate the description of the generator and its screens). The section "CBR and the Repository of Past Cases" provides

* We will hereafter refer to these rich models as *cognitively detailed models* or *thick agents*.

an overview of the repository of past cases and some experimental tools to on how to explore and better use the repository. The section "Conclusions and Next Steps" contains the conclusions.

Architecture and Cycle of the Case-Based Generator

This chapter is an interim progress report on research to make the generator as fully automated and intelligent as we know how. Our goal is to implement the generator as a CBR algorithm; however, not all the requisite capability is easily accomplished. In fact, for a complex domain as in political conflict modeling, it is not clear ahead of time as to how to automate certain steps (see the section "CBR and the Repository of Past Cases"). Our approach is to do this via a spiral engineering process where we have built a first version or a prototype that requires human assistance in some of the reasoning steps. This version can then be circulated to foster dialog about the next steps, a topic we return to in the section "Conclusions and Next Steps."

Conceptually, CBR is commonly described by the CBR cycle as in Figure 17.1—based on Aamodt and Plaza (1994). This cycle consists of four seemingly straightforward sets of steps. However, as just mentioned, we have discovered that we need to modify this to get it to work in our domain, and we need to approach these steps by increasing capability over time.

The retrieval step requires that past cases be organized according to an index scheme that describes the problem space that they are a solution to. This index is most commonly implemented as a graph or a hierarchy of attributes. Once that is created, retrieval is primarily a matter of looking up the nearest neighbor case in an associative attribute graph (e.g., by using a weighted attribute similarity score, Euclidian distance metric). It should be noted that this approach structures the cases into problem–solution pairs. Once the most similar problem is retrieved, it is then possible to reuse the past solution and revise it to fit the needs of the current case.

Solve new problems with the help of a repository of past cases organized within a common indexing structure and via the following:

1. Retrieve similar cases to the problem description.
2. Reuse a solution suggested by a similar case (or portions of solutions from several past cases).
3. Revise or adapt that solution to better fit the new problem if necessary (use scenario-relevant names, rescale parameter values, merge elements from several cases, etc).
4. Retain the new solution once it has been validated. Thus, the repository learns and expands.

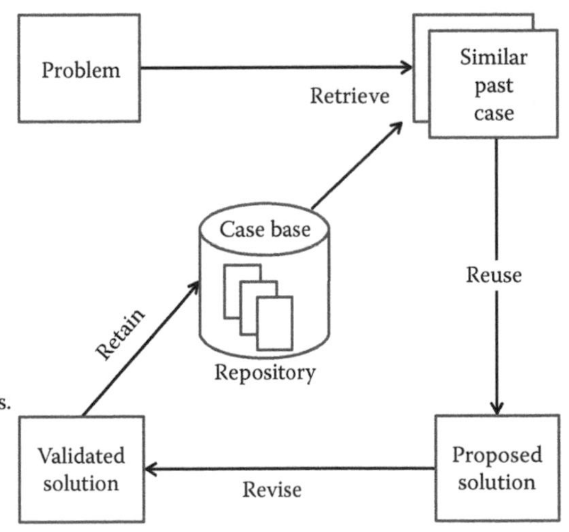

FIGURE 17.1
Basic steps of the CBR cycle (modifications in parentheses).

WHAT IS CASE BASED REASONING?

CBR is based on the simple idea that similar problems have similar solutions. People use this problem-solving technique on a daily basis. The physician tells us that she saw the symptoms before and one course of pills was all that was needed to heal that patient. The mechanic remembers a similar car where he solved the same problem. And we hopefully remember past recipes when we are cooking a new meal. CBR should be considered whenever it is difficult to formulate domain rules and when cases are available. It should also be considered whenever new solutions can be derived from old solutions more easily than from scratch. Many successful commercial applications show the value of this paradigm.

If the past cases are computational models of earlier scenarios (as in this research), then the index must adequately describe the political conflict problems that these models simulate. Thus, we need to have an index or an ontology of the elements of political conflict. In fact, we have a mature and validated questionnaire that serves as the basis for our indexing structure. The repository consists of hundreds of models of past actors, groups, institutions, and related scenario elements that were coded by country desk experts using the StateSim model and its questionnaire. The StateSim model is briefly described in the following.

In order to support many people using StateSim, we derived and validated a questionnaire instrument that collects all the information needed to model a region in the StateSim

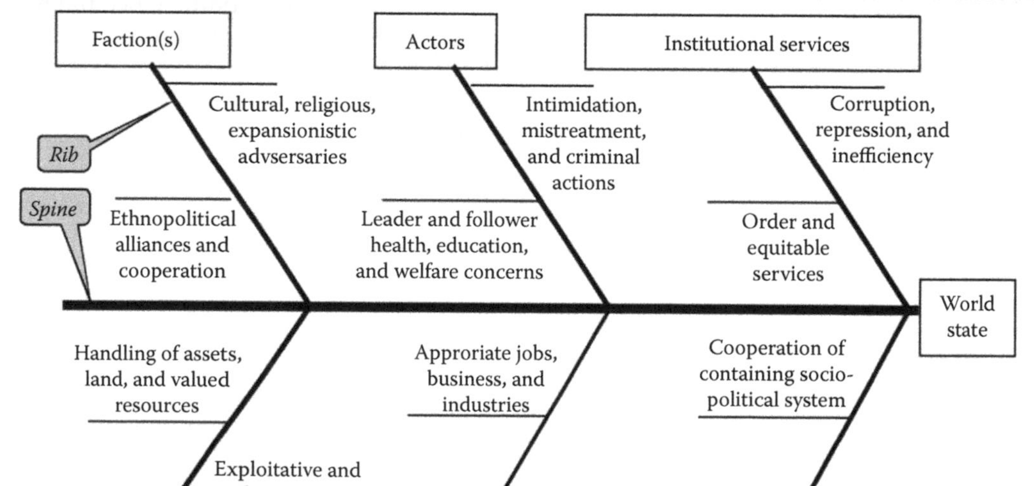

FIGURE 17.2
FAIREST ontology of StateSim parameters and index of the CBR models.

APPLIED EXERCISE

Consider a political scenario you know or which is well documented. Can you readily describe the factions that have the greatest influence, the leader personas, the follower concerns, and what particular resource, economic, and/or public goods and services issues tend to drive the debate? It can be harder than you think to winnow this down to the key variables that need to be quantified to capture the situation at hand. How would a taxonomy like FAIREST help you?

framework. This questionnaire follows the acronym of FAIREST (faction-by-faction actors, institutions, resources, economics, suprasystem, and timelines), and Figure 17.2 shows it in a fishbone diagram. We have found that true country subject matter experts think about the state and its conflicts all the time. What is needed to elicit their mental models and model parameters is largely just a questionnaire instrument. We also use the FAIREST scheme to stimulate a conversation about new idealized designs (and metrics) that seek to mitigate negative factors and promulgate positive ones. As a result, FAIREST serves as the overarching index that describes each model in the repository. It is thus the requisite index needed in the retrieval step of the CBR cycle.

Overall, the FAIREST interview consists of about 80 tables of mostly Likert-style questions, with some free text answers, and some numeric estimates as well. In thinking about using it for the generator, we realized that with the help of the past case repository, we could greatly reduce the questionnaire's length, improve its usability, and thereby speed up the model generation process. Our efforts to do this led to the generator screen walkthrough summarized in the section "The Generator Elicitation Screens." This has been the main accomplishment of this work to date, other than merging all past StateSim cases into the repository.

Background on StateSim

StateSim simulates a state (or cross state or substate) region and important political groups, their ethnic (and other) conflicts, economic and security conditions, political processes, domestic practices, and external influences. StateSim is a model of models that facilitates the codification of alternative theories of factional interaction and the evaluation of policy alternatives. Once it is set up, StateSim executes a game or a conflict scenario in which the factional leader and the follower agents all run autonomously. You are the sole human interacting. You may use a set of actions to influence outcomes and effects (see Figure 17.3) that might occur in the game. Factions are modeled, as in the center of Figure 17.1, where each has a leader, various subfaction leaders or henchmen (e.g., core, fringe, other), a set of starting resources (economy, E; security, S; and politics, P), and a representative setup of thousands of follower agents. A leader agent in a StateSim game will manage his faction's E and S tanks so as to appeal to his followers and to each of the other tribes or factions he wants in his alliance. Each of the leaders of those factions, however, will similarly manage their own E and S assets in trying to keep their subfactions and memberships happy. The followers determine the level of the P tank by voting their membership level (see the section "Factions"). A high P tank means that there are more members to recruit for security missions and/or to train and deploy in economic ventures. So the leaders often find it difficult to move to alignments and positions that are very far from the motivations of their memberships.

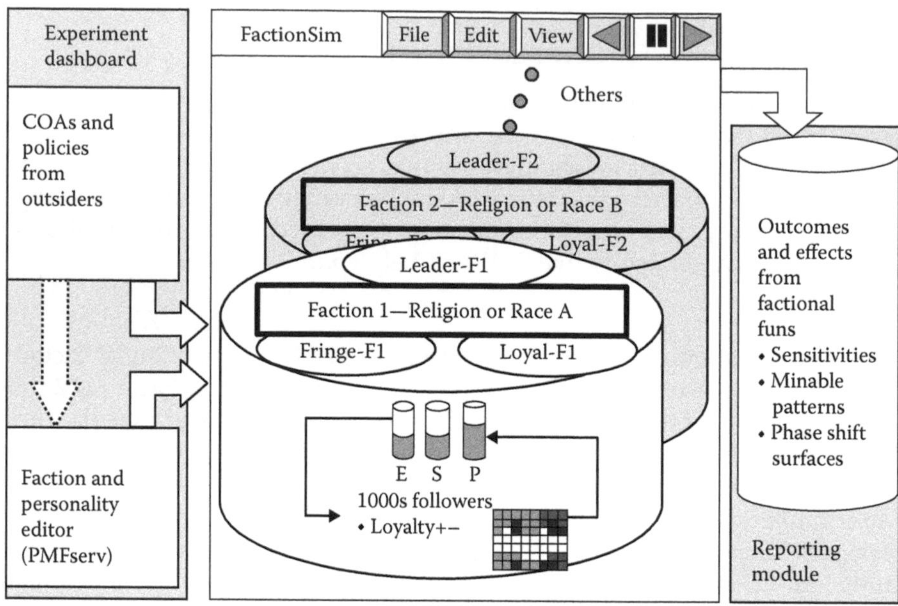

FIGURE 17.3
Models and components that must be synthesized to create a StateSim scenario.

The Generator Elicitation Screens

In order for a StateSim game to work, it is necessary to instantiate the tables of data about the factional leaders and followers and their resources and alignments, norms, and values. This is done by completing the FAIREST taxonomy as described earlier in the section "Architecture and Cycle of the Case-Based Generator." The more accurate the data used, the more realistic the simulation outcomes will be.* Simultaneously, there are a lot of scenarios to consider for any conflict zone, and the easier it is to fill out the FAIREST items and instantiate the StateSim models, the more likely they can be applied in a responsive fashion as conflicts shift and morph during their lifetime. In this section, we will examine the screens that walk the country desk expert through this FAIREST process. In particular, we are using these screens to model the scenario in the use case in the Addendum section.

* StateSim was built under three DARPA programs. DARPA indicated it could predict (>80% accuracy, recall 88%, precision 64%) quarterly events of interest such as coups, rebellions, repression, and intergroup violence. It predicted such events for each quarter from 2006 to 2008 for each of several Pacific rim nations (Bangladesh, Korea, Sri Lanka, Thailand, and Vietnam). It has since been applied to study the impacts of alternate COAs in Afghanistan, Iraq, Nigeria, Palestine, Somalia, and Syria. Aside from the DARPA correlation backcasts, in dozens of other correspondence trials, StateSim has demonstrated better than 80% correlation with the action choices of leaders and followers (insurgencies, rebellions, political repressions, intergroup violence) for numerous countries in Asia, Africa, and across the Middle East, as well as for UK soccer hooligans and US crowd scenes. See, Silverman et al. (2010, 2012) and Bharathy and Silverman (2012). This technology also received a $5000 prize for Best Artificial Intelligence/Pattern of Life in the 2011 Federal Virtual World Competition, and in FY12, it was embedded in an army application (ALOTT) to stimulate the brigade/battalion command staff situation room screens for simulation game exercises aimed at training how to detect and destroy a network of insurgents hiding in a large population of civilians. That latter application has since been deployed to 10 US Army sites.

Since that use case omits suprasystem and timeline, we will only walk through the FAIRE portion of the ontology index.

Factions

The first topic is to find out what factions are involved in the scenario. As the left side of Figure 17.4 shows, the generator presents the user with a taxonomy of all the possible social archetypes that could populate a scenario. In the middle of Figure 17.4, the user selects the archetype and then using the plus (+) sign and right side qualifiers, the user specifies the faction's name and how many of that archetype exists with the scenario. Also, with the pull-downs on the right side, the user selects the general disposition of the group. For instance, there are two groups of refugees: North Razinis and Bursuks. Their disposition is *displaced minority*. The Xanadu military's disposition is semitrained third world army. These dispositions lead to many default values being filled in about the group and the agents in the group. In the section "CBR and the Repository of Past Cases," we will see some repository tools we are working on to help the users to alter these defaults.

Factions are not done until four more screens are filled out as shown in Figure 17.5. StateSim runs a set of multiple games, $G = \{G1, G2, \ldots, Gn\}$ simultaneously. Within a faction, one may observe games between rival leaders, between leaders and followers, and follower on follower. The across-faction games include attempts to cooperate and/or compete with

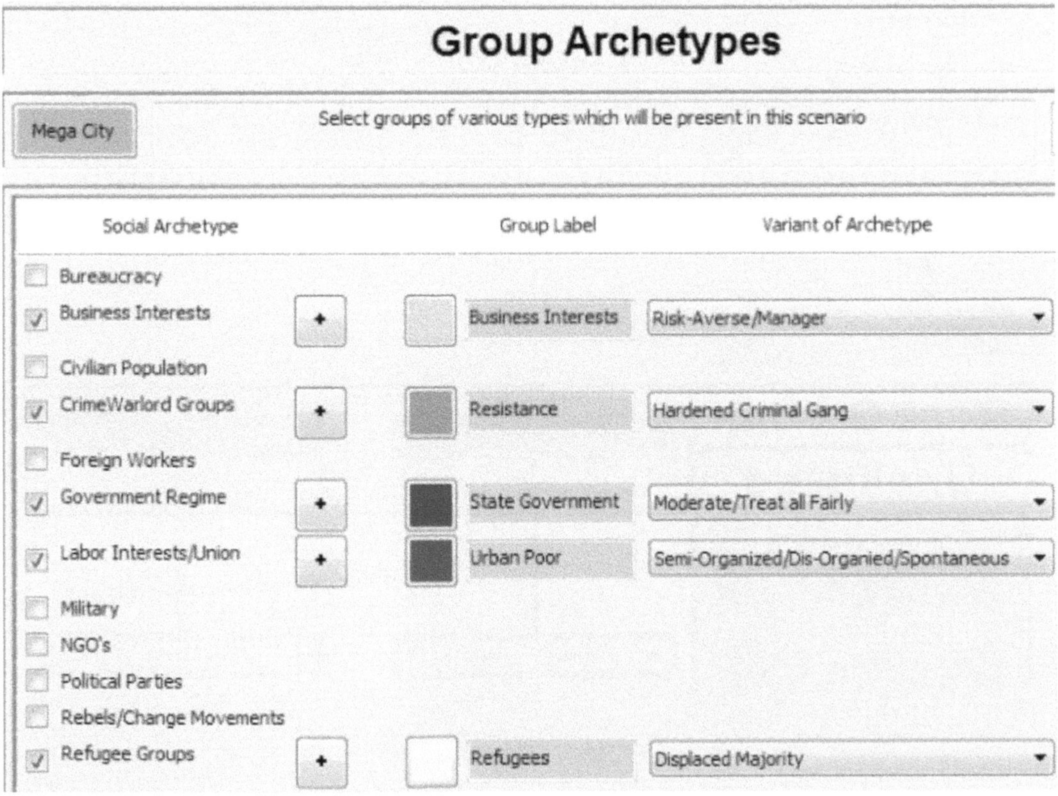

FIGURE 17.4
The scenario is populated with groups drawn from a taxonomy of generic archetypes.

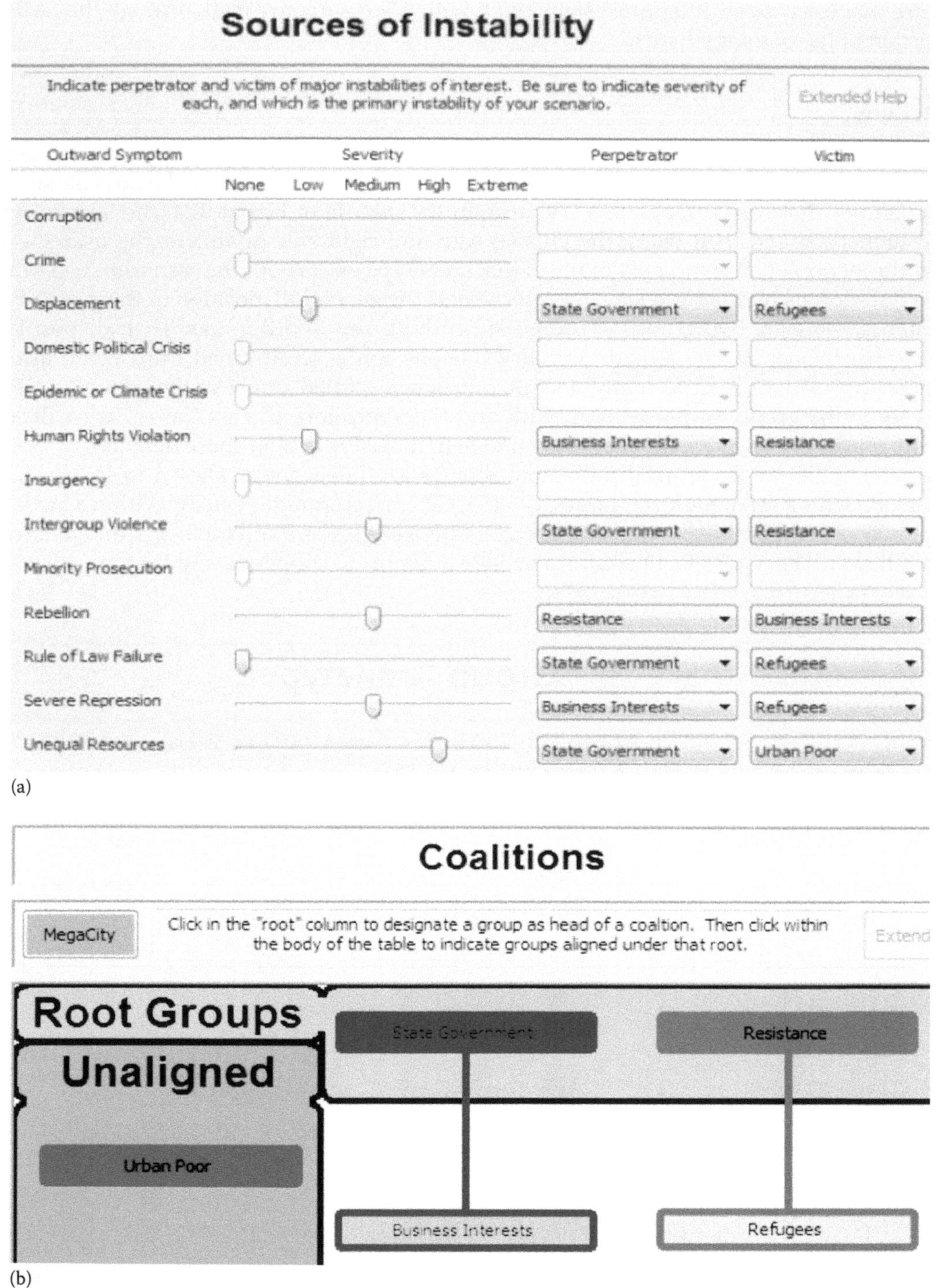

FIGURE 17.5
Clarifying the instabilities, coalitions, sizes, and alignment valences of all the factions: (a) Instabilities—What are the destabilizing issues? (b) Coalitions—What groups lead versus follow? *(Continued)*

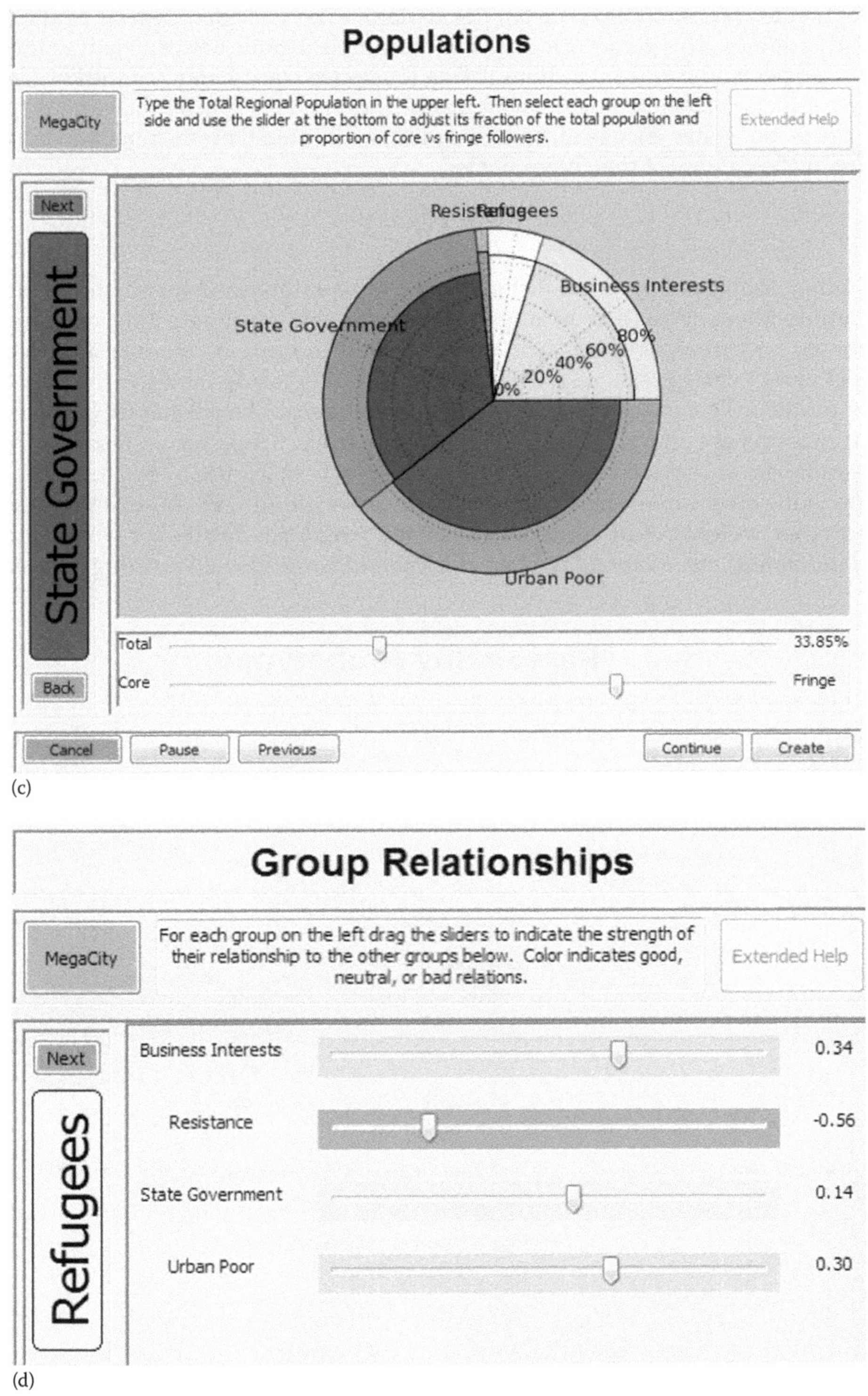

FIGURE 17.5 (CONTINUED)
Clarifying the instabilities, coalitions, sizes, and alignment valences of all the factions: (c) Populace—What is the size of the core and the fringe of each faction? (d) Relations—What are the valences for the group-to-group alignments?

other faction leaders and followers. For discussion's sake, consider these as iterated semi-cooperative games. This game formulation is the simplest game one can analyze involving conflicts between (and within) factions. Using it helps to clarify many of the key elements of these conflicts. Using the screens of Figure 17.5 helps one to specify the various types of games the factions may engage in, how large and committed their membership is, and the intensity or the valence of their alignments.

Actors

StateSim is a tool that allows conflict scenarios to be established in which the factional leader and follower agents all run autonomously; use their groups' assets, resources, and institutions; and freely employ their microdecision making as the situation requires. Macrobehaviors emerge as a result. For this to work, the leader and the follower agents must be profiled. That is, each specific leader and archetypical (core and fringe) follower is profiled in terms of cultural values and personality traits. These are represented through goal, standards, and preference (GSP) trees as shown in Figure 17.6. These are multi-attribute value structures where each tree node is weighted with Bayesian probabilities or importance weights. A preference tree is one's long-term desires for world situations and relations (e.g., no weapons of mass destruction, stop global warming) that may or

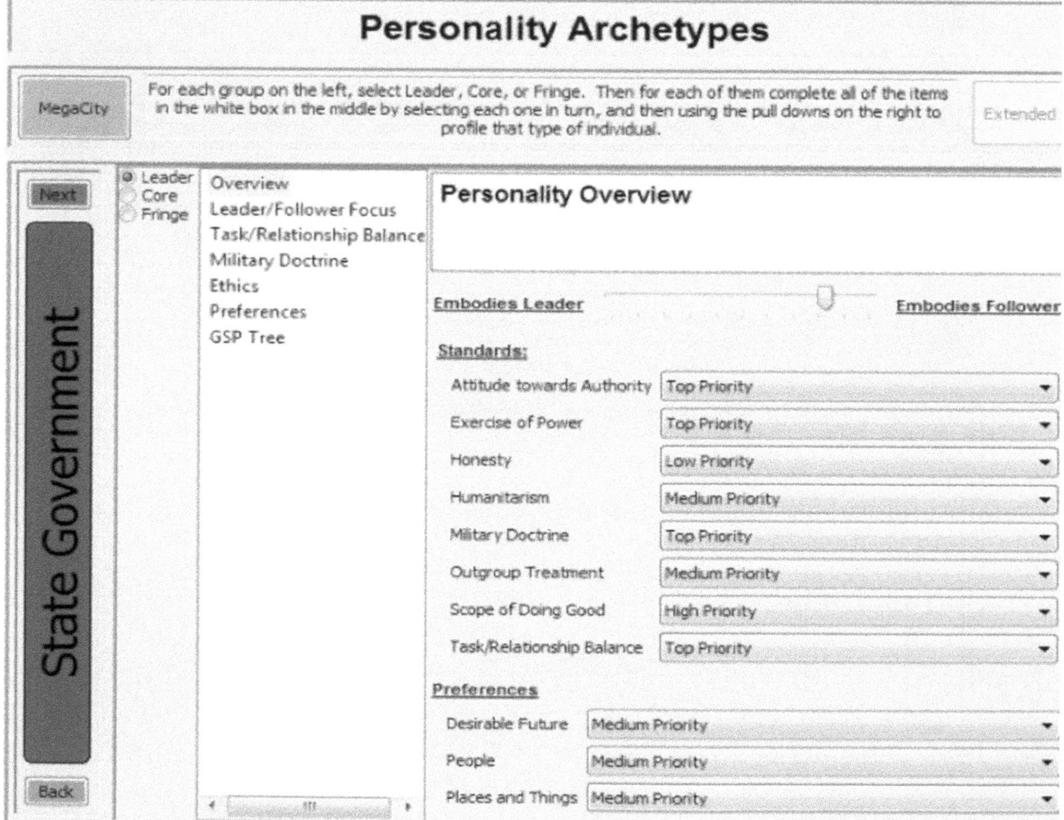

FIGURE 17.6
Leader and follower personalities, norms, and value systems are profiled in the StateSim's GSP tree structure.

may not be achieved in the scope of a scenario. In StateSim agents, this translates into a weighted hierarchy of territories and constituencies (e.g., no assets of leader *X* in resource *Y* of territory *Z*). The standards tree defines the methods an agent is willing to take to attain his/her preferences and what code others should live by as well. Finally, the goal tree holds short-term needs that the agent seeks to satisfy each turn (e.g., vulnerability avoidance, power, rest). The GSP tree is a value model editor that allows one to (1) implement leader and follower profile instruments as nodes on the trees and (2) set the weights on the nodes, which in turn implements a personality profile. Figure 17.6 shows a portion of the screen for specifying the GSP trees of the various leaders and followers in the factions of the scenario. When we get to the repository (the section "CBR and the Repository of Past Cases"), we will see that one can bypass a lot of this step by simply reusing similar characters from past scenarios. For example, a military junta general and an insurgent foot soldier.

Institutions, Resources, and Economics

The economic system currently in StateSim is a mixture of neoclassical and institutional political economy theories. Institutions are used as a mediating force, which control the efficiency (corruption) of certain services and are able to be influenced by groups within a given scenario to shift the equitableness of their service provisions. Political sway may be applied to alter the functioning of the institution, embedding it in a larger political economy system inhabited by groups and their members.

The unit of interaction is the institution as a whole—defined by the interactions between it and the groups in the scenario (Figure 17.7). An institution's primary function is to convert funding into services for groups. Factions, in turn, provide service to the members. Factions, including the government, provide funding and infrastructure usage rights. In turn, each faction has a level of influence over the institution—which it leverages to change the service distribution. Influence cannot only be used to increase favoritism (for one's own group, for example) but can also be used to attempt to promote fairness. The distribution of services is represented as a preferred allotment (as a fraction of the total) toward each group. Institutions are also endowed with a certain level of efficiency. Efficiency is considered the fraction of each dollar that is applied to the service output, as opposed to lost in administration or misuse.

The balance of this section focuses on public institutions, but the StateSim repository also models private and business enterprises that the actors may manage, work at, get goods and services from, and so on. Also, it is possible to substitute more detailed, third-party models of these institutions and enterprises without affecting the ability of StateSim's cognitive agents to interact with them. Thus, the models discussed in this section are defaults, and one can swap in other models without affecting how the actors think through their resource-based, ethnocultural conflicts.

The public institutions currently modeled and well represented in the repository are public works, health, education, legal protections, security, and elections. Public works provide basic needs, such as water and sanitation. Health and education are currently handled by a single institution, which handles healthcare and K–12 schools. Legal protections represent the law enforcement and the courts that enforce laws. Their service is the expectation to protection of full rights under law, as well as to basic human rights. Security carries out defense and attack actions ordered by group leaders. The electoral institution establishes the process by which elections are performed and handles vote counting and announcement of a winner.

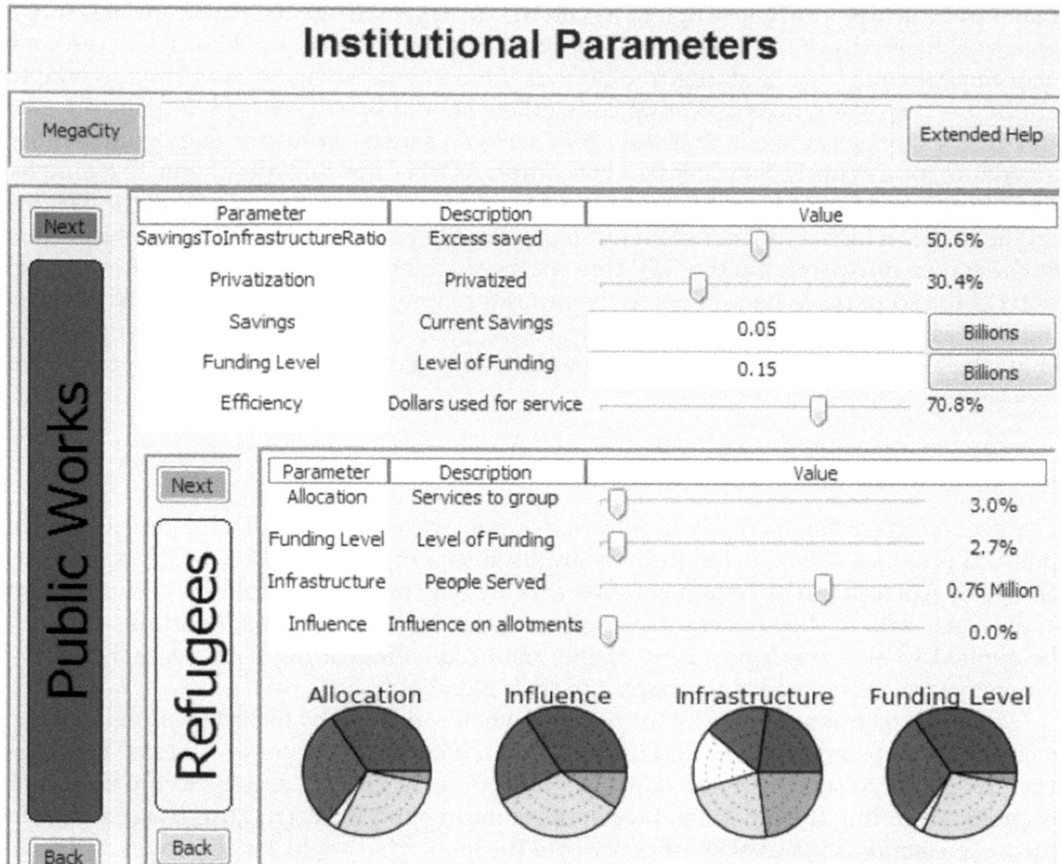

FIGURE 17.7
For each institution (left side), the inner box of sliders need to be set for each faction (near left) to reflect their influence on and funding of the institution as well as their portion of infrastructure and allocations. The outer box shows across factions.

At the macrolevel, this framework makes it fairly straightforward to implement ideas such as the Nobel prize-winning dual sector theory. This argues that a developing nation often includes a small, modern technology sector (faction) run by elites. They exploit a much larger, poor agrarian faction, using them for near-free labor and preventing them from joining the elites. This gives rise to the informal economy faction that provides black market income and jobs and which also may harbor actor intent on chaos (rebellion, insurgency, coup, etc.). Whether or not there is a malicious intent to overthrow the current government and elites, the presence of the informal sector weakens the formal economy (elite faction) by drawing income and taxes away from it and by potentially bribing its institutions and actors to look the other way. We set up many of our country models with these types of factions and the repository thus has these cases.

The followers of each group represent demographics on the order of millions of people. To handle the economic production of each smaller demographic, a stylized Solow growth model is employed (Solow 1956). The specific parameter values of this model are collected in Figure 17.8 and depend on the status of the followers. Each follower's exogenous Solow growth is embedded inside a political economy, which endogenizes the Solow model

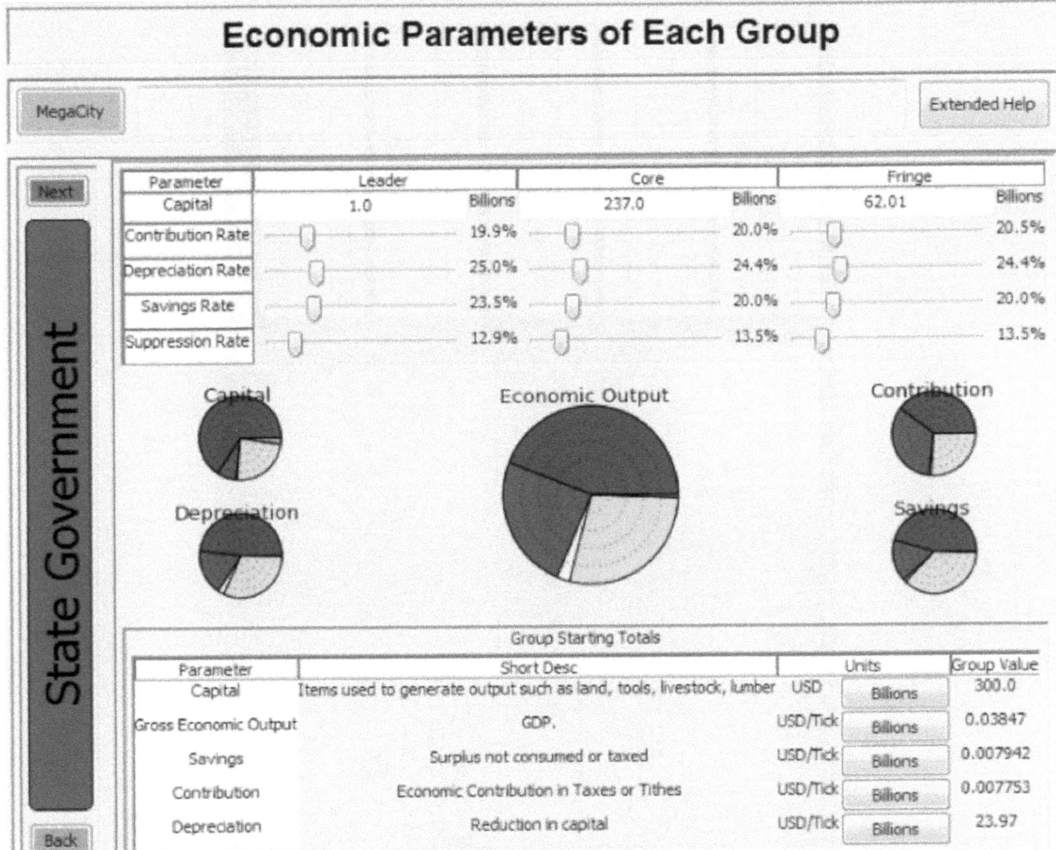

FIGURE 17.8
Third world, developing economies are profiled via these sliders that reflect capital, depreciation, output (GDP), member contributions, and member savings.

parameters. Some parameters remain exogenous, such as savings rate—which is kept constant through time. As savings rates are modeled after the actual demographics in question and the time frame is usually only a few years, fixing the parameter seems reasonable.

Each follower demographics' production depends on their constituency size, capital, education, health, employment level, legal protections, access to basic resources (water, etc.), and level of government repression. These factors parameterize the Solow-type function, in combination with a factor representing technology and exogenous factors, to provide a specific follower's economic output. The economic output of the followers is split into consumption, contribution, and savings. Consumption is lost, for the purposes of this model. Savings are applied to capital, to offset depreciation. Contribution represents taxation, tithing, volunteering, and other methods of contributing to group coffers. Both followers and groups have contributions, with groups contributing to any supergroups they belong to. An individual follower's contributions (tithes, bribes, taxes) are the primary source of growing groups' economy resources.

Using the interview screens of this section, the user eventually converges on a completed scenario description. When this is complete, the user selects a button labeled *finish*, which precipitates a dialog to check for gaps and errors and to confirm that the user is

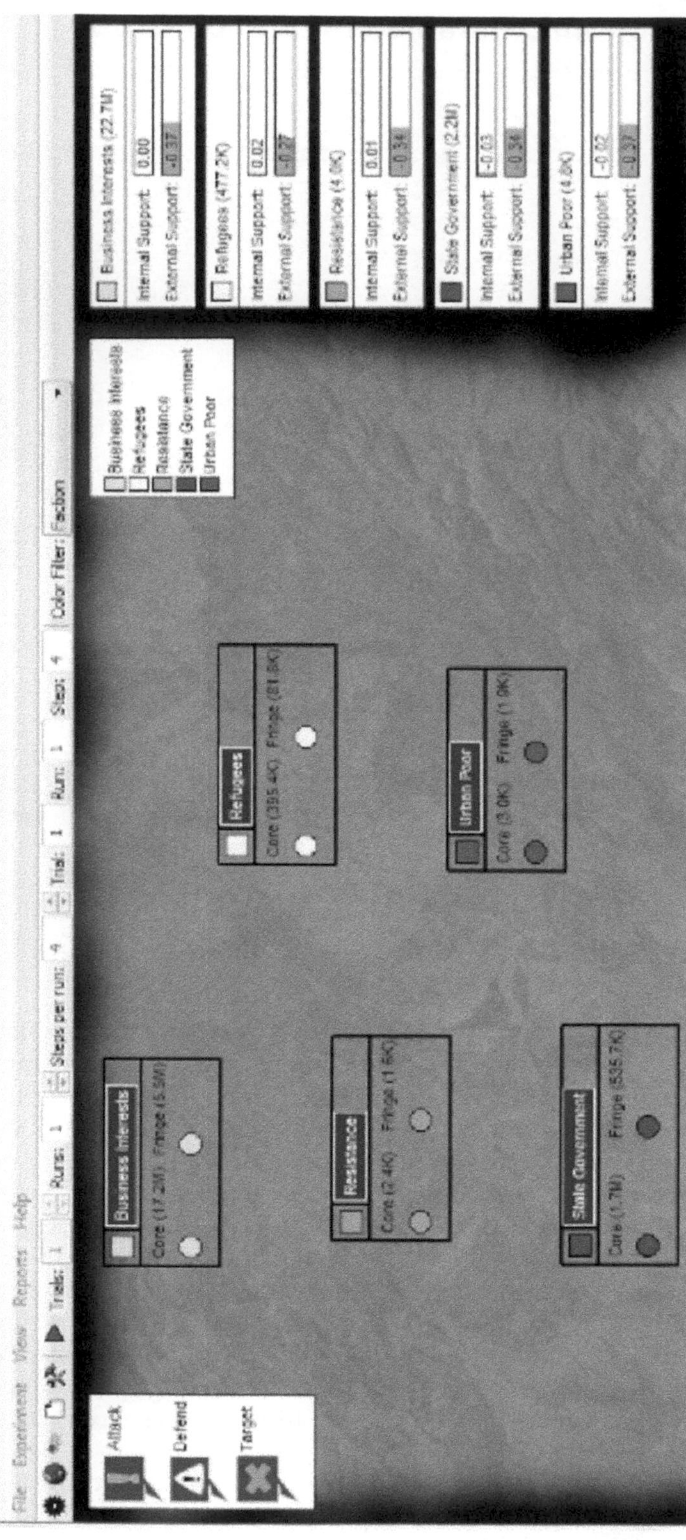

FIGURE 17.9
StateSim model produced for Xanadu.

ready to produce the models. If they confirm this, the generator then produces a working StateSim model out of the scenario description, the selected analogous parts, and the reusable components. Figure 17.9 shows the model it produced for this scenario. This is the standard screen produced for all StateSim models. It shows the groups and the actors on a generic terrain (brown background). The leader agents are small squares in each group box, while the follower archetypes are the small circles with population sizes in parentheses. The level of membership (sentiment) in the various groups is shown as histograms on the right side. At this point, one can run the StateSim to have it play games and evaluate conflict scenarios about the region that has been modeled (as described in earlier the section "Architecture and Cycle of the Case-Based Generator"). We will not run the StateSim in this chapter due to page limits. For more details about the StateSim models, consult the *StateSim Manual* (see Silverman et al. 2010). Examples of StateSim model runs are in the studies by Silverman et al. (2007, 2010) and Bharathy and Silverman (2012).

CBR and the Repository of Past Cases

Filling out the screens of the preceding section for each group, actor, institution, and so on can take up to a day for a small conflict where the country expert knows it well and has all the data easily accessible. This can take several days for a complicated, many-party conflict scenario. If the user is not an expert, this process may take weeks of research and data collection unless one accepts all the default settings that the generator suggests.

With respect to this last point, the generator works on the assumption that generic archetypes can be reused across scenarios (as do all CBR systems). Recall from Figure 17.3 that once the archetypes are selected, many defaults are filled in. For instance, there are three military groups (advanced, third world semitrained, and untrained militias). Selecting one of these qualifiers or dispositions of the military group fills in many of that group's attributes and the personalities and skills of the various group members. The user can inspect these and alter them in the subsequent screens shown in the section "The Generator Elicitation Screens," but there is a lot of autocomplete in this archetype assumption. Likewise, once the user fills in high-level attributes of the institutions, the resources, and economy screens, the generator will automatically scale the remainder of the parameters for the default archetype institutions, resources, and economy variables. That is the point of the CBR cycle described in the section "Architecture and Cycle of the Case-Based Generator." The generator speeds up retrieval through the reuse of the chosen mix of default FAIRE archetypes (factions, actors, institutions, etc.).

A concern of all CBR systems is that if this is done poorly, it can be irritating and even cause delays in editing. Consider the autocomplete and autocorrect features of many word processors and text editing programs. Sometimes they work well and save keystrokes, and other times they block typing the intended word or wording. Thus, how to program the retrieve and revise steps is not always transparent, even for simple domains like editing words. It only gets more involved for editing entire scenarios.

To try and mitigate such concerns, this section reviews the progress we have made thus far in designing and deploying other tools to help users improve how they tune the generator screens. The basic idea is to go beyond the default archetypes and let users directly retrieve and revise elements of past repository cases and elements of different cases that might be helpful if merged together into the current case.

Before discussing those tools, it is worth explaining the repository of past cases in more detail. In the literature, this is referred to as a case base. In our work, it is called the StateSim model repository, or simply, the repository. This repository is organized as a generic library of easily reusable/editable groups, institutions, and actors who are drawn from our prior StateSim models. These are the archetypes that can be quickly edited into a new scenario. Our prior models serve as a starting *empirical* base for a reuse library since this constitutes several hundred models of political actors from around the world (Pacific Rim countries, South Asia, the Middle East, and Africa). These were developed by political scientist SMEs that we hired for various Defense Advanced Research Project Agency (DARPA) projects. As discussed, they used the FAIREST questionnaire (now screens of the section "The Generator Elicitation Screens") to profile all the parameters of these models.

To turn this collection of standalone models into a coherent repository, we first created a structural framework following the FAIREST ontology for organizing the information. We then imported it into the StateSim repository, integrating diverse scenarios from Afghanistan, Iraq, Bangladesh, Thailand, Vietnam, Korea, and Sri Lanka among others. The repository structure contains a faction layer organized by political position, a leader layer organized by personality type, a GSP layer organized by congruence values, an institution layer organized by functionality, and a scenario layer organized by participating group alignments supported by master narratives. We combined all these files into a single large repository file for ease of reference, removed duplicates, and replicated scenarios in the combined file. We also preselected certain past case elements to be the default archetype suggested by a given qualifier (e.g., advanced/trained military, semitrained/third world military, untrained militia). The other cases are still in the repository but are not used by version 1 of the generator.

Additionally, we have begun adding repository tools that might help the user to get more out of the repository. These are not finished, but we explore two tools here (heat maps and spider plots) to raise the idea and to start a dialog about what might be fruitful. Specifically, as a first step, we created an algorithm that computed the congruence (similarity) between repository agents and displayed these in heat maps—a visual display of the extent of the similarity between GSP trees. We have created a functionality capable of generating a heat map from an N-by-N congruence matrix of GSP trees sorted according to the strength of congruence between GSPs (see an example matrix heat map in Figure 17.10). This is for government leaders; other heat maps exist for insurgent leaders, drug lords, elites, etc.

Heat maps are useful for helping the scenario designer to select the reusable past scenario elements (groups, agents, institutions) to retrieve and reuse for a specific purpose in the new scenario. To make use of this, we upgraded it both (1) as an algorithm to search for and output a vector of GSPs for creating flexible lists of comparisons (we also explored expanding this into other dimensions such as social relations, economic and resource parameters, and institutional factors) and (2) into a flexible, interactive system that allows the user to explore what agents are in the repository and to create any comparative matrices they want. The former capability is an attempt to start developing more of the CBR algorithm, but first we wanted to see how users make use of the repository. So, the latter capability provides this capability, although it is not currently wired into the screens of the section "The Generator Elicitation Screens."

Let us illustrate this for an agent that might be needed. Table 17.1 illustrates this process. The left side shows the possible actor archetypes that it found in the repository. The user has selected *military leader*. The right side then displays all the military leaders that the repository knows about and a heat map showing how different they are from each other.

Congruence	Color
1	
0.8 to 0.99	
0.6 to 0.8	
0.4 to 0.6	
0.2 to 0.4	
0.0 to 0.2	

(a)

	HG_Burea	HG_Gov_L	Thai_Gov_	Thai_Bure	Lanka_Go	Thai_Bure	Thai_Gov_	HG_Gov_L	Viet_Cent	Iraq_Bure
HG_BureaucracyApparatus_Leader_GSP_N	1	0.55687	0.55687	0.77437	0.55687	0.77437	0.65096	0.55687	0.70152	0.39796
HG_Gov_Leader_GSP_N	0.55687	1	1	0.50125	1	0.50125	0.68214	1	0.47522	0.19701
Thai_Gov_Leader_GSP	0.55687	1	1	0.50125	1	0.50125	0.68214	1	0.47522	0.19701
Thai_Bureaucracy_Leader_GSP	0.77437	0.50125	0.50125	1	0.50125	1	0.66735	0.50125	0.70971	0.37599
Lanka_Gov_Leader_GSP	0.55687	1	1	0.50125	1	0.50125	0.68214	1	0.47522	0.19701
Thai_Bureaucracy_Leader_GSP_AgConsp	0.77437	0.50125	0.50125	1	0.50125	1	0.66735	0.50125	0.70971	0.37599
Thai_Gov_Leader_GSP_OrigThaks	0.65096	0.68214	0.68214	0.66735	0.68214	0.66735	1	0.68214	0.5244	0.22916
HG_Gov_Leader_GSP	0.55687	1	1	0.50125	1	0.50125	0.68214	1	0.47522	0.19701
Viet_Central_Gov_Leader_GSP	0.70152	0.47522	0.47522	0.70971	0.47522	0.70971	0.5244	0.47522	1	0.4206
Iraq_Bureaucracy_Leader_GSP	0.39796	0.19701	0.19701	0.37599	0.19701	0.37599	0.22916	0.19701	0.4206	1

(b)

FIGURE 17.10
Heat map showing Congruence of Government Leader Agent GSP trees across diverse country leaders developed by different political country desk SMEs.

TABLE 17.1

Repository Evolution

Menu to Select Archetypes 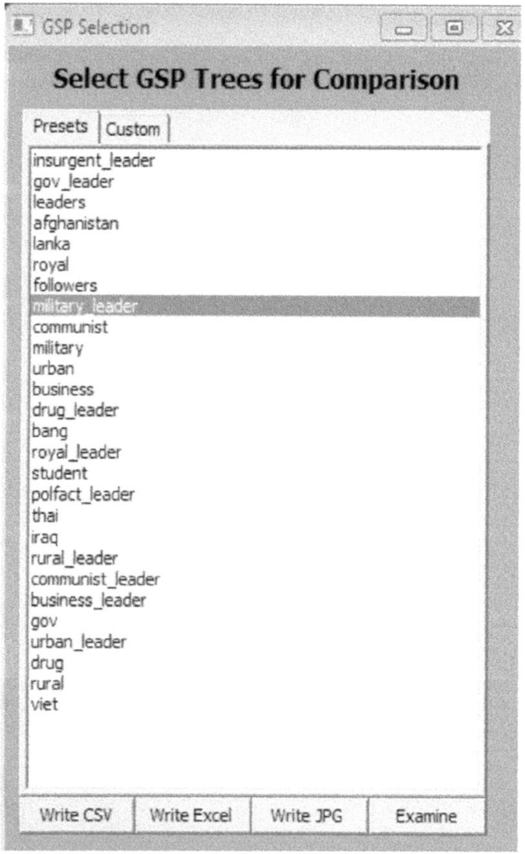	Views to Inspect How Similar or Different Particular Personalities Are That Fall within a Given Archetype 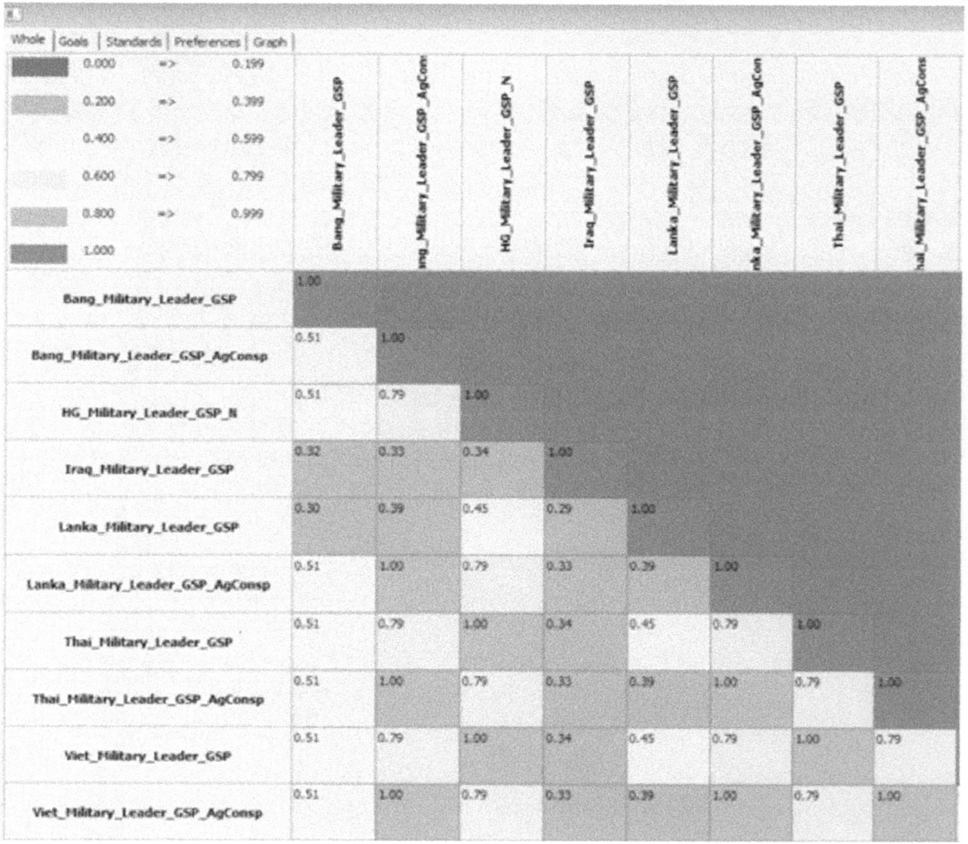

GSP Selection

Select GSP Trees for Comparison

Presets | Custom

insurgent_leader
gov_leader
leaders
afghanistan
lanka
royal
followers
military_leader
communist
military
urban
business
drug_leader
bang
royal_leader
student
polfact_leader
thai
iraq
rural_leader
communist_leader
business_leader
gov
urban_leader
drug
rural
viet

Write CSV | Write Excel | Write JPG | Examine

Whole | Goals | Standards | Preferences | Graph

0.000 => 0.199
0.200 => 0.399
0.400 => 0.599
0.600 => 0.799
0.800 => 0.999
1.000

	Bang_Military_Leader_GSP	ang_Military_Leader_GSP_AgCon	HG_Military_Leader_GSP_N	Iraq_Military_Leader_GSP	Lanka_Military_Leader_GSP	nka_Military_Leader_GSP_AgCon	Thai_Military_Leader_GSP	hai_Military_Leader_GSP_AgCons
Bang_Military_Leader_GSP	1.00							
Bang_Military_Leader_GSP_AgConsp	0.51	1.00						
HG_Military_Leader_GSP_N	0.51	0.79	1.00					
Iraq_Military_Leader_GSP	0.32	0.33	0.34	1.00				
Lanka_Military_Leader_GSP	0.30	0.39	0.45	0.29	1.00			
Lanka_Military_Leader_GSP_AgConsp	0.51	1.00	0.79	0.33	0.39	1.00		
Thai_Military_Leader_GSP	0.51	0.79	1.00	0.34	0.45	0.79	1.00	
Thai_Military_Leader_GSP_AgConsp	0.51	1.00	0.79	0.33	0.39	1.00	0.79	1.00
Viet_Military_Leader_GSP	0.51	0.79	1.00	0.34	0.45	0.79	1.00	0.79
Viet_Military_Leader_GSP_AgConsp	0.51	1.00	0.79	0.33	0.39	1.00	0.79	1.00

Note: We have expanded the spider plot view to help users select agent archetypes to add to their own scenario.

As shown in the top set of tabs on the heat map, the user can directly inspect these differences for the goals, standard, or preferences.

To drill in still further, users can also click and select personality traits and rapidly assemble spider plots (see Figure 17.11) to compare different actors of interest. The repository contains many insurgent leaders, and some are shown in Figure 17.11. The left side of the figure lists just eight specific insurgent leaders from four different countries that the user selected to compare. The right side of the diagram shows a spider plot of the eight insurgent leaders compared in terms of five personality traits. We can see that four of the leaders are fairly similar for two of the traits ((1) be controlling and (2) outgroups are

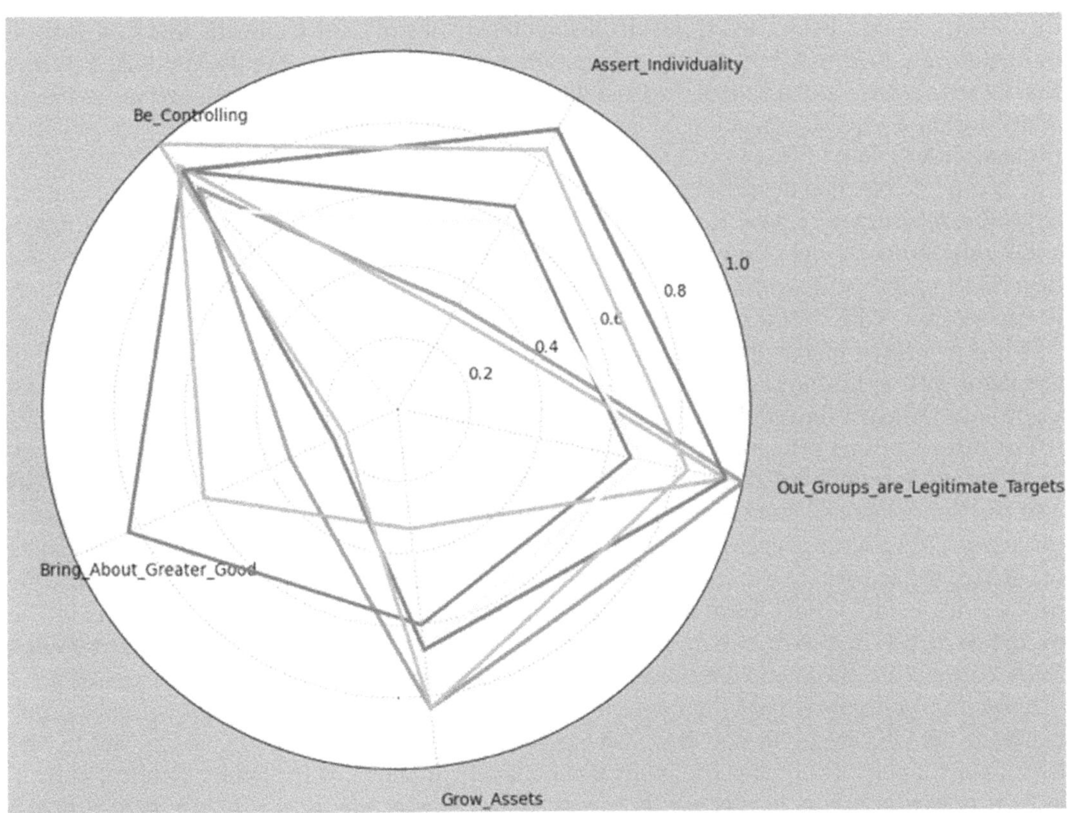

FIGURE 17.11
Spider plot tool showing the importance of five personality traits to eight insurgent leaders in four different countries.

targets), and differ along the others. There are many other traits one can explore in the GSP trees using this tool. In this fashion, the scenario designers tend to find that the spider plots are revealing and informative in helping them to think about what is similar and different about comparable leaders.

Conclusions and Next Steps

Model generation takes a significant amount of time and effort. A valuable approach to scenario editing and model generation is the ability to draw upon experience: to describe problems, reason about their possible impact, and build appropriate simulations by analogy with previous similar widgets. CBR attempts to approximate this ability. If this works well, it would speed up model generation and remove a bottleneck to having political conflict simulations. Such simulations can be invaluable, once validated, for helping policy makers to try out their interventions computationally to see what will be the expected and unexpected effects. Thus, they could make any mistakes in the simulated space and remove these before implementing policies in the real world.

The research presented here has made inroads into producing a rapid CBR generator for political conflict simulations. To do this, the section "Introduction and Purpose" explained the CBR cycle and presented the FAIREST ontology of political scenarios that has shaped our interview tool and that collects parameters for the StateSim set of the models. The section "Architecture and Cycle of the Case-Based Generator" then reviewed our progress on adapting the FAIREST interview into an index of the case base and into a set of screens that the user interacts with to describe the high-level attributes of the new scenario s/he is trying to model. We also described how the CBR algorithm makes use of generic archetypes from the repository of past StateSim models and the interview screens to then automatically generate a model of a new scenario. This is demonstrated for the Xanadu use case described in the Addendum. We further described the repository of past StateSim models in the section "CBR and the Repository of Past Cases."

While we have made progress with this first version of the generator, it falls short of the flexibility of human analogical reasoning, and there are still challenges ahead for an improved version. Two such challenges are discussed in the following.

The most obvious deficiency discussed in this chapter is that the full extent of the repository is not adequately utilized by the generator to date. As described in the section "CBR and the Repository of Past Cases," we had to preset the default archetypes that the generator reuses. While these are drawn from our past StateSim models, they are only a subset of the full repository. The section "CBR and the Repository of Past Cases" showed a couple of illustrative tools (heat maps and spider plots) that help the user see what else is in the repository and as a way to explore alternative analogs to reuse instead of the defaults. Fully connecting these and similar tools to the generator would be a useful next step.

A second concern is that the current version of the generator does not accommodate step 4 of the CBR cycle (update the case base). That is, while the repository can retain any new case that a user creates, the generator cannot make use of the new cases for its analogical reasoning. This is because the preset default cases are all it uses for steps 1 and 2 (retrieve and revise). That too will be changed in the next version. There is research in the CBR community about automatically learning new defaults from past cases and as the case base grows. We will tap into this approach to address this problem.

Megacity Use Case

Analysis of the Megacity Use Case to Fill Generator Screens

Below is our analysis of this book's common use case that we performed in order to extract data to fill out the screens of the generator. The initial versions of those screens are shown in the body of the chapter and further explained here from the viewpoint of the use case. The use case itself, follows this analysis.

Introduction

Our chapter described a rapid scenario generator using a simple and relatively generic five-group scenario as an illustration. It turns out that these five groups occur in the Xanadu megacity scenario as well. To implement that full scenario, we need only to amplify what was already described in the steps and the screens of our chapter. The complete generator approach entails four phases (selector input sources, generator pull phases, model check and experiment setup, scenario outcomes and metrics),* we will be primarily concerned with the second phase (generator pull phase) and later. The generator pull phase is the procedure during which we convert the common conflict picture (a verbal and quantitative description of the situation) to more detailed and organized data that will be later used in the modeling phase. Since the megacity information lacked exact quantitative information, we will also be illustrating ways to estimate the numbers to aid the modeling.

In the following, we will go through the generator pull phase in order to identify major actors, groups, and sources of instability and alliances/coalition strength and intergroup relationships; adapt central issue terms; and select conflict graph position. Then we will attempt to quantify some of the important parameters with reasonable estimates and build the simulation model. Finally, we will discuss possible variants of the generation phase and its applications.

Generator Pull Phase

We are able to first construct (extract) a structured summary of the narrative as the following:

Scenario Name	Megacity Xanadu
Regions	Northern/west/central
Key factions/groups	Directly mentioned in text: state government/Bursuka refugees/North Razini refugees/urban elites/urban workers/rural workers/organized criminal groups/international community; induced from text: business interests group
Description of the tensions and conflicts	• State government does not care about (turns a blind eye to) the refugees. • Local workers treat rural workers as secondary citizens, and refugees as worse. • Organized criminal groups exploit the refugees. • State or government does not care about (turns a blind eye to) organized criminal groups as long as the northern region remains stable.

(Continued)

* A detailed poster is not included in this book chapter due to space constraints, but is available upon request.

Scenario Name	Megacity Xanadu
Institutions	Law/economy/health/education/public works
Central issues	• Lack of institutional provisions and engagements from the state government (limited resources) • Discrimination among the groups • Deteriorating living situation for the refugee groups • Mafia groups exploitation activities upon refugee groups
Group relationships	Can be deduced from the text, estimated in the following table
Instabilities	Unsustainable peace/high likelihood of conflict

From the generator pull phase, we observe the factions in the scenario as in the chart as follows. As we establish the major factions in the scenario, we are then able to fill in the first screen in the generator—group archetypes (Figure 17.3). We first select the social archetypes that exist in the scenario, select the variant of the archetype (induced from provided text), and add subgroups under one particular scenario when necessary.

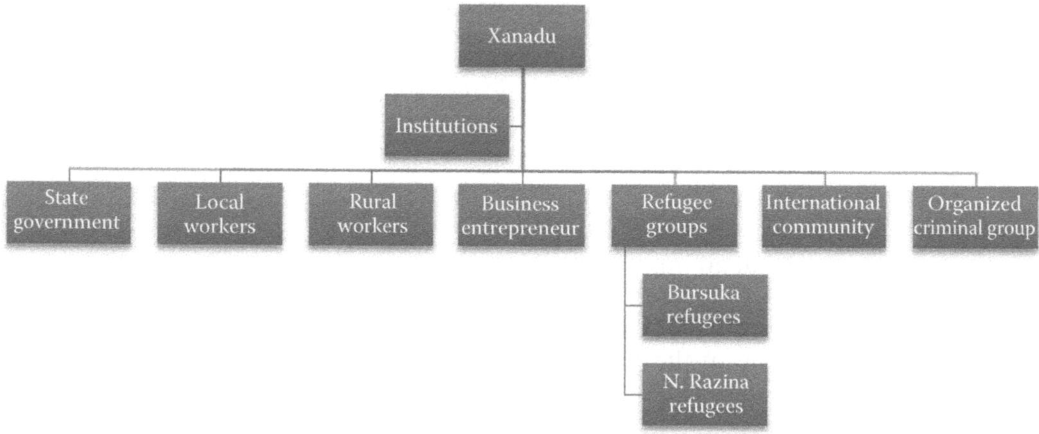

In addition to these directly observed elements of the scenario write-up, we were able to induce a number of parameters that were implied albeit without clear quantification. For example, the StateSim generator models intergroup relationships as ranging from −1 to +1, with 0 being the neutral level. Based on the Xanadu write-up, we roughly estimated the intergroup relationships from the provided narrative as follows:

	State Government	Bursuka Refugees	North Razini Refugees	Urban Workers	Rural Workers	Organized Criminal Group	International Community	Business Interest Group	Urban Elites
State government	1	0.2	0.1	0.8	0.6	0	0.5	0.7	0.9
Bursuka refugees		1	0.5	0.3	0.4	−0.5	0.6	0.4	−0.2
North Razini refugees			1	0.2	0.3	−0.6	0.6	0.3	−0.3
Urban workers				1	0.2	0	0.5	0.6	0.4
Rural workers					1	0	0.5	0.5	0.3

(Continued)

	State Government	Bursuka Refugees	North Razini Refugees	Urban Workers	Rural Workers	Organized Criminal Group	International Community	Business Interest Group	Urban Elites
Organized criminal group						1	−0.5	−0.2	−0.6
International community							1	0.6	0.3
Business interest group								1	0.8
Urban elites									1

We would use the estimated intergroup relationship values from this table to set the strength of the relationship between each group; while not required, the relationship value is assumed here to be symmetrical (A to B is the same as B to A) and ranges from −1 to 1 with −1 being the most negative and 1 the most positive. The input process is done faction by faction. Figure 17.4d in this chapter shows the beginnings of transferring this data to the generator screens.

The text also identified several sources of instability (in the scenario) such as class level, haves versus have nots, criminal element and informal economy, and possibly a religious outgroup. In each case, the scenario text specified three elements—the severity of the issue, the perpetrator, and the victim—although this again was unquantified. Nevertheless, this is sufficient information to allow one to fill in the instabilities and the coalition screens of the generator, and we have given some illustration of that in the body of the chapter (Figure 17.4a and b). In the coalitions screen (Figure 17.4b), we designated the state or the government as the head of one collation, and the business interests as members of this coalition. We also illustrated a resistance, possibly headed by criminal elements with the refugees as a loose affiliate. One could flesh this out still further for the larger set of instabilities and coalition members, although the kernel of it is illustrated in the chapter.

As concerns individual actors, leaders, and followers, the scenario is rather skimpy on detail. However, the generator and the repository are filled with actors from similar scenarios, and one could readily sketch out some hypothetical personalities as in Figure 17.5. For any real-world scenario, there is always more detail available from news reports, social media, expert accounts, and various social science datasets.

The generator also requires estimates of the population size, the core/fringe divides, the GDP size, and the economic resources available to each subgroup. We had very little information about this in the write-up, so we assumed a modest sized population in a third world country. For a lack of any detail, we assumed an 80:20 rule where the haves might hold 80% of the wealth (resources, economy, etc). The have nots might apply this rule again with the second class citizens hold 80% of what is left, and the remaining 20% for the third class refugees. One possible depiction of this type of allocation might be as follows, although many such hypotheses could be explored and entered into the generator as we began to illustrate in Figures 17.4c and 17.7.

Total: 20 Million People Economic: US$40 Billion	**Population Percentage**	**Core/Fringe Ratio**	**Economic Percentage**
State government	34%	3/1	35%
Bursuka refugees	4%	5/1	2%
North Razini refugees	2%	5/1	1%

(Continued)

Total: 20 Million People Economic: US$40 Billion	Population Percentage	Core/Fringe Ratio	Economic Percentage
Urban workers	20%	3/1	15%
Rural workers	20%	4/1	10%
Organized criminal group	1%	6/1	3%
International community	1%	1/1	4%
Business interest group	10%	5/1	15%
Urban elites	8%	3/1	15%

The remaining estimates needed for the generator concern the public goods institutions, their capacities, and their allocations to the various groups of the scenario. In the megacity Xanadu case, most institutions are controlled by the state government, with the exception of religious education (by North Razini refugees) and black market (by organized criminal groups). Figure 17.6 showed the beginnings of this as well as the supplemental material holding more detailed information about each institution's key parameters and its impact on each faction. The allocations are largely done in a way to resemble our estimated economic resources distribution. Also the allocations were done in a way that more resources are allocated to factions that have a positive relationship with the controlling faction of the institution, and less to factions that have a negative relationship with the controlling faction. The parameters of each institution are estimated by reviewing its counterpart in contemporary megacities of developing countries.

Possible Variants in Modeling Phase and Use Case Application

The details of any scenario are open to further fine-tuning and modification on the generator platform—at almost every step along the way, we are able to adjust factors like severity of a certain type of conflict, group relationships, institutional parameters, and so forth. From a simulation perspective, this lets one create alternative hypotheses of the same initial scenario and see how different parameters would affect the outcome of the simulation in later stages. With the megacity Xanadu use case in mind, the following is a list of possible alternative scenarios we may create to conduct a detailed impact analysis:

- We may alter *the nature of relationship* between rural workers and refugee workers to see how that would impact the resource distribution and the economic situation of the refugees.
- We may alter *the initial resource allocation* among different factions to explore its impact on political stability.

TIP

The USAID trains provincial reconstruction teams. These teams are small, joint civilian–military organizations whose mission is to promote governance, security, and reconstruction throughout a given region or country. According to US AID, the only way to get a holistic situational awareness of a region is to focus on the population rather than on the enemy in the scenario. This includes developing an understanding of the population in terms of items like FAIREST as well as their own perceptions as to what are the key issues and how they can be resolved.

- We may adjust *the severity of different sources of instability* to explore their impact on overall stability of the megacity scenario.
- We may adjust *the personality traits* of certain leaders to explore the likelihood of certain events (peace or war, for example).

The central theme of this flexibility in our generator platform is to not only model the most accurate scenario descriptions, but also instantiate a wide range of alternative scenarios to better understand the implications of changes.

Acknowledgments

We gratefully acknowledge the support and guidance of Combating Terrorism Technical Support Office/Pentagon and SoSACorp for the research described here. Thanks also to Qing Du for helping with scenario entry and screen captures. Any errors or misstatements are the responsibility of the authors.

DISCUSSION QUESTIONS

1. Can you describe a modeling domain that is likely to have numerous past cases that could be the basis of a model repository? Describe that domain and explain the structure and the main equations of the models that will likely serve as the past cases.
2. For your repository from question 1: (1) describe the top several layers of a taxonomy of important dimensions, parameters, and possible ranges of values that comprise the models of interest and (2) classify a couple of illustrative models in terms of this taxonomy. (**Note**: *One way to help organize such a taxonomy is to create a class diagram showing object–attribute–value triplets as in the unified modeling language.*)
3. Construct a table comparing the strengths and the weaknesses of applying this repository and taxonomic classification scheme according to the algorithm of Figure 17.1 to generate models in your chosen domain.
4. Is this a useful approach? Can you propose other (better) methods for speeding up the generation of models in your chosen domain? Please elaborate on what added benefits and costs these methods facilitate.

References

Aamodt, A., and Plaza, E. 1994. Case-based reasoning: Foundational issues, methodological variations, and system approaches. *Artificial Intelligence Communications*, 7(1994): 1, 39–52.

Bharathy, G., and Silverman, B. G. 2012. Holistically evaluating agent based social system models. *Simulation: Transactions of the Society for Modeling and Simulation International*, 98(1): 102–135.

Numrich, S. K., and Picucci, P. M. 2012. New challenges: Human, social, cultural, and behavioral modeling. In Tolk, A., *Engineering Principles of Combat Modeling and Distributed Simulation*, 641–667.

Silverman, B. G. et al. 2010, October. User manual for the StateSim viewer (Technical Report). Philadelphia, PA: ACASA/University of Pennsylvania.

Silverman, B. G., Bharathy, G. K., Nye, B., and Eidelson, R. J. 2007, December. Modeling factions for "effects based operations": Part I—Leader and follower behaviors. *Journal of Computational & Mathematical Organization Theory*, 13(4): 379–406.

Silverman, B. G., Pietrocola, D., Nye, B., Weyer, N., Osin, O., and Johnson, D. 2012, Rich socio-cognitive agents for immersive training environments: Case of nonkin village. *JAAMAS*, 24(2): 312–343.

Robert, M., and Solow, R. M. 1956, A contribution to the theory of economic growth. *The Quarterly Journal of Economics*, 70(1): 65–94.

Zacharias, G. L., MacMillan, J., and Van Hemel, S. B. (Eds.). 2008. *Behavior Modeling and Simulation: From Individuals to Societies*. Washington, DC: National Academies Press.

18

Detailed Model Development Case Study: The Peace Game

Matthew J. Powers

CONTENTS

An Introduction to War Gaming 362
The Peace Game 363
Peace Game Scenario Generator 367
  Map 367
  Resources 369
  Players 369
  Regions 370
  Units 372
  Weather 373
Gameplay 375
  Force Movement 376
  Negotiations 376
  Market 381
Conclusion 381
Megacity Use Case 381
  Xanadu Use Case: The Peace Game 381
  Xanadu Map, Resources, and Players 383
  Region Definition 383
  Update Unit Types 386
  Update Players' Units 386
  Weather 390
References 390

ABSTRACT The Peace Game is a flexible war gaming-based software package that captures the complexities of conflict resolution in peace and stability (and other cultural-centric) operations that are underway in regions such as South Sudan and Afghanistan. The Peace Game facilitates multiple players assuming roles as government leaders, military and coalition commanders, NGO representatives, or various militia camps as they attempt to cooperate with or undermine the strategies being implemented by other players. The Peace Game is a manifestation of rigorous, effects-based assessment approaches to operations.

KEY TOPICS

- War gaming generates and explores new ideas or hypotheses.
- A method (unlike methodology) is simply a way of doing something.
- Military leadership relies on war games throughout planning and assessing.
- Epistemology studies the extent of knowledge in complex environments.

An Introduction to War Gaming

> A war game is a warfare model or simulation that does not involve the operations of actual forces, in which the flow of events affects and is affected by decisions made during the course of those events by players representing the opposing sides.[1]

War gaming is a method that harnesses military thinking that requires commanders and decision makers to better understand the environment in which they operate. War gaming aligns with epistemological theory in that practical knowledge emanates from immersing oneself in these modeled environments. War games make no assumption toward how correct an applied strategy is for the projected warfare scenario; rather they facilitate an understanding through capturing operational complexities to include the element of human decision making.[2] That is, a war game's purpose is not to answer any specific question with a point solution. Its purpose is to gain insight into complex questions in order to generate a better analytical focus at the appropriate level of analysis.[3]

War gaming is not designed to predict. War gaming, in and of itself, is not analysis. It is "most productive when used as an organizing and exploratory tool" to inspire ideas that otherwise may have gone unconsidered.[1] The thoughts that cultivate themselves in the minds of the decision makers throughout the gaming process expand beyond the battlefield into the political and behavioral arenas. The US, coalition, and North Atlantic Treaty Organization peacekeeping forces recognize this expanded arena and have used tabletop war gaming as a decades-long mainstay to facilitate COA development during their planning processes. But the nature of tabletop games limits the ability to address the type of what-if scenarios likely to occur in complex cross-cultural peace and stability operational environments. Adjudicating simultaneous human interactions with the myriad of variables that must be considered normally requires panels of SMEs and extensive deliberations for each and every action. Computers blend war gaming with simulation so that player decisions are quickly adjudicated and the modeled environment takes a new shape, all the while calculating numerical and visual results that engage players to make new decisions for the next turn.[4] The course of the gameplay and the computer-enabled sophisticated analysis eventually generates hypotheses, and the elusive *Black Swan*, a highly improbable event that is unpredictable, has enormous effects on the real world and (once it has happened) tricks observers into believing that they would have seen it coming *if only…*[5]

War games are considerably cheaper to execute and assume much less risk than a live scenario. Software war games cost even less (front-end development funding notwithstanding) because they are easily transportable, repeatable, and can often be played over a network with the same benefits of a face-to-face war game. The Peace Game possesses all these benefits.[6]

The Peace Game

This section introduces the Peace Game and refers to the ongoing conflict between Sudan and the Republic of South Sudan (RSS) for contextual understanding. This scenario is for demonstration purposes only and, as such, is only a singular manifestation of infinite possible scenarios. The Peace Game is a simple, practical, stand-alone, process-driven model that harnesses human decisions in an adaptive environment. Simplicity and a common software platform (Java) allows for easy network capability and trivial funding. The gameplay is also simple. Players move forces across the map, manage and trade resources, and negotiate with other players in order to influence civilians or deter potential adversaries while the interface graphically displays the results of player actions. These results are calculated and stored via data-driven algorithms with stochastic elements. The Peace Game combines a war gaming framework with computer simulation to incorporate human decisions in a complex environment. A computer executes the game by shaping the environment according to, and in order to stimulate, player decisions. Modern computing power is amenable to the vast information and data required to reasonably model the nontraditional missions that encompass complex peacekeeping operations.[4] The Peace Game's most important algorithm is that which drives the population. Population behavior, while steeped in historical data, is actually motivated by "less-than-quantifiable subject-social-political issues and behavior" and assumes that human needs (along the lines of religion or ideological motivations) and limited resources are the main drivers behind conflict.[1] These factors, affected by player decisions, influence the population, and the resultant satisfaction level motivates the people to do one of the following[7]:

- Stay in their region.
- Move to the most desirable region within their state.
- Stay in their region but turn to crime, resulting in an immediate decrease in regional stability.
- Move to the region on the map with the highest aggregate levels of resources and stability parameters.
- Move to the region with the most availability of the resource for which the population is most starved.

ALGORITHM DESIGN

During the Peace Game's initial conception, the development team was modeling an unfamiliar, cross-cultural problem of conflict between Sudan and the RSS. The team naively fell into the trap of requiring hard data to model the real world in the form of a playable game. The fear of being wrong impaired imaginative methodology.[8] That fear was vanquished by modeling the unknown in the form of customizable variables such as force structure and capabilities (to name a few), thereby freeing the team to create an interactive environment. Gaining knowledge of culturally interacting environments throughout the gameplay is the main benefit of the Peace Game, not the exhaustive inclusion of data-driven analyses. This is the epistemological heart and soul of the Peace Game.[9]

- Join a terrorist group, thereby motivating terrorist players to decrease stability in regions that are not under their control.
- Die.

The population algorithm acts upon all regions on the map, thereby enabling unconsidered second- and third-order impacts that are in turn affected by the interaction of chance and human decision making in resource-constrained environments. Consider the geopolitical situation of Abyei, wherein the main drivers of this ongoing conflict are both a scarcity (for instance, grazing lands) and an abundance (oil) of natural resources.[10] In 2011, the International Organization for Immigration in Khartoum, Sudan, released *Migration in Sudan: A Country Profile 2011*. The study highlights important resource-centric factors (agriculture) as the main source of livelihood for the large rural population in Southern Sudan and that 30% of the population did not have access to an improved water source.[11] These studies motivate the nondecreasing exponential curve describing the numbers of people that react based on their satisfaction levels. Such a curve seems dramatic, but the exponential shape was chosen to ensure that the players are confronted by the desired decision making situations.[6] The exponential curve is further validated by what is known as the *Malthusian trap*. In his 1798 work "An Essay on the Principle of Population," Thomas R. Malthus formally states the now obvious truth that "population must always be kept down to the level of the means of subsistence." This essay is a baseline for the Peace Game's algorithm. The Malthusian trap is the point in which an exponentially growing population overwhelms the available resources.[12] Further discussion on resource impact on populations can be found in the next section within the subsection "Resources." Figure 18.1 is a visual representation of the population algorithm.

Figure 18.2 is the screen that enables customized population reaction curves according to local attitude toward each resource.

Another important algorithm is that which governs combat between forces, an embellished version of the well-known *Lanchester's Square (Aimed Fire)* law represented as

$$dy/dx = (\alpha/\beta)(x/y). \tag{18.1}$$

The population and the combat equations represent the Peace Game's dice. The outcomes of player decisions (such as combat or attempts at peaceful conflict resolution) are unpredictable due to unintended consequences of both violence and diplomacy. For example, if a battle takes place within a region, the region's security level drops, thereby causing discontent among the population (second-order effect). A follow-up attempt at

POPULATION BEHAVIOR CURVES

The population algorithm assumes an exponential curve to model the number of people that will act according to their satisfaction level. That is, the amount of civilians willing to act dramatically increases when a desired resource set drops below an acceptable threshold. The Peace Game provides default curves that can be modified according to user requirements (see Figure 18.2). The use of an exponential curve is motivated by a similar model used to forecast political instability in Africa where overpopulation and resultant resource drain is identified as one of the most important factors leading to internal conflict.[13]

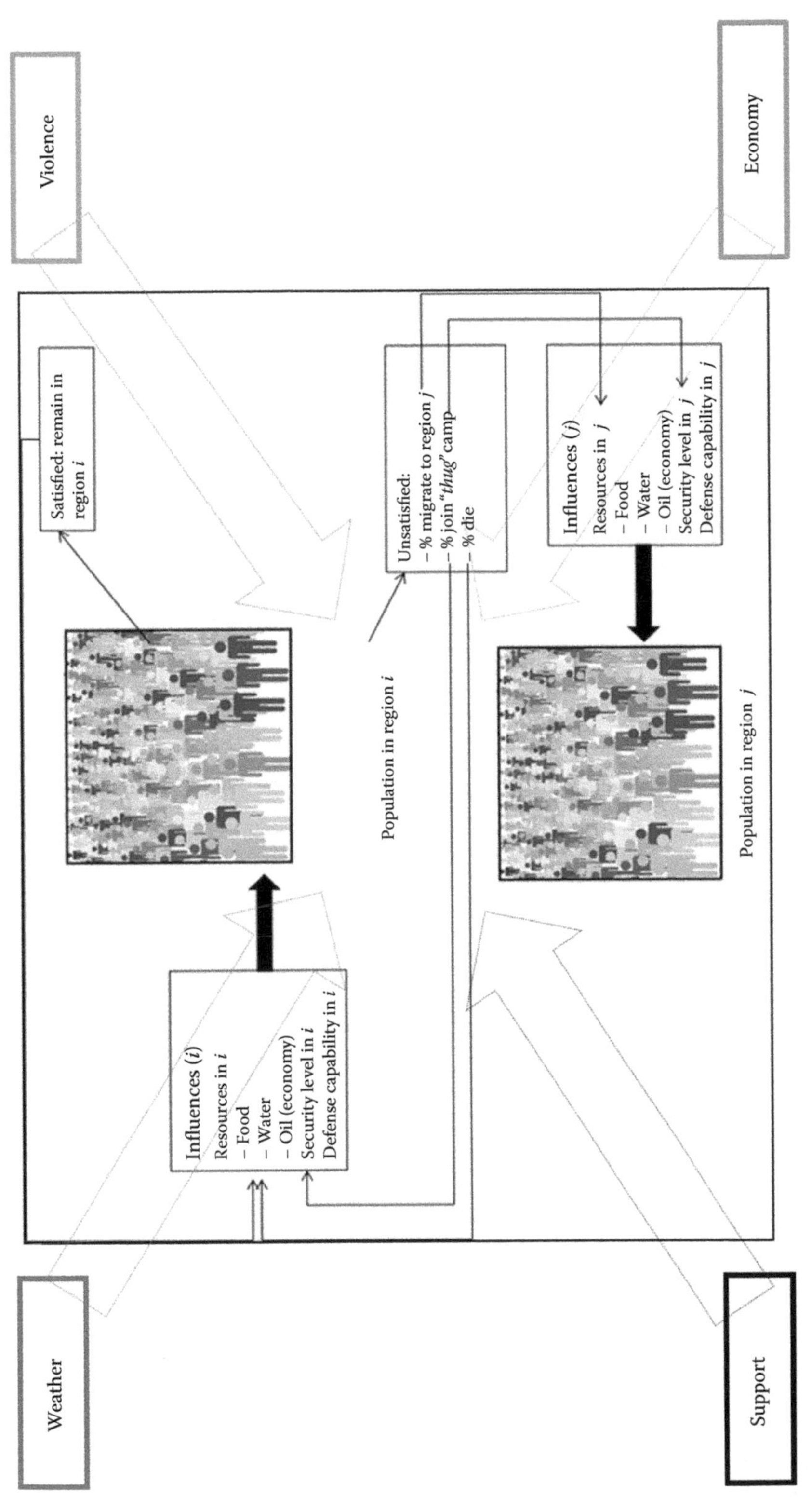

FIGURE 18.1
Depicting algorithm.

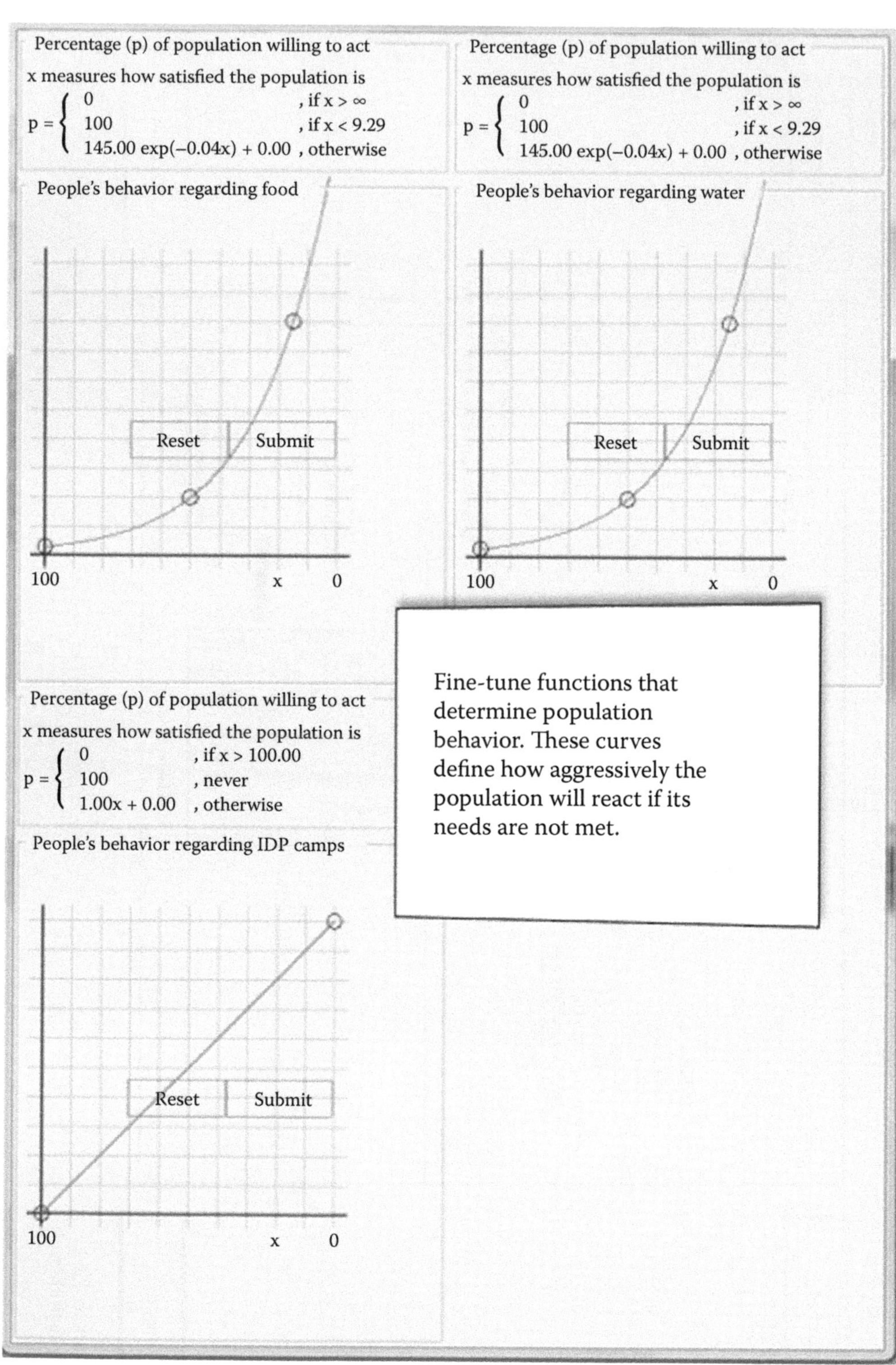

FIGURE 18.2
Depicting exponential curves.

COMBAT EQUATION 18.1

Equation 18.1 defines the change in forces y and x (dy/dx) as a function of the size of force y (y), the size of force x (x), and the attrition coefficients (combat effectiveness values) for x and y (α and β, respectively) using aimed fire (forces are aware of enemy location). This equation is embellished if a force is able to ambush by providing the ambush-enabled force first melee, thereby reducing the unaware force size prior to enacting the aimed fire algorithm in order to account for irregular warfare scenarios. The attrition coefficients (α and β) are generated as random variables along a Gaussian curve, with the mean value of each force's specific curve being a reflection of real-world relative fighting capabilities.

peace through resource and infrastructure development ignites unwelcome immigration according to the population algorithm, resulting in decreased stability (third-order effect). The next sections describe Peace Game scenario generation and gameplay, continuing the context of conflict between Sudan and South Sudan as a case study.

Peace Game Scenario Generator

The scenario generator is the Peace Game's preplay design interface for scenario development. SMEs input parameters according to the Peace Game framework, and these inputs feed the algorithms. Inputs range from big picture to specific details regarding variables that impact the modeled operating environment:

- Map (game board)—The map is where the scenario takes place. It does not need be contiguous or a real location.
- Resources—Resources drive the conflict/competition according to variable levels of importance.
- Players—Players are the game actors and their general capabilities.
- Regions—Regions are the subsets within the game board, each with its own parameters.
- Units (force types)—Units are specific game tokens (each with unique abilities) comprising player forces.
- Weather—Weather variably impacts natural resources throughout the gameplay.

Many of these elements are provided in the scenario generator's first screen (see Figure 18.3).

Map

The map for this case study includes Sudan and the RSS, thus enabling conflict between and within these states.

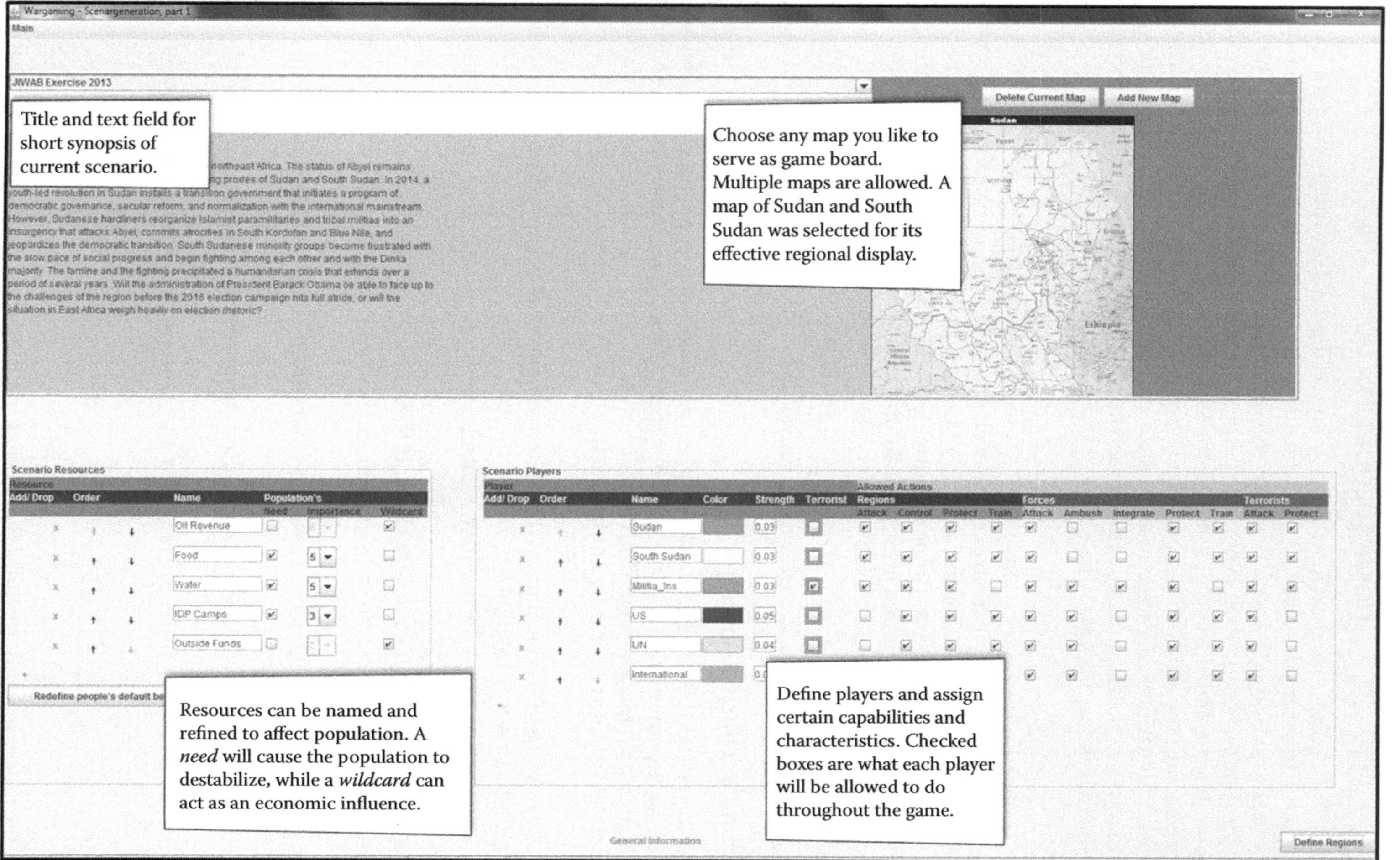

FIGURE 18.3
Scenario generator screenshot 1.

CULTURAL RESOURCES

Game designers are free to label resources along cultural lines. For example, Christianity or Islam represents cultural needs that may exist within regions in Sudan and South Sudan. If a resource (cultural or otherwise) is ranked 4 or 5, the population reacts according to an exponential curve, representing more extreme attitudes toward those cultural parameters. A rank of 1–3 results in a linear function, representing varying degrees of tolerance. Default curves can be manually manipulated if the situation requires a more aggressive attitude.

Resources

The Peace Game defines *resources* as regional variables that drive conflict and competition. These resources need not be physical in nature (such as food, water, or oil) but can represent cultural desires such as education or religion.

A unique, strategically significant resource included in this Sudan/RSS scenario is that of internally displaced people (IDP) camps that can be created as resources on the map throughout the gameplay. An alternate design strategy could be to create IDP camp regions. According to the UN Refugee Agency, about 400,000 new IDPs were registered between January and August 2014 in Sudan alone, and the influx from RSS is expected to continue in 2015.[14] The IDP camps resource is one of many ways to model the IDP/refugee camp effect on the population. By creating an IDP camp resource, modeling flexibility is maximized because US, UN, and international players have the option to develop camps anywhere on the map. Another modeling approach is to create subregions within geographic regions that are defined as IDP/refugee camps, and it is up to the players to manage the supply/demand of those camps. Thanks to the inherent flexibility of the Peace Game, there are often many ways to model the operating environment. However, great strategic care must be taken during the gameplay when considering where to place IDP camps. If regional IDP camps develop enough and the region becomes so desirable that it attracts migrant populations to a point beyond the absorptive capacity of that region, the algorithmic scales tip and stability decreases. This tipping of the scales represents a trap at the end of the Malthusian trap that occurs when attempts are made to solve problems regarding overpopulation and resources. The escape from the Malthusian trap stimulates urban population growth in several ways. Variable birth rates and an exponential population algorithm are used to model a rapid population growth in the Peace Game. A well-nourished population within a given region at the end of a round supports the intensification of the migration processes, allowing for unsustainable levels unless such an effect is anticipated by the acting player. Thus, the escape from the Malthusian trap engenders a rapid growth of population due to both natural increase and migration. This causes tension that, if unchecked, puts the player in a compounded version of a problem thought to have been solved.[13]

Players

The Peace Game defines *players* as the people at the table and the roles they assume in the game. The amount of players and player names depends on scenario requirements. Each player is defined by checking unique parameter combinations:

- Color—Color distinguishes player tokens (units) on the map.
- Strength—Strength is the attrition coefficient that feeds Equation 18.1.

- Terrorist—If checked, this player is defined as terrorist/militia, and its force size can be replenished according to the population algorithm.
- Attack (regions/forces/terrorists)—A player is capable of attacking other regions and/or forces (terrorist or otherwise).
- Control (region only)—A player can wrest regional control from other players.
- Protect (regions/forces/terrorists)—A player can offer protection, causing region/forces (terrorist or otherwise) to temporarily assume the defense capability of the protecting force.
- Train (regions/forces)—A player can increase defense/combat capabilities of regions/forces.
- Ambush (forces only)—Player forces can hide, making them more difficult to detect, and a more advantageous, embellished Equation 18.1 defines their combat.
- Integrate (forces only)—A player cannot only absorb other units into their own, which models the integration of terrorist/militia camps into regular forces as a strategy for conflict resolution, but can also be used to consolidate forces.

There are six players in this Sudan/RSS scenario: Sudan, South Sudan, militia/terrorist, the United States, UN, and an international player (representing an international military/NGO/humanitarian aid collective). Each player is equipped with specific capabilities that must be considered when enacting a strategy. For instance, both Sudan and South Sudan players can attack or protect the militia/terrorist, enabling strategies of a threatening posture or of a negotiation with terrorists in order to obtain their objectives. The militia/terrorist player has a weaker attrition coefficient than the US, UN, and international players, but the ability to ambush/hide and regenerate forces with members of an unsatisfied, desperate population gives that player some strategic advantages. The strategic partnership with militia/terrorist players fits nicely into the real-world and the modeled escalation of tensions between Sudan and RSS. The economic incentive for stability between the neighboring countries is quickly being overwhelmed by desperation due to resource-strapped populations. The Peace Game is capable of modeling the environment and the decisions that will lead these countries into or away from conflict.

Regions

The Peace Game defines *regions* as borders of real estate that overlay the map, each with their own unique parameters. Figure 18.4 shows regional overlays on the map with the Warab region in South Sudan being drawn and parameterized. Scenario designers determine the required number of regions as well the unique parameters for each region:

- Region name
- Control—The player who owns the region at game start. The players enjoy the fruits (resources) of a prosperous region but are also responsible for dealing with regional problems.
- Security—Security is how safe the local population feels. Low security levels feed the population algorithm alongside resource satisfaction levels, so great care must be taken when acting upon a strategy that may adversely affect the safety parameter.

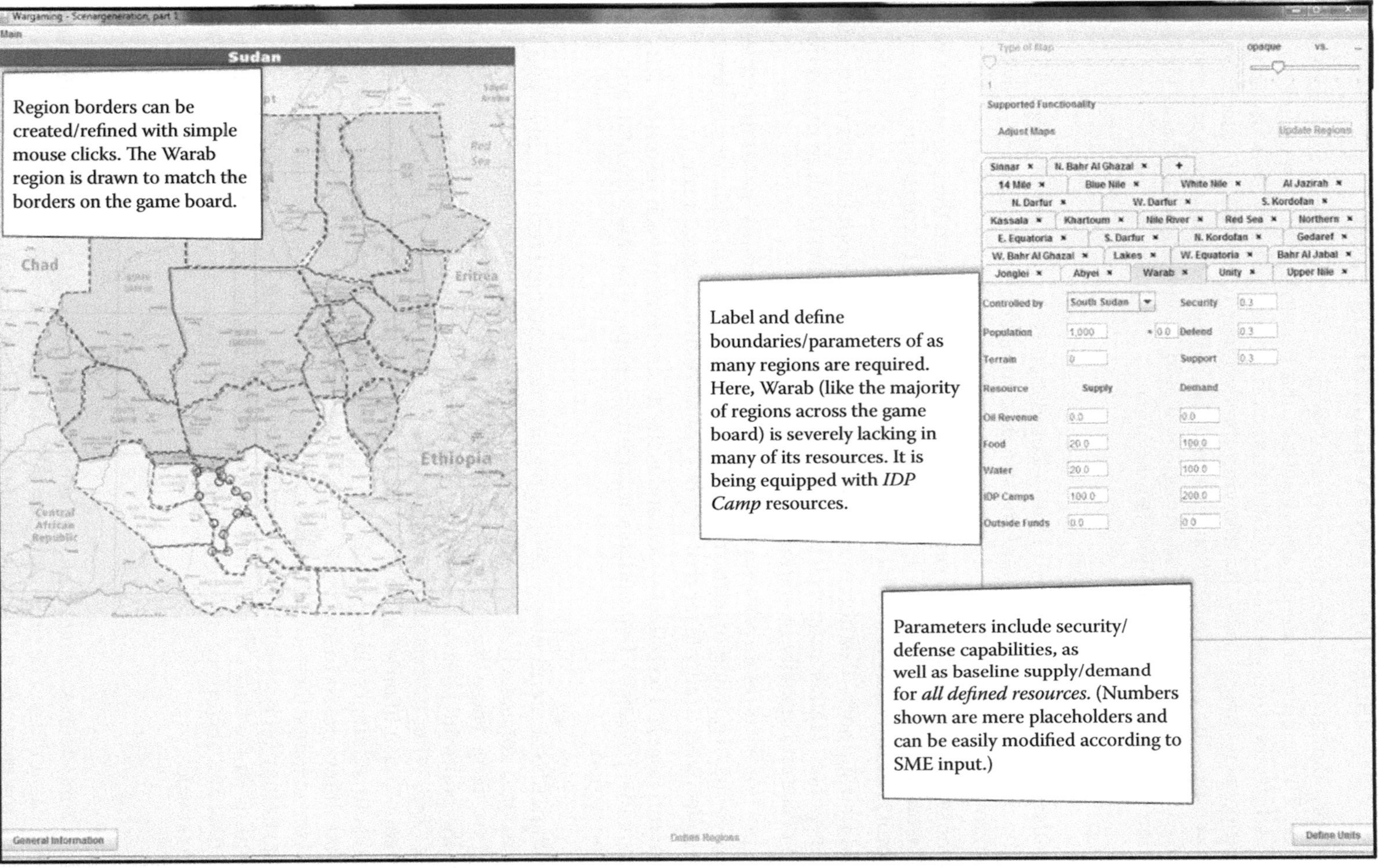

FIGURE 18.4
Scenario generator 2 image.

SUPPLY/DEMAND VERSUS IMPORTANCE

The supply–demand relationship is not analogous to the importance rank for each resource. The supply–demand ratio describes the percentage of unsatisfied population, while the importance rank determines how aggressively that unsatisfied population percentage reacts. By allowing the designer to specify unique supply–demand ratios, the scenario generator accounts for different attitudes toward certain resources. For instance, South Sudan regions would have a higher demand for the Christian resource than that of the Islam resource.

- Population (with optional birth rates)—Population is the number of people living in this region at game start (and the variable number of births occurring within each round).
- Defense capability—Defense capability is the degree to which the population can withstand adverse effects of external attacks (death and decreases in security/stability) or unwelcome, nonlocal force occupation.
- Coalition support—Coalition support is how receptive the population is to outside intervention. Higher support levels for nonlocal intervention result in more effective peace and stability operations.
- Supply (for each resource)—Supply is how much each resource naturally replenishes at the beginning of each round. Resources that exist in a player's region set are collectively available for that player's resource distribution strategy.
- Demand (for each resource)—Demand is the desire level for each resource. This is a dynamic parameter that updates according to population movement in and out of the region.

The Sudan/RSS scenario simply defines regions along recognized state lines and well-known disputed zones such as Abyei and the 14-mile area (the dark regions near the center of Figure 18.4). These regions are particularly interesting because of the dichotomy between the political upheaval and the resource abundance and the resultant population dynamic. Competing tribes and nomads can be exploited as potential tools for competing governments to use in order to strategize a power grab. Interregion overpopulation is a result of resource-hungry citizens and shifting birth rates in and around these regions. The Peace Game allows access to information regarding these factors, but it does not remind (or force) players to juggle these variables when acting upon a strategy. For instance, the South Sudan player may broker a tentative deal with Sudan for control over Abyei and follow-up with an influx of resources to stabilize the region, but if a balance is not struck between resource flow and absorptive capacity, the South Sudan player may have too many mouths feed, so to speak.

Units

The Peace Game defines a *unit* as an element of a force that has its own strengths, weaknesses, and capabilities. The units make up a library that is available to all players so that appropriate forces can be realistically modeled. Units are analogous to chess pieces in a chess game where one player may have an unfair amount of queens while the other player

TIME AND DISTANCE

Time is modeled in the aggregate of the weather, distance, and birth parameters. Regarding weather, if each round represents one month, the designer creates four seasons and allows for each season to last three rounds (therefore, three rounds = three months = one season). Since distance is measured in screen pixels (hence the arbitrary values associated with distance), varying travel/detection radii per round reflect reasonable real-world limitation according to the desired timeframes. Similarly, regional birthrates per round indicate how much time each round represents.

has mostly pawns, some of which are able to move like bishops. Prior to game play, tokens are placed on the map at the location that will be their starting point when the game begins.

A unit's capabilities are based on scenario requirements:

- Combat multiplier (optional)—Combat multiplier is the factor to which a unit increases/decreases player's attrition coefficient (for example, a special forces unit can be defined as 1.5 times as strong as the normal US strength).
- Travel distance—Travel distance is the arbitrary radius defining a unit's ability to move in a round.
- Detection radius (normal)—Detection radius (normal) is how far a unit can see opposing player units.
- Detection radius (hidden)—Detection radius (hidden) is how far a unit can see hidden (ambush capable) units.
- Supply impact (optional)—A unit can permanently increase/decrease a region's per-round supply replenishment for defined resources.
- Combat training (optional)—A unit can train other units and increase their attrition coefficient.
- Defense training (optional)—A unit can train a region and increase its defense capability.

Figure 18.5 illustrates nontraditional units that demonstrate the flexibility of the Peace Game (a nonentity switch that can cut off financial aid and operate an advisor unit). A player with these units among his/her force has the ability to operate as an advisor or to cut off financial aid should the need arise throughout the gameplay.

The inclusion of advisor units in this scenario enables an in-depth cultural understanding of interactions between the security force assistance and security cooperation forces and the Sudan People's Liberation Army (SPLA), the regular army of RSS following the republic's independence in 2011.[15]

Weather

The *weather* parameter defines which resources are impacted by the seasons. For example, if a terrorist supply of a cannabis resource ebbs and flows according to Sudan's weather patterns, the terrorist player may consider a strategy that protects and

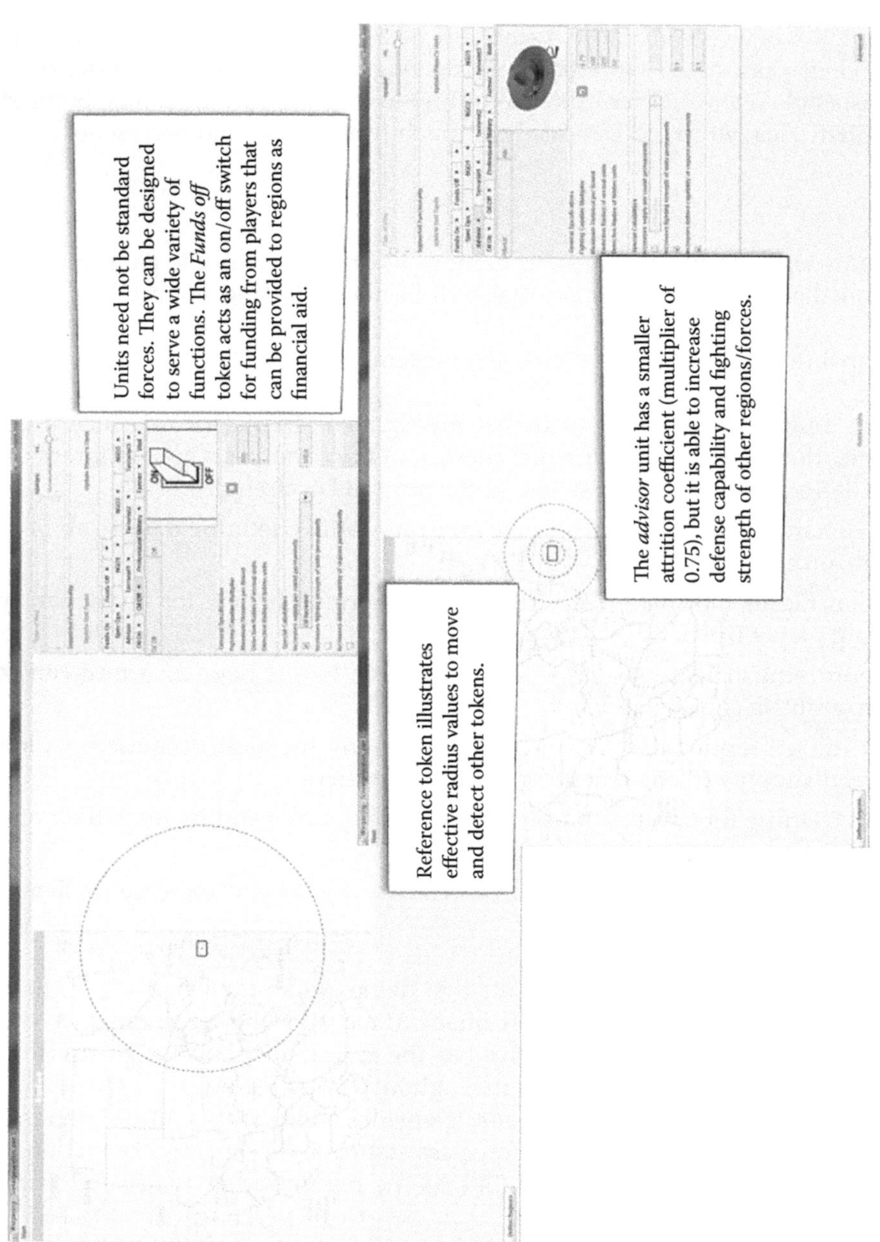

FIGURE 18.5
Scenario generator 3 image.

maximizes the cannabis income. Perhaps the terrorist player must ensure that at least one of its forces is within a striking distance of the cannabis-rich regions during the growing season in order to protect them from invading government forces. This situation aligns with drug-trafficking struggles between the Sudanese government and the illegal drug trade resulting from President al-Bashir's commitment to a drug-free Sudan following the large increase in illicit drug use among segments of the population in 2013. The Peace Game can further model this struggle by designing a demand for cannabis in regions across the map so that a successful counternarcotic strategy results in unintentional displacement and instability due to a portion of unsatisfied and desperate people.[16]

When game designers are satisfied with all the parameters, the scenario is exported and ready to be played. The gameplay is discussed in the next section, continuing the Sudan/RSS contextual case study.

Gameplay

Each game is played as a series of turn-based rounds. When each player completes his/her turn, the round is complete. The game ends when the sponsor or the host decides that the analytic/instructional objectives have been achieved or if a player reaches predetermined objectives (effectively winning the game). The turn order is determined in the scenario generator during player design, but there is no defined order in which a player must make moves and decisions during his/her turn. However, a turn is broken down into three basic phases[7]:

- Phase 1—Force movement: A player moves forces into position and takes action according to player/unit capabilities.
- Phase 2—Negotiations: A player trades resources/negotiates with other players.
- Phase 3—Market: Players manage economic resources to purchase forces/resources.

The algorithms adjudicate the results of player actions, and the complex interactions therein compound to produce unpredictable outcomes. Game moderators (also known as umpires) may explain the results thanks to a familiarity with the formulas, but it is unlikely that they would have been able to accurately predict the outcomes. The responses to the outcomes are up to the players. Open-ended dialogue, trading resources, and negotiations represent the continuous nature of time with regards to the game clock. COA adjudication occurs at the end of each round, at which time the algorithms are implemented. The combination of continuous and discrete time intervals addresses the criticality of time dynamics" intentionally frustrating the players whose strategies are limited by time and available information.[6]

Imperfect information simulates the fog of war, which may induce uninformed decisions. The Peace Game includes a fog of war option that restricts the visibility of external forces so that it is up to each player to manage awareness. Since the players are not allowed perfect knowledge, they are potential victims of their own interpretations of the results of their choices.

Force Movement

The acting player's forces are moved by simply clicking and dragging the token across the map. The map comes to life as each token moves, as seen in Figure 18.6.

The tokens can move and see adversaries as far as their specific radii allow. The shrinking influence circle (colored red during game play) is the trade-off between the geographical displacement and an action's effect. Figure 18.6 shows the South Sudan player's SPLA token moving north through the (highlighted) unity region, en route to the Sudan/RSS border where a regular Sudanese force has appeared in the detection radius. If South Sudan decides to continue north in order to oppose the Sudan force, its influence circle and ability to protect the local inhabitants is minimized. The militia/terrorist player, whose tokens are hidden, can exploit this knowledge if it serves his/her purpose. South Sudan's decision to place the SPLA token in unity means the larger influence circle and regional protection extends to the entire highlighted unity region.

Figure 18.7 is the reaction of the militia/terrorist player to South Sudan's placement of its SPLA token in unity.

The militia/terrorist player's force moves to northern Jonglei, a position close to the SPLA token. This proximity between the tokens indicates potential cooperation between South Sudan and the militia/terrorist player, which may or may not have been explicitly stated aloud during gameplay. This illicit cooperation reflects real-world concerns over corruption in RSS.[17] This particular militia/terrorist token has the ability to increase regional food supplies or decrease defense capabilities, oil, food, water, or IDP camp resources in addition to taking control of, attacking, or protecting a region. A militia/terrorist decision that benefits South Sudan is a further indication of an agreement between the two players, something that other players must keep in mind when dealing with South Sudan and opposing the militia/terrorists.

Negotiations

Force movement and token placement are indications of intent that is witnessed by all players at the table. Negotiations and discussions between players also cue potential adversaries to develop undermining strategies of their own. Human decisions manifest themselves in the deals that are made regarding force movement, resource distribution, and player actions. South Sudan's evident acceptance of the militia/terrorist token in one of its regions (Jonglei) is likely a result of negotiations between these players, indicating that South Sudan may have other than peaceful strategies in mind when it comes to gaining resources and influence. If Sudan attempts to broker a deal with South Sudan regarding oil revenue, and South Sudan adopts an alarming strategy of oil production cessation (as modeled in Figure 18.8), then South Sudan is demonstrating willingness to increase tension between the players in order to obtain its objectives. Such a strategy accurately reflects the real-world tension between Juba and Khartoum over who controls the oil fields in disputed zones and the impasse in negotiations over the financial terms and conditions of oil revenue.

Negotiations include the ability to distribute one's own resources among other players (or their own controlled regions) via the *supply/trade* tab under the *overview* section. Figure 18.9 illustrates this distribution ability, with South Sudan sending 20 food units to Jonglei, despite the region's inhabitation by the militia/terrorist token. The UN player is likely to be displeased with South Sudan's militia/terrorist sanctions and can opt to use its *funds off* token (as seen in Figure 18.8) to cease providing financial aid to South Sudan as long as South Sudan supports militia/terrorist activities.

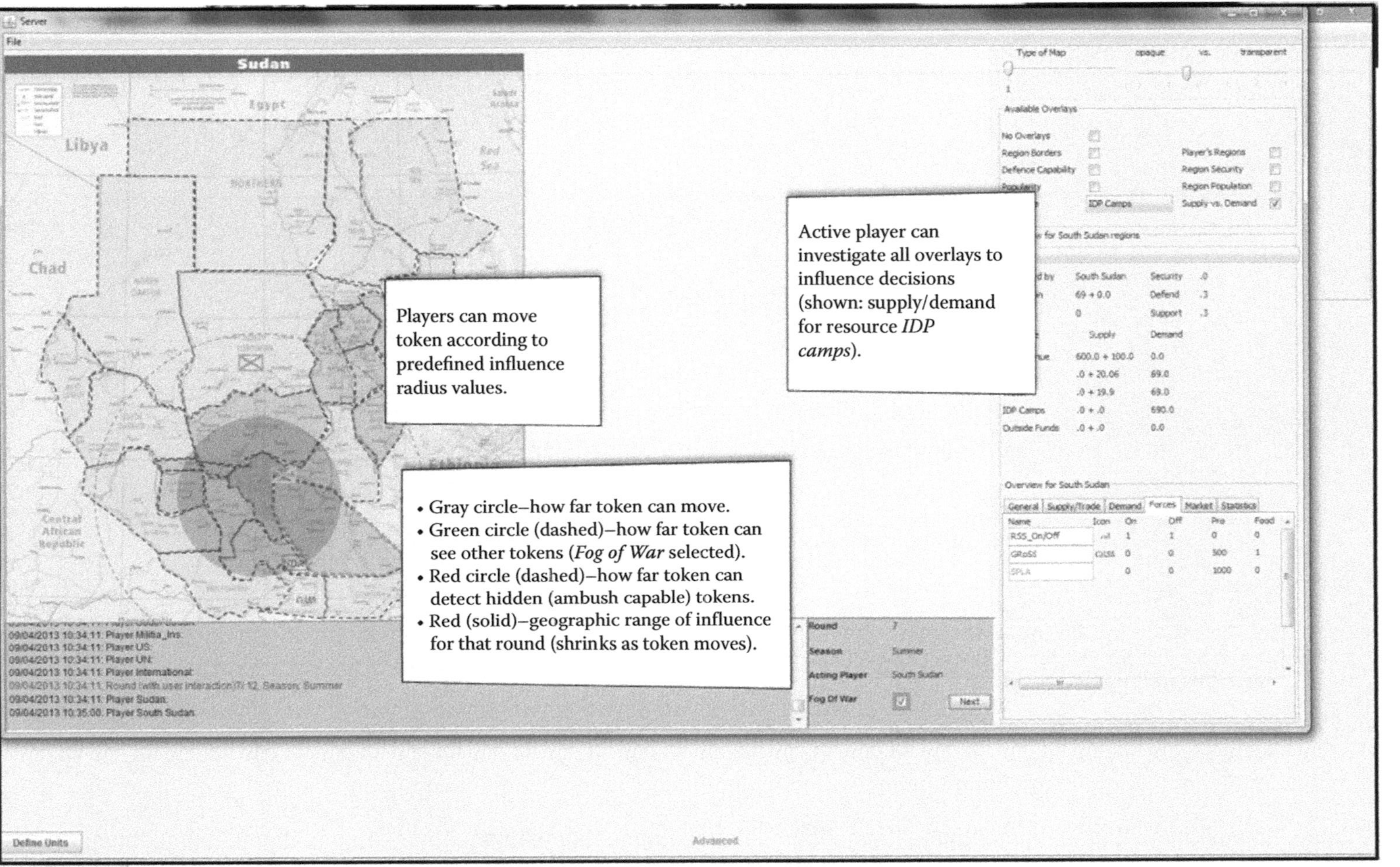

FIGURE 18.6
Gameplay 1 image.

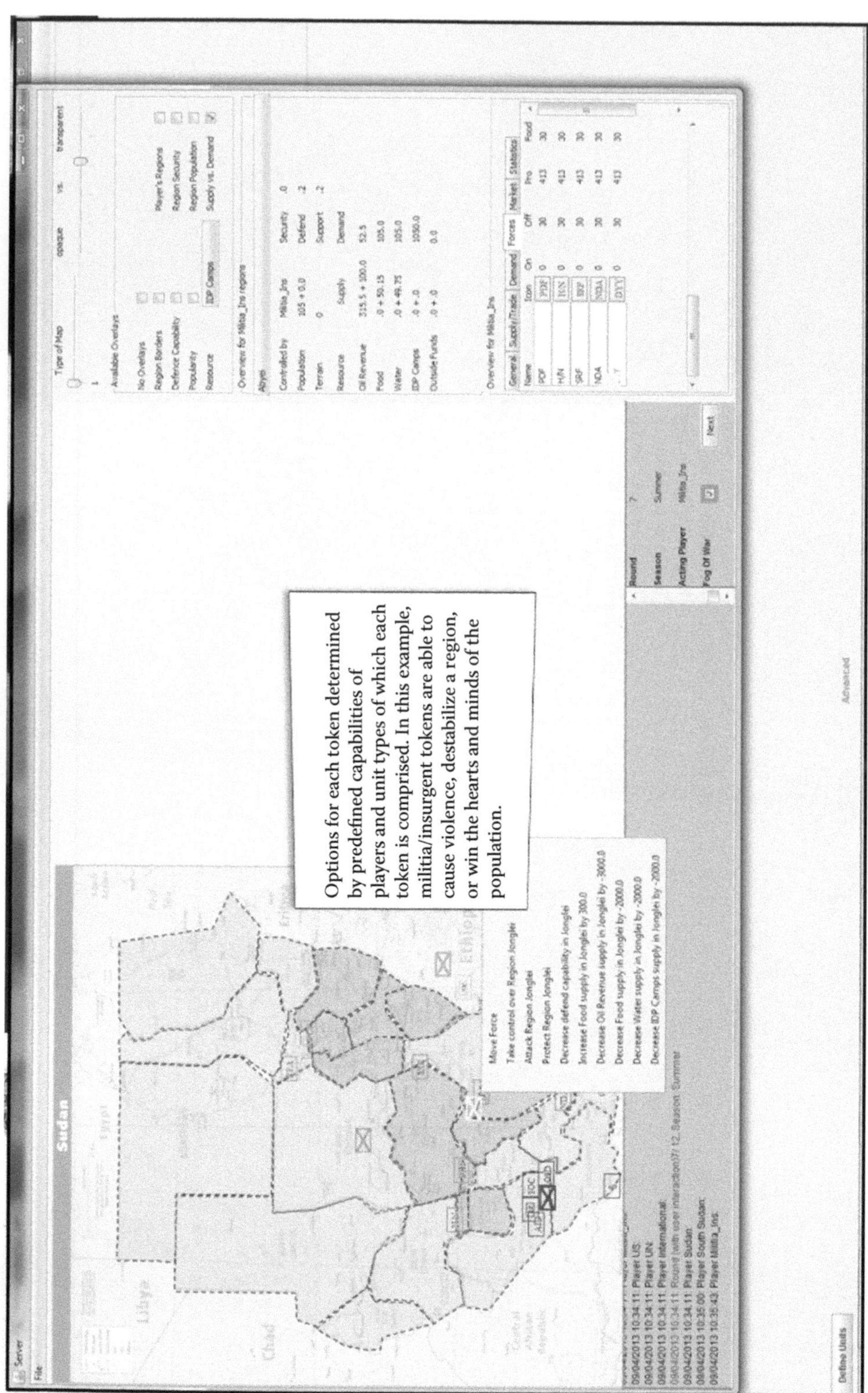

FIGURE 18.7
Gameplay 2 image.

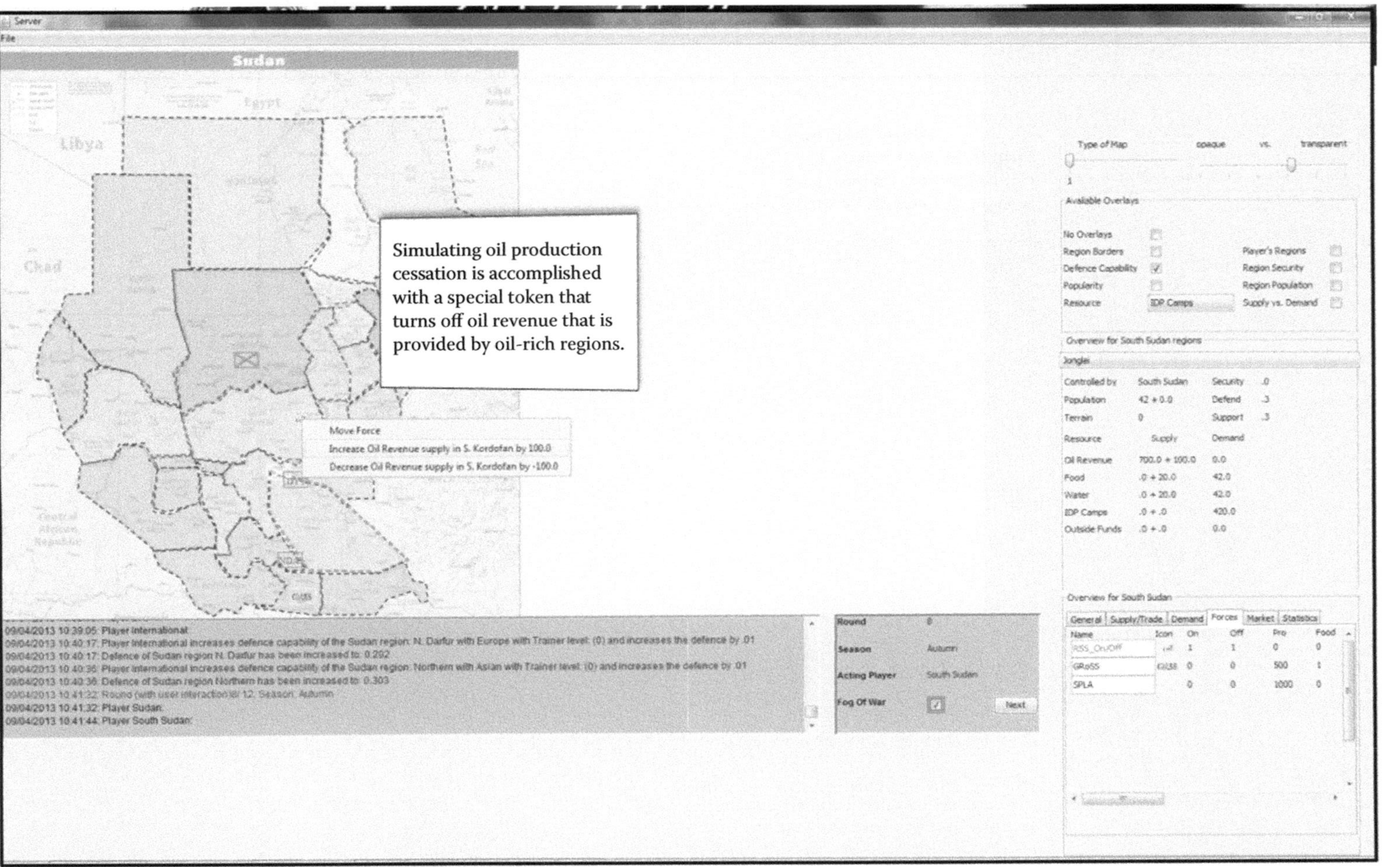

FIGURE 18.8
Gameplay 3 image.

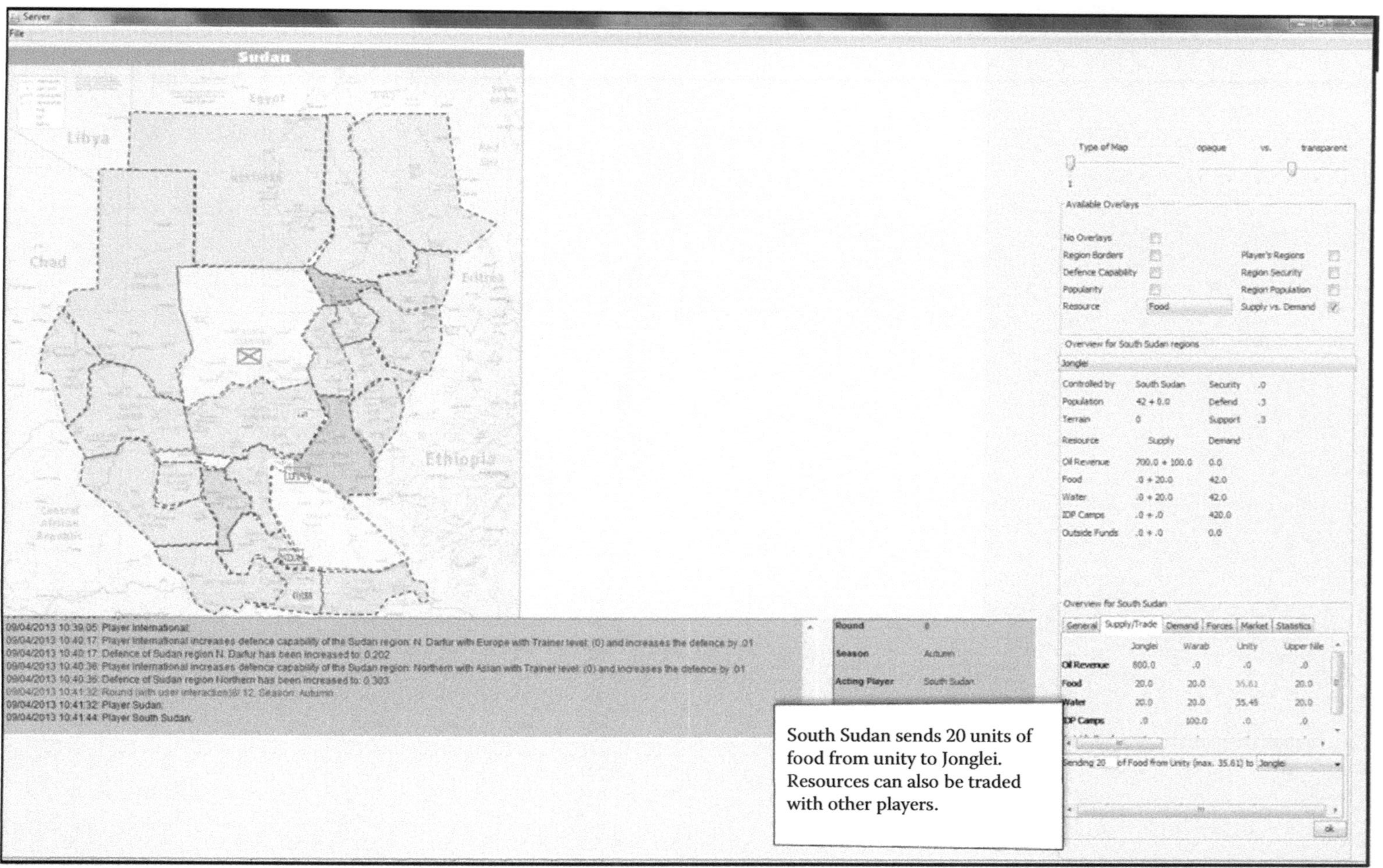

FIGURE 18.9
Gameplay 4 image.

Figure 18.9 also demonstrates the heat map feature of the Peace Game. Jonglei is yellow, illustrating its middle-of-the-spectrum severity regarding the supply/demand for food, as opposed to the green (satisfied) and red (severely lacking) regions. The exact supply/demand ratios are viewable in the *overview* section of the game board.

The supply/demand assessment, mixed with the implemented strategy, drives players to determine ways to spend acquired economic resources (modeled in the Sudan/RSS scenario as oil revenue) in the Peace Game's market.

Market

The *market* tab enables players with the ability to spend economic resources and other resources based on supply/demand and player strategy, as well as forces. South Sudan may not want to send up many more red flags following the UN funding cutoff, so instead of spending resources on forces, which could be used to reinforce its presence along its northern borders along the disputed regions, food is purchased from the market and distributed among the population in one of the hungry regions in South Sudan. Figure 18.10 shows South Sudan purchasing food and sending it to Warab.

The algorithms embedded in the Peace Game adjust the price of each resource (in arbitrary units since scenarios can be designed to represent any place on Earth) according to the overall supply/demand. The fluctuating costs make doing the right thing expensive and possibly infeasible, thus modeling real-world desperation that may lead to unsavory strategies in order to achieve objectives. Further complicating matters is the population algorithm. On the surface, sending food to Warab fixes the immediate hunger problem, but the elevation in Warab's desirability causes population movement into Warab that may cause new tensions if such an influx violates the overall absorptive capacity in that region.

Conclusion

The Peace Game moniker is misleading because it is not a game per se. More accurately, it is an environment onto which unlimited games can be manifested. In software gaming terms, it is a gaming console and a cartridge, where experts can design unique scenarios as needed, effectively allowing them to create their own cartridge. This feature represents multiple learning opportunities for experts and decision makers; designing specific scenarios offers insight into the mechanisms of complex, real-world operating environments, and playing the scenario exposes decision makers to the possible consequences of ignoring those mechanisms. A common critical question of static war games begins with "does this model take [some factor] into consideration?" The Peace Game's framework and inherent flexibility responds to such criticism by facilitating consideration of unlimited factors.

Megacity Use Case

Xanadu Use Case: The Peace Game

The Peace Game's flexibility enables numerous approaches to modeling societies in conflict with reference to economic strife, resources, and illicit actions. The following short

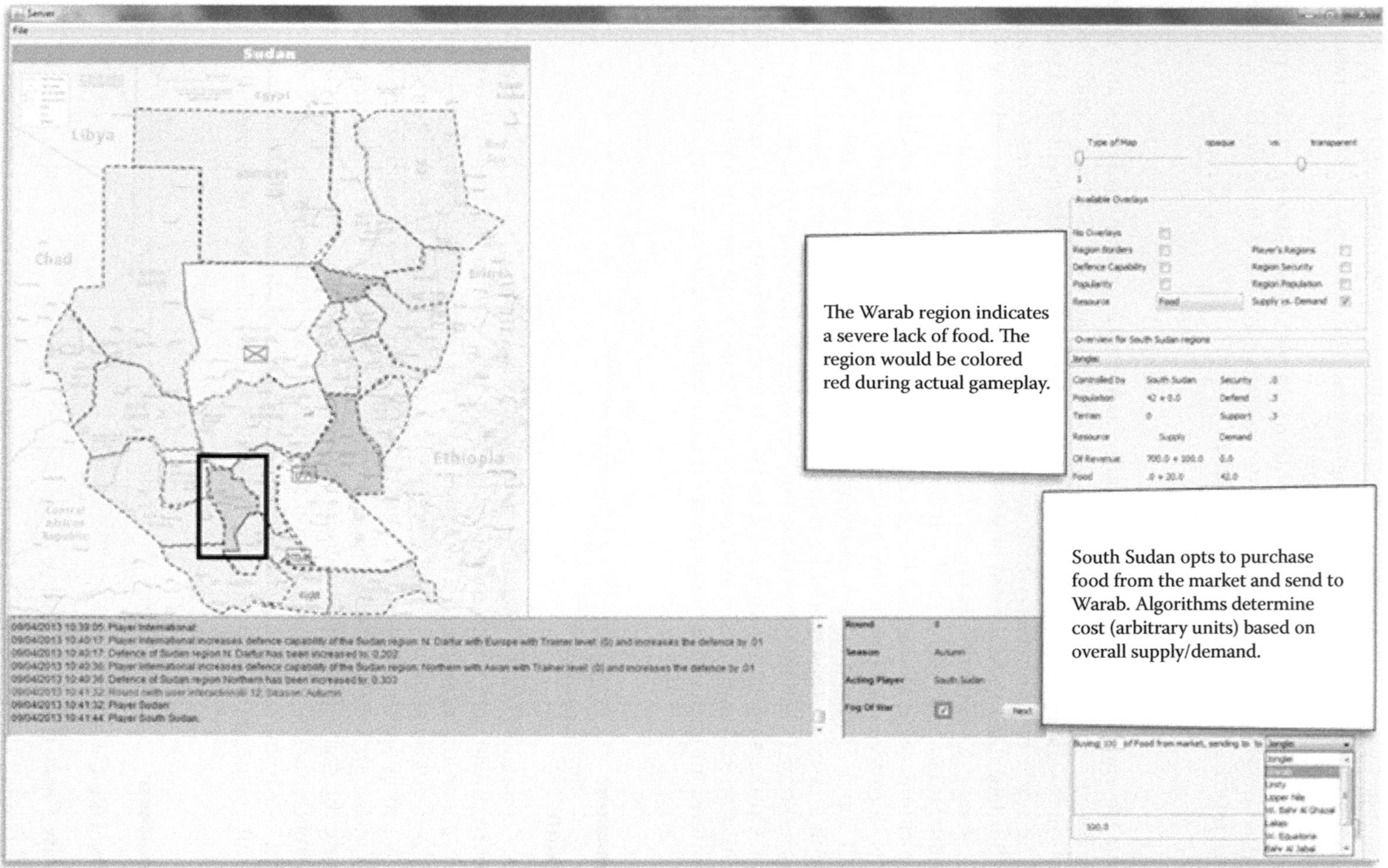

FIGURE 18.10
Gameplay 5 image.

example demonstrates the Peace Game's ability to manifest such a complex (albeit fictional) narrative into a playable game that aids decision making.

Xanadu Map, Resources, and Players

Figure 18.11 displays the big picture considerations for the Xanadu vignette. (Since no Xanadu map has been provided, Figure 18.11 map is of the author's design.)

The scenario is defined by the following:

Map

Dark regions represent megacity Xanadu, broken down into north, south, east, west, and central regions. The light regions to the north and south represent North Razinia and Bursuka, respectively.

Resources

- Economy: Importance—5 (aggressive exponential curve if population is economically unsatisfied).
- Refugee camps: Importance—4 (exponential curve if citizens in need of refugee camps are unsatisfied).
- Sweatshops: Wildcard resource (economic in nature, no associated level of need)—Sweatshops provide means of income to any player in control of a region that produces sweatshops.
- Drugs: Wildcard resource (economic in nature, no associated level of need)—Drugs provide means of income to any player in control of a region that produces drugs.
- Human traffic: Wildcard resource (economic in nature, no associated level of need)—Human traffic provides means of income to any player in control of a region that traffics humans.
- Religion: Importance—4 (exponential curve if citizens with religious needs are unsatisfied).

Players

- Xanadu: It represents the government of Xanadu; It is capable of all actions upon regions, forces, and terrorists.
- N. Razinia: It is the government of North Razinia; it has limited capabilities compared to Xanadu.
- Bursuka: It is the government of Bursuka; it has limited capabilities compared to Xanadu.
- Razinia criminal: It is the criminal element described in narrative.
- Bursuka cleric: It is the religious sect leadership described in scenario.

Region Definition

Point-and-click regions are drawn on the map to define regions, each of which is defined according to scenario parameters (see Figure 18.12).

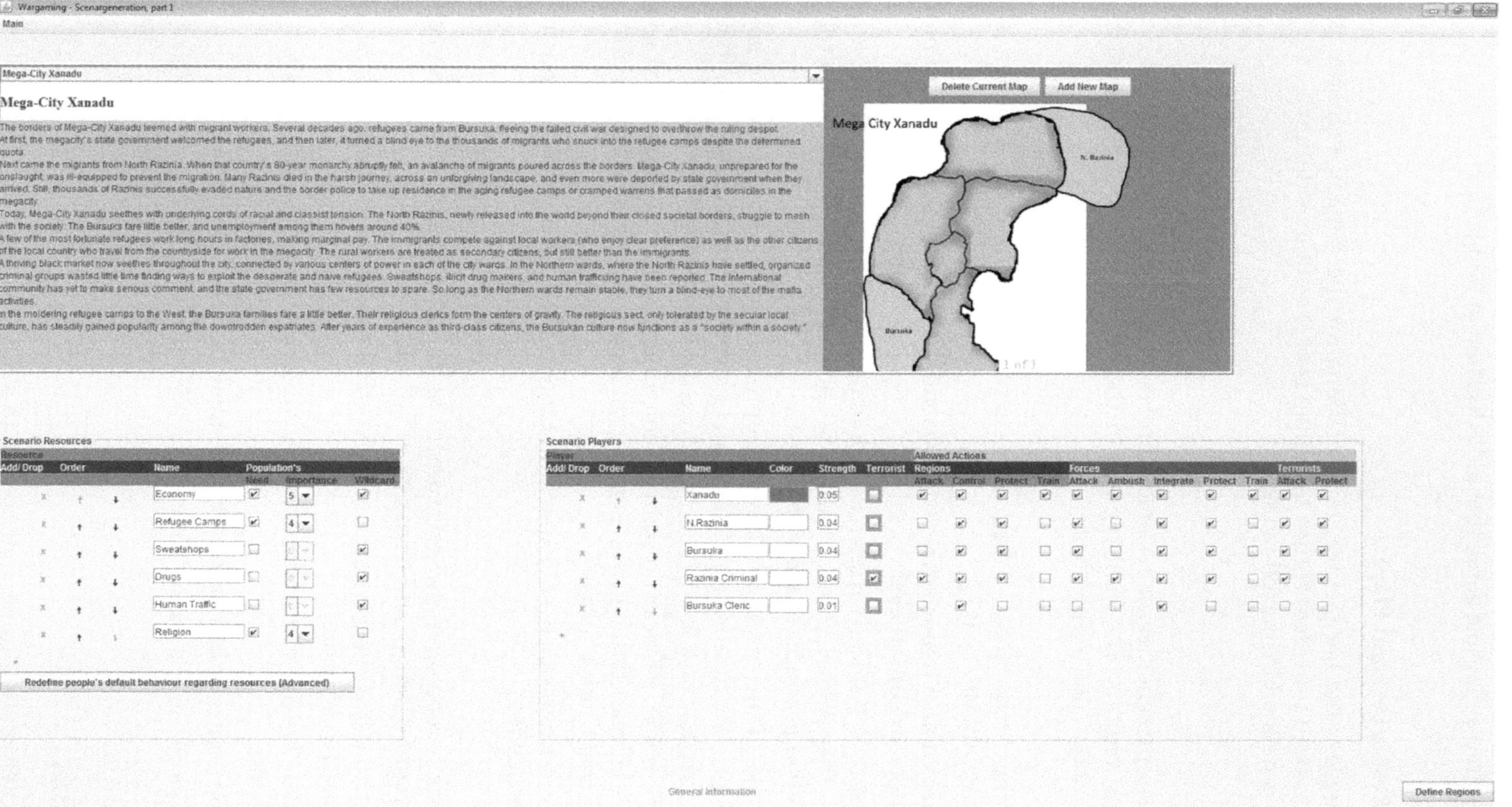

FIGURE 18.11
Peace Game scenario generator (first screen). The scenario is summarized in the gray field. The resources and players are defined with importance levels and capabilities, respectively.

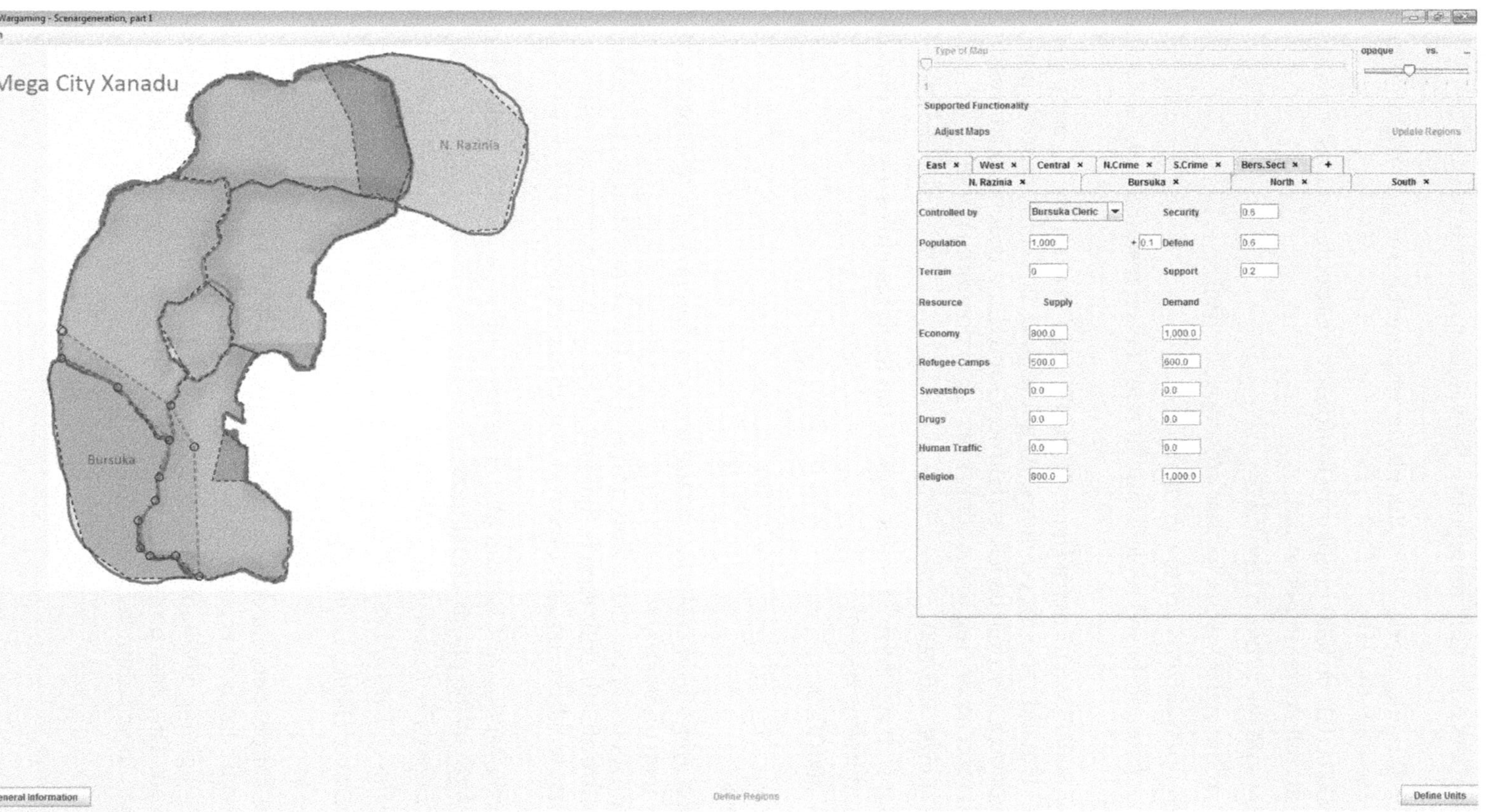

FIGURE 18.12
Peace Game region definition. The dashed line enclosing the region to the north and east of *Bursuka* represents *Bers.Sect*, which is the society within a society described by the scenario narrative. This region, as is the case with all regions, is provided with its own unique player-based color and parameters. The regions are north, south, east, west, central (all referring to the megacity Xanadu), *N.Crime* (criminal-controlled subregion in the northern ward), *S.Crime* (criminal-controlled subregion in the southern ward), *Bursuka*, *N.Razinia*, and *Bers.Sect*.

The following attributes apply to region *Bers.Sect* and are for demonstration purposes only. Scenario designers can easily manipulate the values to fit their needs:

- *Controlled By*: Bursuka cleric—It is the player responsible for this region who is affected by variables therein.
- *Population*: 1000 (+0.1) → 1000 people are uniformly distributed across this region, with 10% birth rate increase per round.
- *Security*: How safe the citizens feel (0.6).
- *Defend*: How robust citizens are to attack/occupation (0.6).
- *Support*: How open citizens are to intervention (0.2).
- *Supply/Demand*: Round-based replenishment rate of each resource versus demand level.
    - Economy: 800/1000 (since *population* = 1000, this region has a one-to-one person/resource demand ratio for economy. At game start, 20% of *Bers.Sect*'s population is economically unsatisfied and, unless something is done to address this, they will react according to economy's level 5 exponential curve).
    - Refugee camps: 500/600.
    - Sweatshops, drugs, human traffic: 0.0/0.0 (no illicit resources are produced or in demand in this region).
    - Religion: 800/1000.

Update Unit Types

Figure 18.13 demonstrates unit creation and the way movement/sight distances are dealt with for arbitrary map distances.

The following attributes apply to the *Boss* unit and are for demonstration purposes only. An important feature of unit definition is that any player can be equipped with any number of any units that will best represent the needs of the scenario:

- *Fighting Capable/Multiplier*: Checked (0.5). The *Boss* unit is capable of combat but at half the strength of the player utilizing it.
- *Maximum Distance Per Round*: 300 (gray circle).
- *Detection Radius of Normal Units*: 400.
- *Detection Radius of Hidden Units*: 200.
- *Special Capabilities*:
    - Increases supply per round permanently: Economy (0.5). *Boss* is able to increase the regional supply of economy resource by 50% if needed.
    - Increases fighting strength/defend capability of units/regions permanently: Unchecked. The *Boss* unit is unable to train other units/region in order to strengthen their parameters.

Update Players' Units

Figure 18.14 shows the *Razinia Criminal* player building and placing force tokens with various units.

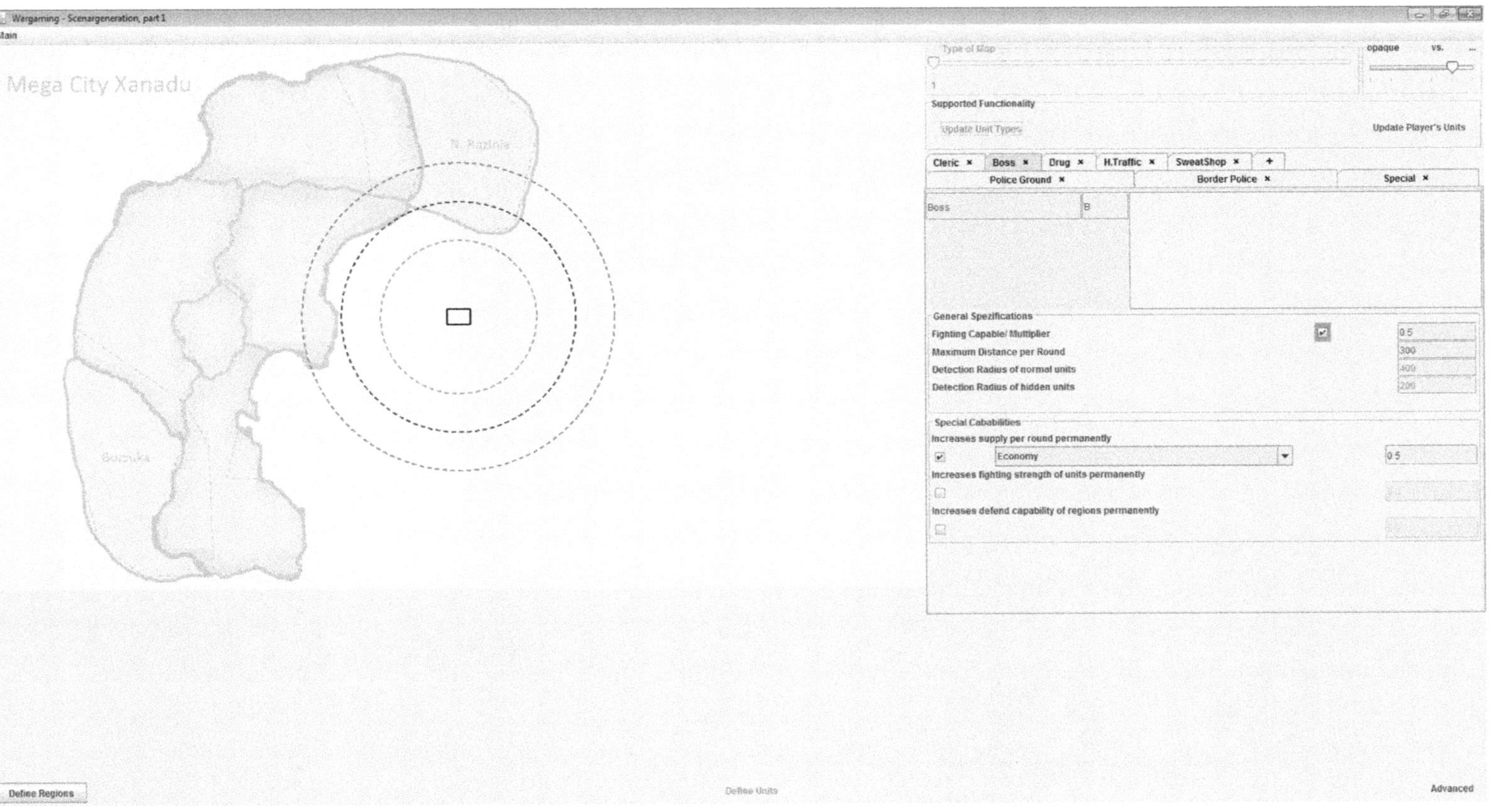

FIGURE 18.13
Peace Game unit definition. Here, a *Boss* unit (a leadership-level unit available to all players) is being parameterized. Since no distance is associated with imported maps, arbitrary distance radius values are displayed on the screen surrounding a reference unit token. In addition to a *Boss* unit, all players can be equipped with *Cleric, Police Ground, Border Police, Special, Drug, Human Traffic,* or *Sweatshop* units.

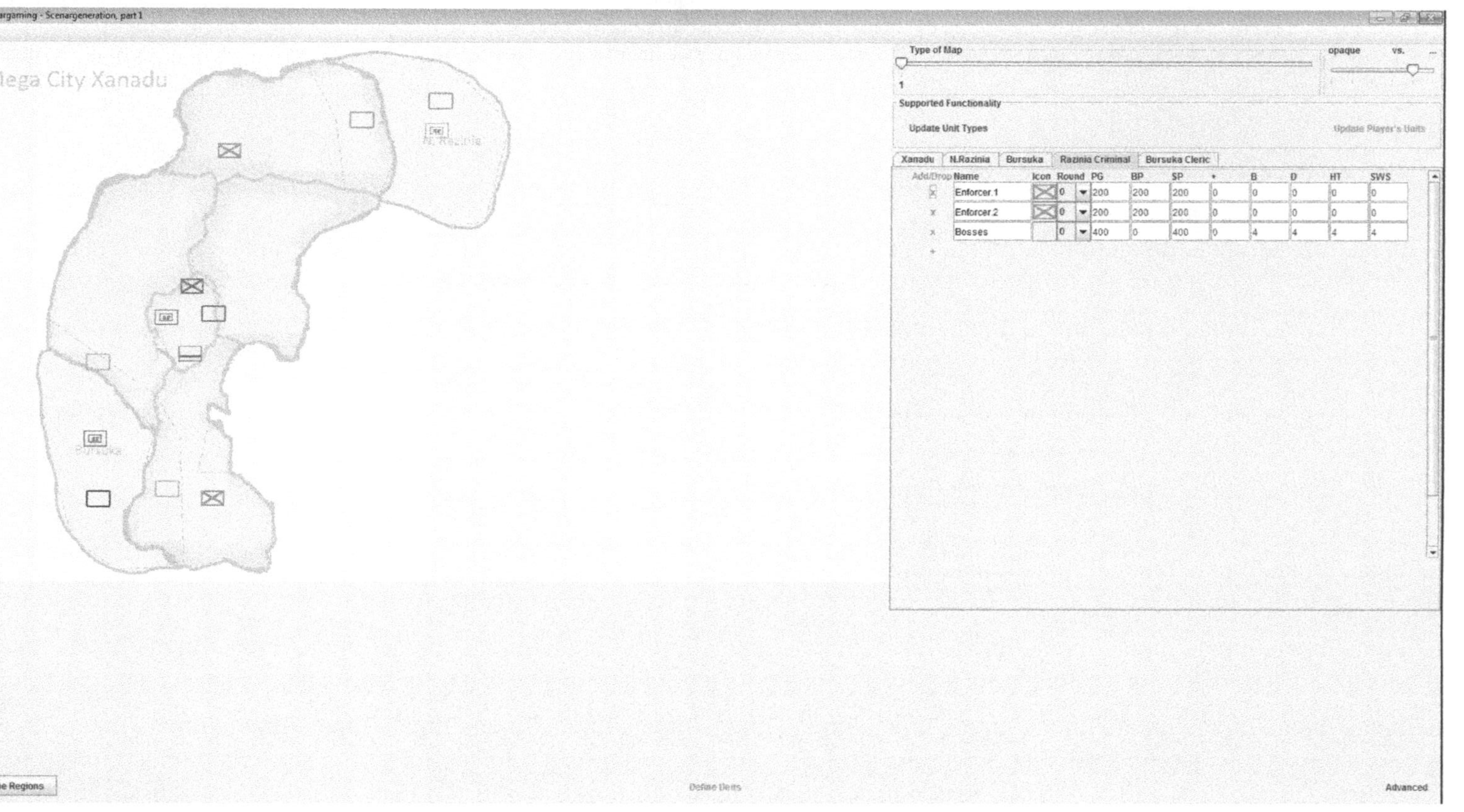

FIGURE 18.14
Peace Game player unit definition. The tokens are placed on the map after they are created and defined on the right side of the screen. Here, the *Razinia Criminal* player has two *Enforcer tokens* (one each in the north/south wards) and a *Bosses* token placed in the *N.Crime* region.

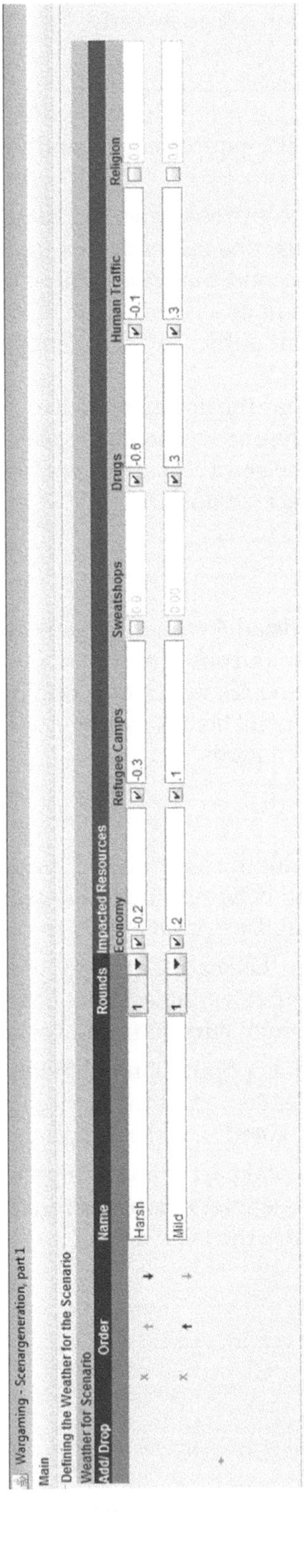

Wargaming - Scenargeneration, part 1

Main

Defining the Weather for the Scenario

Weather for Scenario

Add/ Drop	Order	Name	Rounds	Impacted Resources					
				Economy	Refugee Camps	Sweatshops	Drugs	Human Traffic	Religion
x		Harsh	1	-0.2	-0.3	0.0	-0.6	-0.1	0.0
x		Mild	1	.2	.1	0.00	.3	.3	0.0

FIGURE 18.15
Weather model for Xanadu. The two-type, one-round per pattern (harsh and mild) suggests that each round represents three months.

The following *Razinia Criminal* player forces are made up of various units available in the unit library and are for demonstration purposes only:

- *Enforcer*: *Enforcer* comprises 200 *Police Ground, Border Police,* and *Special* units. While the embedded units are labeled as *police*, the *Razinia Criminal* player is simply constructing force tokens with police-equivalent capabilities as defined in the section "Update Unit Types."
- *Bosses*: *Bosses* comprises 400 *Police Ground* and *Special* units (modeling strong security and combat capabilities that one might expect under a crime boss) as well as four *Boss, Drug, Human Traffic,* and *Sweatshop* units, providing the *Bosses* token with the abilities of the *Boss* unit described in the previous section as well as the ability to establish drug, human traffic, and sweatshop resources in any region.

Since the *Razinia Criminal* player has the ability to control regions and is equipped with forces capable of establishing illicit resource centers, the player is capable of expanding the Razinia-centric criminal empire. The ability to attack regions along with combat-capable forces means the player has the muscle to do so.

Weather

Figure 18.15 demonstrates how modeled weather impacts choice resources. The patterns and the effects are for demonstration purposes only.

A harsh round adversely affects the resources of economy, refugee camps, drugs, and human traffic (negative-associated percentages), while a mild round naturally increases the production in supply-associated regions.

DISCUSSION QUESTIONS

1. Recall that a *Black Swan* is a highly improbable, unpredictable, and significant event that tricks observers into believing that they would have seen it coming *if only…* What are some historical *Black Swan* events that satisfy this description?
2. This chapter demonstrates how the Peace Game can model cultural factors such as Christianity or Islam as a resource variable. What other intangible factors can be modeled through the Peace Game's current framework?
3. User-defined parameters such as population behavior, supply/demand, weather, and force structures/capabilities represent the flexibility of the Peace Game. What other features could be incorporated in order to enhance this flexibility?
4. Player interaction is a fundamental part of any war gaming model. How could one facilitate a gameplay so that these interactions represent the nuances of real-world engagements (for example, under the table deals)?

References

1. Perla, Peter P. *The Art of Wargaming: A Guide for Professionals and Hobbyists.* Naval Institute Press, Anapolis, MD, 1990.
2. Rubel, Robert C. *The Epistemology of War Gaming.* Naval War College, Newport, RI, 2006.

3. Compton, Jon. Toward an epistemology of wargaming: A drunkard's walk. Lecture presented at the Military Operations Research Society (MORS) Brown Bag Lecture Series, Alexandria, VA, 2015, March.
4. Nannini, Christopher J., Jeffrey A. Appleget, and Alejandro S. Hernandez. Game for peace: Progressive education in peace operations. *The Journal of Defense Modeling and Simulation: Applications, Methodology, Technology* 10, no. 3: 283–296, 2013.
5. Perla, Peter P. So a wargamer and a black swan walk into a bar…. Keynote Panel: Connections Wargaming Conference, A Subset of the Third Defense Modeling and Simulation Conference, Orlando, FL, 2008.
6. Rubel, Robert C. *War-Gaming Network-Centric Warfare.* Naval War College, Newport RI, 2001.
7. Powers, Matthew, Jeff Appleget, and Danny Heerlein. The peace game: A software-based model for understanding the complexities of modern conflict at the strategic level. In *Proceedings of the Symposium on Theory of Modeling & Simulation-DEVS Integrative, Society for Computer Simulation International,* p. 12, Tampa, FL, 2014.
8. Bracken, Paul. Net assessment: A practical guide. *Parameters* 36, no. 1: 90, 2006.
9. Powers, Matthew. Being wrong in the right direction. Available at https://www.linkedin.com/pulse/being-wrong-right-direction-matthew-powers?trk=object-title (accessed March 17, 2015), 2015.
10. Lagalice, Clément G. C. Geopolitical situation of Abyei: An ongoing conflict with ethnicities and natural resources as main drivers. PhD dissertation, Università Cattolica del Sacro Cuore Milano, 2011.
11. Al-Tohami, Karar. *Migration in Sudan: A Country Profile 2011.* International Organization for Migration, 2011.
12. Malthus, Thomas Robert. *An essay on the principle of population.* Vol. 1. Cosimo, Inc., 2013.
13. Korotayev, Andrey, Julia Zinkina, Svetlana Kobzeva, Justislav Bozhevolnov, Daria Khaltourina, Artemy Malkov, and Sergey Malkov. A trap at the escape from the trap? Demographic-structural factors of political instability in modern Africa and West Asia. *Cliodynamics: The Journal of Theoretical and Mathematical History* 2, no. 2: 277–303, 2011.
14. The UN Refugee Agency. UNHCR country operations profile—Sudan. Available at http://www.unhcr.org/pages/49e483b76.html (accessed March 19, 2015), 2015.
15. Woodward, Barian A. *Cultural Considerations for Security Cooperation Operations in South Sudan: Understanding the Sudan People's Liberation Army (SPLA).* Marine Corps Command and Staff College, Quantico, VA, 2012.
16. The Sudan Tribune. Sudanese president to chair new anti-drug council. Available at http://www.sudantribune.com/spip.php?article47117 (accessed March 19, 2015), 2013.
17. Voice of America. South Sudan seen as one of the world's most corrupt nations. Available at http://www.voanews.com/content/south-sudan-corruption-transparency-international-annual-index/2544599.html (accessed March 19, 2015), 2014.

Section IV

Applying Sociocultural Models to Gain Insight into Conflict and Instability

In the ideal world, science smoothly progresses through a series of stages, beginning with observations about a phenomenon and ending with a validated theory that not only binds together those observations, but also makes testable predictions about the nature of such future observations across a range of varying conditions. In practice, this scientific method rarely progresses smoothly. Practical challenges abound. These include understanding how to develop and frame a theory about these observations, determining what data should be captured to test the resultant theory, developing methods to capture those data, conducting the appropriate tests to validate (or refute) the theory, and analyzing the results from those tests.

The careful reader will have noticed that this volume is structured, albeit with some license, to provide deeper insights into the practical application of this scientific method to understanding the basis for cross-cultural differences in decision making. The first section of this volume oriented readers to the challenges of establishing the structure or the framework of sociocultural theories and provided readers with insight into critical questions to be addressed during the process of developing these theories. The second section of this volume focused on the unique challenges attendant with capturing and analyzing these data. These include process as well as technical challenges. Yet once these data are collected and analyzed, an equal if not greater challenge lies in using them to validate the original theory and to then implement the validated theory for future applications. While simple experimentation is often the ideal approach to validation, the scope of sociocultural decision making theories and data makes it less than suitable for well-controlled experimentation. The last two sections of this volume address this validation challenge. Specifically, they focus on the development and the application of models and their execution through simulation, to represent (Section III) and to validate theories as well as make predictions and forecast future states based on those theories (Section IV).

There are many ways of defining *modeling, simulation,* and the relationship between them. For the purposes of this section, the following working definition is considered (Cohn, 2007):

Model

- A representation of some aspect of external reality in a program or algorithm
- A tool that assists in making forecasts and predictions for some purpose
- A representation that offers insight about the entity modeled

Simulation

- An artificial situation or environment
- The practice of mimicking some or all the behaviors of one system with a different system
- The experience of a real situation without the associated risks

Together, models, executed through a simulation, provide a unique opportunity to repeatedly test and validate a theory across a wide range of variables and conditions. Their outcome can then be used to not only fine tune theories, but also establish future states extrapolated from those data.

This section focuses on using sociocultural models to represent and understand the drivers of conflict and instability. It explores the challenge of ensuring that the right level of representation is possible—whether it be at the individual, the group, or the larger level. It discusses ways of ensuring that these models can be prevented from being overly biased across different population sampling techniques. Lastly, it provides examples, using real-world data, of how these models can be used to understand how these drivers can influence the future and how this understanding can influence future real-world actions.

An enduring challenge in developing sociocultural models is identifying the right level of granularity—individual, group, social etc—at which the model should be implemented. Chapter 19, authored by Kulzy, Combs, and Fricker, offers the tantalizing potential of models to represent sociocultural behavior at both the individual and social levels. More concretely, it also demonstrates that modeling and simulation tools can be successfully used to evaluate a theoretical model. An equally important challenge with any model that represents human behavior is ensuring that it is not overly biased. Chapter 20, authored by Anderson and Fricker, focuses on the real-world challenge of how to ensure that a model's output accurately reflects the population from which its underlying data were sampled. They describe how a technique known as *raking* can, when combined with known population demographics, can be used to adjust the weights used in a model.

The remaining chapters in this section, Chapters 21–23, focus primarily on applying sociocultural models to real world data and situations. Chapter 21, by Bernard, Backus, Naugel, Jeffers, and Damron, discusses different modeling approaches and techniques for modifying them, using expert opinions. Using a specific modeling approach, the dynamic multiscale assessment tool for integrated cognitive-behavioral actions, applied to a real-world example, they highlight how models can aid in understanding conflict and stability. Chapter 22, by Hall, Blais, and Baird, explores how a different modeling approach, the agent-based multiscale approach, can be used to understand and predict the consequences of a donor nation providing support to a failing or a failed host nation. Using their modeling approach, which incorporates the effects of new technologies like social media, they suggest that resiliency—the ability of a system to adapt to future challenges and opportunities—may be more meaningful than stability—the ability of a system to recalibrate itself in terms of the immediate challenges and opportunities. The last chapter in this section, Chapter 23, by Lofdahl, provides a concrete illustration of how foreign policy decisions may be guided using systems dynamics-based models.

Reference

Cohn, J. (2007). Design and application of modeling and simulation technologies for optimal impact [CD-ROM]. *Proceedings of the 28th Interservice/Industry Training, Simulation and Education Conference, Orlando, FL,* November 29, 2007.

19

Using the Social Framework Model of Trust to Better Understand Trust in Government

Walter W. Kulzy, David J. Y. Combs, and Ronald D. Fricker, Jr.

CONTENTS

Introduction 398
Background 399
Integrative Model of Organizational Trust 400
Components and Definitions of the Integrative Model of Organizational Trust 400
Ability 401
Benevolence 402
Integrity 402
Trust, Risks Taking, Outcomes, and Practical Application 403
Research Hypotheses 404
Methods 404
Sample 405
Measures 405
Analyses 407
Results 410
Discussion 415
Limitations of Current Research 417
Conclusion 418
Megacity Use Case 418
References 419

ABSTRACT We assess the applicability of Mayer, Davis, and Schoorman's (1995) integrative model of organizational trust for modeling citizens' trust in their government using country-level survey data collected in four western Trans-Sahel countries (Burkina Faso, Senegal, Mali, and Nigeria) in 2010. Although the original model focused on trust between individuals, our fundamental supposition is that the model also applies to individuals' trust in an organizational-level entity: government. Our findings also suggest that there are two separate dimensions to the ability and the benevolence/integrity associated with trust in the government, as well as the existence of a new term that we hypothesized is related to government reputation.

KEY TOPICS

- The basis of trust in government at the individual and social levels
- The utility of Mayer, Davis, and Schoorman's (1995) integrative model of organizational trust for understanding citizen trust in government
- Using the R lavaan modeling and simulation package (Rosseel, 2012) to evaluate a theoretical model

Introduction

Multiple social science perspectives state that trust is critical to human interaction. For example, psychologists suggest that "trust is one of the most important components—and perhaps the most essential ingredient—for the development and maintenance of … well-functioning relationships" (Simpson, 2007a, p. 587). Similarly, commentators from the field of international relations have stated that trust within the international system is "the underpinning of all human contact and institutional interaction" (Blind, 2006, p. 3). Organizational management perspectives reach a similar conclusion about the cross-disciplinary importance of trust studies (e.g., Colquitt, Scott, and LePine, 2007). Even American military perspectives from the counterinsurgency battlefields of Iraq and Afghanistan suggest that generating trust between American service members and local populations is more important than kinetics and force. For example, David Kilcullen, the noted Australian counterinsurgency expert, states that trust building in the counterinsurgency context is the military's "true main effort: everything else is secondary: (Kilcullen, 2010, p. 37). As a result of these growing perspectives, researchers have called for more relevant, cross-cultural, macrolevel investigations of trust (e.g., the studies by Bachmann [2011], Li [2011], and Mishra and Mishra [2013]).

This chapter is the first to explore the utility of Mayer, Davis, and Schoorman's (1995) integrative model of organizational trust for understanding citizen trust in government.

RESEARCH IN ACTION: THEORETICAL DEBATE REGARDING THE INTEGRATIVE MODEL OF ORGANIZATIONAL TRUST

It is important to note that there is some debate in the literature regarding the unique contributions of each trustworthiness factor on trust. Some empirical analyses (e.g., the studies by Jarvenpaa, Knoll, and Leidner [1998] and Mayer and Gavin [2005]) have found that integrity and benevolence do not uniquely contribute to trust. For example, Mayer and Gavin (2005), in a study regarding factory workers' trust in leadership, found that benevolence did not uniquely predict trust, while ability and integrity were significant contributors to trust. Mayer and Gavin suggested that this was likely due to multicollinearity among the trustworthiness survey items. On the other hand, Colquitt, Scott, and Lepine's (2007) meta-analysis of dozens of papers on trust found that each trustworthiness factor did have a unique impact on trust.

Mayer, Davis, and Schoorman's (1995) integrative model of organizational trust has been used to explain interpersonal trust outcomes in the business world (Davis et al., 2000), the medical community (Schoorman, Mayer, and Davis, 1996), psychology (Simpson, 2007a,b), and others (e.g., the study by Colquitt, Scott, and LePine, 2007). Despite the overwhelming utility of the Mayer model for understanding interpersonal trust generation in multiple contexts, it has, to our knowledge, never been used to understand citizen trust in the government. This is surprising since Mayer and his colleagues (e.g., in the study by Schoorman, Mayer, and Davis, 2007) claim that their model is robust for understanding an individual's trust in organizations, although they tend to think of trust in business organizations as opposed to governments. As Schoorman, Mayer, and Davis (2007, p. 345) noted, "the 1995 framework is fairly robust across levels of analysis." From their perspective, the same variables that impact trust between people "also affect the extent to which an organization will be trusted" by people (Schoorman, Mayer, and Davis, 2007, p. 345).

This chapter is organized as follows. In the next section, we describe Mayer, Davis, and Schoorman's integrative model of organizational trust and how the elements of the framework can be applied to understand citizen trust in the government. We also discuss existing research on trust in government that, despite not using the Mayer model as an organizing framework, appears to support the Mayer, Davis, and Schoorman concept. Subsequently, we describe our analysis of the survey data from four countries in the Trans-Sahel region of Africa (Senegal, Nigeria, Burkina Faso, and Mali) where we assess whether the integrative model of organizational trust is appropriate for understanding citizen reports of trust in the government. Finally, we conclude with a discussion of our results and recommendations for future research.

Background

Multiple theoretical frameworks, across disciplines, attempt to explain how trust is generated. For example, Wieselquist (1999) propose a trust generation theory based in psychology, often used to explain trust in romantic relationships that describes how trust is generated between two people. Simpson (2007a,b) proposes a dyadic theory of trust generation that focuses on the impact of personality variables across multiple stages of the trust generation process. Hardin (2006) proposes a view based in political philosophy that helps explain how people have developed trust across time. Vigoda-Gadot (2006), in a model of voice orientations, proposes that trust in governance is a function of satisfaction with services, organizational politics, and ethics. Combs, Blincoe, and Vorm (2015) propose a new theoretical framework specifically designed to understand how trust is cross-culturally generated, with an aim toward understanding trust on the battlefield. While each of these trust models is useful, perhaps the most celebrated social science model of trust generation is Mayer, Davis, and Schoorman's now classic integrative model of organizational trust (Mayer, Davis, and Schoorman, 1995). Developed in the mid-1990s to organize a very disjointed social sciences trust literature, Mayer, Davis, and Schoorman's model has now received extensive supporting empirical treatment and, by 2007, has been cited over a thousand times (Schoorman, Mayer, and Davis, 2007).

Integrative Model of Organizational Trust

In 1995, Mayer and his colleagues introduced their integrative model of organizational trust. Trust, from Mayer, Davis, and Schoorman's perspective (1995), is "the willingness of a party to be vulnerable to the actions of another party based on the expectation that the other will perform a particular action important to the trustor, irrespective of the ability to monitor or control that other party." According to their framework, as shown in Figure 19.1, trust is a result of trustworthiness. Trustworthiness, in this case, is a function of the trustor's (the person doing the trusting) perceptions of the trustee's (the person who would be trusted) ability, benevolence, and integrity (each explained in more detail in the next subsection). These components of trustworthiness, along with a person's dispositional propensity to trust, theoretically predict the level of trust a trustor has in a trustee. The Mayer, Davis, and Schoorman model is a feedback loop, where trust is tested in a risk-taking situation, an outcome is observed, and then the trustor's assessment of the trustee is subsequently updated.

Mayer, Davis, and Schoorman's model was originally proposed to explain trust in business relationships. For example, the process in Figure 19.1 was designed to explain when an employee might trust a supervisor. If the employee perceives the supervisor as having sufficient ability, benevolence, and integrity, then the employee is more likely to trust the supervisor and be willing to be vulnerable to his or her direction, leadership, etc.

Components and Definitions of the Integrative Model of Organizational Trust

As noted, Mayer, Davis, and Schoorman (1995) suggest that trustworthiness is made up of ability, benevolence, and integrity. This section characterizes each of these trustworthiness elements and provides research examples indicating that each has an impact on trust in the government. The section concludes with a brief example of how the overall framework could play out regarding trust in government.

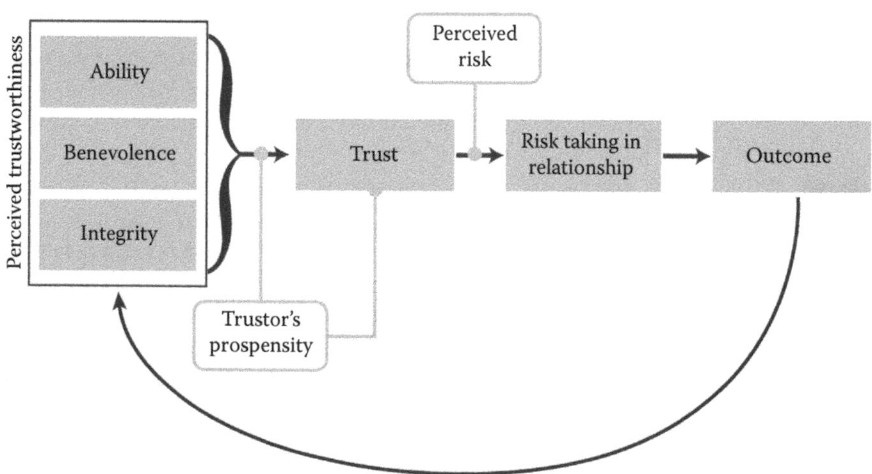

FIGURE 19.1
Mayer, Davis, and Schoorman's (1995) integrative model of organizational trust.

Ability

Mayer, Davis, and Schoorman (1995) define ability as "that group of skills, competencies, and characteristics that enable a party to have influence within some specific domain" (p. 717). Essentially, ability is a trustor's perception of a person's or an organization's ability to successfully complete a task. The impact of this trustworthiness component on trust makes intuitive sense. A trustor would not be willing to make themselves vulnerable to the actions of a trustee if the trustee is incompetent to achieve some relevant task. Of course, as Combs, Blincoe, and Vorm (2015) note, "ability varies from domain to domain and situation to situation." At a personal trust level, an auto mechanic might be very good at repairing domestic cars but have little background with foreign cars. As such, a trustor's sense of a trustee's ability to repair his or her car will clearly vary across situations, as will a trustor's overall trust in the trustee.

From an organizational standpoint, the degree to which a person believes a business or a government is trustworthy, and therefore trusts the business or the government, should also depend on the ability of the organization in question. For example, the trust in an car manufacturer is probably dependent, to a degree, upon the organization's ability to produce a quality product. That is, people probably ask themselves some variant of the question: Can this manufacturer get the job done? Similarly, trust in the government is probably dependent, to a degree, upon the government's ability to produce things like effective security, infrastructure, and economic growth. Like the car manufacturer example, people probably ask themselves some variant of the question: Can this government get the job done?

Multiple research projects, often from the field of political science, have examined the impact of a citizen's perception of government ability on reports of citizen's trust in the government.* For example, Blind (2006) points out that a government's ability to provide security and services impacts the citizens' trust in the government. Blind also states that a sense of government professionalism and competence has much the same effect. Hetherington (1998) similarly points out that a number of perspectives on government trust are based on perceptions of government efficiency:

> Because people are likely to trust things they perceive to be working effectively, the quality of policy outcomes should also help explain trust. Public perceptions of the government's ability to solve problems that are personally most important should have a strong bearing on political trust (p. 794).
>
> Kim (2010) finds "that government performance on the economy, ... the quality of public services, crime, and attention to citizen input are significantly associated with broad public trust in government in both Japan and South Korea" (p. 801).

Similarly, Christensen and Laegreid (2005) conclude that "citizens who are satisfied with specific public services generally have a higher trust in public institutions than citizens who are dissatisfied" (p. 487). Hetherington (2006) also states that "most Americans simply do not think the government is capable of doing the job [referring to a number of policy matters] well enough or fairly enough" (p. 5). In his analysis, Hetherington found that citizen trust in the government was strongly related to the support for government programs such as foreign aid and food stamps.

* It is important to keep in mind that in some cases, the term *trust* is used differently in political science than the way Mayer and his colleauges. use the term. Sometimes the term *trust* in political science more resembles the psychological term *attitude* (e.g., the study by Hetherington [1998]). In other cases, the term *trust* more resembles Mayer and his colleagues trustworthiness construct (e.g., the study by Blind [2006]). Regardless, the usage of the term is similar enough between the political science research and the research by Mayer and his colleagues that we use the concept interchangeably.

Benevolence

Mayer, Davis, and Schoorman (1995) define benevolence as "the extent to which a trustee is believed to want to do good for the trustor" (p. 720). Essentially, benevolence is a trustor's perception that a person or an organization cares about the trustor. A trustor might believe that a trustee has high ability, but if a trustor does not believe that the trustee wants to do good for the trustor, then his or her trust is clearly diminished. As Davis et al. (2000) note, "benevolence represents a positive personal orientation of the trustee to the trustor" (p. 566). They suggest that in a business context, at the individual level, benevolence might manifest for an employee (the trustor) when a manager (the trustee) is willing to go out of his or her way for the employee.

From an organizational standpoint, the degree to which a person or persons believes a business or a government is trustworthy, and therefore trusts the business or government, should also depend on the benevolence of the organization in question. From Schoorman, Mayer, and Davis' perspective (2007), benevolence might not be a major factor in trust in a business since, after all, what business can afford to be truly benevolent? Yet, clearly, the benevolence of a government, the degree to which a trustor believes that a government wants to do good for the citizens, should have an impact on citizen trust. As Vigoda-Gadot (2006) argued,

> ...when citizens perceive bureaucracy as insensitive, feel that it promotes the interests of powerful individuals or groups based on political considerations, and believe that it engages in unfair practices, public attitudes toward democracy may become more cynical. Similarly, citizens may react negatively, either cognitively and/or behaviorally, by reducing their levels of trust and confidence in governance (p. 290).

Thus, if a government seems uncaring about the day-to-day travails of its citizenry, the citizens' willingness to trust that government should be hampered. On the other hand, if a government is perceived as caring and responsive to the needs of its citizens, it makes sense that the citizens might afford the government more trust.

As Blind (2006) notes, in many developed nations where trust is lowest, citizens commonly report that their government does not care about their needs (also see the study by Dalton [2005]). Along these lines, Warren (1999) suggests that governments are better trusted when they take on the interests of their citizens. Warren (2006) also notes that citizen perceptions of government sincerity are helpful in generating government-level trust. Miller (1974) made a similar point when he found that the perception that a government "does not function for [the citizens]" is associated with distrust (p. 951). He also points out that one way to reduce citizens' distrust in the government is for a government to generate policies that are more clearly and obviously responsive to the needs of the citizenry. Miller and Listhaug (1990) came to a similar conclusion in their comparative analysis between Sweden, Norway, and the United States. They found that governments that have a more flexible party system (as opposed to a rigid two-party system) are often more trusted because the citizenry believes that at least some element of the government, even if it is a very small party, cares about their specific needs.

Integrity

The final element of Mayer, Davis, and Schoorman's trustworthiness construct is integrity. Mayer, Davis, and Schoorman (1995) define integrity as "the trustor's perception that the trustee adheres to a set of principles that the trustor finds acceptable" (p. 719). This concept does not imply that the trustor and the trustee have exactly the same set of principles; rather, this definition indicates that the trustor adheres to some consistent moral code and is generally fair and just (e.g., the study by Colquitt, Scott, and LePine [2007]). As Davis et al. (2000) put it, "such factors as consistency, a reputation for honesty, and fairness all contribute to the … perception of integrity" (p. 567). Van Ryzin (2011) found that the "process has a consistently large effect on trust of civil servants" (p. 745), and the beneficial aspects of government processes include fairness, equity, respect, and honesty.

From an organizational standpoint, the degree to which a person or persons believes a business or a government is trustworthy, and therefore trusts the business or government, should also depend on the integrity of the organization in question. From the perspective of a business, the perceptions of integrity are probably based on a company's fulfillment of contractual obligations and a general abiding by the business rules and norms. For a government, the perceptions of integrity are probably based upon similar perceptions. In addition, a sense that a government operates within the rules it has created and avoids corruption probably has an impact on integrity and subsequent trust.

Research on the perceptions of government corruption support the integrative model of organizational trust in the sense that governmental integrity (or lack thereof) affects citizen trust in government. For example, multiple research perspectives note that scandal and corruption have devastating impacts on reports of trust in the government. Warren (2006), for example, notes that corruption is corrosive, undermines democracy, and creates cynicism. Kim (2010) finds "that government performance on … controlling political corruption … [is] significantly associated with broad public trust in government" (p. 801). Research by Morris and Klesner (2010) supports this notion and found that government-level corruption not only diminishes trust in government, but also produces increased corruption, creating a vicious cycle of decreased trust and increased corruption. Morris and Klesner (2010) also point out that the exposure to a political scandal reduces political trust and has even been linked to voter apathy in both the United States (Peters and Welch, 1980) and Mexico (McCann and Dominguez, 1998). Similarly, Dalton (2005) expanded on the pervasiveness of these issues and noted that political scandals over the past several decades have contributed to the decreasing levels of trust in the government found across almost all advanced industrial democracies.

Trust, Risks Taking, Outcomes, and Practical Application

Within the Mayer, Davis, and Schoorman framework, the more trustworthy a trustee is, presumably the more a trustor will be willing to be vulnerable to that person (or organization). In our case, the more trustworthy a government is, the more trust a trustor should have in his or her government in the sense that the trustor should be willing to take on more risk when a risk-taking situation presents itself. For example, a strong trust in a government might manifest itself in risk-taking actions such as general support for a government as well as a practical action such as voting. Presumably, as detailed by the model framework, after a person engages in some kind of risk-taking behavior (such as casting a vote to keep a government in power), the outcome of that risk-taking situation should feedback and update subsequent perceptions of the government's trustworthiness.

A relevant example of this framework of trust generation relates to a recent policy directive of the Obama administration. In June 2012, the US government implemented the Deferred Action for Childhood Arrivals (DACA). Under the DACA, illegal immigrants who were brought into the United States as children, who have graduated from US schools, and who meet other requirements can obtain official deferment of deportation from the US government (US Department of Homeland Security, 2013). But, in order to do so, they must first make themselves very vulnerable by identifying themselves to the government, trusting that the government will honor its promise and not immediately deport them. The degree to which such illegal individuals believe that the American government is trustworthy, that is, possesses ability, benevolence, and integrity, should predict the degree to which they trust the government of the United States and the degree to which they will be willing to take a risk and operate within the bounds of the DACA framework.

Research Hypotheses

This study addresses the following questions: Is the integrative model of organizational trust, applied to citizen trust in their government, empirically supported by our data? If so, is there a consistent model construct across countries? And, in terms of modeling citizen trust in the government, do the benevolence and integrity trustworthiness factors have a unique impact on trust?

To address these questions, we posed the following hypotheses (H):

- H1: The integrative model of organizational trust is an appropriate framework for understanding and modeling citizen trust in the government.
- H2: The ability, integrity, and benevolence trustworthiness factor constructs are country-specific.
- H3: The benevolence and integrity trustworthiness factors each have a unique impact on trust.

Methods

The data for this study comes from surveys independently fielded in Burkina Faso, Mali, Nigeria, and Senegal in November and December of 2010. We selected these nations because of both their geographic proximity in the western Trans-Sahel region of Africa and differences in terms of history, perception of democracy, and adherence to governance. By separately modeling four countries, which represent different governmental, political, and cultural variants (Fricker, Kulzy, and Combs, 2014), we are able to assess how robust the resulting framework is and whether our results are likely to be generalizable or country-specific.

The survey was administered to a representative cross-section of people in each country. The questions are predominantly four- and five-point Likert scale-based and focus on quality of life, governance, politics, security, social tolerance, and opinions about international relations. The surveys were designed and fielded following standard survey principles and procedures (e.g., the studies by Dillman [2006] and Groves et al. [2004]). Upon completion of the instrument design, the surveys underwent extensive pretesting in the field prior to their use in the 2010 data collection effort.

TABLE 19.1

Sample Sizes, Response Rates, and Margins of Error by Country

Country	Initial Sample Size	Final Sample Size	Response Rate	Margin of Error
Burkina Faso	1447	1481	98.0%	2.6%
Mali	2091	1874	89.6%	2.3%
Nigeria	4423	3770	85.2%	1.6%
Senegal	1703	1661	97.5%	2.5%
Aggregate	9664	8786	90.9%	n/a

Sample

Strict probability samples were selected via a stratified, clustered area sampling scheme designed to be nationally representative with a margin of error of 2.6% or less. The target population was adults aged 18 and over. A total of $N = 8786$ responses were obtained; the final sample sizes, the response rates, and the margins of error for each country are shown in Table 19.1.

Data were collected via face-to-face interviews conducted by professional local survey companies using interviewers who were fluent in the local dialect(s) and who were culturally aware and sensitive to local and national customs. With minor exceptions, the surveys asked the same questions, translated into the appropriate languages and dialects, across all four countries.

Measures

Measures were initially derived from the survey questions using an exploratory factor analysis via the R statistical package (R Core Team, 2012). In particular, we used a parallel analysis to determine the number of factors via the `fa.parallel` function of the R `psych` package (Revelle, 2011), and we used the `factanal` function in the base package to derive the factor loadings (using the varimax rotation). The intent of this exploratory modeling is to gain some insight into whether the survey items produce factors that reflect

TIP: USE OF `fa.parallel` TO SUPPORT NUMBER OF FACTORS TO USE IN `factanal` FUNCTION WITHIN R

As Fricker, Kulzy, and Appleget (2012) point out, the determination of the optimal number of factors is counterintuitive; the correct factors are based on supplying the correct number of factors. During this analysis, we called this difficulty the *Goldilocks conundrum*. If we used too few factors, then we would force the observed variables (questions) into factors where they do not belong. If we used too many factors, then we spread out the observed variables and we fail to recognize those that should stick together. Therefore, choosing the number of factors that are just right is necessary to properly characterize the latent variables (factors). In lieu of heuristic methods such as the Kaiser rule and the scree plot, we used a parallel analysis to objectively determine the optimal number.

the Mayer, Davis, and Schoorman trustworthiness constructs—ability, benevolence, and integrity—as well as trust in government and propensity to trust constructs. See the studies by Fricker, Kulzy, and Appleget (2012), Mulaik (2010), and DiStefano, Zhu, and Mîndrilă (2009) for additional discussions of factor analysis.

Tables 19.2 through 19.4 show the resulting factors for Mali, and the detailed results for all four countries, including the factor loadings, are shown in the study by Fricker, Kulzy, and Combs (2014). Table 19.5 shows Guttman's (1945) λ_6 values for all factors and all countries. While Guttman's λ_6 values in Table 19.5 suggest reasonable scale reliability, we note that the fact that the factors that largely replicate across the four countries indicates that the λ_6 values are conservative lower bounds on the actual reliability.

As Tables 19.2 through 19.5 show, and as discussed in more detail by Fricker, Kulzy, and Combs (2014), the factors do in fact align well with the Mayer, Davis, and Schoorman's constructs, although each of the trustworthiness constructs manifests in greater detail than single ability, integrity, and benevolence measures and the specific measures vary some between the countries. Similarly, a clear trustor propensity measure manifests, as do measures of trust in government. These measures were then used as the starting point for structural equation models, discussed in the next section, which we fit to test our hypotheses.

TABLE 19.2

Survey Items That Measure Trust in Government and Trust or Propensity along with Mean and Standard Deviations for Mali Respondents

Measure	Survey Item	Minimum	Maximum	Mean	SD
Trust in government	3. How much trust do you have in President [NAME]?	−2 = No trust at all	+2 = A lot	1.247	1.11
	4. To what extent do you approve of President [NAME]'s overall performance?	−2 = Strongly disapprove	+2 = Strongly approve	1.312	0.91
	5. How much trust do you have in Prime Minister [NAME]?	−2 = No trust at all	+2 = A lot	0.733	1.36
	6. How much trust do you have in your National Assembly?	−2 = No trust at all	+2 = A lot	0.406	1.43
	7. How much trust do you have in your political parties?	−2 = No trust at all	+2 = A lot	−0.175	1.46
	8. How much trust do you have in your armed forces?	−2 = No trust at all	+2 = A lot	1.239	1.11
	9. How much trust do you have in your local police?	−2 = No trust at all	+2 = A lot	0.547	1.40
	10. How much trust do you have in your local government?	−2 = No trust at all	+2 = A lot	0.486	1.35
	11. How much trust do you have in your courts and judges?	−2 = No trust at all	+2 = A lot	−0.292	1.53
Trustor propensity	13. How much do you trust your relatives?	−2 = No trust at all	+2 = A lot	1.770	0.67
	14. How much do you trust people from your ethnic group?	−2 = No trust at all	+2 = A lot	1.429	1.01
	15. How much do you trust people who share your religion?	−2 = No trust at all	+2 = A lot	1.505	0.94
	16. How much do you trust people from other ethnic groups?	−2 = No trust at all	+2 = A lot	0.772	1.35
	17. How much do you trust people who practice a different religion?	−2 = No trust at all	+2 = A lot	0.459	1.47

TABLE 19.3

Survey Items That Measure Integrity and Benevolence along with Mean and Standard Deviations for Mali Respondents

Measure	Survey Item	Minimum	Maximum	Mean	SD
	How much do you agree or disagree with the following statements:				
Free and fair society	18. People of all religions are free to practice their faith.	−2 = Strongly disagree	+2 = Strongly agree	1.623	0.85
	19. People are free to join any organization they want.	−2 = Strongly disagree	+2 = Strongly agree	1.607	0.84
	20. People are free to criticize the government.	−2 = Strongly disagree	+2 = Strongly agree	1.199	1.18
	21. People choose their political leaders through fair elections.	−2 = Strongly disagree	+2 = Strongly agree	0.829	1.41
	22. The media are free to accurately report the news.	−2 = Strongly disagree	+2 = Strongly agree	1.140	1.11
Democracy	1. How satisfied are you with the way democracy works in your country?	−2 = Not at all satisfied	+2 = Very satisfied	0.491	1.28
	2. In your opinion, how stable is democracy in your country?	−2 = Very fragile	+2 = Very stable	0.710	1.17

Analyses

We evaluated Mayer, Davis, and Schoorman's integrative model of organizational trust (Figure 19.1) for explaining citizen trust in the government by fitting structural equation models (SEMs) via the R `lavaan` package (Rosseel, 2012) using standard maximum likelihood estimation with robust standard errors and a Satorra–Bentler scaled test statistic. Before doing so, we assessed the joint significance of ability, integrity, and benevolence, and trustor propensity measures in Tables 19.2 through 19.4 with the trust in government measure(s) for the four countries via baseline multiple regression models (see the study by Kulzy [2012]). These models are similar in approach to the studies by Kim (2010), Vigoda-Gadot (2006), and Christensen and Laegreid (2005) used for modeling trust in the government.

We then evaluated the form of the Mayer, Davis, and Schoorman model in two ways. First, we assessed whether trust in the government is better directly modeled as a function of the measures from Tables 19.3 and 19.4 (and those in the study by Fricker, Kulzy, and Combs [2014] for the other countries) expressed as first-order factors or whether these first-order factors are more appropriately aggregated into second-order factors representing ability, integrity, and benevolence.* The latter would be more consistent with the Mayer, Davis, and Schoorman model of Figure 19.1. Second, we also assessed whether there might be additional significant predictors of trust in the government that are not captured by the

* Here we are using the terminology of Rosseel (2012) and Hair et al. (2009), where first order factors are those factors that are measured directly via survey items while second order factors are functions of the first order factors. Rosseel (2012, p. 6) describes the second order factors as comprising the structural part of the model while the first order factors comprise the measurement part of the model.

TABLE 19.4

Survey Items That Measure Ability along with Mean and Standard Deviations for Mali Respondents

Measure	Survey Item	Minimum	Maximum	Mean	SD
Essential services	32. How would you rate water in your community?	−2 = Poor	+2 = Very good	0.002	1.42
	33. Do you think that water has improved, stayed the same, or gotten worse from a year ago?	−2 = Gotten worse	+2 = Improved	0.528	1.19
	34. How would you rate roads in your community?	−2 = Poor	+2 = Very good	−0.840	1.42
	35. Do you think that roads has improved, stayed the same, or gotten worse from a year ago?	−2 = Gotten worse	+2 = Improved	0.048	1.44
	36. How would you rate electricity in your community?	−2 = Poor	+2 = Very good	−0.703	1.52
	37. Do you think that electricity has improved, stayed the same, or gotten worse from a year ago?	−2 = Gotten worse	+2 = Improved	−0.046	1.37
	38. In your neighborhood, how often do you have electricity?	−2 = Never	+2 = Always	−0.417	1.80
Education and healthcare	28. How would you rate education in your community?	−2 = Poor	+2 = Very good	−0.184	1.38
	29. Do you think that education has improved, stayed the same, or gotten worse from a year ago?	−2 = Gotten worse	+2 = Improved	0.471	1.30
	30. How would you rate health care in your community?	−2 = Poor	+2 = Very good	−0.102	1.37
	31. Do you think that health care has improved, stayed the same, or gotten worse from a year ago?	−2 = Gotten worse	+2 = Improved	0.704	1.25
Security concerns	40. How much do economic problems in the country worry you?	−2 = A lot	+2 = Not at all	1.529	0.96
	41. How much does border security worry you?	−2 = A lot	+2 = Not at all	1.193	1.22
	42. How much do terrorist attacks in neighboring countries worry you?	−2 = A lot	+2 = Not at all	0.859	1.44
	43. How much does the war in Iraq worry you?	−2 = A lot	+2 = Not at all	0.482	1.51
	44. How much does the Palestinian/Israeli crisis worry you?	−2 = A lot	+2 = Not at all	0.526	1.52
	45. How much does the conflict in Afghanistan worry you?	−2 = A lot	+2 = Not at all	0.377	1.53

ability, integrity, and benevolence terms in the Mayer, Davis, and Schoorman model. For example, our survey also contains items about foreign assistance (Table 19.6), as well as items soliciting respondent opinions of other countries (Table 19.7), that coalesce into one or more factors. Testing whether these non-Mayer measures are significant in the models helped us assess whether the integrative model of organizational trust might require additional terms to best model trust in the government.

TABLE 19.5

Guttman's λ_6 Values for Tables 19.2 through 19.4 and Related Factors for all Countries

Measure	Nigeria	Senegal	Mali	Burkina Faso
Trust in government	N/A	N/A	0.70	0.61
• Trust in government agencies	0.77	0.66	N/A	N/A
• Trust in policy makers	0.78	0.63	N/A	N/A
Trustor propensity	0.78	0.66	0.71	0.63
Free and fair society	0.78	0.67	0.71	0.63
Democracy	0.77	0.66	0.72	0.62
Essential services	0.77	0.64	0.72	N/A
• Essential services, current	N/A	N/A	N/A	0.63
• Essential services, changes in	N/A	N/A	N/A	0.66
Electricity	0.78	0.64	N/A	N/A
Roads	0.77	0.64	N/A	N/A
Education and Healthcare	N/A	N/A	0.72	N/A
Security concerns	0.78	0.62	0.74	0.64
Economics and ethnic relations/borders	0.79	N/A	N/A	0.65
Terrorism concerns	N/A	N/A	N/A	0.64

Note: N/A cells indicate factors that are not relevant for the associated country.

TABLE 19.6

Survey Items That Measure Foreign Assistance along with Mean and Standard Deviations for Mali Respondents

Measure	Survey Item	Minimum	Maximum	Mean	SD
Foreign assistance	How much do you think that foreign countries [READ NEXT ITEM] in [SURVEY COUNTRY]?				
	48. Help the military	−2 = Not at all	+2 = A lot	0.813	1.27
	49. Improve public health	−2 = Not at all	+2 = A lot	0.529	1.28
	50. Strengthen the education sector	−2 = Not at all	+2 = A lot	0.355	1.30
	51. Prevent terrorist attacks	−2 = Not at all	+2 = A lot	0.464	1.33
	52. Help reduce corruption	−2 = Not at all	+2 = A lot	−0.212	1.32
	53. Help develop the economy through investment	−2 = Not at all	+2 = A lot	0.474	1.30
	54. Provide humanitarian aid	−2 = Not at all	+2 = A lot	0.637	1.32
	55. Build roads, bridges, and buildings	−2 = Not at all	+2 = A lot	−0.408	1.37

Thus, using the Mali measures to illustrate, we fit the baseline multiple regression model shown in the path diagram of Figure 19.2. Per convention, the squares denote the observed variables (in the form of responses to survey questions), the circles are latent variables (originally derived via exploratory factor analysis), and the arrows denote paths (see the section "Results" for significance tests). In the baseline model, all the factors directly affect the trust in the government variable additively. To simplify this and the subsequent path diagram displays, we suppress the measurement error and disturbance terms as well as labeling the paths with their associated parameters.

TABLE 19.7

Items That Measure Respondent's Opinions of Other Countries

Measure	Survey Item	Minimum	Maximum	Mean	SD
International opinions	Overall, do you have a very favorable, somewhat favorable, somewhat unfavorable, or very unfavorable option of the [READ COUNTRY]?				
	56. United Kingdom	−2 = Very unfavorable	+2 = Very favorable	0.346	1.20
	57. France	−2 = Very unfavorable	+2 = Very favorable	0.695	1.21
	58. United States	−2 = Very unfavorable	+2 = Very favorable	1.122	1.02
	59. China	−2 = Very favorable	+2 = Very unfavorable	1.512	0.84
	60. Iran	−2 = Very favorable	+2 = Very unfavorable	0.319	1.23
	61. Saudi Arabia	−2 = Very favorable	+2 = Very unfavorable	1.152	1.03
	62. Libya	−2 = Very favorable	+2 = Very unfavorable	1.324	0.91
	63. Niger	−2 = Very unfavorable	+2 = Very favorable	0.455	1.23
	64. South Africa	−2 = Very unfavorable	+2 = Very favorable	0.600	1.17

We then compare the baseline model in Figure 19.2 to two alternatives. First, we have a model in Figure 19.3 similar to the baseline model except that trustor propensity also acts as a mediating variable for all the other variables in the model. We refer to this as the *first-order SEM* model formulation, a name that follows from the fact that the model only contains first-order factors. Second, in Figure 19.4, we have the *second-order SEM* model formulation, where this name follows because the model contains both first- and second-order factors. For this model, note that the second-order factors comprising the structural part of the model directly correspond to the Mayer, Davis, and Schoorman model ability and integrity/benevolence constructs of Figure 19.1 as well as a third non-Mayer construct.

Results

Our analysis shows that second-order SEMs are preferred to first-order SEMs and baseline models. When comparing model fits by country, this preference manifests itself in two ways. First, as shown in Table19.8, the fraction of trust in the government variance explained (R^2) is higher, generally by quite a bit, for the second-order SEMs compared to the other model formulations.

Second, Table 19.9 summarizes the key modeling results where, for the purposes of display clarity, we have left out the estimated model parameter values and simply display the significance (or lack thereof) of each of the parameters. (See Tables C-1 to C-4 in the study by Fricker, Kulzy, and Combs [2014] for detailed model results, including p values.)

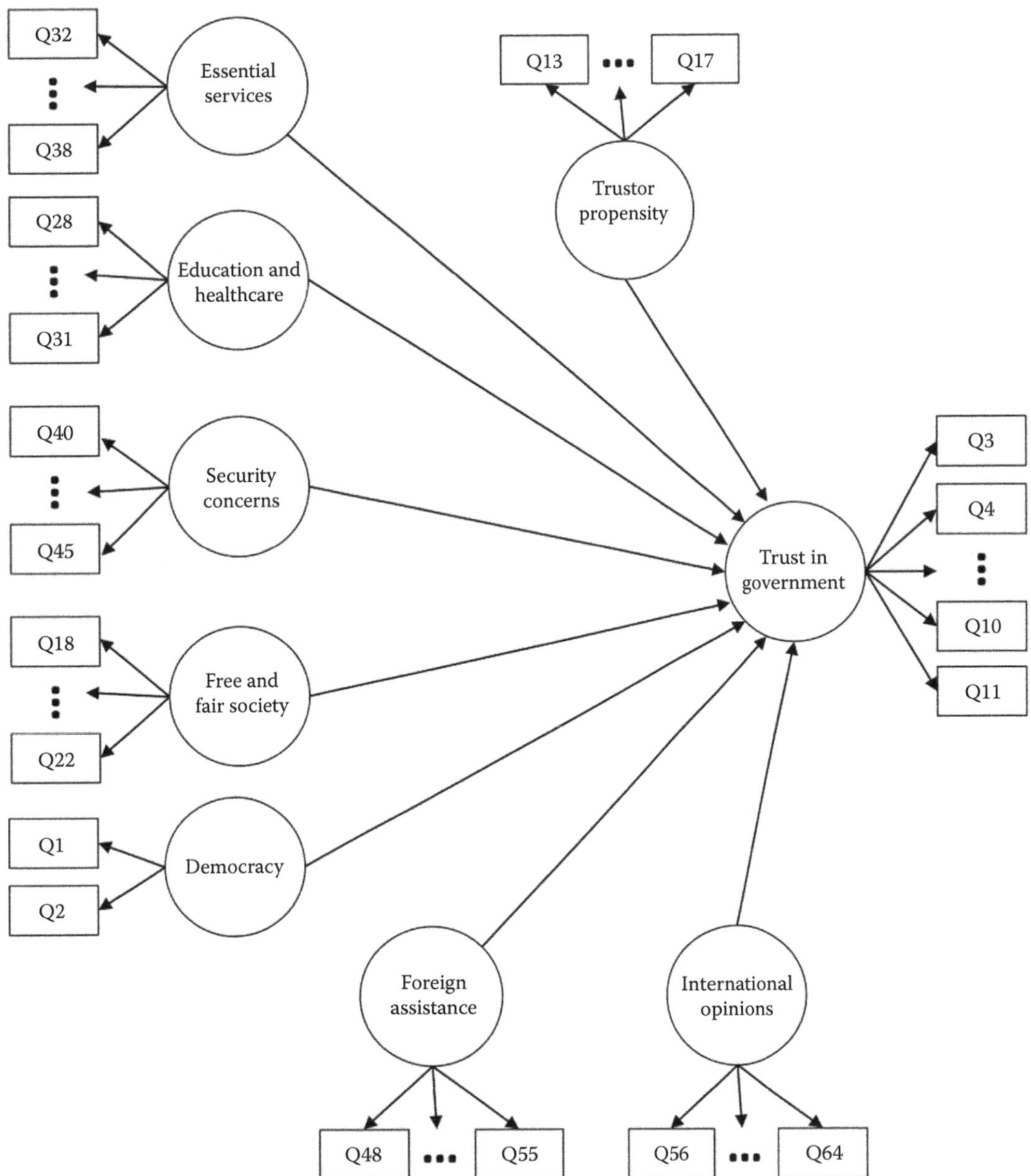

FIGURE 19.2
Baseline model formulation for Mali. Squares denote observed variables (in the form of responses to survey questions), circles are latent variables (derived via exploratory factor analysis), and arrows denote the paths. For display simplicity, only two non-Mayer terms (FA: foreign assistance and IO: International Opinions) are shown, although there were various additional terms that varied by country.

Table 19.9 shows that, broadly speaking, the Mayer, Davis, and Schoorman model formulation is appropriate across all four countries in the sense that virtually all the terms posited by Mayer, Davis, and Schoorman are statistically significant. The only deviations from their theory is that the trustor propensity is not significant in predicting trust in the government in Nigeria, and it does not significantly predict any of the terms in the Senegal

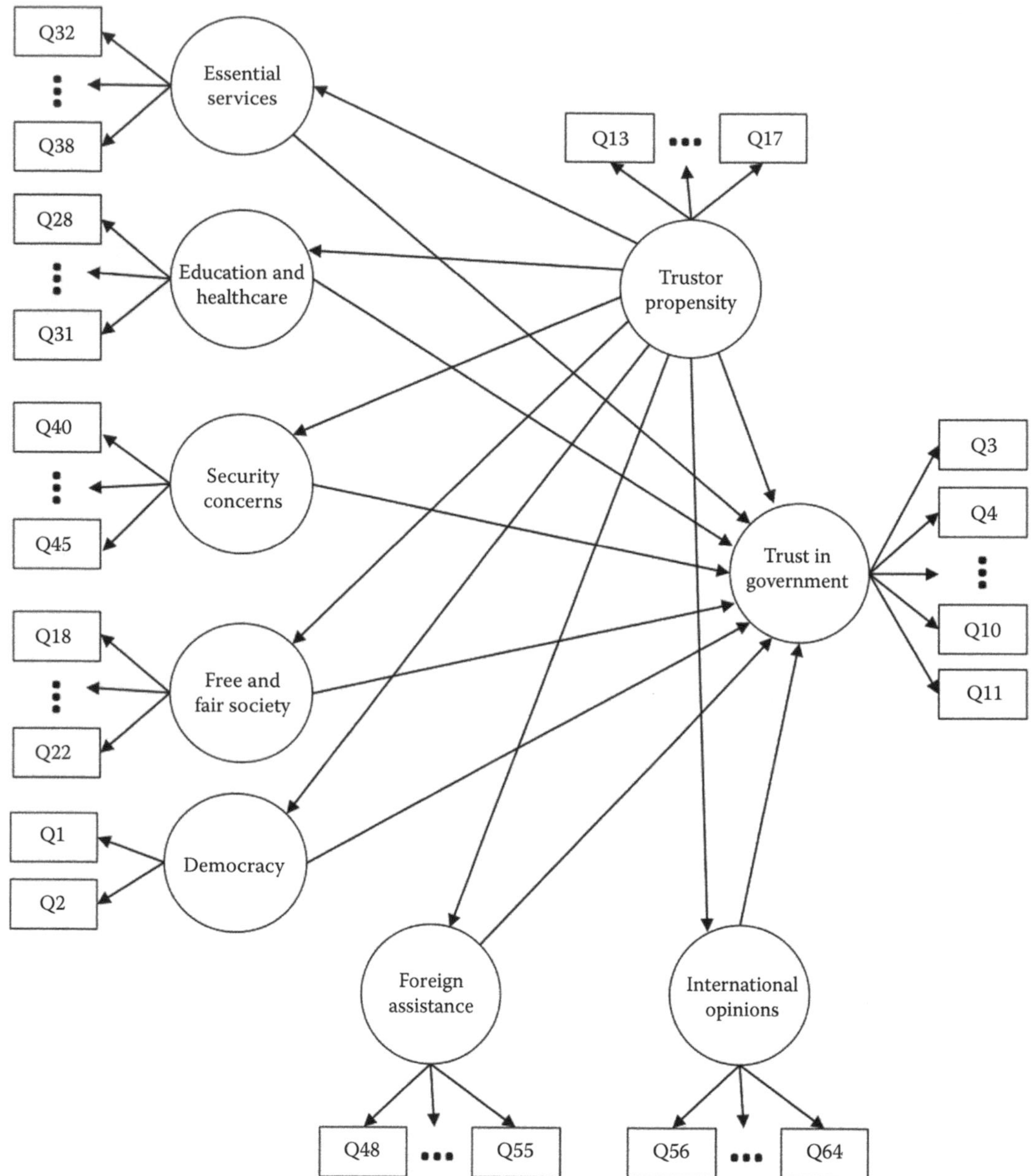

FIGURE 19.3
First-order SEM model formulation for Mali with trustor propensity acting as a moderator variable on the other model variables.

model. Also, Mali and Nigeria have significant non-Mayer terms that are both predictive of trust in the government and often predicted by trustor propensity.

In terms of model fit, the chi-square tests for all four second-order SEM models are significant, although given the survey sample sizes for each country, this is expected (c.f. Table 12.4 of the study by Hair et al. [2009], as well as the discussion in Chapter 13 of the study by Hoyle [2012]). As Table 19.8 shows, for three of the four countries, the models explain from half to three-quarters of the variation in the trust in government outcome

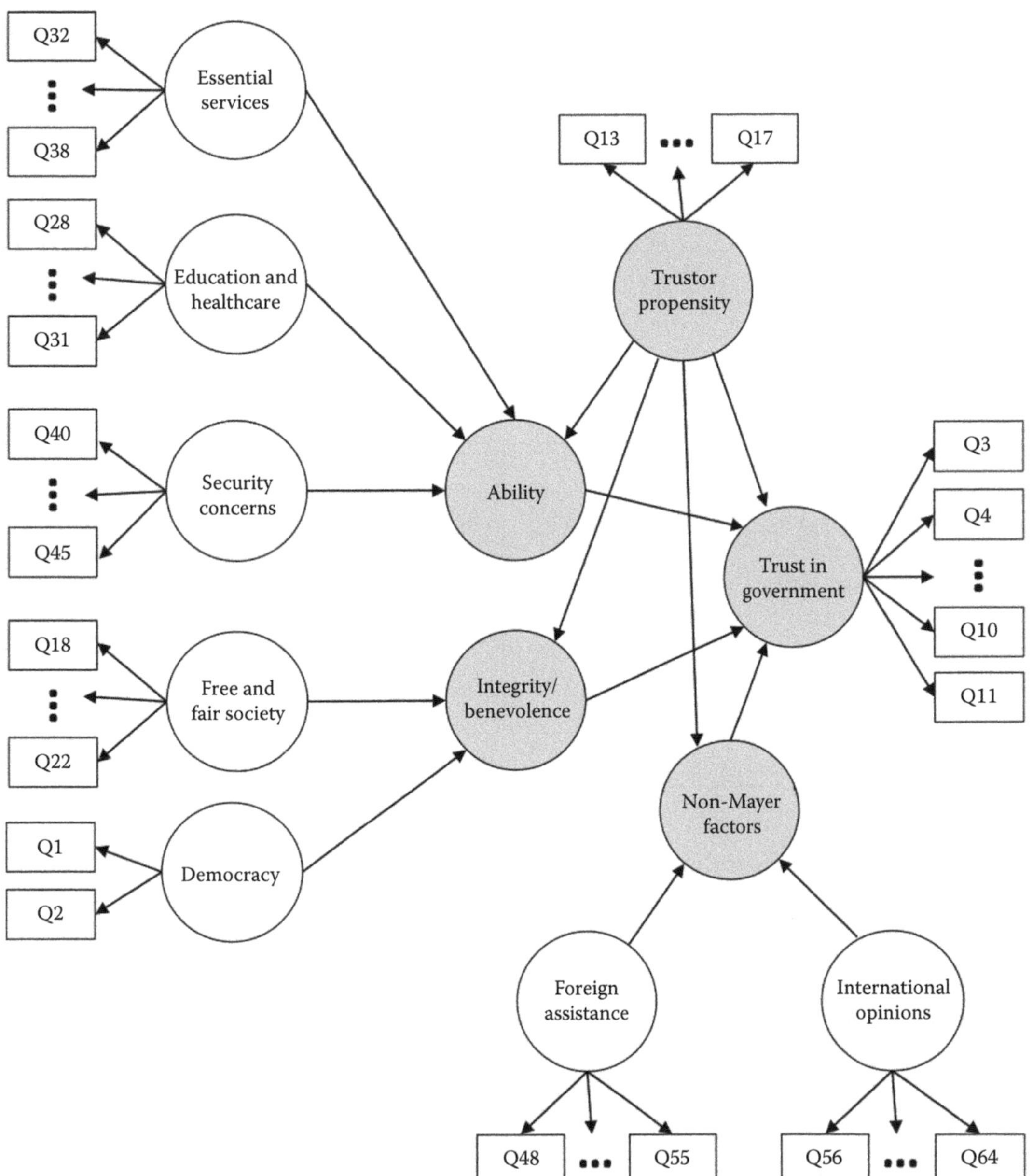

FIGURE 19.4
Second-order SEM model formulation for Mali with second-order factors for ability and integrity/benevolence, per Mayer, Davis, and Schoorman's (1995) integrative model of organizational trust, as well as other non-Mayer terms. The gray circles are those factors that directly affect trust in the government (and the trust in the government term itself).

measure. The comparative fit index is between 0.67 and 0.83 for all four countries (with very similar Tucker–Lewis indices), and the root mean square error of approximation is between 0.057 and 0.072 for all four countries. These are indicative of an adequate model fit but, as we will discuss in the next section, also suggestive that model improvements are possible.

TABLE 19.8

Fraction of Trust in Government Variance Explained (R^2 values) for the Baseline, First-, and Second-Order SEMs by Country

	Burkina Faso	Mali	Nigeria	Senegal
Baseline model	0.430	0.340	0.330	0.230
First-order SEM	0.345	0.390	0.577	0.228
Second-order SEM	0.512	0.685	0.788	0.270

TABLE 19.9

Statistical Significance of the Paths between the Gray Factors in the Second-Order SEM of Figure 19.4

	Burkina Faso (p)	Mali (p)	Nigeria (p)	Senegal (p)
Trust in government				
• Ability	<0.001	<0.001	<0.001	<0.01
• Integrity/benevolence	<0.001	<0.001	<0.001	<0.05
• Trustor propensity	<0.001	<0.05	N/A	N/A
• Other non-Mayer terms	N/A	<0.01	<0.001	N/A
Ability				
• Trustor propensity	<0.001	<0.001	<0.001	N/A
Integrity/benevolence				
• Trustor propensity	<0.001	<0.001	<0.001	N/A
Other non-Mayer terms				
• Trustor propensity	N/A	<0.001	<0.001	N/A

Returning to our hypotheses, we find that our results support two of the three. Specifically, for H1, our results show that the integrative model of organizational trust is an appropriate framework for understanding and modeling citizen trust in the government, at least for the four countries in our analysis. We reach this conclusion based on the fact that our SEM results are consistent with the Mayer, Davis, and Schoorman model for all four countries we evaluated.

For H2, we find that the constructs of the ability, integrity, and benevolence factors in the integrative model of organizational trust are country-specific. What Table 19.9 does not show, but is evident from Appendices B and C of Fricker, Kulzy, and Combs (2014), is the first-order factors that define ability and integrity/benevolence are uniquely specified for each country. For example, the ability in Mali consists of effective education systems and healthcare infrastructure and attracting foreign aid and investment, while in Burkina Faso, the ability is much more broadly defined. This suggests that, while ability, integrity, and benevolence are indeed predictive of trust in the government following the Mayer, Davis, and Schoorman model, just what constitutes ability, integrity, and benevolence is country-specific.

For H3, we were unable to find separate benevolence and integrity trustworthiness factors, each with a unique impact on trust. This may be because integrity and benevolence do not uniquely contribute to trust, as found be Jarvenpaa, Knoll, and Leidner (1998) and Mayer and Gavin (2005), but it could just as well be due to some weakness in our survey items and measures. Hence, we are unable to reach a definitive conclusion for this hypothesis.

Discussion

Our results are consistent with previous research. For example, Safadi and Lombe (2011), Kim (2010), Vigoda-Gadot (2006), Christensen and Laegreid (2005), and Marlowe (2004) all find that governmental ability to provide social services, in one form or another, is associated with increased trust in the government. In terms of our integrity/benevolence construct, Christensen and Laegreid (2005) find, as we do, that "trust in government generally increases according to the level of satisfaction with democracy" (p. 500). Marlowe (2004) also finds a positive association between the perceptions of a democratic system and the trustworthiness of public administrators. Vigoda-Gadot (2006) finds that ethics, as measured in terms of the perceived integrity of civil servants, is a good predictor of trust in governance. Finally, just as we find similarities and differences by country in terms of the specific ability and integrity/benevolence components, Kim (2010) finds that "there are similarities and differences in the factors affecting public trust in Japan and South Korea" (p. 807).

However, our fundamental supposition is that our results support Mayer, Davis, and Schoorman's (1995) model. This is in spite of the fact that the original model was posed as a model of trust between individuals, and we have applied it to individuals' trust in an organizational-level entity: government. In particular, we found that models with second-order latent variables representing ability, benevolence/integrity, and a third non-Mayer term generally fit better models than other model formulations. With the exception of Senegal, the resulting models explained from just over half to three-quarters of the variation in respondent trust in the government. This is substantial for a social science model, although it also suggests that improvements are possible, particularly with respect to better understanding of the non-Mayer term or terms.

Hence, we posit some modifications to the Mayer, Davis, and Schoorman model for modeling trust in the government, in terms of both potential additional factors and a further delineation of what it means for a government to demonstrate ability, benevolence, and integrity. We illustrate these ideas in what we call the integrative model of governmental trust (Figure 19.5), where to begin, we hypothesize that governmental ability has two main

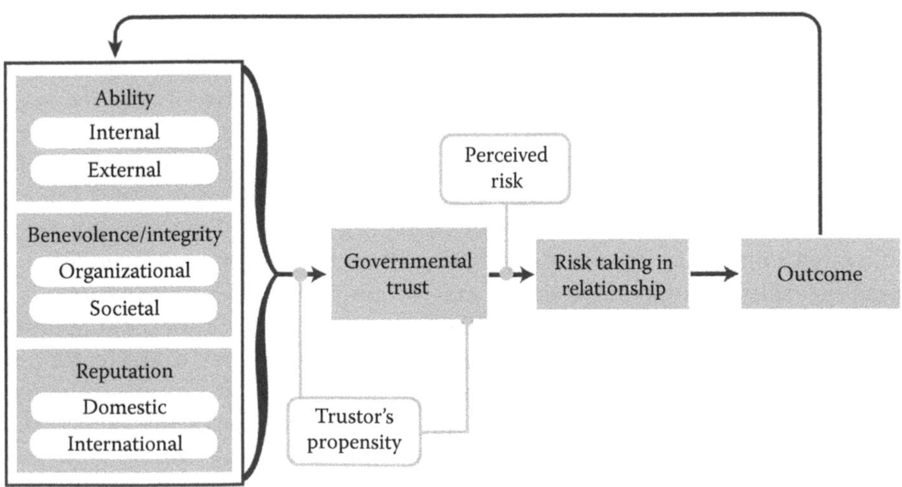

FIGURE 19.5
Integrative model of governmental trust.

facets. The first facet is internal ability, defined as those skills and competencies necessary to provide domestic services desired by citizens, such as having essential infrastructure, effectively managing the country's economy, providing for individual safety and security, etc. The second facet is external ability, defined as those skills and competencies necessary for governing the country within the international community, such as maintaining a national security apparatus, attracting external aid as necessary, etc.

Similarly, we hypothesize that governmental benevolence/integrity also has two main facets. The first facet is organizational benevolence/integrity, which is defined as the extent to which the government adheres to a set of acceptable principles, and it is operated in a manner that is good for the country. For example, the government operates according to democratic principles; it functions in an open and transparent manner. The second facet is societal benevolence/integrity, which is defined as the extent to which the government promotes and advances societal conditions and principles that are good for and desired by the citizenry. For example, the government promotes a society that is free and fair, and peaceful and tolerant.

In addition, our results suggest that there may be a third factor or a set of factors that affect governmental trust. Unfortunately, there were not enough significant variables in our model to fully define it or them, but we hypothesize that they are related to governmental reputation. As with ability and benevolence/integrity, we posit that there may be two facets to governmental reputation. The first facet is the government's international reputation, which is defined as the estimation in which the country and its government is held by the international community, including other countries and governments and the people of other countries. For example, it is the perception of whether the government conducts effective international relations and the country's status among the world's nations. The second facet is the government's domestic reputation, which is defined as the estimation in which the country's citizenry holds the government. This is a reputation for effective governance, as opposed to, for example, whether or not it is actually effective. Whether this facet can be separately measured from ability and benevolence/integrity is an open question.

The role of reputation in trust in the government is potentially an important one because it is likely that most citizens have limited personal interaction with their government. That is, unlike in the Mayer, Davis, and Schoorman interpersonal model in which individuals are likely to have multiple opportunities to iterate around the loop shown in Figure 19.1 and develop trust as a result of direct, repeated interaction, most citizens will likely have much less such interaction with their government. Because of this, we expect that individuals will likely base at least some of their trust in government on the government's reputation as determined via mass media, interaction with family, friends, and colleagues, etc. Furthermore, there are likely to be aspects of ability, benevolence, and integrity that certain citizens are simply unable to observe or experience. In these cases, the only assessment an individual will be able to make, both of how much to trust the government and how much risk is inherent in some action, may be mainly based on reputation.

This additional reputation component of trustworthiness is consistent with some of the popular business management literature. For example, Covey and Merrill (2006) say that trust is based on four cores: integrity, intent, capabilities, and results. The first three directly map to Mayer, Davis, and Schoorman's integrity, benevolence, and ability components of trustworthiness. For the fourth, their results core, Covey and Merrill say it "refers to our track record, our performance, our getting the right things done... when we achieve the results we promised, we establish a positive reputation of performing ... and our reputation precedes us" (p. 55). They go on to say that "Results are the fruits—the visible, tangible, measurable outcomes that are most easily seen and evaluated by others" (p. 56). In this regard, the reputation component in Figure 19.3 slightly differs, as reputation is the result of both directly observable outcomes and unobserved or unobservable outcomes. The latter aspect is important because unobserved outcomes may be correct, actual outcomes, say as related via a responsible news medium, and they may also be incorrect, untrue or distorted outcomes, perhaps reported in other types of mass or social media, or that are spread via rumor, etc. Reputation is also a function of the trustor's viewpoint and disposition.

Of course, reputation may not be the only non-Mayer trustworthiness construct in the integrative model of governmental trust. Van Ryzin (2011) finds that institutional trust depends on the process as much as the outcome. To the extent that Van Ryzin's process construct captures the citizen's observations that the government is ethical and fair, it could be a particular dimension of the Mayer, Davis, and Schoorman benevolence/integrity construct. However, if it is based more on perception than observation, then it may also be a part of the hypothesized reputation trustworthiness construct. And, if it is simply that the process in and of itself is important, then it may be that it is an entirely separate construct. Vigoda-Gadot (2006) finds that trust in the government is a function of satisfaction with services, ethics, and organizational politics. The first two can be mapped onto the Mayer ability and integrity/benevolence trustworthiness constructs. The third, organizational politics, is described by Vigoda-Gadot as "unethical or even immoral or corrupt behavior" (p. 287), and as such, it is a specific dimension within the Mayer integrity/benevolence construct. However, to the extent that governmental trust depends on the alignment of the individual's political views with the government's, there may be a different organizational politics construct that captures this alignment or misalignment as a separate non-Mayer construct. This definition of organizational politics is similar to some of the items in Christensen and Laegreid's (2005) political factors construct.

Limitations of Current Research

There are two limitations of the current research. First, as noted earlier in this chapter, the Mayer, Davis, and Schoorman definition of trust states, in essence, that person A trusts person B when he or she is willing to be vulnerable to the actions of person B. However, the items used to measure trust in the current survey do not explicitly examine a person's willingness to be vulnerable to their government. Instead, the items included are more standard *do you trust person X* items. It is important for future research to include items that tap the *willingness to be vulnerable* element of Mayer, Davis, and Schoorman's model as well as the more standard trust items included in the current survey.

With that said, the items used in this research are consistent with previous research that assessed trust in the government using survey items that asked respondents to directly rate their trust. Examples are as follows:, "Please indicate to what extent you trust the following institutions [the central government, the local government] to operate in the best interests of society" (Kim, 2010). "Below are the names of various institutions, such as the police, the cabinet, the civil service etc. How much trust do you have in each of these institutions?" (Christensen and Laegreid, 2005). "Most civil servants can be trusted to do what is best for the country" (Van Ryzin, 2011). The research by Safadi and Lombe (2011), Van de Walle (2007), Vigoda-Gadot (2006), Rahn and Rudolph (2005), Marlowe (2004), and Chanley, Rudolph, and Rahn (2000) are also based on similar items.

Second, the current survey included only one item that explicitly examined the Mayer concept of benevolence. Future research should aim to more thoroughly examine this construct. Clearly, any government benevolence construct will probably be multifaceted—future work should attempt to examine methods in which governments can demonstrate benevolence to their citizenry and whether citizens perceive those methods as benevolent. Benevolence undoubtedly looks different in different nations with different forms of government. The manner in which a democracy demonstrates benevolence might be by passing laws desired by the populace, while a dictatorship might demonstrate benevolence by unilaterally increasing spending on similarly desired projects. Regardless of the mechanism by which benevolence is demonstrated, the question is do citizens perceive their government as benevolent?

Conclusion

Returning to our model results, our empirical work suggests that the Mayer, Davis, and Schoorman model is an extremely helpful organizational framework for understanding and predicting citizen trust in the government. Of course, citizens of different countries have specific local concerns and needs that impact their trust ratings. The important point here, however, is that the Mayer, Davis, and Schoorman model is robust to capture these nuances. In particular, when thinking about trust, we find that just within our four countries, some separate the trust in their individual leaders from the trust in broader governmental institutions, while others do not distinguish between the two. Similarly, while this research confirms that Mayer, Davis, and Schoorman's categories of ability, benevolence, and integrity are relevant and applicable, it also shows that each of these is also a country-specific construct composed of the things each citizenry judges relevant and important to its society.

In some sense, this brings us full circle back to Combs, Blincoe, and Vorm (2015) assertion that "ability varies from domain to domain and situation to situation," although what we find with trust in the government is that the definition of ability, benevolence, and integrity vary from country to country. This finding thus suggests a series of future research opportunities. One important line of research is further defining survey items from which to derive robust measures of the various aspects of ability, benevolence, integrity, and reputation. Another line of research is further evaluating whether or not reputation is an important aspect of the model and, if so, whether that is the correct characterization of the term. Then, given these results, political scientists, organizational theorists, and international relations experts will likely be interested in comparing and contrasting trust in government models across various countries, societies, and governmental structures.

Megacity Use Case

These results are useful for helping Xanadu understand how to adjust its policies to increase citizen trust in their government. Imagine that the Xanadu policy makers conduct a careful, thoughtful, and rigorous collection of data that permits the fitting of a trust in government SEM that reflects the interests and the needs of its citizenry. Such a model would then become a guide for the government to understand how to best serve its citizens in terms of those policies and services that would improve trust in the government.

The results might, for example, indicate that the Xanadu government should improve its internal ability by providing essential services to the Razini refugees, such as food, shelter, and security, and better health and working conditions for the Bursuka families. Similarly, it might indicate that implementing policies of societal benevolence and integrity, such as policies that decrease racial and class tensions, will assist the government in managing and calming the different sects. Being able to identify and efficiently target the more significant factors will prove valuable to a resource-strapped Xanadu.

Furthermore, fitting a trust in government SEM will not only offer insight into where best to allocate resources, but also offers an assessment by quantifiably measuring the effects the government intends to have on its citizens. For example, it is not hard to imagine that an initial analysis of the model would reveal that Buruska has many concerns, but the one with the greatest influence on trust may be external security. In an effort to instill confidence through governance, the government can leverage this factor and focus the majority of resources toward security. By moving security forces to the border, the government strengthens their benevolence by caring. If they are able to control the refugee flow, then they will also increase their ability factor. The continued use of the model allows the government to better assess if they are influential in moving the needle toward trust.

DISCUSSION QUESTIONS

1. What are the key aspects of Mayer, Davis, and Schoorman's (1995) integrative model of organizational trust for modeling citizens' trust in their government?
2. How effective might this model be at different levels of measurement, using different types of data?
3. In what ways might this model be extended?
4. How would such extensions be validated?

References

Bachmann, R. (2011). At the crossroads: Future directions in trust research. *Journal of Trust Research*, 1(2), 203–213.

Blind, P. K. (2006). Building trust in government in the twenty-first century: Review of literature and emerging issues. *Seventh Global Forum on Reinventing Government Building Trust in Government*, 26–29.

Chanley, V. A., Rudolph, T. J., and Rahn, W. M. (2000). The origins and consequences of public trust in government: A time series analysis. *Public Opinion Quarterly*, 64(3), 239–256.

Christensen, T., and Laegreid, P. (2005). Trust in government: The relative importance of service satisfaction, political factors, and demography. *Public Performance and Management Review*, 28(4), 487–511.

Colquitt, J. A., Scott, B. S., and LePine, J. A. (2007). Trust, trustworthiness, and trust propensity: A meta-analytic test of their unique relationships with risk taking and job performance. *Journal of Applied Psychology*, 92(4), 909–927.

Combs, D. J. Y., Blincoe, S, and Vorm, E. (2015). Trust, attitudes, and social influence: The cross cultural social psychology of counterinsurgency. *Department of Defense Manual.*

Covey, S. M. R., and Merrill, R. R. (2006). *The Speed of Trust: The One Thing That Changes Everything*. New York: Free Press.

Dalton, R. J. (2005). The social transformation of trust in government. *International Review of Sociology*, 15(1), 133–154.

Davis, J. H., Schoorman, F. D., Mayer, R. C., and Tan, H. H. (2000). The trusted general manager and business unit performance: Empirical evidence of a competitive advantage. *Strategic Management Journal*, 21, 563–576.

Dillman, D. A. (2006). *Mail and Internet Surveys: The Tailored Design Method 2007 Update with New Internet, Visual, and Mixed-Mode Guide*. New York: John Wiley & Sons.

DiStefano, C., Zhu, M., and Mîndrilă, D. (2009). Understanding and using factor scores: Considerations for the applied researcher. *Practical Assessment, Research and Evaluation*, 14(20), 1–11.

Fricker, Jr., R. D., Kulzy, W. W., and Appleget, J. A. (2012). From data to information: Using factor analysis with survey data. *Phalanx*, 45, 30–34.

Fricker Jr, R. D., Kulzy, W. W., and Combs, D. J. Y. (2014). *The Integrative Model of Organizational Trust as a Framework for Understanding Trust in Government*. Monterey, CA: Naval Postgraduate School, Technical Report NPS-OR-14-003.

Groves, R. M., Fowler, Jr., F. J., Couper, M. P., Lepkowski, J. M, Singer, E., and Tourangeau, R. (2004). *Survey Methodology* (Second ed.). New York: Wiley-Interscience.

Guttman, L. (1945). A basis for analyzing test-retest reliability. *Psychometrika*, 10(4), 255–282.

Hair, Jr., J. F., Black, W. C., Babin, B. J., and Anderson, R. E. (2009). *Multivariate Data Analysis* (Seventh ed.). Upper Saddle River, NJ: Prentice Hall.

Hardin, R. (2006). *Trust*. Cambridge: Polity.

Hetherington, M. J. (1998). The political relevance of political trust. *American Political Science Review*, 92(4), 791–808.

Hetherington, M. J. (2006). *Why Trust Matters: Declining Political Trust and the Demise of American Liberalism*. Princeton, NJ: Princeton University Press.

Hoyle, R. H. (2012). *Handbook of Structural Equation Modeling*. New York: The Guilford Press.

Jarvenpaa, S. L., Knoll, K., and Leidner, D. E. (1998). Is anybody out there? Antecedents of trust in global virtual teams. *Journal of Management Information Systems*, 14, 29–64.

Kilcullen, D. (2010). *Counterinsurgency*. Oxford: Oxford University Press.

Kim, S. (2010). Public trust in government in Japan and Korea: Does the rise of critical citizens matter? *Public Administration Review*, 70(5), 801–810.

Kulzy, W. (2012). Modeling indigenous population attitudes in support of irregular warfare analysis. Master of Science thesis. Monterey, CA: Naval Postgraduate School.

Li, P. P. (2011). The rigour–relevance balance for engaged scholarship: New frame and new agenda for trust research and beyond. *Journal of Trust Research*, 1(1), 1–21.

Marlowe, J. (2004). Part of the solution or cogs in the system? The origins and consequences of trust in public administrators. *Public Integrity*, 6(2), 93–113.

Mayer, R. C., and Gavin, M. B. (2005). Trust in management and performance: Who minds the shop while the employees watch the boss? *Academy of Management Journal*, 48(5), 874–888.

Mayer, R. C., Davis, J. H., and Schoorman, F. D. (1995). An integrative model of organizational trust. *Academy of Management Review*, 20(3), 709–734.

McCann, J. A., and Domínguez, J. I. (1998). Mexicans react to political corruption and electoral fraud: An assessment of public opinion and voting behaviour. *Electoral Studies*, 17(4), 483–504.

Miller, A. H. (1974). Political issues and trust in government, 1964–1970. *American Political Science Review*, 68(3), 951–972.

Miller, A. H., and Listhaug, O. (1990). Political parties and confidence in government: A comparison of Norway, Sweden and the United States. British *Journal of Political Science*, 20(3), 357–386.

Mishra, A. K., and Mishra, K. E. (2013). The research on trust in leadership: The need for context. *Journal of Trust Research*, 3(1), 59–69.

Morris, S. D., and Klesner, J. L. (2010). Corruption and trust: Theoretical considerations and evidence from Mexico. *Comparative Political Studies*, 43(10), 1258–1285.

Mulaik, S. A. (2010). *Foundations of Factor Analysis* (Second ed.). New York: CRC Press.

Peters, J. G., and Welch, S. (1980). The effect of charges of corruption on voting behaviour in congressional elections. *American Political Science Review*, 74(3), 697–708.

R Core Team. (2012). *R: A Language and Environment for Statistical Computing*. Vienna: R Foundation for Statistical Computing.

Rahn, W. M., and Rudolph, T. J. (2005). A tale of political trust in American cities. *Public Opinion Quarterly*, 69(4), 530–560.

Revelle, W. (2011). Psych: Procedures for personality and psychological research (R package version 1.0-51). Evanston: Northwestern University.

Rosseel, Y. (2012). lavaan: An R package for structural equation modelling. *Journal of Statistical Software*, 48(2), 1–36.

Safadi, N. S., and Lombe, M. (2011). Exploring the relationship between trust in government and the provision of social services in countries of the global south: The case of Palestine. *Journal of Social Service Research*, 37, 403–411.

Schoorman, F. D., Mayer, R. C., and Davis, J. H. (1996). Empowerment in veterinary clinics: The role of trust in delegation. Paper presented at the *11th annual meeting of the Society for Industrial and Organizational Psychology, San Diego, CA.*

Schoorman, F. D., Mayer, R. C., and Davis, J. H. (2007). An integrative model of organizational trust: Past, present, and future. *Academy of Management Review*, 32(2), 344–354.

Simpson, J. A. (2007a). Foundations of interpersonal trust. In Higgins, T., and Kruglanski, A. (Eds.), *Social Psychology: Handbook of Basic Principles* (pp. 587–605). New York: The Guilford Press.

Simpson, J. A. (2007b). Psychological foundations of trust. *Current Directions in Psychological Science*, 16, 264–268.

US Department of Homeland Security. (2013). Deferred action for childhood arrivals. Retrieved from http://www.dhs.gov/deferred-action-childhood-arrivals.

Van de Walle, S. (2007). Determinants of confidence in the civil service: An international comparison. *Research in Public Policy Analysis and Management*, 16, 171–201.

Van Ryzin, G. G. (2011). Outcomes, process, and trust of civil servants. *Journal of Public Administration Research and Theory*, 21(4), 745–760.

Vigoda-Gadot, E. (2006). Citizens' perceptions of politics and ethics in public administration: A five-year national study of their relationship to satisfaction with services, trust in governance, and voice orientations. *Journal of Public Administration Research and Theory*, 17(2), 285–305.

Warren, M. E. (1999). *Democracy and Trust*. Cambridge: Cambridge University Press.

Warren, M. E. (2006). Democracy and deceit: Regulating appearances of corruption. *American Journal of Political Science*, 50(1), 160–74.

Wieselquist, J., Rusbult, C. E., Foster, C. A., and Agnew, C. R. (1999). Commitment, pro-relationship behaviour, and trust in close relationships. *Journal of Personality and Social Psychology*, 77(5), 942–966.

20

Understanding Public Opinion toward Violent Extremists

Lewis A. Anderson and Ronald D. Fricker, Jr.

CONTENTS

Extensions to the Use Case 424
Introduction 424
Model-Based versus Design-Based Inference 425
A Word on Complex Survey Sampling 426
The Way It Should Be 427
The Raking Algorithm 428
An Illustrative (but Actual) Application 431
Applying the Raking Technique 435
Evaluating Changes in Public Opinion toward VEOs 436
Public Trust in VEOs 439
Religious Differences Supporting VEOs 440
Age Group Differences Supporting VEOs 441
Conclusions 444
Megacities Use Case 446
References 447

ABSTRACT Properly representing a population using survey data leaves analysts with a list of challenges. This chapter describes in detail how to work through one of the most common challenges: how to properly represent the population after the survey has been fielded and the data have been collected. By using a poststratification technique called iterative proportional fitting, or raking, in conjunction with known population demographics, the sampling weights can be adjusted for each individual respondent. Doing so essentially makes some respondents (those undersampled) represent more than one person, while others (those oversampled) will represent less than a person, forcing the data to represent the known demographics.

KEY TOPICS

- Defining the difference between design-based and model-based inference
- The underpinning of survey analysis: The Horvitz–Thompson estimator
- Modeling survey data through poststratification
- Using iterative proportional fitting (raking) as a poststratification technique

Extensions to the Use Case

Joe is a recently deployed military operations research (OR) analyst, eager to showcase his analytical talents. He is also deploying on the heels of one of the most monumental terrorism crises in the host nation's history.

This particular crisis started when a highly militarized violent extremist organization (VEO) entered a heavy populated coastal city by boat, taking over the city hall in one swift evening engagement. For the next 20 days, gun battles ensued between this VEO and the government's armed forces, forcing over 130,000 residents to flee the area. Cornered and outnumbered, the terrorists then took to storming other communities, taking residents hostage and burning homes. According to reports, over 200 people were killed in the crisis, and while most casualties were terrorists, several friendly soldiers, police, and civilians were also killed. While the people in the country, where he is deployed to, are subjected to tyranny, insurgency, and terrorism almost daily, this crisis took terrorism to a level not seen before in this city.

"About three years ago, the local government contracted a professional polling company to gain access to several of the conflict affected areas in this country," Joe's boss informs him. "Twice a year, with some oversight from us, this polling company creates, pretests, and fields a survey questionnaire." Joe's boss continues, "The latest survey wave just came in, and what I need you to do is tell me if this latest conflict changed the public's attitude toward the VEOs operating in these areas."

Joe went back to his desk with a few things on his mind. "Easy day, I just have to compare the public's perception of VEOs from the new survey to previous surveys," he thought. As he turned on his computer he thought, "But this report has to be an inference to the entire population, not just the sample right?" As he sat down he quietly said, "I wonder how well the polling company matched the survey sample to the population demographics?"

Introduction

Joe's situation could be an actual scenario, and in today's global fight against indirect and asymmetric warfare, surveys have become an important tool for military planners and OR analysts. Surveys help commanders and policy makers gain valuable insight into the performance of an organization being trained and equipped by US forces and/or a population's attitudes and opinions toward various groups. After all, in most missions today, host-nation public support is vital for a military unit's success. Furthermore, a good survey analysis could queue key leaders to possible on-ramps for negotiations with VEOs.

Joe undoubtedly has a daunting task, facing a huge data set with thousands of observations (respondents) and hundreds of variables (questions). With the time crunch Joe faces, a simple univariate analysis, comparing the latest survey responses to previous survey responses by quickly computing tables and graphs, is the most likely course of action (COA) he and most analysts would take. With that said, it is important not to underestimate the challenge this type of analysis poses. It is not uncommon for surveys, like the one Joe is analyzing, to contain over 150 questions where the analyst is responsible for deciding which specific survey questions will answer the objective, and in so doing, the results could be fraught with human error (Fricker and Kulzy, 2015).

Nevertheless, the goal of any survey analysis is to estimate something about the entire population from a selected sample. In this aspect, the inference solely depends on the sample being representative of the population. Therefore, Joe's last question to himself, "I wonder how well the polling company matched the survey sample to the population demographics?" is an important question that must be addressed before any analysis should take place; but, unfortunately, it is one that most analysts might overlook. So, if the sample demographics are not representative of the known population demographics, all an analyst can truly say about his/her results from the survey data is that they are representative of the sample, not the population. This chapter will show how to appropriately draw inferences from the sample by essentially forcing the sample demographics to match the known population demographics using a model.

Model-Based versus Design-Based Inference

There are two approaches for conducting statistical inference from a sample of data to a population. The most widely used is the *model-based inference*, where the generalization from sample to population is based on a mathematical probability model. From this model, which is based on assumptions about how the data are collected, conclusions can then be drawn about the entire population. For example, the hypothesis tests taught in introductory statistics courses, such as *t*-tests and analysis of variance, are model-based; they assume the data, or the statistic of interest, follow a particular probability distribution, and, typically, the data are independent and identically distributed. Therefore, the inferential results are only as accurate as the probability model (Lumley, 2010).

In contrast, the *design-based inference*, which is typically used with surveys, is based on the notion that the data (people in this case) are collected according to a known and carefully controlled sampling scheme. The methodology used to collect the survey sample does not typically result in independent and identically distributed data, but because all the information about the complex sampling scheme can be known, the appropriate mathematical calculations can be done that allow an inference from sample to population without appealing to a model.

A WORD ON COMPLEX SURVEY SAMPLING, PART 1

The probability for selecting a person to take a particular survey is just the product of the probability at each stage of the sampling process. For example, a person's selection probability might be the probability of selecting the respondent's neighborhood times the probability of selecting the particular block within the neighborhood times the probability the respondent's house is selected times the probability of selecting the respondent from the adults within the respondent's household. Using this sampling method, it is easy to see that each person could have a different probability of being selected to take the survey.

A Word on Complex Survey Sampling

The fundamental statistical underpinning of any survey is the probability sample. In an ideal surveying world, data would be collected via a simple random sampling (SRS) where a model-based inferential analysis would be perfectly appropriate. However, real-world constraints, driven by numerous things like a limited survey budget and time, require the use of complex sampling.

One commonly used complex sampling technique is stratified sampling, which involves sampling by population strata—uniquely identifiable, nonoverlapping groups within the population. By using this technique, important subgroups within the population, such as high-poverty neighborhoods or ethnic minorities, may be oversampled. These subgroups then have a higher probability of being selected to take the survey and thus would be over-represented in the sample.

A popular complex sampling technique used in face-to-face interviews is cluster sampling, in which the respondents are sampled in groups. With this technique, it is more efficient to transport an interviewer to one location where he or she interviews a number of individuals located in close geographic proximity. However, because of this geographic proximity, cluster sampling can, and often does, result in a correlation between observations within groups. That is, people who live or work near each other tend to be more alike, and thus answer the survey questions more similarly, than people chosen at random throughout the population.

Furthermore, in most large-scale, real-world surveys, samples are drawn using both stratification and clustering. For example, a population is first split into strata from which either a single-stage or a multistage cluster sampling is then conducted.

Each complex survey sampling design has its own advantages (and disadvantages), and typically, the sampling approach for any given survey is uniquely designed to meet the objectives and the constraints of that survey effort. For example, as compared to a SRS, a stratified sample can produce smaller margins of error. Additionally, if the estimates of the population parameters are desired for subgroups of the population, then these subgroups can be identified as strata, drawing down the cost of the survey (Lohr, 1999).

Just imagine, for example, a complex survey sample where rural areas were oversampled and where the respondent responses differ among those who live in urban areas from those who live in rural areas. If not accounted for, this would result in a rural bias in the survey results. The key takeaway is that a simple model-based inference is usually the wrong way to analyze survey data arising from complex sampling. For example, except in special circumstances, stratification will result in a sample that is not representative of the population, in the sense that the various subgroups will be over- or under-represented, and without accounting for this in the analysis, the resulting point estimates will be incorrect. Similarly, except for special circumstances, the use of cluster sampling will result in correlated survey responses that, if not appropriately accounted for in the analysis, will result in incorrectly estimated margins of error (and thus incorrect confidence intervals

A WORD ON COMPLEX SURVEY SAMPLING, PART 2

The details of complex survey sampling fill books. Interested readers should consult one or more of the standard texts, including Lohr (1999), Cochran (1977), and Kish (1995) to name a few.

[CIs] and hypothesis tests). Accounting for such complications in the analysis makes a design-based model the appropriate inferential method.

The Way It Should Be

Regardless of the sampling methodology used, a survey sample should be drawn from the population probabilistically, and in order to use a design-based model, the sampling methodology must meet two specific properties:

1. Every individual (i) in the population must have a nonzero probability (p_i) of ending up in the sample. With the SRS and some special complex sample designs, these probabilities are equal, but in general, they could be different for every sampled individual.
2. The probability (p_i) must be known for every individual in the sample.

Since the sample is unlikely to be representative of the entire population, *sampling weights* are used to account for these design-based inequalities. That is, each respondent in the final sample is given a sampling weight (w_i) that is the inverse of the sampling probability: $w_i = 1/p_i$. These weights, perhaps further adjusted to account for other aspects of the survey process (e.g., unequal response rates), are then used in the *Horvitz–Thompson estimator* (Horvitz and Thompson 1952) to do the appropriate inference from sample to population.

For example, if one is interested in using survey data to estimate the (unobserved) population mean response to question j (shown as: μ_j), then the Horvitz–Thompson estimator for a sample size n is

$$\hat{\mu}_j = \frac{\sum_{i=1}^{n} w_i x_{ij}}{\sum_{i=1}^{n} w_i}.$$

Intuitively, the weights are the number of individuals in the population that the respondent represents, where those with a small selection probability represent more of the population and those with larger selection probabilities represent fewer people. In the extreme case of $p_i = 1$, person i has a weight of $w_i = 1$ and thus only represents him or herself.

THE WAY IT SHOULD BE

Introduced in 1952 by Daniel G. Horvitz and Donovan J. Thompson, the Horvitz–Thompson estimator is the foundation for all survey analysis regardless of the sampling methodology. It is important to note that these weights and the Horvitz–Thompson estimator only provide for correct point estimation. Also required, although beyond the scope of this chapter, is the correct estimation of standard errors. For those interested, see the study by Lohr (1999) for an introduction to estimating standard errors for complex sample designs.

So, any large-scale survey, specifically those conducted by a professional research company, should provide sampling weights along with documentation on how the weights were calculated in the analysis file. These weights must be used in the survey analysis, and this will likely require special software designed for survey analysis, particularly in order to calculate the correct standard errors. Finally, and most importantly, any survey using complex sampling needs sampling weights (provided or calculated) in order to conduct the correct analysis.

But what if, even with the provided weights, the survey estimates of the known population characteristics are off? Or, perhaps even worse, what if the weights are not provided at all? These are serious but real issues, both of which have been observed by the authors with surveys conducted under DoD contracts. Fortunately, there exist several established post-stratification techniques, such as raking (Lumley, 2010), to estimate or modify the sampling weights in order to ensure that the survey estimates match known population characteristics.

The Raking Algorithm

Raking (otherwise known as iterative proportional fitting, sample balancing, or raking ratio estimation) is a model-based approach for adjusting the sampling weights of the sample data based on known population totals (usually from a census). Essentially, raking forces the survey totals to match the known population totals by assigning a weight to each respondent, making an inference to the entire population, not just to the sample, possible. However, the actual algorithm involves repeatedly estimating weights across each set of variables in turn until the weights converge and stop changing.

	Female	Male	Survey totals	Census totals
Rural	20	17	37	57
Urban	39	24	63	43
Survey totals	59	41		⇑
Census totals	49	51	⇐	Target

	Female	Male	Survey totals	Census totals
Rural	20(1.2978722)	17(1.8262812)	57	57
Urban	39(0.5908348)	24(0.8313842)	43	43
Survey totals	49	51		⇑
Census totals	49	51	⇐	Target

FIGURE 20.1
Marginal distributions with survey totals being raked to match the census totals by giving each respondent a sampling weight. For example, each female from a rural locale has a sampling weight of 1.2978722.

Figure 20.1 is a hypothetical example of the raking algorithm applied to two demographic variables each with two categories, Gender (female or male) and Locale (rural or urban). Notice that the survey totals for each category do not match the census totals for the population, which is the case for most complex surveys. Using raking, the weights (shown in the lower right table of Figure 20.1 in parenthesis) are calculated so the weighted survey totals match the census totals. For example, since urban locations were oversampled, the individuals from these locations are given a weight smaller than one, while individuals in the rural locations, that were undersampled, are given weights greater than one. Similarly, females were oversampled, so their weights are smaller than their male counterparts who were undersampled.

As shown, Figure 20.1 is a fairly easy example when there are only two variables with two categories, but now imagine raking on four to six variables with multiple categories. Thankfully, there is software available that does this procedure in a matter of seconds.

The most logical and popular survey variables to use with raking are demographics (i.e., gender, age, religion) for two reasons. First, population totals for demographics are often available, for example, from census data, and second, survey responses to these types of variables typically have a low nonresponse rate.

But in order to converge, a few rules prior to raking must be considered:

1. The population totals for each raking variable need to be the same. For example, one raking variable may be pulled from a 2008 census and another from a 2012 census. Since it is unlikely that the 2008 population is equal to the 2012 population, it is necessary to determine and fix a population total and then ratio-adjust any variables that do not sum to the fixed population total.
2. Careful consideration also needs to be taken to ensure that the raking variables in the survey sample and the known population are measured by the same criteria. For example, the survey sample may measure the socioeconomic status by tangible items around the respondent's household, such as the number of television sets, whereas the known (census) socioeconomic status directly comes from the respondent's household income.

Before raking, one will most likely need to do some data cleaning or recoding as well. The majority of recoding for survey analysis is generally done for basically three reasons (depending on the complexity of the analysis):

1. There must be a naming consistency between variable categories of each survey wave. For example, when comparing two survey waves, one variable category may be named differently but measure the same criteria. Or perhaps, for whatever reason, respondents were given the choice of *Don't Know*, but in another survey wave, they were given the choice *Refused to Answer*. However, one thing that is more commonly seen in data recoding is that one survey wave has a respondent answer in all caps *EXCELLENT*, and another survey wave has the response in lower case *excellent*.
2. Some variable categories may need collapsing due to too many categories. Doing so reduces the variance of the raked weights as well as spurious significant results (type I errors) when conducting statistical tests such as chi-square tests or *t*-tests. Of note, the chi-square test (difference in distributions) is preferred over the *t*-test (difference in means) with survey demographics simply because the *t*-test requires survey responses to be numerical. However, for both tests, as a simple rule, variables with more than five categories should be considered for collapsing. For

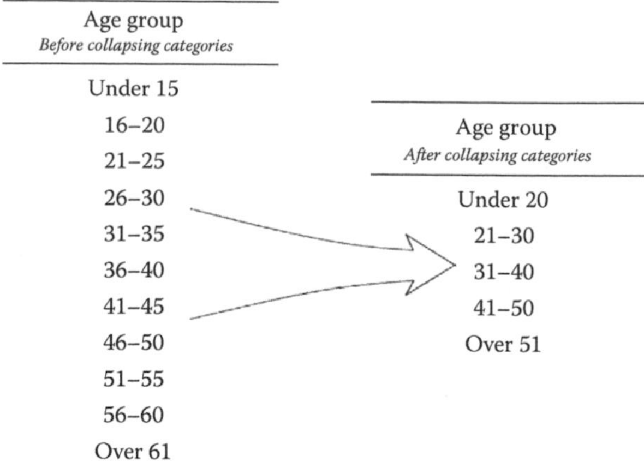

FIGURE 20.2
Collapsing all variables to five or less categories will reduce the number of type I errors when conducting statistical tests such as the chi-square test or the *t*-test.

example, the age group variable shown in Figure 20.2 had 11 original categories broken down in five-year increments. By breaking this variable down into 10-year increments, it is reduced to just 5 categories.

3. Some variable categories may need collapsing due to too few responses. Once the second recommendation is done, this rule almost weeds itself out. However, as a general rule, in order to reduce type I errors, when conducting a chi-square test for homogeneity between surveys (which is how we discover if there is a difference between survey responses), each variable category in each survey wave must have an expected count of at least five (Wackerly et al., 2007). The expected count for a given cell is calculated by multiplying the number of observations in the associated row by the number of observations in the associated column, then dividing by the total sample size.

For example, in Table 20.1, survey waves I and II were combined (1000 total observations) to add statistical power, and compared to survey wave III (500 observations). To find the expected count for class AB in waves I/II, take the row count for waves I/II—class AB (1000), multiply this by class AB's column count (13), and divide by the total number of observations (1500) to get 8.67. This is above the five expected count per cell general rule, but doing the same thing for class AB of wave III gives an expected count of only 4.33. Therefore, this variable should be collapsed into three categories, shown in Table 20.2.

Once the variables are recoded to ensure consistency between the waves, no variable has more than five categories, and no category has an expected count less than five, the survey demographics between waves can be accurately compared.

TABLE 20.1

Number of Socioeconomic Counts for Waves I/II and Wave III before Collapsing the Categories

Survey	Class AB	Class C1	Class C2	Class D	Class E
Waves I/II	11	4	38	440	507
Wave III	2	6	27	280	185

TABLE 20.2

Number of Socioeconomic Counts for Waves I/II and Wave III after Collapsing the Categories

Survey	Class ABC	Class D	Class E
Waves I/II	53	440	507
Wave III	35	280	185

An Illustrative (but Actual) Application

At the beginning of this chapter, Joe, a deployed military operations analyst, was given the task of comparing survey waves following a high profile terrorist event in the host country where he is deployed, to determine if this event changed the public's attitude toward the VEOs operating in these areas. This research is perfect for this survey data since the survey's purpose was to gain insight into population attitudes and opinions about various governments, military organizations, and VEOs, as well as assessing the population's perceived quality of life and views of various programs.

Overall, 500 respondents were sampled in each survey wave, and the respondents answered 144 questions, of which 15 were related to demographics. To add statistical power, the data from two survey waves fielded before a particular event were combined (consisting of 1000 respondents—waves I/II) and compared with the data from a survey wave fielded after the event (consisting of 500 respondents—wave III).

Joe's boss told him that the survey itself was contracted out to a commercial entity. After looking into the provided survey data file, Joe summarizes the sampling scheme used to collect the data:

- The data were collected using complex survey sampling (multistage clustering).
- Sampling weights were provided by the survey company but seem to inappropriately capture the variation in sampling probabilities (i.e., they seem too uniform).
- Certain geographic locations within the area of interest were purposely avoided.
- Adjacent locations from previous survey waves were used in subsequent survey waves.

There are some serious sampling issues mentioned here, particularly the last two. Since certain locations were purposely eliminated from the sampling scheme, the probability of selecting individuals in those locations is zero. And, since some locations were purposely selected (i.e., with probability 1), not every stage of the sampling process was random. Both these issues make traditional design-based survey analysis via the Horvitz–Thompson

APPLYING THE RAKING TECHNIQUE

Most survey questions are asked using a Likert scale (i.e., better now to worse now as in Figure 20.3). The best way to visualize this type of data is by using stacked bar plots like in Figure 20.3 or divergent stacked bar plots as in Figure 20.4. It is the analyst's job to know what will be both visually appealing and informative to the client.

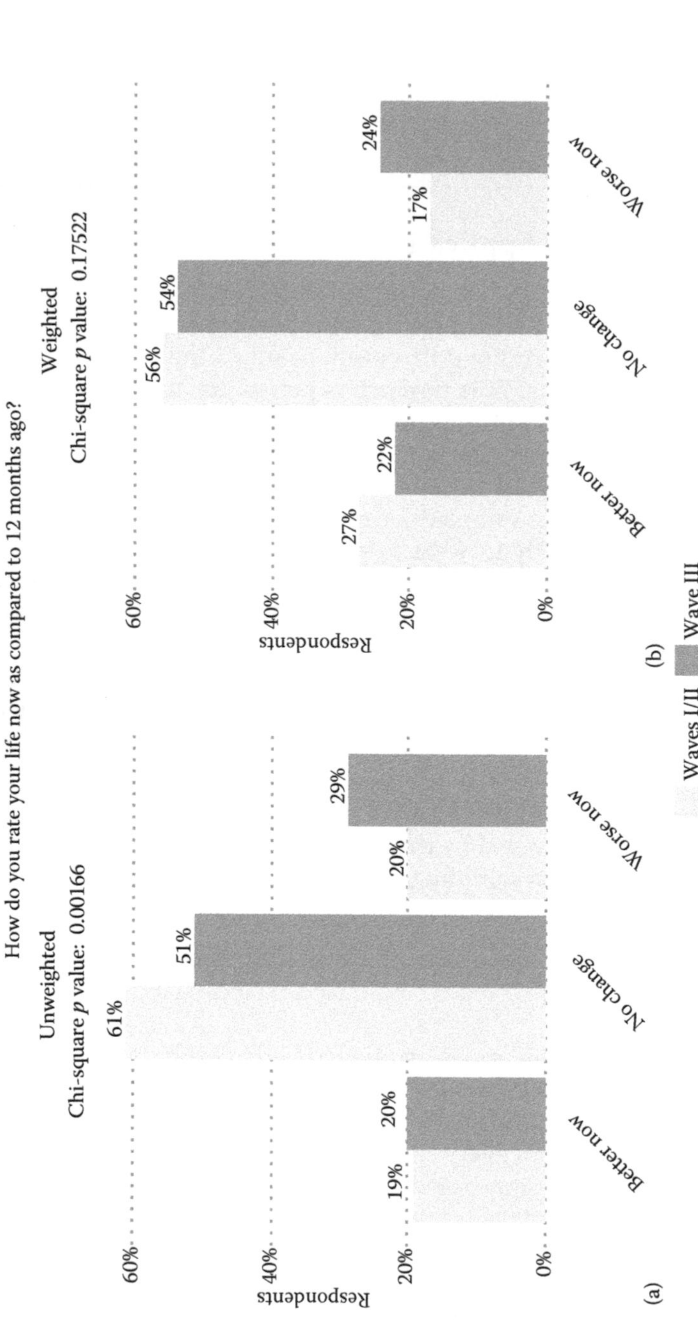

FIGURE 20.3
The unweighted data seem to show that the respondents are reporting that (a) their lives are worse now compared to a year ago. However, the difference is not statistically significant for the (b) weighted data.

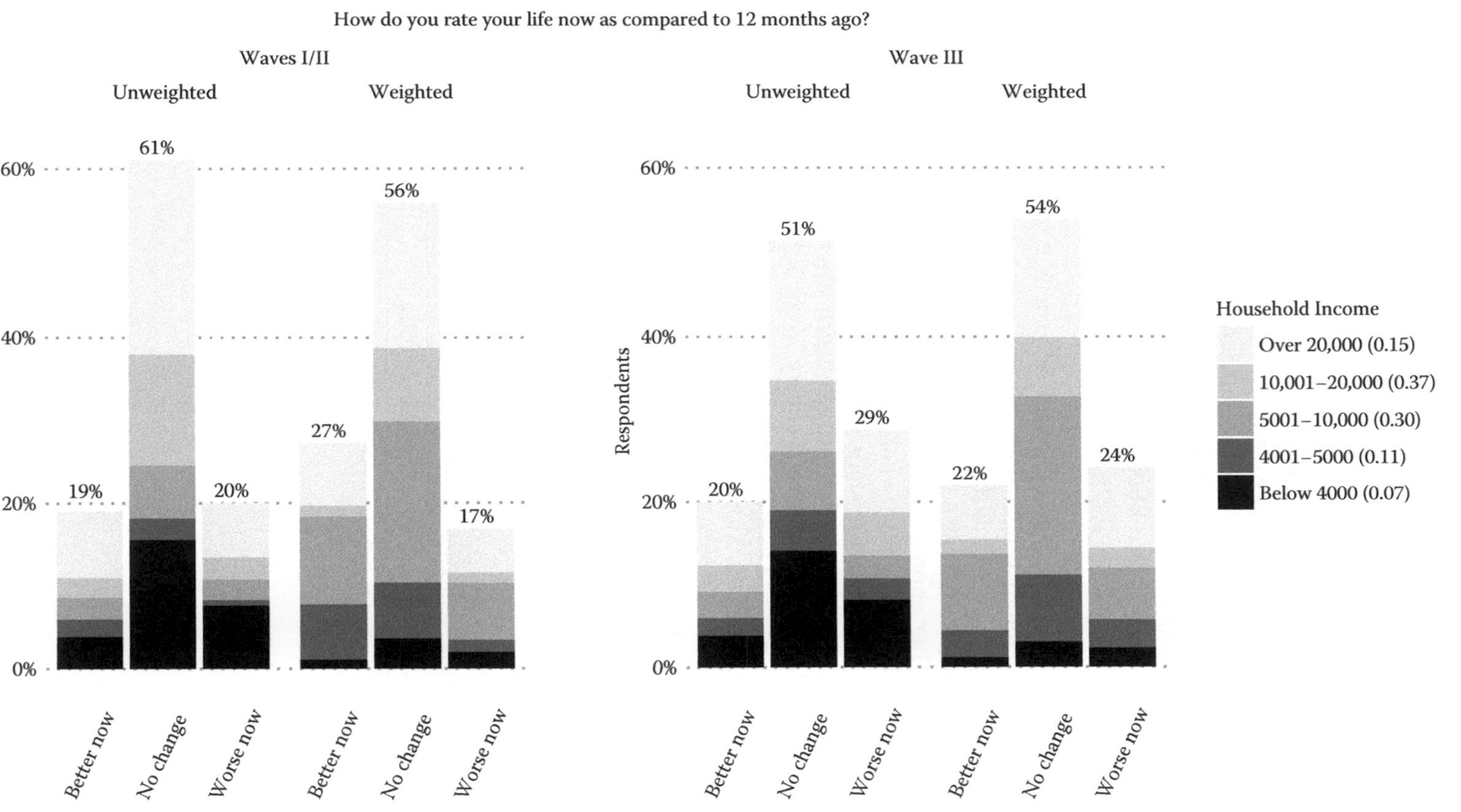

FIGURE 20.4
This stacked bar plot shows how respondents belonging to each household income category answered the question, "How do you rate your life now as compared to 12 months ago?" (a) waves I/II and (b) wave III.

estimator problematic, and this will unfortunately contribute to making the resulting samples unrepresentative of the population.

Lastly, the company claims to have done a multistage cluster sampling scheme. Here, the municipalities were selected first, then the neighborhoods, then the spots within the neighborhoods, then the households, and then the individuals within each household. Therefore, it makes sense that the survey research company provided sampling weights. However, a respondent's sampling weight, according to the survey research company, is one of two values. And being that each respondent's probability of being selected to take the survey is the product of the probability at each stage of this sampling scheme, each respondent can theoretically have a different probability of being selected. This little fact leads to a safe assumption that the provided sampling weights are incorrect, but with a just little analysis on the sampling weights and the demographics, this assumption is verified.

To prove that the provided weights did not correct for this unrepresentativeness, Table 20.3 shows chi-square test p values comparing the distribution of the demographic characteristics between waves I/II and wave III data. The *Unweighted* column shows the results from chi-square tests conducted on the raw data, while the *Weighted* column shows the results after appropriately using the survey company's weights in the chi-square calculation (see the study by Lohr [1999] for discussion).

Table 20.3 shows that there are statistical differences at the 0.05 significance level for socioeconomic status and personal identity between waves I/II and III. Had the sampling been consistently done wave to wave, it is very likely (although not guaranteed) that the sample demographics would be stable between surveys. That is, there would be no statistically significant differences. More importantly, if the sampling weights were correctly calculated, they should have corrected the demographic differences between the waves as well. What Table 20.3 shows is that neither the raw data nor the (contractor) weighted data in at least one of the waves (if not all of them) are not representative in at least these two demographic dimensions, which is an issue that must be addressed before conducting any analysis.

TABLE 20.3

Statistical Differences between Waves I/II and Wave III Using the Chi-Square Test at the 0.05 Significance Level

Demographics	Unweighted p Value	Weighted p Value
Age	0.12110	0.11096
Locale	0.60073	0.61613
Socio-economic	0.00294	0.00288
Gender	1.00000	1.00000
Age group	0.80292	0.79262
Marital status	0.25264	0.24203
Education level	0.79494	0.78393
Work status	0.97935	0.98152
Occupation	0.31626	0.32301
Home owner	0.07428	0.07456
Income	0.25629	0.24393
Religion	0.07076	0.06933
Ethnicity	0.63281	0.63348
Personal identity	0.01127	0.01162

Applying the Raking Technique

It is important to realize at this point that since the survey sample was collected using a complex survey sampling methodology, there will most likely be demographics that are either over- or undersampled. Therefore, a designed-based analysis conducted on the unweighted survey data is unlikely to be representative of the larger population.

This is where raking comes into the analysis. As mentioned earlier, the first step to raking is finding known population totals. In this case, the known population totals came from a 2010 census, where five census demographics, shown in Figure 20.5, were, category for category, comparable to, and consistent, with the survey demographics.

Using the survey package in R (Lumley, 2010) with the calibrate function, the census totals for each variable were used to calculate weights, where the procedure is separately performed for waves I/II and wave III. The result, as seen in the table on the right side of Figure 20.5, show the weighted data now match one another and the census data quite closely across each variable category. Likewise, performing the chi-square test on the demographics now produces no statistically significant differences between waves I/II and wave III. The result is that the weighted data are more representative of the population.

A key takeaway is that, in some instances, the weighted survey data (with raked sampling weights) change the analysis results. For example, as shown in Figure 20.6, the raw (i.e., unweighted) data for the question "how do you rate your life now as compared to 12 months ago?" is statistically different between waves I/II and III, with almost 1 in 10 people shifting their view toward the *worse now* category. This seems to show that a substantial portion of the population believe that their lives are getting worse. However, the weighted data show no statistically significant difference between surveys, and the people, for the most part, have the same view of their lives now as compared to 12 months ago.

To give some insight into how the sampling weights adjust the responses, this question is broken down in Figure 20.7 by the income demographic. Here, note that those with lower incomes are over-represented in the sample and are thus disproportionately affecting the results compared to the population. The weighting corrects for this imbalance in the sample.

Now, it may be that those with lower incomes do rate their lives as worse now than a year ago. And that may be a critical issue to address. But when doing the correct inference from sample to population, using the raked weights, there is no significant change over the entire population.

Furthermore, although not shown here, there are instances in these data where the weighted sampling design does not change the analytical outcome for other survey questions, and the same conclusion would be drawn regardless of whether the sampling weights were used or not. Even so, the percentages of each category within the question will be different when using sampling weights in the analysis.

Finally, a few things about these sampling weights (obtained through raking) need to be considered. First, if the sample demographics grossly differ from the actual population demographics (i.e., from a census), then expect the variance between the raked sampling weights to be large. A simple examination between the maximum and minimum weights can give the analyst a quick variance snapshot on the sampling weights. For example, with a simple random sample, the weights are the same for all respondents, and thus, the variance of the weights is zero. From there, larger variances indicate greater differences among

	Waves I/II	Wave III	Census
Locale			
Rural	0.54	0.56	0.47
Urban	0.46	0.44	0.53
Age group			
Under 30	0.26	0.29	0.37
30–39	0.26	0.25	0.24
40–49	0.22	0.22	0.18
50–59	0.14	0.12	0.12
Over 60	0.12	0.12	0.09
Marital status			
Married	0.78	0.75	0.51
Single	0.17	0.18	0.45
Widow	0.05	0.07	0.04
Ethnicity			
A	0.20	0.19	0.07
B	0.13	0.15	0.28
C	0.06	0.05	0.16
D	0.48	0.53	0.03
E	0.15	0.12	0.46
Income			
Below 4000	0.27	0.26	0.07
4001–5000	0.19	0.17	0.11
5001–10,000	0.38	0.34	0.30
10,001–20,000	0.12	0.13	0.37
Over 20,000	0.05	0.10	0.15

	Waves I/II	Wave III	Data
Locale			
Rural	0.49	0.44	0.47
Urban	0.51	0.56	0.53
Age group			
Under 30	0.34	0.35	0.37
30–39	0.26	0.25	0.24
40–49	0.19	0.17	0.18
50–59	0.12	0.13	0.12
Over 60	0.09	0.09	0.09
Marital status			
Married	0.48	0.51	0.51
Single	0.47	0.45	0.45
Widow	0.05	0.04	0.04
Ethnicity			
A	0.08	0.07	0.07
B	0.27	0.29	0.28
C	0.16	0.15	0.16
D	0.04	0.05	0.03
E	0.47	0.46	0.46
Income			
Below 4000	0.07	0.07	0.07
4001–5000	0.11	0.10	0.11
5001–10,000	0.31	0.30	0.30
10,001–20,000	0.37	0.37	0.37
Over 20,000	0.15	0.14	0.15

FIGURE 20.5
Unweighted Waves I/II and III demographics on the left and weighted demographics, after the implementation of the raking algorithm, are on the right. In the left image, the highlighted survey demographics significantly deviate from the population demographics. Observe how, after raking, the weighted survey sample demographics in the right image closely resemble the census demographics.

the weights. There are a few general sampling weight size guidelines accepted by the survey analysis community. The most generally accepted guideline is that the weights should never be larger than five times the mean weight (Debell and Krosnick, 2009). For example, if the mean weight for 500 respondents was 1.0, then any weight larger than 5.0 would be truncated or simply capped with a weight of 5.0, but small weights should never be truncated "because large weight values increase the potential for outliers to affect analyses and are more likely to inflate variance, while small values do not have these consequences" (Debell and Krosnick, 2009). Of course, a statistical software such as R is required to do this across multiple raking variables, and it goes without saying that these truncated cases need to be documented in the analysis results.

Evaluating Changes in Public Opinion toward VEOs

Now there are two important assumptions that have to be made before analyzing the survey data. The first assumption is that the sampling weights obtained through raking are now appropriate for all the survey questions, and the second assumption is that inference using the obtained sampling weights represent the entire population and not just the sample.

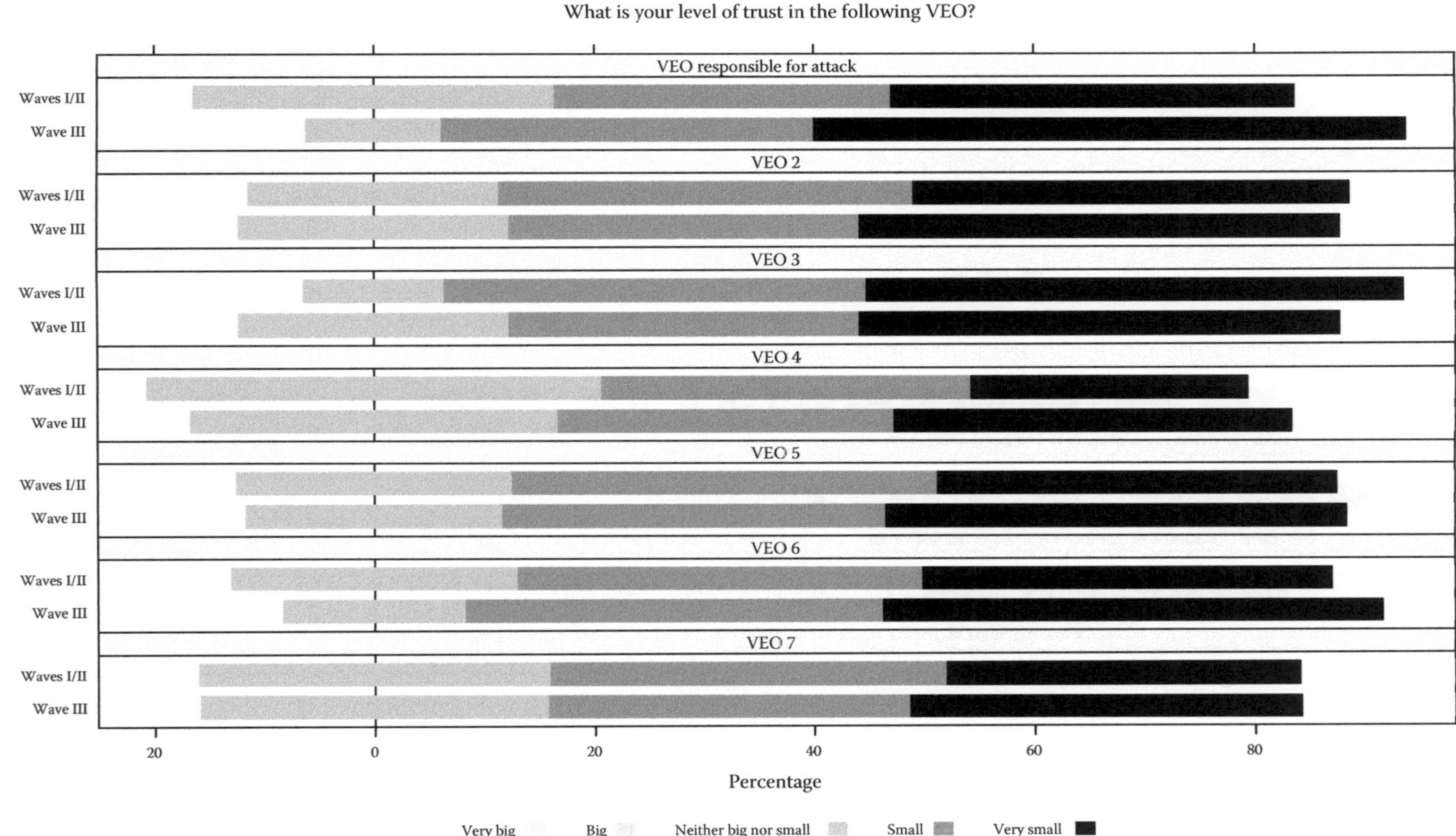

FIGURE 20.6
VEO trust level percentage comparison between survey waves.

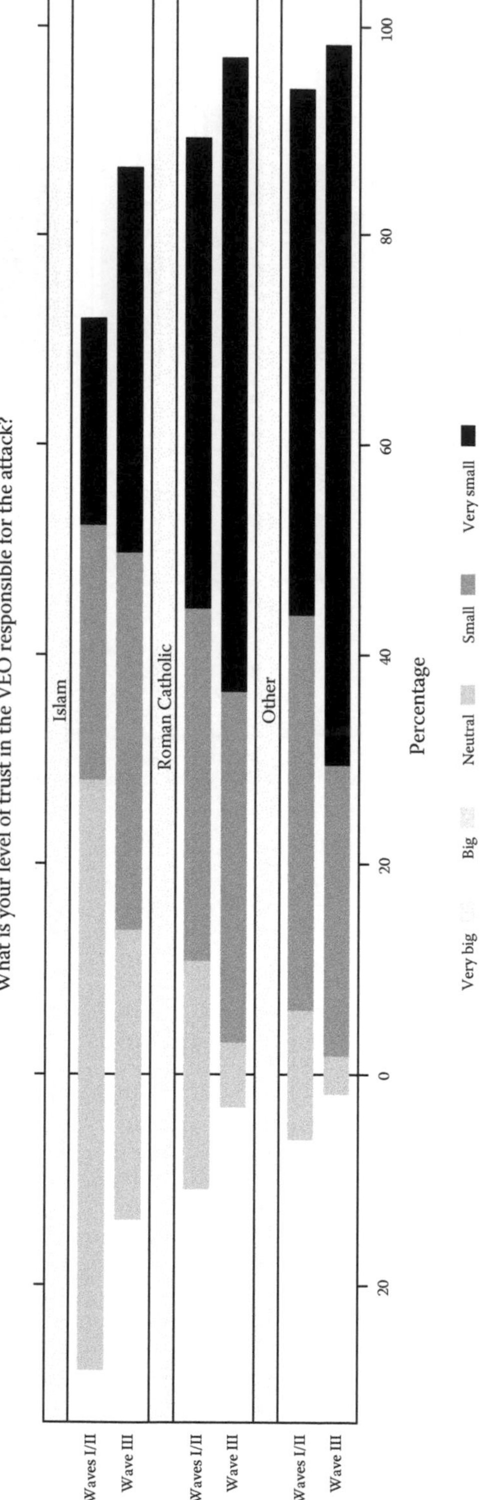

FIGURE 20.7
The VEO's level of trust appears to have shifted toward *Very Small* from waves I/II to wave III for all religions.

There are two statistical tests commonly used in this research, the chi-square test and the *t*-test. The chi-square test used in this analysis tests how the extent of which the observed distribution of data from waves I/II fits with the distribution of data from wave III. This test is specifically designed for categorical data such as survey data. The two-sample *t*-test used in this analysis is a two-sided *t*-test. It determines whether the means from waves I/II and wave III are statistically different. Both tests, the chi-square test and two-sided *t*-test, are usually measured at either a 0.05 or 0.1 significance level.

There are several ways to present survey data, but the most important rule of thumb is not to inundate the client with over 100 bar graphs or a graphic for each survey question. Probably the most important part of being an analyst is knowing the client's preferred presentation style. In any case, revisiting the initial question is a good start: "Did the latest conflict change the public's attitude toward the VEOs operating in these areas?" In this instance, a broad overview of the data (comparing waves I/II to wave III) is done first to highlight any trends in the public perceptions of a particular VEO. Then, if warranted, the public perceptions are broken down into population demographics to highlight where the key group's (i.e., religious factions) perceptions toward VEOs lay.

Public Trust in VEOs

In both survey waves, respondents were asked, "What is your level of trust in the following VEO?" This question is rated on a five-point scale, ranging from *Very Big* to *Very Small*. Figure 20.6 compares this level of trust between survey waves for the seven VEOs operating in the region of interest. As shown in Figure 20.6, VEO trust has never been strong since there are no respondents who answered either *Big* or *Very Big*. Fortunately, with the exception of VEO 3, it appears that the respondents' overall level of trust in each VEO has gone down from waves I/II to wave III. For example, after the hostage crisis situation, the level of trust for VEO 4 and 6 dropped from *Neither Big nor Small* to *Very Small* by over 10%. However, one concern, when looking at Figure 20.6, is that it appears that the overall population is more trusting of VEO 3, with 1 in 10 attitudes shifting in that direction.

Table 20.4 is the first of such table where the statistical tests discussed at the beginning of this chapter were run on the survey design model. These numbers show that,

TABLE 20.4

Statistical Difference in the Level of Trust for the VEO of Interest between Waves I/II and Wave III

	p Value	
VEO	**Chi-Square Test**	**Two-Sided *t*-Test**
VEO of interest	0.0001	0.0026
VEO 2	0.7689	0.8776
VEO 3	0.1291	0.2357
VEO 4	0.2711	0.1643
VEO 5	0.7834	0.5715
VEO 6	0.1603	0.0727
VEO 7	0.8311	0.7401

while the level of trust previously appeared to be different for all VEOs between survey waves in Figure 20.6, the fact is that the level of trust for the VEO responsible for the attack is the only statistically significant VEO after the hostage crisis situation. This is an interesting insight, considering that this was the only VEO involved in the hostage crisis situation.

The percentages provide insight into why this VEO is statistically different between survey waves. Specifically, the population's trust in this VEO dramatically shifted from *Neither Big nor Small* to *Very Small* by about 22%. This is undoubtedly good news for the local government and the organizations vested in the population's security. Furthermore, all survey demographics have at least one category showing a statistical difference between survey waves when comparing this VEO's trust level. That said, of all the demographics, religion and age group show the largest changes in the overall attitude toward this VEO's trust.

Religious Differences Supporting VEOs

As mentioned earlier, certain variables need to be recoded because there are just too many categories. The issue is that too many categories, and categories with low response rates, will likely create an unnecessary variance, producing spurious statistical significance or more type I errors, thus skewing results. Here, the religion demographic was recoded into three categories. The two largest categories, Islam and Roman Catholic, were not recoded. However, 27 other religions were recoded into one category simply named *Other*.

Therefore, after recoding the religion categories, the graphics can be produced, and the statistical tests can be ran between waves I/II and wave III. Figure 20.7 shows a broad overview of the population's trust in the VEO responsible for the attack by religion and between survey waves. While it is somewhat difficult to draw percentages off Figure 20.7, it does provide some insight, especially when coupled with Table 20.5. For example, it appears from Figure 20.7 that across all religions, the population's trust has shifted toward *Very Small*. Table 20.5 supports this fact for Muslims and Catholics by showing

TABLE 20.5

Statistical Difference in the Level of Trust for the VEO of Interest between Waves I/II and Wave III by Religion

	p Value	
Religion	**Chi-Square Test**	**Two-Sided *t*-Test**
Islam	0.0426	0.0086
Roman Catholic	0.0029	0.0439
Other	0.4549	0.9343

that these religions are statistically different between waves I/II and wave III. However, *Others* show no statistical difference between waves I/II and wave III, although, from Figure 20.7, there does appear to be a shift in attitudes toward a *Very Small* trust in the VEO of interest.

As previously seen in Figure 20.6, the population's level of trust in the VEO responsible for the attack has never been strong, regardless of religious denomination, since no respondents said that they have neither a *Big* nor a *Very Big* trust in this VEO. But it is also difficult to draw percentages off Figure 20.7; therefore, the data specifically for Muslims and Catholics (the two religions statistically different between survey waves) are plotted with percentages in Figure 20.8.

As seen from Figure 20.8, the Muslim's *Neither Big nor Small* opinion toward this VEO dropped from 56% to 27% after the hostage crisis situation, with all this trust shifting toward the *Very Small* category. It should be noted, however, that while this level of trust dropped dramatically, Muslims still hold over a 20% higher *Neither Big nor Small* level of trust for this VEO than any other religion. For Catholics, the results after the hostage crisis situation show approximately a 16% shift to the right, that is, toward a *Very Small* trust in the VEO responsible for the attack.

Age Group Differences Supporting VEOs

The other interesting demographic when comparing trust levels of the VEO responsible for the attack is age group. As previously discussed, ages less than 30 were combined into one category, with age groups being collapsed by 10-year intervals afterward. The respondents over 60 years old were combined into one category to finish the age group recoding.

Figure 20.9 shows all age groups trending toward a less trusting view of the VEO of interest. Although it is not represented in Figure 20.9, a total of one out of five people shifted his or her opinion from *Neither Big nor Small* to the right, with the youngest population, under 30, holding about half those individuals. The statistical tests in Table 20.6 demonstrate this fact by showing that all the age groups younger than 50 years old are statistically different between survey waves.

Although age groups 50–59 and over 60 are not statistically different between waves I/II and wave III, Figure 20.10 lays out the distribution of these two age groups. It is evident from this figure that these age groups are still trending toward a less trusting view of the VEO responsible for the attack. Although small, the age group 50–59 had about a 7% drop in the *Neither Big nor Small* trust for the VEO, with almost all these people skipping the *Small* category and immediately jumping into the *Very Small* category. The change in the over 60 age group is even more pronounced with one in five people who were either in the *Neither Big nor Small* or the *Small* category in waves I/II going to the *Very Small* category in wave III.

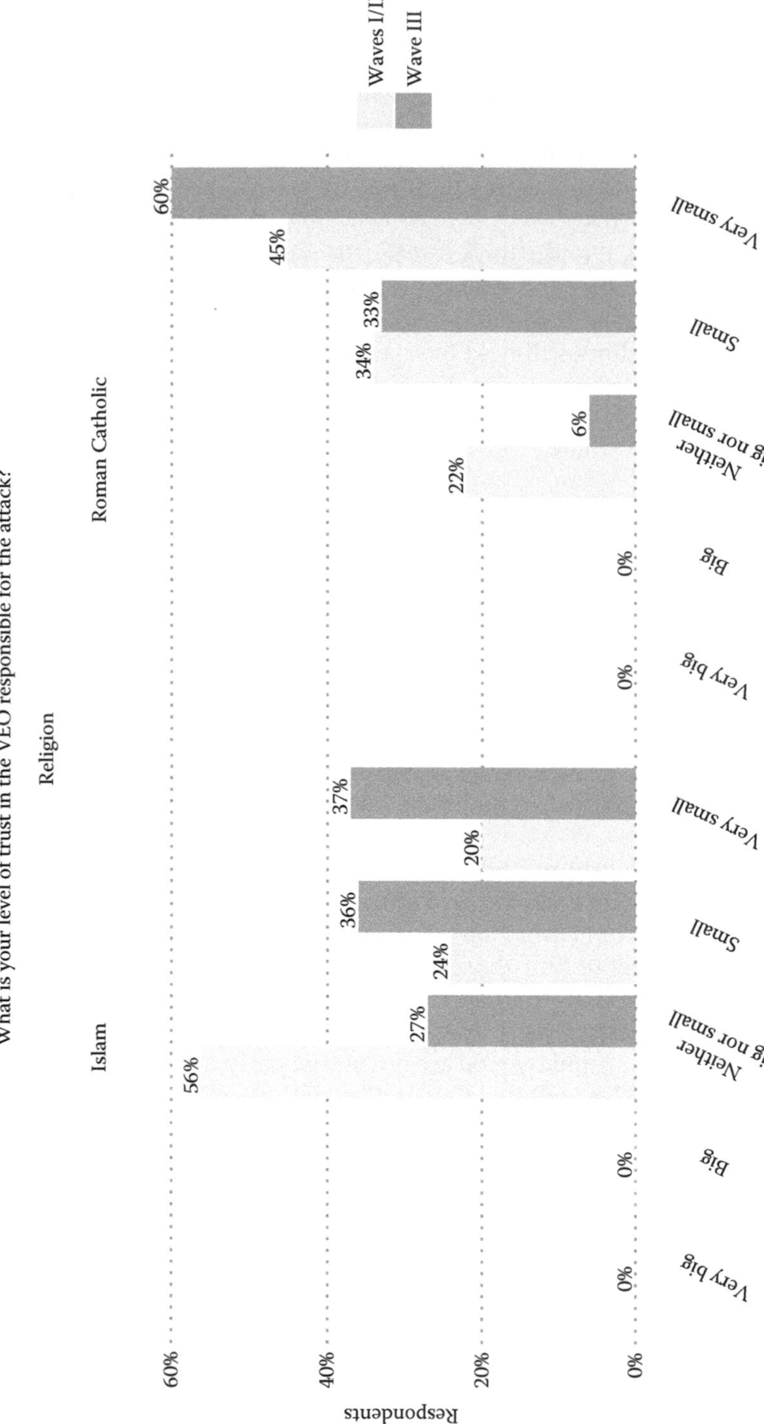

FIGURE 20.8
With percentages added to the plot, it is easy to see that the level of trust for the VEO has shifted from *Neither Big nor Small* to *Very Small* for both religions.

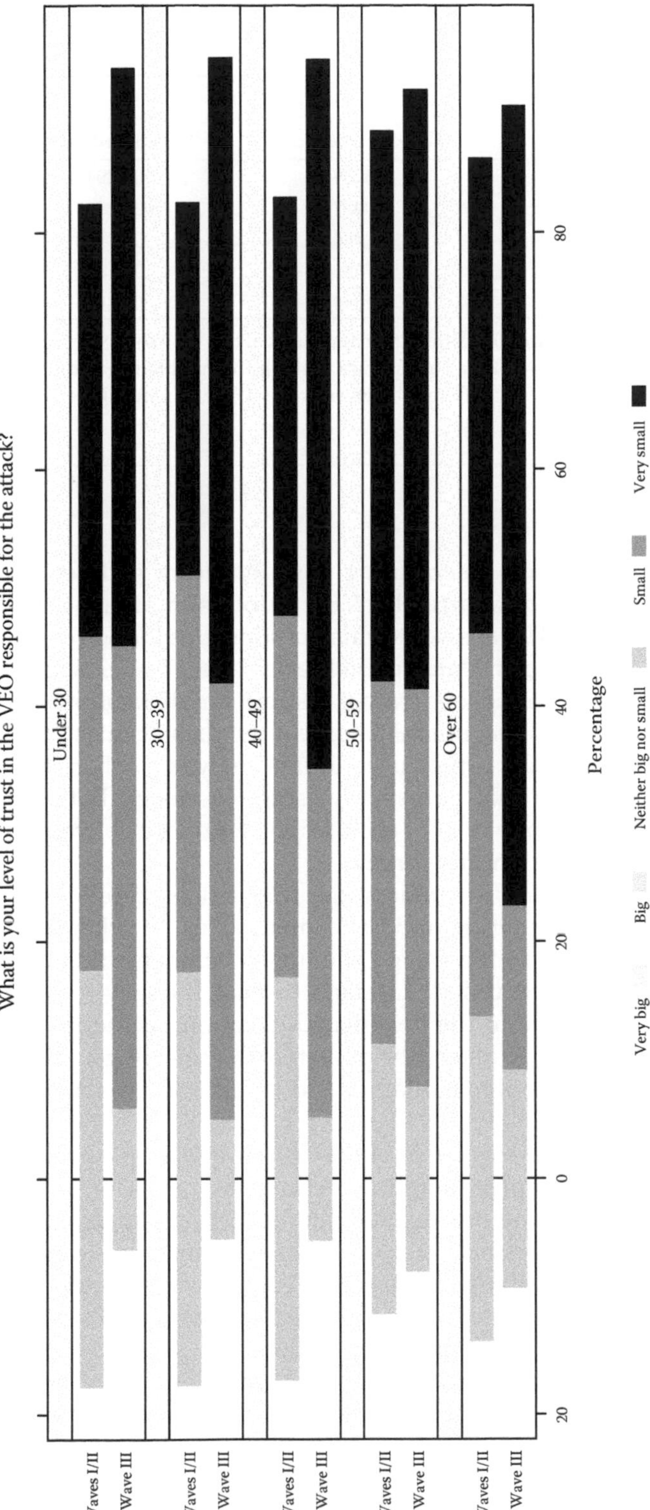

FIGURE 20.9
The VEO's level of trust appears to have shifted toward *Very Small* from waves I/II to wave III throughout all age groups.

TABLE 20.6

All Age Groups Younger than 50 Years Old are Statistically Different between Survey Waves

	p Value	
Age Group	**Chi-Square Test**	**Two-Sided *t*-Test**
Under 30	0.0259	0.0294
30–39	0.0077	0.0286
40–49	0.0200	0.0040
50–59	0.7688	0.5107
Over 60	0.1089	0.5857

Note: From Figure 20.9, although it appears that the VEO's level of trust has decreased for all age groups, only the highlighted age groups are statistically different between waves I/II and wave III.

Conclusions

The earlier analysis focused on determining any significant differences in population attitudes and opinions toward a particular VEO before and after a crisis that the VEO was responsible for starting. The answer to this question was found by using a univariate analysis on key survey questions. Unfortunately, there were issues with the survey sampling. In complex survey sampling, one cannot assume simple random sampling. Instead, analysis must account for the design used to collect the data such as complex multistage clustering, as is the case with the survey used in this analysis. Earlier, there was a discussion, in some detail, on how to analyze a survey using this design, but in short, in order to draw inference on the entire population, each respondent needs a weight. However, this presented three major issues with this survey.

The clustering details needed to produce the unequal probabilities and weights for respondents are unknown, rendering traditional analysis via the Horvitz–Thompson estimator impossible. Furthermore, after wave I, the survey company did not randomly select regions at the primary sampling unit (PSU) stage. Instead, the survey company purposely eliminated certain regions due to the heightened situation, or the inaccessibility to the residents in that region. Doing so makes the probability of selecting a person in that region zero; this, in turn, makes the weight of that individual infinite, thus making design-based survey analysis impossible.

The survey sample is not representative of the city's population according to a 2010 census. Most surveys using a complex sampling methodology over- or undersample certain demographics. In particular, the unweighted survey data under-represents single individuals, who are under 30 years old, in households with incomes over $10,000 a month, and from an urban region. Similarly, the unweighted data over-represents older, married, poorer, and rural residents. Therefore, any univariate analysis conducted on the raw survey data is not representative of the larger population.

The survey company did provide weights with the raw data; however, these weights do not appropriately account for observed demographic differences and thus do not appropriately account for the sampling differences between the survey samples and the population. This is important because, if the survey sampling had been consistently done wave to wave, the sample demographics would be consistent from wave to wave.

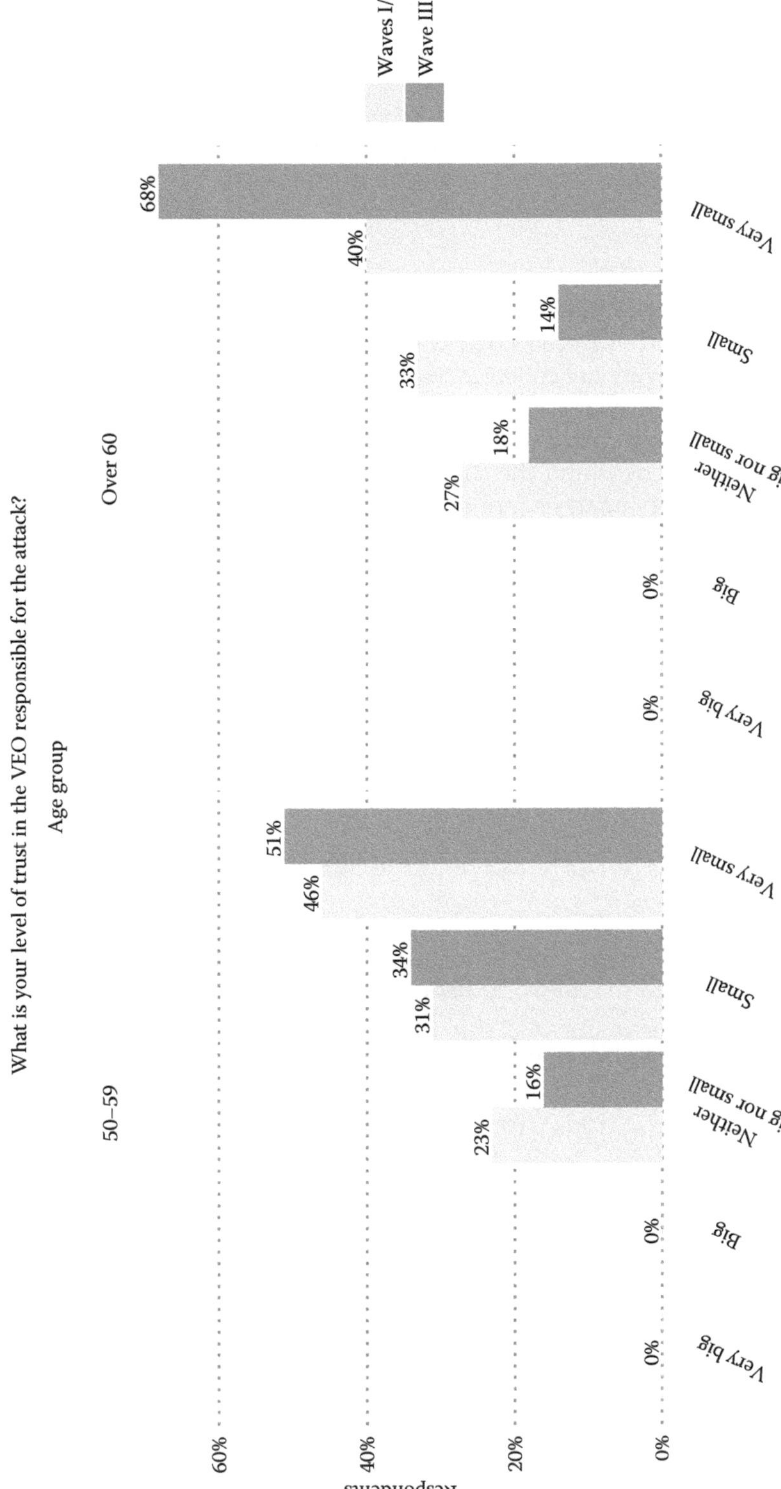

FIGURE 20.10
Even though the 50–59 and over 60 statistical tests showed no significance between waves I/II and wave III, the VEO's level of trust among these age groups did shift toward *Very Small*.

CONCLUSION

The poststratification technique used in this analysis is not the only established poststratification technique. The entire concept of poststratification is to make the survey sample more representative of the population; therefore, any technique used to do this will theoretically make the analysis stronger. Sensitivity analysis could be conducted on several of these techniques, comparing results, and determining which technique is best suited for a particular survey.

Had the survey company executed (and documented) an appropriate complex sampling scheme, this analysis could have proceeded using standard survey analysis techniques. However, because the data were not collected using a random design, a model-based set of weights had to be derived in order to proceed with the analysis. In doing so, this analysis is forced to assume that the weights derived via raking, on a set of demographics common between the population's census and the survey data, are appropriate. Ultimately, this is unknowable, but a sensitivity analysis (not done in this chapter) would show that most of the resulting conclusions are not particularly sensitive to the choice of weights.

According to Lumley (2010), even if there is a large bias in the survey, such as an over- or an undersampling of certain demographics, raking will make the sample more representative of the population. The point is, analysts need to compare the survey demographics between survey waves and, to the extent it is available, census demographics. Then, since the tools are available, any issues that arise need to be corrected. If not, all that can be said about the results is that they are representative of the sample, not of the population.

Megacities Use Case

The following chapter sections describe how the previous poststratification methods and univariate analysis is applied to real-world data:

- Evaluating changes in public opinion toward VEOs
- Public trust in VEOs
- Religious differences supporting VEOs
- Age group differences supporting VEOs

DISCUSSION QUESTIONS

1. Given several waves of survey data, what would be the first step you would take in as the analyst?
2. If you could not find census data or another gold standard representative of the population demographics, would you tell your client the results represent the population? How would you explain your reasoning to the client?
3. What is the difference between model-based and design-based inferences?
4. What does poststratification attempt to achieve and is it always necessary?

References

Cochran, W. G. (1977) *Sampling Techniques*. Third Edition, John Wiley & Sons, New York.

DeBell, M., and Krosnick, J. A. (2009) Computing weights for American national election study survey data. Retrieved March 26, 2014, www.electionstudies.org/resources/papers/nes012427.pdf.

Fricker, R. D. Jr., and Kulzy, W. W. (2015, June) The survey process: With an emphasis on survey data analysis. *Phalanx Magazine, Military Operations Research Society*, 48(2), 50–56.

Horvitz, D. G., and Thompson, D. J. (1952) A generalization of sampling without replacement from a finite universe. *Journal of the American Statistical Association*, 47, 663–685.

Kish, L. (1995) *Survey Sampling*. John Wiley & Sons, New York.

Lohr, S. L. (1999) *Sampling: Design and Analysis*, Second Edition. Duxbury Press, Pacific Grove, CA.

Lumley, T. (2010) *Complex Surveys: A Guide to Analysis Using R*. Wiley, New York.

Wackerly, D. D., Mendenhall, W., and Scheaffer, R. L. (2007) *Mathematical Statistics with Applications*. Cengage Learning, Stamford, CT.

21

Modeling Sociocultural Influences on Decision Making: Assessing Conflict and Stability

Michael L. Bernard, George A. Backus, Asmeret Bier Naugle, Robert F. Jeffers, and Regan W. Damron

CONTENTS

Introduction 450
  Computational Modeling: System Dynamics, Agent-Based, and Hybrid Models 450
  Theoretical Orientations of Computational Modeling 451
Modeling Example: Hybrid Computational Models 451
  Methodology 452
    Problem Question to be Modeled 452
    Knowledge Acquisition Process 453
    Accounting for SME Bias 454
    Macrolevel Theoretical Representation 454
    Microlevel Theoretical Representation 457
  Computational Structure of DYMATICA 458
    The Process of Representing Behavior 460
  DYMATICA in Action: Assessing Environmental, Geopolitical, and Societal Factors Influencing Unrest and Government Support 462
    Entities and Decisions 463
    The Model 463
    Scenario Conclusions 467
General Conclusions 467
Megacities Use Case: Applying DYMATICA to the Megacities Narrative 468
  Entities and Decisions 468
    Potential System Behavior: Overextension of Government Resources 469
    Potential System Behavior: Sociocultural Conflict 469
References 470

ABSTRACT This chapter describes a theory-based analytical capability to enable analysts to better assess the influence of events on groups interacting within a country or a region. These events can include changes in policy, human-made or natural disasters, war, or other changes in environmental/economic conditions. This chapter includes a discussion of (1) the underlying psychological, social, and economic theories that are synthesized within its structure; (2) the inclusion of data and expert opinion into the modeling structure; (3) the methods used to computationally instantiate theories and data/opinion; (4) the types of assessments that are generated; and (5) the implications of these assessments in comparison to current events.

KEY TOPICS

- General types of computational modeling: System dynamics, agent-based, and hybrid models
- Introduction to a hybrid modeling example: dynamic multiscale assessment tool for integrated cognitive behavioral actions (DYMATICA) computational model
- Macro- and microlevel decision theory representation within DYMATICA
- Computational structure of DYMATICA and the process of representing behavior
- Applying DYMATICA to a real-world example and the Megacities narrative

Introduction

> Simply stated, the lesson of the last decade is that failing to understand the human dimension of conflict is too costly in lives, resources, and political will for the nation to bear. Once a conflict commences, it is already too late to begin the process of learning about the population and its politics. The optimal condition is for our leaders to have the ability to influence budding conflicts "left of bang*," that is, *before* tensions turn violent. Left of bang, policy options are more numerous, costs of engagement are lower, and information flows more freely to more actors. After a conflict begins, options decrease markedly, the policy costs rise rapidly, and information becomes scarce and expensive.... A new concept should seek to explain how populations understand their reality, why they choose either to support or resist their governments, how they organize themselves socially and politically, and why and how their beliefs transform over time. (Flynn et al., 2012, pp. 13–14)

A common obstacle associated with the effort to better assess potential behaviors within different societies is the shear difficulty in comprehending the population's dynamic nature, particularly over time and considering feedback effects. Obtaining better insight with regard to how populations understand their reality and responding accordingly is particularly difficult. These obstacles are becoming more problematic as small groups, and even individuals, are increasingly able to harm large segments of society. Thus, determining what is actually achievable in obtaining better insight into how societies perceive their environment, make decisions, and ultimately behave is becoming increasingly important. Efforts to achieve this insight have employed a number of qualitative and quantitative modeling techniques. This chapter discusses a quantitative approach to simulating human behavior in order to better assess various conflict- and stability-focused scenarios. More specifically, this chapter focuses on the development of methods to quantifiably address prevailing dynamics and unintended higher-order consequences of events on populations over time. It is not, however, focused on predicting specific individual or group behaviors at some particular time.

Computational Modeling: System Dynamics, Agent-Based, and Hybrid Models

To most effectively understand conflict and stability, models assessing sociocultural influences on behaviors should incorporate causality, dynamic tendencies, and some

* More formally known as the *shaping* or *preoperational* phase before a conflict.

specification of decision making processes. System dynamics and agent-based modeling techniques can incorporate these characteristics to address issues relating to conflict and stability. System dynamics models (Forrester 1961; Sterman 2000; Ford 2009) use difference equations, determined by causal structure, to dynamically simulate systems of interest. They are generally constructed using stock and flow systems (described in the following) with carefully specified mathematical relationships between variables. System dynamics models are generally not strongly driven by exogenous factors, although exogenous factors do often play a role. The models account for feedback dynamics and time delays in the system.

While system dynamics models tend to be most useful for broad-level, aggregated systems assessment, agent-based models (Macy and Willer 2002; De Marchi and Page 2014) are useful for simulating interactions between individuals and assessing complex emergent group behaviors over time. In comparison to system dynamics, agent-based models tend to use simpler descriptions of decision making processes, but more complex networks of individuals. They also allow for probabilistic interaction, while system dynamics models are generally deterministic.

System dynamics and agent-based models have some similarities that make them particularly applicable to sociocultural modeling, creating opportunities for a hybrid simulation. Both types of models can be used to simulate decision making processes where interactions between people or groups are fundamental to the simulation outcomes. While agent-based models tend to focus on individual behaviors and system dynamics models tend to focus on aggregate behaviors, there are no hard rules delineating the granularity for each type of model. In some cases, combining system dynamics and agent-based techniques can improve simulations by incorporating individual and emergent behavior with structural and group dynamics.

Theoretical Orientations of Computational Modeling

The model's theoretical orientation is every bit as critical as the computational approach in determining the assessment's overall direction and scope. Robust theories that have been developed and empirically tested across social, religious, and political boundaries are found within anthropological, economic, political science, psychological, and sociological disciplines. Incorporating a theory within computational models can enable the theories and the data to be interrogated in a much more precise manner. Since each discipline typically has a different focus, integrating multiple theories across these foci potentially allows the model to address a wider range of behaviors. For instance, anthropology addresses cultural behavior aspects, economics (or more precisely, behavioral economics) addresses how decision making affects behavioral choices, psychology addresses the mechanisms and the processes underlying decision making, political science addresses how individual preferences are aggregated and expressed under various institutional and systemic constraints, and sociology addresses the processes underlying group interaction and potential conflict within societies.

Modeling Example: Hybrid Computational Models

An example of a computational model that has incorporated system dynamics and agent-based approaches, as well as several theoretical methodologies, is the DYMATICA modeling effort within Sandia National Laboratories'* Behavioral Influence Assessment program. The objective of this hybrid modeling approach is to have greater decision and behavioral insight by providing a modeling structure that represents interactions among groups within societies. The basis for the approach is to assess the dynamics and the key psychosocial, geopolitical, and economic processes underlying how people make decisions and express behaviors over time. Included in these simulations are behaviors that affect the decision making of others, creating complex feedback loops within and between individuals and groups. Each simulated behavior is a function of individual psychosocial characteristics (described in the following) along with environmental and group dynamic factors. This effort's goal is to inform planning processes and decisions that minimize the likelihood of undesirable consequences by providing a more systematic (and thus, replicable) analysis of group perceptions, intentions, and behaviors, as well as the environmental variables that bear upon their interactions (Bernard et al. 2014; Bier and Bernard 2014). To help achieve this objective, this program is developing models that synthesize data-supported, behavioral economic, political, psychological, and sociological human behavior theories. This synthesis is further supported by an independent, theory-based analytical assessment of historical socioeconomic data. As it is designed to inform rather than predict, DYMATICA focuses on examining likely dynamic repercussions of actions, not on generating specific behavior estimates. The result is a unified framework that connects the multiple scales of human behavior (from individual to societal interactions) to the external (geopolitical, physical, and socioeconomic) world.

Computationally, the DYMATICA structure consists of a modeling framework, model simulators, and an analysis approach. The current structure allows for an assessment of models across different domains (i.e., different countries, groups, individuals, and scenarios of interest). For example, Figure 21.1 shows a simplified conceptual representation of a hypothetical DYMATICA structure that involves the modeling of two interacting groups and several leaders. Exogenous inputs to the model (e.g., global economic factors and general population support) influence the dynamic interactions within and between the entities. Each simulated behavior is a function of psychological characteristics along with environmental and group dynamic factors. This enables the assessment of group behaviors as the groups react to other's perceptions and world conditions.

Methodology

Problem Question to be Modeled

The modeling process begins with a problem question to be addressed. Modeling a problem helps to bound the model, focus the list of auxiliary questions, and define project tasks. The scope of the problem will help to determine the level of detail in the model; a very broadly defined problem will usually result in a high-level model, whereas a more targeted problem can lead to a more detailed model. The time horizon, the groups and/or individuals to be modeled, and the major variables of interest are typically identified at this point. The general scope of the model structure is determined based on the overarching question(s). At this stage, the process of generating possible subquestions begins. Vetting the subquestions with an analyst consumer helps to further refine this structure.

* Sandia National Laboratories is a multiprogram laboratory managed and operated by Sandia Corporation, a wholly owned subsidiary of Lockheed Martin Corporation, for the US Department of Energy's National Nuclear Security Administration under contract DE-AC04-94AL85000. This effort is supported by the DoD and internal Laboratory Directed Research and Development funds.

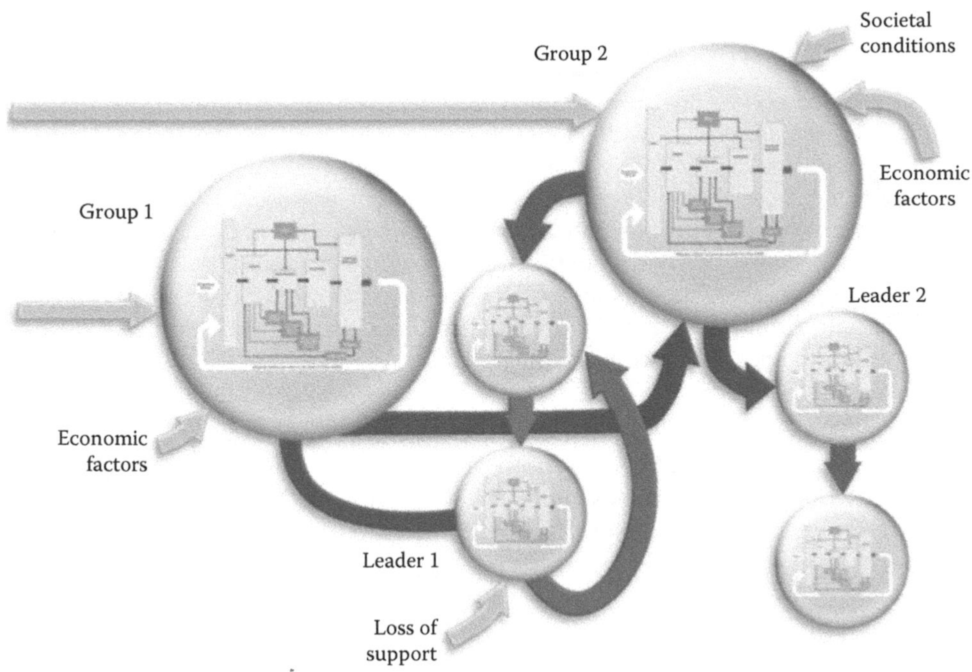

FIGURE 21.1
Conceptual diagram of the full systems view of a hypothetical application of DYMATICA.

Knowledge Acquisition Process

The process of populating the model begins at a macrolevel. At this level, we seek to understand the essential human influencers in the system and begin to identify the fundamental dynamics within and between entities. Subsystem structures may also be included to represent noncognitive processes, such as economic or resource dynamics within the broader system. The specific expressions pertaining to each influencer, and what behaviors each influencer can invoke, are determined through the use of SME guidance and available data. SMEs can hypothesize relevant concepts and abstract notions that may not be apparent from the available data. Analytical methods can also enable the estimation of how the hypothesized behaviors could occur based on the knowledge of a group's behavior in other circumstances.

As the model is developed, SME guidance, data, and report information are used to add specificity to the structure. From the overall question, it is determined what should be modeled at a more macro, social-systems level and which entities (individuals, groups) should be modeled at a more detailed or microlevel. Macroinformation typically pertains to societal systems that include economic and behavioral trends, whereas microinformation pertains to a group's or an individual's decision making process.

The microinformation is documented via the DYMATICA knowledge structure. The knowledge structure is consistent with specific, decision making psychosocial theories. It is organized in a manner that characterizes the decision processes of individuals and acts as a scaffolding for the organization of psychosocial processes underlying decision making. The intent of this structure is to record the elicited SME knowledge content with respect to a modeled group of individuals. Specifically, the structure can capture information such as cognitive perceptions, motivations, norms, and potential behaviors. The

KNOWLEDGE ACQUISITION

The process of knowledge acquisition generally includes the initial elicitation of knowledge from an SME, as well as the explication, the coding, and the presentation of knowledge (Cooke 2007). These methods can, however, vary in accordance with the specific needs of the task and the degree of rigor imposed on them. Currently, there are numerous techniques and tools to help facilitate the elicitation and the structuring of knowledge within expert-type systems—particularly those that focus on technical processes, such as mechanical troubleshooting (Shaw and Woodward 1990). These techniques range from unstructured and structured interviews to development of ontologies and entity-attribute grids.

SME information is then treated as hypothesis generators, having partial knowledge of the entire system. These hypotheses are tested within the model. Hypotheses that are not falsified are retained as components of uncertainty for potential behaviors and interventions impact.

Accounting for SME Bias

Dealing with SME bias will always be a challenge. Each SME brings with them educational, cultural, and political biases that can influence their interpretations of a situation, a society, a group, and/or a person. This cannot only provide important insight, but also has the potential to skew the elicited information. The knowledge structure enables comparison and integration of views from a variety of SMEs in a common format. This is important for incorporating multiple SME views regarding the decision making of a specific entity. Because the output from each SME can be compared to that of all other SMEs, the structure can be used to assess similarities and differences between SME opinions and to examine how knowledge from multiple experts converges. We typically want multiple SMEs that have different cross-cultural expertise to include multiple points of view. Each SME knowledge structure is then treated as a separate instantiation within the DYMATICA model. The spread of simulated outcomes will consequently cover all elicited opinions, where common and unique views can be separately assessed for consistency with actual behaviors as they unfold. Unique views that do not contribute to the assessment's accuracy can then be devalued. It is also possible that unique views can be highly consistent with certain observed behaviors. If so, greater weight can be placed on these views. This process is accomplished through the use of quantitative techniques such as sensitivity analysis and uncertainty quantification, which can be designed to compare the accuracy of each SME's knowledge contributions across time. When data cannot identify the most appropriate SME perspectives, the ensemble of perspectives can generate an uncertainty envelope as an output. The analysis of intervention options then utilizes that envelope to determine confidence in those intervention's outcomes.*

Macrolevel Theoretical Representation

The DYMATICA structure rests upon a synthesis of psychosocial theories that can be (1) integrated into a representation of behavior, (2) translated into a set of computational

* A greater description of the DYMATICA knowledge structure is discussed by Bernard (2015).

equations, and (3) instantiated, tested, and verified using accessible data. The intent is to use robust theories that apply across behaviors and societies to describe innate human characteristics. While no theory has been empirically tested across all cultures, certain theories have been used across a large number of societies and seem to explain some general human tendencies.

In parallel with psychosocial theories (described in the following), a set of behavioral economic theories—also extensively evaluated with experimental and historical data—have been incorporated into the general DYMATICA framework. It should not be too surprising that the behavioral economic theories mesh well with the included psychosocial theories, since behavioral economics fundamentally describes how people make choices. The theory set describes how behaviors are derived from decisions and how choices are made to determine those decisions (see Figure 21.2 for a conceptual illustration of the representation of decisions within DYMATICA). For instance, the prospect theory suggests that decision makers negatively weigh potential losses more strongly than they positively weigh comparable gains (Kahneman and Tversky 1979). The human response to perceived risk lends itself to behavioral responses within computational models. Also, DYMATICA uses a flexible version of the (psychologically framed) qualitative choice theory (QCT) developed by McFadden (1984). This is used to quantitatively capture how individuals make behavioral choices based on how they weigh information, tastes, beliefs, and preferences.

The robust parameterization of QCT is often based on data readily obtainable in the field. Other techniques can further determine the correct functional representation of the QCT utility formulation for the problem at hand (Keeney and Raiffa 1993). A key part of the decision process is the filtering of information and the extent to which experience biases the decision process. At a group level, the probabilistic nature leads to a mean value response because a random variation in one direction by an individual is balanced by the reverse variation of another individual. The enduring population (society) aspects dominate group behaviors. The decision process' transient and stable components are identified by cointegration (also Granger causality) methods pioneered by Granger (1969). These same methods also ascertain the filtering and the delayed response processes associated with information perception and behavior. These methods and others are summarized by Backus and Glass (2006). Moreover, they can integrate disparate perspectives and information, qualitative as well as quantitative, into analysis and decision support systems. They are also compatible with orthodox macroeconomic assumptions and used for all manner of choices (including those associated with security).

PROSPECT THEORY

The prospect theory is a descriptive model of decision making that focuses on factors that help determine the degree of risk one is willing to take. Specifically, how one frames a situation can affect the degree of perceived loss and/or gain one is willing to take. These decisions are evaluated using heuristics regarding the potential value of losses and gains rather than the final outcome. Thus, a greater weighting of behavioral tendencies might be placed on groups seeking to avoiding losses (such as in war, finances, etc.) than pursing greater gains (Kahneman and Tversky 1979).

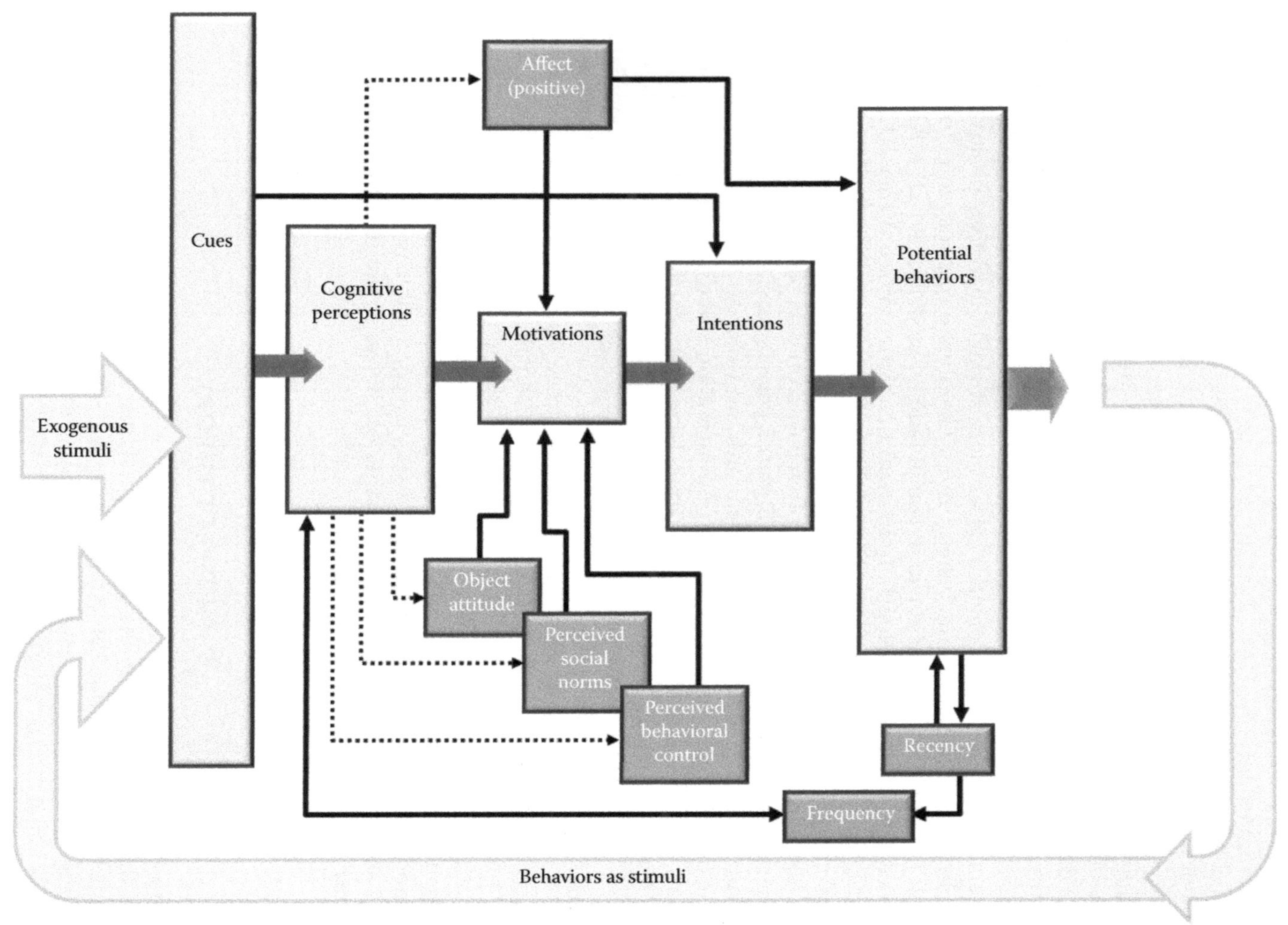

FIGURE 21.2
Conceptual diagram of the psychosocial (decision) model.

Microlevel Theoretical Representation

In order to understand and predict our environments, humans attempt to find patterns in stimuli. If relevant, stimuli can be perceived as cues to particular cognitive perceptions regarding one's current environment. Cue patterns may be associated with an existing or a potential situation or state. The notion that, in many circumstances, can trigger a cognitive perception without the need for extensive deliberation has been proposed by researchers such as Klein et al. (1993) in their model of recognition-primed decision making. This process is represented in Figure 21.2.

Individual differences in culture and experiences may cause the same stimuli to be interpreted differently, thus stimulating different cognitive perceptions within individuals (Wisniewski 1995). For example, groups that share similar cultures and/or religious backgrounds may generally have common cognitive perceptions that differ from other, less similar, groups. These concepts involve the representation of environmental cues and relevant knowledge in a manner that accommodates conceptual pattern recognition.

In DYMATICA, a cognitive perception is considered to be an estimate of some attribute or state in an environment that may affect an existing attitude. When a cognitive perception rises to full consciousness, it will be compared to a cognitive perception template that is stored in long-term memory. These templates store semantic self and environmental perceptions and serve to categorize one's perceptions of one's world (Markman 1999).

Cognitive perceptions may stimulate other cognitive processes such as emotional reactions (which we more broadly characterize as positive and negative valences), attitudes, expectations associated with perceived social norms, and perceived behavioral control over potential behaviors. These processes may help stimulate a motivation to perform some specific type of behavior. If the motivation is high enough, it can stimulate an intention or a set of intentions to perform some type of behavior (Fishbein and Ajzen 1975). The intention to perform a specific behavior is typically a function of what is perceived to be actionable. Thus, upon assessing the environment, intentions that are not perceived to be attainable will lose strength while intentions that are perceived to be attainable will gain strength. Moreover, the valence associated with an emotion (low to high positive, low to high negative) can mediate both the selection and the degree of risk one is willing to take associated with those behaviors (Bernard and Smith, 2006). The actual behavior that is realized is a function of the intent, the associated emotional strength, and the perceived environmental conditions indicating that a behavior is indeed actionable. Additional factors that can affect the likelihood of a behavior being realized include how often and how recently that behavior has been previously acted upon. That is, previous behaviors tend to be good predictors of future behaviors (Bagozzi and Kimmel 1995).

With respect to attitudes, cognitive perceptions are thought to be associations or linkages that people establish between attributes of attitude objects (Fishbein and Ajzen 1975). An attitude can be thought of as a general and relatively enduring evaluative response to an attitude object, where an attitude object can be a person, a group, an issue, or a concept (Petty and Wegener, 1997). This evaluative response generally has some degree of favor or disfavor, approach or avoidance, or attraction or aversion toward that object (Ajzen 1991; 2005). This is expressed with differences in emotional direction that can be "bifurcated into positive and negative evaluations" (Eagly and Chaiken 1993, p. 4). A general theoretical model supporting the notion that attitudes and intent are highly associated with behavior is the theory of planned behavior (TPB; Ajzen 1991). The TPB postulates a process in which behaviors are influenced by (1) current attitudes toward a specific behavior, the (2) subjective norms associated with acting out that behavior, and (3) perception that carrying out this behavior

A NOTE ON CORRELATIONS

The strength of correlations between cognitive elements has shown considerable variability across a number of studies (Ajzen 2005). However, a major factor causing this is the difference in the granularity in the measurement between these elements. This is particularly true for attitudes, since attitudes are often not directed toward a specific behavior and, thus, may be very broad in their level of specificity, while the measure of behaviors may be very specific. Intentions, however, are typically concerned with a specific behavior and, therefore, are generally more consistent in their level of specificity with behaviors.

is within the person's control. The combination of these factors forms a behavioral intention state, which can serve to drive that person's actual behavior (Ajzen 1991; Madden, Ellen, and Ajzen 1992). It is asserted in the TPB that an individual's intention captures the factors that influence some type of behavior, which is indicative of one's effort to perform that behavior.

The success of the TPB has led researchers, such as Perugini and Bagozzi (2001), to extend this theory in an attempt to explain a larger percentage of variance associated with behavior. Perugini and Bagozzi's model of goal-directed behavior (MGB) includes predictor variables of attitudes, positive and negative emotion, subjective norms, and perceived behavioral control, which drive desires. Desires in their model drive intentions.* As with the TPB, the perceived behavioral control also mediates behavior. In addition, they include the frequency and the recency of past behavior, which mediates potential behaviors. Perugini and Bagozzi (2001) argue that individuals take into account both their attitudes and emotion regarding potential achievement or failure with respect to a sought after goal. That is, an attitude is an "evaluative response toward an object or act that, once learned, is triggered automatically" (p. 82). The processes underlying emotion, on the other hand, are "more dynamic and entail self-regulation in response to feedback" (p. 82). As discussed earlier, adding these predictor variables should, in many circumstances, explain a higher percentage of the variance associated with behavior. If so, MGB would further broaden and enhance the TPB by providing greater predictability of behaviors.

Computational Structure of DYMATICA

A DYMATICA model uses system dynamics to simulate interactions between cognitive entities in the context of a problem of interest (see Figure 21.3). These models consist of cognitive model sectors, which use environmental cues to determine behaviors for each cognitive entity of interest, and world model sectors, which include all noncognitive elements in the simulation (such as economics, resources, or population growth). An assessment begins with a scenario, often including an initiating event associated with certain cues. Cognitive entities interpret these cues as cognitive perceptions, determined by linear weighted sums of cues with coefficients based on the beliefs of each entity. Entities form expectations about their world based on their cognitive perceptions. The difference between expectations and perceptions is called discordance. Discordance is the normalized difference between perceptions and expectations.

* Note that in the MGB, Perugini and Bagozzi chose the more specific term *desire* instead of *motivation*. DYMATICA uses the more hierarchically expansive term *motivation* to represent a broader range of drives.

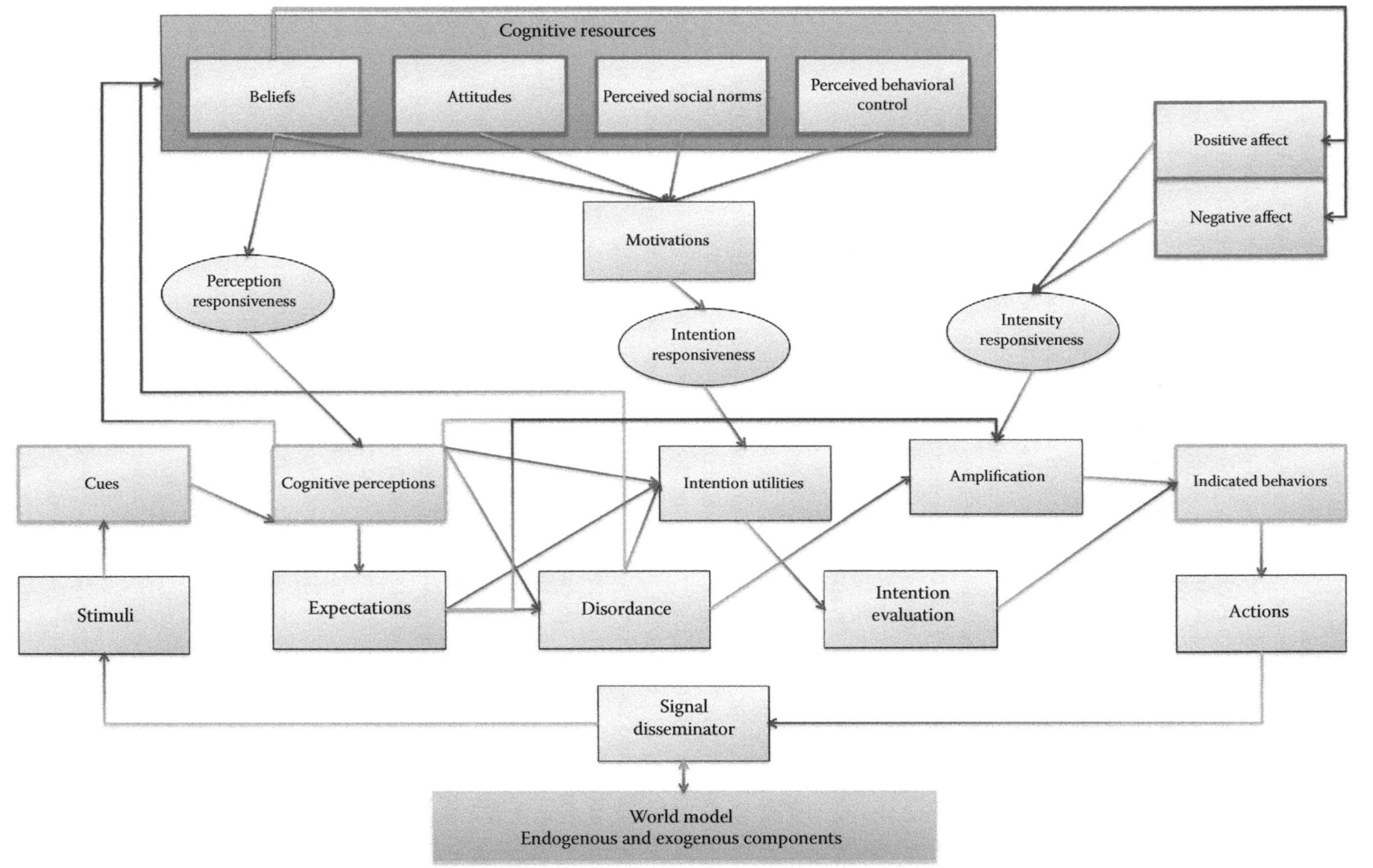

FIGURE 21.3
Overview of the DYMATICA structure.

Each entity calculates an intention utility, or perceived benefit of taking the corresponding action, for each potential behavioral choice. Intention utilities are linear weighted sums of cognitive perceptions, expectations, and discordance, with weights determined by the entities' cognitive resources (perceptions, attitudes, perceived social norms, and perceived behavioral control). These weights are determined by SMEs, literature, or other data and may be different for each entity. The model uses QCT (McFadden 1982) to select the intentions that each entity will pursue. This is based on a multinomial logic function that determines either the probability of selecting a particular behavior from a set (for individuals) or the fraction of people that will select that behavior (for groups). In situations where emotion affects the magnitude of an intention, the model determines the amplification using a linear weighted sum of perceptions, expectations, and discordance. Weights for amplification equations are based on positive and negative emotions and are determined by SMEs, literature reviews, or other data. Intention evaluations are multiplied by amplification to determine the indicated behaviors of each entity. Actions, or physical realizations of behaviors, are delayed versions of these indicated behaviors. Both actions and world model outputs (which can also depend on actions) can act as cues for cognitive entities in subsequent time steps.

The Process of Representing Behavior

Representing potential behavioral responses and counterresponses is first achieved through causal loop diagramming, which causally relates the interactions included in the model. The causal loop diagram is then mapped to a stock-and-flow diagram that explicitly details the information flow and physical quantities through the system. A key feature is the designation of stocks that represent the accumulation of information, experience, monetary, or physical quantities. These stocks, or state variables, largely characterize a system's nature and its responses. The change in the value of stocks over time is the differential part of the differential equation approach to computational modeling, which the system dynamics paradigm approximates. The theory's exact mathematical expression is anchored in the accumulation of flow into and out of the stocks. The mathematical expression of the flows comes from a causal interpretation of the theory into the mathematical language. The data determine the parameters that control the progression of the simulated values through time. Rigorous statistical techniques determine the appropriate parameters and the uncertainty associated with their use. This uncertainty can later define the confidence in the results of an intervention analysis.

As discussed earlier, populating the structure with information is achieved via the DYMATICA knowledge structure. For example, circled formulas in the rightmost image in Figure 21.4 are populated with information, via the knowledge structure, shown on the left in Figure 21.4. As these structures are developed, more detailed domain information is increasingly used to populate the models and to help ensure that the macrolevel and microlevel structures are consistent. That is, this information can be used to further strengthen the overall systems structure. The quantitative data comes from such things as survey polls, economic output reports and projections, and demographics, all of which provide useful information pertaining to perceptions, attitudes, behaviors, and trends.

This approach to modeling is made possible by assuming a fixed set of potential behaviors embodied in a representation of the individual or group. The representation contains the preferences and the personality characteristics pertinent to the relevant decision making. While the magnitude of interactions may change, the model

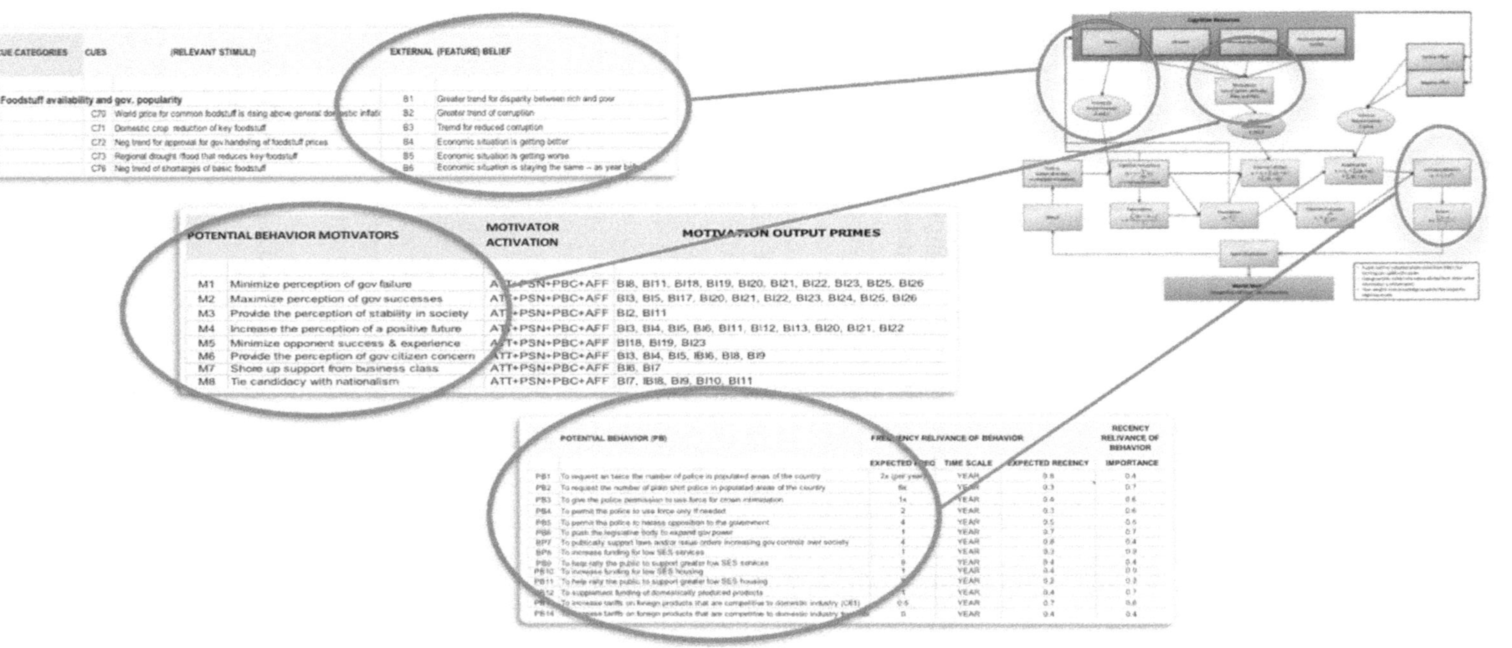

FIGURE 21.4
Example of a knowledge structure showing cognitive perception, motivation, and potential behavior information.

does not produce new paths of cognition. All potential interactions are determined via initial model parameterization. Over time, the simulation will be less applicable in that the modeled individuals or groups will be responding to a new set of cues. This will require updating the parameters within the models, either exogenously or endogenously.

Although the relationships between the elements within the model are indeed fixed due to their derivation from the knowledge structure, it is possible to vary world model (system-level) inputs over theoretically or empirically appropriate ranges and store the model results at each value (or combination of values, if referencing more than one input variable). This enables end users with little modeling and simulation experience (e.g., analysts) to adjust the levels of the input variables and evaluate the changes in the model results without having access to the infrastructure required for a series of computationally intensive model runs. Being able to adapt the model output allows the users to reflect real-world changes as they unfold and/or to analyze different potential scenarios germane to the research question, even as the underlying model mechanics remain static.

DYMATICA in Action: Assessing Environmental, Geopolitical, and Societal Factors Influencing Unrest and Government Support

An illustrative DYMATICA model assessment is presented in this section. In this example, the model focuses on potential geopolitical tensions arising out of the construction of the Grand Ethiopian Renaissance Dam (GERD) in Ethiopia. Recently, the construction of the GERD has generated tensions between Egypt and Ethiopia over the control of the Nile River in northern Africa. These tensions have the potential to exacerbate existing stresses within Egypt, leading to greater internal conflict.* DYMATICA was used to assess how tensions within Egypt might evolve from the impacts of the GERD. The interplay between four parties (the Egyptian regime, the military elite, a more militant Islamist population, and the general population) over an upcoming 10-year period was addressed. Similar to what happened before the Arab Spring, the rise of food prices, which was a strong driver for the unrest leading up to the Arab Spring events, is hypothesized to produce economic stress that could be driven by the GERD, albeit with different political undertones. With this hypothesis in mind, DYMATICA was applied to lend insight to the following questions:

1. How might unrest manifest within Egypt's progress? What party would receive the brunt of the blame, given the different assumptions about the impact of the GERD?
2. How do different Egyptian economic policies and factors (such as use of food and energy subsidies, military spending, and external food or energy price shocks) affect the population's response to the GERD?
3. How do Egyptian noneconomic policies such as suppression of unrest and government messaging influence how the population will respond to GERD's impact?

* In March 2015, an agreement was signed between Egypt, Sudan, and Ethiopia, providing a base for cooperation in the region regarding the GERD.

MODELING AND REPRESENTING ENTITIES

This model represents macrolevel societal interactions as well as microlevel sociopolitical entities that are inclusive of various groups within a society. These entities can be thought of as existing within a causal structure where decisions of certain entities define the strength of relationships. By determining the relationships between the entities, a list of potential behaviors, motivations for these behaviors, and cues that would activate these motivations was generated. This set of potential behaviors, motivations, and cues, along with how they are related to one another is the core cognitive input that defines this modeling structure.

Entities and Decisions

The following set of cognitive entities was chosen for simulation as they pertain to the set of questions presented earlier. These entities were represented using a cognitive behavioral modeling structure described earlier.

1. The regime: The regime represents a set of high-ranking government officials (which include former military officers). The regime is primarily interested in improving the Egyptian economy, decreasing the perception of instability by internal and external parties, and maintaining power. The regime is assumed to have enough influence to decrease GERD's impacts through diplomacy.
2. The military elite: The military elite represent high-ranking officials within the Egyptian military and military-connected business owners and executives who are motivated to keep their wealthy and powerful positions.
3. The Islamists: The Islamist represents the Egyptian population motivated to increase conservative Islam within the Egyptian government.
4. The general population: In essence, the general population is everyone else. Both the general population and the Islamists are quite poor by western standards. Beliefs include a high sense of pride and the belief that Egypt remains the regional leader. Motivations stem from the distrust of corrupt regimes and the distrust of colonial influencers, balanced with a buildup of fatigue from recent instability and economic hardships.

The Model

Cognitive entities interact not only with each other in DYMATICA, but also with the surrounding world. For this effort, the world model represented the effect of food and energy subsidies on finances, government income dependencies, and balance of protestors and enforcement in the streets. Varying the scenarios to the world model provided insight into how internal and external economic facets may influence future behavior within Egypt. These inputs include the size and timing of GERD's impact on agricultural production; the size and timing of GERD's impact on energy production; the price elasticity of food and energy supply; the external food or energy price shocks; and the efficiency of food and energy subsidies.

Figure 21.5 shows the multiple feedback loops affecting how population discontent can lead to unrest and opposition toward the current regime. Of particular note are the

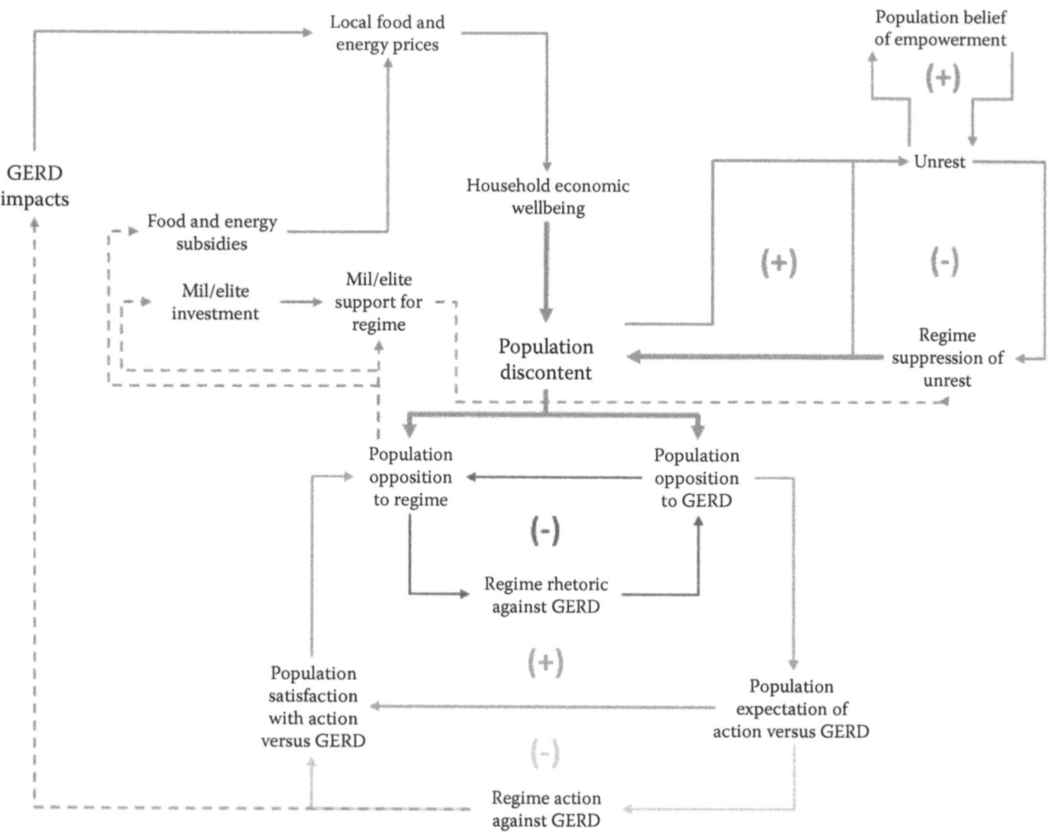

FIGURE 21.5
Causal loop diagram of the key hypothesized relationships associated with unrest in Egypt.

positive and negative loops toward the right of the diagram. The positive loop most closely situated to population discontent is referred to as the suppression loop, in which the government's decision to suppress unrest directly leads to more population dissatisfaction and more unrest. This loop is dependent on the mechanisms used in suppression, as well as the population's reaction to these methods. Even if the suppression is successful at first, the empowerment loop (in the upper right-hand corner) can lead to a long-term escalation of unrest. The suppression loop affects how successful governments will be at quelling unrest in the short term, while the empowerment loop determines the amount of time a regime has to fix the underlying conditions that lead to unrest.

Household well-being, calculated using perceived income inequality, is a direct driver of population discontent. Cognitive entities compare their own discretionary income to that of others in the model. Discretionary income for Islamists and the general population (which are economically identical) is shown in Figure 21.6. At the beginning of GERD's potential impact, Egyptians receive a large reduction (nearly 20%) in their discretionary income, which begins to recover as the regime increases food and energy subsidies. However, the regime cannot afford these subsidies for the duration of the disruption, and the population's discretionary income begins to decline again. The military elite (not shown) experience almost no impact to their discretionary income.

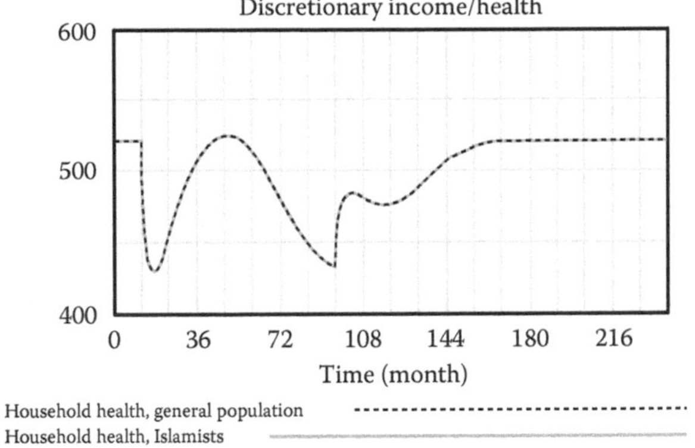

FIGURE 21.6
Discretionary income for the Islamists and the general population.

In the initial assessment, the regime is attempting to recover the economy (thus improving discretionary income) by making new investments in material supporting the military elite, but that will take time to manifest. These investments may include new resource development such as natural gas or mineral deposits or development of military strength to improve competitive stature in the region. In the meantime, they increase food and energy subsidies and attempt to negatively direct public opinion toward GERD and away from the regime. This economic and emotional balancing act largely drives their success in avoiding large protests and the aforementioned unrest tipping point. Figure 21.7 shows how the regime blames GERD, and through the use of government messaging, how this affects the population's and the military elite's views of GERD. The graph (Figure 21.7a) can be interpreted as the relative amount of effort the regime is putting into blaming one cause versus another for hardships. The graph (Figure 21.7b) represents what entities are saying about GERD, where higher than 0.5 means that they are saying negative things, and lower than 0.5 means that they are saying positive things. Notably, the Islamists respond much faster to the messaging. This is because they are assumed to be a more cohesive group in which opinion can spread more quickly, as opposed to the larger, more diverse general population, and the military elite who are less affected by messaging.

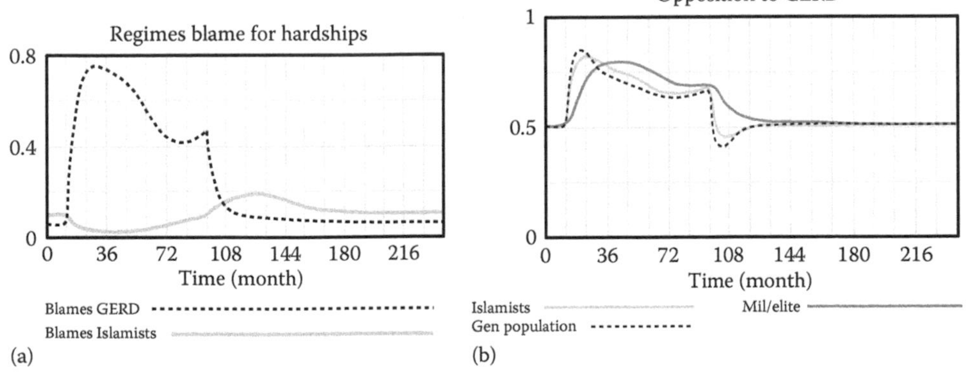

FIGURE 21.7
Percentage of (a) the propaganda activity by the regime and (b) the opposition to GERD by different entities.

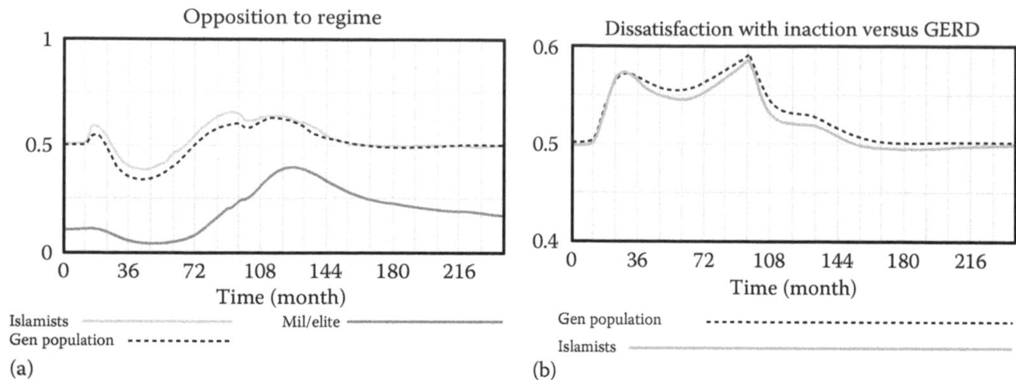

FIGURE 21.8
(a) Negative communications directed toward the regime and (b) dissatisfaction with the regime's inaction against GERD.

While the population is worse off and perceives a comparative disadvantage to the military elite, their anger toward GERD offsets much of the dissatisfaction that they would otherwise have toward the regime. Figure 21.8a shows the strength of negative opinions by the cognitive entities. Because the subsidies and the opinion against GERD take time to manifest, negative communications about the regime increase for the first year of impacts but are slowly replaced by relatively positive opinions as groups direct their anger toward GERD. By the fourth year of impacts, as the regime can no longer afford subsidies, and as the dissatisfaction toward the regime's inaction against GERD builds, opinion shifts against the regime. The dissatisfaction with inaction component is shown in Figure 21.8b. The slight dip around month 60 is because the regime is taking diplomatic action against GERD's use, but soon thereafter, the population becomes again dissatisfied, as the regime's actions do not completely eliminate GERD's effects. A slow decline in market food and energy prices is the result of the regime's diplomatic action.

The actual number of protestors in the model is a function of the population's opposition to the regime and the regime's ability to suppress and deter unrest. The regime can successfully suppress and deter more unrest if they have military support. Figure 21.9a shows the funding that the regime is funneling to the military elite. In the very initial stages of

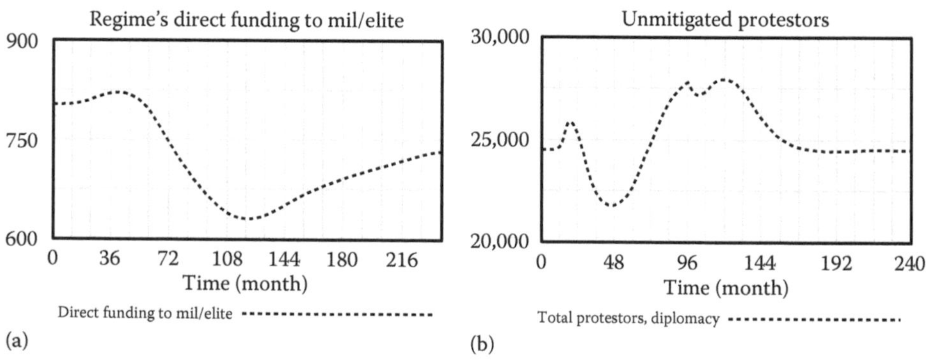

FIGURE 21.9
(a) The funding of military and elite programs and (b) the number of actual protestors. Defunding of the military elite could lead to their refusing to support the regime in their suppression of unrest.

impacts, the regime is trying to increase this funding as an investment in the recovery. As subsidies become too expensive, the regime chooses to defund the military elite programs. This results in the regime's decay of military elite support, as shown by the green curve of Figure 21.8. This decreasing support is another potential source for an unrest tipping point but does not become a problem in the base case simulation, where diplomacy is at least marginally effective at decreasing GERD's impacts.

The amount of unrest declines well after the dam's impacts are over—in fact, the unrest peaks right as GERD is almost full. This dynamic is due to the slippage of food and energy subsidies, the resulting decrease in discretionary income for the public, and the use of suppression by the government during the upswing in protests. All this serves to show that unrest can be a heavily lagging indicator of actual problems, and that the regime's balance between sentiment management and funding decisions is critical to understand when assessing an external shock such as GERD.

Scenario Conclusions

This example scenario is meant show the relationships that could affect the unrest in Egypt between important groups during the period in which GERD is refilling. Some early insights include the following:

1. The highest unrest was found well after GERD's impacts were over. This counterintuitive finding is largely due to the long delays in the system. Structurally, unrest is buffered by the regime's ability to decrease the economic hardship of the population and also by its ability to spread propaganda against GERD.
2. Propaganda against GERD and food/energy subsides only delay unrest. The best way to prevent unrest is to prevent or mitigate GERD's impacts. Effective diplomacy is one strategy that was found to prevent extreme unrest in the model.
3. External shocks to food prices (not related to GERD) would have a significant effect on the popular support for the regime. A price shock occurring before GERD filling period has the potential to be more disruptive than a price shock after this period.

General Conclusions

The DYMATICA framework is based on principles that can encompass a large number of entities with any number of alternative decisions, and with any level of interrelationship complexity. Because of these characteristics, DYMATICA is well suited to examining questions relating to how world model (system-level) variables and/or interventions affect the aggregate expression of individual or group behavioral outcomes. There are many potential use cases for this approach, to name but a few:

1. Estimation of the potential effects of proposed interventions (including unintended consequences) that seek to alter behavior by affecting the cognitive domain (e.g., information operations, deterrence scenarios, proposed engagement options to arrest or avoid conflict escalation) before they are executed

2. Modeling of latent instability within a society (or group of societies) by modeling tensions between groups (in terms of their beliefs and behaviors) and simulating to understand what types of trigger events could ignite conflict
3. Definition of the temporal sequence and the timing of steps in more complex interventions to achieve the highest likelihood of success (however that is defined)
4. Exploration regarding what sets of system-level scenarios might lead to virtuous or vicious cycles (situations in which behaviors create positive [vicious] or negative [virtuous]* feedback into the system and cause or prevent behaviors of the same type)

All this becomes especially useful in light of the realization that hybrid warfare has returned to prominence following Russia's recent annexation of Crimea. Indeed, joint military doctrine is evolving as of this writing to embrace the human aspects of military operations (HAMO), which "describes how the Joint Force identifies, understands, anticipates, and influences relevant individuals, groups, and populations" (HAMO Working Group 2015, p. 3). As the US military comes to grips with these challenges and begins to build them into its training and doctrine, this type of modeling and simulation becomes more obviously and immediately impactful, particularly when "operations will frequently depend on the application of unique capabilities to identify relevant actors in the environment and influence their decision making and behavior in a manner consistent with the desired state" (HAMO Working Group 2015).

Megacities Use Case: Applying DYMATICA to the Megacities Narrative

In addressing the fictional Megacities narrative, the DYMATICA approach could be applied to lend insight to questions, such as the following:

1. How might unrest within Xanadu progress and who would it be directed toward given the interactions between the different immigrant/nonimmigrant populations?
2. How do different Xanadu economic policies and factors such as rise in crime and lack of jobs affect the population's response to the immigrant population?
3. How do Xanadu noneconomic policies toward immigrant populations affect unrest and influence how the Xanadu population will respond to immigrant population over the next 10 years?

Entities and Decisions

A DYMATICA model would simulate system interactions relevant to the questions of interest. The model would incorporate behavioral tendencies associated with the relevant cultures, including differences in belief structures, attitudes, norms, motivations, and intensity of expressed emotions. A macrolevel systems model would first be developed to represent the societal-level interactions. At the microlevel, where decision making would be represented, potential entities could include the following:

* Positive feedback reinforces behaviors that attempt to force the system to depart dramatically from the norm. Negative feedback reinforces behaviors that attempt to force the system back to the norm.

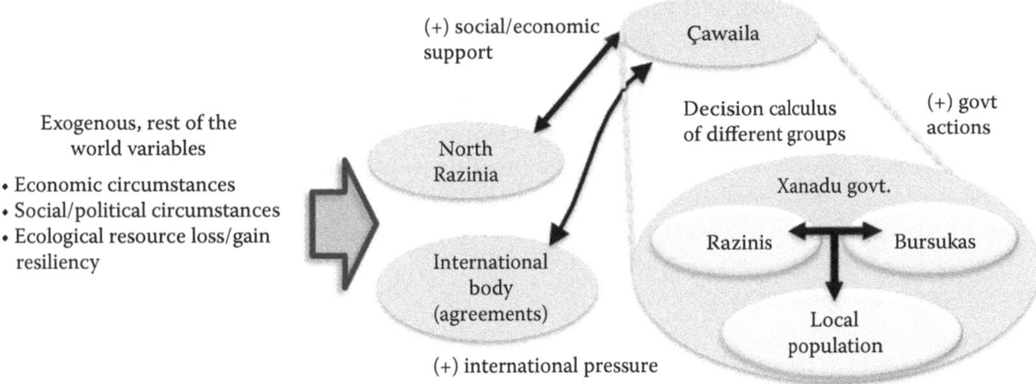

FIGURE 21.10
Example of a simplified DYMATICA model.

1. The Xanadu city government: The Xanadu city government is assumed to comprise former military individuals; the government is primarily interested in improving Xanadu economy, decreasing the perception of instability by internal and external parties, and maintaining power.
2. The Bursuka refugee population: The Bursuka refugee population is assumed to comprise disparate, poor individuals from a different cultural and ethnic background than the Xanadu population.
3. The Razinia migrant population: The Razinia migrant population is assumed to comprise disparate, poor individuals from a different cultural and ethnic background than the Xanadu population.
4. The general Xanadu population: The general population is quite poor by western standards. Their beliefs include a high sense of former pride and the belief that Xanadu remains a leading city.

By determining the relationships between the entities, a list of potential behaviors, motivations for these behaviors, and cues that would activate these motivations would be generated (see Figure 21.10 as an example).

Potential System Behavior: Overextension of Government Resources

Within DYMATICA models, exogenous variables can be used to stimulate the model. For example, the influx of large populations within Xanadu would heavily strain the Xanadu resources. This includes housing, policing, and medical resources (to name a few). Each variable would need to be included in the model, which could be adjusted to assess how increasing or decreasing each resource would affect the stability within Xanadu over time.

Potential System Behavior: Sociocultural Conflict

Another potential behavior of interest is the interaction between the Bursuka refugees, the Razinia migrants, the general Xanadu population, and the Xanadu city government. Potential conflict between the Bursukas and Razinis for resources, as well as segments of the Xanadu population would be high. Each population's sociocultural characteristics would need to be modeled, including how each population interacts with each other. What is important is each population's perceived plight relative to others. Local unemployment, crime, and political influence can have a strong effect on the perceived status and the well-being of each population. Adjusting these variables to assess the potential impact on the overall Xanadu society, over time, could provide important insight regarding how the government should prioritize its resources. The insights from this assessment could provide a discussion forum detailing relationships responsible for the unrest within Xanadu—and could ultimately help explore the policy options required to lower future unrest.

DISCUSSION QUESTIONS

1. What are the differences between system dynamics and agent-based models?
2. How could hybrid computational models potentially increase the utility of sociocultural assessments?
3. What are the macro- and microlevel decision theories represented within DYMATICA?
4. Generally, how does DYMATICA represent decision making and behavior?

References

Ajzen, I. 1991. The theory of planned behavior. *Organizational Behavior and Human Decision Processes.* 50: 179–211.

Ajzen, I. 2005. *Attitudes, Personality, and Behavior* (Second Ed.). New York: McGraw-Hill.

Backus, G. A., and Glass, R. J. 2006. An agent-based model component to a framework for the analysis of terrorist-group dynamics. Sandia Technical Report, SAND2006-0860P. Albuquerque, NM: Sandia National Laboratories.

Bagozzi, R. P., and Kimmel, S. K. 1995. A comparison of leading theories for the prediction of goal-directed behaviours. *British Journal of Social Psychology.* 34: 437–461.

Bernard, M. L. 2015. Developing a capability to elicit and structure psychosocial decision information within computational models. *Proceedings of the 6th International Conference on Applied Human Factors and Ergonomics.* http://www.ahfe2016.org.

Bernard, M. L., and Smith, B. 2006. The effects of emotional states and traits on risky decision making. Sandia Technical Report, SAND2006-7642. Albuquerque, NM: Sandia National Laboratories.

Bernard, M. L., Backus, G. A., and Bier, A. B. 2014. Behavioral influence assessment (BIA): A multi-scale system to assess dynamic behaviors within groups and societies across time. *Proceedings of the 5th International Conference on Applied Human Factors and Ergonomics.* 5: 161.

Bier, A. B., and Bernard, M. L. 2014. Validating a hybrid cognitive-system dynamics model of team interaction. *Proceedings of the 5th International Conference on Applied Human Factors and Ergonomics.* 5: 209.

Bier, A., Bernard, M. L., Backus, G., Glickman, M., and Verzi S. 2011. Using a hybrid cognitive-system dynamics model to anticipate the influence of events and actions on human behaviors. *29th International Conference of the System Dynamics.* http://www.ahfe2016.org.

De Marchi, S., and Page, S. E. 2014. Agent-based models. *Annual Review of Political Science.* 17: 1–20.

Cooke, N. J. 2007. Knowledge elicitation. In *Handbook of Applied Cognition.* ed. Durso, F. T. Hoboken, NJ: John Wiley & Sons.

Eagly, A. H., and Chaiken, S. 1993. *The Psychology of Attitudes.* Fort Worth, TX: Harcourt Brace Jovanovich.

Fishbein, M., and Ajzen, I. 1975. *Belief, Attitude, Intention, and Behaviour: An Introduction to Theory and Research*. Reading, MA: Addison-Wesley.

Flynn, M. T., Sisco, J., and Ellis, D. C. 2012. "Left of Bang": The value of sociocultural analysis in today's environment. *Prism: A Journal of the Center for Complex Operations*. 3(4): 12–21.

Ford, A. 2009. *Modeling the Environment*. Washington, DC: Island Press.

Forrester, J. W. 1961. *Industrial Dynamics*. Portland, OR: Productivity Press.

Granger, C. 1981. Some properties of time series data and their use in econometric model specification. *Journal of Econometrics*. 16: 121–130.

Human Aspects of Military Operations (HAMO) Working Group. 2015. Joint concept for human aspects of military operations (JC-HAMO) Draft v. 0.26, dated February 24, 2015. Washington, DC: Joint Staff, United States Department of Defense.

Kahneman, D., and Tversky, A. 1979. Prospect theory: An analysis of decision under risk. *Econometrica: Journal of the Econometric Society*. 1: 263–291.

Keeney, R. L., and Raiffa, H. 1993. *Decisions with Multiple Objectives: Preferences and Value Trade-Offs*. Cambridge University Press.

Klein, G. A., Orasanu, J., Calderwood, R., and Zsambok, C. E. (Eds.). 1993. *Decision Making in Action: Models and Methods*. Norwood, NJ: Ablex Publishing Corporation.

Macy, M. W., and Willer, R. 2002. From factors to actors: Computational sociology and agent-based modeling. *Annual Review of Sociology*. 1: 143–166.

Madden, T. J., Ellen, P. S., and Ajzen, I. 1992. A comparison of the theory of planned behaviour and the theory of reasoned action. *Personality and Social Psychology Bulletin*. 18: 3–9.

Markman, A. B. 1999. *Knowledge Representation. Mahwah*, NJ: Erlbaum.

McFadden, D. 1982. Qualitative response models. In *Advances in Econometrics*. Ed. Hildenbrand, W. New York: Cambridge University Press.

McFadden, D. 1984. Econometric analysis of qualitative response models. In *Handbook of Econometrics, Volume II*. North Holland, Amsterdam: Elsevier Science Publishers BV.

Perugini, M., and Bagozzi, R. P. 2001. The role of desires and anticipated emotions in goal-directed behaviours: Broadening and deepening the theory of planned behavior. *British Journal of Social Psychology*. 40: 79–98.

Petty, R. E., and Wegener, D. T. 1997. Attitude change: Multiple roles for persuasion variables. In *Handbook of Social Psychology*, ed. D. Gilbert, S. Fiske, G. Lindzey. 4th ed. New York: McGraw-Hill.

Shaw, M. L. G., and Woodward, J. B. 1990. Modeling expert knowledge. *Knowledge Acquisition*. 2: 179–206.

Sterman J. D. 2000. *Business dynamics: Systems Thinking and Modeling for a Complex World*. New York: Irwin/McGraw-Hill.

Wisniewski, E. J. 1995. Prior knowledge and functionally relevant features in concept learning. *Journal of Experimental Psychology: Learning Memory & Cognition*. 21: 449–468.

22

Modeling Social System Resiliency: An Agent-Based Multiscale Approach

Steven B. Hall, Curtis L. Blais, and Ryan G. Baird*

CONTENTS

Introduction 474
The Foundations of Social Identity Group Birth, Growth, Transcendence, and Dissolution 478
The Coevolutionary Ethnogenesis of Social Identity 481
Shared Perceptions of Threat and Betrayal 481
Shared Thoughts of Values and Narratives 483
Shared Intentions to Create a Better Situation 485
Conditions Underlying SIG Growth in Potential or Capacity to Act 485
Social Resiliency as a Multiscale Phenomenon 487
The Adaptive Cycle Process 487
Multiscale Influences on Resiliency 488
Megacity Use Case 489
Situation Summary 489
Substantive Intuitions on How the Situation Might Evolve 489
Where an Operational Optimization Model Would Help 490
Overt Crackdown on Organized Crime 491
Education, Legal, and Economic Support for Razini Population 491
Deportation Program Applied to All Illegal Immigrants 491
Acceptance of Foreign Aid and Foreign Workers to Help Administer That Aid 492
Reintegration of the Bursuka 492
References 493

ABSTRACT This chapter describes a fundamental challenge often faced by donor agents when they attempt to provide support to a failing or failed host nation during its attempt to recover autonomous operations. That challenge is fundamentally defined by the complex relationship between the operations that are conducted in support of the host nation recovery process and the emergence of that nation's innate capacity to resiliently absorb and/or exploit unanticipated challenges and novel opportunities. The role of complex adaptive systems modeling in understanding and anticipating the influence of such operations on the emergence of this resiliency is discussed and illustrated with key components of a foundational modeling architecture.

* Disclaimer: The views and the opinions expressed in this chapter are those of the authors and do not necessarily reflect the official policy or the position of any agency of the US government.

KEY TOPICS

- Resiliency is often more important than stability in sustaining autonomy.
- Stability results from managing the growth of potential within multiple sectors.
- Ethnogenetic models provide a framework for understanding stability dynamics.
- Social media technologies influence the spatial–temporal patterns of ethnogenesis.
- Multiscale interactions between stable and unstable systems yield resiliency.
- Multiagent complex adaptive system (CAS) models can capture all these influences on nation-state resiliency.

Introduction

The US foreign policy, when it comes to providing support to disaster recovery, humanitarian assistance, and reconstruction operations, is often framed and justified in terms of the assumed value of enhancing the stability of the host nation, region, and/or global world order (Army, 2014). Yet that focus has recently come under fire as being too blunt an instrument (Renn and Diehl, 2015). We argue here that both our immediate goal and our longer-term interests, in providing support to such assistance operations, would often be better served by a more nuanced intention to enhance resiliency.

Stability and resiliency are both a function of a system's capacity to adapt to its operational environment, yet there is an important and far-reaching distinction as well, which we highlight and advocate in the following as the cornerstone to more effective assistance operations.

Stability (Gilboa and Matsui, 1991) refers to a system's capacity to orient toward and adjust to the immediate challenges and opportunities of its operational environment so as to retain the capacity (i.e., the potential) to execute core functionality. Within social identity groups (SIGs), such as nation-states, given their essential reliance on collective action as the fundamental response mechanism, the measure of an assistance operation's success is often conceptualized and quantified in terms of the *duration of the peace* (Renn and Diehl, 2015), i.e., the maintenance of a capacity to collectively act, achieved by the assistance operation.

Given that stability is optimally achieved by an SIG's adaptation to the existing structure of the challenge(s) and opportunity(ies) of its operational environment, resilience is, in contrast, achieved by adapting to the knowable nature of the dynamic changes in that challenge/opportunity structure. Consequently, resiliency is quantified in terms of the *potential maintained over the (predictable) evaluation period.* That is, resilience is a measure of the estimated likelihood that the SIG will retain the capacity to stabilize in the context of the environmental situations that are likely to arise. In the fullest sense, resiliency is a measure of the SIG's capacity to both minimize the impact of potential threats and maximize the impact of potential opportunities to its stabilization capacity.

If the stability challenge is condensed down to "is the current environment defined by one large and/or coordinated challenge/opportunity or by many smaller uncoordinated challenges/opportunities?" then the resiliency challenge can, in contrast, be condensed down to "how often and how quickly might the operational environment switch from being fundamentally cohesive to fundamentally noncohesive or even swarming (i.e., self-organizing and

WHAT ARE THE CHALLENGES AND OPPORTUNITIES FACED BY AN SIG?

Situations that constitute SIG challenges and opportunities fall into two natural kinds: firstly, those situations with the capacity to modify the SIG's core sustainment functionality (e.g., processes that exchange essential materiel, energy and/or information, with the SIG's external environment along with processes that maintain the group's internal cohesion) and, secondarily, those situations with the capacity to impact the SIG's ability to develop and maintain stability enhancing relations with external entities such as economic collaborations and identity transforming symbioses. This second category includes impacts to what we label the *neurotic degree* of the SIG.

potentially self-adaptive) in nature." That is, resiliency fundamentally addresses a system's capacity to adapt its stabilization strategy in response to the nature of the uncertain but characterizable changing of the structure of the operational environment.

While in principle a system (e.g., an SIG) can be known to operate stably in the system's current operational environment, the inherent uncertainty of the system's future operational environment, especially when that environment is strongly influenced by other cognitive agents, can only be evaluated, at best, against a (generally nonlinear) model of the likely evolution(s) of that environment. An implication of this inherent (and desirable) uncertainty, in the evolution of the environment, is that an optimally resilient system is often nonoptimally stable in a particular situation and may, at times, intentionally partially destabilize itself in order to maintain resiliency. This topic will be explored further throughout this chapter.

While, at least conceptually, measuring both the stability and the resiliency of an SIG is a tractable endeavor, the task of enhancing an SIG's resiliency can be fundamentally more challenging than enhancing its stability. That challenge is, at least in part, a consequence of an innate drive within humans to both feel and (narratively) understand how they belong to their social group (Koltko-Rivera, 2006).

Optimizing an SIG's stability requires only that the individual constituents are each capable of adopting a state-oriented, narratively anchored belief regarding what their SIG needs to achieve in redistributing decision making authority (along with a consensual mechanism for arbitrating differences; e.g., majority rules) in order for both the individual's need to belong and the collective's need for stability to be satisfied. Optimizing an SIG's resiliency, however, demands the individual to become socially process-oriented, i.e., focused on how they prefer social change to occur. Such a focus is both considerably less concrete than a preference for a specific preferred power structure and more fluid due to the stochastic and nonlinear nature of the models on which they depend. Moreover such process-oriented convictions diminish the effectiveness of the traditional reified narrative in satisfying the need to belong and consequently trigger the risk aversion response that is dominant in humans (Kahneman and Tversky, 1979). Narratives, founded in mutually recognized values, provide for the role differentiation that in enlightened and postenlightened society define legitimate group membership. Thus, a focus on social resiliency motivates a fundamental transformation in the socially constructed notions of identity and belonging.

Several dimensions of the definition of resiliency offered earlier need some additional explication to be operationally comprehensible.

Firstly, it should be noted that the structure of the SIG's resource allocation authority pragmatically determines to what extent the SIG will be managed hierarchically (and consequently authoritatively) versus horizontally (and consequently market-based). Centralized control, with its capacity to relatively rapidly command large-scale reallocations of the SIG's collective resources, is preferable when the encountered challenge/opportunity approximates the scale of the SIG, while decentralized control offers an advantage when encountering threats/opportunities of smaller scale but with greater agility. The bottom line is that the costs associated with making market-based transactions relative to the costs associated with maintaining a hierarchical decision structure, in a particular challenge/opportunity environment, determines the optimal power distribution stabilization strategy (Coase, 1937).

Secondly, conceptually, a focus on optimizing adaptability accepts the premise that managed *instability* can successfully create opportunities for competitors, which promise enhanced overall survival fitness, within the ever-evolving operational environment, to get a toehold in what would otherwise be a temporally optimized exclusive monopoly. If managed well, such a focus on maximizing the adaptability yields a higher mean retention of stability than an exclusive focus on stability with its temporally narrower focus. Consequently, a highly adaptive SIG may not be as efficient as it could be in procuring resources from its environment, but it is more likely to survive; which is presumably our ultimate objective in providing assistance.

The remainder of this chapter is dedicated to outlining the fundamental building blocks that are required to model the nonlinear and uncertain behaviors discussed earlier and to consequently measure the potential influence of assistance operations on both the reestablishment/reconstruction of immediate nation-state stability and their longer-term influence on adaptability.

Part of the challenge we face squarely in this domain is the irrational bias we have for stability (Kahneman and Tversky, 1979; Tversky and Kahneman, 1981). Not unlike forest management, where the resiliency of the forest is often achieved by letting the groves burn, social resiliency is not synonymous with social stability; yet unexamined, our risk-averse nature inclines us toward wanting stability even at the cost of resiliency. So the first challenge we must confront is how much and what kind of diversity do we need in order to optimize global social resiliency.

A theoretically grounded answer to this question would provide much needed guidance in terms of when a change in a specific SIG's (e.g., nation's) state (whether it be a revolution, secession, federation, or annexation) is more likely to serve the global resiliency objective than would the maintenance of the status quo.

But independent of what specific objective is called for in any particular social context, minimizing the loss of more potential than is necessary to achieve the desired end state is taken as self-evidently an objective. We may need to let the grove burn in order to open up new venues for innovators but we do not want to burn more than necessary or stifle the recovery.

Assuming that we know whether the global system needs more or less, the same or different kinds of diversity, the question becomes "how do we know if the global system is on a trajectory toward achieving that desired end state versus in need of some revectoring?" That is, when and how do we judiciously conduct interventions with the intent to nudge the conditions toward greater resiliency? That question can be decomposed into three largely independent questions: When are we about to create (1) more diversity, (2) less diversity, and (3) a different kind of diversity? We will turn to these questions next.

Before doing so, we must first elucidate some key terminology used in this discussion:

- Agent: An agent is an autonomous entity possessing perceptual and cognitive capabilities that are, in part, defined by its unique space–time trajectory and behavioral capabilities that are fundamentally motivated by self-preservation and/or identity growth.
- SIG: An SIG is an agent composed of agents who share a mutually known set of values and that either mutually recognize the value of the unique history (role) of their fellow constituents or trust in their collectivist commitment.
- Value: A value is a contingently persistent belief regarding the desirability or the undesirability of a defined state or process that influences agent choice.
- Self: A self is the set of agents that an agent will sacrifice managed resources in order to maintain.
- Other: An other is a nonself agent that is engaged, by the self, if at all, on a boundedly rational self-interested economic basis.
- Narrative: A narrative is a set of chronologically ordered choices made by an agent, motivated by its values, along with the knowledge of their subsequent consequences.

NOTE: These definitions earlier assume that humans, as cognitive social agents, are essentially dynamic in nature. That is, we are composed of constantly evolving multiple identities that are defined by, potentially noncompatible, values whose behavioral salience is subject to both environmental influences and evolving roles that are constantly being renegotiated to reduce social distance in pursuit and/or maintenance of higher-level identities (Turner and Reynolds, 2010).

We begin this conceptual modeling discussion of diversity change by proposing an abstract characterization of the triggering condition for each of the three defined categories of change. These notions will be elaborated on in subsequent sections.

Much of the traditional literature on insurrection has focused on a shared perception of the state as acting unfairly (in one form or the other) as the compelling instigator of rebellion (Gurr, 1970; Merton, 1938).

We propose that a more consistent (Kendall, 2011) (and compelling) characterization can be found in a subtle refinement of that shared perception as being of the state's betrayal. Postulating perceived betrayal as the fundamental motivator denigrates the role of any perception of unfairness (in the distribution of collective action costs/benefits) to providing evidence of this more onerous charge. Numerous examples from the annals of history (Durschmied, 2013; Pons, 2014; Serge, 2011) support John Gottman's claim that it is betrayal, the "noxious invader, arriving with great stealth … (which) lies at the heart of every failing relationship" (Gottman and Silver, 2012), that encompasses our international and intranational relations as well.

Perhaps the reason that the perception of betrayal is so often linked to resolute revolutionary movements can be traced back to its capacity to trigger innate behaviors that are resistant to the deterrent influences of rational self-interest. Social psychologists have compellingly demonstrated (cross-culturally) the willingness of people to individually sacrifice to punish agents that exploit the inherent trust that underlies collective action (Güth, Schmittberger, and Schwarze, 1982). *Traitors* seem to evoke intentions in us to meaningfully modify the offending behavior that can trump our core self-survival-oriented needs. And with that shared sense of betrayal (of the understood social contract between a nation and its state), the state's legitimacy (from which its authority, in part, derives) begins eroding, further exacerbating any revolutionary or secessionist sources of instability. In the next section, we will deconstruct how the constituents of an SIG come to collectively believe that they have been betrayed.

Meanwhile we posit that annexations and federations are most often triggered by quite different underlying motivations. A collective intention to annex an other is proposed to be contingent on the collective perception of an economic advantage in usurping authority over the other's managed resources relative to the estimated gains anticipated from collaborating with them, while the federation of the self with an other is proposed to be contingent on the perception of a mutual reification of the respective roles of the self and the other that has yielded shared value(s) and consequently constitutes a new shared identity (Beitz, 2013).

The next section begins this discussion by characterizing a framework in which to conceptualize the process of insurrection, its role in SIG formation, and the processes that underlie counterinsurgency and/or reunification/(re)federation outcomes.

The Foundations of Social Identity Group Birth, Growth, Transcendence, and Dissolution

We have said that stability is a measure of how well matched the distribution of resource allocation decision making is to the current state of the operational environment's challenge/opportunity dynamics, while resiliency is a measure of how effective the SIG's processes are in modifying that distribution to maximize potential over an uncertain but characterizable future. In both cases, the SIG's potential to harness collective resources to minimize the impact of shared challenges while maximizing the advantage of shared opportunities underwrites the key measure of effectiveness.

One of the inherent struggles that human SIGs confront in achieving resiliency is postulated to be a consequence of the reluctance of their constituents to embrace the notion of an uncertain future due to it adverse implications for their sense of belonging. But another obstacle also stands in the way. The social and cognitive processes that mediate the emergence of both shared identities and their optimal stability response to the current environment are subject to limitations and biases that all but ensure that there will be systematic mismatches between the social identities that form and the set of individuals that truly share environmental influences, as well as mismatches between the structure of their stability response and the operational dynamics of the environment. Understanding these irrational, but not unpredictable, biases along with the near ubiquitous human reluctance to embrace uncertainty facilitates our endeavor to model the inevitable tension that contributes to social identity dynamics; thus, we start the discussion here.

To understand how any idea comes to be taken as a self-evidently true or good notion by a collective set of agents is where we must start, for while most such mutually shared ideas do not trigger the emergence of a new shared social identity or how it ought to collect and spend its resources, the ones that do (described in the following) are subject to the same constraints that any idea must overcome in order to be shared.

Most new ideas, adopted by individual agents, are self-perceived as being implications of ideas that they believe are already accepted as true or good by an SIG to which they belong. Sometimes these implications are purely logical or phenomenological in origin, but more often, they represent the application of existing beliefs to new situations. This would be a straightforward process to model if we humans all maintained a consistent set of beliefs about the world and how we should respond to it, but alas we do not.

Unfortunately, for modeling purposes, the future, except within the fairly limited domain of the physical sciences, is complex and, much like the weather, difficult for anyone to accurately predict over the temporal horizons of interest to humans. Consequently we humans tend to maintain multiple hypotheses (sometimes reified by different SIGs) about what is happening in the world and what we should do about it. And, to further complicate the picture, what we monitor in the world and consequently perceive is biased toward confirming what we believe is happening and supportive of what we are considering doing about it. Which is all to say that predicting the beliefs that are, in practice, adopted by individuals requires the modeling of a self-fulfilling feedback loop that turns out to be exacerbated by social reification.*

The impact of social concurrence on the perceived reality with which a proposition is held, in conjunction with a variety of intrinsic human cognitive limitations and biases, creates an opportunity for SIGs to collectively converge on symbolically mediated narratives (i.e., value-laden models) regarding the state of their shared operational environment and the effectiveness of their response to it that are at odds with reasoned analysis. We elaborate on two specific relevant classes of these irrational biases as follows: *cognitive biases* and *social biases*.

The sources that we characterize as cognitive biases are arguably traceable to predispositions that have evolutionarily served our species' survival interests. They include biases referred to as risk aversion (Kahneman and Tversky, 1979), comparative optimism (Shepperd, Carroll, Grace, and Terry, 2002), narrative reconstruction (McAdams, 2001), and hedonic adaptation (Lieberman, Ochsner, Gilbert, and Schacter, 2001) among others. The critical takeaways include (1) most people make decisions that are nonoptimally conservative, given their current value system, (2) most people have little self-understanding of how their value system automatically adjusts to lived (especially irrevocable) experience, and (3) most people reevaluate the propitiousness of their decisions by their perception of the propitiousness of the final outcome while ignoring any pain/suffering or pleasure/reward involved in getting there. The full range of human cognitive biases remains an active area of research,† but in principle, modeling their impact is reasonably simple and straightforward, after they have been identified, as is evaluating the influence that any assistance operation might have on making any collective decision more rational.

The second class of irrational biases are collectively called here the *social biases* due to their origination in the complex relationship between our drive to sustain our own physical existence and our drive for social belonging, i.e., to find meaning in relationship to something other (Koltko-Rivera, 2006). These biases range from the Golden Rule' (Narvaez, 2010) bias (most people will not directly hurt a member of their own SIG even if it would be in the overall interest of their SIG to do so) to Dunbar's number (Dunbar, 2010) and optimal distinctiveness (Smaldino, Pickett, Sherman, and Schank, 2012) (which both argue for innate limitations/preferences for SIG size and/or role/identity differentiation within the group) to the conformity biases first clearly elucidated by Asch (1951).

---

* Social reification of our ideas is hypothesized to be an innate process underlying what has often been referred to as the *social construction of reality* (Berger and Luckmann, 1966; Bredenlöw, 2001). The belief that an experience is shared or at least sharable bestows in and of itself a sense of reality to the thought. It is arguably what is at the core or our infatuation with empiricism and our horror with solipsism. It is, in simple terms, what distinguishes a dream from reality.

† See https://en.wikipedia.org/wiki/List_of_cognitive_biases.

The conformity experiments of Soloman Asch (1951) and more recently the modeling work of the bounded confidence theorists (e.g., Deffuant, Amblard, Weisbuch, and Faure [2002] and Hegselmann and Krause [2002]) demonstrate not only that the more resolutely an agent holds a belief, the less susceptible they are to the influence of others, but also, more tantalizingly, that if we know the social network structure and the initial distributions of belief resoluteness, we can predict whether the network will converge on a shared conviction on the truth or the value status of any particular proposition or whether the network will balkanize into multiple independent subnets of belief/value conviction.

The social reifying influences referenced earlier in conjunction with the work of social identity and self-categorization theorist (e.g., Turner, Oakes, Haslam, and McGarty [1994]) offer the promise of being able to model and predict which subnetworks of agents within a larger network will emerge as sharing a social identity and how those social identities will likely expand and shrink over time.

It is in regard to these dynamically evolving SIGs that we want to first associate a stability/rebellion model and then a model of resiliency. The work of theorists such as Lustick, Miodownik, and Eidelson (2004) on the evolution of secessionist social identities and Davis and O'Mahony (2013) on the evolution of violent/terrorist social identities provides a framework for the predictive modeling of the multiscale social stability that provides the basis for making the instability management assessment at the heart of resiliency evaluation.

One of the more challenging aspects of developing a valid model of SIG potential (i.e., stability) growth lies in the complex dynamic that governs whether the drive to sustain autonomy (via control of the operational environment) or the drive to belong (via the dynamic relationship maintenance) will dominate in any particular context for any particular agent. Where the agents of an SIG are on this continuum between the desire to control and the desire to relate influences two important aspects of SIG behavior. First, as the mean desire to control increases, so does the mean resoluteness or conviction with which agents hold their beliefs and values and consequently, as modeled by the bounded confidence theorists, the likelihood of social balkanization. Similarly, as the mean desire to relate increases, so does the probability that previously distinct SIGs will adopt a shared identity.

And it is fear that plays the critical role here in determining the relative saliency of these competing drives. Fear (perceived threats to survival) and anxiety (perceived threats to belonging) tend to evoke habitual responses in humans that tap into their wisdom (i.e., knowledge gained through experience). Evolutionarily, these habitual responses serve our immediate survival/self-sustainment interests but at the expense of relationship formation and maintenance. Such invocations of wisdom tend to fixate the belonging response (O'Connell, 1980) onto a desire to control social distance; which is, in a felicitous and stable relationship, always in dynamic flux.

Moreover, it is in these dynamic social distance relationships that the identity flexibility, at the heart of social resiliency, is most likely to emerge. Thus, somewhat paradoxically, while the goal-oriented behavior of the fearful crowd is easier to model (and consequently predict), it is destined to be less resilient than its nonfearful counterpart simply because it is more likely to become fixated on the immediate challenge/opportunity dynamics and thus become focused on achieving stability (at the cost of resiliency).

In summary, understanding how fear influences the relative strength of the self-survival versus the social belonging drives and, consequently the distribution of belief/value convictions, facilitates the predictive modeling of how SIGs self-organize into semi-independent constituent SIGs that each exhibit the capacity to motivate resource expenditures, among their members, intended to sustain the group. Moreover, understanding these relationships facilitates the modeling of the relative strengths of these identifications that individual agents maintain with each of the potentially many subdominant constituent SIGs to which they belong.

The next section deepens our discussion of the necessary preconditions for the emergence of rebellious (revolution and/or secessionist) social identities.

The Coevolutionary Ethnogenesis of Social Identity

The process involved in a group of people coming to collectively think of themselves as a people, thus enabling their capacity to deliberately act in their own collective interest, is only in part fueled by our innate social inclinations. Social groups take on an identity of their own (i.e., become an SIG) only as a consequence of their collective perception of an alien social group that is perceived as other, i.e., that behaves in a way that is collectively judged to be untrustworthy.

Traditionally (i.e., before the widespread availability of technology-mediated communication systems), most SIGs emerged from the interaction of social groups that were in geographical proximity to each other but whose unique histories solidified discrepant value systems, which in turn yielded mutually unpredictable and untrustworthy behavior. In such traditional environments, following the emergence of a defined identity, the innate tendency for individuals to be indoctrinated into the dominant identity of their social neighborhood made the prediction of an individual's SIG membership a relative straight forward function of determining what geographically defined boundary the individual lived within.

In the modern era of increased demographic mobility and urbanization, the boundaries that define the in-group versus the out-group status of a particular agent vis-à-vis a particular SIG are increasingly nongeographically defined but instead rely on explicit expressions of sacred value attachments, behaviors that are consistent with such attachments, and/or superficial signs that suggest such attachments (e.g., race, gender, age, and custom).

This diminishing of our ability to pragmatically assume SIG membership has effectively returned us to a more primordial state in which social identities are once again much more commonly emerging from preconscious social groups and/or being actively renegotiated among existing SIGs. This renewed focus on the perception of intersocial group value conflicts, and, even more saliently, on the perception of an other's incapacity to act trustworthily, independent of value differences, as the dominant ethnogenetic forces is where we turn next in our quest to understand (and model) the default trajectory of a social network's distribution of decision-making authority. The next three subsections define the memes that precondition the emergence of a new social identity.

Shared Perceptions of Threat and Betrayal

Agents, acting in their own selfish survival interests, generally find value in collaborating with other agents only when they share a common challenge/opportunity. The cost of participating in any such collective actions must yield what is perceived to be an investment profit, after discounting the expected returns over the life of the investment (i.e., possess what is called in economics an attractive net present value [Magni, 2009]), for each of the participants or they will not long participate, and the collective action will fail.

These challenges/opportunities (Hardin, 1968) arise almost inevitably when a social group is confronted with limited common pool resources (e.g., water, pasture lands, fishing hatcheries) and can arise as a consequence of either the incidental behavior of others (e.g., migration patterns and pollution generation) or their intentional actions (e.g., threats of war, evangelizing, and offerings of alliance), but in each case, the impacted agents must continue to perceive that the required taxation is yielding them a personal profit not just theoretically but in reality. The state may have taken on a legitimate collective action intention (i.e., that should have produced, for its constituents, a positive return on investment), but mismanagement, including ineffective deterrence of would-be free riders (Ostrom, 1997) can be sufficient for the SIG to reevaluate the legitimacy of the action and/or the capability of the state.

Moreover, not all shared challenges/opportunities can be, even at best, pragmatically addressed via collective action in such a way as to yield equal benefits to all members of the SIG. The nonuniform nature of the impact and/or the nonuniform benefit of the collective action often results in the members living within the frontier either realizing less profit from the collective action or being subsidized by those not living on the frontier. Both cases present a narrative challenge to the state in legitimizing how the final resolution is fair to all concerned. Failure to compellingly formulate such a narrative increases the risk that the state will come to be seen, by at least some constituencies within the SIG, as not acting in its interests even if they continue to reap a profit from the collective action (c.f. *relative deprivation* [Gurr, 1970]).

But the sense of betrayal that we have argued earlier that lies behind the most stalwart and persistent insurrectionist movements represents a special kind of threat beyond simple unfair or unprofitable treatment. Unlike most kinds of threats, which are perceived to challenge the survival of the self, betrayal is perceived as a threat to the foundations of social identity and collective action, i.e., to the possibility of identity formation, maintenance, and growth. Social psychologists and cognitive neuroscientists have compellingly demonstrated this phenomenon via experiments such as the ultimatum game (Nowak, Page, and Sigmund, 2000) in which humans are shown to be inclined, even when interacting with strangers, to impose disciplinary actions on unfair resource allocation decision makers, even when it is personally costly for them to do so. The key take away is that among agents that are assumed to share a SIG, *the perception of an intention to exploit the trust of another is sufficient to trigger a selfless intention to deter the exploiting agent from forming such intentions in the future.*

The ubiquity of betrayal as the most enduring of insurrectionist motivations can be more readily understood when its resistance to extinction is recognized. Betrayal can be portrayed as a kind of moral violation and as such provides an opportunity for meaningful behavior that is not subject to the typical economically oriented self-oriented profit-loss analyses that can thwart insurrections motivated by the desire to remove an incompetent state or even an unfair state. In summary, a sense of betrayal is triggered when the following conditions are perceived:

1. Agents *A*, *B*, and *C* are all mutually known to each other to be constituents of a common SIG.
2. *A* expresses an intention to *B* with the expectation of influencing *B*'s behavior (i.e., *A* makes a commitment to *B*).
3. *B* consequently acts in such a way as to benefit *A* based on its trust in *A*'s expressed intention.
4. *A* fails to deliver on the commitment made while reaping the benefit of *B*'s actions.
5. *B* (or *C*) concludes that *A* did not sincerely hold the commitment intention when it was expressed (i.e., *A* intentionally exploited *B*'s trust).

A couple of quick notes should be highlighted here: (1) Third-party agents (i.e., who are neither betraying nor betrayed) have a tendency to be pulled into the fight. (2) Desperation and/or relative deprivation are not, in themselves, sufficient to trigger a sense of betrayal, but they can increase the plausibility of such a narrative, especially if not effectively countered with a narrative that refutes these conditions as evidence of insincerity.

Shared Thoughts of Values and Narratives

While an individual agent can perceive/feel that they are being threatened (or betrayed), the emergence of a behaviorally relevant SIG requires the agents to mutually share this sense of threat/betrayal within a context of what constitutes a legitimate group membership. And that legitimacy is fundamentally based on a consensus regarding what the SIG represents or stands for vis-à-vis agents outside the SIG. This observation importantly implies that the values that are held by an SIG are always in relation to those that are other, as if to say "we value *X* … relative to them," and consequently the identity of an SIG evolves as a function of the others with whom they interact and the nature of what is perceived to be their values. In this context, it is useful to distinguish two kinds of SIGs: SIGs that assume a superior stance with regard to out-groups and SIGs that assume a role stance (Guimond, 2005; Rentfrow, 2010).

SIGs that regard others as morally (or ethically) inferior to themselves are generally (Lefebvre, 2001) limited to antagonistic relations with others. They adopt value systems that are predicated on an assumed understanding of the other and a consequent ability to predict their interactional behaviors, and these behaviors are inevitably seen as lacking in social and/or survival competency if for no other reason than because of their predictability and exploitability. And in a kind of self-fulfilling prophecy, that kind of mechanistic characterization of the other, with rare exception, does in fact trigger a predictably reifying antagonistic response (e.g., the study by Bruneau and Saxe [2012]). Perhaps more surprisingly, the members of such superior SIGs report a mean level of happiness that is less than that of the members of the role stance SIGs that are discussed next (Diener and Biswas-Diener, 2009). New SIGs, born out of the betrayal being discussed here, generally begin as members of this class of SIGs.

Role stance SIGs, unlike superior stance SIGs, are defined by their willingness to negotiate role specializations with other SIGs. Arguably, the source of the unique characteristics of role stance SIGs is a narrative formalization of their potential value to others as being contextual in nature, i.e., that the source of these values is derived from their unique historical (space–time) experience.

Such SIGs present themselves as offering a potentially valuable role within an inter-SIG collaboration (or within common encompassing SIGs) as opposed to claiming a uniquely important role, as do superior stance SIGs. Initial encounters between role stance SIGS are defined by a concurrent search for reifiable roles that offer potential pragmatic value to the other while honoring their sacred values (i.e., values that are self-perceived to provide SIG coherency). Pragmatically, this dance requires a considerable level of inter-SIG social skill involving a nuanced blend of an understanding of the need to issue and respond to boundary warnings (signals that indicate that sacred values are being threatened) with an understanding of how to offer choices with multiple acceptable responses. Roughly speaking, we can distinguish two kinds of role structuring that can be negotiated by agents within such multiagent role stance SIGs.

The first kind involves the assumption of homogenous roles by most of the agents within the SIG. Hierarchically structured SIGs are often of this form as are most social animals

THE POTENTIAL POWER OF INTER-SIG NEGOTIATIONS

In November 2015, students at the University of Missouri began protesting perceived racial inequality. The subsequent unwillingness of the administration to discuss their concerns escalated the protests and eventually precipitated the resignation of the university system's president and the chancellor of the Columbia campus. One can speculate on the sense of betrayal expressed by the students (together with sympathetic faculty) and the different kinds of SIGs—superior stance and role stance—operating in the situation. Perhaps the administration could have averted much of the outcome by quickly moving into negotiated role specializations to give voice to and establish avenues for action with the other SIG(s) involved.

and social insects. The assumed roles of constituent (commoditized) agents are relatively static and context insensitive. Decision making here tends to have a democratic quality to it (King and Cowlishaw, 2009).

The second kind of role structuring involves a stabilization of heterogeneous roles taken on by some or all the agents within the SIG. In such, SIGs decision making is considerably more nonlinear, and as such, each constituent role can assume a position in which it is instrumental in determining collective behavior (Michelena, Jeanson, Deneubourg, and Sibbald, 2010).

The resiliency benefits of such heterogeneous role SIGs relative to homogeneous role SIGs can be understood as being a function of the increased access to a broader range of actionable experience and thus the possibility of generating emergent intelligence that exceeds the aggregate intelligence of the individual agents. We note here that the collective perception of emergent intelligence is a necessary condition for identity transcendence; the process by which constituents of two SIGs, in economic collaboration with each other, develop emotional bonds to a persistent encompassing SIG.

What is key to this ethnogenesis argument is that the kind of SIG that emerges from a perceived betrayal and its likely interactional trajectory with the perceived traitor depends on the values that come to be consecrated as shared within the social group. If these values are effectively in essence "they are evil to have abused our trust," then a confrontational relation is destined to evolve, but if what emerges is something closer to "they know not what they do," then a reconciliation is much more likely to occur, which may in turn yield an even more resilient SIG characterized by an emergent intelligence.

THE BIOLOGY OF EMOTIONAL ATTACHMENT

Oxytocin and vasopressin are thought to be involved in the emotional attachment process between individual humans. One current model suggests that social isolation leads to stress and consequently to the release of cortisol. Successful social interactions increase the amount of oxytocin, which reduces stress and stress hormones and reinforces the behavior linkage between the individuals. Weak interpersonal bonds have been identified as the primary conduit of information flow within social networks (Granovetter, 2005).

Shared Intentions to Create a Better Situation

The final collective cognitive state that underlies the emergence of a new SIG is the belief that something meaningful can be done by us in response to the betrayal that we are facing. In many ways, like a fish in water, the betrayal cannot be clearly seen until the possibility of it being otherwise is collectively conceptualized. Thus, the betrayal, the betrayed, and a meaningful response all in some sense come into collective existence at once. This coevolutionary development of the preconditioning memes of insurrection provides the fertile grounds in which an emergence of the initial, superior stance stage of SIG development often emerges (most often with the assistance of a charismatic leader) in righteous insurrectionist furor.

When we reference the role of a shared meaningful action/response meme in the SIG ethnogenesis process, we intend the word to imply a shared conviction that the result of the intended action will be regarded by the self and the other, when it is fully processed, to constitute a permanent change in how the (social) world is perceived and consequently how the SIG will henceforth be behaviorally governed, e.g., all people have a right to freedom. What is being implicitly proposed in such meaningful revolutionary action is something worth fighting for and worth sacrificing for as opposed to something that seems like a profitable endeavor now but might change in the future. Such values assume a sacred position in the SIG's belief structure and play an important role in maintaining the SIG integrity and coherency both during the insurrection and in its aftermath if and when successful.

Conditions Underlying SIG Growth in Potential or Capacity to Act

Once an SIG comes into existence, the first challenge it confronts is to increase its potential or capacity to act in the service of the collective interests of its constituents. This need is often notably salient, given the adversity in which most SIGs are born. The United States Institute of Peace/Peacekeeping and Stability Operations Institute (USIP/PKSOI)'s *Guiding Principles for Stabilization and Reconstruction* (Cole and Hsu 2009) provides a starting point for conceptualizing the end states that promote the growth of such potential.

THE NEURAL UNDERPINNINGS OF SOCIAL PERCEPTIONS, THOUGHTS, AND INTENTIONS

An intriguing new suggestion is beginning to take form that one of the influences on whether any particular meme will be adopted by a given social group has to do with whether it resonates within that group at a frequency that facilitates the entire group seeing a repeating pattern at roughly the same time. Some of the support for this notion comes from work within the cognitive neuroscience community that is beginning to clarify the role of brainwaves in managing cognition within the brain (Fries, 2015). Modeled social networks, like neural networks, are prone to generate traveling waves of excitation that are, in the social network context, mediated by refractory periods determined by the emotional half-life response to the memes that travel through them. Evidence of such social resonance of the three memes described in the section "The Coevolutionary Ethnogenesis of Social Identity" has been found to presage a number of historical revolutionary movements.

THE ESSENTIAL CAPABILITIES OF A LIVING SYSTEM

The living system theory (Miller, 1978) represents an attempt to succinctly characterize what constitutes a living system. Miller argues that such systems are inherently hierarchical and multiscale and that they are composed of a variety of processing capabilities. These capabilities include the ability to process matter-energy and information. Defined subcapabilities within the matter-energy category include the ability to ingest, distribute, convert, produce, extrude, motor, and support. Defined subcapabilities within the information category include input transducing, internal transducing, channelizing, timing, decoding, associating, remembering, deciding, encoding, and output transducing. Reassuringly what the USIP/PKSOI defines as necessary for stable states is consistent with what Miller claims are the necessary capabilities for any living systems.

The USIP's characterized end states include a well-defined domain in which constituents can expect to operate free of coercive influences (a safe and secure environment), where life-sustaining goods and services are universally available (sustainable economy), where collective responses to shared threats and opportunities can be predictably coordinated (stable governance), where constituents are committed to maintaining the SIG (social well-being), and where effective mechanisms are in place to ensure that the decision authority remains collectivist-oriented (rule of law).

In the absence of external interventions, these five end states tend to positively feedback on each other such that their growths and declines (along with the consequent SIG's growth potential) remain highly correlated until either an inadequate or a superfluous measure of potential precipitates a reorganization of the SIG.

The interpretation of inadequate and superfluous potential is a function of the SIG's environmental context in concert with its intrinsic willingness to sacrifice. An inadequate potential is fundamentally a function of a collective constituent assessment that the cost of maintaining the SIG is no longer worth the profits being reaped in terms of increased survival probability and/or in a sense of belonging. The interpretation of a superfluous potential is, however, most often not a consequence of a collective assessment but rather a collective blindness to the fragility that results from overoptimization. SIGs in overoptimized states of fragility are generally destined to either disintegrate in a failed attempt to effectively deal with a novel situation or manage to enter into a mutually supportive symbiotic (identity) relation with one or more external SIGs.

The two critical challenges that any external agent must confront in attempting to support the revitalization of growth potential in another SIG are (1) that any positive development that is delivered successfully carries with it a potential to negatively impact the emerging social identity that underlies the SIG's foundational social well-being and (2) in an overexuberant drive to help the SIG reestablish its growth potential, the donor agent may unwittingly put the host nation on a path to becoming overoptimized.

The contextually sensitive positive and negative feedback dependencies operating among the development of the five end states have proven to be sufficiently complicated that they challenge the inherent limitations of human cognitive processing to reliably calculate the probable outcome of any proposed intervention. Computerized multiagent system dynamics models are now starting to be developed that address this limitation.

The second challenge is more deeply rooted in our inherent preference for certainty and stability even in the face of constant social change, and it is to this challenge that we turn in greater depth in the next section.

Social Resiliency as a Multiscale Phenomenon

In the section "The Foundations of SIG Birth, Growth, Transcendence, and Dissolution," we explored the conceptual modeling foundations required for predicting a state's stability potential or capacity development and the influence of operations intended to support that development. Often, however, the ultimate objective is not so much to maximize the capacity of a state but rather its resiliency. A state operating at peak efficiency that collapses the first time anything goes wrong is rarely what anyone is striving to achieve.

But the relationship between capacity and resiliency is not immediately self-evident. Fortunately, resiliency has long been a topic of interest within the ecosystems community, and it is there that we now turn for insight. A conceptual framework called *panarchy* (Gunderson and Holling, 2002) serves as our foundation here.

Panarchy theory begins by characterizing the inherent cyclic fragility that is characteristic of living systems and then goes on to describe how and why the linking of multiscale living systems works to enhance the overall system resiliency. We simply outline the discussion in the sections below.

The Adaptive Cycle Process

The adaptive cycle process, see Figure 22.1, described by Gunderson and Holling (2002), fundamentally describes what happens in a natural or a social ecosystem when survival of the fittest plays itself out.

The process naturally evolves to a high potential state (K) where only the most efficient (and effectively networked) have survived. Unfortunately, this state is optimized only over a narrow spatial–temporal range that is, in part, driven by the harshness of the environment. The inevitable outcome is an overall increase in the fragility of the ecosystem to reasonably rare and/or novel events. The collapse of the ecosystem is predictably destabilizing but can ultimately produce an increase in fitness. Following the collapse, new kinds of agents (innovators/SIGs) have a chance to take root in a less competitive environment, and new kinds of collaborative relations form, and often the reconstructed system is more

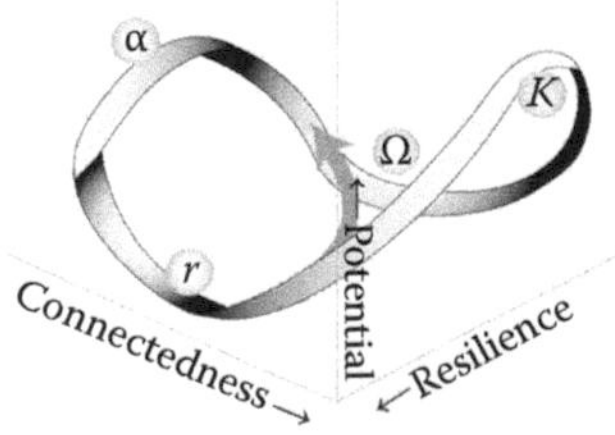

FIGURE 22.1
The adaptive cycle process. r, growth or exploitation; K, conservation; Ω, collapse or realease; α, reorganization.

attuned to its operational environment than its collapsed predecessor. This improved adaptation to current conditions is at the root of increased resiliency.

The potential problem comes about in the collapse process. If the ecosystem does not get enough support, the loss of potential can be so great as to drown out any eventual benefit of being better adapted to meet the current circumstances. The next section describes the key multiscale influences that determine whether these periodic collapses collectively have a net positive or a net negative impact on the overall system resiliency.

Multiscale Influences on Resiliency

The first observation worth noting is that resilient ecological systems are composed of discretely multiscale subecosystems. The lowest level adaptive cycles are both small in spatial extent and fast in their adaptive cycling. Higher-level subsystems are both increasingly larger in spatial extent and slower in adaptive cycling, see Figure 22.2.

The second observation of note is that how fast (and how efficiently) any particular adaptive cycle traverses its state space depends on the state of the adaptive cycles both above and below it. And in particular, the collapse tends to be contagious working up in scale, while how long and severe the loss of resources is, during the collapse, tends to be influenced top-down. In general, resiliency is maximized when an ecosystem's superordinate ecosystem has high potential during its collapse, to aid in its recovery and minimize potential loss, but if the superordinate has too much potential (i.e., is overly optimized), then the ecosystem's collapse may lead to the collapse of the superordinate as well, magnifying the potential loss and limiting any recovery support. The bottom line is that the negative impact on the overall multiscale (i.e., panarchic) system's resiliency, generated by maximizing the stability potential of any particular SIG, *increases* as a function of the proximity of the SIG's spatial–temporal extent to the most encompassing SIG of the panarchy (i.e., it becomes increasingly important to avoid potential collapse [maximized stability] as you approach the top-level SIG).

The challenge, confronting the resiliency growth manager, is to ensure that the groves occasionally burn while ensuring that the whole forest does not go up in smoke. And the

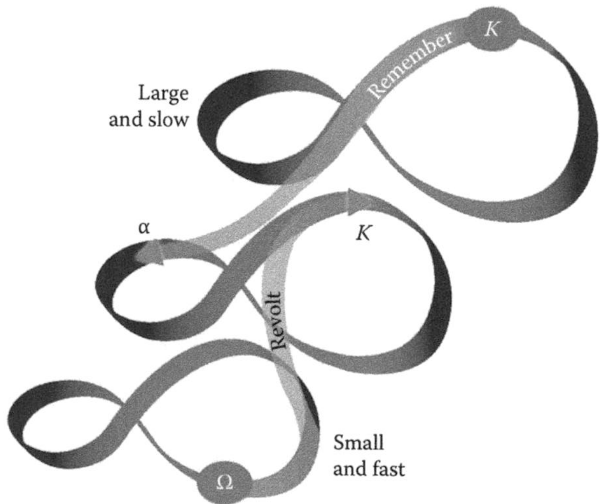

FIGURE 22.2
The panarchic resiliency model.

challenge, for the developer of a model-based tool supporting resiliency growth management, is to demonstrate compellingly how and where support operations should be conducted to yield this panarchic behavior.

Multiagent based complex adaptive system (CAS) models have been identified as offering the most promising foundation for capturing the context (including geographical) sensitivity, nonlinear behavior, and self-adaptive, self-organizing emergent behavior that characterize this social resiliency. The interested reader is referred to the National Research Council for an extended discussion on alternative social behavior modeling approaches and their relative strengths and weaknesses (Zacharias, MacMillan, and Hemel, 2008).

Megacity Use Case

Situation Summary

Xanadu, a megacity, is divided into wards. Organized crime elements exist in each ward and are linked together. The city is seething with tension and has become recognized as in danger of destabilizing. Discrimination against rural inhabitants (Rurals) as well as refugees is growing. Governance is secular with, at best, tolerance of any politically motivated clergy.

The Bursuka, a refugee group, are living in camps in the western wards. They are not well integrated with their neighbors. With a high unemployment rate, they are considered third class citizens. Operating largely as their own SIG, religious clerics are beginning to play a politically relevant state role among them.

The North Razinis, another refugee group, are living in the northern wards. They are also not well integrated with their neighbors and enjoy even less social status than the Bursuka. They are being actively and obviously exploited by organized crime elements. Many are living in Xanadu without formal documentation.

Çawaila, the state within which Xanadu resides, can offer little in the way of assistance to Xanadu. Overall, the urbanized population is growing. That growth offers both promise (in terms of building state legitimacy, social well-being, and a sustainable economy) as well as despair (in terms of increased social unrest, polarization, and loss of ecological sustainability).

The international community is at risk of growing wary of the unmanaged organized crime-based trafficking and the threat to stability. Economic investments may soon begin to dry up.

Potentially developing conditions that threaten to exacerbate the stability situation include increasing accessibility to the social media by refugee populations, a falling mean standard of living for everyone as the population pressures continue to mount, growing strength of the organized crime elements, and increased exploitation of refugee populations.

Substantive Intuitions on How the Situation Might Evolve

Substantive intuitions often serve to guide operational responses in such crisis situations. One of the key roles of modeling and simulation-based tools of social stability and resilience is to provide a check on those intuitions. Thus, we begin this discussion with an intuition of what may happen in Xanadu.

The prevailing tolerance of organized crime activity precipitates an unprecedented growth of the syndicate with a consequent direct and indirect net negative influence on local and state potentials. This growth, initially fueled by the easily exploited Razini and later spread to encompass the underemployed Bursuka, now threatens to drive the entire city into warring crime factions where no endogenously initiated meaningful collective activity is possible. Both Çawaila and the international community grow more anxious as they foresee a need to step in as a variety of self-interested SIGs begin crystalizing within Xanadu.

The mayor of Xanadu is aware of this mounting external pressure to resolve the problem. Foreign investment is beginning to dry up as investors, weary of future development risks, start taking a wait and see attitude. International organizations have begun hinting that foreign aid might be available to Çawaila/Xanadu if the various restrictions on how it is to be spent are accepted. Those demands are locally perceived as requiring a substantial loss in the autonomy for both Çawaila and Xanadu, and the mayor delays in taking action.

Meanwhile, Bursuka clerics begin demanding a more just distribution of resources, consistent with scripture. It is a move that is perceived as a governance threat by both Xanadu and Çawaila itself. Xanadu overreacts (in part due to the religious tone of the demands) losing further legitimacy and further alienating both the Bursuka and now the Razini.

Simultaneously, an increasingly discriminatory stance toward the *have nots* in general, fueled in part by the overall decline in the standard of living, triggers a demand for greater income equality among the Rurals. The resulting protests engender further economic hardships for the local government (already economically stressed) along with a further erosion of legitimacy. The lack of an effective response incites the Rurals into a dialogue with the Bursuka on a generalized antiestablishment Bursuka rebellion.

The Razini, a vulnerable population and not naturally inclined to rebel on their own, are collectively simmering from their exploitation at the hands of the crime syndicates and from their loss of dignity as the once proud citizens of a monarchy. Hints that they might be drawn into the Bursuka rebellion have been responded to with threats to enact a mass undocumented worker expulsion program.

Xanadu finally falls into a full-blown revolutionary (secessionist) violence, when the crime syndicates, out of what can best be characterized as frustration with the profit-taking constraints imposed by Xanadu, opportunistically fund rebels to protest. The ultimate result is a long-term descent into deprivation and violence that becomes a breeding ground for violent global system discontents.

Where an Operational Optimization Model Would Help

The role a fully operationalized stability, humanitarian, and reconstruction operations optimization model (SHROOM) would play in this analysis is to first confirm that the current trajectory is indeed as dire as the intuition suggests, i.e., will the failure spiral of the five end states continue? Or will some steady (albeit less than desirable) state be achieved? Often, as noted earlier, our intuitions fail to fully appreciate the human capacity to both accommodate to new situations and find innovative solutions within them, and consequently the situation may not be as dire as we imagine.

Consequently, the utilization of SHROOM begins by helping us to assess the existence or the plausibility of an emergence of one or more sets of politically relevant agents within Xanadu that have (1) assumed a superior stance with regard to the state, (2) begun resonating with the precursors to betrayal, and (3) showed the capacity to modulate their information flow so as to accommodate other vulnerable agents and engage them in an emergent insurgent identity.

Given that SHROOM confirms the dire substantively intuitive prognosis, the further role of SHROOM would be to support an exploration of (1) what kinds of external support would be both welcomed by Xanadu/Çawaila and would be most effective in preventing the collapse or, barring the identification of a viable solution of this type, (2) what strategy/operations would be most effective in the military occupation of Çawaila/Xanadu.

In this scenario, we discuss five potential strategic/operational responses that the operationalized SHROOM could help assess using the theoretically grounded insights provided by the social/biological sciences combined with the lesson-learned insights yielded from past operational experience.

In the analysis of each of these strategic responses, the fundamental use of SHROOM involves the search of an operational design space in which available resources are dynamically reallocated to alternative operations (categorized by their primary sector/end state development objective) assigned to various defined regions and at various scales. The objective function in this search being, as described earlier, not the direct influence on the growth of potential at the various scales but rather the emergent resiliency of the state.

Overt Crackdown on Organized Crime

We begin here by assuming that there is a demonstrable risk in cracking down too aggressively on the crime syndicates, lest an unmanageable popular backlash be triggered as a function of the syndicate's current role in providing public goods to some constituencies within Xanadu.

SHROOM could then be used to systematically explore how the optimal mix between cracking down on the crime syndicates and directly providing goods/services to these underserved constituencies should evolve optimally over time and then assess whether this strategy would be sufficient to stop the looming anarchy.

Education, Legal, and Economic Support for Razini Population

We begin here by assuming that the existence of an alienated and abused Razini refugee population plays a critical role in both establishing the local/global discontent that pressurizes the insurrection and that their participation was key to it becoming unmanageable.

SHROOM could then be used to explore systematically if an amnesty program, granting citizenship rights, could be combined with an integration aid package expeditiously enough such that the international organized crime concerns were addressed leaving investment revenues intact and the Razini willing to grant enough legitimacy to the state so as to resist joining the rebellion.

Deportation Program Applied to All Illegal Immigrants

We begin here by assuming that the rebellion would not occur if the very existence of the Razini in Xanadu was not a given.

SHROOM could then be used to systematically explore the consequences of deporting the Razini back to their home country. Foreign opinion backlash and possible additional loss of investment funds would be systematically evaluated against the potential benefits of having no vulnerable population for the syndicates to prey on and an immediate rise in the mean Xanadu standard of living as a consequence of no longer carrying the burden of supporting the Razini.

Acceptance of Foreign Aid and Foreign Workers to Help Administer That Aid

We begin here by assuming that the offered foreign aid, if accepted, would be sufficient to prevent the rebellion, barring any consequences not currently foreseen.

SHROOM could then be used to explore systematically how Xanadu's conceding to the conditions of foreign aid, with its contingent of foreign workers and administrators, would in fact influence the stability/resiliency of the Xanadu/Çawaila and if any unanticipated effects might undercut the value that those funds would generate.

Reintegration of the Bursuka

We begin here by assuming that the clerically led Bursuka protests and the state's subsequent overreaction to them are instrumental in fueling the unmanageable rebellion that follows.

SHROOM could then be used to explore systematically if an explicit program of power sharing could rectify the alienation and the governance delegitimacy experienced by the Bursuka, due to local governance weakness. SHROOM could assess whether, given the plausible timelines required for implementation, such a power sharing arrangement (especially given the Bursuka's current resistance to secular decision making) with its consequently more equitable/legitimate distribution of goods/service, would likely happen fast enough to address the growing sense of exploitation that is emerging collectively among the Bursuka.

SHROOM would not only allow the examination of all these scenarios, and in turn compare their success or failure against one another, but also allow the analyst/planner to compare timelines of success/failure. Finally, it would allow for combinations of these interventions to be examined, which would allow decision makers the ability to truly examine causally complex solutions that sociotechnical systems in trouble usually require.

DISCUSSION QUESTIONS

1. How does *resiliency* differ from *stability* and why are humans innately more inclined to pursue one over the other?
2. What kinds of systematic mistakes do humans make when deciding how to allocate resources.
3. Explain how the five USIP-defined end states of a high potential nation-state are developmentally linked.
4. Why does a focus on maximizing social stability sometimes work counter to the objective of creating social resiliency?
5. Within a multiscale living system, such as an SIG, where does maximized stability represent the greatest risk to resiliency?
6. Why do many theorists in this domain believe that multiagent complex adaptive systems modeling techniques offer the most promising approach to modeling social resiliency?

References

Army. (2014). FM 3-24 MCWP 3-33.5: Insurgencies and countering insurgencies: Counterinsurgency Washington, DC: US Army, 2006. Retrieved from http://scholar.google.com/scholar?hl=enan dbtnG=Searchandq=intitle:FM+3-24+MCWP+3-33.5#2.

Asch, S. E. (1951). The effects of group pressure upon the modification and distortion of judgments. *Group Leadership and Men: Research in Human Relations*, 177–190. Oxford, UK: Carnegie Press, ix, 293 pp.

Beitz, C. R. (2013). From practice to theory. *Constellations*, 20: 27–37. doi: 10.1111/cons.12019.

Berger, P. L., and Luckmann, T. (1966). *The Social Construction of Reality*. Harmondsworth (Original: Doubleday, Garden City, N.Y.): Penguin.

Bredenlöw, T. (2001). An invitation to social construction. Kenneth J. Gergen; Sage Publications, Thousands Oaks, CA, 1999, 248 pp. 17 (4), 519–522.

Bruneau, E. G., and Saxe, R. (2012). The power of being heard: The benefits of "perspective-giving" in the context of intergroup conflict. *Journal of Experimental Social Psychology*, 48(4), 855–866.

Coase, R. H. (1937). The nature of the firm. *Economica*, 4(16), 386–405.

Cole, B., and Hsu, E. (2009). *Guiding Principles for Stabilization and Reconstruction*. Washington, DC: United States Institute of Peace Press.

Davis, P. K., and O'Mahony, A. (2013). A computational model of public support for insurgency and terrorism. Retrieved from http://www.rand.org/pubs/technical_reports/TR1220.html.

Deffuant, G., Amblard, F., Weisbuch, G., and Faure, T. (2002). How can extremism prevail? A study based on the relative agreement interaction model. *Journal of Artificial Societies and Social Simulation*, 5(4), 1–26.

Diener, E., and Biswas-Diener, R. (2009). *Happiness: Unlocking the Mysteries of Psychological Wealth. Happiness: Unlocking the Mysteries of Psychological Wealth*. Hoboken, NJ: Blackwell Publishing Ltd.

Dunbar, R. I. M. (2010). Coevolution of neocortical size, group size and language in humans. *Behavioral and Brain Sciences*, 16(04), 681.

Durschmied, E. (2013). *Whisper of the Blade: Revolutions, Mayhem, Betrayal, Glory and Death* (Vol. 5). London: Hodder and Stoughton. Retrieved from https://books.google.com/books/about/Whisper_of_the_Blade.html?id=BvE5AgAAQBAJandpgis=1.

Fries, P. (2015). Rhythms for cognition: Communication through coherence. *Neuron*, 88(1), 220–235.

Gilboa, I., and Matsui, A. (1991). Notes and comments: Social stability and equilibrium. *Econometrica*, 59(3), 859–867.

Gottman, J., and Silver, N. (2012). What makes love last? How to build trust and avoid betrayal. Retrieved from http://www.barnesandnoble.com/w/what-makes-love-last-john-gottman/1111759392.

Granovetter, M. (2005). The impact of social structure on economic outcomes. *Journal of Economic Perspectives*, 19(1), 33–50.

Guimond, S. (2005). *Social Comparison and Social Psychology*. Cambridge: Cambridge University Press.

Gunderson, L. H., and Holling, C. S. (2002). *Panarchy: Understanding Transformations in Systems of Humans and Nature*. Washington, DC: Island.

Gurr, T. R. (1970). *Why Men Rebel*. Princeton, NJ: Princeton University Press.

Güth, W., Schmittberger, R., and Schwarze, B. (1982). An experimental analysis of ultimatum bargaining. *Journal of Economic Behavior and Organization*, 3(4), 367–388.

Hardin, G. (1968). The Tragedy of the commons. *Science*, 162(June), 1243–1248.

Hegselmann, R., and Krause, U. (2002). Opinion dynamics and bounded confidence. *Simulation*, 5(3), 2.

Kahneman, D., and Tversky, A. (1979). Prospect theory: An analysis of decision under risk. *Econometrica: Journal of the Econometric Society*, 47(2), 263–292. Retrieved from http://www.jstor.org/stable/1914185.

Kendall, D. (2011). Sociology in our times: The essentials. Retrieved from http://books.google.com/books?id=MtcnWRJ8nVwCandpgis=1.

King, A. J., and Cowlishaw, G. (2009). Leaders, followers and group decision making. *Communicative and Integrative Biology*, 2(2), 147–50. Retrieved from http://www.pubmedcentral.nih.gov/articlerender.fcgi?artid=2686370andtool=pmcentrezandrendertype=abstract.

Koltko-Rivera, M. E. (2006). Rediscovering the later version of Maslow's hierarchy of needs: Self-transcendence and opportunities for theory, research, and unification. *Review of General Psychology*, 10, 302–317.

Lefebvre, V. A. (2001). Algebra of conscience. Retrieved from http://www.amazon.co.uk/dp/0792371216.

Lieberman, M. D., Ochsner, K. N., Gilbert, D. T., and Schacter, D. L. (2001). Do amnesics exhibit cognitive dissonance reduction. *Psychological Science*, 12(2), 135–140. Retrieved from http://www.scn.ucla.edu/pdf/Amnesic.published.pdf.

Lustick, I. S., Miodownik, D., and Eidelson, R. J. (2004). Secessionism in multicultural states: Does sharing power prevent or encourage it? *American Political Science Review*, 98(2), 209–229.

Magni, C. A. (2009). Investment decisions, net present value and bounded rationality. *Quantitative Finance*, 9(8), 967–979.

McAdams, D. P. (2001). The psychology of life stories. *Review of General Psychology*, 5(2), 100–122.

Merton, R. K. (1938). Social structure and anomie. *American Sociological Review*, 3(5), 672.

Michelena, P., Jeanson, R., Deneubourg, J.-L., and Sibbald, A. M. (2010). Personality and collective decision making in foraging herbivores. *Proceedings. Biological Sciences/The Royal Society*, 277(1684), 1093–1099.

Miller, J. (1978). *Living Systems*. New York: McGraw-Hill.

Narvaez, D. (2010). The neuroscience of fair play: Why we (usually) follow the Golden Rule. *Journal of Moral Education*, 39(1), 113–116.

Nowak, M. A, Page, K. M., and Sigmund, K. (2000). Fairness versus reason in the ultimatum game. *Science (New York)*, 289(5485), 1773–1775.

O'Connell, A. N. (1980). Karen Horney: Theorist in psychoanalysis and feminine psychology. *Psychology of Women Quarterly*, 5, 81–93.

Ostrom, E. (1997). A behavioral approach to the rational choice theory of collective action: Presidential address, American Political Science Association. *American Political Science Association*, 92(1), 1–22. Retrieved from http://www.jstor.org/stable/2585925.

Pons, S. (2014). *The Global Revolution: A History of International Communism 1917–1991*. Oxford: Oxford University Press. Retrieved from https://books.google.com/books?id=Es9LBAAAQBAJandpgis=1.

Renn, D., and Diehl, P. F. (2015). *Pay Me Now or Pay Me Later PKSOI Papers*. Carlisle, PA: United States Army War College Press. Retrieved from http://www.joomag.com/magazine/pay-me-now-or-pay-me-later-pksoi-papers/0329991001445629588?short.

Rentfrow, P. J. (2010). Statewide differences in personality: Toward a psychological geography of the United States. *The American Psychologist*, 65(6), 548–558.

Serge, V. (2011). *Witness to the German Revolution*. Chicago, IL: Haymarket Books. Retrieved from https://books.google.com/books?id=a99G_WHaP0ICandpgis=1.

Shepperd, J. A, Carroll, P., Grace, J., and Terry, M. (2002). Exploring the causes of comparative optimism. *Psychologica Belgica*, 42(1–2), 65–98. Retrieved from http://search.ebscohost.com/login.aspx?direct=trueanddb=psyhandAN=2002-06881-007andsite=ehost-live\nshepperd@ufl.edu.

Smaldino, P., Pickett, C., Sherman, J., and Schank, J. (2012). An agent-based model of social identity dynamics. … *Societies and Social* …. Retrieved from http://jasss.soc.surrey.ac.uk/15/4/7.html.

Turner, J. C., and Reynolds, K. J. (2010). The story of social identity. Postmes, T. & Branscombe, N. (Eds). *Rediscovering Social Identity: Core Sources*. Psychology Press.

Turner, J. C., Oakes, P. J., Haslam, S. A., and McGarty, C. (1994). Self and collective: Cognition and social context. *Personality and Social Psychology Bulletin*, 20(5), 454–463.

Tversky, A., and Kahneman, D. (1981). The framing of decisions and the psychology of choice. *Science*, 211(4481), 453–458.

Zacharias, G. L., MacMillan, J., and Van Hemel, S. B. (2008). Behavioral modeling and simulation: From individuals to societies. National Research Council, National Academies Press.

23

Applying Modeling and Simulation to Foreign Policy: An Afghan Example

Corey Lofdahl

CONTENTS

Introduction 496
Reviewing the US Experience in Afghanistan 496
Governance Competition Model 498
Applying the Governance Competition Model to Iraq 504
Conclusion 508
References 509

ABSTRACT Foreign policy as implemented by the US military in the early twenty-first century has not achieved the results that initially motivated its interventions in other countries. Senior decision makers and their supporting analysts must account for the relevant policy factors and their complex interactions to improve their policy results. They must better assess the costs, the benefits, and the complex consequences of proposed policies and interventions. Modeling and simulation can aid foreign policy analysis by leveraging computers to help decision makers and their analysts to organize the myriad factors and understand the complex interactions associated with interventions in foreign social systems. Computer-based modeling and simulation accomplishes this by specifying, quantifying, and integrating causal relationships and available data into a unified analytic framework. The resulting simulation allows analysts to perform what-if scenario analyses that explore and characterize the short- and long-term consequences of potential policies and interventions. Examples are provided from the US experience in Afghanistan through a system dynamics simulation of local decision making with an emphasis on the network effects of proximity and compatibility associated with implementing democracy. Lessons learned from Afghanistan are then applied to the strategy formulation issues currently facing the US foreign policy community regarding ISIS.

KEY TOPICS

- The US experience was initially driven by the imposition of Western style democracy, which was eventually supplemented with the more locally oriented counterinsurgency program of the village stability operations (VSO) and the Afghan Local Police (ALP).
- The impact of VSO/ALP is then examined in terms of network effects, which comprise proximity and compatibility.

- A computer-based simulation is created that specifies and quantifies the network effects, which is then used to examine the history of the US intervention in Afghanistan.
- The computer simulation is then respecified to examine current US policy toward ISIS.

Introduction

The US foreign policy community is currently monitoring the expansion of ISIS and is considering its military options for engaging with it and reducing its growth (e.g., the study by Kenner and Jakes [2015], Rand and Heras [2015], and Wood [2015]). US interventions in Iraq and Afghanistan over the past decade, however, give ample reason for the foreign policy community to pause and reflect before committing to another military campaign because both of these interventions have not delivered the beneficial results that were originally envisioned. In the case of Iraq, the creation of ISIS itself can be traced to the unintended consequence from the US invasion in 2003 and the toppling of Iraq's leader, Saddam Hussein. For Afghanistan, over a decade of effort in what is now America's longest running war has not removed the Taliban from the country nor has it resulted in a stable and functioning democratic government (Economist 2014). This mismatch between envisioned results and actual outcomes is variously described as "unintended consequences," "counterintuitive behavior of social systems" (Forrester 1971), or "policy resistance" (Sterman 2000). The US DoD, understanding its core responsibility involves intervening in such social systems, sought to address and understand them through effects-based operations (EBO) (Deptula 2001). However, the EBO analytic enterprise did not ultimately prove successful in helping the military to achieve success in comprehending the complexities of their foreign interventions (Mattis 2008). Like EBO, the system dynamics (SD) simulation methodology was designed to analyze complex social systems including direct, indirect, and cascading consequences (Davis 2001), and it has been demonstrated to be useful in analyzing foreign interventions (Lofdahl 2010).

This study performs a retrospective analysis by applying SD to the case of the US intervention in Afghanistan, and it does so in three parts. First, the background behind the Afghan intervention is reviewed in terms of (1) the initial reasoning behind and the initial goals of the intervention, (2) what actually occurred as a result of the intervention, and (3) the response by the United States to unexpected events during the intervention. Second, an SD-based governance competition model (GCM) is developed that considers the cultural factors in an effort to explain the intervention dynamics developed in the first background section. Third, the GCM-based analytic structure created in the first two sections will be applied to the problem of strategy creation and intervention formulation as the United States considers its policy options regarding ISIS. The study concludes by identifying follow-on research steps based on the analysis performed herein.

Reviewing the US Experience in Afghanistan

With the Afghanistan war effort currently focused on concluding the extended intervention and bringing the international troops home, it is instructive to review the reasons why the United States went there in the first place. Afghanistan was invaded by the United States soon after September 11, 2001, because Al Qaeda used the country as a staging area for the passenger aircraft-based attacks on New York and Washington (Naylor 2005). The primary policy goals for the United States were twofold. First, the invasion was designed to prevent Afghanistan from being used as a base for more attacks against the United States, which required driving the Taliban out of power and out of the country (*Washington Post* 2001). Second, the United States additionally sought to remake, rebuild, and aid Afghanistan. Said President George W. Bush,

> At the same time (the US drives out the Taliban), the oppressed people of Afghanistan will know the generosity of America and our allies. As we strike military targets, we will also drop food, medicine, and supplies to the starving and suffering men and women and children of Afghanistan. (*Washington Post* 2001)

In the parlance of the American military, both kinetic security and nonkinetic governance and development COAs would be used to engage and shape Afghanistan during the intervention. Part of the nonkinetic policy plan included standing up a new, American-style democratic government to replace the ousted Taliban. Policy critics wondered if the Middle East would prove an appropriate and hospitable location to implement American-style democracy, but one of the key goals of the US intervention was to implement democracies in both in Iraq and Afghanistan. The intellectual foundation for this policy goal was articulated most clearly and influentially by Natan Sharansky who visited President George W. Bush in November 2004 (Milbank 2004; Stern 2005). Sharansky and Dermer (2004) provided intellectual support for Bush's policy of activist intervention in foreign affairs and criticized the notion that democracy is culturally contingent and potentially unsuitable for Middle Eastern countries. In other words, Middle East countries deserve the benefits of democracy just as much as the Western, developed countries. This foreign policy understanding of activist intervention in the Middle East to reduce the threat of terror through the military to establish democracies in the Middle Ease via aid caused President Bush to say "I felt like his (Sharansky's) book confirmed what I believe." (Sharansky and Dermer 2004).

Democratic elections were implemented by the United States, and Hamid Karzai was elected the first president of GIRoA in October 2004. The security for the election was very good as the Taliban had been routed and removed from Afghanistan in late 2001 (Naylor 2005). However, a democratically elected Afghan government backed by the militarily superior United States and the international coalition partners did not lead to a stable Afghanistan nor a Taliban that remained defeated over the long term. While US forces fought and implemented economic and governance improvements as directed by the newly formulated counterinsurgency (COIN) doctrine (US Army 2006), the Taliban, rather than contracting, instead expanded its influence as indicated by the growth of shadow governors (Flynn 2009). Taliban shadow governors act like standard, GIRoA governors, but in a parallel and competitive manner and, like their legitimate counterparts, are supported by force. In 2005, 11 of 34 Afghan provinces had shadow governors, but in 2009,

A democratically elected Afghan government backed by the militarily superior United States and the international coalition partners did not lead to a stable Afghanistan nor a Taliban that remained defeated over the long term.

that number has expanded to 33 with only the capital Kabul not having one. This expansion of political influence shows that the Taliban did not cease to exist after their defeat in 2001 but instead escaped to Pakistan, regrouped, reorganized, refitted, and then reentered Afghanistan (BBC 2011). While the expansion of Taliban shadow governance is only a single indicator, it is nevertheless significant and shows that a key indicator of the fledgling democratic GIRoA was moving in a counterintuitive, counterproductive, and problematic direction.

This Taliban expansion dynamic indicated that something fundamental about the US intervention in Afghanistan was going wrong and that the intervention itself might be based on faulty assumptions. Moreover, the briefing slide from which the Taliban expansion evidence was extracted is entitled "Time is Running Out" (Flynn 2009), which indicates that time itself was a critical factor in the US intervention. This and other analyses of the Afghan campaign's progress led to a reevaluation and fundamental redirection as the COIN strategy dictated less emphasis on purely military operations and more focus on the population. This redirection took the form of VSO/ALP that started in August 2010. VSO/ALP recommended that US forces move away from Afghanistan's center of political power in Kabul, leave their fortified and culturally insulating military bases, and instead interact with the Afghan locals in precisely those rural areas where the Taliban had been expanding (Saum-Manning 2012).

As the US intervention in Afghanistan winds down, multiple critiques and analyses have been published with two specifically mentioned here. First, the intelligence community has argued that policy makers exhibit a persistent tendency to overestimate their ability to change other cultures through international interventions: "American decision makers have shown a need for help in isolating and understanding the complexity, weight, and relevance of culture as they consider foreign policy initiatives" (Johnson and Berrett 2011, p. 1). Given the complexity of foreign cultures and their attendant social systems, it makes sense to represent and specify them in a manner that supports their analysis and improves the likelihood of policy success. Kilcullen (2010, p. 192) provides a primer on how to implement COIN campaigns and ends his study by stating the need for a systems model of insurgency, which recognizes the need for a more realistic, structured, and nuanced understanding of complex social systems and policies designed to influence them. For this study, computer-based simulation generally, and SD specifically, is used to order and analyze the complexity, the weight, and the relevance of culture for policy decision makers; an example of which is developed in the following section.

Governance Competition Model

The US policy in Afghanistan was initially driven by the priorities of senior US decision makers rather than those of the Afghans themselves. Implementing democracy based on a Western theory, regardless of the priorities or the history of Afghans, provides a clear example of policy overreach. Moreover, even if the Afghans did want democracy implemented in their country, it remains to be seen whether or not it is practical, workable, affordable, and likely to result in policy success. Given that the use of computer-based tools to help inform senior decision makers is even today extremely limited, it is usual for policy discussions to focus on only a small subset of factors known to be of interest to the decision makers. It is comparatively difficult to examine multiple, culturally informed, and interrelated decision factors from the perspective of the affected population even if such studies result in policies that have a higher likelihood of success. However, focusing on a country's local population is exactly what is recommended by the Army's COIN strategy (US Army 2006), which stresses protecting the population and treating it as the prize, the ultimate goal of the intervention, the campaign, the mission, or the operation.

While focusing on only a small number of decision criteria may help to convince policy makers, they may only have a tangential relationship to the concerns of the impacted population. Behavioral scientists have for decades recognized that multiple factors impact decision making including Arrow (1958), Kahnemann and Tversky (1979), Simon (1983), and Klein (1998), among others, who contributed to transitioning our understanding of decision making from spare, abstract, and artificial to an increasingly realistic, complex, and naturalistic process. With regard to this study, we seek to depict the decision making process of local Afghans who increasingly accepted the presence of the Taliban relative to GIRoA in their rural provinces and districts between 2005 and 2009. Two driving decision factors emerge from the American policy goals described in the section "Reviewing the US Experience in Afghanistan": (1) security, which indicates the ability of Afghans to travel safely and live their lives without threat of violence, and (2) democracy, which indicates governance quality based on the idea that governments selected by the people will be more answerable and attentive to their concerns and needs.

However, there are additional factors that bear on policy success. For this study, we concentrate on network effects that consist of presence and compatibility factors. *Presence* is an indicator of the physical proximity that is necessary to interact with a population and implement a policy. Regarding Afghanistan, the initial centralization of the US campaign when it built up GIRoA in Kabul worked against the goal of bringing good governance to all the people throughout the country regardless of whether the government was democratically elected or not. The Taliban, in contrast, lived among and interacted with the Afghans in the more rural, less densely populated, and poorer provinces and districts. Moreover, the Taliban also exhibited increased compatibility with the local Afghans as compared with the GIRoA representatives from Kabul. The VSO/ALP program was designed to address these shortcomings by training an ALP force comprised of locals and connecting them to the Kabul-based GIRoA government (Saum-Manning 2012). The Taliban, in contrast with Americans, was not considered by local Afghans to be a foreign occupying force even though they had spent time in Pakistan. The Taliban were ethnically and culturally Afghan and easily blended with the local population, even though they sometimes brutalized and extorted them, but such are the harsh realities of Afghanistan in the early twenty-first century.

Finally, there is matter of financing, which impacts the long-term stability and perceived reliability of Afghan governance. Recall that Afghanistan has suffered through more than three decades of war: 10 years with the Soviet Union, 10 years of civil war that resulted in the Taliban achieving power, and 10 years of the Americans and their international partners attempting to remove the Taliban. The Afghans who have survived the turmoil this long are finely attuned to the factors that impact their local security, and they carefully consider the information available to them before committing to a government or cause.

Network effects, and the proximity and compatibility factors they imply, provide a causal explanation behind the cultural inertia that makes changing a country hard.

GIRoA is supported by the Americans who come from very far away and are not likely to stay, so the Afghans reason that they do not have a permanent presence and may choose to return home and cut them loose. The Taliban, in contrast, are supported by Pakistan, Afghanistan's eastern neighbor, and so will be there forever. Moreover, the more than 300,000 Afghan security forces trained by the United States and their international partners are expensive to maintain and require large and continuing amounts of foreign aid (Chandrasekaran 2012). The Taliban, in contrast, are comparatively frugal to place in the field, which once again argues for their longevity and relative permanence.

From 2005 to 2009, the Taliban expanded throughout Afghanistan as the United States sought to support and establish a democratically elected government, which indicates a competition between the Taliban and the GIRoA. As we consider how this cultural competition might be modeled, we look for the dynamics in which purportedly superior products, concepts, or ideas end up losing to other, seemingly inferior alternatives. Network effects that embody the presence and the compatibility of cultural factors are quite separate and far removed from their theoretical attractiveness or hypothesized benefits (Sterman 2000, Section 10.8). Specifically, network effects take into consideration the impact of an installed base of products, concepts, or policies and help to identify the key features that impact its attractiveness. For Afghanistan, that installed base takes the form of the country's social and political culture, which is resistant to change. Network effects, and the proximity and compatibility factors they imply, provide a causal explanation behind the cultural inertia that makes changing a country hard (Johnson and Berrett 2011). Network effects also help to explain the difficulties associated with attempting to transform a culture too quickly or radically—for example, democratizing a Middle East country as proposed by President Bush (Sharansky and Dermer 2004). Figure 23.1 shows a GCM that articulates, specifies, integrates, and quantifies these network effects.

The weight and the relevance of cultural competition are specified through five competitive factors: (1) security, (2) democracy, (3) presence, (4) compatibility, and (5) funding. First, security is clearly a relevant factor as the government that is best able to provide a stable environment in which people can live and work will be deemed superior by its population. Second, democracy has already been discussed at length, but for this analysis, let us assume that it is indeed a superior form of government. Therefore, the government that is more democratic will be preferred by the Afghans. Third, the network factor of physical presence or proximity is considered. Barfield (2010) argues that the Afghan government has for decades been characterized by high degrees of centralization in Kabul, but it is also separate and insulated from the population. This is a different situation from the United States in which the impact of Washington, D.C., is felt throughout the country, and functional working relationships between the capital and the rest of the country are well established and assumed. This means that for the rural Afghan provinces, there is only a tenuous understanding of and relationship with GIRoA in Kabul. This also means that the Taliban shadow governors who establish themselves in the remote rural Afghan provinces can attain and exercise political power with relatively little interference from Kabul. Fourth, the related network factor of cultural compatibility indicates that the Taliban, who have usually been born and raised in Afghanistan, are often less foreign than those from Kabul, which is especially important in an intensely tribal country like Afghanistan. Fifth, long-term government funding for GIRoA and from Kabul for security remains uncertain, which directly contributes to a lack of confidence from the population. The Taliban insurgency—because they are local, lightly equipped, and comparatively efficient—can sustain themselves almost indefinitely, while the much more expensive GIRoA forces may exist only so long as the international funding does not run out.

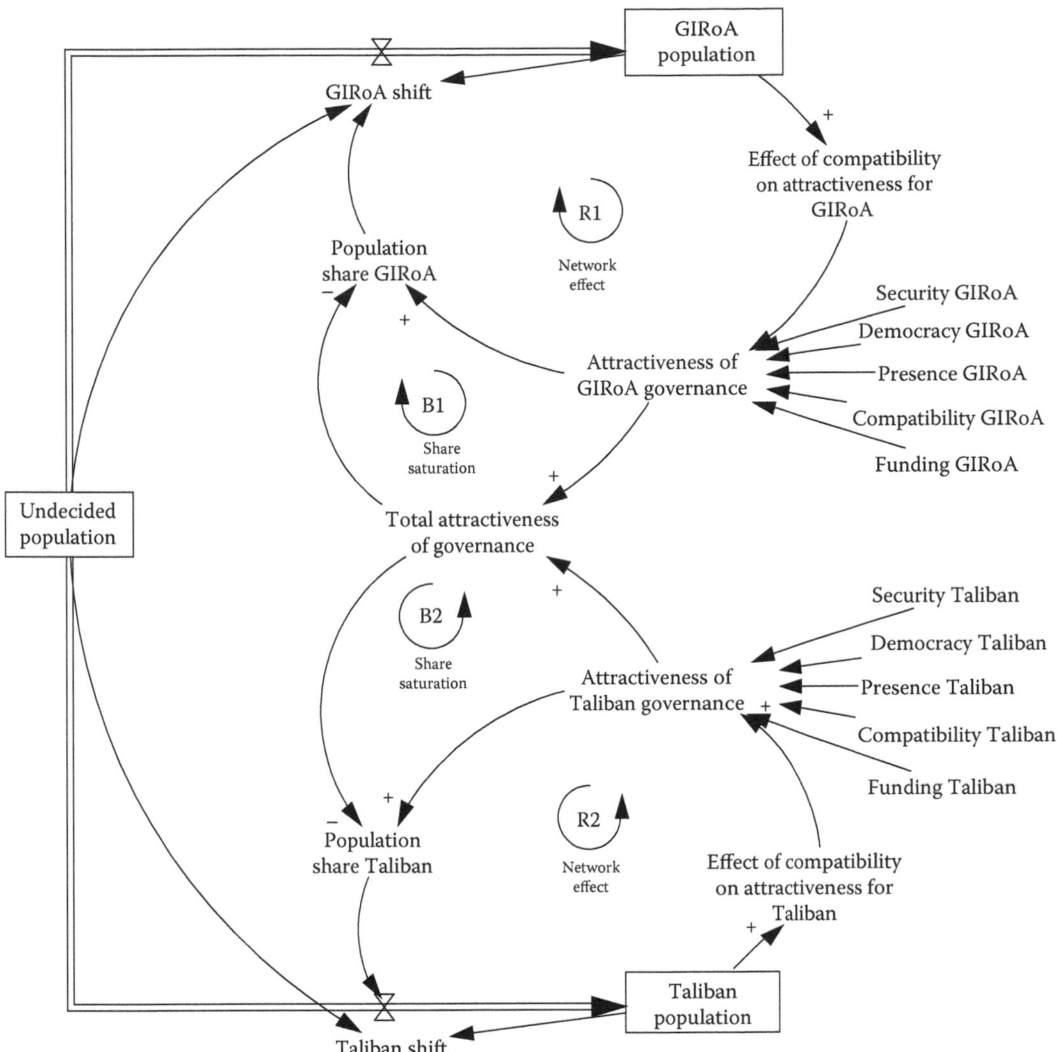

FIGURE 23.1
GCM for Afghanistan.

These five factors combine into an overall attractiveness score for both GIRoA and the Taliban that, in turn, impacts their respective share of popular support. The GCM's three-stock representation of population with individual representations for each of the two competing forms of governance and a third, undecided population has become a common computational feature of COIN simulation models (Lofdahl 2010). The influences on these three stocks form two competing reinforcing feedback loops, labeled R1 and R2 in Figure 23.1, that are driven by the network effects that, in turn, determine the relative attractiveness of Afghan governance. That is, the number of Afghan locals who collectively accept a form of governance increases based on the attractiveness of that form of governance and its compatibility with the local population. Phrased differently, the GCM model describes how the very popularity of a government makes it even more popular, hence the reinforcing, self-referential nature of the causal relationships.

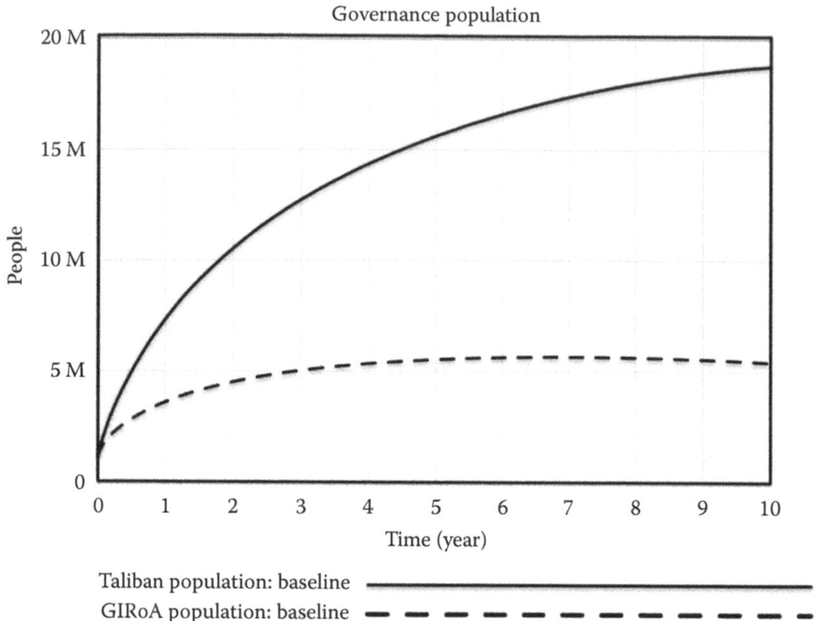

FIGURE 23.2
GCM dynamics from 2002 (year 0) to 2012 (year 10).

Figure 23.2 shows the dynamic consequences that derive from the GCM shown in Figure 23.1. If the five factors combine to make the Taliban more attractive than GIRoA—for example, if the network effects of presence and compatibility are more highly valued by the Afghans than democracy—then the Taliban will attract support from more of the country's population. This dynamic has spatial consequences as well because, as the populations in the rural areas feel disconnected from GIRoA, they will turn to the Taliban alternative as depicted temporally in Figure 23.2.

Experienced military commanders recognize that no campaign plan survives the first contact with the enemy, and the US experience in Afghanistan is no exception. US commanders recognized that the dynamics depicted in Figure 23.2 were moving in the Taliban's favor, so they undertook a corrective action in the form of COIN generally, and VSO/ALP specifically, which directly addressed the network effects of presence and compatibility. That is, they moved US forces out of Kabul and into the rural areas through VSO, which increased their presence, impact, and effectiveness. The coalition forces trained local Afghans in the rural areas to defend themselves through ALP, which increased the cultural compatibility with the Afghan locals. A sensitivity analysis of the resulting GCM model dynamics is provided with the implementation of VSO/ALP shown at two-year increments for populations that support the Taliban (Figure 23.3) and GIRoA (Figure 23.4). Figures 23.3 and 23.4 demonstrate the time dependence of campaign planning, that the time at which a policy action is undertaken can have an important impact on its overall effectiveness. For VSO/ALP, the earlier it is implemented, the more effective it will be. In actuality, VSO/ALP was implemented late in the campaign in 2010, approximately eight years after the invasion, which worked against the program's overall effectiveness.

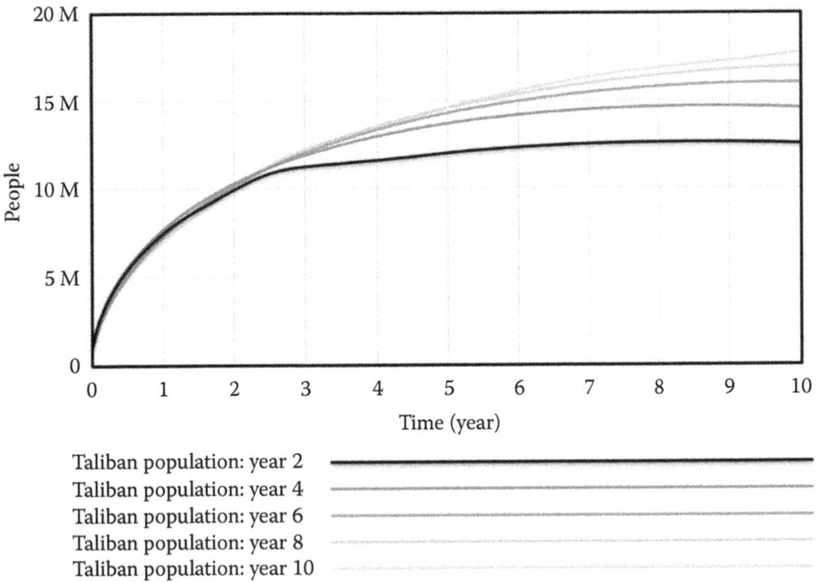

FIGURE 23.3
GCM-based sensitivity analysis of Taliban population for implementing VSO/ALP for 2004 (year 2), 2006 (year 4), 2008 (year 6), and 2010 (year 8).

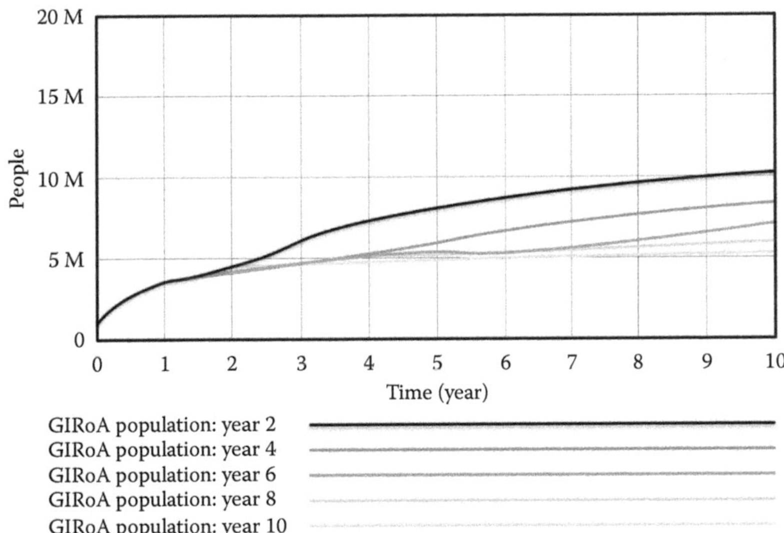

FIGURE 23.4
GCM-based sensitivity analysis of GIRoA population for implementing VSO/ALP in 2004 (year 2), 2006 (year 4), 2008 (year 6), and 2010 (year 8).

Applying the Governance Competition Model to Iraq

Iraq was the second of America's twenty-first century interventions that began in March 2003, mere months after the invasion of Afghanistan. While the initial invasion of Iraq was considered successful, ill-considered attempts at radical and rapid governmental changes—that is, De-Ba'athification—and ineffective attempts at governance after the fall of Saddam Hussein gave rise to an insurgency (Frontline 2014a). American attempts at establishing a democratic government that included both Shia and Sunnis supported by an army not driven by sectarian competition and infighting have proven unsuccessful. The demonstrated inability of Shias and Sunnis to share power in Iraq has significantly contributed to the rise of ISIS, also called the Islamic State in the Levant (Frontline 2014b). In the fall of 2014, President Obama announced somewhat controversially that the United States did not have a strategy to counter the rise of ISIS (Carter, Shoichet, and Alkhshali 2014). While President Obama and DC-style politics were justifiably criticized for this absence of strategy, it should be noted that deriving a successful strategy for addressing a policy problem as difficult as ISIS remains an open question. The GCM developed in this study will be redirected to structure an analysis to provide an initial foundation for an ISIS strategy that will increase the probability of policy success as compared with more circumscribed, traditional, and nonmodel-based approaches that have been demonstrated to be problematic.

The ISIS portion of the study begins by outlining the basic parameters of America's potential strategy. First, the United States no longer has the luxury of sending 100,000 troops into theatre again as there is no longer the budget nor the political appetite for doing so on the part of the president nor the American people. Nevertheless, doing nothing is not an option as indicated by the significant territorial gains made by ISIS, especially in the Sunni-populated regions of eastern Syria and western Iraq, including Mosul (Chulov 2014). This expansion is especially worrisome as Al-Qaeda was able to carry out the 9/11 attacks without the resources or the territory currently controlled by ISIS. As part of their propaganda campaign, ISIS has published a map showing the proposed, desired, or projected expansion of ISIS that conceptually corresponds to the expansion of the Taliban in Afghanistan (Hall 2014). While it is unlikely that ISIS will expand to the extent portrayed, it is worth considering that American policy makers would have quickly dismissed the possibility of Taliban expansion that actually occurred after 2004. This means that the matter of possible ISIS expansion should be taken seriously and actively countered to ensure that it does not, in fact, occur. The question then becomes how can the United States implement a strategy that effectively impacts and impedes ISIS, but does so at an acceptable cost?

Figure 23.5 shows the GCM modified to support the analysis of ISIS with the fundamental competition analyzed between the Federal Government of Iraq (FGoI) and ISIS. Moreover, both FGoI and ISIS receive external support. For FGoI, external support comes from geographically proximate Shia countries, primarily Iran, and controversially, the United States (Youssef 2015) because Iran and the United States have traditionally been enemies. For ISIS, the initial support came from other Sunni countries, including Saudi Arabia and Qatar, and Iraqis living in Jordan (Windrem 2014; Frontline 2014b). However, the fundamental drivers underlying this model are the local decision makers in the Sunni areas of Iraq who must decide whether to place their allegiance in FGoI or ISIS. The factors that drive their individual decisions are specified, quantified, and compared in this section with the goal of creating an analytic foundation that can support a more locally informed and effective American strategy for countering ISIS expansion.

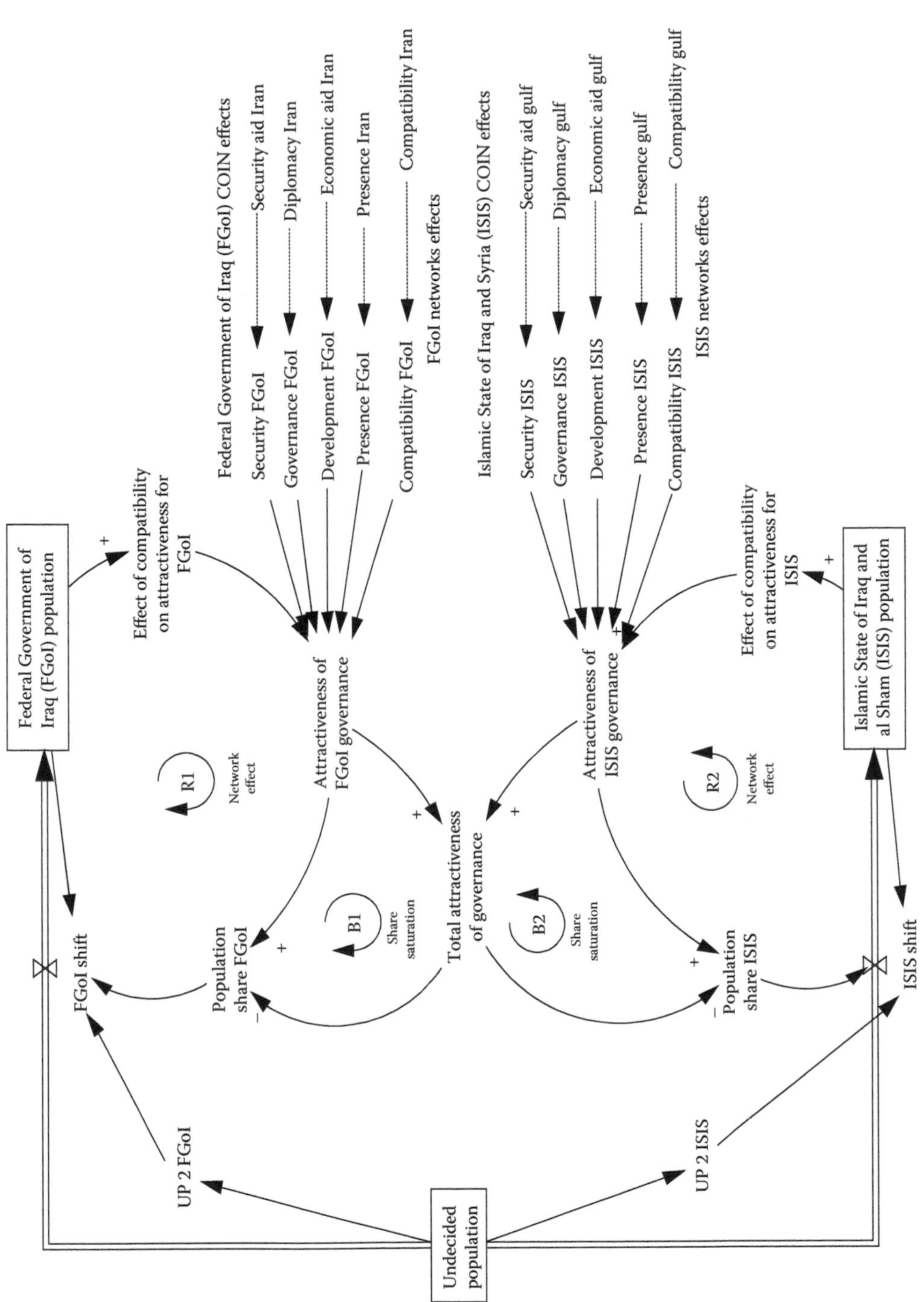

FIGURE 23.5
GCM modified for ISIS.

First, there exists a security competition between FGoI and ISIS, with the latter demonstrating a convincing operational advantage over the former, especially in the traditional Sunni heartland of western Iraq. During the summer of 2014, ISIS took over surprising amounts of Iraq's traditional Sunni areas causing much concern in the international community (Chulov 2014). The Iraqi army proved no match for ISIS in battle and frequently fled before it. The Iraqi army members were poorly led in battle as many of the most capable Sunni officers had been fired by the FGoI and replaced with lesser trained Shia political appointees. The Shia members of the Iraqi army who were captured by ISIS were frequently killed. Moreover, the stated goal of ISIS to take over much of the Middle East and North Africa could no longer be comfortably ignored, especially as their presence on social media was just as potent as on the battlefield (Hall 2014). ISIS expanded so quickly and widely that they were soon found on the outskirts of Baghdad, and Shia reinforcements were called in from Iran to help protect Iraq's capital (Youssef 2015). How this unlikely coalition of Iraq, Iran, and the United States will play out in confronting and countering ISIS remains unclear, but ISIS appears to be a significant threat for the foreseeable future.

Second, comparing the commitment to democracy between FGoI and ISIS provides a clear and stark contrast. FGoI is a democratically elected government that was created explicitly by and in the image of the United States, while ISIS explicitly rejects elections and democracy (Wood 2015). In reviewing the impact of democracy specifically and governance generally, we should recall the original US policy context in which democracy was the centerpiece, which is to say that democracy was implemented with too much optimism due to its perceived ability to remake, improve, and pacify countries. After a decade of war in Afghanistan and Iraq, as well as the Arab Spring countries, this has not proven to be the case: democracy has not remade these countries, and this failure was not due to lack of effort or insufficient resources. Specifically, democracy was viewed positively by US policy makers because of its perceived ability to calm the sectarianism that has, from the American perspective, bedeviled the Middle East. In Afghanistan, democratic elections did not prevent the return and the spread of the Taliban, and in Iraq, democratic elections did not cause Sunnis and Shia to put aside their past differences and work together in a cooperative fashion to build a better Iraq. In fact, Iraq's then president, the Shiite Nouri al-Maliki, used his leadership position to persecute Sunnis in 2012 and 2013, which helped fuel the rise of ISIS (Frontline 2014b). The explicit rejection of democracy by Iraq's Sunni population provides further evidence that democratic governance proves problematic in the Middle East. Democracy will ultimately be evaluated by its utility in an intergroup competition context—that is, by its ability to aid group survival—as depicted by the GCM (Wilson and Wilson 2008) rather than the enthusiasm with which it is implemented.

Third, the ongoing conflict or competition between Shia-dominated FGoI and Sunni-dominated ISIS has spatial consequences as ISIS expands throughout western Iraq and eastern Syria. However, these macrolevel regional effects are motivated and driven by the lower-level network effects of physical presence and cultural compatibility, which are, in turn, determined by interpersonal relationships and individual decisions. That is, the individuals in Iraq make their allegiance decisions based on which organization can best provide them the necessities of life, starting with security. Many Sunnis have decided to reject FGoI and support ISIS. These decisions are certainly subject to variation and are not unanimous, but the practical effect of this aggregate political dynamic is that the Shia and the Sunni population in Iraq, which were formerly mixed, are geographically separating. FGoI generally and Maliki specifically purged his government of Sunnis and attacked Sunni areas. ISIS attacked the Iraqi army in response; it now controls large areas of Iraq, and it has killed whatever Shia it could capture. From a network effects perspective, the shared presence between Shias and Sunnis is decreasing as FGoI and ISIS align and consolidate themselves in their respective regions of control.

Fourth, the network effects of culture operationalize themselves as compatibility. When policy makers seek to influence a region, how compatible are the proposed interventions with what the local decision makers already know? The primary policy proposed by the United States, implementing democracy, has not proven empirically compatible with Middle East decision makers insofar as it was envisioned to ease tensions among groups in conflict and lead to improved governance, which has not proven the case in either Afghanistan or Iraq. Local decision makers instead make their allegiance choices based on regional tribal membership and relationships rather than international notions and norms of democratically elected governments. So extreme is this tribal separation that currently, the areas controlled by ISIS are virtual no-go areas for anybody who is not recognized as a member of the Sunni community, including journalists (Harkin 2014). It is possible that a program like VSO/ALP (Saum-Manning 2012) that recognizes and leverages the strengths of local relationships and tribal culture could be implemented in Iraq to build trust between the Shia and Sunni communities, but it is not clear that the United States is willing to commit the time and the resources necessary to make this happen. Moreover, it would have been more effective to have supported such institutions while the Sunnis and the Shias were living and working together rather than attempt to reunite them after they have violently split themselves apart.

Fifth, local decision makers choose the organizations to which they give their allegiance—be it Shia-dominated FGoI or Sunni-dominated ISIS—and this consideration is informed by the perceived longevity and stability of the organization, which is driven, in turn, by its ability to generate revenues and garner support. ISIS initially obtained financial help from Sunnis in other Gulf states, such as Saudi Arabia, Qatar, and Jordan, when they were attacked by Maliki's Iraqi army (Windrem 2014). Now that ISIS controls a significant territory (Chulov 2014), it is able to generate its own revenue streams. Just how significant, stable, and varied these sources of revenue are remain matters for further study. So too, the FGoI also depends on both internal support and external aid for its revenues. For their security, FGoI depends on both the United States for weapons and air support and Iran for fighters to supplement their army as they contend with ISIS. Note that the traditional enmity between the United States and Iran leads to some fractious diplomacy and unusual coalitions (Youssef 2015).

As the United States looks to create a strategy to address, confront, and eventually defeat ISIS, it must realistically assess the situation on the ground from the perspective of local decision makers. The strategy must also take into consideration the fact that the United States will be both unwilling and unable to commit significant numbers of troops in the foreseeable future. However, the United States does have options that center on the use of special operations forces (Robinson 2013) and information operations (Joint Chiefs of Staff 2014) for strategic effect, which supplements and extends the US military's traditional air superiority. With smaller and more subtle levels of effort centered on supporting and training local forces to defend themselves rather than have the United States do it for them—the traditional mission of the US Army Special Forces—the time frame of such operations will need to be extended so that they can achieve strategic impacts over many years rather than execute quick operations such as changing regimes. Moreover, the Russians have successfully supplemented their special operations forces with information operations that influence whole populations as demonstrated in Ukraine (Darczewska 2014). If the United States expects to compete on the world stage with the likes of ISIS and Russia, then it too will have to incorporate information operations with its special operations.

Conclusion

This study has explored the American experience in Afghanistan in terms of the policy goal of implementing democracy as well as the network effects of presence and compatibility. These network effects have been placed within a preference structure that emphasizes the local decision makers rather than the political pressures and desires of US policy makers in Washington, D.C. These local preferences were placed within a GCM that accounted for the macrolevel relationships that drive policy outcomes, such as the number of people who adopt a mode of governance or exhibit certain cultural behaviors, which change individual incentives by making that type of governance or those behaviors more attractive. This model-based study also helps to specify and reveal the underlying factors that explain why societies are not so easily shaped and remade as policy makers might think (Johnson and Berrett 2011). Based on the US experiences in Afghanistan and Iraq, this study counsels caution, carefulness, and circumspection when intervening in and crafting policies for interventions in other countries.

Walt (2015) similarly counsels caution by arguing first that the policy community lacks a detailed and sophisticated understanding of societies whose histories and cultures are significantly different from the United States. In this study, we have addressed this cultural shortcoming by specifying a preference structure that captures the particularities of culture without making claims of spanning all its detailed complexity. Second, Walt (2015) notes that the military is an imprecise instrument with which to implement sensitive cultural interventions and one that, due to the inherent complexity associated with culture, is likely to lead to unintended consequences (Forrester 1971). For this study, the American policy goal of transferring and implementing democracy as defined and recommended by Sharansky and Dermer (2004) was pursued by President Bush (Stern 2005) with the best of intentions, but his interventions did not produce the security and stability outcomes originally envisioned. The frequency and seriousness of unintended policy consequences argues for the increased utilization of special operations forces as they are more highly trained and thus able to operate in the full-spectrum COIN operations that span security, governance, and development (Robinson 2013). Additionally, as special operations forces learn to operate more strategically over longer periods, there will be more opportunity to readjust and correct campaigns when they deviate from expectations.

In conclusion, the systems-based foundation provided by this study can be further developed in two significant ways. First, the preference structures provided for Afghanistan (Figure 23.1) and Iraq (Figure 23.5) are not fully developed because the relevant decision factors remain independent of each other. There are surely dependencies among security, democracy, presence, compatibility, and funding that are not currently represented by the GCM and that merit further study. The prose descriptions provided in this study give the first indications of possible relationships among these factors and may possibly indicate additional decision factors for inclusion in future iterations of the GCM. Second, the details of the network effects are obscured by the aggregation inherent in SD simulation. While the macrolevel dynamics are captured by the GCM, it makes sense to explicitly model the actual networks as described by Rahmandad and Sterman (2008). Developing the network aspects of the GCM would also allow space to be explicitly represented by a geographic information system (Lofdahl 2002). Representing the networks and displaying the model's results in geographic context would also allow for the spatial expansion processes (O'Loughlin et al 1998) at the heart of Taliban and ISIS growth to be explicitly modeled and examined.

References

Arrow, Kenneth. 1958. Utilities, attitudes, choices: A review note. *Econometrica* 26(1): 1–23.

Barfield, Thomas. 2010. *Afghanistan: A Cultural and Political History*. Princeton, NJ: Princeton.

BBC. 2011, October 26. *Secret Pakistan: Double Cross*. London: British Broadcasting Corporation (BBC) video documentary.

Carter, Chelsea J., Catherine E. Shoichet, and Hamdi Alkhshali. 2014, September 4. Obama on ISIS in Syria: "We don't have a strategy yet." *CNN, U.S. Online Edition*.

Chandrasekaran, Rajiv. 2012, October 20. Afghan ecurity force's rapid expansion comes at a cost as readiness lags. *Washington Post*.

Chulov, Martin. 2014, June 10. ISIS insurgents seize control of Iraqi city of Mosul. *The Guardian*.

Darczewska, Jolanta. 2014. The anatomy of Russian information warfare: The Crimean operation: A case study. *Point of View 42*. Warsaw: Center for Eastern Studies (OSW).

Davis, Paul K. 2001. *Effects-Based Operations (EBO): A Grand Challenge for the Analytical Community*. Santa Monica, CA: RAND Corporation.

Deptula, David A. 2001. *Effects Based Operations: Change in the Nature of Warfare*. Arlington, VA: Aerospace Education Foundation.

Economist. 2014, September 22. Divide and rule: Afghanistan's disputed election. *The Economist*.

Flynn, Michael. 2009, December 22. *State of the Insurgency, Trends, Intentions and Objectives*. Director of Intelligence, International Security Assistance Force (ISAF), US Forces, Kabul, Afghanistan.

Forrester, Jay W. 1971, January. Counterintuitive behavior of social systems. *Technology Review* 73.

Frontline. 2014a, July 29. Losing Iraq. Frontline video documentary. *WGBH Boston*. Boston: WGBH Boston.

Frontline. 2014b, October 28. The rise of ISIS. Frontline video documentary. *WGBH Boston featuring Martin Smith*. Boston: WGBH Boston.

Hall, John. 2014, July 1. The ISIS map of the world: Militants outline chilling five-year plan for global domination as they declare formation of caliphate and change their name to the Islamic State. *Daily Mail*.

Harkin, James. 2014, May. Evaporated. *Vanity Fair*.

Joint Chiefs of Staff. 2014, November 20. *Information Operations*. Joint Publication (JP) 3-13 (Change 1). Washington, DC: Joint Chiefs of Staff (JCOS).

Johnson, Jeannie L. and Matthew T. Berrett. 2011, June. Cultural topography: A new research tool for intelligence analysis. *Studies in Intelligence* 55(2): 2.

Kahneman, Daniel and Amos Tversky. 1979. Prospect theory: An analysis of decision under risk. *Econometrica* 47(2): 263.

Kenner, David and Lara Jakes. 2015, February 13. The Islamic state inches toward a showdown with the U.S. Marine Corps. *Foreign Policy*.

Kilcullen, David. 2010. *Counterinsurgency*. New York: Oxford.

Klein, Gary. 1998. *Sources of Power: How People Make Decisions*. Cambridge, MA: MIT Press.

Lofdahl, Corey. 2002. *Environmental Impacts of Globalization and Trade: A Systems Study*. Cambridge, MA: MIT Press.

Lofdahl, Corey. 2010. Governance and society. In Kott, Alexander and Gary Citrenbaum eds., *Estimating Impact: A Handbook of Computational Methods and Models for Anticipating Economic, Social, Political, and Security Effects in International Interventions*. New York: Springer-Verlag.

Mattis, James N. 2008, Autumn. USJFCOM commander's guidance for effects-based operations. *Parameters*.

Milbank, Dana. 2004, November 23. An Israeli hawk accepts the President's invitation. *Washington Post*.

Naylor, Sean. 2005. *Not a Good Day to Die: The Untold Story of Operation Anaconda*. New York: Penguin.

O'Loughlin, John, Michael D. Ward, Corey L. Lofdahl, Jordin S. Cohen, David S. Brown, David Reilly, Kristian S. Gleditsch, and Michael Shin. 1998. The diffusion of democracy, 1946–1994. *Annals of the American Association of Geographers* 88(4): 545–574.

Rahmandad, Hazir and John D. Sterman. 2008, May. Heterogeneity and network structure in the dynamics of diffusion: Comparing agent-based and differential equation models. *Management Science* 54(5): 998–1014.

Rand, Dafna H. and Nicholas A. Heras. 2015, March 16. Iraq's Sunni reawakening: How to defeat ISIS and save the country. *Foreign Affairs*.

Robinson, Linda. 2013, April. The future of U.S. Special Operations Forces. Council Special Report No. 66. New York: Council on Foreign Relations.

Saum-Manning, Lisa. 2012, December. VSO/ALP: Comparing past and current challenges to Afghan local defense. *Small Wars Journal*.

Sharansky, Natan and Ron Dermer. 2004. *The Case for Democracy: The Power of Freedom to Overcome Tyranny and Terror*. New York: Public Affairs Press.

Simon, Herbert A. 1983. *Reason in Human Affairs*. Stanford, CA: Stanford University Press.

Sterman, John D. 2000. *Business Dynamics: Systems Thinking and Modeling for a Complex World*. Boston: McGraw-Hill.

Stern, Yoav. 2005, February 22. Want to know what Bush thinks? Read Sharansky. *Haaretz*.

US Army. 2006, December. *Counterinsurgency* (FM3-24). Field Manual. Washington, DC: Headquarters, Department of the Army.

Walt, Stephen M. 2015, January 23. If only we'd just spent more blood and treasure in Yemen: Five lessons from misadventures in the Middle East that Washington just can't seem to learn. *Foreign Policy*.

Washington Post. 2001, October 7. Bush announces strikes against Taliban. *Washington Post*.

Wilson, David Sloan and Edward O. Wilson. 2008, September–October. Evolution "For the Good of the Group." *American Scientist* 96: 380–389.

Windrem, Robert. 2014, September 21. Who's funding ISIS? Wealthy Gulf "angel investors," officials say. *NBC News*.

Wood, Graeme. 2015, March. What ISIS really wants. *The Atlantic*.

Youssef, Nancy A. 2015, March 25. U.S. backs Iran with airstrikes against ISIS. *The Daily Beast*.

Subject Index

Page numbers followed by f, t, and n indicate figures, tables, and footnotes, respectively.

A

Ability, defined, 401
ABMs, *see* Agent-based models (ABMs)
Academic achievement tests, 89
Access
  data, to disconnected communities, 154, 156t
  data ownership and, 154
  limited, data collection and, 111
Accidents
  cognitive biases and, 104–106, 105f
  cultural differences role in (case studies)
    Korean Air Flight 801 disaster, 106–107
    Union Carbide's Bhopal chemical plant accident, 107
  cultural differences role in (case study), 106–107
Adaptive cycle process, social resiliency, 487–488, 487f
Advanced Progressive Matrices (APM), 89
Adverse events
  descriptive statistics by district, 251–252, 251t, 252f
  GIS-based analysis and visualization, 252–254, 253t–254t
  mapping, *vs.* population density, 254–255
AF1, 116–117, 121, 122, 128; *see also* High-risk ethnography (HRE); Sixty Percent Mission (case study)
Afghanistan
  GCM for, 498–503, 501f, 502f
  population survey (case study); *see also* Population surveys
    factor analysis (results), 231–232
    history, 230–231
    indicators deriving from factors, 232–233, 232t, 240t–242t
  regions of, 248, 249f
  unstable sociopolitical environment in (case study), 146–147
  US experience in, 496–498
  US interventions in, 496, 498
  war in (2004–2010), geographical modeling, 243, 246–247
    researching design and methodology, 247–251
    results and discussion, 251–259
Afghan Security Forces Fund, 245
Age group differences, VEOs and, 441, 443f, 444t
Agent, defined, 477
Agent-based models (ABMs)
  computational modeling, 450–451
  populations, 19–20
Aggregation
  defined, 298
  uncertainty, 298
Al-Assad, Bashar, 191, 192
Aleatory (statistical) uncertainty, 293
Algorithmic mind, 87
Al Qaeda, 29, 37, 197, 497
Althusser, Louis, 36
Ambiguity, 168
American College Test (ACT), 89
American Revolution, 35, 36
Anomaly detection
  entropy-based, 214–217
    drill down of anomalous/suspicious behavior, 216–217, 216f–217f
    information measures detection in time using KL divergence, 215–216
  probabilistic semantics, 214–217
Anxiety, 168
APM, *see* Advanced Progressive Matrices (APM)
Application program interface (API), 49
Arabic discourse, in-group/out-group factors in, 191t
Arabic script, hashtags in, 50
Arab Spring of 2011, 63
ArcMap 10.2 software, ESRI, 255
Aristotle, 32
Army and Marine Corps, 270
Assertional knowledge (ABox), 208, 211
Assortative mixing
  ERGMs, 71–72, 72f
  opinion spread modeling and, 71–74, 73t

Athena simulation
challenge for, 21
expeditionary modeling, 20–21; *see also* Expeditionary modeling (ExMo)
Attention-direction devices, 190
Attitude, 401n
Audiences attributes
behavioral habits, 48
consumption habits, 48
social media messaging and, 48
Augmentation, of social media-based dissent, 46–47
Authority, 190

B

BAM, *see* Binary agreement model (BAM)
Bayesian fusion, uncertainty reduction using, 303–304, 304f
megacity use case (fictional), 309–312, 310f–311f, 310t–311t
Bayesian networks (BNs) leverage probabilistic measures, 293
Bayes rule, 303–304
Beer for Data, 146
Behavioral habits, of audiences
social media messaging and, 48
Behavioral Influence Assessment program (Sandia National Laboratories), 452, 452n
Behavior science
cross-cultural research in, 164–165; *see also* Non-WEIRD societies; WEIRD societies
Belief function, 294
Belief mass function, 293–294
Benevolence, 402
Betrayal, shared perceptions of, 481–483
Biases
cognitive, 479
accidents and, 104–106, 105f
case studies, 106–107
decision making and, 103–108
loss aversion, 105, 106
overview, 103–104
power distance, 106–107
cognitive reflection and, 91–92
SME, 454
social, 479
Bidirectional causality, 319
CCM for, 322–323
Big data, 205, 206, 267
Big data tools, 16
Binary agreement model (BAM), 62; *see also* Naming game-based (NG) model
Black Swan, 362
Blogs, 48
Boko Haram, 12, 181
violence by, forecasting, 198–199, 198f–199f
Boston Marathon bombings (2013)
social media use during, 44
Boston Tea Party, 35, 36, 37, 38
Bottom-up processing, narrative comprehension, 33–34
Bounded awareness, 104
Bounded rationality; *see also* Cognitive biases
overview, 103–104
Boyle, Susan, 50
Brown, Brooks, 182
Buddy Media, 48
Bush, George H. W., 38
Bush, George W., 38, 497
Bystander responsibility
data collection from disconnected communities and, 152, 156t

C

Call data records (CDRs), in NetSense data, 71, 77
Capability-based assessment (CBA), 270
Capacity and resiliency, relation between, 487
Case-based reasoning (CBR) algorithm, 335, 336
applied exercise, 339
architecture and cycle, 337–339, 337f–338f
described, 338
megacity use case (fictional), 355–359
and repository of past cases, 349–354, 351f–353f
StateSim model; *see also* StateSim model
background, 339, 340f
generator elicitation screens, 340–349
overview, 338–339, 338f
Causality, 315, 316; *see also* Causal/predictive relationships
Causal/predictive relationships, 315–333
challenges of complex patterns, 318–319
bidirectional causality, 319
qualitative features and effects, 319
temporal offsets, 318–319
convergent cross mapping (CCM), for bidirectional causality, 322–323
dynamic time warping (DTW), for uneven temporal relationships, 321–322
Granger causality, 321

megacities use case (fictional), 332–333
model validation of poverty and violent conflict (case study), 325–331, 326f, 328f–330f
overview, 315–318
qualitative feature-based analysis, 323–325, 324f
validation methodology, 319–320
Cause-and-effect relationship, 39
CBA, *see* Capability-based assessment (CBA)
CBR algorithm, *see* Case-based reasoning (CBR) algorithm
CCM, *see* Convergent cross mapping (CCM)
Center for Disease Control, 44
Cheney, Dick, 38
Chi-square test, 412, 429, 430, 434, 435, 439
Choices
values and, 38
CIA Fact Book, 222
CIs, *see* Confidence intervals (CIs)
City; *see also* Megacity(ies); Urban areas
described, 12–13
Civil Information Management Data Processing System (CIM-DPS), 128
Class background
opinion spread modeling and, 74–75, 74t
Classical logic, 214
Classical set theory
uncertainty modeling and, 293
Cluster sampling, 426
Coca-Cola, 48
Coevolution
of models, data, and operating environment/mission, 11, 11f
network/opinion, 76–80, 78f–81f
Coevolutionary ethnogenesis, of social identity, 481–485; *see also* Social identity groups (SIGs)
shared intentions to create better situation, 485
shared perceptions of threat and betrayal, 481–483
shared thoughts of values and narratives, 483–484
Cognitive biases, 479
accidents and, 104–106, 105f
case studies
aviation, 106–107
Korean Air Flight 801 disaster, 106–107
Union Carbide's Bhopal chemical plant accident, 107
decision making and, 103–108
loss aversion, 105, 106
overview, 103–104
power distance, 106–107
Cognitively detailed models, 336n
Cognitive narratology, 34
Cognitive perceptions
hybrid computational models and, 457
Cognitive reflection; *see also* Cognitive reflection test (CRT)
and biases, 91–92
as cognitive style, 90
cultural differences in, 92–97, 93t–94t, 95f–96f
culture and, 85–97
dual process theories, 85–86
type 1 process, 86–89
type 2 process, 86–89
intelligence and, 89–92
measure of, 88–89
religious beliefs and, 90–91
and resistance to irrational beliefs, 90–91
Cognitive reflection test (CRT)
CIs for, 94, 95f–96f
cognitive biases and, 91–92
cognitive styles and, 90
descriptive statistics, 92, 93t–94t
intelligence tests and, 89
and intuitive thinking measures, 90
meta-analysis, 92, 94–97, 95f–96f
overview, 85, 88
problems, 88–89
purpose of, 88
Cognitive schema
narrative structural forms as, 32–33
Cognitive styles, 87
cognitive reflection as, 90
CRT and, 90
indicator, 90, 90n
measure of, 88
modes of processing, 86
COIN strategy, 497, 498
Collectivism *vs.* individualism, as Hofstede's cultural dimension, 169–170, 170f
Common Rule, 125, 125n
Communication
defined, 295
SCC of military operations and, 274, 275t
uncertainty and, 295
Communication barriers
to data collection, 111
Communities, disconnected, *see* Disconnected communities
Community analysis, 51–52
Comparative optimism, 479

Complex adaptive system (CAS) models, 489
Complexity, 315
  of sociocultural systems, 318–319; *see also* Causal/predictive relationships
    bidirectional causality, 319
    qualitative features and effects, 319
    temporal offsets, 318–319
Complex survey sampling, 425, 426–427
Comprehension
  described, 31
  narrative as mode of, 31
Comprehension, narrative
  bottom-up processing and data/schema mismatch, 33–34
  structural forms, 32–33
Computational modeling
  agent-based modeling, 450–451
  hybrid models, 450–451, 450–452, 453f; *see also* Hybrid computational models
    cognitive perceptions, 457
    knowledge acquisition process, 453–454
    macrolevel theoretical representation, 454–455
    methodology, 452–458
    microlevel theoretical representation, 456f, 457–458
    problem question, 452
    SME bias, 454
  system dynamics, 450–451
  theoretical orientations of, 451
Computational models
  ExMo approach, 8
Conceptual idea level (culture), 1, 3, 167
Conditional entropy, 214, 216–217
  uncertainty reduction and management, 301–302
Confidence intervals (CIs), 94, 426–427
  for CRT scores, 94, 95f–96f
Conflict(s)
  conceptualization of, 192–193, 193f
  detection, semantic processing of mobile-collected data and, 211–212
  dynamics of, 192–193, 193f, 194f
  "gray zone" conflicts, 267
  poverty and, relationships between, 325–331, 326f, 328f–330f
  reasoning-based methods for, 305–307
    first-order logic, 305–306
    Markov logic networks, 306–307
  violent, validation model of (case study), 325–331, 326f, 328f–330f; *see also* Causal/predictive relationships
Conjunctive normal form (CNF), 306
Connectedness, 7
Consumption habits, of audiences
  social media messaging and, 48
Context and implicit relationships, data analysis and, 113
Convergent cross mapping (CCM), 317, 325, 327, 329, 329f
  for bidirectional causality, 322–323
Cooperative defaults, 92
Coordination
  SCC of military operations and, 274, 275t
Correlations, strength of, 458
Counterresponses, 460
Courses of action (COAs), 8, 9, 11, 12, 181, 424
Critical data
  expeditionary modeling, 13
Cross-cultural narrative research project, 33
Cross-cultural social science research, 164–165; *see also* Non-WEIRD societies; WEIRD societies
Cross entropy, 215–216
Crowdsourcing, 143
CRT, *see* Cognitive reflection test (CRT)
#CtrlSec campaign, 46
Cultural awareness
  SCC of military operations and, 274, 275t
Cultural differences
  in cognitive reflection, 92–97, 93t–94t, 95f–96f
  role in accidents (case studies)
    Korean Air Flight 801 disaster, 106–107
    Union Carbide's Bhopal chemical plant accident, 107
Cultural resources
  Peace Game, 369
Cultural ties
  intercultural ties, 74
  intracultural, 74
  measurement and, 74
Culture cycle, 1–2; *see also* Culture(s)
  information processing within, 177–178
  Markus and Conner's framework, 166–168, 167f
    conceptual ideas level, 167
    individualism–collectivism within, 171–174, 172f–174f
    individual level, 166
    institution level, 166
    integration of with Hofstede dimensions, 171
    interaction level, 166
  megacity use case (fictional), 176–178

Culture(s); *see also* Culture cycle
cognitive reflection and, 85–97
biases, 91–92
as cognitive style, 90
cultural differences in, 92–97, 93t–94t, 95f–96f
dual process theories, 85–89
intelligence and, 89–92
measure of, 88–89
resistance to irrational beliefs, 90–91
type 1 process, 86–89
type 2 process, 86–89
conceptual idea level, 1, 3, 167
defined, 166
Hofstede dimensions, 168–171
individualism–collectivism, 169–170, 170f
integration with culture cycle, 171
masculinity–femininity distinction, 169
newer dimensions, 170–171
power distance, 168
uncertainty avoidance, 168–169
individual level, 1, 2, 166
institution level, 1, 3, 166
interaction level, 1, 2–3, 166
levels of, 1–2, 165
opinion spread in empirical networks and, 70–81
class background, 75, 75t, 76f
megacity and Notre Dame NetSense data, 70–71
network/opinion coevolution, 76–80, 78f–81f
race and ethnicity, 74–75, 74t
sociodemographic differences impacts on activity and assortative mixing, 71–74, 72f, 73t
CultureSim, 18–19
Cyberattacks, 267

D

Darkness, of urban areas, 8
DARPA program, 340n
Data
described, 183
emic perspective in, 181–201
example, 185t
future research, 199–200
importance, for forecasting, 198–199, 198f–199f
interpretation, 184–185, 185t
megacity use case (fictional), 200–201
overview, 181–182
problem of inference, 183–184
text analytic tools for violence forecasting, 195–198, 196f, 197f
threat narrative, 185–194
researcher/analyst relationship to, 182–183
Data analysis, 112–114
challenges
context and implicit relationships, 113
data handling, 113
inadvertent bias, 113
latent *vs.* observable variables, 113
noisy data, 113
entropy, 114
GIS approaches, 114
mobile-collected data (semantic processing), 211–214
conflict detection, 211–212
inconsistency detection, 212–213
paraconsistency reasoners, 213–214
parallel planning process (HRE), 121–122
Data Analysis Using Regression and Multilevel/Hierarchical Models, 224
Database
relational solutions, 206
semantic store-based solutions, 206; *see also* Semantic knowledge management
Data collection, 111–112
challenges, 139
communication barriers, 111
ethical ambiguity, 112
limited access, 111
noisy data, 112
population vulnerability, 111
from disconnected communities, *see* Disconnected communities
emic perspective, *see* Emic perspective
expeditionary modeling, 12–18, 17f
Lagos (as megacity example), 12–18, 16t
Nigeria (as megacity example), 12
PMESII-ASCOPE factors, 13, 14t
high-risk ethnography, 112, 115–130; *see also* High-risk ethnography (HRE)
mobile
semantic knowledge management and, 209–211, 209f–210f
semantic knowledge management, *see* Semantic knowledge management
WEIRD populations, 112
Data handling, 113
Data management
SCD, 125–128
Data ownership, 154, 156t

Data preprocessing
uncertainty propagation and, 297
Data privacy, 154–155, 156t
Data processing
uncertainty propagation and, 297
Data/schema mismatch, narrative comprehension, 33–34
Data sensitivity, 125
HRE data, 125
Data-to-decisions process
uncertainty propagation in, 296f, 297
Data uncertainties, 292
measurement errors and, 292
Dawkins, Richard, 45
De-Ba'athification, 504
Deceptive information
entropy-based detection, 214–215
Decision making
cognitive biases and, 103–108; *see also* Cognitive biases
narrative and, 37–38
shared component of, 39
Default interventionist approaches, 86
Defaults
cooperative, 92
defined, 92, 92n
uncooperative, 92
Defense Advanced Research Project Agency (DARPA) projects, 350
Deferred Action for Childhood Arrivals (DACA), 403
Dempster–Shafer (DS) theory, 293–294
Department of Defense (DoD), 125n, 164, 246, 496
Common Rule, 125, 125n
HSCB modeling program, 247
Department of Defense Directive (DODD), 125n
Description logics (DL), 208
query for inconsistencies, 212–213
Design-based inference
vs. model-based inference, population surveys, 425
Design-based model, population surveys, 427
Differentiation, defined, 189
Digital Globe, 142–143
Digital media, 43; *see also* Social media
Digital subscriber line (DSL), 48
Diplomatic, information, military, and economic (DIME) actions, 20, 22
Disconnected communities, 136–159
connection with
need for, 137–138
startup principles, 157
ethical concerns, 150–155
bystander responsibility, 152
data ownership and access, 154
emerging framework, 155t–156t
exploitation, 152–153
poor data quality and validation, 153
poor security and safety, 151
privacy, 154–155
unintended/negative influence, 151–152
implementation challenges, 138–149, 138t
emerging framework, 149–150, 150t
harsh physical environments, 144–146
limited resources, 138–140
other sociocultural differences, 148–149
poor infrastructure, 140–144
unstable sociopolitical environment, 146–147
megacity use case (fictional), 157–159
harsh physical environment, 158
limited resources, 157–158
other sociocultural differences, 159
poor infrastructure, 158
unstable sociopolitical environment, 158–159
overview, 136–137
Discordance, 458
Displaced minority, 341
Dissent, social media-based
suppression and augmentation, 46–47
Distributed Common Ground System (DCGS), 128
DL, *see* Description logics (DL)
DOTMLPF, 281
DTW, *see* Dynamic time warping (DTW)
Dual processes, 86
Dual process theories, cognitive reflection, 85–86
algorithmic mind, 87
overview, 85–86
reflective mind, 87
type 1 process, 86–89
type 2 process, 86–89
Dual systems, 86
Dual types of thinking, 86
Duration of the peace, 474
DYMATICA model, 450, 452, 454
behavioral responses, representing, 460–462, 461f
computational structure of, 458–462, 459f
environmental, geopolitical, and societal factors assessment, 462–467
entities and decisions, 463
representing entities, 463
scenario conclusions, 467

macrolevel theoretical representation, 454, 455
megacities use case (fictional), 468–470, 469f
entities and decisions, 468–470
overextension of government resources, 469
sociocultural conflict, 469–470
microlevel theoretical representation, 456f, 457–458
Dynamicity
urban areas, 8
Dynamic multiscale assessment tool for integrated cognitive behavioral actions (DYMATICA) computational model, *see* DYMATICA model
Dynamic time warping (DTW), 317, 325, 327, 328f
advantages, 322
ForwardDTW, 322
for uneven temporal relationships, 321–322

E

Early model-based event recognition using surrogates (EMBERS) system, 19
Ebola outbreak of 2014–2015
social media use during, 44
Economic system, in StateSim models, 345–349, 346f–348f
Effects-based operations (EBO), 496
Egyptian revolution (2011), 44
E-mail, 48
Emic perspective, 181–201; *see also* Etic perspective
data, described, 183
example, 185t
forecasting
human behavior, 187–188
importance for, 198–199, 198f–199f
future perspectives, 199–200
interpretation, 184–185, 185t
megacity use case (fictional), 200–201
overview, 181–182
problem of inference, 183–184
researcher/analyst relationship to data, 182–183
text analytic tools for violence forecasting, 195–198, 196f, 197f
threat narrative, 185–194
assessment, 186–188
conflict dynamics, 192–193, 193f, 194f
defined, 185–186
forecasting methodologies, 190–192, 191t
human behavior forecasting and, 187–188
in search of, 188–189
Emotional attachment process, 484
Empirical networks
culture and gender impact on opinion spread in, 70–81
class background, 75, 75t, 76f
megacity and Notre Dame NetSense data, 70–71
network/opinion coevolution, 76–80, 78f–81f
race and ethnicity, 74–75, 74t
sociodemographic differences impacts on activity and assortative mixing, 71–74, 72f, 73t
Entity extraction, 188
Entropy, 207
anomaly detection based on, 214–217
drill down of anomalous/suspicious behavior, 216–217, 216f–217f
information measures detection in time using KL divergence, 215–216
basic, 214
calculations, 214, 215
conditional, 214, 216–217
cross, 215–216
defined, 114, 207
KL divergence, 214
maximal, 215
uncertainty reduction and management, 299–300, 299t
Epistemic (systematic) uncertainty, 293
Erdõs-Rényi (ER) networks, 66, 67f, 69
Error bar calculation, 237
Ethical ambiguity
data collection and, 112
Ethical considerations
disconnected communities, 150–155
bystander responsibility, 152
data ownership and access, 154
emerging framework, 155t–156t
exploitation, 152–153
poor data quality and validation, 153
poor security and safety, 151
privacy, 154–155
unintended/negative influence, 151–152
Ethnicity, *see* Race and ethnicity
Ethnogenesis, coevolutionary, of social identity, 481–485; *see also* Social identity groups (SIGs)
shared intentions to create better situation, 485
shared perceptions of threat and betrayal, 481–483
shared thoughts of values and narratives, 483–484

Etic perspective, 185; *see also* Emic perspective
example, 185t
ETL, *see* Extract–transfer–load (ETL) process
Euromaidan movement, 44
Event coding, for forecasting, 188
Evidence theory
uncertainty modeling and, 293–294
Evidentiality, 190
ExMo, *see* Expeditionary modeling (ExMo)
Expectations *vs.* perceptions, 458
Expeditionary modeling (ExMo), 5–22
challenges, 18
coevolution of models, data, and operating environment/mission, 11, 11f
computational models, 8
critical data, 13
human resources, 10
overview, 5–7
requirements (data), 12–18
CEWS coded news event data, 16, 17f
Lagos (as megacity example), 12–18, 16t
Nigeria (as megacity example), 12
PMESII-ASCOPE factors, 14t
requirements (models)
Athena (example), 20–21
multimodel composition, 21–22, 21f
populations, modeling, 19–20
urban environments modeling, 18–19
scenario, 11–12
statistical models, 8–9
urban operation challenges, 7–8, 7f
vision of, 8–12, 9f, 11f
Xanadu, as fictional megacity use case, 11–12
Exploitation
of vulnerable communities, data collection and, 152–153, 156t
Exponential random graph models (ERGMs), 71–72, 72f
Extract–transfer–load (ETL) process, 206
semantic, 206

F

Facebook, 44, 48, 49, 54; *see also* Social media
ISIS campaigns, 46
peak activity times, 51
in social movements, 44
use for misinformation and propaganda purposes, 45–46
use in Ebola outbreak of 2014–2015, 44
use in Nepal earthquake 2015, 44
Facebook egonetwork, TM for, 69, 69f–70f
Facebook's Graph Search, 207
Face-to-face interviews, 426
Factanal function, integrative model of organizational trust, 405
Factor analysis
population surveys, 223–224, 224f
Afghanistan survey (case study), 230–232
comparison of analysis, 226t
enduring questions, 225
error bar calculation, 237
example of rankings, 229t
key areas, 231
Mann–Whitney test, 228–229, 233
mean respondent display, 235
methodology, 225–230, 226t–229t
MICh tool, 232–234
normal distribution, 230
output, 227t
recoded questions for, 226–227, 227t
respondent scores, 228, 228t
"Fanta for the Funny" campaign (Coca-Cola), 48
Fa.parallel function, integrative model of organizational trust, 405
Federal Government of Iraq (FGoI), 504, 506, 507
Femininity *vs.* masculinity, as Hofstede's cultural dimension, 169
FGoI, *see* Federal Government of Iraq (FGoI)
50 Cent Party, 46
First-order logic, 305–306
First-order SEM model formulation, 410
Forced labor victims, limited resources for, 138–140
Force movement, Peace Game, 376, 377f, 378f
Forecasting, 187
Boko Haram violence, 198–199, 198f–199f
emic perspective and, 198–199, 198f–199f
event coding for, 188
human behavior, 187–188
methodologies, 190–192
model-based, 20
social identity-based methodology, 190–192
techniques for, 188
violence, text analytic tools for, 195–198, 196f, 197f
idea density, 195
integrative complexity, 195
keyness, 195–196
sentiment/affect of in-group toward outgroups, 196
Foreign Military Studies Office (FMSO), 117
Foreign policy, 495–507
governance competition model (GCM), 498–503

for Afghanistan, 498–503, 501f, 502f
for Iraq, 504–507, 505f
overview, 495–496
US experience in Afghanistan and, 496–498
VSO/ALP program and, 498, 499, 502, 503f, 507
ForwardDTW, 322
Four-valued logic, 214
Frank, Scot, 148
Freytag, Gustav, 32
Freytag's triangle/Freytag's pyramid, 32
From wave, 228–229
Fuzzy measure theory
uncertainty modeling and, 294
Fuzzy set theory
uncertainty modeling and, 294

G

Gameplay, Peace Game, 375–381; *see also* Peace Game
force movement, 376, 377f, 378f
market tab, 381, 382f
negotiations, 376, 379f, 380f, 381
Gaps and mitigation measures, SCC, 278–282, 279t–281t
Gatov, Vasily, 46
GCM, *see* Governance competition model (GCM)
Gender differences
in network ties, 73
opinion spread in empirical networks and, 70–81
class background, 75, 75t, 76f
megacity and Notre Dame NetSense data, 70–71
network/opinion coevolution, 76–80, 78f–81f
race and ethnicity, 74–75, 74t
sociodemographic differences impacts on activity and assortative mixing, 71–74, 72f, 73t
General Motors (GM), 29
Generator elicitation screens, StateSim models, 340–349
actors, 344–345, 344f
factions, 341–344, 341f–343f
institutions, resources, and economics, 345–349, 346f–348f
Geographical modeling
background, 245–247
HSCB modeling and, 246–247
megacity use case (fictional), 263–264
nation-building, 245–246
overview, 243
researching design and methodology
data set description, 247–249, 249f, 249t, 250f
point density analysis, 251
study area, 247, 248f
results and discussion
descriptive statistics of adverse events by district, 251–252, 251t, 252f, 253t–254t, 255f
GIS-based analysis and visualization of adverse events, 252–258, 256f–262f
terrorism, described, 244
in terrorism studies, 245
Geographic information systems (GISs), 9; *see also* Geographical modeling
ExMo approach, 9, 10
HSCB modeling and, 246–247
point density analysis, 251
Gephi, 52
#GhostSec campaign, 46
Gill, Kohl, 139, 140
Girkin, Igor, 45
GIRoA, 497–498, 499, 500–502, 503f
GISs, *see* Geographic information systems (GISs)
Global events, social media in, 43–47; *see also* Social media
augmentation via, 46–47
for misinformation, 45–46
for propaganda purposes, 45–46
in social movements, 44
for suppression of dissent, 46–47
Globalization, 5, 267
Globe Study, 172
Google AdWords, 49, 55
Google Flu Trends, 19
Google's Knowledge Graph, 207
Google's Project Loon, 143–144
Governance competition model (GCM), 498–503
for Afghanistan, 498–503, 501f, 502f
for Iraq, 504–507, 505f
Government Accountability Office (GAO), U.S., 222
Grand Ethiopian Renaissance Dam (GERD), Ethiopia, 462
Granger causality, 317, 325, 326–327, 326f
for temporal offsets in causal/predictive relationships, 321
Graph data technology, 207
"Gray zone" conflicts, 267
G-test, 195–196
Guiding Principles for Stabilization and Reconstruction (USIP/PKSOI), 485

H

Harris, Eric, 182
Harsh physical environments
 data collection from disconnected communities and, 144–146
 discussion of, 145
 key principles, 146
 in LRA-affected regions, 144–145
 megacity use case (fictional), 158
Hashtags
 in Arabic script, 50
 co-opting, 50
 value on social media platforms, 49–50
Health monitoring
 parallel planning process (HRE), 123
Heuer, Richards, 184
High-risk environments
 and need for HRE, 118
High-risk ethnographer, 119
High-risk ethnography (HRE), 112, 115–130
 concepts, 119–120
 data sensitivity, 125
 experience, 120t
 high-risk environments and need for, 118
 HRE team
  role of, 118–119
  security and integration of, 120–123
 megacity use case (fictional), 130–132
 methods, 119–120, 120t
 overview, 117
 parallel planning process, 120–123
  data analysis, 121–122
  data sets, 122
  health monitoring, 123
  logistics monitoring, 123
  overwatch activities, 122–123
  primary data, 122
  RDPA, 122
  secondary data, 122
  tertiary data, 122
  threat monitoring, 123
 practical experience, 122
 primary data, 124t, 125
 research parameters, 120t
 secondary data, 124t, 125
 Sixty Percent Mission (case study), 121, 128–130
 social complexity data, 123–128
  information and, 123–125
  management of, 125–128
  overview, 123
 tertiary data, 124t, 125
 vs. traditional social science
  experience, 120t
  research parameters, 120t
Hitler, Adolf, 38
Hofstede's culture dimensions, 168–171
 individualism–collectivism, 169–170, 170f
 integration with culture cycle, 171
 masculinity–femininity distinction, 169
 megacity use case (fictional), 176–178
 newer dimensions, 170–171
 power distance, 168
 uncertainty avoidance, 168–169
Homophily, 62, 73
Hootsuite, 51
Horvitz–Thompson estimator, 427
Hot deck imputation method, 224
HRE, *see* High-risk ethnography (HRE)
HRE team; *see also* High-risk ethnography (HRE)
 role of, 118–119
 security and integration (parallel planning process), 120–123
  data analysis, 121–122
  overwatch activities, 122–123
HSCB modeling, *see* Human social culture behavior (HSCB) modeling
Human behavior forecasting, 187–188
Human behavior modeling (HBM), 336
Human resources
 expeditionary modeling (ExMo), 10
Human social cultural behavioral models, 10
Human social culture behavior (HSCB) modeling, 244
 GIS modeling and, 246–247
Human Terrain System (HTS), 116, 117
Human Terrain Team (HTT), 116
Hurricane Katrina (August 2005), 6
Hussein, Saddam, 33, 38, 496, 504
Hybrid computational models, 450–452, 453f
 cognitive perceptions, 457
 DYMATICA
  assessment, 462–467
  megacities use case (fictional), 468–470, 469f
  structure, 458–462, 459f
 knowledge acquisition process, 453–454
 macrolevel theoretical representation, 454–455
 methodology, 452–458
 microlevel theoretical representation, 456f, 457–458
 objective, 452
 problem question, 452
 SME bias, 454

I

ICEWS, *see* Integrated crisis early warning system (ICEWS)
Idea density, 195
Ideological functions, and narrative, 36–37
Ideology
  defined, 36
  narrative and, 35–38
    decision making and, 37–38
    ideological functions, 36–37
    master narratives, 35–36
    values and, 37–38
Impact assessment
  SCC of military operations and, 276t, 277
Implementation challenges, disconnected communities, 138–149, 138t
  emerging framework, 149–150, 150t
  harsh physical environments, 144–146
    discussion of, 145
    key principles, 146
    in LRA-affected regions, 144–145
    megacity use case (fictional), 158
  limited resources, 138–140
    discussion on, 140
    key principles, 142
    megacity use case (fictional), 157–158
    victims of forced labor, 138–140
  poor infrastructure, 140–144
    discussion on, 143–144
    key principles, 144
    megacity use case (fictional), 158
    in Nepal after 2015 earthquakes (case study), 140, 141–144
  sociocultural differences, 148–149
    key principles, 149
    megacity use case (fictional), 159
    in nomadic communities (Himalayan Plateau, China), case study, 148–149
  unstable sociopolitical environment, 146–147
    in Afghanistan (case study), 146–147
    discussion of, 147
    key principles, 147
    megacity use case (fictional), 158–159
Inadvertent bias
  data analysis and, 113
Inconsistencies, identification and resolution
  for uncertainty reduction and management, 304–307
    first-order logic, 305–306
    Markov logic networks, 306–307
    reasoning-based methods, 305–307
    source trust weighting, 304–305
Inconsistent information detection
  DL query for, 212–213
  semantic processing of mobile-collected data and, 212–213
Independence *vs.* interdependence, 171, 172
Individualism–collectivism
  within culture cycle, 171–174, 172f–174f
  as Hofstede's cultural dimension, 169–170, 170f
Individualized thresholds, dynamics with, 68–69, 69f–70f
Individual level, of culture, 1, 2, 166
Indulgence *vs.* restraint, as Hofstede's cultural dimension, 170–171
Inference, problem of, 183–184
Influencers identification, social media and, 52
Information
  HRE data and, 123–125, 124t
  SCD and, 123–125
  vs. uncertainty, 300
Information measures detection, in time using KL divergence, 215–216
Information processing
  within culture cycle, 177–178
  uncertainty propagation and, 296f, 297
Information theoretic methods
  for reduction and management uncertainty, 299–303
    conditional entropy, 301–302
    entropy, 299–300, 299t
    joint entropy, 300–301, 301t
    mutual information, 302–303
Infrastructure, poor, *see* Poor infrastructure
Instagram, 48
Institution level, culture, 1, 3, 166
Institutions parameters, StateSim models, 345–349, 346f–348f
Integrated crisis early warning system (ICEWS), 16, 16n
  statistical models of populations, 19
Integration, defined, 189
Integrative complexity, 195
  defined, 189
Integrative model of organizational trust, 397–419, 400f; *see also* Trust
  ability, 401
  analyses, 406t–410t, 407–410, 411f–413f
  background, 399
  benevolence, 402
  components, 400–403
  defined, 400, 401
  discussion, 415–418
  factanal function, 405

fa.parallel function, 405
integrity, 402–403
limitations of current research, 417–418
measures, 405–406, 405t
megacity use case (fictional), 418–419
methods, 404
overview, 398–399
research hypotheses, 404
results, 410–414, 414t
risks taking, outcomes, and practical application, 403–404
sample, 405, 405t
theoretical debate, 398
Integrity, 402–403
Intelligence
cognitive reflection and, 89–92
biases, 91–92
as cognitive style, 90
resistance to irrational beliefs and, 90–91
Intelligence, surveillance, and reconnaissance (ISR), 15
Intelligence preparation of the battlespace (IPB) process, 8, 8n
Intelligence tests
circumstances for, 89
CRT and, 89
Intensifiers, 190
Intentions
neural foundation of, 485
shared, 485
Interaction level, culture, 1, 2–3, 166
Intercultural ties, 74
Interdependence *vs.* independence, 171, 172
International Organization of Migrants, 154
Internet memes, 45
Inter-SIG negotiations, power of, 484
Intracultural ties, 74
Intuitive thinking, 88
CRT and measures of, 90
Invisible Children, 144, 145
IPB process, *see* Intelligence preparation of the battlespace (IPB) process
Iraq
GCM for, 504–507, 505f
US interventions in, 496
Iraq War, 34–35
Irrational beliefs, cognitive reflection and resistance to, 90–91
Irregular warfare, defined, 246
ISIS, *see* Islamic State of Iraq and al-Sham (ISIS)
Islamic State of Iraq and al-Sham (ISIS), 29, 37, 45–46, 50, 52, 181, 267, 496, 507
emic and etic interpretations from, 185, 185t
social media campaigns, 46
Telegram as communication platform for, 54
Iterative proportional fitting, *see* Raking algorithm

J

Japanese ISIS-chan movement, 46–47
Joint entropy
uncertainty reduction and management, 300–301, 301t
Joint force land component commander (JFLCC), 11, 12
Joint Operations, 270
Joint Strategic Capabilities Plan, 2010, 272
Joint Task Force Rapid Relief, 11
Joint Warfighting Center, 272

K

Karzai, Hamid, 497
Kathmandu Living Labs (KLL), 142–143
Kerry, John, 50
Keyness, 195–197
Knowledge
defined, 295
Knowledge acquisition, 454
hybrid computational models, 453–454
Knowledge base, 208
Knowledge refinement
uncertainty propagation and, 297–298
Korean Air Flight 801 disaster, 106–107
Kullback–Leibler (KL) divergence, 214
information measures detection in time using, 215–216

L

Labor trafficking and violations, 139
LaborVoices, 139, 154
challenges for, 139–140
obstacles to data collection, 139
SmartLine, 139–140
Lagos
ExMo data requirements and, 12–18, 16t
PMESII data from open sources, 15, 16t
Lanchester's Square (Aimed Fire) law, 364
Language form element, 190
Large Knowledge Collider platform, 213
Latent *vs.* observable variables, data analysis and, 113
Learning in Graphical Models, 304
Length, social media messaging, 48–49

Lexicalization, 190
Likert scale, 223
Limited access
  data collection and, 111
Limited resources
  data collection from disconnected communities and, 138–140
    megacity use case (fictional), 157–158
  discussion on, 140
  key principles, 142
  for victims of forced labor, 138–140
Littoralization, 7
Living system, capabilities of, 486
Logistics monitoring
  parallel planning process (HRE), 123
Long-term–short-term orientation, as Hofstede's cultural dimension, 170–171
Lord's Resistance Army (LRA)-affected regions
  harsh physical environments in (case study), 144–145
  poor security and safety in, 151
Loss aversion, 105, 106

M

Macrolevel theoretical representation, hybrid computational models, 454–455
Malthusian trap, 364
Mann–Whitney test, 228–229, 233
Map, Peace Game, 367
Market tab, Peace Game, 381, 382f
Markov logic networks (MLN), 306–307
Markus and Conner's culture cycle, 166–168, 167f
  conceptual ideas level, 167
  individualism–collectivism within, 171–174, 172f–174f
  individual level, 166
  information processing within, 177–178
  institution level, 166
  integration of with Hofstede dimensions, 171
  interaction level, 166
  megacity use case (fictional), 176–178
Masculinity–femininity distinction, as Hofstede's cultural dimension, 169
Master narrative, 33, 35–36, 38
Master Narratives of Islamist Extremism, 30, 35
Master Rationality Motive Scale (MRMS), 90
MBTI (Myers–Briggs type indicator), 90, 90n
McCain, John, 29, 30
Measurement errors, data uncertainties and, 292
Measure of indicator change (MICh) tool
  population surveys, 233–234
Media environment
  understanding before messaging, 47–48
Megacity(ies); *see also* Urban areas
  connectedness, 7
  defined, 5, xxvii
  ExMo approach, *see* Expeditionary modeling (ExMo)
  Lagos as example, ExMo and, 12–18, 16t
  littoralization, 7
  Nigeria as example, ExMo and, 12
  Notre Dame NetSense data, 70–71
  overview, 5–6
  population density
    current perspective, 5–6
    future perspective, 5–6
  scale of operations in, 6
  statistics, 5–6
  urbanization, 7
  urban operation challenges, 7–8, 7f
  Xanadu (as fictional use case), *see* Xanadu (fictional megacity use case)
Meme, 45
  defined, 45
Message length, in social media, 48–49
Messaging, social media, 47–53
  audience attributes (behavioral habits, consumption habits, etc.), 48
  hashtags, 49–50
  media environment and, 47–48
  message length, 48–49
  narrative analysis, 52–53
  social network analysis, 51–52
  timing for optimum exposure, 51
  user roles within network, 52
  volume of posts, 51
Metamodels, 22
MGB, *see* Model of goal-directed behavior (MGB)
Microlevel theoretical representation, hybrid computational models, 456f, 457–458
Migration in Sudan: A Country Profile 2011, 364
Military operations, 269
  capability-based assessment (CBA), 270
  gaps and mitigation measures, 278–282, 279t–282
  megacity use case (fictional), 282–284
    peacekeeping operations, 282–284
  missions, 270–273
    operational phases, 271–272, 271t
  missions by phase description, 272–273
  overview, 270
  sociocultural capabilities, 267, 273–277
    categories, 274–277, 275t–276t

defined, 273–274, 273t
by operational phase, 277–278, 278t
MINERVA Initiative, 164n
Misinformation operations
social media and, 45–46
Missions, military operations, 270–273
operational phases, 271–272, 271t
by phase description, 272–273
shape phase, 271–272, 271t
Mobile data collection
data analysis and fusion
conflict detection, 211–212
inconsistency detection, 212–213
paraconsistency reasoners, 213–214
semantic knowledge management and, 209–211, 209f–210f
Model analyst's toolkit (MAT), 325
Model-based forecasting, 20
Model-based inference
vs. design-based inference, population surveys, 425
Model/modeling; *see also specific models*
defined, 394
and simulation, relation between, 394
Model of goal-directed behavior (MGB), 458, 458n
Modes of processing, 86
MRMS (Master Rationality Motive Scale), 90
Mujahideen, 31
Multiscale influences, on resiliency, 488–489, 488f
Mutual information
uncertainty reduction and management, 302–303
Myers–Briggs type indicator (MBTI), 90, 90n
MySpace, 48

N

Naming game-based (NG) model
opinion dynamics, 61, 62–66, 64f, 65f
opinion evolution process, 76
tipping point, 62–63, 64f
waning commitment, 63, 65–66, 65f
Narrative, defined, 477
Narrative analysis
social media messaging and, 52–53
Narrative arc, 31
Narrative coherence, 34
Narrative comprehension
bottom-up processing, 33–34
data/schema mismatch, 33–34
structural forms, 32–33
Narrative Comprehension and Film, 31
Narrative fidelity, 34, 35
Narrative Landmines: Rumors, Islamist Extremism and the Struggle for Strategic Influence, 34
Narrative reconstruction, 479
Narrative(s); *see also* Threat narrative
decision making and, 37–38
shared component, 39
defined, 28, 30
described, 30
dual logics of, 30
ideological functions and, 36–37
ideology and, 35–38
importance, 28
master, 33, 35–36
misinterpretation, 29
as mode of comprehension, 31
overview, 28
rationality, 34–35
role in self-concept, 38
shared thoughts of, 483–484
sociocultural artifact, 30
values and, 37–38
National Aeronautics and Space Administration (NASA), 304–305
National Air and Space Intelligence Center, 182, 190
National Intelligence, Open Source Center, 245
Nation-building, geographical modeling and, 245–246
Natural disasters; *see also specific case studies*
social media use and, 44
Naval Postgraduate School, 222
Naxalites, 196
Naxalite violence, 196
NBC Meet the Press, 38
Near past occurrence window, concept of, 124
Need for cognition (NFC) scale, 90
Negotiations, Peace Game, 376, 379f, 380f, 381
Nepal earthquake 2015
poor infrastructure after, 140, 141–144
social media use during, 44
NetSense study (University of Notre Dame), 70–71, 72f, 79
CDRs in, 71, 77
Network/opinion coevolution, 76–80, 78f–81f
Network ties, gender distinctions in, 73
NFC (need for cognition) scale, 90
NG model, *see* Naming game-based (NG) model
Nigeria
ExMo data requirements and, 12

Nin, Anaïs, 184
9/11 attacks, 245, 497, 504
NodeXL, 52
Noisy data, 112, 113
Nomadic communities in Himalayan Plateau, China (case study), 148–149
Nominalization, 190
Non-Internet users (global), 140, 141f; *see also* Disconnected communities
Non-WEIRD societies, 163, 165; *see also* WEIRD societies
  cross-cultural social science research, 164–165
  vs. WEIRD societies, 163, 165
    cross-cultural social science research, 165
Notre Dame NetSense data, 70–71, 72f
  CDRs in, 71, 77
  megacity and, 70–71

O

Obama, Barack, 29, 37, 504
Objective indicators, 222–223
Objective uncertainty, 293
Observable *vs.* latent variables, data analysis and, 113
Observer perspective, 185; *see also* Emic perspective; Etic perspective
#OccupyBoston, 47
#OccupyDenver, 47
Occupy Wall Street movement, 47
OED (One Earth Designs), 148–149
One Earth Designs (OED), 148–149
Ontology, 208
Open source data, 15
OpenStreetMap, 142
OpenTable (online restaurant reservations), 19
Operational phases, military operations, 271–272, 271t
  sociocultural capabilities by, 277–278, 277f, 278t
Operations research (OR) analyst, 424
Opinion, 61–81
  culture and gender impact on spread of, in empirical networks, 70–81
    class background, 75, 75t, 76f
    megacity and Notre Dame NetSense data, 70–71
    network/opinion coevolution, 76–80, 78f–81f
    race and ethnicity, 74–75, 74t
    sociodemographic differences impacts on activity and assortative mixing, 71–74, 72f, 73t
  naming game-based (NG) model, 61, 62–66, 64f, 65f
  overview, 61–62
  public, toward VEOs, 436–439
  threshold model (TM), 61, 66–70
    individualized thresholds, 68–69, 69f–70f
    tipping points, 66–68, 67f
Opinion mining, *see* Sentiment analysis
#OPISIS campaign, 46
Organizational trust, integrative model of, *see* Integrative model of organizational trust
Other, defined, 477
Outrage culture, 47
Overwatch activities, parallel planning process (HRE), 122–123
  health monitoring, 123
  logistics monitoring, 123
  threat monitoring, 123
OWL API standard, 212
OWL-DL, 208
Oxytocin, 484

P

PageRank algorithm, 304
Panarchy theory, 487
Paraconsistency reasoners, 212, 213–214
Parallel planning process, HRE, 120–123
  data analysis, 121–122
  data sets, 122
  overwatch activities, 122–123
    health monitoring, 123
    logistics monitoring, 123
    threat monitoring, 123
  primary data, 122
  RDPA, 122
  secondary data, 122
  tertiary data, 122
Pashto, 191
Peace Game, 361, 363–367
  algorithm design, 363, 365f
  exponential curves, 364, 366f
  factors, 363–364
  gameplay, 375–381
    force movement, 376, 377f, 378f
    market tab, 381, 382f
    negotiations, 376, 379f, 380f, 381
  megacity use case (fictional), 381–390
    map, 383
    players, 383
    players' units, 386, 388f, 390
    region definition, 383, 385f, 386

resources, 383
unit types, 386, 387f
weather, 389f, 390
population algorithm, 364, 366f
scenario generator, 367–375, 368f
cultural resources, 369
map, 367
players, 369–370
regions, 370–372, 371f
resources, 369
units, 372–373
weather, 372f, 373–375
Sudan/RSS scenario, 363, 369, 370, 372, 375
Peak activity times, social media messaging, 51
People's War Group (PWG), 196–197, 196f
Perceptions *vs.* expectations, 458
Pinterest, 48
Planning
SCC of military operations, 274, 275t
Plausibility function, 294
Players, Peace Game, 369–370
PMESII-ASCOPE factors
expeditionary modeling, 13, 14t
Point density analysis
geographical modeling, 251
Political, military, economic, social, infrastructure, and information (PMESII) variable, 13, 14t
Athena simulation and, 20
data for Lagos from open sources, 15, 16t
Polling, 15
Poor data quality and validation
disconnected communities and, 151, 156t
Poor infrastructure
data collection from disconnected communities and, 140–144
discussion on, 143–144
key principles, 144
megacity use case (fictional), 158
in Nepal after 2015 earthquakes (case study), 140, 141–144
Poor security and safety
data collection from disconnected communities and, 151, 155t
Population algorithm, Peace Game, 364, 366f
Population growth, megatrends of, 7
Population(s)
ABMs, 19–20
engaging, SCC of military operations and, 274, 275t–276t, 277
expeditionary modeling, 19–20; *see also* Expeditionary modeling (ExMo)
megacities
current perspective, 5–6
future perspective, 5–6
model-based forecasting, 20
statistical models, 19
system dynamics models, 19
Population surveys
Afghanistan survey (case study)
factor analysis (results), 231–232
history, 230–231
indicators deriving from factors, 232–233, 232t, 240t–242t
analytics suite for displaying measured change
error bar calculation, 237
mean respondent display, 235, 235f
proportionality of indicator support display, 236–237, 236f
spatiotemporal indicator change, 233–235, 233f–234f
application (real world example), 431–434, 432f–433f, 434t
background, 222–223
complex sampling, 425, 426–427
design-based model, 427
enduring questions, 225
goal of, 425
Horvitz–Thompson estimator, 427
megacity use case (fictional), 238–239
methodology, 225–230, 226t–229t
model-based *vs.* design-based inference, 425
overview, 221–222
raking algorithm, 428–430, 428f, 430f, 430t–431t
application, 435–436, 436f
megacities use case (fictional), 446
sampling methodology, 427–428
sampling weights, 427, 428
simple random sampling (SRS), 426
statistical primer, 223–224, 224f
factor analysis, 223–224, 224f; *see also* Factor analysis
missing data, 223, 224
VEOs
age group differences supporting, 441, 442f–443f, 444t
changes in public opinion toward, 436–439, 437f–438f
public trust in, 439–440, 439t
religious differences supporting, 440–441, 440t
Population vulnerability
data collection and, 111

Poverty
and conflict, relationships between, 325–331, 326f, 328f–330f
validation model of (case study), 325–331, 326f, 328f–330f; *see also* Causal/predictive relationships
Power distance
defined, 106, 107, 168
as Hofstede's cultural dimension, 168
Korean Air Flight 801 disaster, 106–107
Primary data, 122
HRE data, 124t, 125
Privacy, data, 154–155, 156t
Probabilistic semantics
for anomaly detection, 214–217
Probability of the subset, 293
Probability theory, uncertainty modeling and, 293
Problem question, hybrid computational models, 452
Propaganda purposes, social media for, 45–46
Propagation, of uncertainty, 295–298, 296f; *see also* Uncertainty(ies)
data preprocessing and, 297
data processing and, 297
in data-to-decisions process, 296f, 297
information processing and, 296f, 297
knowledge refinement and, 297–298
Propositional idea density (PD), 195
Prospect theory, 455
Public goods game, 92, 92n
Public misinformation
social media for, 45–46, 47
Public opinion, toward VEOs, 436–439
Public trust, toward VEOs, 437f–438f, 439–440, 439t
Pulp Fiction (film), 33

Q

QCT, *see* Qualitative choice theory (QCT)
QuakeMap.org, 142, 143
Qualitative choice theory (QCT), 455, 460
Qualitative feature-based analysis, 323–325, 324f
Quotations, 190
Qzone, 48

R

Race and ethnicity
opinion spread modeling and, 74–75, 74t
Raking algorithm, 428–430, 428f, 430f, 430t–431t
application, 435–436, 436f
real world example, 431–434, 432f–433f, 434t
megacities use case (fictional), 446
rules, 429
Raking ratio estimation, *see* Raking algorithm
Rana Plaza factory complex collapse in April'13 (Bangladesh), 139
RAND Corporation, 273
2011 study, 8
study on unstable sociopolitical environment in Afghanistan (2012), 146–147
Rapid scenario generator, 335
Rational–experiential inventory (REI), 90, 90n
Rationality, narrative, 34–35
Raven Matrices, 89
Reachback, 119n
Reasoning-based methods
for uncertainty reduction and management, 305–307
first-order logic, 305–306
Markov logic networks, 306–307
Recombinant data processing and analysis (RDPA), 122
ReelSEO, 48
Reflective mind, 87
Regions, Peace Game, 370–372, 371f
REI (rational–experiential inventory), 90, 90n
Relational database solutions, 206
disadvantages, 206
Religious beliefs, cognitive reflection and, 90–91
Religious differences, VEOs and, 440–441, 440t
Research
cross-cultural social science, 164–165
semantic and entropy techniques (future perspectives), 217–218
Researcher/analyst, relationship to data, 182–183
Resiliency, 474–475
capacity and, relation between, 487
defined, 474
of SIG, 475, 478
social
adaptive cycle process, 487–488, 487f
multiscale influences, 488–489, 488f
The Resolve LRA Crisis Initiative, 144, 145
Resource description framework (RDF), 208
Resources
limited
data collection from disconnected communities and, 138–140
discussion on, 140

key principles, 142
megacity use case (fictional), 157–158
for victims of forced labor, 138–140
Peace Game, 369
StateSim models, 345–349, 346f–348f
Restraint *vs.* indulgence, as Hofstede's cultural dimension, 170–171
Risk aversion, 479
Risk-taking actions, trust and, 403–404
R lavaan package, 407
Rough set theory
uncertainty modeling and, 294
Royal Dutch Airlines (KLM) Flight 4805 disaster, 105–106
Russert, Tim, 38
Russia–Ukraine–Crimea conflict, 267

S

Sadat, Anwar, 36, 37, 38, 39
Safavids, 52
Safety considerations
data collection from disconnected communities and, 151, 155t
in LRA-affected regions, 151
Sample balancing, *see* Raking algorithm
Sampling methodology, population surveys, 427–428
Sampling weights, 427, 428
Sandia National Laboratories' Behavioral Influence Assessment program, 452, 452n
SAS Text Miner, 53
Satorra–Bentler scaled test, 407
SCC, *see* Sociocultural capabilities (SCC)
SCD, *see* Social complexity data (SCD)
Scenario generator, Peace Game, 367–375, 368f
cultural resources, 369
map, 367
players, 369–370
regions, 370–372, 371f
resources, 369
units, 372–373
weather, 372f, 373–375
Scholastic Aptitude Test (SAT), 89
Schweitzer, Col. Martin P., 116, 117, 121, 129; *see also* Sixty percent mission
SCI (suspicious, conflicting, inconsistent) data interpreter, 211
Secondary data, 122
HRE data, 124t, 125
Second-order SEM model formulation, 410
Security
data collection from disconnected communities and, 151, 155t
in LRA-affected regions, 151
Self, defined, 477
Self-concept, 38
narrative systems role in, 38
The Selfish Gene, 45
Semantic ETL, 206
Semantic knowledge management, 207–209
data analysis and fusion
conflict detection, 211–212
inconsistency detection, 212–213
paraconsistency reasoners, 213–214
future research, 217–218
megacity use case (fictional), 219–220
mobile data collection to support, 209–211, 209f–210f
overview, 207
probabilistic semantics for anomaly detection
entropy-based, 214–217
workflow, 211
Semantic knowledge representation, 207–209
Semantic reasoners, 206, 208, 209, 211, 212
Semantic store-based solutions
advantages, 206
Sentiment analysis, 188
Shape phase, military operations, 271–272, 271t
Sharia, 37
Shias, 504, 507
SIGs, *see* Social identity groups (SIGs)
Simple random sampling (SRS), 426
Simulation
defined, 394
and models, relation between, 394
Sixty Percent Mission (case study), 112, 121, 128–130; *see also* High-risk ethnography (HRE)
Small-scale societies, 165
Smart cards, 276
SmartLine, 139–140
SMART system, *see* Strategic measuring and reporting tool (SMART) system
SME bias, 454
SnapChat, 54
Sociability effects, 71
Social biases, 479
Social complexity data (SCD)
HRE and, 123–128
information and, 123–125
management of, 125–128

overview, 123
research context timeline concept, 126f
storage framework, 127f
Social construction of reality, 479n
Social identity, 188–189; *see also* Social identity groups (SIGs)
coevolutionary ethnogenesis of, 481–485
shared intentions to create better situation, 485
shared perceptions of threat and betrayal, 481–483
shared thoughts of values and narratives, 483–484
forecasting methodologies and, 190–192
Social identity groups (SIGs), 474
challenges and opportunities, 475, 480
coevolutionary ethnogenesis of social identity, 481–485
shared intentions to create better situation, 485
shared perceptions of threat and betrayal, 481–483
shared thoughts of values and narratives, 483–484
cognitive biases, 479
defined, 477
foundations of, 478–487
growth, conditions underlying, 485–487
inter-SIG negotiations, power of, 484
megacity use case (fictional), 489–492
neurotic degree of, 475
resiliency of, 475, 478
resource allocation authority, 476
social biases, 479
social resiliency
adaptive cycle process, 487–488, 487f
multiscale influences, 488–489, 488f
stability of, 475
stability/rebellion model, 480
Social identity theory, 188
Social media, 43–55; *see also* Facebook; Twitter
augmentation via, 46–47
experts, 47
growing use in global events, 43–47
influencers identification, 52
megacity Xanadu (as case study, fictional), 54–55
messaging, 47–53
audience attributes (behavioral habits, consumption habits, etc.), 48
hashtags, 49–50
media environment and, 47–48
message length, 48–49
narrative analysis, 52–53
social network analysis, 51–52
timing for optimum exposure, 51
user roles within network, 52
volume of posts, 51
for misinformation, 45–46
overview, 43
for propaganda purposes, 45–46
suppression via, 46–47
use in Ebola outbreak of 2014–2015, 44
use in social movements, 44
Social movements
social media use in, 44
Social network analysis, 51–52
user roles in, 52
Social network mapping, 51
Social perceptions, neural foundation of, 485
Social reification, 479n
Social resiliency
adaptive cycle process, 487–488, 487f
multiscale influences, 488–489, 488f
Social science, cross-cultural research in, 164–165; *see also* Non-WEIRD societies; WEIRD societies
Social structure, 270
Sociocultural artifact
narrative, 30
Sociocultural capabilities (SCC)
categories, 274–277, 275t–276t
coordination and communication, 274, 275t
detailed cultural awareness, 274, 275t
engaging the population, 274, 275t–276t, 277
impact assessment, 276t, 277
planning, 274, 275t
defined, 273, 273t
gaps and mitigation measures, 278–282, 279t–281t
intersections, by mission activities, 277, 277f
military operations, 267, 273–277
megacity use case (fictional), 282–284
by operational phase, 277–278, 278t
Sociocultural differences
data collection from disconnected communities and, 148–149
key principles, 149
megacity use case (fictional), 159
in nomadic communities (Himalayan Plateau, China), case study, 148–149
Sociocultural models
causal/predictive relationships in, *see* Causal/predictive relationships, in sociocultural models

practical benefits, 307
uncertainty reduction and management in, 298–304
information theoretic methods, 299–303
using Bayesian fusion, 303–304, 304f
Sociodemographic differences
ERGMs, 71–72, 72f
impacts on opinion spread, 71–74, 72f, 73t
NetSense within-study network, 71–72, 72f
Solow growth model, 346–347
SolSource, 148, 149
Source trust weighting, 304–305
SPARQL-DL, 208
SPARQL query, 212
SRS, *see* Simple random sampling (SRS)
Stability
defined, 474
of SIG, 475
Stability, humanitarian, and reconstruction operations optimization model (SHROOM), 490–492
Star Wars (film), 39
2012/2013 State of the World's Cities Report, xxvii
StateSim models, 335, 340n
background, 339, 340f
economic system in, 345–349, 346f–348f
generator elicitation screens, 340–349
actors, 344–345, 344f
factions, 341–344, 341f–343f
institutions, resources, and economics, 345–349, 346f–348f
overview, 338–339, 338f
Statistical models
ABMs, 19–20
ExMo approach, 8–9
ABMs, 19–20
populations, modeling, 19
Strategic measuring and reporting tool (SMART) system, 222, 223
Structural equation models (SEMs), 407, 410
Subjective indicators, 222, 223
Subjective uncertainty, 293
Subject matter experts (SMEs), 267–268, 460; *see also* SME bias
Sudan/RSS scenario, Peace Game and, 363, 369, 370, 372, 375; *see also* Peace Game
Sunnis, 504, 506, 507
Supply–demand relationship, 372
Suppression, of social media-based dissent, 46–47
Survey wave, 231
Suspicious, conflicting, inconsistent (SCI) data interpreter, 211
Suspicious behavior/information
entropy-based detection, 214–215, 216–217, 216f–217f
Swarm, defined, 46
Synthesio, 49
Sysomos, 49
System dynamics (SD) models, 19, 450–451, 496
populations, 19

T

Target audience
attributes of, social media messaging and, 48
behavioral habits, 48
consumption habits, 48
Tea Act of 1773, 37
Telegram, 54
Temporal offsets, 318–319
in causal/predictive relationships, Granger causality for, 321
Tensions
urban areas, 8
Terminological knowledge (TBox), 208, 211
Terminology, social media messaging, 49
Terrorism
described, 244
geographical aspects, 244; *see also* Geographical modeling
geographical modeling in studies of, 245
Terrorist incidents; *see also specific entries*
analysis, with GIS, 244; *see also* Geographical modeling
Tertiary data, 122
HRE data, 124t, 125
Text analysis methods, 188
Theoretical orientations, computational modeling, 451
Theoretical representation, hybrid computational models
macrolevel, 454–455
microlevel, 456f, 457–458
Theory of planned behavior (TPB), 457–458
Thick agents, 336n
Thoughts, neural foundation of, 485
Threat, shared perceptions of, 481–483
Threat monitoring
parallel planning process (HRE), 123
Threat narrative, 185–194; *see also* Narrative(s)
assessment, 186–188
conflict dynamics, 192–193, 193f, 194f
defined, 185–186

forecasting methodologies, 190–192, 191t
human behavior forecasting and, 187–188
in search of emic perspective, 188–189
Three Block War, 6
Threshold fraction (ϕ), 66
Threshold model (TM)
for Facebook egonetwork, 69, 69f–70f
individualized thresholds, 68–69, 69f–70f
opinion dynamics, 61, 66–70
tipping points in, 66–68, 67f
Thresholds, individualized, 68–69, 69f–70f
Ties
cultural, measurement and, 74
intercultural, 74
intracultural, 74
network, gender distinctions in, 73
Timing
social media messaging and, 51
Tipping point
in NG model, 62–63, 64f
in threshold models, 66–68, 67f
TM, *see* Threshold model (TM)
TPB, *see* Theory of planned behavior (TPB)
Track Social, 48
Traditional ethnography; *see also* High-risk ethnography (HRE)
vs. HRE
experience, 120t
research parameters, 120t
Traffic models, 18
Traitors, 477–478
Trolls, 46
Trust, 290, 398, 401n; *see also* Uncertainty(ies)
aspects of, 290
integrative model of organizational, 397–419, 400f
ability, 401
analyses, 406t–410t, 407–410, 411f–413f
background, 399
benevolence, 402
components, 400–403
defined, 400, 401
discussion, 415–418
integrity, 402–403
limitations of current research, 417–418
measures, 405–406, 405t
megacity use case (fictional), 418–419
methods, 404
overview, 398–399
research hypotheses, 404
results, 410–414, 414t
risks taking, outcomes, and practical application, 403–404
sample, 405, 405t
theoretical debate, 398
public, toward VEOs, 437f–438f, 439–440, 439t
and uncertainty, relationship between, 290, 298
T-test, 429
Tumblr, 48
TweetDeck, 51
Twitter, 44, 48, 49, 142, 305; *see also* Social media
ISIS supporters on, 46
peak activity times, 51
in social movements, 44
use for misinformation and propaganda purposes, 45–46
use in Ebola outbreak of 2014–2015, 44
value of hashtags, 49–50
Type 1 process, dual process theories, 86–89
Type 2 process, dual process theories, 86–89

U

Ukraine's Right Sector, 45
Ummah, 31
Uncertainty avoidance
defined, 168
as Hofstede's cultural dimension, 168–169
Uncertainty(ies); *see also* Trust
aggregation and, 298
aleatory (statistical), 293
categories, 292
characterizing, 291–298
by communication mechanisms, 295
data, 292
defined, 291
epistemic (systematic), 293
megacity use case (fictional), 308–312
crisis management scenario, 308–309
crisis response management application architecture (example), 309
using Bayesian fusion, 309–312, 310f–311f, 310t–311t
objective, 293
overview, 289–290
propagation, 295–298, 296f
data preprocessing, 297
data processing, 297
in data-to-decisions process, 296f, 297
information processing, 296f, 297
knowledge refinement, 297–298
reduction and management, 298–307
inconsistencies, identifying and resolving, 304–307
information theoretic methods, 299–303

in sociocultural models, 298–304
using Bayesian fusion, 303–304, 304f
representation, 292–294
sociocultural applications and, 289–290
subjective, 293
trust and, relationship between, 290, 298
as untrustworthy, 290
vs. information, 300
Uncertainty modeling, 291
belief function and, 294
belief mass function and, 293–294
classical set theory and, 293
evidence theory and, 293–294
fuzzy measure theory and, 294
fuzzy set theory and, 294
methods, 293–294
plausibility function and, 294
probability theory and, 293
rough set theory and, 294
Uncooperative defaults, 92
Uneven temporal relationships, DTW for, 321–322
Unintended/negative influence, data collection from disconnected communities and, 151–152, 155t
Union Carbide's Bhopal chemical plant accident, 107
United States
experience in Afghanistan, review of, 496–498
interventions
in Afghanistan, 496, 498
in Iraq, 496
United States Africa Command (AFRICOM), 222, 223
United States Institute of Peace/Peacekeeping and Stability Operations Institute (USIP/PKSOI), 485–486
Units, Peace Game, 372–373
Universal Joint Task List, 2002, 272
Unstable sociopolitical environment
in Afghanistan (case study), 146–147
data collection from disconnected communities and, 146–147
discussion of, 147
key principles, 147
megacity use case (fictional), 158–159
Urban areas; *see also* Megacity(ies)
darkness of, 8
described, 13
dynamicity, 8
operation challenges, 7–8, 7f
potential tensions, 8
Urban design models, 18–19
CultureSim, 18–19
UrbanResolve, 18–19
UrbanSim system, 18
Urban environments
expeditionary modeling, 18–19; *see also* Expeditionary modeling (ExMo)
traffic models, 18
Urbanization, 5, 7
operation challenges, 7–8, 7f
Urban operation challenges, 7–8, 7f
UrbanResolve, 18–19
UrbanSim system, 18
US Air Force Research Laboratory, 182, 190, 191
Users, roles in social network analysis, 52
US foreign policy community, 496; *see also* Foreign policy

V

Validation
for causal analysis in sociocultural models, 317, 319–320; *see also* Causal/predictive relationships
model of poverty and violent conflict (case study), 325–331, 326f, 328f–330f
Value-laden narrative system, 38
Value(s)
choices and, 38
defined, 477
narrative and, 37–38
shared thoughts of, 483–484
Vasopressin, 484
VEO, *see* Violent extremist organization (VEO)
Verification and validation (V&V), 317
Victims of forced labor, limited resources for, 138–140
Video length
on YouTube, 48–49
Vienna Matrix Test (VMT), 89
Vine, 48
Violence
by Boko Haram, forecasting, 198–199, 198f–199f
conflict, validation model of (case study), 325–331, 326f, 328f–330f; *see also* Causal/predictive relationships
forecasting, text analytic tools, 195–198, 196f, 197f
idea density, 195
integrative complexity, 195

keyness, 195–196
sentiment/affect of in-group toward out-groups, 196
Violent extremist organization (VEO), 424
age group differences supporting, 441, 442f–443f, 444t
changes in public opinion toward, 436–439, 437f–438f
public trust in, 439–440, 439t
religious differences supporting, 440–441, 440t
Vision
ExMo approach, 8–12, 9f, 11f
VK, 45, 48
Vocabulary diversity, 195
Vocabulary tests, 89
Volume, of social media posts, 51
VSO/ALP program, 498, 499, 502, 503f, 507

W

Waning commitment, NG model, 63, 65–66, 65f
War gaming, 8, 11, 12; *see also* Peace Game
described, 362
Warner, Dave, 146–147
War on Islam, 31
Weather, Peace Game, 372f, 373–375
Web ontology language (OWL), 208
WeChat, 48, 54
Wechsler Abbreviated Scale of Intelligence (WASI), 89
Wechsler's intelligence tests, 89
WEIRD societies, 163, 164, 165; *see also* Non-WEIRD societies
cross-cultural social science research, 164–165
future research, implications for, 175
Hofstede dimensions, 168–171
individualism–collectivism, 169–170, 170f
masculinity–femininity distinction, 169
newer dimensions, 170–171
power distance, 168
uncertainty avoidance, 168–169
integration of culture cycle and Hofstede dimensions, 171
Markus and Conner's culture cycle framework, 166–168, 167f
conceptual ideas level, 167
individualism–collectivism within, 171–174, 172f–174f
individual level, 166
institution level, 166
interaction level, 166
research gap, addressing, 166–174
vs. non-WEIRD societies, 163, 165
cross-cultural social science research, 164–165
Western, educated, industrialized, rich, democratic (WEIRD) societies, *see* WEIRD societies
WhatsApp, 54
Wonderlic Personnel Test (WPT), 89
World Bank, 222, 325
World Health Organization, 44
Worldview
defined, 189
dimensions of, 184
Worldwide incidents tracking system (WITS), 248

X

Xanadu (fictional megacity use case), xxvii–xxviii
causal/predictive relationships, 332–333
CBR algorithm, 355–359
disconnected communities, 157–159
harsh physical environment, 158
limited resources, 157–158
other sociocultural differences, 159
poor infrastructure, 158
unstable sociopolitical environment, 158–159
DYMATICA model, 468–470, 469f
entities and decisions, 468–470
overextension of government resources, 469
sociocultural conflict, 469–470
emic perspective, 200–201
expeditionary modeling scenario, 11–12
geographical modeling, 263–264
high-risk ethnography (HRE), 130–132
Hofstede's culture dimensions, 176–178
integrative model of organizational trust, 332–333
Markus and Conner's culture cycle, 176–178
information processing within, 177–178
military operations, 282–284
peacekeeping operations, 282–284
network/opinion coevolution, 77–80
Notre Dame NetSense data, 70–71
Peace Game, 381–390
map, 383
players, 383

players' units, 386, 388f, 390
region definition, 383, 385f, 386
resources, 383
unit types, 386, 387f
weather, 389f, 390
population surveys, 238–239
raking algorithm, 446
semantic and entropy techniques, 219–220
social identity groups (SIGs), 489–492
SHROOM, 490–492
situation summary, 489
substantive intuitions, 489–490
social media, 54–55
uncertainty reduction and management, 308–312
crisis management scenario, 308–309
crisis response management application architecture (example), 309
using Bayesian fusion, 309–312, 310f–311f, 310t–311t

Y

Yanukovych, Viktor, 44
Yarosh, Dmytro, 45
YouTube, 46, 48, 198
video length, 48–49

Author Index

Page numbers followed by n indicate notes.

A

Aamodt, A., 337
Abbe, A., 276
Abrams, D., 188
Aczél, B., 94, 95
Adamson, W. G., 18
Adolphs, R., xxiii
Agar, M., 185
Agger, K., 144
Agnew, C. R., 399
Ahram, T., 248
Ajzen, I., 457, 458
Akiyama, E., 94, 95
Al-Tohami, K., 364
Albaity, M., 91, 94
Alcorn, B., 19
Aldrich, J. H., 61
Alexander, V., 28
Alford, E., 48
Alger, H., 29
Alkhshali, H., 504
Alter, A. L., 89, 93, 95, 97
Altmann, S., 92
Amblard, F., 480
Ames, D., 185
An, N., 94, 95
Anderson, D., 54
Anderson, L. A., 394, 423–446
Anderson, R. E., 407, 412
Anselin, L., 246
Appelbaum, R. P., 246
Appleget, J. A., 114, 221–242, 362, 363, 375, 405, 406
Apps, P., 146
Arackaparambil, C., 218
Aral, S., 71
Argote, L., xxiii
Aristotle, 32
Arrow, K., 499
Asch, S. E., 479, 480
Auletta-Young, C., 144
Ayyub, B. M., 291

B

Babin, B. J., 407, 412
Bacallao-Pino, L., 47
Bachmann, R., 398
Backus, G. A., 394, 449–470
Baezconde-Garbanati, L. A., 40
Bagozzi, R. P., 457, 458
Bahgat, K., 246
Bailey, M., xxvii
Baird, R. G., 394, 473–492
Bakdash, J. Z., 2, 61–81
Baker, T., 190, 191, 196
Baker-Brown, G., 189, 195
Ballard, E. J., 189, 195
Ballová Mikušková, E., 94, 95, 96, 97
Bandura, A., 189, 191
Barajas, L., 137, 143, 147
Barfield, T., 500
Barnett, J., 325
Baronchelli, A., 61, 62
Barr, N., 90, 93, 95
Barrat, A., 61, 62
Barraza, J. A., 28
Bateson, G., xxiii
Bauer, J., 208
Bazerman, M. H., 104
Beavin, L. E., 28
Beck, R. A., 245
Beitz, C. R., 478
Benedictus, L., 6
Benjamin, P., 268, 289–312
Benoit, T. St., 112, 115–132
Berger, P. L., 479n
Bernard, M. L., 394, 449–470
Bernardi, D. L., 28, 34
Berrebi, C., 245
Berrett, M. T., 498, 500, 508
Best, D. L., 165
Bharathy, G. K., 268, 335–359
Bhatnagar, D., 276
Bhattacharjee, Y., 246
Bhuvanagiri, L., 218

Bier, A. B., 452
Biswas-Diener, R., 483
Black, J. S., 276
Black, W. C., 407, 412
Blair, E., 44
Blais, C. L., 394, 473–492
Blincoe, S., 112, 163–178, 399, 401, 418
Blind, P. K., 398, 401, 402
Bluck, S., 189, 195
Bobashev, G. V., 20
Böhnke, J. R., 246
Boiney, J., 10
Boisot, M., 295
Boland, J. E., 165, 177
Bolt, R. K., 189
Bolzendahl, C. I., 77
Bond, M., 168
Bonham-Carter, G. F., 251
Bonilla-Silva, E., 77
Booth, R. J., 197
Borum, R., 189
Bosch-Domènech, A., 93
Box, G. P., 321
Bozhevolnov, J., 364, 369
Bozkurt, H., 248
Brachman, R. J., 209, 297
Bracken, P., 363
Brady, H. E., 316
Brafman, O., 105
Brafman, R., 105
Braithwaite, A., 325
Braithwaite, E., 245
Brañas-Garza, P., 93, 94
Branigan, E., 31, 32, 33
Bratus, S., 218
Bredenlöw, T., 479n
Brett, J. M., 276
Brin, S., 304
Brislin, R. W., 276
Brock, T. C., 28
Brody, J., 218
Bronowski, J., 184
Brown, D. E., 165, 245
Brown, D. S., 508
Brown, R. A., 182
Bruneau, E. G., 483
Bryden, R., 108
Buccino, G., 40
Buchtel, E. E., 86, 87
Bullock, B., 15
Burke, K., 31
Burke, P., 189
Burns, N., 89, 91
Burns, W. J., 93, 95
Burstein, P., 61
Byman, D., 275

C

Caglayan, A., 114, 205–220, 218
Caglioti, E., 61, 62
Cahalane, D. J., 68, 69
Çakit, E., 114, 243–264
Calderwood, R., 457
Calvert, B., 146, 147
Camerer, C., xxiii
Campitelli, G., 94, 95
Canals, A., 295
Caplan, J. M., 245
Cappa, S. F., 40
Card, B., 137, 143, 147
Carley, K., 77
Carpenter, J., 93, 95
Carroll, P., 479
Carter, C. J., 504
Carter, M. A., 112, 115–132
Cassani, L., 114, 205–220
Casstevens, R., 268, 315–333
Castellano, C., 61
Castelló, X., 62
Cavojová, V., 85–98
Centola, D., 66, 68
Chaiken, S., 457
Chamberlain, J., 268, 315–333
Chamberlain, R. G., 20
Chan, P., 322
Chandrasekar, C., 245
Chandrasekaran, R., 500
Chanley, V. A., 417
Chapman, G. B., 89
Chatterjee, J. S., 40
Chen, L., 276
Cheong, P. H., 28, 34
Cheyne, J. A., 89, 90, 93, 95, 98
Child, T. B., 246
Choucri, N., 19
Christensen, T., 401, 407, 415, 417
Christiansen, J., 22
Chua, H. F., 165, 177
Chulov, M., 504, 506, 507
Chung, C. K., 197
Ciavarelli, A. P., 108
Claes, L., 87
Clancy, J., 246
Coase, R. H., 476
Cochran, W. G., 426

Cohen, J. S., 394, 508
Cohn, J. V., xxv
Cokely, E. T., 89, 92
Cole, B., 485
Collier, D., 316
Collier, P., 325
Collins, J. J., 245
Colquitt, J. A., 398, 399, 403
Combs, D. J. Y., xxv, 112, 163–178, 394, 397–419
Compton, J., 362
Conant, E., 52
Conner, A., 1–2, 3, 112, 163, 164, 166–168, 170, 171, 172, 173, 174, 176, 177, 178
Conover, W., 228, 229
Cooke, N. J., 454
Cools, E., 90n
Corbin, J. M., 191
Corcoran, J., 245
Corgnet, B., 93, 95
Corman, S. R., 30, 31, 32, 33, 35–36, 37, 40
Cosenza, V., 48
Couper, M. P., 404
Cover, T. M., 298
Covey, S. M. R., 416
Covington, M., 192, 195, 196, 197
Cowlishaw, G., 484
Cragin, W., 144
Crossett, C., 246
Cruz, M., 136
Cushner, K., 276
Cutter, S. L., 245

D

Dall'Asta, L. A., 61, 62
Dalton, J., 245
Dalton, R. J., 402, 403
Damron, R. W., 394, 449–470
Danin, R., 46
Danna, M., 40
Darczewska, J., 507
Dasandi, N., 325
Davenport, T. H., 125
Davis, J. H., 397, 398, 399, 400, 401, 402, 403, 406, 407, 408, 410, 411, 413, 414, 415, 416, 417, 418, 419
Davis, P. K., 182, 480, 496
Davis, R., 276
De Marchi, S., 451
De Neys, W., 93, 95
de Vries, B., 189, 195
DeBell, M., 436
Deffuant, G., 480
Dehghani, M., xxiii
Dekker, S., 109
Delen, D., 298
Delfabbro, P., 89, 91
Demirci, S., 245
Dempster, A. P., 393
Deneubourg, J.-L., 484
Denning, D., 244
Deptula, D. A., 496
Dermer, R., 497, 500, 508
Desai, K., 139, 140
Design, B., 137, 143, 147
Dessai, V. M., 108
Dettmer, J., 45
Devichand, M., 45
Dieckmann, N. F., 93, 95
Diehl, P. F., 474
Diener, E., 483
Dillman, D. A., 404
DiStefano, C., 406
Ditto, P., 90, 91, 93
Dixon, R., xxvii
Djankov, S., 325
Dolph, J., 22
Domingos, P., 305, 306
Dominguez, J. I., 403
Donohue, K., 92, 93, 95
Douglass, F., 29
Downward, G. S., 148
Doyle, C., 2, 61–81
Drapeau, M. D., 246
Dudeková, K., 89, 94, 97
Dugan, L., 245
Dunbar, R. I. M., xxiii, 479
Duquette, W. H., 20
Durschmied, E., 477
Dusen, E. V., 151
Dyer, G., 146
Dyer, N., 276

E

Eagly, A. H., 457
Edison, T., 29
Egeth, J. D., 19
Eguíluz, V. M., 66
Eidelson, R. J., 336, 349, 480
Ekhtiari, H., xxiii
Eldivan, I. S., 245
Ellen, P. S., 458
Elliott, V. L., 325
Ellis, D. C., 8, 450
Emenheiser, J., 62, 63

Engel, P., 54
Epley, N., 89, 93, 95, 97
Epstein, J. M., 20
Epstein, S., 86, 90n
Erraguntla, M., 298
Errickson, K. W., 32
Espín, A. M., 93
Evans, J. S., 86, 87
Evans, M., 12
Eyre, R. N., 89, 93, 95, 97
Eyre, S., 66

F

Fairlie, B. M., 2, 43–56
Falk, A., 92
Fang, J., 213
Farrelly, D., 87
Farry, M., 15
Faure, T., 480
Fazio, F., 40
Feaver, P., 61
Fehr, E., xxiii
Felici, M., 61, 62
Felix, K. M., 6
Feltz, A., 92
Fenstermacher, L., 113, 181–201
Fernbach, P. M., 93, 95
Fishbein, M., 457
Fisher, M., 47
Fisher, W., 34, 35
Fisk, M., 31, 40
Flin, R., 108
Flynn, M. T., 450, 497, 498
Forbus, K., xxiii
Ford, A., 451
Forrest, A., 142
Forrester, J. W., 451, 496, 508
Fortunato, S., 61
Fosgaard, T. R., 92
Foster, C. A., 399
Foster, D., 10
Foster, J., 136
Fotheringham, S., 251
Fowler, Jr., F. J., 404
Francis, M. E., 197
Frank, L. B., 40
Frank, M. G., 189, 196
Frankish, K., 86
Franklin, B., 29
Frederick, S., 88, 89, 90, 91, 93, 95, 97
Freeman, H., xxv
Freeman, J. R., 321
Freytag, G., 32
Fricker, R. D., Jr., 223, 224, 226, 394, 397–416, 423–446
Fries, P., 485
Fugelsang, J., 89, 90, 93, 95, 98
Furlan, S., 89, 90, 93, 94, 95
Furlow, R. B., 32

G

Galehouse, D., 63, 64
Gallese, V., 40
Gallos, L. K., 66
Ganguly, S., 218
Garces, M., 19
García-Muñoz, T., 94
Garris, C. P., 112, 163–178
Gavin, M. B., 398, 414
Geeraerts, L., 93, 95
Geiler, W., 20
Gelfand, M. J., 276
Gelman, A., 224
Gelman, R., 89
Gelpi, C., 61
Gentner, D., xxiii
Gervais, W. M., 91, 98
Geuss, M., 142
Giacalone, R., 168
Gilbert, D. T., 46, 479
Gilboa, I., 474
Gill, K., 139, 140, 154
Gladwell, M., 63
Glass, R. J., 455
Gleditsch, K. S., 508
Gleeson, J. P., 68, 69
Glenn, R. W., 275
Glimcher, P. W., xxiii
Godoy, A., 87
Goedecke, D. M., 20
Goldsmith, D., 19
Goldthorpe, J. H., 316, 319, 320, 321
Gonzalez, R. H., 93, 94, 95
Goodall, B., 36, 37
Goodall, H. L., Jr., 30, 33, 35–36
Goodchild, M. F., 246
Gottman, J., 477
Grace, J., 479
Graffeo, C., 112, 115–132
Graham, J., 90, 91, 93
Graham, M., 93, 95
Granger, C. W. J., 315, 317, 321, 322, 325, 326, 327, 329, 331, 332, 455
Granovetter, M., 61, 66, 68, 484

Green, M. C., 28
Greene, J. D., 89, 91
Greiner, L. H., 195
Groll, E., 50, 54
Groves, R. M., 404
Grusky, D. B., 77
Guimond, S., 483
Gulick, L. M., 276
Gunderson, L. H., 487
Gupta, M. M., 291
Gurr, T. R., 477, 482
Gustin, S., 44
Güth, W., 477
Guthrie, A., 50
Guttman, L., 406, 409

H

Hachen, D. S., 71, 79
Hagmayer, Y., 87
Haidt, J., 90, 91, 93
Hair, Jr., J. F., 407, 412
Hall, D. L., 308
Hall, J., 504, 506
Hall, S. B., 394, 473–492
Halpin, S. M., 276
Halverson, J. R., 30, 33, 35–36
Hanák, R., 85–98
Handley, S. J., 87
Hanges, P., 168
Hansen, L. G., 92
Hardin, G., 481
Hardin, R., 399
Hardisty, D. J., 91
Harkin, J., 507
Harley, C., 87
Harrell, F. E., 318
Harris, K., 50
Harris, M., xxvii, 185
Harthorn, B. H., 246
Hartman, A., 190, 191, 196
Haslam, S. A., 480
Havlin, S., 66
Heerlein, D., 363, 375
Hegre, H, 325
Hegselmann, R., 480
Heine, S. J., 88, 164, 165
Helmreich, R. L., 108
Hemel, V. S. B., 489
Hendrex, D., xxvii
Henrich, J., 88, 164, 165
Henrion, M., 291
Hepner, G. F., 245
Heras, N. A., 496
Herman, J. L., 276
Herman, L., 36
Hernandez, A. S., 362, 363
Hetherington, M. J., 401
Heuer, R. J., 124, 184
Hicken, A., 19
Hill, A. V., 92, 93, 95
Hill, J., 224
Hill, K., 154
Hilleary, C., 44
Hingst, R. G., 3, 5–23
Hirshleifer, J., 290
Hitzler, P., 214
Hodges, S. D., 97
Hoeffler, A., 325
Hoffman, M. A., 3, 5–23, 15
Hoffman, R. F., xxiii
Hofstede, G. J., 106, 107, 112, 163, 164, 166, 167, 168–171, 173, 175, 176, 178
Hogg, M. A., 188
Holland, R. W., 37–38
Holling, C. S., 487
Holmes-Eber, P., 275, 276
Hoppe, E. I., 91
Horridge, M., 208
Horvitz, D. G., 423, 427, 431, 444
Hosgood, H. D., 148
House, R., 172
Howard, R., 46
Hoyle, H., 245
Hoyle, R. H., 412
Hsu, E., 485
Hu, J., 276
Hu, W., 148
Huang, Z., 213
Hudson, D., 325
Hunter, G. J., 308
Huo, Y. J., 276
Hwang, H., 189, 196

I

Ikei, M., 173
Iñiguez, G., 66
Iyer, R., 90, 91, 93

J

Jager, S. M., 276
Jakes, L., 496
James, G., 92, 93, 95
Jandora, J., 276

Jarvenpaa, S. L., 398, 414
Jeanson, R., 484
Jeffers, R. F., 394, 449–470
Jeltsen, M., 45
Jenkins, G. M., 321
Jentsch, F., 124
Johnson, D., 340n
Johnson, J. L., 498, 500, 508
Johnson, L., 50
Johnson, S. D., 245
Jordan, M., 304
Jøsang, A., 290
Juanchich, M., 87, 94, 95, 96, 97

K

Kahan, D. M., 91, 93, 95
Kahneman, D., 86, 87, 103, 455, 475, 476, 479, 499
Kain, Eric, 47
Kalish, Y., 71
Kameda, N., 276
Kamrany, N. M., 246
Kang, C. S., 112, 135–160
Karampourniotis, P. D., 2, 61–81
Karsai, M., 66
Karwowski, W., 104, 105, 114, 243–264
Kaski, K., 66
Keebler, J. R., 124
Keeney, R. L., 455
Kelley, C. M., 89
Kelly, J., 144
Kempe, D., 66
Kemper, S. J., 195
Kendall, D., 477
Kennedy, L. W., 245
Kenner, D., 496
Kertész, J., 66
Kettler, B. P., 3, 5–23
Khaltourina, D., 364, 369
Kikas, R., 66
Kilcullen, D., 6, 7, 13, 164, 398, 498
Kim, S., 401, 403, 407, 415, 417
Kimmel, S. K., 457
King, A. J., 484
King, C., 118
Kirby, M., 62, 63
Kish, L., 426
Kiss, H. J., 94, 95
Kitayama, S., 167, 170, 171, 172, 173, 174
Kitsak, M., 66
Kjeldsen, S., 275
Klaaren, K. J., 97
Klein, G. A., xxiii, 457, 499
Klein, G. L., 19
Kleinberg, J., 66
Klesner, J. L., 403
Knoll, K., 398, 414
Kobzeva, S., 364, 369
Koehler, D. J., 89, 90, 92, 93, 95, 98
Koehler, J., 246
Kohlhase, J. E., xxvii
Kohn, D., 182
Koleva, S., 90, 91, 93
Koltko-Rivera, M. E., 187, 189, 475, 479
Kondo, M., 276
Koneèný, M., 89, 94, 97
Korniss, G., 2, 61–81
Korotayev, A., 364, 369
Kostovicová, L., 89, 94, 97
Krause, U., 480
Krishnan, R., xxiii
Krosnick, J. A., 436
Krulak, C. C., 6
Kujal, P., 93, 95
Kulzy, W. W., 223, 224, 226, 394, 397–419, 424
Kumar, M. V., 245
Kurtz, C. F., 119, 187
Kusterer, D. J., 91
Kuznar, L. A., 113, 181–201

L

L'Herrou, T, 66, 68
Labollita, M., 94, 95
Laegreid, P., 401, 407, 415, 417
LaFleur, S. J., 97
LaFree, G. L., 245
Lagalice, C. G. C., 364
Lakdawalla, D., 245
Lall, A., 218
Landis, D., 276
Larrick, R., 276
Latane, B., 66, 68
Lautenschlager, J., 22
Lawrence, T. E., 267
Lee, B. A., 285n
Lee, C. S. G., 294
Lee, G., 248
Lee, K., 51
Lee, T. Y., 45
Leedom, D. K., xxiii
Lefebvre, V. A., 483
LeGree, L., 275, 276
Leiber, N., 139, 140
Leidner, D. E., 398, 414

Leiter, M., 51
LePine, J. A., 398, 399, 403
Lepkowski, J. M., 404
Lett, J., 185
Leung, K., 168, 185, 276
Levesque, H. J., 297
Levey, B., 188
Li, J., 276
Li, P. P., 398
Liang, D. W., xxiii
Liberali, J. M., 89, 90, 93, 94, 95
Lickel, B., 185
Lieberman, M. D., 479
Liljeros, S., 66
Lim, C., 62
Lin, C. T., 294
Lind, E. A., 276
Lisle, D. J., 97
Listhaug, O., 402
Liu, H., 245
Liu, S., 71
Lizardo, O., 2, 61–81
Llinas, J., 308
Lofdahl, C., 395, 495–508
Lohr, S. L., 426, 427, 434
Lombardo, T., 144
Lombe, M., 415, 417
Loreto, V., 61, 62
Loshin, D., 154
Louis, R. St., 93, 95
Lu, Q., 62
Lucie, M., 49
Luckmann, T., 479n
Lumley, T., 425, 428, 435, 446
Lundry, C., 28, 32, 34
Luo, J. , 276
Lusher, D., 71
Lustick, I. S., 317, 480

M

Ma, M., 44
Ma, Y., 214
Macannuco, D., 19
Maclin, B., 144
MacMillan, J., 336, 489
Macy, M. W., 66, 451
Madden, T. J., 458
Maddux, J., 114, 221–242
Madnick, S. E., 19
Madrigal, A., 44
Magni, C. A., 481
Maheswaran, D., xxiv
Maier, F., 214
Makse, H. A., 66
Malkov, A., 364, 369
Malkov, S., 364, 369
Malthus, T. R., 364
Markesbery, W. R., 195
Markman, A. B., 457
Markus, H., 1–2, 3, 112, 163, 164, 166–168, 170, 171, 172, 173, 174, 176, 177, 178
Marlowe, J., 415, 417
Masi, A., 46
Masuda, T., 177
Matsui, A., 474
Matsumoto, D., 189, 196
Mattis, J. N., 496
Mayer, R. C., 298, 397, 398, 399, 400, 401, 402, 403, 406, 407, 408, 410, 411, 413, 414, 415, 416, 417, 418, 419
McAdams, D. P., 479
McCann, J. A., 403
McCrae, R. R., 165
McFadden, D., 455, 460
McGarty, C., 480
McInnes, W., 48
McNemar, Q., 165
Mearns, K., 108
Medby, J. J., 275
Medina, R. M., 245, 246
Melin, N., xxvii
Mendenhall, W., 430
Mendoza, M. T., 276
Meng, L., 71
Merrill, R. R., 416
Merritt, A., 108
Merton, R. K., 477
Mertz, C. K., 93, 95
Michelena, P., 484
Mignone, B. K., 246
Mikusinski, P., 248
Milbank, D., 497
Miller, A. H., 402
Miller, C., 45
Miller, J., 486
Mills, A. J., 119
Mills, J. H., 119
Mîndrilă, D., 406
Minkov, 164, 168, 169, 170, 173, 176
Miodownik, D., 480
Mishra, A. K., 398
Mishra, K. E., 398
Mistree, D., 19
Moon, B., xxiii

Moon, W. H., 185
Moore, D. A., 104
Moran, A., 290
Moran, M. B., 40
Moreland, R. L., xxiii
Morgan, D. L., 275
Morgan, M. G., 291
Moritz, B. B., 92, 93, 95
Morone, F., 66
Morris, M. W., 185, 276
Morris, S. D., 403
Morrison, J. B., 19
Mortimer, J. A., 195
Moss, K., 49
Moyer, D., 18, 19
Muchnik, L., 66, 71
Mulaik, S. A., 406
Munck, G. L., 316
Munoz, A., 146, 147
Murata, A., 3, 103–109
Murphy, K. P., 303
Murphy, S. T., 40
Myaskovsky, L., xxiii
Myers, C., 322
Myers, D. J., 77

N

Nakamura, T., 104, 105
Nannini, C. J., 362, 363
Nardi, D., 209
Narvaez, D., 479
Nassr, W., 50
Naugle, A. B., 394, 449–470
Naylor, S., 497
Neal Reilly, S., 268, 315–333
Neal Reilly, W. S., 325
Neill, J., 225
Newstead, S. E., 87
Nguyen, T., 63, 64
Nicholson, D., 124
Nisbett, R. E., 88, 165, 177
Noel, H., 76, 77, 78
Nofsinger, J. R., 91, 93, 95
Nordvall, A., 94, 95
Norenzayan, A., 86, 87, 88, 91, 98, 164, 165
Novitskiy, N., 93, 95
Nowak, M. A., 482
Null, C., 151
Numrich, S. K., 336
Nye, B., 336, 340n, 349
Nyhan, B., 76, 77, 78

O

O'Brien, S., 9, 21, 188, 192, 195, 196, 197, 198
O'Connell, A. N., 480
O'Connor, P., 108
O'Loughlin, J., 508
O'Mahony, A., 480
Oakes, P. J., 480
Obrecht, N. A., 89
Ochsner, K. N., 479
Oechssler, J. J., 91, 93, 95
Ogihara, M., 218
Ohbuchi, K., 276
Okumura, S., 276
Oldham, C., 200
Olive, J. L., 185
Olson, P., 54
Olszewski, R. T., 324
Omoregie, H., 174
Oppenheimer, D. M., 89, 93, 95, 97
Orasanu, J., 457
Osin, O., 340n
Over, D. E., 86
Owen, B., 182
Owen, K., 15

P

Pacini, R., 90n
Packer, G., 12, 18
Page, K. M., 482
Page, L., 304
Page, S. E., 451
Palfi, B., 94, 95
Pardo, S. T., 89, 90, 93, 94, 95
Pariente, W., 151
Parsia, B., 208
Pattison, P., 71
Paulo, D., 66
Pennebaker, J. W., 197
Pennycook, G., 89, 90, 93, 95, 98
Perani, D., 40
Perla, P. P., 362, 363
Perry, W. L., 182, 268, 269–285
Perugini, M., 458
Peters, E., 93, 95
Peters, J. G., 403
Peterson, R. B., 276
Petty, R. E., 457
Pfautz, J., 15
Pfeffer, A., 325
Pickett, C., 479
Picucci, P. M., 336

Pierson, B., 19
Pietrocola, D., 340n
Pike, K. L., 185
Piovesan, M., 92
Plaza, E., 337
Plutchik, R., 193
Poellabauer, C., 71
Polat, E., 245
Poldrack, R. A., xxiii
Pons, S., 477
Pool, R., 246
Porter, D., 93, 95
Porter, L. W., 276
Powers, M. J., 268, 361–390
Presti, S. L., 290
Pyszczynski, T., 189

Q

Quillin, B., 136

R

Rabiner, L., 322
Rahman, M., 91, 94
Rahmandad, H., 508
Rahn, W. M., 417
Raiffa, H., 455
Ramachandran, S., 268, 289–312
Ramakrishnan, N., 19
Ramautar, J., 93, 95
Rand, D. G., 89, 91
Rand, D. H., 496
Razmyar, S., 93, 95
Reason, J., 107
Reeve, C. L., 93, 95
Reifler, J., 61
Reilly, D., 508
Reilly, N., 45
Renfro, C. L., 188
Renn, D., 474
Rentfrow, P. J., 483
Revelle, W., 405
Rey, B., 51
Reyna, V. F., 89, 90, 93, 94, 95
Reynal-Querol, M., 325
Reynolds, K. J., 477
Richardson, M., 305
Ries, E., 157
Riley, J. G., 290
Rizzolatti, G., 40
Robert, M., 346, 347
Roberts, K. H., 108
Robins, G., 71
Robinson, L., 507, 508
Robinson, S. L., 290
Rodriguez-Lara, I., 94, 95
Rogerson, P., 251
Roider, A., 91, 93, 95
Romano, B., 20
Rosa-García, A., 94, 95
Roshan, P., 182
Ross, S., 145
Rosseel, Y., 398, 407
Ruan, Z., 66
Rubel, R. C., 362, 364, 375
Rudolph, T. J., 417
Rusbult, C. E., 399
Rusnak, D. M., 245
Russel, S., 306
Russo, R., xxvii
Ruston, S. W., 3, 27–41

S

Saccuman, M. C., 40
Sachdeva, S., xxiii
Safadi, N. S., 415, 417
Salerno, J. J., 20
Salmoni, B. A., 275, 276
Salvador, S., 322
Sambanis, N., 325
Sampat, R., 45
Sanborn, J. K., 6
Sarmah, S., 44
Satterfield, J. M., 196
Sattler, U., 208
Saunders, C., 86
Saxe, R., 483
Schacter, D. L., 479
Schank, J., 479
Schatz, S., xxv
Scheaffer, R. L., 430
Schellekens, P., 136
Schmittberger, R., 477
Schmitz, P. W., 91, 93, 95
Schmorrow, D., 19
Schooler, J. W., 97
Schoorman, F. D., 397, 398, 399, 400, 401, 402, 403, 406, 407, 408, 410, 411, 413, 414, 415, 416, 417, 418, 419
Schwarze, B., 477
Sciarini, L. W., 124
Scifo, P., 40
Scott, B. S., 398, 399, 403
Sears, D. O., 165

Seawright, J., 316
Seddon, M., 46
Sekar, V., 218
Seli, P., 89, 90, 98
Seow, W. J., 148
Serge, V., 477
Shachter, R., 291
Shafer, G. A., 293
Shakarian, P., 66
Shaley, O., 275
Shannon, C. E., 298
Sharansky, N., 497, 500, 508
Sharp, K. T., 61
Shavitt, S., xxiv
Shaw, M. L. G., 454
Sheer, V. C., 276
Shellman, S. M., 188, 192, 193, 194, 195, 196, 197, 198
Shenhav, A., 89, 91
Shepperd, J. A., 479
Sherman, J., 479
Shi, X., 94, 95
Shimada, J., 276
Shin, M., 508
Shoichet, C. E., 504
Shroder, J., 245
Shube, J. N., 93, 95
Shubina, A., 218
Sibbald, A. M., 484
Siebeneck, L. K., 245
Siegel, E., 125
Siegel, M. D., 19
Sigmund, K., 482
Silver, N., 9, 477
Silverman, B. G., 268, 335–359
Simon, H. A., 499
Simpson, J. A., 398, 399
Simunich, K. L., 22
Sindelar, D., 46
Singer, E., 404
Singh, P. P., 2, 61–81, 245
Sirota, M., 87, 94, 95, 96, 97
Sisco, J., 8, 450
Slack, J., 54
Sliva, A., 268, 315–333
Sloman, S. A., 93, 95
Smaldino, P., 479
Smith, B., 457
Snijders, T. A. B., 79, 81
Snowden, D. J., 119, 187
Snowdon, D. A., 195
Sohn, H., 107
Solon, O., 45
Solow, R. M., 346, 347
Sorvino, C., 146
Sreenivasan, S., 62, 63, 64, 65, 66, 67, 68, 69, 70
St. Clair, C., 191
Stanley, H. E., 66
Stanovich, K. E., 86, 87, 89, 90, 93, 95, 97, 98
Stanton, J., 247
Steels, L., 61, 62
Steglich, C. E. G., 79, 81
Stein, L. M., 89, 90, 93, 94, 95
Stenmark, T., 276
Stephan, W. G., 188
Stephenson, A., 191
Sterbenz, C., 46
Sterman J. D., 451, 496, 500, 508
Sterman, J. D., 508
Stern, Y., 497, 508
Storey, J., 36
Strauss, A., 191
Striegel, A., 71
Suedfeld, P., 189, 195
Suen, I. S., 245
Sugihara, G., 323
Sullivan, J., 276
Sun, D. Q., 268, 335–359
Sundararajan, A., 71
Sunstein, C. R., 92n
Swisher , R., 112, 115–132
Sydelko, P., 22
Szaszi, B., 94, 95
Szollosi, A., 94, 95
Szymanski, B. K., 2, 61–81

T

Tajfel, H., 188
Takahashi, Y., 276
Taleb, N. N., 187
Tan, H. H., 399, 402, 403
Tardos, E., 66
Tarnoff, C., 245
Terris, E. T., 28
Terry, M., 479
Tetlock, P. E., 8, 189, 195
Tettamanti, M., 40
Thaler, R. H., 92n
Themnér, L., 325
Thomas, J. A., 298
Thompson, D. J., 423, 427, 431, 444
Thompson, W., 248
Thomson, K. S., 89
Thurlow, A., 119
Toman, P., 190, 191, 196

Toplak, M. E., 89, 90, 93, 95
Torrens, P. M., 18
Tourangeau, R., 404
Townsend, S. S. M., 174
Trafton, A., 48
Trethewey, A., 36, 37
Tubin, M. R., 317
Turner, J. C., 188, 477, 480
Tusler, M., 93, 95
Tversky, A., 86, 87, 103, 455, 475, 476, 479, 499
Tyler, T. R., 276

U

Uchida, Y., 174
Ucko, D. H., 164
Uganec, C., 48
Ulfarsson, G. F., 18

V

Vadakkeveedu, K., 268, 289–312
Van de Bunt, G. G., 79, 81
Van de Walle, S., 417
van den Bercken, J., 87
Van den Broeck, H., 90n
Van Dijk, T. A., 190, 191
van Harmelen, F., 213
Van Hemel, S. B., 336
Van Ryzin, G. G., 403, 417
Vandiver, J., 144
Varma, A., 91, 93, 95
Vermeulen, R., 148
Verplanken, B., 37–38
Vervaeck, B., 36
Vigoda-Gadot, E., 399, 402, 407, 415, 417
Voorhies, P., 182
Vorm, E. S., 112, 163–178, 399, 401, 418

W

Wachowicz, M., 308
Wackerly, D. D., 430
Waddell, P., 18
Wagemans, J., 93, 95
Waldram, H., 50
Wallensteen, P., 325
Walt, S. M., 508
Waltz, E., 20
Wan, K., 168
Wang, C., 71, 79
Ward, A., 45
Ward, M. D., 508
Ware, J., 245
Warren, M. E., 402, 403
Watson, S., 290
Watts, D. J., 61, 66
Webb, J., 245
Webb, W. A., 306
Weber, E. U., 91
Weeden, K. A., 77
Wegener, D. T., 457
Weick, K. E., 119
Weisbuch, G., 480
Wekstein, D. R., 195
Welch, S., 403
Weller, J. A., 93, 95
Welsh, M., 89, 91
Werth, L. N., 40
West, R. F., 89, 90, 93, 95
Weyer, N., 268, 335–359, 340n
Wieselquist, J., 399
Willer, R., 451
Williams, J. E., 165
Williams, K. Y., 276
Willis, J. W., 185
Wilson, D. S., 506
Wilson, E. O., 506
Wilson, I., 245
Wilson, T. D., 97, 98
Windrem, R., 504, 507
Wisniewski, E. J., 457
Witteman, C., 87
Wolf, J., 93, 95
Wong, F. D., 6
Wood, G., 496, 506
Woods, L., 51
Woodward, B. A., 373
Woodward, J. B., 454
Wright, H., 87
Wunderle, W. D., 276

X

Xie, J., 62, 63
Xie, M., 245
Xu, J., 218

Y

Yager, M., 190, 191, 194, 198
Yamada, I., 245
Yan, G., 218
Yeung, D., 182
Yin, R. K., 185
Young, J. K., 188

Youssef, N. A., 504, 506, 507
Yu, F., 20

Z

Zacharias, G. L., 336, 489
Zak, P., 28
Zara, C., 47
Zarrella, D., 49
Zhang, H., 218
Zhang, W., 62
Zhao, N., 40
Zhu, M., 406
Zinkina, J., 364, 369
Zinman, J., 151
Zsambok, C. E., 457
Zürcher, C., 246
Zwane, A. P., 151

Milton Keynes UK
Ingram Content Group UK Ltd.
UKHW052027071024
449327UK00027B/2464